Applying Calculus in Economic and Life Sciences

by
Dennis Sentilles

University of Wisconsin-Madison
Department of Mathematics

Brooks/Cole Publishing Company

I(T)P An International Thomson Publishing Company

Boston * Albany * Bonn * Cincinnati * Detroit * London * Madrid * Melbourne * Mexico City
New York * Paris * San Francisco * Singapore * Tokyo * Toronto * Washington

COPYRIGHT © 1997 by Brooks/Cole Publishing Company
A Division of International Thomson Publishing Inc.
I(T)P The ITP logo is a registered trademark under license.

Printed in the United States of America

For more information, contact Brooks/Cole Publishing Company, 511 Forest Lodge Road, Pacific Grove, CA 93950, or electronically at http://www.thomson.com/brookscole.html

International Thomson Publishing Europe
Berkshire House 168-173
High Holborn
London, WC1V 7AA, England

International Thomson Editores
Campos Eliseos 385, Piso 7
Col. Polanco
11560 México D.F. México

Thomas Nelson Australia
102 Dodds Street
South Melbourne 3205
Victoria, Australia

International Thomson Publishing Asia
221 Henderson Road
#05-10 Henderson Building
Singapore 0315

Nelson Canada
1120 Birchmount Road
Scarborough, Ontario
Canada M1K 5G4

International Thomson Publishing Japan
Hirakawacho Kyowa Building, 3F
2-2-1 Hirakawacho
Chiyoda-ku, Tokyo 102, Japan

International Thomson Publishing GmbH
Königswinterer Strasse 418
53227 Bonn, Germany

International Thomson Publishing Southern Africa
Building 18, Constantia Park
240 Old Pretoria Road
Halfway House, 1685 South Africa

All rights reserved. No part of this work covered by the copyright hereon may be reproduced or used in any form or by any means—graphic, electronic, or mechanical, including photocopying, recording, taping, or information storage and retrieval systems—without the written permission of the publisher.

ISBN 0-534-49725-X

The Adaptable Courseware Program consists of products and additions to existing Brooks/Cole Publishing Company products that are produced from camera-ready copy. Peer review, class testing, and accuracy are primarily the responsibility of the author(s).

Contents

Preface   viii
List of Applications   xii
What Is Applied Calculus?   1

Chapter 0   **Functions**   2

        Real Systems as Mathematical Functions   2
0.1   Mathematics and Systems Modeling   3
0.2   Functions and Function Notation   13
0.3   Linear Functions and Graphs   22
0.4   Graphs of Nonlinear Functions   30
0.5   Chain Processes and New Functions from Old   40
0.6   Describing and Measuring Change   48
        Chapter 0 Summary   54
        Chapter 0 Summary Exercises   55

Chapter 1   **Limits and the Derivative**   57

        On the Threshold of Calculus   57
1.1   Approximating the Tangent to a Curve   58
1.2   The Limit of a Function   66
1.3   Continuous Functions and the Limit Properties   74
1.4   The Derivative of a Function   86
1.5   The Derivative as Rate of Change   94
        Chapter 1 Summary   103
        Chapter 1 Summary Exercises   104

Chapter 2   **The Derivative**   106

       State and Change-of-State   106

2.1   The Derivative of a Function: Basic Operations   107

2.2   The Chain Rule   112

2.3   Differentiation of Products and Quotients   118

2.4   Applications to Graphing Functions   126

2.5   Maximum and Minimum Values of a Function   137

     A. The First Derivative Test for Extrema   140

     B. The Second Derivative Test for Extrema and Inflection Points   142

     C. The Maximum and Minimum Values of a Continuous Function on an Interval   148

2.6   Optimization   154

     A. Optimum Numerical States   157

     B. Optimum Geometric States   158

     C. Optimum Allocation of Resources   160

     D. Optimum Profit and Revenue   164

     Chapter 2 Summary   174

     Chapter 2 Summary Exercises   175

Chapter 3   **Related Rates and the Differential**   178

       Further Applications of the Derivative   178

3.1   Implicit Differentiation   179

3.2   Related Rates   185

3.3   The Differential of a Function   195

     Chapter 3 Summary   203

     Chapter 3 Summary Exercises   204

Chapter 4   **Integration**   205

       The Whole as the Sum of Continually Varying Parts   205

4.1   Integration and the Area under a Curve   206

4.2   The Fundamental Theorem of Calculus   218

4.3   The Indefinite Integral and Integration by Substitution   225

4.4   Applications of the Indefinite Integral   233

4.5   Applications of the Definite Integral   242

     A. The Area of the Region between Two Curves   243

        B. The Average Value of a Function 244
        C. The Volume of a Solid 246
        D. Producers', Consumers', and Polluters' Surplus 249
        E. Accumulation in the Environment 253
        Chapter 4 Summary 257
        Chapter 4 Summary Exercises 258

Chapter 5   The Exponential and Logarithmic Functions 261

        Two Special Functions 261
5.1 The Exponential Function and Its Derivative 262
5.2 The Calculus of the Exponential Function 276
5.3 The Natural Logarithm 285
5.4 The Calculus of the Logarithmic Function 297
5.5 Exponential Growth and Decay 303
        Chapter 5 Summary 320
        Chapter 5 Summary Exercises 320

Chapter 6   Functions of Several Variables 322

        Three Dimensions and Beyond 322
6.1 Functions of Several Variables and Their Graphs 323
6.2 Partial Differentiation 331
6.3 Optimization 341
6.4 Linear Regression by Least Squares 351
6.5 Constrained Optimization and Lagrange Multipliers 360
6.6 The Differential 370
6.7 The Double Integral 378
        Chapter 6 Summary 386
        Chapter 6 Summary Exercises 388

Chapter 7   Applied Integration 390

        Accumulation and Concentration 390
7.1 Integration by Parts 391
7.2 Density and Accumulation 399
7.3 Flow Rate Through a Compartment 409
7.4 Arrival and Accumulation Processes 415

	7.5	Approximate Integration 424
		Chapter 7 Summary 432
		Chapter 7 Summary Exercises 433

Chapter 8

Differential Equations   435

	Laws of Change in a Dynamical System 435
8.1	Differential Equations and Their Solutions 436
8.2	Growth in a Limited Environment 445
8.3	Evolution of Time-Autonomous Systems 455
8.4	Solution by Separation of Variables 465
8.5	Applications to Compartment-Mixing Processes 473
	A. Applications to Business and Economics 475
	B. Applications to Natural Systems 479
8.6	Euler's Method of Approximate Solution 489
8.7	Compartment-Mixing Processes with Variable Rates 496
8.8	Two-Compartment Mixing and Linear Systems 502
	Chapter 8 Summary 511
	Chapter 8 Summary Exercises 512

Chapter 9

Probability   514

	Continuously Distributed Random Variables 514
9.1	Improper Integrals 515
9.2	The Probability Distribution of a Random Variable 523
9.3	The Probability Density Function of a Continuous Random Variable 533
9.4	The Expectation and Variance of a Random Variable 541
9.5	The Normal Distribution 551
	Chapter 9 Summary 560
	Chapter 9 Summary Exercises 561

Chapter 10

Sequences and Series   563

	Discrete Change 563
10.1	Sequences 564
10.2	First-Order Difference Equations 570
10.3	Infinite Series 582

	10.4	Power Series 593
		Chapter 10 Summary 600
		Chapter 10 Summary Exercises 602

Chapter 11 Trigonometric Functions 604

Repetitive Behavior: Periodic Systems 604

11.1	The Radian Measure of an Angle 605
11.2	The Basic Trigonometric Functions 610
11.3	Differentiation and Integration of Trigonometric Functions 622
11.4	Applications of Trigonometric Functions 629

Chapter 11 Summary 638

Chapter 11 Summary Exercises 639

Appendix Further Development 641

A.1	Continuity and Differentiation 641
A.2	Functions with Equal Derivatives and L'Hôpital's Rule 644
A.3	Logarithmic Scaling and Best-Fit Polynomial Curves 650
A.4	Hubbert's Curve: A Model for Resource Depletion 656
A.5	The Newton-Raphson Algorithm 660

Answers 665

Index 699

Preface

> Every mathematician has a particular way of thinking about mathematics, but rarely makes it explicit. Yet such perspectives, in any particular field, can be of great value to non-experts who must apply the results . . . or want to use them.
>
> Morris W. Hirsch

Applying Calculus in the Economic and Life Sciences is a class-tested, two-semester text addressed to students in business, economics, the life and social sciences, and other applied fields. Too often these students enter an applied calculus course poorly prepared, and what they learn is merely mechanical manipulation—without the understanding necessary to apply calculus to their own fields.

My goal is for students to complete the course in applied calculus with a "reading knowledge" of the subject, with the ability to comprehend work in their own field that uses calculus, and with a good sense for when and how calculus is an appropriate tool. The text is not mathematically rigorous, but it strives to be mathematically correct as it builds concepts devoted to meaningful applications—applications that give information, via mathematics, about systems outside mathematics.

While the full range of drill for the derivative and integral and all other topics is included, the book concentrates on the following essential goals:
- Making explicit the reasoning process in applications so students can build problem-solving skills.
- Helping students learn to move from descriptive language to mathematical formulation *and back again*—for I believe that students' ability to understand a solution is directly related to their ability to explain in English what has been done.
- Making notation and mathematical form an object as real as the microbe or the balance sheet students already know.
- Finding meaning and representation in the limit, derivative, and integral.
- Weaving the topics of calculus into a course of whole cloth.

Organization

Initially I try to achieve these goals through naive, everyday examples from students' experience or imagination, seeking meaning rather than depth or rigor and avoiding unappreciated conceptual extremes. These examples help students refresh algebra skills and, more importantly, establish that the tools of calculus have external (and internal) meaning.

In the early chapters I initiate a handy paradigm of the calculus, drawn from the language of dynamical systems, which students immediately understand and which helps them see the course whole: If we could make a motion picture of an evolving system, whether the system is in business or economics or the natural world, we would obtain a filmstrip recording its changing states over time. A function f is our filmstrip. Its range is the collection of images recorded on the filmstrip. Each function value represents a single state—a single photo—of the evolving system. Its derivative displays the film in motion. And $f'(t)$ is Newton and Leibniz's incredibly useful and precise measure of its change-of-state at the moment t. This metaphor is reinforced through selected optional exercises and is required only in the solution of related-rate problems, where it has proved to be very effective.

The concepts of the derivative and the integral are initially developed solely in the context of familiar algebraic functions by Chapter 4. They are then applied in Chapter 5 to the study and use of the comparatively difficult exponential and logarithmic functions. The overall content is otherwise fairly standard, and, with the exception that integration by parts is needed in Chapters 8 and 9, the remaining chapters may be covered in any order (as the flowchart shows).

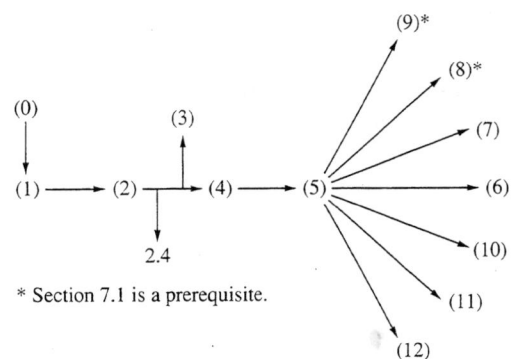

* Section 7.1 is a prerequisite.

Presentation

The presentation was often guided by memories of my own student days, when of course I could understand every word of a poem but perhaps not comprehend the poem and often needed an unpoetic prose version before I could find the poet's meaning. Thinking mathematically is a special form of cognition that students do not necessarily share with us, and leading students through every step does not ensure that they understand the solution. A prose version, an analogy, a holistic impression—however unmathematical—can help build student enthusiasm for the subject and help them see the algebraic details of solution as small steps in a meaningful whole. Indeed, like other instructors, I have found that when students work in a meaningful context, they often succeed at the very same algebra they might miss otherwise. More generally, I believe that this incredibly versatile and powerful subject does and will continue to offer most when presented in a conceptual framework for application, rather than as a list of rote procedures to be learned.

Features

Math modeling A back-and-forth interpretation between descriptive language and mathematical expression is used consistently to help students learn how to construct meaningful mathematical representations of external structures and, conversely, to draw external conclusions out of mathematical expression. This style directly encourages problem-solving skills.

Chain rule before the product and quotient rules This frequently used and most important rule of differentiation is more concise and lends itself more readily to applied interpretation than either the product rule or the quotient rule; presenting it first gives students an immediate sense of learning something new and useful.

Early coverage and continued application of the definite integral The Fundamental Theorem, substitution, and the basic nature of application of both the definite and indefinite integral are first studied and practiced using familiar algebraic functions and only then brought to bear upon the calculus of the log and exponential functions. Continued application of the definite integral throughout the text is tied to a single, clearly delineated theme I call "AP: The Fundamental Principle of Application of the Definite Integral."

The exponential function as the invariant of calculus This opening analogy with the number 1 of arithmetic emphasizes the importance and basic simplicity of the exponential function in calculus and helps students immediately link conceptual insight with important applications.

Compartment-mixing models Students tend to be most enthusiastic about the differential equation and integral applications of Chapters 7 and 8, where a unifying compartment-mixing model is used for all of these applications. Students quickly see how to apply this model in their own fields. The practical nature of the integral models in Chapter 7 is also made plain by using these models with the trapezoidal rule when only incomplete information (data) is available.

Integration by parts Using Reimann-Stieltjes notation makes integration by parts a simpler symbol-driven procedure in which students need only to choose one part.

Well-correlated exercises and examples To address the frequent student complaint that "the problems are not like the examples," worked examples of medium-level difficulty are consistently employed.

Brief algebra and geometry background reviews These are located immediately before selected problem sets as they are specifically needed rather than being grouped in an isolated appendix.

BASIC computer programs These somewhat interactive programs are available on disk and keyed to particular exercises. At the discretion of the instructor, they can be used to enrich selected topics, and they allow students to solve more interesting applications—these programs take on the burden of calculation without eliminating the need for students to understand the mathematical and applied meaning of the program output.

Further development A brief appendix of more substantial theory and application is included for instructors who wish to enrich the course and take students farther into the background and utility of certain topics without intruding on section-by-section development. Included are the mean value theorem and L'Hôpital's rule, Newton-Raphson root approximation, continuity and differentiation, resource depletion, and nonlinear least squares techniques.

General Proof of what students regard as obvious is avoided; instead, formulas are interpreted for their "reasonableness." Essential ideas are often previewed two or three times, and in each exercise set, drill and rote practice problems precede the core exercises. Text formatting is used to highlight the shortest possible list of core theorems, rules, and formulas that each section, and ultimately the whole course, turns on. Key equations are set in boldface type. In cases where *form* rather than content is what students should recognize, the word itself is emphasized. Finally, every effort is made to provide readable, complete coverage of each topic to free more classroom time for discussion and problem solving, where the most rapid learning takes place.

Supplements

The following supplemental material is available to instructors without charge:

- A supplemental chapter covering Matrices and Linear Programming
- An Instructor's Manual providing worked-out solutions to even-numbered exercises
- A Student's Solutions Manual (by Kevin Evens, Columbia College) providing worked-out solutions to odd-numbered exercises
- A Test Item Booklet providing sample test, quiz, or worksheet questions with corresponding solutions
- A computerized testing system for the IBM PC and compatible computers
- A computer applications disk for the IBM PC and compatibles and for computers in the Apple II series

Acknowledgments

An undertaking of this scope could not succeed without the effort and care of many individuals. I would particularly like to thank Paul Burcham, who did not live to see the book's completion, for his advice and encouragement when the book was in its embryonic stage, and Barbara Maring, for her absolute commitment to the worth of this project. I also want to thank the students who have taken the course from me over many years using various other texts. Their term projects and special assignments have furnished the raw material for many of the better applications outside mathematics, where my expertise is quite limited. Kevin Evens (Columbia College) has patiently worked all the exercises and reworked many, and he has class-tested the text as well.

My appreciation also goes to my editors at Wadsworth—Jim Harrison, Kevin Howat, and Barbara Holland—who have patiently and with much understanding shepherded the text through its various passages. In the book's production I have been delighted to find the end of my efforts in the experienced hands of the "pros" at Wadsworth, especially Sandra Craig.

Finally, I want to express my thanks for the corrections and contributions of the reviewers and problem checkers. The reviewers were Bob C. Denton, Orange Coast Community College; John A. Beachy, Northern Illinois University; Gary L. Ebert, University of Delaware; Don Edmondson, University of Texas at Austin; Joe S. Evans, Middle Tennessee State University; Albert Fadell, State University of New York at Buffalo; Howard Frisinger, Colorado State University; Warren B. Gordon, City University of New York, Bernard Baruch College; John T. Gresser, Bowling Green State University; Rutger Hangelbroek, Western Illinois University; John Haverhals, Bradley University; Stephen Kuhn, University of Tennessee at Chattanooga; Peter Morris, Pennsylvania State University; Barbara Osofsky, Rutgers University; and Wayne Powell, Oklahoma State University. The problem checkers were Stephen Kuhn, University of Tennessee at Chattanooga; Jerry Metzger, University of North Dakota; Karl Seydel, a student at Standford University; Henry Smith, University of New Orleans; Clive Taylor, Mission College; and Ellen T. Wood, Stephen F. Austin State University.

Dennis Sentilles
Professor of Mathematics

List of Applications

Business and Economics

Accumulated savings, 309, 417, 418, 431
Advertising allocation, 369
Advertising and customer retention, 434
Advertising and declining sales, 440
Affluence and income, 655
Allometric relationship, 294
Annuity, 486
Assets and stock values, 444, 454
Average costs, 45
Average production, 255, 385
Breakdowns, 369
Break-even analysis, 47
Cash flow, 98, 105, 638
Cash flow accumulation, 217
Cobb-Douglas production functions, 338, 340, 364
Comparative rate of sales, 102
Compound interest, 275
Consumption-indifference curve, 194
Continuous compounding, 306
Cost function, 27
Cost of advertising, 241
Cost of returns, 241
Crop losses, 312, 318
Crop yields, 562
Currency exchange, 486, 508
Customer retention, 519, 522
Customer surge, 281, 284
Defective product losses, 413, 414, 431, 434, 638
Demand function, 45
Demand rates, 125
Doubling time and half-life, 309
Efficient allocation, 348, 350, 368, 370
Exponential decay, 307
Exports/imports, 358, 359
Flow rate through a compartment, 411
Growth comparisons, 54
Growth of investments, 281, 284
Growth that "feeds on itself," 276
Inflation, 12, 308, 317, 321, 424, 501
Interest rates, 42
Investment, 423
Investment growth, 464, 475, 485, 512, 573, 580
Investment returns, 133, 137
Investment strategy, 454
Licensing polluters, 252
Loan cost, 317
Loans, 477, 486, 582
Long-run production, 522
Manufacturing time, 550
Marginal cost, 96
Marginal cost, marginal revenue, 167
Marginal production/optimization, 366, 369
Marginal profit, 96
Marginal yields, 235, 240
Market models, 58
Market saturation, 66
Market share, 465, 471
Maximum efficiency, 346, 350
Moving average, 225
Multiplier effect, 585, 592
Oil production, 358
Optimal design, 138, 160, 170, 176, 177
Optimal inventory, 162, 171, 176, 177
Optimal revenue, 166, 171, 172, 173, 176, 177
Optimum allocation of resources, 150, 171
Optimum geometric states, 158, 169
Optimum growth yield, 164, 172, 176, 177
Peak customer rate, 657
Peak harvest, 659
Present value, 313, 318
Present value analysis, 422
Price equilibrium, 37
Producer, consumer, and polluter surplus, 256
Production demand on resources, 550
Production efficiency, 350, 455
Product reliability, 546, 556, 560
Product satisfaction, 389
Rate competition, 65
Rate of sales per advertising, 96
Reliability, 548, 550
Rents, 423
Resource allocation, 478, 486, 495, 501, 602
Revenue/cost, 340
Revenue functions, 46
Revenue rates, 126
Sales accumulation, 257
Sales growth, 318, 454
Sales projections, 92, 104
Supply and demand, 37-38
Surge-dissipation curve, 280
Synergistic sales, 185
Tangent projection, 94
Value, 398
Worker output, 414

Environmental Sciences

Accumulated fish kill, 282, 284
Accumulated rainfall, 409, 424

Accumulation in the environment, 241, 253, 257
Algae spread, 5
Antibiotic food additives, 444
Doubling time and half-life, 309
Energy costs, 30
Energy efficiency, 159
Environmental impact, 29, 501, 602
Environmental management, 414
Environmental spreading, 235
Expanding areas, 100
Exponential decay, 307
Fertilizer pollution, 430
Flow rate through a compartment, 411
Greenhouse effect, 487
Growth that "feeds on itself," 276
Heat loss, 257, 337
Licensing polluters, 252
Nitrogen exchange, 505
Optimal design, 138, 160, 170, 176, 177
Optimum allocation of resources, 150, 171
Optimum geometric states, 158, 169
Ozone loss, 311
Pesticide accumulation, 215
Pollutant dispersal and accumulation, 245, 664
Producer, consumer, and polluter surplus, 256
Radioactive decay, 137, 315, 319, 321
Sedimentation, 480
Surge-dissipation curve, 280
Toxicity and the environment, 441, 486, 495, 498, 513
Water resources, 414

Life Sciences

Aging, 359
Alcohol consumption, 444, 501
Bacterial growth, 318
Cardiovascular capacity/rate, 374, 411, 434
Cell division, 13
Cholesterol accumulation, 408
Disease control, 431
DNA hybridization, 472
Drug accumulation, 423
Drug dosage, 296
Drug metabolization, 317, 488, 499, 509, 512, 577, 581
Epidemics, 52, 450
Food chains, 118
Genetic weakness, 487
Glucose metabolization, 454, 513
Gompertz growth functions, 442
Heat stress, 359
Hormonal secretion, 509
Hormonal trigger, 284
Insulin levels, 424
Insulin treatment, 488
Ion exchange, 482, 510
Memory, 487
Mytosis, 264
Natural growth, 465
Passive diffusion, 473, 507
Perception of a stimulus, 290
Population accumulation, 216–217
Predator-prey interaction, 195
Resource harvesting, 257
Sound, 297
Tomography, 408
Tumor growth, 487
Wind chill, 359

Social Sciences

Arrival times, 550
Birth and death processes, 423, 424
Cultural transformation, 424
Discovery/consumption, 657
Earnings satisfaction, 296
Epidemics, 580
Learning curve, 29, 454, 462
Loan cost, 317
Malthusian growth, 319
Meeting times, 532, 556, 560
Military costs, 319
Population accumulation, 419, 431, 432, 433, 434, 638
Population distribution, 402, 403, 408
Population dynamics, 465, 487, 509, 575, 580
Resource depletion, 318
Spread of a rumor, 241, 454, 472
Time restraints/allocation, 12, 48, 56
Television viewing, 358
Wealth and happiness, 296
Weber-Fechner law, 472
Workplace tedium, 6, 21, 126

General

Air exchange, 481
Arrival times, 550
Autocatalytic reactions, 454
Average rate of change, 63, 95
Average value, 247
Back taxes, 423
Braking efficiency, 378
Chain processes, 42, 47
Chemical reaction, 472
Closeness, 73
Contact lens cleaning, 487
Cooling, 294
Density, 401, 403
Design choices, 4, 16, 21
Differential approximation and estimation, 199
Distribution, 401, 403
Efficient allocation, 348, 350, 368, 370
Entropy, 303
Error estimation, 377
Exponential change, 305
Frequency data, 540
Growth that "feeds on itself," 276
Heating and cooling, 446, 453
Hitting a tennis ball, 192, 202
H-test for Lagrangian optimization, 363
Imperfect models, 7
Inherent error, 197, 202, 373, 377
Lagrangian control, 376
Load-bearing strength, 388
Logistic growth and decay, 449, 461
Measuring change, 65
Optimum numerical states, 157, 169
Order versus complexity, 340
Peak loads, 414
Periodic models, 621, 622
Physical distribution, 404, 406, 431
Predicting world records, 656
Prize winnings, 424, 434
Random numbers, 532
Rate in less rate out analysis, 475–479
Rates for chain processes, 114, 118
Rate versus state, 94
Related rates, 185–194
Relative motion, 111
Relativistic mass, 126
Resource allocation, 478, 486, 495, 501, 602
Solar efficiency, 370
Speed yields distance, 237
State, 22, 64
Surge-dissipation curve, 280
System reliability, 378
Tennis, 103
Two-dimensional accumulation, 389
Vibrating spring, 633
Volume, 246, 256

What Is Applied Calculus?

Calculus is the mathematics of change. It is both a method for measuring change and a language for describing change.

Little can be learned in a fixed, frozen situation: It is outside the web of life. Think of your favorite activity, whether it is tennis, skiing, music, or some other interest. How do you come to know it better? Besides much practice, you try changes in your approach and notice the outcome. Similarly, a biologist learns more about an organism by changing its environment and observing the outcome. A business firm changes its advertising and measures the result through sales. Applied calculus gives mathematical form to such a common experience.

Much of your experience in mathematics has been learning the mechanics of abstract operations. If you approach calculus that way—as just a series of problems that you have to learn how to do—you will miss much of what this subject is: a language and concept and method for exact description and then precise computation.

The two main processes of calculus are differentiation and integration. We first encounter these in Chapters 2 and 4 and then use them throughout the book. **Differentiation** computes the rate at which change is occurring, given information about how much is present. **Integration** computes how much is present, given the rate at which change is occurring. Differentiation and integration are the reverse of each other. Together, they describe and measure change in a variety of applications.

Both differentiation and integration rest on one idea: the concept of **limit value.** This unique idea separates calculus from all previous mathematics you have learned. It is the mathematical basis for a theme that unites the calculus: Compute something complex, represented by the "limit value," *by approximation* by something simpler. Ironically, such a powerful approach to applying mathematics rests on an idea that is found only in pure mathematics. There seems to be no ready analogue of the concept of limit value in any other subject. Since the limit concept is both subtle and technical, we must wait until Chapter 1 before giving it any useful meaning.

Remember the day you first got the keys to the family car? Suddenly you could do a lot more and a lot more easily. The variety of experience you could sample increased exponentially. In the world of mathematics and its uses calculus is like that. With it you can and will understand a lot more and a lot more easily.

Chapter 0

Functions

Real Systems as Mathematical Functions

In our application of calculus to the economic and life sciences, we will consider a wide variety of uses. We can unify our approach to these applications by viewing each as the study of some relationship between particular parts of a system, such as a production system, an ecosystem, one of the human body's physiological systems, or the changing state of some part of a system over time. We apply calculus in each case by creating a mathematical model of the relationship we wish to study. The basic model of calculus is the mathematical concept of a function, which we use to represent the relationship between particular variables within a system where one variable depends on the others. A function is a kind of motion picture film of a changing system expressed in symbols that allows us to see any particular state of the system and, when coupled with calculus, allows us to see the system in motion.

0.1 Mathematics and Systems Modeling

This prologue to calculus uses mathematics you have learned in earlier courses to build mathematical models of simple systems.

In earlier algebra and geometry courses, you have seen that mathematics may be studied and learned in abstraction—apart from applications to the external world. While we can and sometimes will study calculus in the same way, calculus is an immensely powerful applied tool precisely because its concepts closely reflect processes at work in the external world. Thus we will study calculus as mathematics immersed in its applications, with steady insight given by one to the other. This overview relates your past experience in algebra and geometry to this new subject. It will help you begin to learn how to frame real-world problems so that you can solve them with calculus. The review and exercises at the end of this section can refresh your memory and sharpen your skills.

We are able to apply mathematics to the external world whenever we can *represent* a real-world problem in symbolic form and then manipulate these symbols by general rules. Such a symbolic representation is called a **mathematical model.** As we will see in this chapter, the basic model used in calculus is called a function. A function model is built upon the familiar idea of using a letter to represent a **variable,** any element of the problem that is initially undetermined. A model built upon variables then becomes a statement of the problem in symbolic form that may be manipulated mathematically. Each of the examples in this section builds a model of a structure found in the external world.

We can think of a mathematical model of the external world as describing a **system,** a collection of interrelated elements that we naturally regard as a whole. For example, a commercial business that derives profit from human and material resources is a system. The human body is a system. Since the world is complex, one system may lie within another. A business exists within the larger economic system. A corn crop growing in nutritious soil is a system within the larger natural ecosystem.

As we will learn, calculus is the mathematics of change. Ultimately we want to derive a model of the change in a system from a model of what the system is like in one fixed but general instance. Figure 0.1 suggests one of the dynamics of changing prices within our economic system. As the Federal Reserve Bank raises interest rates, fewer houses are built, and this in turn may cause the price of lumber to fall. In Chapter 1, we will introduce the derivative of a function and use it to represent the relationship between the condition, or state, a system is in and how that state is changing.

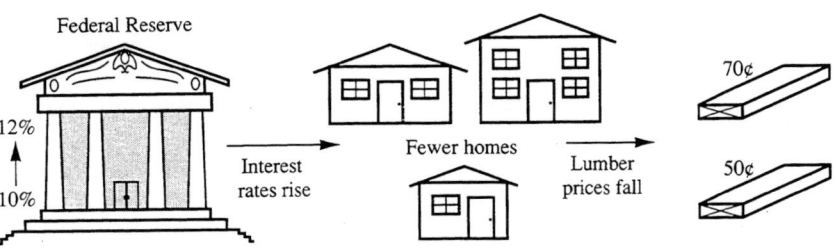

Figure 0.1

Although a system may be complex, we can often learn something important about it by focusing narrowly, say, on only two variables in the system, where one *depends on* the other. A manufacturer can focus on how the number of machines used affects profit and can ignore labor costs. A farmer can focus on how the size of a corn crop depends on the amount of nitrogen in the soil, ignoring potash and phosphate. A doctor can measure a patient's pulse 10 minutes after giving a sedative and ignore body temperature.

This last example also suggests that a system can be thought of as an observable activity coupled with the abstract concept of *time*. A manufacturer is concerned with the return on investment in equipment before the next loan payment is due. Agribusiness is interested in day-to-day crop yields during harvest. A doctor, having given a sedative, needs to know what quantity of it remains in the patient's body t hours later.

We are hardly ready to study such intricate systems. Instead we begin with instances we can readily imagine and see how a mathematical model can be built on the introduction of a letter to represent a key variable in the system. Our first example uses both algebraic and geometric reasoning.

EXAMPLE 1

A manufacturer wishes to make an open-topped box from a 10-by-10-in. piece of cardboard by cutting a square from each corner and bending up the resulting sides to form a box, as illustrated in Figure 0.2.

a. Describe a mathematical model of how the size of the cut at each corner determines the volume of the resulting box.

b. Determine the volume of the boxes that result from a cut of 2 in. and from a cut of 3 in.

Solution

a. There are many ways to build such a box. It could be tall and thin or short and flat or something more reasonable.

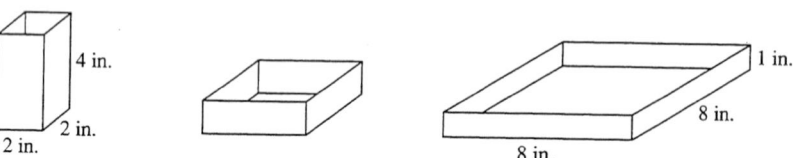

Aside from appearance, the volume of the resulting box can vary considerably. Our tall, thin box has a volume of $2 \cdot 2 \cdot 4 = 16$ in.^{3}; our short, flat box has a volume of $8 \cdot 8 \cdot 1 = 64$ in.^{3} and thus could hold considerably more.

The manufacturer needs to think in terms of the correspondence between the possible size of the cut-out corners and the resulting volume of the box. Since each of the four cut-out corners is to be a square (and the length of each side of a square is the same), let us use the letter x to denote the length of the side of an arbitrarily sized cut-out. (Note that x can represent any number between 0 and 5.) We may then imagine the construction process as represented by Figure 0.2.

Figure 0.2

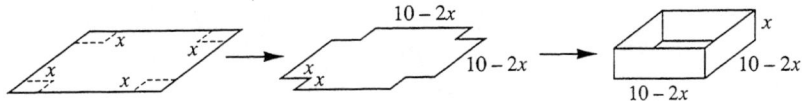

Since the volume of a box is the product of its length, width, and height, the resulting volume for a cut of length x is given by

$$(10 - 2x)(10 - 2x)(x) = (100 - 40x + 4x^2)x$$
$$= 100x - 40x^2 + 4x^3$$

The correspondence

$$x \rightarrow 4x^3 - 40x^2 + 100x$$

is a mathematical model for constructing the box and knowing the resulting volume, where we read the "$\rightarrow$" as "corresponds to."

b. Therefore a cut of $x = 2$ in. results in a box of volume

$$4 \cdot 2^3 - 40 \cdot 2^2 + 100 \cdot 2 = 32 - 160 + 200 = 72 \text{ in.}^3$$

and a cut of $x = 3$ in. results in a volume of

$$4 \cdot 3^3 - 40 \cdot 3^2 + 100 \cdot 3 = 108 - 360 + 300 = 48 \text{ in.}^3 \blacksquare$$

The mathematical model in Example 1 illustrates change as an inherent part of a dependency relationship. The model introduces a letter to represent the **independent variable**, that variable subject to choice. In Example 1, the independent variable is the depth of the cut in each corner. The volume of the resulting box is the **dependent variable** because it depends on the length of the cut-out corner. The formula in this example offers the manufacturer a model of the whole range of variability of volume dependent upon length. In Chapter 2, we use this form as an entity in itself to find the particular length x that yields a box of maximum volume, thus making the most efficient use of material.

Our next example looks at a different sort of change, **exponential change,** where a quantity increases or decreases by an amount that is proportional to the amount already present. Such change is prevalent in both the life and economic sciences. Example 2 considers the growing business of fish farming for profit, where nutrients introduced into a pond can lead to the growth of algae on the pond's surface. These simple organisms reproduce exponentially whenever food and space are available. The algae in turn inhibit the transmission of sunlight to the subsurface organisms needed for a healthy ecological balance in the pond.

EXAMPLE 2

A fish farmer notices that 2 sq yd of algae have appeared on one of the farm's 512-sq-yd ponds by the end of the first week of June. By the end of the second week algae cover 4 sq yd, and by the end of the third week 8.

a. Formulate the apparent correspondence between the number of weeks since June 1 and the amount of surface area covered by algae.

b. How many square yards will be covered by the end of the fifth week if this formula is a good model for algae growth?

c. If nothing is done to eliminate the algae, in how many weeks will one-fourth of the pond be covered, according to this model?

Solution

a. Let w denote the number of weeks since June 1, represented by $w = 1$. At that time there are $2 = 2^1$ sq yd of algae growth. After 2 weeks there are $4 = 2^2$ sq yd of growth, and after 3 weeks there are $8 = 2(4) = 2(2^2) = 2^3$ sq yd of growth. From the evidence, we conclude that at the end of week w there will be 2^w sq yd of algae on the pond.

b. After $w = 5$ weeks, there will be $2^5 = 32$ sq yd of algae.

c. Since the pond covers 512 sq yd, one-fourth of the pond equals $\frac{512}{4} = 128$ sq yd. This much of the pond will be covered by algae when $2^w = 128$. Arithmetic with powers of 2 shows you that w must be 7. Thus by approximately the end of the third week of July one-fourth of the pond will be covered.

The correspondence $w \to 2^w$ relates the time variable w to the number of square yards of algae present at time "w," thus describing the state of this system at that time. Notice that the variable w appears as an exponent in the model, giving this model the name exponential growth. Figure 0.3 provides a natural picture of this mathematical model and shows the weekly doubling of the algae cover.

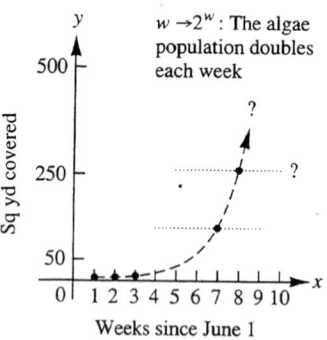

Figure 0.3

We will see later that calculus is essential to the study of systems exhibiting exponential change over time and that *the* exponential function, written e^x, which we will study in Chapter 5, gives us a universal model for such systems.

Our next example models time demands in the familiar scientific process of testing against a pair of individuals. This model also uses the concept of **proportional reasoning**; recall that a quantity A is directly proportional to a quantity B if $A = kB$ for some constant k.

EXAMPLE 3

The fruit fly *Drosophila melanogaster* has long been a favorite subject of biological experiments because these insects rapidly reproduce successive generations, and external characteristics observable with a hand lens often closely

6   Chapter 0   Functions

reflect internal genetic structure. Suppose a student is testing the reaction of *Drosophila* to a specific chemical thought to be especially toxic to individual *Drosophila* that are *recessive* for a certain inherited trait. The student is given a population of 100 *Drosophila*. Suppose this population satisfies Mendel's laws of inheritance exactly, so that one-fourth of the population exhibits the recessive trait while the remaining three-fourths are either hybrid or dominant. Her task is to extract a *pair* of individuals from the population, where only one individual in the pair is recessive, and then to expose the pair to the chemical and observe the consequences. For statistical validity the experiment must be repeated 20 times. For each repetition the student must select individual *Drosophila* and examine these for the recessive or nonrecessive trait until an appropriate pair is found.

a. How many *Drosophila* in the initial population are recessive and how many are not?

b. What is the initial ratio of nonrecessive to recessive individuals?

c. Describe the correspondence between the number of experiments the student has run and the ratio of nonrecessive to recessive individuals remaining in the population. What is this ratio after the experiment has been repeated five times?

d. To repeat the experiment the student must select another pair. Assume that the amount of time it takes to find an appropriate pair is *directly proportional* to the ratio of nonrecessive to recessive individuals remaining. She observes that it takes 10 minutes to isolate the 6th pair. Under this assumption how long might she expect to take to isolate the 16th pair? The 20th pair?

Solution

a. The student can expect to find $\frac{1}{4}(100) = 25$ recessive individuals, and therefore 75 nonrecessives, in the population.

b. The initial ratio is $\frac{75}{25} = 3$; recessive individuals are initially outnumbered 3-to-1 in the population.

c. Suppose that the experiment has been completed "n" times. Then $25 - n$ recessive individuals remain along with $75 - n$ nonrecessives. The correspondence

$$n \to \left(\frac{75 - n}{25 - n}\right)$$

represents the ratio of nonrecessives to recessives after n experiments.

After five repetitions this ratio is $(75 - 5)/(25 - 5) = 3.5$.

d. Let us denote by T the amount of time it takes to isolate a pair after n experiments.

To say that the time needed to isolate a pair is directly proportional to the ratio of recessives to nonrecessives is to say that the isolation time T equals a multiple, say, k, of the existing ratio. We can express this English statement with an algebraic statement:

$$T = k\left(\frac{75 - n}{25 - n}\right)$$

where k is the constant of proportionality. Because it takes 10 minutes to isolate the 6th pair (after 5 pairs have been chosen), we know that $T = 10$ when $n = 5$. Therefore

$$10 = k\left(\frac{75-5}{25-5}\right) = k\left(\frac{70}{20}\right)$$

and hence $k = \left(\frac{2}{7}\right)(10) = \frac{20}{7}$.

If we assume that the time it takes to isolate another pair remains directly proportional to the existing ratio, even as that ratio changes, we know that after the 15th experiment it will take

$$T = \left(\frac{20}{7}\right)\left(\frac{75-15}{25-15}\right) = \left(\frac{20}{7}\right)\left(\frac{60}{10}\right) \simeq 17.14 \text{ min}$$

to isolate a 16th pair.

Similarly it will take

$$T = \left(\frac{20}{7}\right)\left(\frac{75-19}{25-19}\right) \simeq 26.67 \text{ min}$$

to isolate a 20th pair.

Figure 0.4 illustrates the relationships

$$n \to \frac{75-n}{25-n} \quad \text{and} \quad T = \frac{20}{7}\left(\frac{75-n}{25-n}\right)$$

As the number n that has been selected approaches 25 and few recessive *Drosophila* remain, the isolation time T grows larger and larger since it is directly proportional to the ratio of nonrecessives to recessives [Figure 0.4(b)]. Notice that the model reasonably reflects the tedium of such work, however interesting the experiment itself might be; such a model can be used to study workplace psychology.

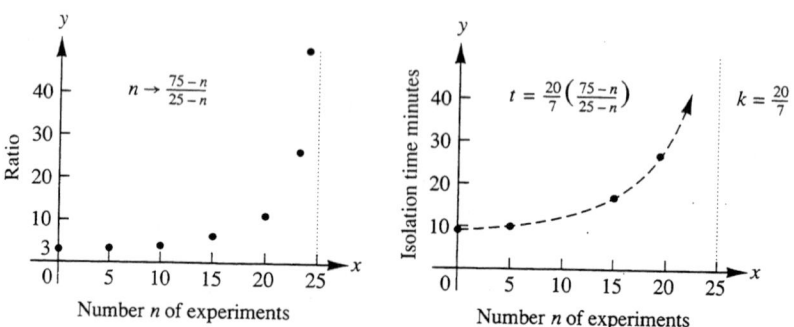

Figure 0.4

Each of the solutions in the preceding examples gives us a mathematical model of the process at hand. A mathematical model is the mathematician's simulation device. A simulator attempts to replicate the real world in a controlled way so that it can be experienced in simple form. Such simulation is highly useful, but it also has its limitations. If an aspect of reality is too subtle, the simulator may ignore it entirely. For example, an automobile driving simulator does not attempt to simulate the difference between cornering in a station wagon and cornering in a sports car. A mathematical model that simulates a real system can also be subject to limitations, as the next example shows.

EXAMPLE 4

A water skier is being pulled along on water skis at a constant horizontal speed of 50 ft/sec and then up an inclined ramp with a slope of $\frac{1}{5}$—a vertical rise of 1 ft for each 5 ft of horizontal run—as in Figure 0.5.

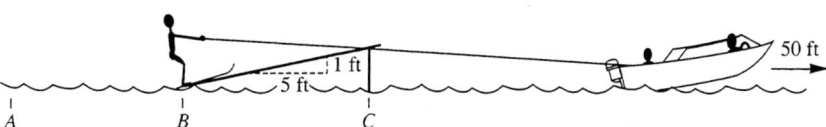

Figure 0.5

Describe how the skier's *vertical* speed depends on his or her position between points A and C, ignoring the angle of the tow rope, the roughness of the water, and the skier's vertical speed at point B.

Solution At any position x between points A and B, the skier is not moving vertically. Thus the skier's vertical speed is 0.

At a position x between B and C, the skier is moving horizontally at a speed of 50 ft/sec along a ramp with a rise/run ratio of $\frac{1}{5}$. This causes the skier to move vertically at a speed of $\frac{1}{5}(50) = 10$ ft/sec.

The relationship between position x and vertical speed is given by

$$x \to \begin{cases} 0 & \text{if } x \text{ is between } A \text{ and } B \\ 10 & \text{if } x \text{ is between } B \text{ and } C \end{cases}$$

Under our assumptions we do not specify the skier's speed *at* point B. ∎

This example shows that a mathematical model need not be restricted to a single algebraic formula as in Examples 1, 2, and 3. It also shows that such a model can be built on "idealizing" assumptions that necessarily place limitations on the model. In this model we are uncertain of the vertical speed *at* point B and so model the speed at points other than B. This model is used in Chapter 1 to clarify the key idea in calculus of a "limit value."

In each of our examples, the formula that we derive serves as a mathematical model of a whole system and also makes us aware of the dynamic, changing nature of a system. It is very useful to distinguish between the *system* itself, which is represented by the whole model, and one condition, or *state*, of the system, represented by one value produced by the model. In Example 1, many boxes (resulting states) are possible, and our formula (the model) has many distinct values. On the other hand, in Example 4, the skier is in only two states—on the water or on the ramp—and these are reflected by the two values allowed by the model.

We can think of the distinction between state and system in another way. If we could make a motion picture of a changing system, we would have a filmstrip representing the system as a whole. Each frame of the filmstrip is a single photo representing one state of the system. A mathematical model corresponds to the whole filmstrip; one value produced by the model corresponds to a single frame. We will soon see that the central tool of the calculus, the derivative of a function, will allow us to show the system in motion—even when no photographic images of the system can possibly be made!

The examples in this section are a first step in learning to think of the external world in terms of mathematical models. You may find it useful to keep our filmstrip analogy in mind as you learn to solve problems that begin with information about both the state of a system and how that state is changing.

Background Review 0.1

The area of a rectangle is

$$l \times w$$

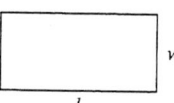

and the volume of a box is

$$(\text{cross-sectional area}) \times h = l \times w \times h$$

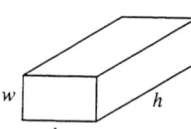

In a right triangle, we always have

$$a^2 + b^2 = c^2$$

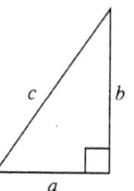

The area A and circumference C of a circle of radius r are

$$A = \pi r^2 \quad \text{and} \quad C = 2\pi r$$

respectively.

The slope of a straight line is the numerical ratio of its rise to its run:

$$S = \frac{\text{rise}}{\text{run}}$$

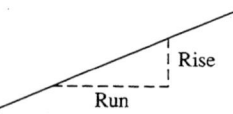

---

10 Chapter 0 Functions

Exercises 0.1

1. Repeat Example 1 for a piece of cardboard measuring 10 in. wide by 20 in. long: Model the volume of the box resulting from a cut of x in. on each corner. What is the resulting volume when $x = 1, 2, 4,$ or $\sqrt{2}$ in., respectively?

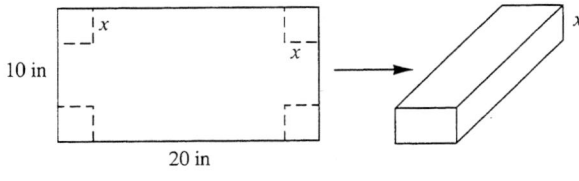

Figure 0.6

2. Two automobiles leave Colby, Kansas, at noon. One automobile is driving west toward Denver at 50 mph, and the other is driving east toward Kansas City at 55 mph (Figure 0.7).

   a. How far apart are they at three o'clock?

   b. What condition is this system in at six o'clock? That is, how far apart are the cars at six o'clock?

   c. Describe the distance between the cars at any time t after departure ($t = 0$ is noon) by a formula using the variable t.

3. Exponential growth can be loosely described as "growth that feeds on itself." For this reason a system undergoing exponential growth can change rapidly. Suppose that the fish farmer in Example 2 decides to wait until the pond is half-covered with algae before treating it. How much time would remain at this point before the pond would be completely covered?

4. Consider Figure 0.8. Describe the length of arc subtended by the angle θ in terms of the number of degrees in the angle. [*Hint:* Think in terms of arc length A in proportion to the whole circle, which has length 2π. Ask yourself: How long is the arc when θ is 90°? When θ is 120°?]

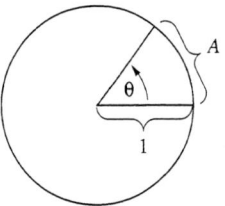

Figure 0.8

5. a. Referring to Example 3, how long might we expect it to take the student to isolate the 24th pair? The 25th pair?

   b. A second student who is asked to repeat the experiments of Example 3 with a second collection of 100 *Drosophila* finds that it takes 14 min to isolate the 6th pair. How much time might the second student expect to take to isolate the 10th, 15th, 20th, and 25th pairs if isolation time is always directly proportional to the remaining ratio?

   c. How do both models mathematically reflect the physical impossibility of isolating a 26th pair?

6. Two cars leave an intersection at the same time. One drives north at 30 mph and the other drives west at 40 mph (Figure 0.9).

   a. Describe the distance between the cars t hr after departure. [*Hint:* Use a fact about right triangles and that distance equals rate times time.]

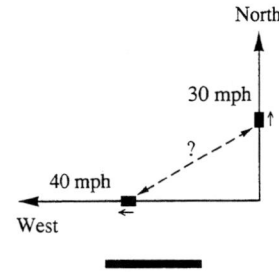

Figure 0.9

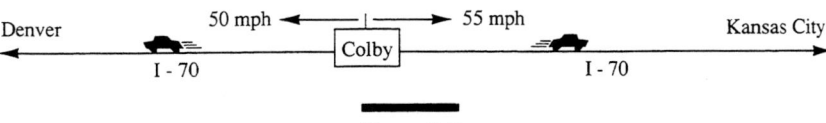

Figure 0.7

b. What is the state of this system at $t = 3$ hr; that is, what is the distance between the two cars?

7. A box initially contains 60 red balls and 40 white balls. Each minute someone removes 2 red balls and 1 white ball. Describe the correspondence between the time t of removal and the proportion of red balls to white balls remaining in the box at time t.

a. What is the initial ratio at time $t = 0$?

b. What is the ratio t min later?

c. What is the condition of this system after 5 min, 15 min, and 25 min, as expressed by this ratio?

[*Caution:* This exercise differs from Example 3 in that here we assume a constant removal time of 1 minute.]

8. Suppose that you are walking along a horizontal sidewalk and then step up onto a platform that is 1 ft high. Describe the correspondence between where you are and your vertical position (height) above the horizontal, omitting the point B from consideration.

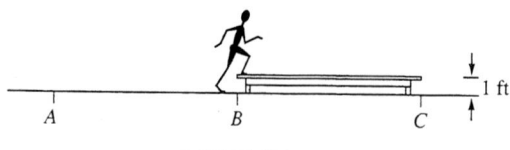

Figure 0.10

9. A ladder 5 m long resting against a wall and on an oil-slicked surface at the bottom suddenly starts to slide away from the wall.

a. Describe the correspondence between how far the foot of the ladder is from the wall and the height of the top of the ladder along the wall. [*Hint:* Introduce two letters to represent the position of the two ends of the ladder. Then think in terms not of movement but of a single photo of this changing system.]

b. What is the vertical state of this system when the bottom of the ladder is 3 ft from the wall? That is, how high is the top of the ladder?

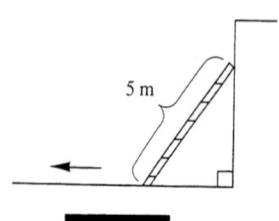

Figure 0.11

10. The Full-Bloom Nursery has set aside a portion of its property, valued at $1,000, for the cultivation of a new flowering shrub that costs $15 to grow to selling height. It sells the shrub for $75.

a. Describe the correspondence between the number of shrubs to be sold and the cost to produce them.

b. Describe the relationship between the number sold and the total profit.

c. How many plants must be sold to break even?

d. How many plants must be sold to generate a $500 profit?

11. Suppose that you walk along a sidewalk, step up and over a curb, and then walk up a hillside with slope $\frac{1}{5}$. Suppose that while doing all this you manage to maintain a constant horizontal speed of 6 ft/sec. With this information you can only guess at your vertical speed as you step over the curb, but it should be different from your vertical speed either before or after the curb. Make such a guess. Describe this "system" by a mathematical model, like Example 4, which models your *vertical speed* as your position changes.

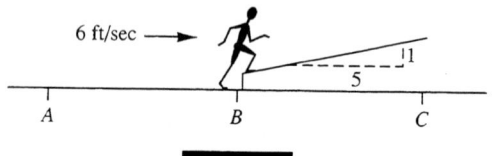

Figure 0.12

Challenge Problems

12. If inflation of the currency is 10% per year, then an item costing $1.00 today will cost $1.10 one year from now. That $1.10 item will cost $1.21 one year later; that is, $1.10 + 0.10(1.10) = (1.10)^2$. Describe this correspondence between time and the future cost of an item costing $1.00 today.

13. At the State Fair the following game can be played. You are given a box with 50 bolts and 30 nuts. You are to search for a nut and a bolt, fasten the two together, put the piece aside, and repeat the process. If your search and assembly time exceeds 20 sec, then the game is over and the prize you win is determined by the number of pieces you have fastened together. If you manage to put 30 pieces together, you win a round-the-world cruise; if you fasten only 10 pieces, you win a rubber duck. The game's operator knows that it takes the average player $5/R^2$ sec to complete the search and assembly, where the ratio of nuts to bolts remaining in the box is R.

How many seconds will the average player take for the first assembly? The second assembly? The fifth asembly? Can the average player expect to win more than a rubber duck? What of the above-average player who takes $4/R^2$ sec? (There are intermediate prizes.)

0.2 Functions and Function Notation

In this section we define a function and learn to use function notation.

Functions play the role in calculus that numbers play in arithmetic and letters play in algebra. A function is the basic object that we work with every day in this course. In applications of calculus, functions represent the whole relationship between two (or more) variables as a single entity.

Definition

A **function** is a relationship f that assigns to each number x in a given set of real numbers A exactly one other number called the *function value of f at x*. We then say that f is a function *defined on the set A*.

The following example illustrates a simple relationship that can be represented by such a real-valued function of a single real variable.

EXAMPLE 1

A virus invades a cell at time $t = 0$. Its DNA uses the cell's metabolism to construct replicas of itself. At the end of one minute the cell contains two viral particles. At the end of the second minute each of these two has produced another viral particle; there are now four. This process continues, and the number of viral particles doubles each minute thereafter. Describe the relationship between the number of minutes since the virus invaded the cell and the population of viral particles within the cell as a function called P, and identify the function value of P at each number in the set on which P is defined.

Solution Let the set A consist of the numbers 0, 1, 2, 3, ... Number each successive minute since the virus first invaded the cell, beginning with $t = 0$. At time $t = 0$ there are $2^0 = 1$ viral particles in the cell. After $t = 1$ minute there are $2^1 = 2$ viral particles. After $t = 2$ minutes there are $2^2 = 4$; after 3, there are $2^3 = 8$. Since the number of viral particles doubles each minute, the desired function P is the relationship $t \to 2^t$, which assigns to each t in the set A the number 2^t of viral particles present in the tth minute. The number 2^t is the function value of P at t. ∎

Notice that we have used the letters t and P rather than x and f in Example 1. Such letters remind us of the topic at hand. We used them here to emphasize that what matters is the relationship itself, not what it is called.

A principal utility of functions lies in *function notation*, or the symbols we use to represent a function value.

Definition   Let f be a function defined on a set A. The **function value** of f at x is denoted by

$$f(x)$$

and read "f of x." The set A is called the *domain* of f; the collection of all function values $f(x)$ is then called the *range* of the function f.

The symbol "$f(x)$" is our name for the number that corresponds to the number x in the relationship f; the function value $f(x)$ is also called the **image** of f at x. Notice from the definition of a function that the function value $f(x)$ must always be a real number.

We will discuss the domain and range of a function in more detail shortly. For the moment, think of the **domain** of f as the set of values of the variable x that can be used in expressing the relationship f. In Example 1, the domain of P is the set of numbers $0, 1, 2, 3, \ldots$.

The **range** of f is the set of real numbers that the relationship f assigns to numbers in its domain. Some numbers in the range of P are 1, 2, 4, and 8 because these numbers are assigned to the numbers 0, 1, 2, and 3 through the relationship $t \to 2^t$.

Let us relate this notation to the solution to Example 1. We can summarize that example and its solution by the abstract statement $P(t) = 2^t$. The symbol $P(t)$ represents the number of viral particles after t minutes, and the equation $P(t) = 2^t$ is a brief way of writing the English sentence "the population of viral particles after t minutes is 2^t."

By using function notation, we can represent a function directly with an algebraic formula.

EXAMPLE 2

Let f be a function whose image at the number x is $f(x) = x^2 - 2x + 3$ for any x in the set A of all real numbers.

a. Find the function values $f(0)$, $f(1)$, $f(2)$, $f(-3)$, $f(\sqrt{5})$, $f(\frac{1}{2})$, $f(\pi)$, $f(a)$, and $f(a + 1)$.

b. Discuss why the function given by $g(z) = z^2 - 2z + 3$ for any real number z is exactly the same function as f.

Solution

a. From the given formula for f, the number

$$3 = 0^2 - 2(0) + 3$$

is the value that the function f assigns to $x = 0$; thus $f(0) = 3$. The remaining function values are

$$f(1) = 1^2 - 2 \cdot 1 + 3 = 2$$
$$f(2) = 2^2 - 2 \cdot 2 + 3 = 3$$

$$f(-3) = (-3)^2 - 2(-3) + 3 = 18$$
$$f(\sqrt{5}) = (\sqrt{5})^2 - 2\sqrt{5} + 3 = 8 - 2\sqrt{5}$$
$$f\left(\frac{1}{2}\right) = \left(\frac{1}{2}\right)^2 - 2\left(\frac{1}{2}\right) + 3 = \frac{9}{4}$$
$$f(\pi) = (\pi)^2 - 2(\pi) + 3 = \pi^2 - 2\pi + 3$$
$$f(a) = a^2 - 2a + 3$$

and
$$f(a + 1) = (a + 1)^2 - 2(a + 1) + 3$$
$$= a^2 + 2a + 1 - 2a - 2 + 3 = a^2 + 2$$

b. The function g is the same function as f (despite its different "clothing"), because it describes the same relationship between numbers as f. For example,
$$g(-3) = (-3)^2 - 2(-3) + 3 = 18 = f(-3) \;\blacksquare$$

We see from this example that the formula defining the function f is a means for finding its function values. Thus f may be thought of as a machine with input x and output $f(x)$; the interior workings of this machine consist of the formula defining f (see Figure 0.13).

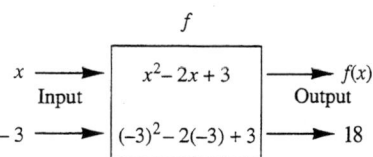

Figure 0.13

Notice also in Example 2 that a function is a dependency relation in a particular direction. The function value $f(x)$ depends on the given value of x. The relationship f is in the direction from x to $f(x)$, as Figure 0.13 illustrates.

The Domain of a Function

If f is a function given by the equation
$$f(x) = \frac{2}{x - 1}$$
there can be no number corresponding to $x = 1$ because the formula $2/(x - 1)$ cannot be evaluated at 1, for we cannot divide by 0. We agree in such a case that the domain of this function is the set A of all numbers except 1.

For a second example, consider the function $g(x) = \sqrt{4 - x^2}$. If we try to sample the values of this expression—say, at $x = 0$, $x = \pm 1$, $x = \pm 2$, and $x = \pm 3$—we obtain a real number only when $-2 \le x \le +2$. Thus the domain of g is the interval of numbers $[-2, 2]$ (see Background Review 0.2 on inequalities, intervals of numbers). Combining these two examples, if h is the function

$$h(x) = \frac{\sqrt{4 - x^2}}{x - 1}$$

the domain of h is all numbers between and including -2 and 2, but excluding 1, to insure that the numerator is a real number and the denominator is not zero.

When we say *all* numbers in these examples, we mean to include not only integers like 1, 2, -2, 0, and so on, but also all fractions and other real numbers such as $\sqrt{2}$ and $\pi/3$. Unless otherwise specified in a particular case, the *domain* of a function defined directly by a single formula *consists of all those real numbers for which the formula yields a real number by ordinary arithmetic.*

In an application this is not necessarily so, and we will use the context of the application to help us determine an appropriate domain for the function, as in our next example.

EXAMPLE 3

John Handy has a 50-ft length of fencing and intends to bend it and then join the ends to form a rectangular enclosure for his garden.
a. Find a formula for the function representing the area of the resulting rectangle as a function of John's choice for the width of his garden.
b. What is the resulting area for a width of 10 ft? 12 ft?

Solution

a. Let w denote the width chosen for the garden. Since the available length of fencing material is 50 ft, w can be any number between 0 and 25 ft (although either of these two extremes will produce a garden of no area!).

We wish to find a formula for, let us say, $A(w)$, the area of the rectangular garden that results from a choice of w ft for its width.

Let s denote the length of the side of the garden. The area A of the garden is then given by $A = ws$. However, this formula depends on two variables and so does not answer the question above.

Let us think further. The garden will have a perimeter equal to 50 ft of fencing. Therefore (see Figure 0.14)

$$w + s + w + s = 50$$

or

$$2w + 2s = 50$$

so that

$$s = \frac{50 - 2w}{2} = 25 - w$$

Consequently, $A = ws = w(25 - w)$, and a formula defining the function A is given by

$$A(w) = w(25 - w) = 25w - w^2$$

with domain $0 \leq w \leq 25$.

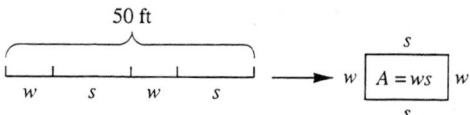

Figure 0.14

b. If $w = 10$ then the area of the garden will be $A(10) = 150$ ft^2. Similarly, $A(12) = 156$ ft^2. ∎

In Chapter 2 we will use the function A to determine the particular width w that yields a garden of *maximum* area.

The Range of a Function

Recall that the range of a function f is the set of all function values $f(x)$ for x in the domain of f.

EXAMPLE 4

Let $f(x) = x^2 + 1$ and suppose that the domain of f is taken to be all numbers between 0 and 2. Describe the range of f.

Solution If x can have any value between 0 and 2, then x^2 can have any value between 0 and 4. Therefore $x^2 + 1$ has any value between 1 and 5. The range of the function is the interval of all numbers between 1 and 5. ∎

The range of a function is not always so easy to describe.

EXAMPLE 5

Let $f(x) = (x^2 + 5)/(x - 2)$ with domain all $x \neq 2$. Is the number -2 in the range of f?

Solution It is difficult to describe the entire range of this function. Here we are only asking whether the particular number -2 is in its range. The number -2 will be in the range if we can find a number x such that $f(x) = -2$.

According to the formula defining f, this last equation is the same as the equation

$$\frac{x^2 + 5}{x - 2} = -2$$

Solving for x we have

$$x^2 + 5 = -2(x - 2)$$

or

$$x^2 + 5 = -2x + 4$$

or

$$x^2 + 2x + 1 = 0$$

or

$$(x + 1)^2 = 0$$

This equation has the solution $x = -1$. Observe that indeed $f(-1) = -2$. Thus -2 is in the range of f and is the value of f at $x = -1$. ∎

The difficulty in describing the range of a function is compensated for by its utility. In Example 3 the range of the function A represents the areas of all possible rectangles. We will learn how to describe the range of many functions in Chapter 2.

Functions Defined Piecewise

To understand the fundamental concept of a limit value (Chapter 1), we must deal with functions defined in an "irregular" way. We begin with an example of the useful notion of a **constant function,** wherein we see that all function values of a constant function are the same: Its range has only one element, and its domain can be any set we wish.

EXAMPLE 6

Let $f(t)$ be the temperature in degrees Fahrenheit at which pure water freezes on day t. Write a formula for f.

Solution For all time t, $f(t) = 32$. ∎

We will also need a function that is "piecewise constant." Such a function has different constant values on each of two (or more) different intervals in its domain.

EXAMPLE 7

Imagine a person walking along a sidewalk to a curb and then up a hill with slope $\frac{1}{5}$ (see Figure 0.15). The sidewalk and hill each account for 20 ft of horizontal distance. The walker deliberately maintains a constant horizontal speed of 6 ft/sec and at the curb steps up with a vertical speed of 4 ft/sec. Describe the person's *vertical* speed as a function V of the person's position x measured from the left end of the sidewalk.

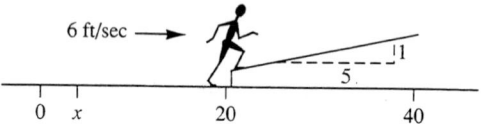

Figure 0.15

Solution On the flat sidewalk the individual's vertical speed is 0. Thus $V(x) = 0$ if x is between 0 and 20. At 20 ft the individual steps up quickly with a given vertical speed of 4 ft/sec. Thus $V(20) = 4$. Finally, since the hillside has slope $\frac{1}{5}$ and the walker is moving horizontally at 6 ft/sec, the individual's vertical speed on the hillside must be $\frac{6}{5}$ ft/sec. Thus $V(x) = \frac{6}{5}$ when x is between 20 and 40. We then write V in the following standard notation:

$$V(x) = \begin{cases} 0 & 0 \leq x < 20 \\ 4 & x = 20 \\ \frac{6}{5} & 20 < x \leq 40 \end{cases}$$ ∎

Example 7 describes a function whose domain is all x such that $0 \leq x \leq 40$ and whose function values cannot be given by a single algebraic formula, in contrast to earlier examples. The range of this function has exactly three elements: 0, 4, and $\frac{6}{5}$.

This function is an idealized model of physical reality and is intended to help you relate the need to define a function in an irregular way with something you can easily imagine. This example will also be used in Chapter 1 to help you to understand the key point about the limit value of a function.

Functions defined piecewise, as in Example 7, can also be given directly. If f is the function defined by

$$f(x) = \begin{cases} \dfrac{x^2 + 1}{x - 1} & \text{if } x \neq 1 \\ 3 & \text{if } x = 1 \end{cases}$$

we are to understand this to mean the following: The domain A of f is all real numbers, including 1. For $x \neq 1$, we find function values in the usual way:

$$f(0) = \frac{0^2 + 1}{0 - 1} = -1 \qquad f(2) = \frac{2^2 + 1}{2 - 1} = 5 \qquad f\left(\frac{1}{2}\right) = \frac{\left(\frac{1}{2}\right)^2 + 1}{\frac{1}{2} - 1} = \frac{-10}{4} = -\frac{5}{2}$$

But—and this is the point of Example 7—we are to understand from the given format defining f that $f(1) = 3$; this value has *no relation* to the algebraic formula given for the definition of $f(x)$ when $x \neq 1$.

Applications of functions as mathematical models of systems outside mathematics will be more meaningful if you carefully distinguish between a function and its function values. It may help to think of a *function* as a roll of motion picture film showing the whole course of a changing system. One *function value* is like one frame of the film and represents one state of the system. The range of the function is the complete collection of all the possible states of the system, represented by the full set of individual frames of the film showing its differing states.

Background Review 0.2

A horizontal line is used to represent the real number system. On this line a unit distance is arbitrarily chosen and measured to the right of a point 0 representing zero.

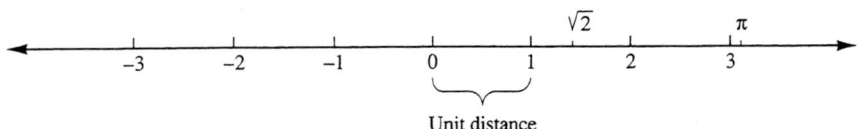

All other real numbers are represented as points on this line. Whole numbers are whole multiples of the entire unit. Rational numbers a/b are a multiples of the fractional part $1/b$ of the unit; 5/2 is five halves. The placement of irrational numbers, such as $\sqrt{2}$ or π, which cannot be a fraction, must be estimated: $\sqrt{2} \simeq 1.4$, $\pi \simeq 3.14$.

Every point on the line has some real number measurement associated with it.

For real numbers a and b we write

$$a \le b \quad \text{"a less than or equal to b"}$$

to mean that, in the drawing, a lies to the left of b, and may equal b, whereas

$$a < b \quad \text{"a less than b"}$$

means a lies strictly to the left of b and $a \ne b$.

The *absolute value* $|a|$ of a number a is its distance (as a *positive* number) from 0; thus

$$|2| = 2 \qquad \left|-\tfrac{3}{2}\right| = \tfrac{3}{2}$$

In general then
$$|a| = \begin{cases} a & \text{if } a \ge 0 \\ -a & \text{if } a < 0 \end{cases}$$

With two numbers a and b, one also associates four *intervals* of numbers:

The closed interval $[a, b]$, which contains a and b and all numbers x between a and b.

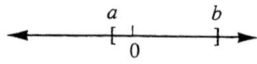

The open interval (a, b),

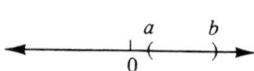

which contains all numbers x strictly between a and b, that is, all x such that $a < x$ and $x < b$ (written $a < x < b$). Notice that an open interval contains no largest or smallest number.

The half-open/half-closed intervals, which are denoted

$[a, b)$

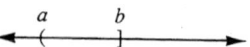

$(a, b]$

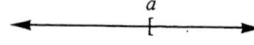

The real number line has no end in either direction. We denote intervals without end by

$[a, \infty)$

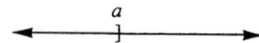

or

$(-\infty, a]$

or by (a, ∞) or $(-\infty, a)$; the symbol "[" ("(") means that we include (exclude) the endpoint a in the interval. The symbol ∞ is called *infinity*. It is not a number and only reminds us that the real number line is unending.

Exercises 0.2

Compute the function value of each function given in Exercises 1–8 at each of the numbers 1, 0, -1, 2, $\sqrt{2}$, $\frac{1}{2}$, a, $a + 1$, and $a + h$.

1. $f(x) = 2x + 1$
2. $f(x) = 7 - x$
3. $g(t) = t^2 - 1$
4. $f(u) = \dfrac{u^2 - 4}{u + 2}$
5. $h(x) = \dfrac{x + 1}{x - 3}$
6. $g(r) = 13$
7. $k(p) = (p + 1)^2$
8. $r(x) = |x - 3|$

Describe the domain of each function in Exercises 9–14.

9. $f(x) = 1 - x^2$
10. $g(t) = \sqrt{1 - t^2}$
11. $h(w) = \dfrac{w - 1}{w + 1}$
12. $k(z) = \dfrac{1}{z}$
13. $f(t) = \dfrac{1}{t(t + 1)(t - 2)}$
14. $z(t) = \dfrac{1}{1 + |t|}$

15. Find five distinct numbers in the range of the function in Example 3.

16. Describe the range of $f(x) = x^2 + 1$ if we take its domain to be all real numbers.

Determine whether the number 4 is in the range of the functions in Exercises 17–24. If so, what number (or numbers) is 4 a function value of? [*Hint:* Refer to Example 5.] You may need to use the quadratic formula for the roots of a quadratic equation $ax^2 + bx + c = 0$. These roots are given by

$$x = \dfrac{-b \pm \sqrt{b^2 - 4ac}}{2a}$$

17. $f(x) = 2 - 2x$
18. $g(t) = t^2 + 3t$
19. $h(z) = \dfrac{z + 1}{z - 1}$
20. $l(p) = \sqrt{p - 1}$
21. $k(s) = (s + 1)(s + 2)$
22. $f(t) = \sqrt{1 - t^2}$
23. $h(w) = \dfrac{w^2 - 4}{w - 2}$
24. $z(t) = |t - 2|$

Compute $f(-1)$, $f(0)$, $f\left(\tfrac{3}{2}\right)$, $f(2)$, $f(\sqrt{2})$, $f(a)$, and $f(a + 1)$ (where $0 < a \leq \tfrac{1}{2}$) in Exercises 25 and 26.

25. $f = \begin{cases} x + 1 & x \leq 0 \\ 3 & 0 < x \leq 1 \\ x - 1 & 1 < x \end{cases}$

26. $f(x) = \begin{cases} \dfrac{x^2 - 1}{x - 1} & x < 1 \\ 2 & x = 1 \\ 3x + 1 & x > 1 \end{cases}$

27. Suppose that $P(x) = 4{,}000\left(1 - \dfrac{1}{x}\right)$ is the profit in dollars earned by the Ace Manufacturing Company from the sale of x (thousands) of its spindles. Thus if 2,000 spindles are sold, the profit is $P(2) = 4{,}000\left(1 - \tfrac{1}{2}\right) = \$2{,}000$. How many must be sold to earn \$3,000? To earn \$3,800? To earn \$5,000?

28. A company that produces breakfast cereal wishes to sell its product in boxes that hold 64 in.³ of cereal when full. The company decides that a box with a square face would distinguish it from the boxes of competing products. Let x denote the length of the side of this square face, and let w denote the width of the box. Note then that $x^2 w = 64$, since the box must have a volume of 64 in.³. Find a formula for a function S of x representing the *total surface area* of the box. Be sure that your solution contains only one variable.

29. A box contains 40 nuts, 40 washers, and 40 bolts at time $t = 0$. Once each minute a worker reaches into the box and removes 1 nut, 1 bolt, and 2 washers and uses these to join two pieces moving down the assembly line. Let $f(t)$ denote the ratio of nuts, bolts, and washers in the box to the number of washers only, t minutes later.

   a. Describe this function representing the worker's increasing frustration at finding 2 washers for each nut and bolt pair.

   b. What is $f(0)$? $f(10)$? $f(15)$? Interpret these numbers in words.

   c. What is the domain of this function? How is the fact that there is no mathematically valid $f(20)$ in accord with physical reality?

30. Why does the formula $f(x) = \pm\sqrt{1 - x^2}$ not define a function? Look very closely at the definition of function.

31. Let $P(t)$ denote the number of bicycles produced by a manufacturing firm by the tth day of the year. For example, $P(1)$ represents the number produced by the end of the first day. In similar English, express the meaning of the following:

   a. $P(2)$
   b. $P(10)$
   c. $P(a)$
   d. $P(5 + h)$ where $h = 3$
   e. $P(29 + h) - P(29)$
   f. $P(30)$
   g. $P(365)$
   h. $P(a + 3)$

i. $P(30) - P(29)$ **j.** $\dfrac{P(29 + 5) - P(29)}{5}$

32. A person is pulled on water skis onto a ramp with a slope of $\frac{1}{6}$. Suppose that the momentum of the boat maintains the skier's *horizontal* speed at 30 ft/sec (Figure 0.16). Describe the skier's *vertical* speed as a function of the skier's position x between the points A and C, omitting the point B from the domain of this function (and ignoring the roughness of the water and angle of the tow rope).

33. If you drop a rock from the top of a 100-ft-high building, its height $h(t)$ above the ground t sec later (ignoring air resistance) will be given by

$$h(t) = 100 - 16t^2 \quad \text{with domain} \quad 0 \le t \le 2.5$$

Think of the function h as a motion picture film of the rock's fall, where each function value $h(t)$ corresponds to a single frame of the film—that frame taken at time t. That is, each value $h(t)$ shows you where the rock is, in the sense of telling you the height of the rock, at time t. Thus the function h models the falling rock "system," and each function value tells you the state of the system at time t.

a. What is the state of this system at time $t = 0$; $t = 1$; $t = 2$? That is, where is the rock at these times?

b. When does the rock hit the ground? [*Hint:* Imagine what the film would show at that moment—it would show $h(t) = 0$! Which "frame" is this?]

c. When was the rock "in the state of being" 36 ft above the ground?

In Chapter 1, we will introduce the *derivative* of a function; this derived function will play the role of showing the film in motion and telling us the change-of-state of this system—the speed at which the rock is falling at any time.

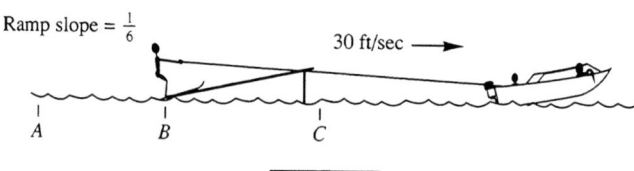

Figure 0.16

0.3 Linear Functions and Graphs

The purpose of this section is to describe the relationship between linear functions and the equation of a straight line in terms of the graph of a function.

The graph of a function f is a line or curve that will allow us to picture the relationship f apart from the algebraic details defining f. We will draw such a graph in the two-dimensional plane. One dimension allows us to represent the domain of the function and the other allows us to represent its range. The ability to regard a function as a picture—its graph—and to draw basic graphs rather quickly is extremely useful.

Cartesian Coordinates

Recall that the Cartesian, or rectangular, coordinate system consists of two lines drawn at a right angle, as in Figure 0.17. Each line represents the real number line. The point where these two lines meet is called the **origin** of the coordinate system. The horizontal line is most often called the **x-axis**; the vertical line is called the **y-axis**. Any location P in this rectangular coordinate system is uniquely specified by two measurements a and b, called its **coordinates**. The first coordinate a of P is measured horizontally along the x-axis and is the directed distance of P from the y-axis; the second coordinate b is measured vertically along the y-axis. The location P is then called a *point* in the coordinate system with numerical coordinates (a, b). Conversely, any given pair of numbers (c, d) defines a single point Q in the coordinate system; to locate Q, simply measure c units horizontally and then d units vertically. Note that the origin has coordinates $(0, 0)$.

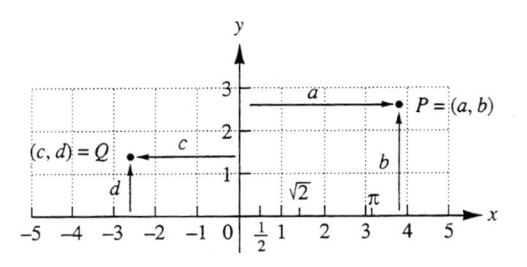

Figure 0.17

The Graph of a Function

Section 0.2 suggests that a function is essentially a collection of pairs $(x, f(x))$ of a variable x and its corresponding value $f(x)$.

Definition

The **graph of a function** f is the collection of all points $(x, f(x))$ pictured in a rectangular coordinate system.

When drawing this picture—or **plotting the graph**—of a function, we measure domain values x horizontally and range values $f(x)$ vertically. The graph of a function can typically be a drawing such as Figure 0.18.

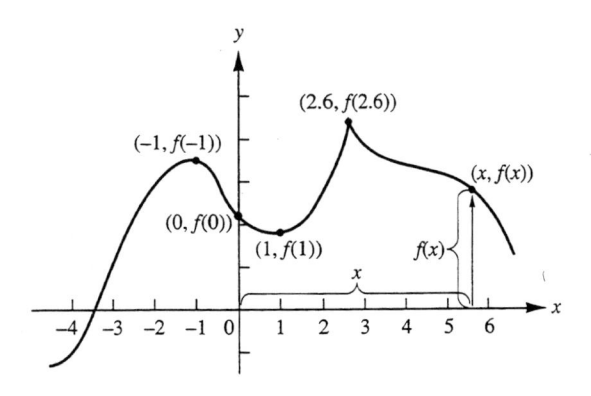

Figure 0.18

Remark. The graph of a function gives us some idea of its range. The values in the range are precisely the second coordinate values $f(x)$—each measured

0.3 Linear Functions and Graphs

in the vertical direction—of each point $(x, f(x))$ on the graph. Note this again: $f(x)$ *is the (directed) distance to the graph measured from the point* x *on the* x-*axis.*

Since we denote points in an x–y coordinate system as pairs (x, y), we commonly write $y = f(x)$ and call this equation the function f. We then speak of graphing the equation $y = f(x)$. For example, we now think of the function $f(x) = 2x + 1$ equally well as the equation $y = 2x + 1$. The graph of f consists of all points $(x, y) = (x, 2x + 1) = (x, f(x))$ drawn in the coordinate plane.

You may recall from algebra that such an equation has a straight line as its graph. In this section we study such functions and their graphs and then graph more general functions in Section 0.4.

Linear Functions and the Slope of a Straight Line

A **linear function** is a function f whose graph is a (nonvertical) straight line. A linear function can always be put into the particular algebraic form $f(x) = mx + b$. To see this let us begin with a straight line and derive its equation.

Recall that two points in space determine a unique straight line. Let us relate this to the idea of the **slope** of a straight line, a numerical measure of how steep the line is.

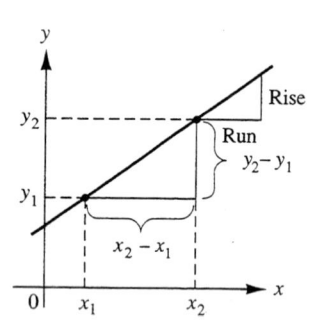

Figure 0.19

In Figure 0.19 the slope S of the straight line L is the ratio of its vertical "rise" to its horizontal "run." That is,

$$S = \frac{\text{rise}}{\text{run}}$$

To compute this number, recall from plane geometry that just as any two points on L determine L uniquely, the slope S may be computed along any part of L and is always the same. Thus, in the notation of Figure 0.19,

$$S = \frac{y_2 - y_1}{x_2 - x_1} \qquad x_1 \neq x_2$$

Before going further let us note two special cases of this formula. If the line is horizontal, then $S = 0$, and vice versa. If the line is vertical, we cannot compute S (because $x_2 - x_1 = 0$), and we do not assign a numerical slope to a vertical line but simply refer to it as a vertical line. The slope formula yields two algebraic forms of the equation of a straight line.

24 Chapter 0 Functions

The Slope–Intercept Form of the Equation of a Straight Line

Consider Figure 0.20. If we know the height b at which the line L intercepts the y-axis and if we know the slope $S = m$ of the line, then we know where every other point (x, y) is on the line, since, computing the slope between $(0, b)$ and any other point (x, y) on L, we must have

$$m = \frac{y - b}{x - 0} = \frac{y - b}{x}$$

or
$$y = mx + b \qquad (1)$$

This equation is called the **slope–intercept form** of the equation of a straight line. Notice that the slope m is precisely how much the graph rises (or falls) in a distance of one unit.

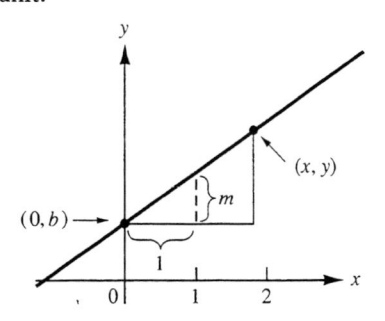

Figure 0.20

EXAMPLE 1

Write the equation of the straight line with y-intercept $b = 2$ and slope $m = -\frac{1}{2}$.

Solution The equation is $y = -\frac{1}{2}x + 2$, by substitution in the general formula $y = mx + b$ (1). Notice that a line with a negative slope falls as x moves to the right (Figure 0.21).

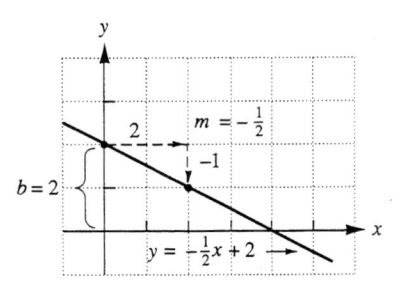

Figure 0.21

Conversely, it can be shown that an equation of the form $y = mx + b$ must be the equation of a straight line. Moreover, this straight line has slope m and intercept b, for, using the two simplest values of x, when

$$x = 0 \quad \text{then} \quad y = m \cdot 0 + b = b$$

and when
$$x = 1 \quad \text{then} \quad y = m \cdot 1 + b = m + b$$

0.3 Linear Functions and Graphs

so that y changes from b to $m + b$ when x rises by one unit. Therefore the straight line has slope

$$\frac{(m + b) - b}{1} = m$$

as illustrated in Figure 0.22.

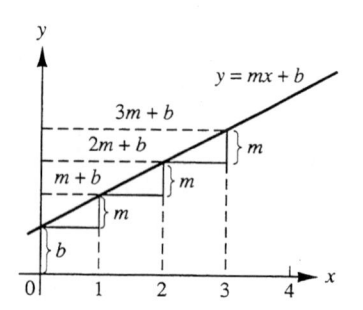

Figure 0.22

EXAMPLE 2

Find the slope and intercept of the equation $3y + 2 = 6x - 1$.

Solution We begin by solving this equation for y by writing

$$3y = 6x - 1 - 2 = 6x - 3$$

or

$$y = \frac{6x - 3}{3} = 2x - 1$$

The equation $y = 2x - 1$ is the equation of a straight line with slope 2 and intercept -1. We leave it to the reader to graph this equation. ∎

The Point–Slope Form of the Equation of a Straight Line

Let us consider a second way to obtain the equation of a straight line. This time we imagine that we know the slope S of a line and that we know one point $P = (x_1, y_1)$ on the line (see Figure 0.23).

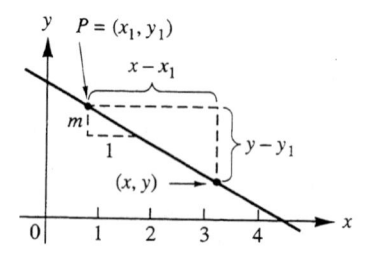

Figure 0.23

If (x, y) denotes *any other* point on the line, we must have

$$m = \frac{y - y_1}{x - x_1}$$

26   Chapter 0   Functions

or
$$y - y_1 = m(x - x_1) \qquad (2)$$

This equation is known as the **point–slope form** of the equation of a straight line.

EXAMPLE 3
Find the equation of the line with slope $m = \frac{3}{2}$ passing through the point $P = (-2, 1)$.

Solution By substitution in Formula 2, the equation must be
$$y - 1 = \frac{3}{2}(x - (-2))$$

or
$$y - 1 = \frac{3}{2}(x + 2) = \frac{3x}{2} + 3$$

or
$$y = \frac{3}{2}x + 4 \quad \blacksquare$$

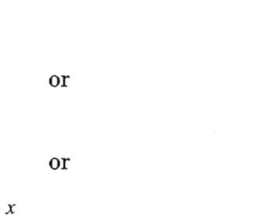

This last equation is the slope–intercept form of the equation of the same line (see Figure 0.24). Thus we may obtain the slope–intercept form of the line from the point–slope form.

Thus the graph of a straight line (that is not a vertical line) always has equation $y = mx + b$, and a linear function, whose graph is such a line, can be written as $f(x) = mx + b$. Such functions arise easily in applications.

Figure 0.24

EXAMPLE 4
The Fresh Food Gardening Company invests $1,500 in a greenhouse to produce fresh lettuce in late winter for the local market. It costs $30 to grow and sell one 50-lb case of lettuce.

a. Describe the total cost to the Fresh Food Company to produce x cases of lettuce as a linear function, and draw its graph.

b. What is the cost of producing five cases of lettuce?

Solution

a. Let $C(x)$ denote the cost to produce x cases of lettuce.

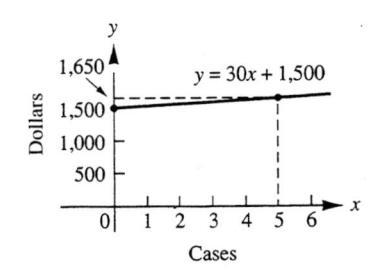

Figure 0.25

Since there is an initial cost of $1,500 to produce zero cases of lettuce and an additional cost of $30 to produce each additional case, we can represent production costs as a straight line with intercept $b = 1{,}500$ and slope $m = 30$.

Accordingly, this graph (see Figure 0.25) has equation $y = 30x + 1{,}500$. Therefore $C(x) = 30x + 1{,}500$.

b. The cost to produce five cases is $C(5) = 30(5) + 1{,}500 = 1{,}650$ dollars. ∎

Background Review 0.3

Whenever you perform multiplication or its inverse (division) on a sum (or difference), you must use the **distributive law**:

$$a(b + c) = ab + ac$$

If you do not use this law, you will make an error. For example,

$$\frac{2x + 6}{2} = \frac{2(x + 3)}{2} = x + 3$$

is correct, and the distributive law is used in going to the intermediate step. *Do not* be tempted instead to write

$$\frac{\cancel{2}x + 6}{\cancel{2}} = x + 6$$

which is *incorrect*; here the distributive law is not used, and division (cancellation of the number 2) is performed on a sum $2x + 6$.

Exercises 0.3

Find the equation for each straight line described in Exercises 1–7, and graph each line.

1. The slope is -2, the intercept is 3.
2. The slope is $\frac{1}{10}$, the intercept is -1.
3. The line has slope -1 and contains the point $(-1, 1)$.
4. The line has slope $-\frac{3}{2}$ and contains $(0, 3)$.
5. The line contains the two points $(-1, 1)$ and $(4, 1)$. [*Hint:* Compute the slope of the line.]
6. The line contains the two points $(1, 2)$ and $(4, 1)$.
7. The line contains the two points $(-1, -1)$ and $(3, 1)$.

Find the slope and y-intercept of each of the lines given by the equations in Exercises 8–15. Graph these lines.

8. $y + 2 = x$
9. $y = 3(x - 1)$
10. $y + 1 = 2x - 1$
11. $y + \frac{1}{2}x = 2$
12. $\frac{y}{4} = \frac{x - 6}{2}$
13. $y - 2 = 0$
14. $x + 1 = 3$
15. $2y = \dfrac{4x - 12}{2}$

28   Chapter 0   Functions

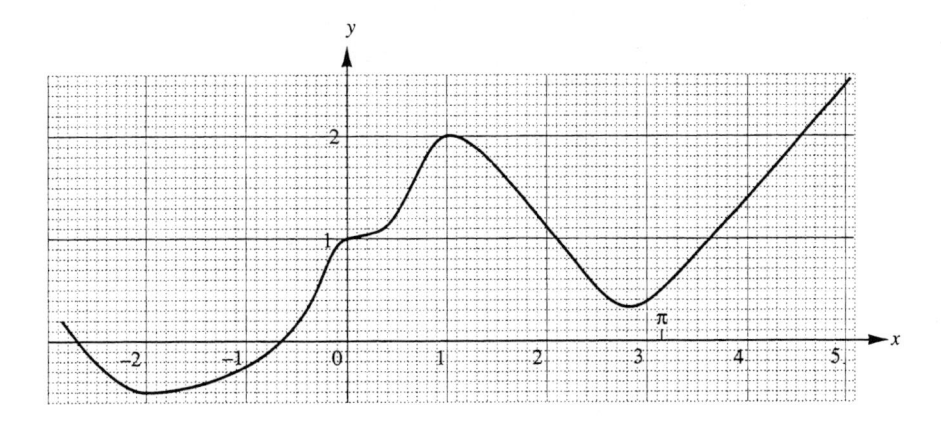

Figure 0.26

A constant function $f(x) = C$ is a linear function whose graph has zero slope. Graph the constant functions in Exercises 16 and 17.

16. $f(x) = 2$ **17.** $g(x) = -3$

18. This exercise asks about properties of a very general graph. What are the values of $f(0), f(-2), f(1), f\left(\frac{5}{2}\right)$, $f(\pi)$, and $f(x + h)$, where $x = 2$ and $h = \frac{1}{10}$, for the graph in Figure 0.26? What is the value of $[f(3 + 2) - f(3)]/2$? Is $f(0) < f(1)$? Is $f(1) < f(3)$? For how many x's is $f(x) = \frac{3}{2}$? Where is this graph rising (or falling) the slowest?

19. Suppose that you have a given amount of material to learn for an upcoming test. We denote this amount by 1, for one unit of material. Let $f(t)$ denote the amount of material you know after t hours of study. If you are learning new material at a rate directly proportional to how much you know and how much you do not know, it turns out that (by sophisticated methods we will later study) your learning curve looks like the graph in Figure 0.27.

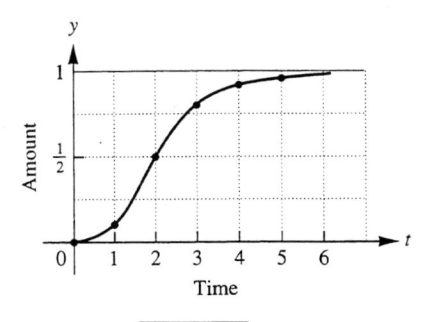

Figure 0.27

How much of the material have you learned at the end of 1 hr? 2 hr? 4 hr? 5 hr? At what time do you appear to be learning material the fastest? What does the shape of the learning curve suggest about the rate at which you learn new material?

20. Observe that parallel lines have the same slope. Find the equation of the straight line parallel to $y + x - 2 = 2x + 1$ with intercept -1.

21. The Strung Manufacturing Company finds that it costs $22 to make one tennis racket. But before even a single racket is made, the company has a fixed cost of $30,000 for buildings and manufacturing equipment. Find a formula for $C(x)$, the total cost of manufacturing x tennis rackets. Interpret the numbers 22 and 30,000 with respect to the graph of $y = C(x)$.

Exercises 22–25 explore the use of linear functions in proportional reasoning.

A quantity y is said to be **directly proportional** to a quantity x if y is a constant multiple of x; that is, $y = kx$ for some constant k called the **proportionality constant**. (For example, at a constant speed r, the distance traveled d is directly proportional to time in motion t; indeed $d = rt$ and the speed r is the proportionality constant.) Since $y = kx$, directly proportional quantities have a linear relation whose graph is a straight line with slope m being the proportionality constant k.

22. Let P be the number of people floating down the Colorado River in rubber rafts over a holiday weekend. Let T be the amount of plastic trash floating in the river. Express the equation $T = kP$ in a sentence using the word *proportional*.

0.3 Linear Functions and Graphs

23. Let P be the initial amount deposited in a bank account that pays 8% simple interest per year, and let A be the amount added to the account 1 year later. Then A is proportional to the amount P initially in the account. Express this as an equation using A, P, and 0.08. What is the proportionality constant?

24. In general, population growth is directly proportional to population size. If P_0 is the population of a large city and P_1 is the population 1 year later, then $P_1 - P_0$ is the growth in population. Express the first sentence in this exercise as an equation.

25. The temperature in a department store is maintained at 80° during the Christmas shopping season. The outside temperature can vary between 0° and 40° during this period. The store must charge 5 cents extra on each $100 item to cover its heating cost for each difference in degree between the constant interior temperature of the store and the average exterior temperature during this period.

   a. Let $H(x)$ be the extra cost on a $100 item if the average outside temperature is x degrees. Show that this heating cost is proportional to the temperature difference $80 - x$.
   b. Graph $y = H(x)$.
   c. What are the values $H(10)$, $H(25)$, and $H(40)$, and what do they mean to a shopper hoping for a breath of cool air?

0.4 Graphs of Nonlinear Functions

The purpose of this section is to gain familiarity with the most frequently seen graphs of nonlinear functions.

The graph of a nonlinear function $y = f(x)$ cannot be a straight line. Nonlinear functions can be given by virtually any kind of algebraic formula, and their graphs can take many forms. A certain few nonlinear functions, however, occur repeatedly as handy examples of basic ideas. We begin the section with the graphs of these and then see how these graphs may be used to represent more complex nonlinear functions.

Six Basic Nonlinear Graphs

Each of the functions and graphs that we study in Examples 1 through 6 has the simplest algebraic and geometric form of the class of formulas and curves to which it belongs. The first of these is the simplest quadratic function, $y = f(x) = x^2$, whose graph has the geometric form called a **parabola**.

EXAMPLE 1

Graph the equation $y = f(x) = x^2$.

Solution We begin by plotting a few points from the table of values in Figure 0.28 and assume regularity, or a smoothly changing curve, between these

x	x^2
0	0
1	1
-1	1
2	4
-2	4
$\frac{1}{2}$	$\frac{1}{4}$
$-\frac{1}{2}$	$\frac{1}{4}$

Figure 0.28

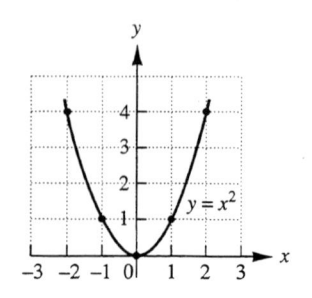

points. Because the graph is a continually changing curve, it cannot be fully determined by plotting any finite number of points. ∎

Our second example is the **absolute value function** $y = f(x) = |x|$. Notice that the continually changing graph of this function has an abrupt change in slope at the point $x = 0$ (see Figure 0.29).

EXAMPLE 2

Graph $y = f(x)$, where $f(x) = |x|$.

Solution

x	$|x|$
0	0
1	1
−1	1
2	2
−2	2

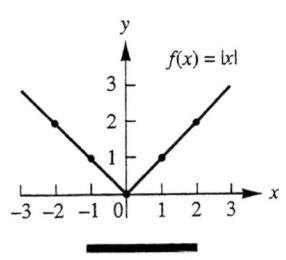

Figure 0.29

Our third example is the simplest cubic function $y = f(x) = x^3$. Notice how its graph seems to rise continually as x moves to the right, except at $x = 0$, where the graph appears to level off "for an instant" (see Figure 0.30).

EXAMPLE 3

Graph $y = x^3$. That is, graph the function $f(x) = x^3$.

Solution

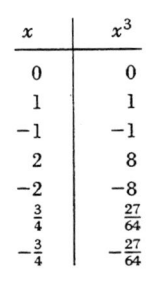

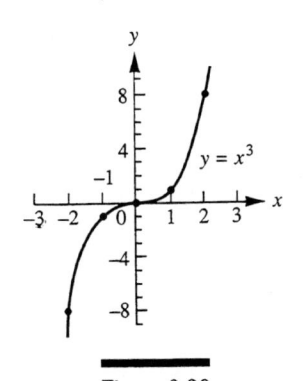

x	x^3
0	0
1	1
−1	−1
2	8
−2	−8
$\frac{3}{4}$	$\frac{27}{64}$
$-\frac{3}{4}$	$-\frac{27}{64}$

Figure 0.30

Note that for convenience we use a different scale on the y-axis.

The **hyperbola** is another basic geometric shape. A **hyperbola** is the graph of the function $y = f(x) = 1/x$ that expresses algebraic division in its simplest form (see Figure 0.31).

EXAMPLE 4

Graph $f(x) = 1/x$, $x \neq 0$.

Solution

x	$\dfrac{1}{x}$
1	1
-1	-1
2	$\frac{1}{2}$
-2	$-\frac{1}{2}$
$\frac{1}{3}$	3
$-\frac{1}{3}$	-3
3	$\frac{1}{3}$
-3	$-\frac{1}{3}$
$\frac{1}{4}$	4
$-\frac{1}{4}$	-4

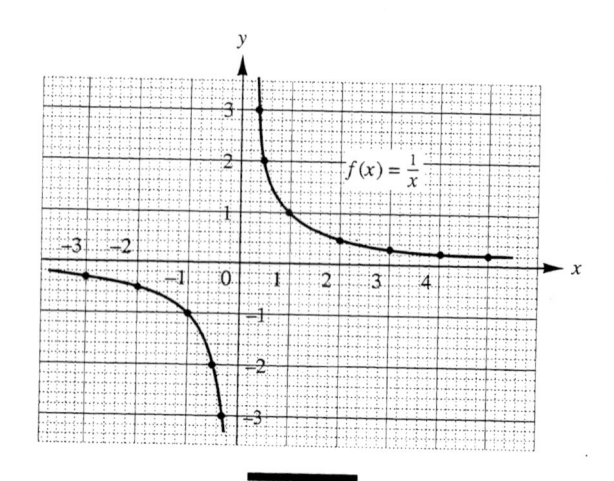

Figure 0.31

The graph in Figure 0.31 also illustrates the concept of an *asymptote*. The graph of $y = 1/x$ curves always closer to, but never coincides with, the straight line $y = 0$ (the x-axis). The line $y = 0$ is called a **horizontal asymptote** for the graph of $y = 1/x$. The same is true for the vertical line $x = 0$ (the y-axis); as x approaches zero the graph of $y = 1/x$ curves closer and closer to the y-axis but never coincides with it. The line $x = 0$ is called a **vertical asymptote** for the graph of $y = 1/x$.

Our next function illustrates the simplest algebraic form of *exponential growth*. Its graph is the **exponential growth curve** that we first encountered in Example 2 in Section 0.1. This curve is neither a parabola nor a hyperbola (see Figure 0.32).

EXAMPLE 5

Graph $E(x) = 2^x$.

Solution Recall that $a^0 = 1$ and $a^{-n} = 1/a^n$ if $a \neq 0$. We use these facts to construct the table of values in Figure 0.32.

x	2^x
0	1
1	2
-1	$\frac{1}{2}$
2	4
-2	$\frac{1}{4}$
3	8
-3	$\frac{1}{8}$

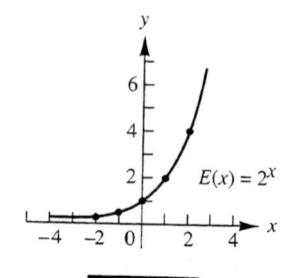

Figure 0.32

32 Chapter 0 Functions

In Example 5, the x-axis is again a horizontal asymptote for the curve $y = 2^x$, since this curve approaches ever closer to but never coincides with the horizontal straight line $y = 0$. Unlike the graph in Example 4, the graph of $y = 2^x$ has *no* vertical asymptote, because it does not approach any vertical line but continues to rise rapidly as it slowly moves to the right.

Our next example is that of a piecewise constant function that also has a different function value (from its two constant values) at exactly one point. This example shows how we graph a function that is defined in an irregular way by more than a single algebraic formula (see Figure 0.33).

EXAMPLE 6

Graph

$$V(x) = \begin{cases} 0 & 0 \leq x < 20 \\ 4 & x = 20 \\ \frac{6}{5} & 20 < x \leq 40 \end{cases}$$

Solution The formula for V already gives the table of points $(x, V(x))$, and we plot the graph directly from it. Note the "open" circle indicating the absence of a point at the end of each horizontal line. ∎

Figure 0.33

Our final example combines the idea of Example 6 with an algebraic simplification to show the graph of a function that is linear except for one point. This function does not, however, appear to be linear at first.

EXAMPLE 7

Graph

$$f(x) = \begin{cases} \dfrac{x^2 - 4}{x - 2} & x \neq 2 \\ 1 & x = 2 \end{cases}$$

Solution When $x \neq 2$,

$$\frac{x^2 - 4}{x - 2} = \frac{(x - 2)(x + 2)}{x - 2} = x + 2$$

We use this simpler formula to determine the table of values in Figure 0.34 for $x \neq 2$. For $x = 2$, we are given that $f(2) = 1$. Note that this graph has a gap at height 4.

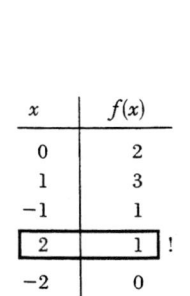

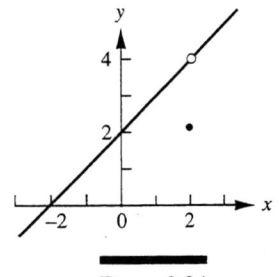

x	$f(x)$
0	2
1	3
−1	1
2	1 !
−2	0
3	5
−3	−1

Figure 0.34

0.4 Graphs of Nonlinear Functions   33

Graphs by Translation and Reflection

Transposition of a graph is a geometric idea that we want to learn to recognize algebraically. Consider graphing the three functions

$$f(x) = x^2 + 1 \qquad g(x) = -x^2 \qquad \text{and} \qquad h(x) = (x - 1)^2$$

in relation to the graph of $b(x) = x^2$. We wish to see that each of these graphs is only a simple geometric translation or reflection of the basic graph of $b(x) = x^2$ shown again in Figure 0.35.

x	x^2
0	0
1	1
-1	1
2	4
-2	4
$\frac{1}{2}$	$\frac{1}{4}$
$-\frac{1}{2}$	$\frac{1}{4}$

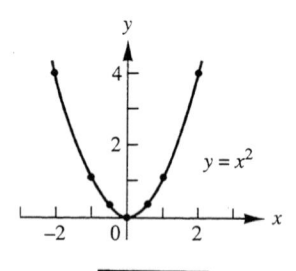

Figure 0.35

The formula $f(x) = x^2 + 1$ merely *adds the number* 1 to each value of the basic graph $b(x) = x^2$ (see Figure 0.36). (The graph in Figure 0.36 is a **translation** of the basic graph in Figure 0.35.)

x	x^2	$x^2 + 1$
0	0	1
1	1	2
-1	1	2
2	4	5
$\frac{1}{2}$	$\frac{1}{4}$	$\frac{5}{4}$
$-\frac{1}{2}$	$\frac{1}{4}$	$\frac{5}{4}$

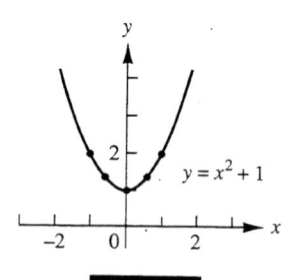

Figure 0.36

The formula $g(x) = -x^2$ merely *reverses the sign* of the basic graph $b(x) = x^2$, giving the graph of g in Figure 0.37. (The graph in Figure 0.37 is a **reflection** of the basic graph in Figure 0.35.)

x	$-x^2$
0	0
1	-1
-1	-1
2	-4
-2	-4
$\frac{1}{2}$	$-\frac{1}{4}$
$-\frac{1}{2}$	$-\frac{1}{4}$

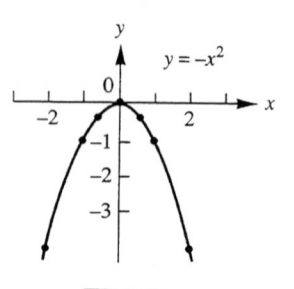

Figure 0.37

34   Chapter 0   Functions

Finally, and less obviously, the formula $h(x) = (x - 1)^2$ moves the "center of symmetry" of $y = x^2$ from the line $x = 0$ to the line $x = 1$ (see Figure 0.38).

x	$x - 1$	$(x - 1)^2$
0	−1	1
1	0	0
2	1	1
3	2	4
−1	−2	4
−2	−3	9

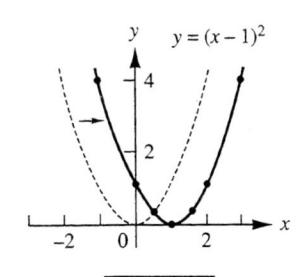

Figure 0.38

You may wish to compute a table of values and plot a few points for these three examples to further verify that Figures 0.36–0.38 are indeed their graphs.

Putting these three ideas together, we can easily graph the apparently more complicated function $k(x) = -(x - 1)^2 + 1$ (see Figure 0.39) by first translating our basic graph one unit to the right, reflecting it about the x-axis, and then raising the graph one unit.

x	$-(x - 1)^2 + 1$
0	0
1	1
2	0
−1	−3
3	−3

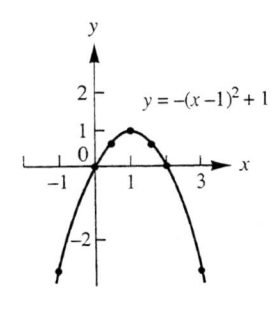

Figure 0.39

Geometrically, the center of symmetry is now the line $x = 1$, and the basic graph is inverted and raised by one unit. Algebraically, this corresponds to the minus sign and the addition of 1 to every value of $(x - 1)^2$.

Using similar reasoning and the graph of $b(x) = 1/x$ (Figure 0.31), you can see with some effort that the graph of

$$f(x) = \frac{1}{x - 2} - 1 \qquad x \neq 2$$

is that shown in Figure 0.40.

In Figure 0.40, the line $y = -1$ is now a horizontal asymptote for the graph; the line $x = 2$ is now a vertical asymptote. Every value $b(x) = 1/x$ is shifted two units to the right and reduced by one unit; notice how this is indicated by the algebraic formula for f.

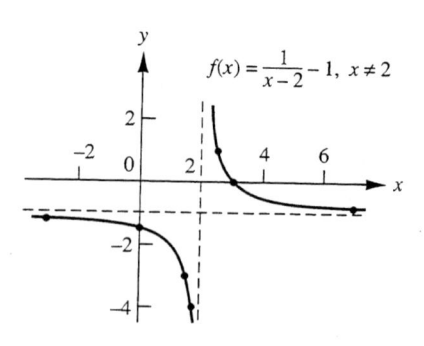

Figure 0.40

These examples indicate that the shape and position of certain graphs can be seen in their algebraic form. The next example shows how to put a quadratic function into the needed form.

EXAMPLE 8

Graph $y = x^2 - 4x + 3$.

Solution We need to rewrite this equation in the form

$$y = \pm(x - A)^2 + B$$

and apply the ideas of translation, reflection, and symmetry to obtain the graph. We obtain this form by *completing the square*. Note that $(x \pm a)^2 = x^2 \pm 2ax + a^2$, so that the constant term a^2 is always the square of one-half the middle term—that is, $a^2 = [\frac{1}{2}(\pm 2a)]^2$. Thus we rewrite $x^2 - 4x$ into a perfect square by noting that $\frac{1}{2}(-4) = -2$ and proceeding as follows:

$$y = x^2 - 4x + 3 = x^2 - 4x + (-2)^2 - (-2)^2 + 3 \quad \text{Add and subtract } (-2)^2$$
$$= (x - 2)^2 - 4 + 3$$
$$= (x - 2)^2 - 1$$

(The number -2 is one-half the middle coefficient -4.) The graph of this equation is seen in Figure 0.41 to be centered about the line $x = 2$ with all values of $b(x) = x^2$ reduced by 1. ∎

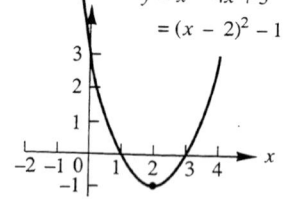

Figure 0.41

Example 8 illustrates the fact that the graph of any function $f(x) = ax^2 + bx + c$ is a transformation of the graph of the basic parabola $b(x) = x^2$.

Graphs That Do Not Represent Functions

Not every graph is the graph of a function, nor does every equation define a function. This significant point is due to a detail in the definition of a function—namely, that with each x is associated a *single $f(x)$*—and is important in applications, because a system undergoing change is in only one state or condition at any one moment. [Returning to an earlier analogy, if we think of a function as a motion picture film of a changing system and a function value as a single frame, then on one frame we see only one picture: On frame x we see only one picture $f(x)$.] Geometrically, a graph is the graph of a function if

36   Chapter 0   Functions

it meets the *vertical-line test*: Any vertical line hits the graph at no more than one point. The graph in Figure 0.42(a) does not meet this requirement, because *two y values* (not *a single y* value) are associated with most x values. Likewise, the equation

$$x^2 + y^2 = 4$$

or equivalently

$$y = \pm\sqrt{4 - x^2}$$

does not associate a single value of y with a given x. If we graph this equation we obtain Figure 0.42(b). This is the graph of a circle of radius 4 and center at (0, 0). Thus a circle is not the graph of a function.

x	$y = \pm\sqrt{4-x^2}$
0	± 2
1	$\pm\sqrt{3}$
-1	$\pm\sqrt{3}$
2	0
-2	0

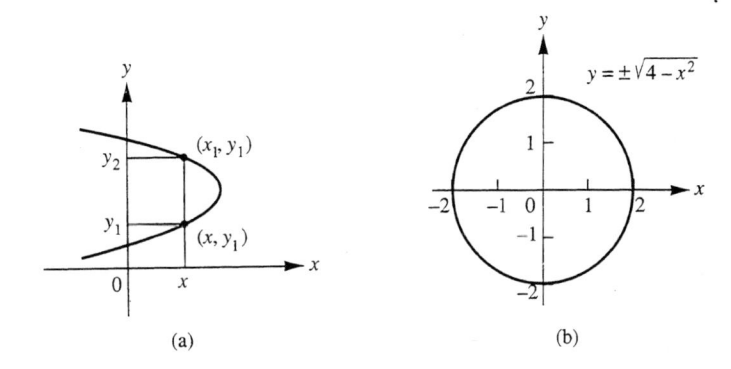

Figure 0.42

An Application to Economics

In basic economic theory, as the price P of a commodity rises, two things happen: (1) The supply S of the commodity rises, and (2) the demand D for the commodity falls. Thus if we view supply and demand as functions of price P, the graphs of S and D might appear as in Figure 0.43.

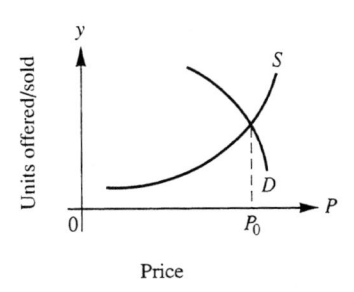

Figure 0.43

The price P_0 where these graphs cross is known as the *equilibrium price*, where supply equals demand, or in the language of functions, where $S(P) = D(P)$. That is, the particular number P_0 is a solution of the equation

$$S(P) = D(P)$$

0.4 Graphs of Nonlinear Functions

Thus, if
$$S(P) = 2P^2 + 3P + 1 \quad \text{and} \quad D(P) = -2P^2 + 5P + 7$$
the equilibrium price is where
$$2P^2 + 3P + 1 = -2P^2 + 5P + 7$$
or
$$4P^2 - 2P - 6 = 0$$
or
$$2P^2 - P - 3 = 0$$
or
$$(2P - 3)(P + 1) = 0$$

The equation then has two solutions: $P_0 = \frac{3}{2}$ and $P_0 = -1$. Only $P_0 = \frac{3}{2} = \$1.50$ has economic meaning and is the equilibrium price in this system.

Background Review 0.4

The quadratic formula is used to solve any second-degree (quadratic) equation $ax^2 + bx + c = 0$. The roots, or solutions, of this equation are given by

$$x = \frac{-b \pm \sqrt{b^2 - 4ac}}{2a} \quad \text{if } a \neq 0$$

Any quadratic expression in the form $x^2 + (a + b)x + ab$ may be factored as $(x + a)(x + b)$

Thus $x^2 + 3x - 4 = (x + 4)(x - 1)$ since
$$x^2 + 3x - 4 = x^2 + (4 + (-1))x + 4(-1)$$

In general, a function of the form
$$f(x) = a_n x^n + a_{n-1} x^{n-1} + \cdots + a_1 x + a_0 \quad a_n \neq 0$$
is called a **polynomial** (function) of degree n and usually cannot be easily factored or graphed.

Exercises 0.4

Graph each of the functions in Exercises 1–18 by plotting a sufficient number of points on the graph.

1. $f(x) = 2x + 1$
2. $g(x) = 2x^2$
3. $f(x) = -|x|$
4. $k(x) = |2x|$
5. $h(t) = |t - 2|$
6. $f(t) = -t^3$
7. $w(x) = \frac{1}{x^2}$
8. $f(x) = x^4$
9. $s(x) = \sqrt{x}$
10. $f(u) = \sqrt{u} - 2$
11. $y = 3^x$
12. $f(x) = \frac{1}{2}^x$

38 Chapter 0 Functions

13. $y = x + \dfrac{1}{x}$

14. $y = 1 + x^{1/2}$ (recall that $x^{1/2} = \sqrt{x}$)

15. $y = x^{2/3}$ (recall that $x^{2/3} = \sqrt[3]{x^2}$)

16. $y = (5 - x)(x - 3)$ 17. $y = x(x - 1)(x + 2)$

18. $f(x) = \dfrac{x^2}{2^x}$

19. Graph
$$f(x) = \begin{cases} 1 & 0 \le x \le 2 \\ x - 1 & 2 < x \le 3 \\ -(x - 3)^2 + 2 & 3 < x \le 4 \end{cases}$$

20. Graph the three functions:
  a. $f(x) = x + 2$
  b. $g(x) = \begin{cases} \dfrac{x^2 - 4}{x - 2} & x \ne 2 \\ 6 & x = 2 \end{cases}$
  c. $h(x) = \begin{cases} \dfrac{x^2 - 4}{x - 2} & x \ne 2 \\ 4 & x = 2 \end{cases}$

  Which two of these three functions are the same function?

21. Graph the function $V(x) = (10 - 2x)(10 - 2x)(x)$, $0 \le x \le 5$, of Example 1 of Section 0.1. What does the graph tell you about where to cut the cardboard so as to achieve the greatest volume for the resulting box?

22. Graph the function of Exercise 27, Section 0.2.

23. Graph the function $f(t) = (60 - 2t)/(20 - t)$ of Exercise 29, Section 0.2. Identify its horizontal and vertical asymptotes.

24. Try to graph $y = x + (1/x)$, not by plotting points, but rather by thinking of the graph as the *sum* of the y-values of the graphs of $y = x$ and $y = 1/x$. Do this by first graphing the latter two and then graphing $y = x + (1/x)$ on the same axis system by "visually adding" the two graphs.

Draw each set of related graphs in Exercises 25–32 on the same axis system. Identify any horizontal or vertical asymptotes that you find.

25. $b(x) = x^2$ $y = -x^2 + 2$ $y = (x + 1)^2 + 1$
26. $b(x) = x^2$ $y = 3x^2$ $y = \tfrac{1}{3}(x - 1)^2 + 1$
27. $b(x) = \dfrac{1}{x}$ $y = \dfrac{1}{x + 1}$ $y = \dfrac{2}{x + 1} + 3$
28. $b(x) = |x|$ $y = |x - 2|$ $y = -|x + 1| + 3$
29. $b(x) = x^3$ $y = -x^3$ $y = (x - 2)^3 + 1$
30. $b(x) = x^3$ $y = 2x^3$ $y = -(x + 3)^3$
31. $b(x) = 2^x$ $y = 2^{x-1}$ $y = 2^{x+1} - 3$
32. $b(x) = 2^x$ $y = 3(2^x)$ $y = \tfrac{1}{3}(2^x)$

In Exercises 33 and 34, write the equation of a parabola with the given graph.

33.

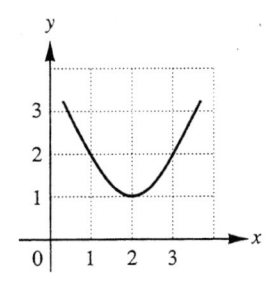

Figure 0.44

34.

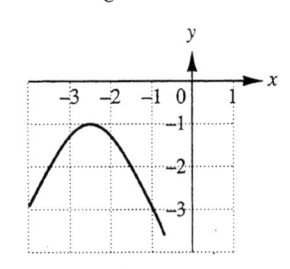

Figure 0.45

If $y = f(x)$ and $y = g(x)$ are two functions, their graphs will cross (intersect) each other at any point x where $f(x) = g(x)$, since then both graphs have the same height at x. Thus you find the points of intersection of two graphs by solving for x in the equation $f(x) = g(x)$.

Find the points of intersection of the two graphs given by the equations in Exercises 35–42. You may need to use the quadratic formula in some cases.

35. $y = -x^2 + 4$ $y = x^2 - 4$
36. $y = \tfrac{1}{2}x + 1$ $y = x^2$
37. $y = \dfrac{1}{x}$ $y = x + 3$
38. $y = -x^2 + x - 1$ $y = x^2 + 2x + 3$
39. $y^2 + x^2 = 1$ $y = 2x$
40. $y^2 + x^2 = 1$ $y = x - 1$
41. $y = x^3$ $y = x^2 - 3x$
    [*Hint:* Factor an x in the resulting equation.]
42. $y = \sqrt{x} + 1$ $y = -x + 3$
    [*Hint:* Let $x = z^2$.]

0.4 Graphs of Nonlinear Functions

43. You need a calculator with a $\boxed{y^x}$ key to do this exercise. In Example 5, we sketched a graph of the function

$$y = 2^x$$

From algebra you already know how to calculate its values when x is a rational number; for example, $2^1 = 2$, $2^2 = 4$, $2^5 = 32$, $2^{-3} = \left(\frac{1}{2}\right)^3 = \frac{1}{8}$, $2^{1/2} = \sqrt{2} \approx 1.414$, $2^{1/3} = \sqrt[3]{2} \approx 1.26$. The graph in Example 5, however, indicates that 2^x has a value for *any* real number x. It is technically difficult to define numbers like $2^{\sqrt{2}}$ and 2^π; we will not do so. Use your calculator to evaluate the following powers of 2, and compare the results with the value indicated by the graph in Figure 0.32: $2^{\sqrt{2}}$, $2^{\pi/3}$, $2^{0.012}$, $2^{-0.99}$, $2^{-0.00001}$.

Computer Application Problems

Use the BASIC program FUNVAL1 to graph the functions in Exercises 44–46 on the indicated interval. This program will give you values of the function at $N + 1$ equally spaced points in the interval.

44. $f(x) = 2x^4 + 0.7x - 1.5$ on $[-2, 2]$

45. $f(x) = \dfrac{x^3}{3} - \dfrac{7x^2}{8} + \dfrac{5x}{8} + 1$ on $[0, 2]$

46. $f(x) = \left(1 + \dfrac{3}{x}\right)^x$ on $(0, 1]$

Note that in BASIC a product $a \cdot b$ is entered as a*b and a power x^n is entered as x∧n.

0.5 Chain Processes and New Functions from Old

The purpose of this section is to show how functions are combined to represent an overall relationship in terms of its constituent smaller parts.

One thing leads to another, and that to yet another. When a little rain falls in the morning, some students take umbrellas to class, and Lost-and-Found has more business the following day. If interest rates on home mortgage loans begin to rise, fewer homes are built and then lumber prices fall. Soon thereafter Idaho sawmills begin to shut down, and laid-off workers elect a new representative to Congress, who then calls for the Federal Reserve Board to lower interest rates.

These examples indicate that an overall relationship within a larger system is made up of smaller, more direct relationships. Moreover, the representative from Idaho who wishes to affect interest rates needs to know not only that interest rates have risen and lumber prices fallen but also how the dependence works itself out through the *intervening process* of home construction. Calculus needs a similar mechanism. This mechanism, called the **composition of functions**, is a way of combining functions and allows us to construct an overall relationship from a chain of relationships, where each relationship leads into the next.

Definition

Let f be a function whose domain includes the range of another function g. The *composite of f with g*, denoted by $f \circ g$, is the function whose value at x is given by the equation

$$(f \circ g)(x) = f(g(x))$$

This concept is a bit abstract. It defines a new function $f \circ g$ from two given functions f and g. This new function is definitely *not* the product of f and g, as the next example shows.

EXAMPLE 1
Let $f(x) = x^2 + x + 1$ and $g(x) = x - 3$.
a. Compute $f(g(4))$ and $f(g(\sqrt{2}))$.
b. Find a formula for $f \circ g$.

Solution
a. Since $g(4) = 4 - 3 = 1$, then
$$f(g(4)) = [g(4)]^2 + g(4) + 1 = (1)^2 + 1 + 1 = 3$$
Similarly,
$$\begin{aligned} f(g(\sqrt{2})) &= [g(\sqrt{2})]^2 + g(\sqrt{2}) + 1 \\ &= (\sqrt{2} - 3)^2 + (\sqrt{2} - 3) + 1 \\ &= 2 - 6\sqrt{2} + 9 + \sqrt{2} - 2 = 9 - 5\sqrt{2} \end{aligned}$$
b. $(f \circ g)(x) = f(g(x)) = [g(x)]^2 + g(x) + 1 = (x - 3)^2 + (x - 3) + 1$
$= x^2 - 6x + 9 + x - 2 = x^2 - 5x + 7$ ∎

That is, to find a formula for the composite function $f \circ g$, we only need to substitute $g(x)$ (and then its formula) wherever the variable appears in the formula for f.

In Example 1, once we have a general formula for $f \circ g$ from Solution (b), we can use it to obtain the values in Solution (a). For example, from Solution (b)
$$(f \circ g)(4) = 4^2 - 5 \cdot 4 + 7 = 23 - 20 = 3$$
We can also compute the composite function $g \circ f$ in Example 1. The result, however, may surprise you.

EXAMPLE 2
Show that for the functions given in Example 1
$$f \circ g \neq g \circ f$$

Solution We have already seen that $(f \circ g)(x) = x^2 - 5x + 7$. Now $(g \circ f)(x) = g(f(x)) = f(x) - 3 = (x^2 + x + 1) - 3 = x^2 + x - 2$. Since $x^2 - 5x + 7 \neq x^2 + x - 2$, then $f \circ g \neq g \circ f$. ∎

This example is the general rule. Never expect that $f \circ g = g \circ f$. Always assume that $f \circ g \neq g \circ f$.

An Application of Composite Functions

EXAMPLE 3

Suppose that

$$g(r) = \frac{50 - 2r}{15}$$

is the number (in millions) of homes built when the interest rate on home loans is $r\%$

and

$$f(x) = \frac{x + 5}{10}$$

is the price per board foot of Douglas fir lumber when x millions of homes are built (one board foot is a piece of lumber 12 by 12 by 1 in. thick).

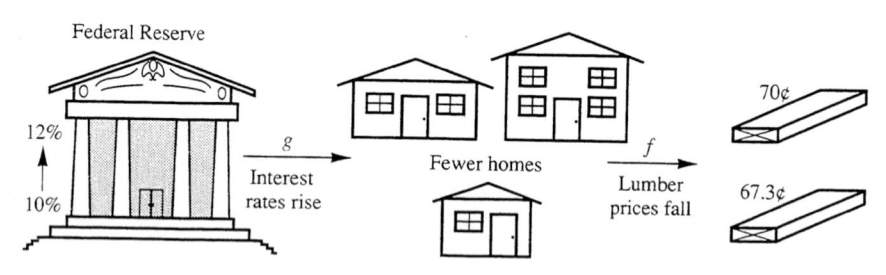

Figure 0.46

a. When the interest rate is $r = 10\%$, how many millions of homes are built? At this level of home building, what is the price of lumber?
b. What is the price of lumber when the interest rate is $r = 10\%$? Relate this answer to $f(g(10))$.
c. Write a formula for $f \circ g$. What does $f \circ g$ represent?
d. What is the price of lumber when the interest rate is $r = 12\%$?

Solution

a. When the interest rate r is 10%, there will be

$$g(10) = \frac{50 - 2(10)}{15} = 2 \text{ million homes built}$$

When $x = 2$ million homes are built, the price of lumber is

$$f(2) = \frac{2 + 5}{10} = \frac{7}{10} = \$0.70 \text{ per board foot}$$

b. From (a), at $r = 10\%$, 2 million homes are built and at this level of production the price of lumber is $0.70 per board foot. This is exactly the number $f(g(10))$.

c. By definition

$$(f \circ g)(r) = f(g(r)) = \frac{g(r) + 5}{10}$$

$$= \frac{[(50 - 2r)/15] + 5}{10} = \frac{125 - 2r}{150}$$

Using (a) and (b) as a guide, we see that $f \circ g$ is the function that represents the dependency of lumber prices upon interest rates.

d. Using (c), the lumber price at $r = 12$ is

$$(f \circ g)(12) = \frac{125 - 2(12)}{150} \approx \$0.673 \text{ per board foot} \blacksquare$$

This example illustrates a key point: The overall direct dependence of lumber prices on interest rates is given by the formula found for $f \circ g$. We could then let $L(r) = (f \circ g)(r)$ denote this newly formulated overall relationship. A diagram like that in Figure 0.47 can be used to illustrate such a composite, or chain, relationship.

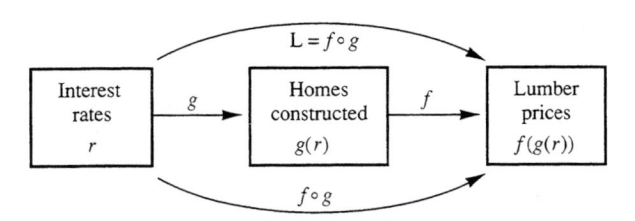

Figure 0.47

The three functions f, g, and $L = f \circ g$ relate all the variables at issue. The composite function $L = f \circ g$ is the direct relationship between the *beginning* and *end* of the chain; this chain begins in the domain of g and ends in the range of f.

Simplification via Composition

The composition of functions will also repeatedly be used to break down complicated formulas into simpler parts, that is, to decompose an overall relationship into a chain of simpler relationships.

EXAMPLE 4

Write $h(x) = \sqrt{(x+3)^2 + 1}$ as a composition of functions in two different ways.

Solution We can write $h(x) = f(g(x))$ where $f(x) = \sqrt{x^2 + 1}$ and $g(x) = x + 3$, since

$$f(g(x)) = \sqrt{g(x)^2 + 1} = \sqrt{(x+3)^2 + 1}$$

We can also write $h(x) = p(q(x))$ where $p(x) = \sqrt{x}$ and $q(x) = (x+3)^2 + 1$. Thus the decomposition of h as a composite function is not unique; the advantage of one decomposition over another will become clear in use. $\blacksquare$

Digression. A decomposition of the kind we have seen in Example 4 is closely related to the subtle yet essential concept of mathematical *form*. An untaught awareness of form, as distinct from detail, and as an entity in itself, is often found among those who are "good" at algebra. *Form* is that property of an algebraic/functional expression that remains when all particulars are gone. This key concept is not easy to convey or grasp. You might think of form as akin to "a footprint in the sand," conveying a message of substance and direction but containing nothing in itself. The solution to the next example is no more than a recognition of abstract form.

EXAMPLE 5

Write $h(x) = 2^{x^2-1}$ as a composition of functions.

Solution   We can write $h(x) = f(g(x))$ where $f(x) = 2^x$ and $g(x) = x^2 - 1$, since $f(g(x)) = 2^{g(x)} = 2^{x^2-1}$. ∎

The Domain of a Composite Function

Let us consider a brief example of a technical detail involved in the composition of functions. If $f(x) = 1/(1 - x)$ and $g(x) = 2x + 3$, then $f(g(-1))$ has no meaning, for $g(-1) = 1$ and $f(1)$ has no meaning: The domain of f does not contain $g(1)$. For any $x \neq -1$, we can compute $f(g(x))$ as before:

$$f(g(x)) = \frac{1}{1 - g(x)} = \frac{1}{1 - (2x + 3)} = \frac{1}{-2x - 2} = \frac{1}{-2(x + 1)}$$

Thus the domain of $f \circ g$ is not necessarily the same as g. On the other hand,

$$g(f(x)) = 2f(x) + 3 = 2\left(\frac{1}{1 - x}\right) + 3 = \frac{2 + 3(1 - x)}{1 - x} = \frac{5 - 3x}{1 - x}$$

can be computed for any value of x in the domain of f, since the domain of g contains every value $f(x)$ of f.

New Functions from Old via Ordinary Arithmetic

We conclude this section with some easier ways of forming new functions from old. Here we define the sum $f + g$, difference $f - g$, product $f \cdot g$, and quotient f/g of functions f and g by

$$(f + g)(x) = f(x) + g(x)$$

$$(f - g)(x) = f(x) - g(x)$$

$$(f \cdot g)(x) = f(x)g(x)$$

$$\left(\frac{f}{g}\right)(x) = \frac{f(x)}{g(x)} \quad \text{where } g(x) \neq 0$$

EXAMPLE 6

Find $f + g, f - g, f \cdot g$, and f/g for $f(x) = x + 1$ and $g(x) = x^2 - 1$.

Solution

$$(f + g)(x) = (x + 1) + (x^2 - 1) = x^2 + x$$
$$(f - g)(x) = (x + 1) - (x^2 - 1) = x - x^2 + 2$$
$$(f \cdot g)(x) = (x + 1)(x^2 - 1) = x^3 + x^2 - x - 1$$

and $\left(\dfrac{f}{g}\right)(x) = \dfrac{x + 1}{x^2 - 1} = \dfrac{x + 1}{(x + 1)(x - 1)} = \dfrac{1}{x - 1}$ if $x \neq \pm 1$ ∎

The sum, difference, product, and quotient of functions can have applied meaning if the functions themselves do. For example, if

$$C(x) = \text{total cost (in dollars) to manufacture } x \text{ items}$$

and $S(x) = $ total sales (in dollars) of these x items

then $(S - C)(x) = S(x) - C(x) = $ sales $-$ cost $=$ profit

On the other hand

$$(S + C)(x) = S(x) + C(x)$$

while having (dollar) meaning, is not commonly useful. Let us consider two further examples from economics.

EXAMPLE 7

Suppose that $C(x) = 0.02x^2 + 5x + 1{,}000$ is the cost to manufacture x items. Represent the average cost (per item) to produce x items by a function.

Solution Consider first a particular case. The cost to produce, say, 10 items is $C(10) = 0.02(10)^2 + 5(10) + 1{,}000 = \$1{,}052$. Therefore the average cost per item of these 10 items is $\$1{,}052/10 = \$105.20$.

This example tells us that the average cost function must be a quotient: total cost divided by the number of items produced. Let $A(x)$ be the average cost to produce x items. The function A is given by the formula

$$A(x) = \frac{C(x)}{x} = \frac{0.02x^2 + 5x + 1{,}000}{x} = 0.02x + 5 + \frac{1{,}000}{x} \quad ∎$$

A **demand function** expresses the relationship between the price of an item and the demand by consumers for it. Economists treat this relationship in a special way for reasons that become apparent in Example 8. In economics a demand function p is defined as follows: Let

$$p(x) = \text{price of an item when } x \text{ items are produced}$$

(Note that p is not the number of items demanded at a certain price but rather the reverse relationship.) Since the price of an item tends to fall as more items become available, the graph of a demand function often looks like the graph in Figure 0.48.

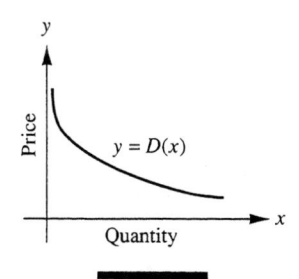

Figure 0.48

EXAMPLE 8

Suppose that
$$p(x) = -0.03x + 20$$
is the price (demanded) when x items are available for sale. Find a function representing the total revenue from the production and sale of x items.

Solution Let us think first in specific terms. When $x = 300$ items are available, the price of each item will be $p(300) = -0.03(300) + 20 = \$11$. Therefore the total revenue from the sale of these 300 items is $\$11 \times 300 = \$3,300$. That is,

$$\text{Revenue} = \text{unit price} \times \text{quantity sold}$$

Let $R(x)$ represent the total revenue from the sale of x items. Since $p(x)$ is the price per item when x items are sold, then $R(x)$ is a product of two functions:

$$R(x) = x \cdot p(x) = x(-0.03x + 20) = -0.03x^2 + 20x \quad \blacksquare$$

Background Review 0.5

Let a, b, c, and d be real numbers. Then

1. $\dfrac{a}{b} + \dfrac{c}{b} = \dfrac{1}{b}(a + c) = \dfrac{a + c}{b}$, $b \neq 0$

2. $\dfrac{a}{b} + \dfrac{c}{d} = \dfrac{ad + bc}{bd}$

   $\dfrac{a}{b} + 1 = \dfrac{a}{b} + \dfrac{b}{b} = \dfrac{a + b}{b}$, $b, d \neq 0$

3. $\dfrac{a/b}{c/b} = \dfrac{a}{c}$, $b, c \neq 0$

4. $\dfrac{a/b}{c/d} = \dfrac{a}{b} \cdot \dfrac{d}{c} = \dfrac{ad}{bc}$, $b, c, d \neq 0$

5. $a^2 - b^2 = (a - b)(a + b)$

   Note: $\dfrac{a^2 - 1}{a} \neq a - 1$ but $\dfrac{a^2 - a}{a} = a - 1$

Exercises 0.5

Compute (a) $(f \circ g)(x)$ and (b) $(g \circ f)(x)$, and then compute (c) $f(g(2))$ and (d) $f(g(a + 1))$, where defined, for each pair of functions in Exercises 1–6.

1. $f(x) = 2x + 1$ $g(x) = 1 - x^2$
2. $f(x) = \sqrt{x - 1}$ $g(x) = x^2 - 1$, $x \geq 0$
3. $f(x) = x^2$ $g(x) = \sqrt{x - 1}$
4. $f(x) = \log_{10} x$ $g(x) = |x - 1|$
5. $f(x) = \dfrac{x + 1}{x - 1}$ $g(x) = \dfrac{1}{x}$
6. $f(x) = x\sqrt{x + 1}$ $g(x) = x^2$

Write each function $h(x)$ in Exercises 7–12 as a composition $f(g(x))$. Is your answer the only possible answer?

7. $h(x) = 2(x + 1)^{13}$
8. $h(x) = \sqrt{x - 1} + 1$
9. $h(x) = ((x - 1)^2 + 1)^{1/3}$
10. $h(x) = (\sqrt{x} + 1)^7$
11. $h(x) = 2^{(x-1)^2}$
12. $h(x) = 2\sqrt{x + 3} + 5$

For each pair of functions in Exercises 13–17, compute (a) $f + g$ and (b) f/g in simplest algebraic form.

13. $f(x) = \dfrac{1}{x}$ \qquad $g(x) = x$

14. $f(x) = x^2 - 1$ \qquad $g(x) = x - 1$

15. $f(x) = \dfrac{x}{1 + x}$ \qquad $g(x) = \dfrac{1 - x}{1 + x}$

16. $f(x) = \dfrac{x - 1}{x + 1}$ \qquad $g(x) = \dfrac{2}{(x + 1)^2}$

17. $f(x) = \dfrac{-x}{x + 2}$ \qquad $g(x) = \dfrac{x}{x + 1}$

18. Let $f(x) = x^2$ and $g(x) = (x/2) - 1$. Graph the functions f, g, $f \circ g$, and $g \circ f$ on the same coordinate system.

19. Suppose that $B(x)$ is the number of bass living in a lake that supports x bluegill and that $g(w)$ is the number of bluegill in the lake when the water spider population of the lake is w (thousands). If $B(x) = x/20$ and $g(w) = w + 10{,}000$, how many bass are there when the lake supports 50,000 water spiders? When there are 100,000 spiders? When there are s spiders? When there are $s + h$ spiders? Interpret both $B(g(w))$ and $g(B(x))$.

20. A manufacturer of basketballs finds that in t hours of operation its plant produces $B(t) = 2t^2 - t$ hundred basketballs and that the cost to produce x hundreds of basketballs is $C(x) = 40 + 600x$ dollars. How much does it cost to run the plant for 2 hours? Find the formula, as a composite function, of production cost as a function of the number of hours the plant is operating. How much does it cost to run the plant for 4 hours? For $t + 2$ hours? Does $B(C(x))$ have any real-world meaning?

21. Suppose that while playing tennis, you observe that you win the point 2/3 of the time that you return the ball to the deep backhand corner of the court. Thus $w(x) = (2/3)x$ is the number of points you win on x such returns. On the other hand, suppose that you only succeed in placing the ball in the deep backhand corner on $p(t) = 3t/4$ of the times t you attempt such a return (hitting out of court otherwise). If you try 12 deep backhand returns, how many points should you win? On 16 such tries? On z tries? On $z + 4$ tries? Express what $w(p(t))$ represents in words. What does $p(w(x))$ represent?

22. Let $E(y) = $ the per capita beef consumption by Americans when beef costs y dollars per pound, and let $B(x) = $ the cost of beef (per pound) when corn costs x dollars per bushel. What does $E \circ B$ represent? What does $B \circ E$ represent? Are these functions equal?

23. Suppose that $C(x) = (x^2/100) + 3x + 900$ is the cost to manufacture x items.

   a. What does it cost to produce nothing at all; that is, what are the fixed costs before production begins?
   b. What is the average cost to produce 20 items?
   c. Define a function that describes the average cost to produce x items. Use this function to answer part (b) again.

24. Specify two functions f and g so that f/g represents the proportion of nuts to nuts and bolts in a box initially containing 75 nuts and 95 bolts, from which 1 nut and 1 bolt are removed each minute.

Challenge Problems

25. Suppose that the demand function for x items produced in Exercise 23 is
$$p(x) = \dfrac{-x}{50} + 24$$

   a. What is the total revenue for the sale of 20 items?
   b. Define a function that describes the total revenue from the sale of x items.
   c. Using the cost function in Exercise 23, define a function that represents the total profit from the sale of x items.
   d. At what level of production does the manufacturer "break even" (that is, revenue equals cost)?
   e. Define two functions, one representing average cost per item and the other average revenue per item, for the production and sale of x items. At what levels of production are these equal?

26. Suppose that an assembly-line robot has 1 min to remove 1 rivet and 1 washer from a box that initially contains 50 rivets and 75 washers. When the ratio of rivets to washers is R, the robot needs $3/R^2$ sec to find the next rivet and washer. Model this system by a composite function $f \circ g$ relating the time of removal to the time taken to find the next rivet and washer. Diagram this system and give explicit formulas and the domains of f and g. When will this system "crash"—that is, when will the time of removal exceed the 1 min allowed to find the next pair?

0.6 Describing and Measuring Change

The purpose of this section is to learn to translate ordinary English language into mathematical language and vice versa. In it we discuss average rate of change and average amount present.

Calculus was invented independently in the seventeenth century by Isaac Newton and Gottfried Leibniz, who used it to answer questions going back over 2,000 years. Today calculus, using their same concepts, answers questions that these mathematicians never imagined. This is possible because mathematics is a language and structure for thought that exists independently of its uses. In this respect, mathematical language is like ordinary language. For example, you can "go to a movie," or you can "go to Paris." The word *go* in these two statements is an applied abstraction, applicable to diverse circumstances. The language of mathematics—$f(t)$, $f(t + h)$, $f(3 + 0.2)$, and so on—though less familiar, is no more abstract and is equally applicable to diverse circumstances. In this section you will learn how to relate the concepts and computations of calculus to its applications via a study of the language and meaning of function notation. This is not a difficult topic, but it is an important one. It may be compared to the driver learning how to read a road map before departing on the journey.

Function Notation as Language: Average Rate of Change

We will discuss the language and uses of function notation in the context of measuring change. We begin by learning how to measure change in a situation where change is very slow.

Foresters studying the growth of southern yellow pine timber in a test plot are interested in the total amount of lumber available from the test plot, as a whole, over time. Let $L(t)$ denote the amount of lumber (in board feet) in the entire test plot at time t. The graph of the function L might look like the graph in Figure 0.49, since as the trees grow they shade each other and compete for nutrients, slowing overall growth.

The graph in Figure 0.49 could equally well model bacterial growth in a closed environment where growth is rapid once the bacteria are established but

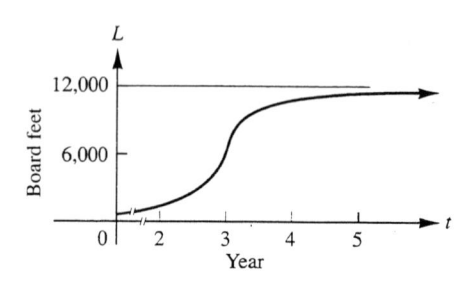

Figure 0.49

then slows as the bacterial population approaches the maximum population the environment can carry. In Chapter 8, we will see that this graph also models the spread of an infectious disease, a rumor in a population, or the sales growth of a new product; all of these rapidly increase and then level off near some saturation level. Such a model, which finds a general description and method that can describe a variety of similar phenomena, is one of the goals of mathematics.

In this discussion we do not suppose that we have an actual formula for $L(t)$, since a particular formula is irrelevant. We wish to study two things:

1. How computations expressed in function notation can be interpreted in ordinary language.
2. How statements in ordinary language can be expressed in function notation.

We illustrate this through a series of questions and answers. These questions deal with various states of this lumber production system—how much lumber is available at time t, represented by the function value $L(t)$—and with changes in these amounts.

EXAMPLE 1

a. How much lumber is available in the test plot in the fourth year?

   Solution This is represented by $L(4)$ which, from the graph, is approximately 10,800 board feet.

b. How much growth occurs between the second and fifth years?

   Solution $L(5) - L(2)$: (amount in year 5) minus (amount in year 2). The actual number, as indicated by the graph, is

   $$L(5) - L(2) = 11{,}500 - 1{,}500 = 10{,}000 \text{ board feet}$$

c. What does $[L(5) - L(2)]/3$ represent? Compute its value.

   Solution This is the average rate of growth per year in lumber in years 2 through 5. The numerator in this fraction measures the amount of growth—the change in total board feet available—in a three-year period. Division by 3 averages this growth over the three-year period. From Figure 0.50 this number appears to be

   $$\frac{11{,}500 - 1{,}500}{3} = \frac{10{,}000}{3} \approx 3{,}333 \text{ board feet per year}$$

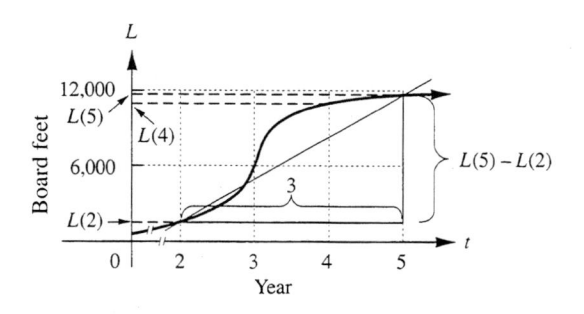

Figure 0.50

0.6 Describing and Measuring Change

Notice that (b) and (c) can be visualized on the graph in Figure 0.50. It is significant that the number $[L(5) - L(2)]/3$ is the *slope* of the line S.

d. What does $[L(3)]/3$ represent?

Solution Nothing special. This is the amount of lumber available in the third year divided by 3, the number of years. $[L(3) - L(0)]/3$, on the other hand, represents, again, the average rate of growth *per year* over the first 3 years.

e. What does $[L(2 + 3) - L(2)]/3$ represent?

Solution The same quantity as (c) because, after all, $L(2 + 3)$ and $L(5)$ mean the same thing.

f. How much growth occurs in the first 6 months of the second year?

Solution $L(2.5) - L(2)$, or $L(2 + 0.5) - L(2)$. This is the amount at time 2 subtracted from the amount available one-half year later, at time $t = 2 + 0.5 = 2.5$ years.

g. How much growth occurs in the first day of the second year?

Solution As in (f), the answer must be $L[2 + (1/365)] - L(2)$.

h. What is the *average rate of growth* in the first 6 months of the second year? In the first day of that year?

Solution As in (e), the computations are

$$\frac{L(2 + 0.5) - L(2)}{0.5} = \frac{2{,}700 - 1{,}500}{0.5} = 2{,}400 \text{ board feet per year}$$

and

$$\frac{L\left(2 + \frac{1}{365}\right) - L(2)}{\frac{1}{365}}$$

The numerical value of this quotient is too small to be seen on the graph in Figure 0.50. In each case we have an *amount* of growth divided by the *time* span in which the growth occurred.

i. What does $L(2 + h) - L(2)$ represent, where $h > 0$?

Solution The amount of growth between the beginning of year 2 and the later time $2 + h$.

j. What does $[L(2 + h) - L(2)]/h$ measure?

Solution The amount of growth (the change in the amount of lumber available) between time 2 and time $2 + h$ divided by the amount of time h that has elapsed between these two moments. Like (c) and (h), this computes the average rate of growth per year *over the time interval* from 2 to $2 + h$. Note that $h = (2 + h) - 2$.

k. How can the average rate of growth between an arbitrary time t and a later time $t + h$ be computed?

Solution As

$$\frac{L(t + h) - L(t)}{h} = \frac{\text{change in amount}}{\text{elapsed time}} \quad \blacksquare$$

Average Amount Present and the Area Beneath a Graph

There is a tendency to confuse the idea of *average rate* of growth or change with the idea of *average amount* present. For example, suppose that this class meets three times this week and the instructor notices that on each day there is the same number, say 30, of students present. Then the *average change* in the number of students present is 0. But the *average number* of students present is 30, since

$$\frac{30 + 30 + 30}{3} = 30$$

We conclude this section with the function-based language of *average amount present*.

EXAMPLE 2

Again, recall the graph of lumber available in the forester's test plot in year t (see Figure 0.51).

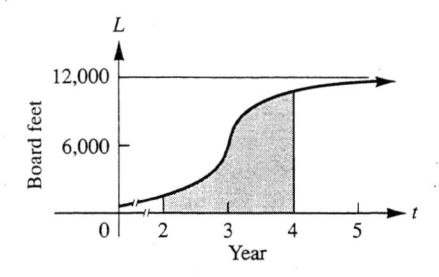

Figure 0.51

a. What is the average amount of lumber available at the beginning of years 2 through 4 inclusive?

b. What does

$$\frac{L(2) + L(2.5) + L(3) + L(3.5) + L(4) + L(4.5)}{6} =$$

$$\frac{1{,}500 + 2{,}700 + 6{,}000 + 10{,}000 + 11{,}000 + 11{,}250}{6} = 7{,}075$$

represent?

Solution

a. Recall that the average of a collection of n numbers $a_1, a_2, \ldots, a_n$ is $A = (a_1 + a_2 + \cdots + a_n)/n$. (For example, the average of the four numbers 2, 7, 10, and 13 is $(2 + 7 + 10 + 13)/4 = 8$.) Since $L(2)$, $L(3)$, and $L(4)$ are the amounts of lumber available at these times, the average amount available over this period is, from Figure 0.52(a),

$$\frac{L(2) + L(3) + L(4)}{3} = \frac{1{,}500 + 6{,}000 + 11{,}000}{3} \approx 6{,}166 \text{ board feet}$$

because $L(t)$ is the amount available at the beginning of each year.

b. Here the foresters have visited the test plot at six-month intervals for three years and have measured how much lumber is available at each visit. Then they have averaged these six amounts over the three-year period. This average available is different from what they would find if they visited the test plot only at yearly intervals—namely, $[L(2) + L(3) + L(4)]/3 = 6,166$. This difference is shown on the graph in Figure 0.52. The answer in (a) is one-third of the area of the three large rectangles of width one year in Figure 0.52(a). The answer in (b) is one-third of the six small rectangles of width one-half year in Figure 0.52(b).

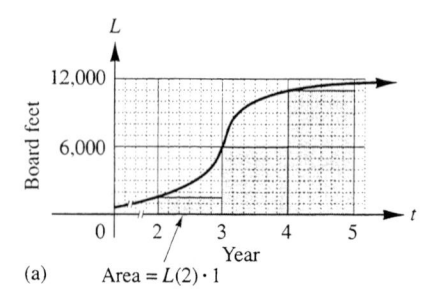

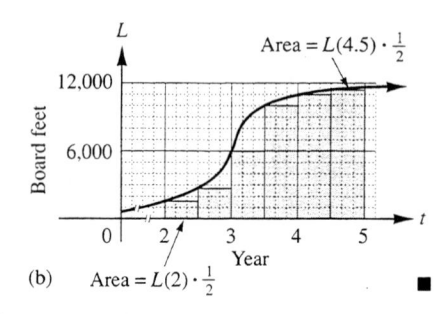

Figure 0.52

(a) Area = $L(2) \cdot 1$

(b) Area = $L(2) \cdot \frac{1}{2}$

Area = $L(4.5) \cdot \frac{1}{2}$

Figure 0.52 suggests that the "true" average amount of lumber available in years 2 through 5 is one-third of the exact area beneath the graph of L between year 2 and year 5. The gaps between the rectangular pieces and the curve of L itself represent the foresters' ignorance of the true average amount if they visit the test plot at only six-month or yearly intervals. They would find less error if they measured more often, but they cannot visit the test plot at every instant in time and actually measure the true average. Integration, which we study in Chapter 4, enables us to compute a true average amount.

Exercises 0.6

It is unlikely that foresters would ever actually consider the kinds of equations addressed in this section in the way that we did, since time and tree growth in a forest are so slow. In Exercises 1–7, we consider the spread of a highly contagious flu virus in a school of 900 students. Let $f(t)$ denote the number of students who have caught the flu by the end of day t after it first appears, where a "day" means an 8 A.M. to 4 P.M. school day (see Figure 0.53), and day t "begins" at $t - 1$.

1. Write in ordinary language, using as few mathematical words as possible, what the following computations represent.

   a. $f(4.5)$

   b. $f(4) - f(2)$

   c. $\dfrac{f(6) - f(4)}{2}$

   d. $\dfrac{f(5)}{5}$

   e. $f\left(4 + \dfrac{1}{24}\right) - f(4)$

   f. $\dfrac{f(4 + h) - f(4)}{h}, \; 0 < h < \dfrac{1}{24}$

   g. $\dfrac{f(4 + h) - f(4)}{h}, \; -\dfrac{1}{24} < h < 0$

   h. $f(t + h) - f(t), \; h > 0$

2. Draw a graph that illustrates the geometrical meaning of the computations in parts (b), (c), (e), (f), (g), and (h) of Exercise 1. (See questions (c), (d), and (g) in Example 1.)

3. On which days does the flu

   a. Appear to be spreading fastest?

   b. Appear to be spreading slowest?

   c. Appear to be spreading at equal rates?

52   Chapter 0   Functions

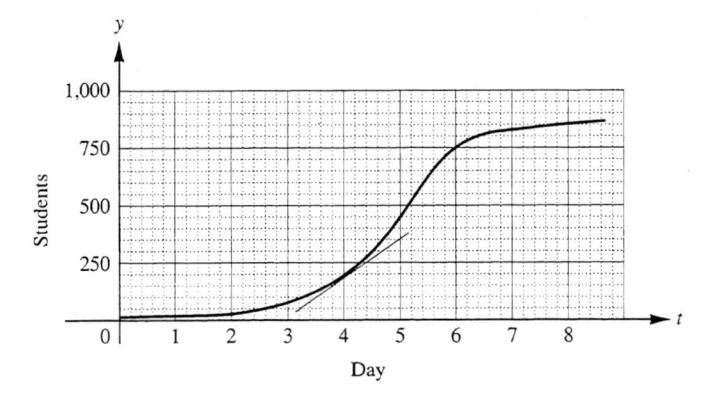

Figure 0.53

4. **a.** What is your best estimate of how fast this flu is spreading on day 4?
   b. When does the flu appear to be spreading at the rate of 150 students per day? (This is hard to pin down exactly!)
   c. Suppose that public health authorities could begin an immediate vaccination program against this flu in which 150 students per day could be inoculated. What would be the likely result if the program is instituted on day 2; on day 4? Illustrate your answer by a graph.

5. Write down, in function notation, the following quantities.
   a. The number of new cases of flu between days 5 and 6.
   b. The average number of new cases per day in the first 6 days.
   c. The average number of new cases per day over days 5 and 6.
   d. The average number of new cases per day between day t and day $t + 1$.
   e. The average number of new cases per day over the morning of the fifth day.

6. Based on the graph, estimate the number answers to questions (a), (b), (c), and (e) of Exercise 5.

7. This question is about average amount present rather than average rate of change: A public health official calls the school's principal each morning of days 4, 5, and 6 and asks how many students are ill with the flu. At noon on day 6 the health official reports the average number of students with the flu to the town's newspaper. [*Remark.* Day 1 begins at $t = 0$, and so on.]

   a. In function notation, what number would appear in the newspaper?
   b. Based on the graph, estimate this number.
   c. The health official decides to call the school twice a day (at the beginning and middle of each day) and report the average number of cases over days 4, 5, and 6. What number, in function notation, would he report?
   d. Based on the graph, estimate this number.
   e. What is your best estimate of the true average number of students with the flu over days 4, 5, and 6? Can you relate this estimate to a certain area below the graph in Figure 0.53?

8. This exercise begins with information about speed (or rate) and asks about the distance covered. Recall that

   $$\text{Distance} = \text{rate} \times \text{time}$$

   Suppose that you are riding a bicycle down the road at a constant rate. Suddenly a gust of wind comes up from behind and for 3 sec increases your speed. The graph of your velocity $V(t)$ at time t might look like the graph in Figure 0.54.

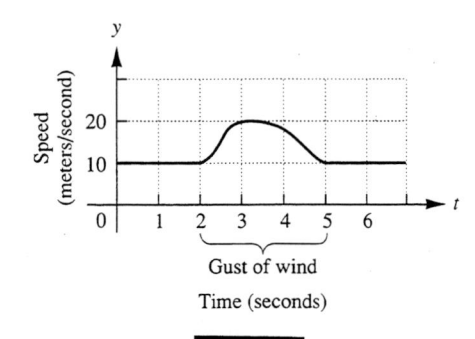

Figure 0.54

0.6 Describing and Measuring Change

Answer each question (a)–(f) in ordinary language, and try to indicate an area in Figure 0.54 that illustrates your answer.

a. How far do you travel in the first 2 sec?

b. What does $V(1) \cdot \frac{1}{2}$ represent in terms of distance traveled?

c. Does $V(2) \cdot \frac{1}{2}$ represent a distance that you actually travel in the half-second beyond time 2?

d. Do you travel a longer or shorter distance between times 0 and 2 than between times 2 and 4?

e. What does

$$V(2) \cdot \tfrac{1}{2} + V(\tfrac{5}{2}) \cdot \tfrac{1}{2} + V(3) \cdot \tfrac{1}{2} +$$
$$V(\tfrac{7}{2}) \cdot \tfrac{1}{2} + V(4) \cdot \tfrac{1}{2} + V(\tfrac{9}{2}) \cdot \tfrac{1}{2}$$

represent?

f. Considering parts (a) and (e) and the area beneath the graph, estimate how far you actually traveled between time 2 and time 5. Estimate the error between how far you actually traveled and the number in part (e).

9. "Auto sales have risen 30% in April over March of this year, an increase twice that of a year ago." Let us examine this statement in terms of function notation.

First recall that a percentage is a proportion of a whole. Thus if a basket contains 20 lb of apples and oranges and if 8 lb of these consist of oranges, then the basket contains 40% oranges. Shown in Figure 0.55 is a hypothetical graph of auto sales in each month over a $1\frac{1}{2}$-year period. Notice that months are numbered, with "0" being January. Use this graph to answer the following questions.

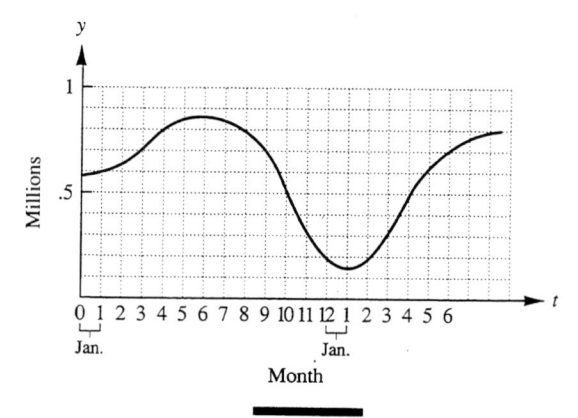

Figure 0.55

a. What is the level of sales in April of the first year; of the second year? What is the level of sales in March of each year? How much did sales change in each year between March and April? What is the percentage change in sales from March to April, relative to the total number of sales in March, in each year? In which years were sales largest?

b. If $S(x)$ = total sales in month x, explain in ordinary language what $S(6) - S(5)$ represents. What does

$$\frac{S(6) - S(5)}{S(5)}$$

represent?

c. If x is any month, what does

$$\frac{S(x+1) - S(x)}{S(x)}$$

represent?

Chapter 0 Summary

1. Applied calculus is the study of relationships between variables that depend on each other within a system.

2. A relationship is expressed in mathematical form as a function. A function serves as a mathematical model of the system under study. The formula and notation $f(x)$ for a function are used to discuss and compute particular values in the *range* of the relationship as these values depend on a value x in the *domain* of the function. The number $f(x)$ represents a state or condition of the system that the function f models.

3. The graph of a function is an alternate way to view a function as a whole. The simplest graph is that of a *linear function* $f(x) = mx + b$. You should know the graphs of the *nonlinear functions* $y = x^2$, $|x|$, and $1/x$ and how they may be translated and reflected.

4. The composition of functions has two uses. It represents a chain relationship, and it is also used to represent the abstract *form* of functional expressions.

5. The expression $f(x + h) - f(x)$ measures the change in function values between x and at $x + h$. Hence $[f(x + h) - f(x)]/h$ measures the average change between these values since $h = (x + h) - x$ is the change in the variable x.

Chapter 0 Summary Exercises

In Exercises 1–6, evaluate the given function at the numbers 2, -1, $\sqrt{3}$, $1 - h$, and a^2.

1. $f(x) = (x + 1)^2$
2. $g(t) = \dfrac{2t}{3t + 1}$
3. $h(u) = \dfrac{u^2 - 1}{u^2 + 1}$
4. $f(z) = z\sqrt{z + 1}$
5. $k(x) = x^3 - 5$
6. $f(x) = \dfrac{x + 1}{2x}$

Describe the domain of each function in Exercises 7–12.

7. $f(t) = t\sqrt{t - 1}$
8. $g(x) = \dfrac{x}{x^2 + 1}$
9. $h(z) = \dfrac{z}{z^2 - z - 2}$
10. $f(x) = \sqrt{\dfrac{1}{x - 1}}$
11. $f(x) = \begin{cases} \dfrac{\sqrt{x - 1}}{x - 2} & x \ne 2 \\ 1 & x = 2 \end{cases}$
12. $g(x) = \begin{cases} \dfrac{x - 2}{x^2 - 1} & x \ne \pm 1 \\ 3 & x = -1 \end{cases}$

Determine whether $y = 4$ is in the range of each function in Exercises 13–18. If so, find the value of the independent variable that yields the function value 4.

13. $f(x) = 3x + 1$
14. $g(t) = t^2 - 5$
15. $h(u) = \dfrac{u + 1}{u - 1}$
16. $f(t) = \dfrac{t^2 - 4}{t - 2}$
17. $k(z) = z^3 + 12$
18. $h(x) = -|x - 4|$

Let $f(x) = (x + 1)/(x - 1)$ and $g(x) = 2/x$. Compute the value of the compositions in Exercises 19–26.

19. $f(g(1))$
20. $g(f(3))$
21. $f(g(x))$
22. $g(f(t))$
23. $f(g(2)) + 3$
24. $g(f(2 + 1))$
25. $f(g(a^2))$
26. $g(f(a + h))$

Give the slope, x-intercept, y-intercept, and one other point on the graph of each of the equations in Exercises 27–30. Graph each equation.

27. $y = 2x - 3$
28. $3y = -9x - 3$
29. $x = 3y + 4$
30. $y - 1 = 3 - \tfrac{1}{2}x$

Write the equation of the straight line given by the information in Exercises 31–36. Graph each equation.

31. Slope -2; y-intercept 1
32. x-intercept 3; y-intercept 2
33. Slope $\sqrt{2}$; $(\sqrt{2}, 4)$ on the line
34. $(-1, 1)$ and $\left(1, \tfrac{1}{2}\right)$ on the line
35. Slope $\tfrac{1}{2}$; x-intercept 2
36. Slope $\tfrac{1}{2}$; y-intercept $-\tfrac{3}{4}$

Graph each function in Exercises 37–42.

37. $f(x) = (x - 1)^2$
38. $g(x) = \left(\tfrac{1}{2}\right)^x$
39. $h(t) = t - \dfrac{1}{t}$
40. $f(x) = 2|x|$
41. $g(t) = (t - 2)^2 - 3$
42. $h(z) = -1$

43. Find all points of intersection for the graphs of $f(x) = -x + 1$ and $g(t) = t^2 - 1$.

44. Find all points of intersection for the graphs of $h(z) = 2z$ and $f(t) = 1/(t + 1)$.

Write each expression in Exercises 45–48 as a composition $f \circ g$ in some way. Specify f and g separately.

45. $\sqrt{1 - x^2}$

46. $\dfrac{3}{t^2 + 5}$

47. $(8 + x^2)^3$

48. 2^{x^2+1}

Write each expression in Exercises 49–52 as a product $f \cdot g$ or a quotient f/g of two functions f and g in some way; specify f and g.

49. $(x + 1)\sqrt{1 + 2x}$

50. $\dfrac{t^2 - 1}{t + 2}$

51. $(x^2 + 1)\left(\dfrac{3}{x + 2}\right)$

52. $2^x 3^{x+1}$

53. The loading dock at a shoe factory initially contains 35 boxes of tennis shoes and 60 boxes of jogging shoes. Each minute a conveyor belt delivers 5 boxes of tennis shoes and 3 boxes of jogging shoes to the dock. Express the ratio of boxes of tennis shoes to boxes of jogging shoes 5 minutes later, assuming that none are removed.

54. Refer to Exercise 53. Now suppose that each minute a worker loads 4 boxes of tennis shoes and 6 boxes of jogging shoes onto a truck. Express the ratio of boxes of tennis shoes to the total number of boxes of shoes at the loading dock t minutes later.

55. At time t (minutes) there are $2^{0.03t}$ yeast organisms in a rising loaf of french bread. The yeast organisms produce both alcohol and carbon dioxide in the ratio of $(1 + 50x)/(1 + x)$ μL of CO_2 per μL of alcohol for each x thousand yeast organisms. Express this ratio as a function of time t. Estimate the ratio of CO_2 to alcohol at the end of 2 hours of rising.

56. Profits depend on sales, which in turn depend on advertising. A company earns $P(x) = [25 - (x/200)]x$ dollars for the sale of x units, and $U(y) = 800[1 - 5{,}000/(y + 5{,}000)]$ is the number of units sold for an advertising expenditure of y dollars. What profit accrues from spending $5,000 on advertising? From spending y dollars on advertising?

57. Let $R(x)$ be the total revenue accruing to a company from the sale of x thousand units. Using this notation, express the following:

   a. Total revenue from the sale of 3,000 units.

   b. Average revenue per unit from the sale of 5,000 units.

   c. Average revenue per thousand units from the sale of 5,000 units.

   d. Average revenue per thousand units due to the sale of 3,000 units after 2,000 have already been sold.

58. Let $f(t)$ denote the number of snow tires sold in the first t days after the first snow of the season. Write in clear, ordinary language what each of the following computations represents.

   a. $f(4)$

   b. $\dfrac{f(5)}{5}$

   c. $f(5) - f(2)$

   d. $\dfrac{f(5) - f(2)}{3}$

   e. $f\!\left(3 + \tfrac{1}{4}\right) - f(3)$

   f. $\dfrac{f\!\left(3 + \tfrac{1}{4}\right) - f(3)}{\tfrac{1}{4}}$

56 Chapter 0 Functions

Chapter 1

Limits and the Derivative

On the Threshold of Calculus

The concept of a *limit value* allows us to enter a whole new domain of mathematics called *mathematical analysis*, and separates analysis from algebra, geometry, and earlier mathematics as surely and sharply as the surface of a smooth pond separates life above it from life below it. As you will see, calculus is not a new or more difficult form of algebra. In algebra, equality is the rule. In calculus, we approach problems through inequality and approximation—with approximation taken to a "limit." Calculus organizes this approach into concepts and methods that can often be given as equations, much as in algebra. The basis for these is the idea of the limit value of a function. We first use the limit concept to define the computational tool of the calculus, the *derivative of a function*, beginning with approximations of a particular form that permit us to go beyond a static relationship among variables and to estimate change. The limit of these approximations defines the derivative and allows us to precisely measure the dynamics of change in the external world.

1.1 Approximating the Tangent to a Curve

How can we compute an exact rate of change at a fixed instant? In this section we begin to answer this question by approximation *of the exact rate of change by the simpler concept of average rate of change.*

A common underlying theme in calculus recurs in a variety of particular forms: compute by approximation. More specifically, we wish to measure a quantity that is in some sense complex—the area of a region with a curved boundary or the slope of the line tangent to a curve—and we try to approximate this number by measuring a quantity that is much simpler—the total area of a group of rectangles that almost cover the region or the slope of a line that meets the curve at two points that are very close to each other. Once you are able to approach a computation in this way, calculus becomes a remarkable tool of vast application. In Section 1.2, we will study the key concept that allows us to organize approximation into a set of exact rules and procedures. In this section we will demonstrate how computation by approximation is used in a specific case. We will compute the exact rate of change of a curve at a point by approximation by its average rate of change over an interval.

Approximating the Rate of Growth of a Curve

Consider a local market that reached a saturation level of 10,000 unit sales over a five-month period and in which total sales $U(t)$ by time t grew month by month as indicated in Figure 1.1. At first sales grew slowly; then they increased more rapidly until about half the market was reached. After that, sales again grew more slowly; finally, sales rose at an even slower rate as the saturation level was approached. We want to compare and compute *rates* of growth at various times; these rates directly affect cash flow into the company. The attempt to compute the exact rate at a particular time will lead us to the concept of the derivative of a function, introduced in Section 1.4, after we have learned about limit values. Here we take only the first step in computing such a rate.

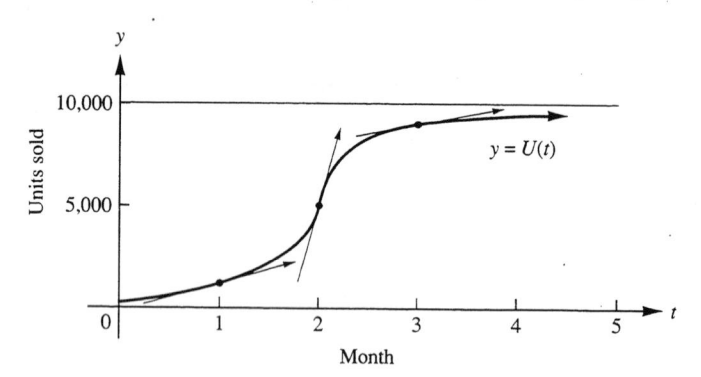

Figure 1.1

In Figure 1.1, the rate of sales growth appears to be fastest at the start of month 2 and appears to be faster at the end of 2.5 months than at the end of 1 month. Let us consider month 2. When we say that growth is fastest at this time, we mean that the graph (of total growth) is rising fastest at that moment. This is not, however, a precise enough description of the rate of growth at month 2. How can we make more precise the observation from Figure 1.1 that growth is faster at month 2 than, say, at month 3? And how can we measure how fast growth is at these two times?

Our intuitive sense of "rate of growth" suggests that the rate of growth at month 2 or month 3 coincides with the direction of the graph at that moment, as measured by the rise/run ratio, or slope, of a straight line drawn in exactly the direction of the graph at that time. We indicate these straight lines in Figure 1.2 and call these the tangent lines to the graph at time 2 and time 3, respectively. The **tangent line** to a curve at a point is the best straight-line approximation we can make to the curve at that point.

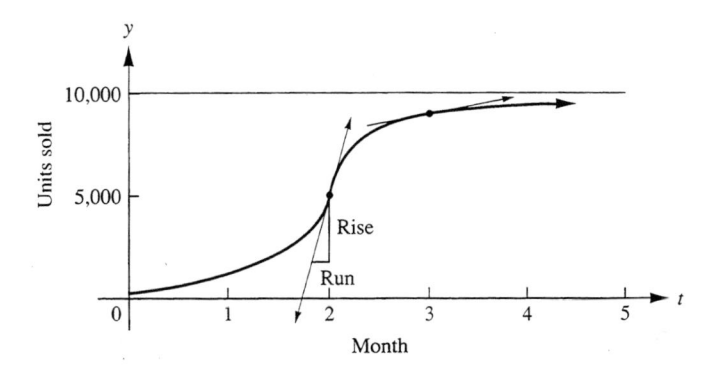

Figure 1.2

How can we compute or measure the slope of a line tangent to a curve? In fact we cannot even draw it precisely, since the width of a pencil mark obscures part of the curve. This tangent and its slope is the complex, exact answer that we need, but we can only approximate it by something simpler. First, we measure total sales by time $t = 2$, $U(2)$, and (a half-month later) $U(2 + 0.5)$. Then we calculate

$$\frac{U(2 + 0.5) - U(2)}{0.5} \simeq \frac{8{,}250 - 5{,}000}{0.5} = 6{,}500$$

This number is the *average* growth over the two-week period beginning at month 2, and it is also the slope m of the line through the two points indicated in Figure 1.3. Such a "two-point line" is called a **secant line**.

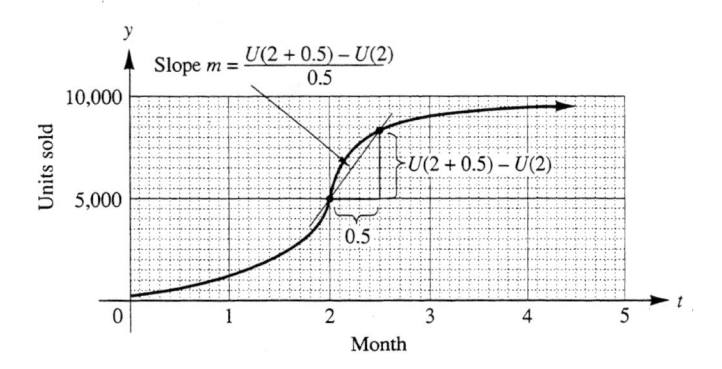

Figure 1.3

1.1 Approximating the Tangent to a Curve

This approximation is not accurate enough, however, because this calculation does not answer the original question, How fast is growth *at* month 2? Moreover, using a half-month period for an approximation at other times could give an erroneous impression. Recall our earlier observation that the growth rate appears to be faster at month 2.5 than at month 1. But the calculations

$$m_1 = \frac{U(1 + .05) - U(1)}{0.5} \approx 1{,}500 \text{ units/month}$$

and $$m_2 = \frac{U(2.5 + 0.5) - U(2.5)}{0.5} \approx 1{,}500 \text{ units/month}$$

give approximately the same answer. This is indicated by the slope of the secant lines in Figure 1.4. The time shift 0.5 (2 weeks) is too long a time to measure the growth rate at time $t = 1$ versus the growth rate at time $t = 2.5$.

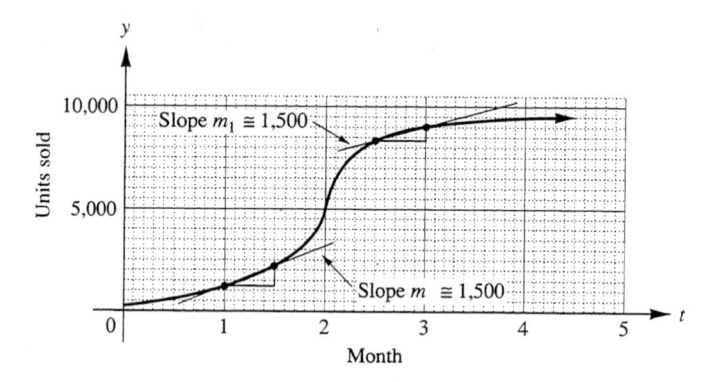

Figure 1.4

In Figure 1.5, we would like to measure the slope of the tangent line T at the point $(2, U(2))$, thus measuring the rate of growth *at that moment*. Such a measurement can be made only through approximation, not directly. We can measure total sales by month 2 and at some time just before or just beyond month 2—say, at time $2 + h$ (where h may be negative or positive). For example, we could measure total sales 1 week rather than 2 weeks later, using $h = 0.25$ (one-fourth of a month). We could then average growth over this 1-week period by computing

$$\frac{U(2 + 0.25) - U(2)}{0.25} \approx \frac{7{,}500 - 5{,}000}{0.25} = 10{,}000 \text{ units/month}$$

This calculation would give a better approximation to the rate of growth at month 2; notice that the secant line $S_{0.25}$ more nearly matches the desired tangent line.

Nevertheless, this calculation is still not good enough. The secant line is not the tangent line, and the average growth

$$\frac{U(2 + h) - U(2)}{h} \text{ sales/month}$$

where h is 0.5 (2 weeks later) or 0.25 (1 week later) is not the rate of growth at the instant when month 2 begins.

60   Chapter 1   Limits and the Derivative

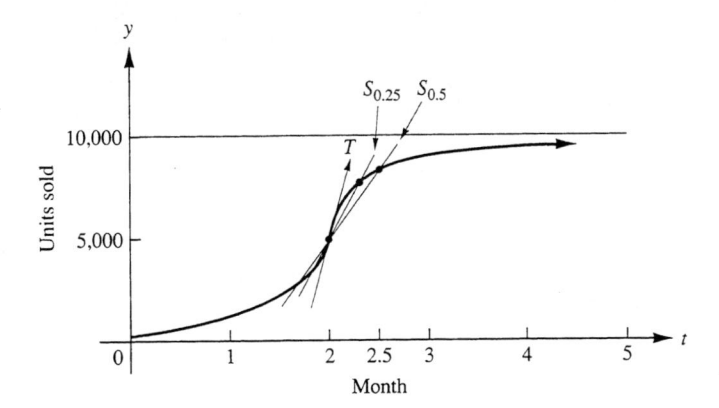

Figure 1.5

We could tirelessly measure total sales at month 2 and 1 day later $\left(h = \frac{1}{30}\right)$ and obtain a better approximation, more nearly the true rate at month 2 but not the exact value. An even better approximation could be obtained by measuring sales at the beginning of month 2 and 1 hour later $\left(h = \frac{1}{720}\right)$ or even only a half-hour $\left(h = \frac{1}{1,440}\right)$ later, but we would still not get an exact rate of growth or the exact slope of the tangent. Any amount of time h that elapses beyond the moment month 2 begins introduces some error and gives only an average rate of growth $[U(2 + h) - U(2)]/h$ and a secant line S only approximating the tangent line T. (In Figure 1.6, we have enlarged the scale, but imagine that the time gap h is very small.)

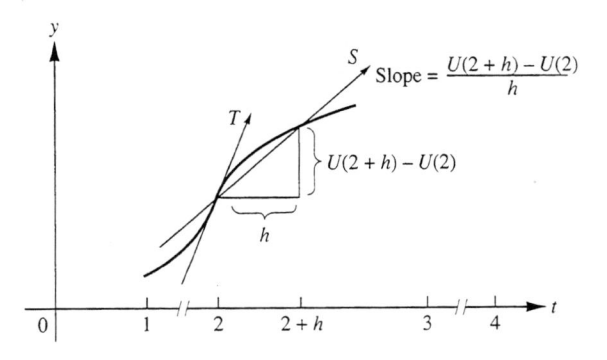

Figure 1.6

Apparently, we have to measure the total sales growth "instantaneously," which is physically impossible. Or, we must evaluate the approximation formula $[U(2 + h) - U(2)]/h$ at $h = 0$, which is mathematically impossible, since $[U(2 + 0) - U(2)]/0 = 0/0$ is meaningless.

Calculus will avoid such repetitive measurements and estimates of slope. If we can write an algebraic formula for the function $y = U(t)$, we can use calculus to find the exact rate of growth and the slope of the tangent T directly. To reach this point and understand the needed technique, we next make some specific calculations of the sort just discussed, using a particular function and its algebraic formula.

1.1 Approximating the Tangent to a Curve

Calculating the Complex by the Simpler

Let us consider the familiar function $f(t) = t^2$, $t \geq 0$, and study its rates of growth. Note from Figure 1.7 that the graph of f rises more steeply as t increases in value. We will try to compute how fast the graph is rising at $t = 1$, or alternatively, to compute the slope of the tangent line T to the graph at the point $(1, f(1)) = (1, 1^2) = (1, 1)$. The drawing in Figure 1.7 indicates that this slope is approximately 2. But we cannot be sure, since any slope arrived at visually may not be exactly right.

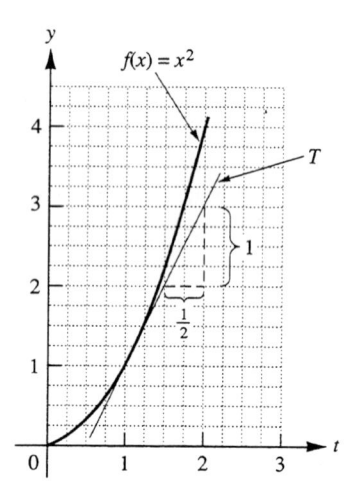

Figure 1.7

Applying the ideas discussed earlier in this section, let us compute the slopes of secant lines that hit the graph at the two points $(1, f(1))$ and $(1 + h, f(1 + h))$, where h is a small positive but nonzero number. The slope of this secant line is $[f(1 + h) - f(1)]/h$, as Figure 1.8 indicates. [*Note:* We can use a small negative number h as well; see Exercise 1(d).]

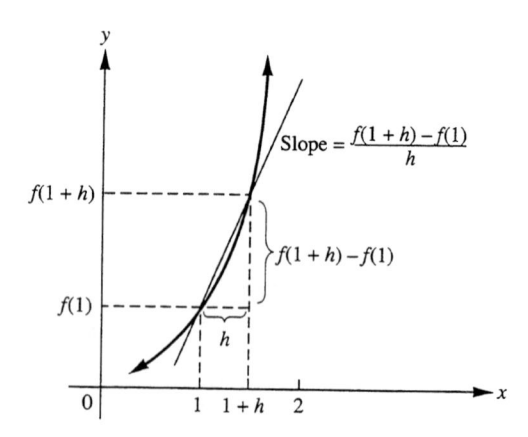

Figure 1.8

We can make this computation by preparing a table (see Table 1.1). The trend, indicated by the right-hand column of the table, is clear. The closer the point $1 + h$ is to the point 1, the closer the slope of the two-point secant line is to 2, the slope of the apparent tangent. When 1 and $1 + h$ are only 0.0005 unit apart, the slope of the two-point line differs from the apparent slope of the tangent by only 0.0004.

Table 1.1

h	$1 + h$	$f(1 + h) = (1 + h)^2$	$f(1)$	$f(1 + h) - f(1)$	$[f(1 + h) - f(1)]/h$
0.5	1.5	2.25	1	1.25	2.5
0.25	1.25	1.5625	1	0.5625	2.25
0.10	1.1	1.2100	1	0.21	2.1
0.05	1.05	1.1025	1	0.1025	2.05
0.005	1.005	1.010025	1	0.010025	2.005
0.0005	1.0005	1.0010002	1	0.0010002	2.0004

Table 1.1 was prepared with the aid of a calculator. There is an easier way to compute the number $[f(1 + h) - f(1)]/h$ in column 6. Recalling that $f(x) = x^2$, we see that (for $h \neq 0$)

$$\frac{f(1 + h) - f(1)}{h} = \frac{(1 + h)^2 - (1)^2}{h} = \frac{1 + 2h + h^2 - 1}{h} = \frac{2h + h^2}{h} = 2 + h \quad (1)$$

Hence even without a calculator we can make a table to any desired degree of accuracy (see Table 1.2). Notice that line 6 in Table 1.2, relying as it does on Equation 1, reveals an unsuspected roundoff error of 0.0001 in Table 1.1. Thus Equation 1 is an easier and more accurate method of approximating the slope of the tangent and the rate of change of $f(x) = x^2$ at $x = 1$.

Table 1.2

h	$1 + h$	$[f(1 + h) - f(1)]/h = 2 + h$
0.5	1.5	2.5
0.25	1.25	2.25
0.10	1.10	2.1
0.05	1.05	2.05
0.005	1.005	2.005
0.0005	1.0005	2.0005
0.00000005	1.00000005	2.00000005

Such a computation can be used to find a formula that approximates the slope of the tangent T at an *arbitrary* point $(x, f(x))$ on the graph of $f(x) = x^2$. For, using a second point $(x + h, f(x + h))$ as indicated in Figure 1.9, we can calculate the slope of the approximating secant as

$$\frac{f(x + h) - f(x)}{h} = \frac{(x + h)^2 - x^2}{h}$$

$$= \frac{x^2 + 2xh + h^2 - x^2}{h} = \frac{2xh + h^2}{h} = 2x + h$$

1.1 Approximating the Tangent to a Curve

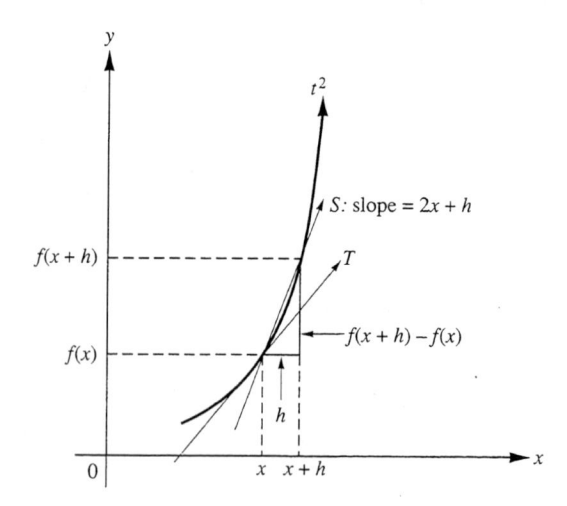

Figure 1.9

Thus, for example, the slope of the tangent to the graph at $x = 3$ is approximately $2(3) + h = 6 + h$. This is the exact slope of an approximating secant line between the points $(3, 9)$ and $(3 + h, (3 + h)^2)$.

For the moment this is the best we can do in a reasoned way. To find a formula for the exact slope of the tangent line, we must learn about limit values in Section 1.2.

Exercises 1.1

1. Draw a graph of $f(x) = x^2$ on 10×10 mesh-per-inch graph paper. Then draw the best tangent line T you can at the point $(2, f(2))$.

   a. Calculate the slope of T using the mesh on the paper.

   b. Approximate T by drawing a secant line through the points $(2, f(2))$ and $(2 + h, f(2 + h))$, where $h = 1, 0.5, 0.2,$ and 0.1, and measure the slope of each secant line.

   c. Compute $[f(2 + h) - f(2)]/h$ for $h = 1, 0.5, 0.2,$ and 0.1.

   d. Repeat parts (b) and (c) for $h = -1, -0.5, -0.2,$ and -0.1.

   e. Repeat parts (a), (b), (c), and (d) at the point $(3, f(3))$. [*Hint:* Replace 2 by 3 in the drawings and calculations.]

2. Repeat parts (a), (b), and (c) of Exercise 1 for

   a. The graph of $f(x) = x^3$, replacing $(2, f(2))$ by $(1, f(1))$.

   b. The graph of $f(x) = 1/x$ at $(2, f(2))$.

   (See Figures 0.18 and 0.19.)

3. You may ask, "If I know how much I have of some quantity, why should I want to know the rate at which the quantity is changing?" Consider the following rough model of the discovery and consumption of a natural resource (such as oil). Let $D(t)$ be the amount discovered in year t and let $C(t)$ be the amount used, or consumed, in year t. Since people do not even think to use a resource until after it is found, there is always a time lag T_L between discovery and consumption. Over a long enough period of time the graphs of C and D could look something like those in

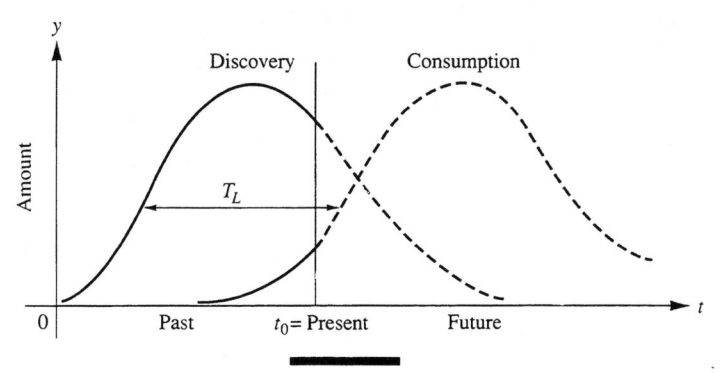

Figure 1.10

Figure 1.10. Graphs like these provide an overview of the growth and decline, but society as a whole tends to see only short intervals of time near the present.

a. Which is greater at time t_0 (the present), $D(t_0)$ or $C(t_0)$?

b. If observers only considered the amounts $D(t)$ and $C(t)$ over an interval of time near t_0, what could they conclude? Answer in ordinary language.

c. If observers considered instead the (average) rate of change of D and C near time t_0, what could they conclude?

d. In light of your responses to (b) and (c), is it possible that the same set of data can elicit the contradictory conclusions of "satisfaction" and "concern"? Try to phrase your response using the words *state* and *rate*.

4. On a sheet of 10 × 10 mesh-per-inch graph paper, draw the graphs of $y = x^2$ and $y = x^3$ for $0 \leq x \leq 1$, representing this interval by at least 5 inches of line. Notice that the graph of $y = x^2$ is never below the graph of $y = x^3$, but that both graphs start at (0, 0) and end at (1, 1). At which point x does the graph of $y = x^3$ begin to increase faster than the graph of $y = x^2$? (Estimate as closely as you can.)

5. A pro football team has heard of a hot linebacker prospect playing at Obscure University. They call one of their scouts in the Obscure area and ask for some film on him. The scout has a two-frame-per-second movie camera and films the linebacker on short 40-yard dashes down the field. The head coach receives the film and, by counting the yard markers on the football field as they appear in each frame of the film, can get some idea of how fast his prospect really is. In doing this the coach creates the chart in Table 1.3.

Table 1.3

Frame	Time (sec)	Yard Line
0	0	0
1	$\frac{1}{2}$	4
2	1	12
3	$\frac{3}{2}$	18
4	2	23
5	$\frac{5}{2}$	28
6	3	32
7	$\frac{7}{2}$	36
8	4	39

a. What is the linebacker's average speed in the first 4 sec; the first 2 sec; the first second; the first half-second? In the first half-second after he has already run 18 yd? After he has run 36 yd?

b. Draw a graph of Table 1.3, and indicate on it the geometric representation of your answers in part (a).

6. The coach in Exercise 5 realizes that he needs better information on this prospect, so he sends his scout a four-frame-per-second movie camera. In return the scout sends back film from which Table 1.4 is constructed.

Table 1.4

Frame	Time (sec)	Yard Line
0	0	0
1	$\frac{1}{4}$	1
2	$\frac{1}{2}$	4
3	$\frac{3}{4}$	$7\frac{1}{2}$
4	1	12
5	$\frac{5}{4}$	15
6	$\frac{3}{2}$	18
7	$\frac{7}{4}$	21
8	2	23

a. What is the linebacker's average speed in the first second? In the first $\frac{3}{4}$ sec? In the first $\frac{1}{4}$ sec? Between $\frac{1}{4}$ and $\frac{3}{4}$ sec? Between the $\frac{5}{4}$- and 2-sec points?

b. Draw a graph of the information in Table 1.4.

c. Having no lack of financial resources, the coach sends his scout an eight-frame-per-second camera and receives film in return. Estimate, by drawing a graph, what his third roll of film will show.

7. A sales curve $y = S(t)$ of the shape shown and discussed in Figure 1.1 is a good model of how sales might grow in a limited market. Imagine going back in time to $t = 0$ when sales began, and suppose that the manufacturer had assumed that the growth of sales would take such a shape but did not know what the eventual market saturation level would prove to be. How would a knowledge of the *rate* of sales growth have allowed the manufacturer to predict this saturation level long before it was reached? [*Hint:* What characteristic of the rate of growth, for a curve of this shape, coincides with the sales level that is one-half market saturation?]

8. If you throw a ball straight up into the air, it rises to some height and then begins to fall back to the ground. Let $s(t)$ be the height of the ball at time t. The graph of s could look like that shown in Figure 1.11.

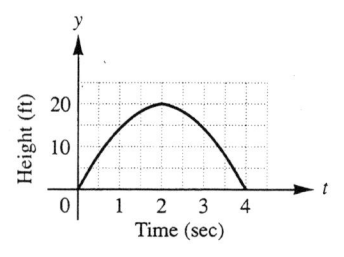

Figure 1.11

a. What is the average speed of the ball in the first 2 sec? In the time period from 1 to 2 sec?

b. At what time does its average speed appear to be less than 5 ft/sec?

c. How fast does it appear to be going at time $t = 2$?

d. Relate each of your answers above to the geometry of secant and tangent lines to the graph in Figure 1.11.

Computer Application Problems

Use the BASIC program TANGENT to estimate the slope of the tangent line to each of the graphs in Exercises 9–12 at the indicated points a. Indicate the tangent and its slope on a graph.

9. $f(x) = x^9$ at $a = 0$, $a = 0.5$, $a = 1$, and $a = 1.5$.
10. $g(x) = 1/x$ at $a = 1, 2$, and 3.
11. $E(x) = 2^x$ at $a = 0, 1$, and 2.
12. $f(x) = \sqrt{x}$ at $a = 1$ and 4. (Enter $\sqrt{x}$ as "x∧0.5" in the program.)

1.2 The Limit of a Function

The purpose of this section is to introduce the limit value of a function at a point.

Calculus, and with it the concept of limit value, grew out of attempts to understand motion as the rate of change of position over time. Thus we might say that calculus developed from efforts to understand time as a measure.

Imagine an arrow flying toward a target, as in Figure 1.12. Its position is continually changing over time. In contrast, say, to the successive bounces of a basketball, where each bounce is separated in our mind from the next, we would find it nearly impossible to separate the smoothly changing successive positions of the arrow in its flight. The roots of calculus lie in this question of Greek antiquity: *When* does the arrow hit its target? Is there a *last* moment just before the arrow hits? Is there a *first* moment after it hits? The question is not one of our eyes seeing the impact but of our conception of time itself. Can we conceive of the *very next* moment after this moment? No. Between any two moments we might notice, there was another moment between the two. If the arrow has not yet reached the target, it can certainly move a little closer—for example, half the remaining distance—without yet hitting the target.

Figure 1.12

We call a series of definite, distinguishable events, one clearly following another (like the successive bounces of a basketball), a **sequence.** In contrast, a collection of events, where one event blends smoothly into another (like the changing positions of an arrow in flight), is called a **continuum.** Our concept of time itself is a continuum, with one moment blending smoothly into another.

Figure 1.13

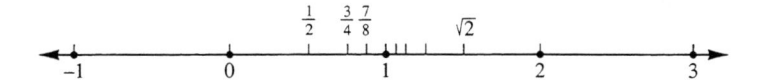

The real number system is a continuum, and we commonly illustrate it this way, as in Figure 1.13. While the whole numbers—1, 2, 3, and so on—form a sequence, the collection of *all* real numbers, consisting of the positive and negative whole numbers along with 0 and all fractions and infinite decimals, blend smoothly into each other, like moments in time or the changing positions of an arrow in flight. For example, suppose we ask, "Is there a *next* real number after the number 1?" If I answer, "1.0001," you need only reply, "But 1.00001 is closer" to contradict me. If we ask, "Is there a *last* real number before the number 1?", you might reply "0.999999." Then I need only say, "But 0.9999999 is closer still" to contradict you. We might think of the number "1" as the moment when the arrow hits the target. Until the arrow is there, it can always get a little closer, however close it is. The idea of a continuum like the real number system is that, while we can always find a real number as close to a given number as we like, we can then equally well find another number closer still. This image of something always approaching but never actually reaching a final destination is the key to understanding the meaning of a limit value in calculus.

Let us now define the limit of a function. Imagine that f represents a function whose domain is a set of real numbers and that b represents a fixed real number. Suppose further that $a < b < c$ and that the domain of f includes the open interval (a, c) except possibly the number b (see Figure 1.14).

1.2 The Limit of a Function   67

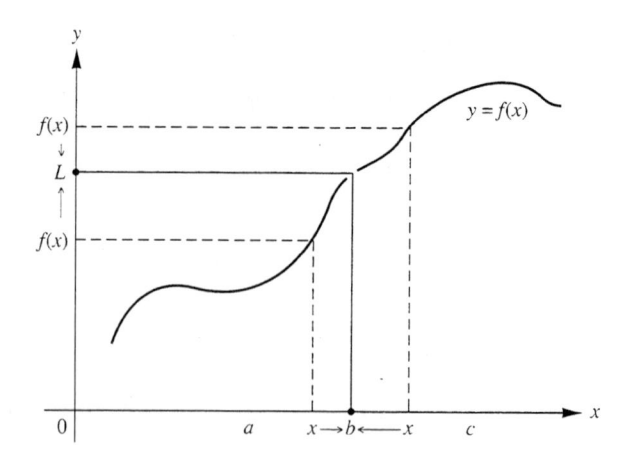

Figure 1.14

Definition

The **limit** of $f(x)$, as x approaches b, is the number L—written as

$$\lim_{x \to b} f(x) = L$$

if the following is true:

> The function values $f(x)$ are as near to the number L as we demand, if x is near enough to b, but never exactly equal to b

The notation $\lim_{x \to b} f(x) = L$ is read "the limit of $f(x)$, as x approaches b, is L."

Alternatively (and in practice this is how we find the limit of a function), the number L is the *single* real number that best approximates all the numbers $f(x)$ for numbers x near, but unequal to, the number b.

The two key points about the concept of limit value nicely match the example of the arrow approaching its target. Interpret the point b as the time when the arrow strikes the target and $f(x)$ as the height of the tip of the arrow at time x (see Figure 1.15). If we cannot know exactly when the arrow strikes, we can admit this by only asking in the limit that x be near, *but never equal to, b*. Secondly, notice that the height of an arrow in flight may never actually equal the height of its final point of impact on the target. In the limit definition the number L plays the role of this point of impact, and we can have $\lim_{x \to b} f(x) = L$ without $f(x)$ ever being equal to L. That is, the point of impact of a flying arrow is its "limiting position" and may never actually be attained in flight. This willingness to forsake equality of x with b and $f(x)$ with L and accept only approximation—but approximation as near as we want—is what makes calculus the unique mathematical tool that it is.

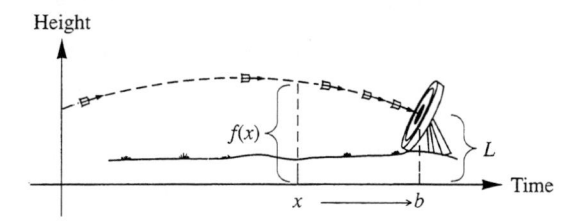

Figure 1.15

Let us see how the limit definition is used in practice for particular functions.

EXAMPLE 1

Let $f(x) = 2x - 1$ and $b = 2$. Find $\lim_{x \to 2} f(x)$.

Solution We have $\lim_{x \to 2}(2x - 1) = 3$ because if x almost equals 2, $2x - 1$ almost equals $2 \cdot 2 - 1 = 3$. This is indicated in Figure 1.16.

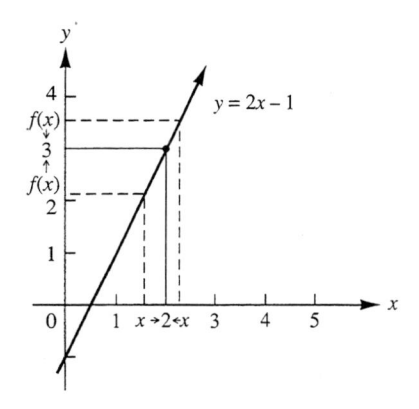

Figure 1.16

This first example is not very interesting, because the limit value is the function value, $f(2) = 3$. The next example brings up an essential point and, as we will soon see, deals with the kind of limit that must be found in order to compute an exact rate of change.

EXAMPLE 2

Let $f(x) = (x^2 - 9)/(x - 3)$ and $b = 3$. Find

$$\lim_{x \to 3} \frac{x^2 - 9}{x - 3}$$

(Note that 3 is not in the domain of this function.)

Solution If x is near 3, then both $x - 3$ and $x^2 - 9$ are near 0. Is their ratio $(x^2 - 9)/(x - 3)$ near 1? No. Remember that we are interested only in values of $f(x)$ when x is near, but not equal to, 3. Notice that if $x \neq 3$, then

$$f(x) = \frac{x^2 - 9}{x - 3} = \frac{(x - 3)(x + 3)}{x - 3} = x + 3$$

1.2 The Limit of a Function

(Since $x \neq 3$, then $x - 3 \neq 0$, and we may legitimately divide by $x - 3$.) Thus when x is near, but unequal to, 3, $f(x) = x + 3$, and therefore $f(x)$ is near $3 + 3 = 6$. That is,

$$\lim_{x \to 3} \frac{x^2 - 9}{x - 3} = 6$$

We indicate this in Figure 1.17, showing the graph of f as the straight line $y = x + 3$ for any $x \neq 3$.

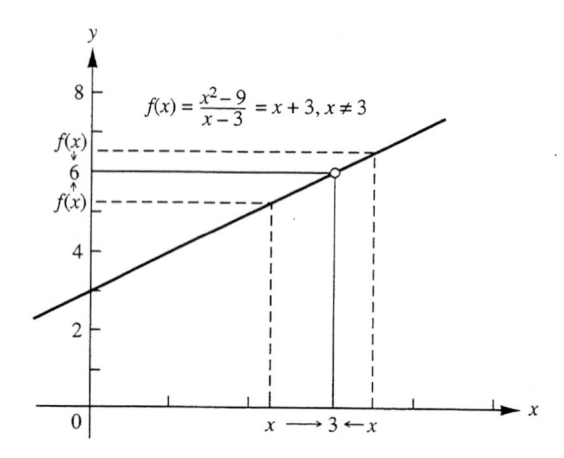

Figure 1.17

Figure 1.17 also suggests that we may interpret the number $\lim_{x \to b} f(x)$ as the answer to the question "As x gets arbitrarily close to the number b from either direction (but never equals b), to what number are the values of $f(x)$ inexorably headed, whether these values ever reach this number or not?" ∎

EXAMPLE 3

Not every function has a limit! Recall the earlier example of a person walking along a sidewalk, up a curb, and then up a hill (see Figure 1.18).

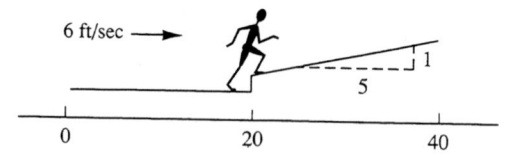

Figure 1.18

Assume that the person maintains a constant horizontal speed of 6 ft/sec and that his or her vertical speed upon stepping over the curb is 4 ft/sec. The person's vertical speed is given by the function

$$V(x) = \begin{cases} 0 & 0 \leq x < 20 \\ 4 & x = 20 \\ \frac{6}{5} & 40 > x > 20 \end{cases}$$

What is the $\lim_{x \to 20} V(x)$?

Solution First, if x is near 20 but $x \neq 20$, then $V(x)$ is either 0 or $\frac{6}{5}$. So the limit can only be decided by these two values; the fact that $V(20) = 4$ plays no role at all. To find the limit we might at first be tempted to compromise and average the two values of 0 and $\frac{6}{5}$ and say that the limit is $\frac{3}{5}$. But this cannot be right because of the definition of limit. This requires that the limit L be as close as possible to the values $V(x)$. Certainly there are other numbers closer than $\frac{3}{5}$ to $\frac{6}{5}$ or to 0. But there is no *single* number as close as we can imagine to both 0 and $\frac{6}{5}$, and so

$$\lim_{x \to 20} V(x)$$

does not exist (see Figure 1.19).

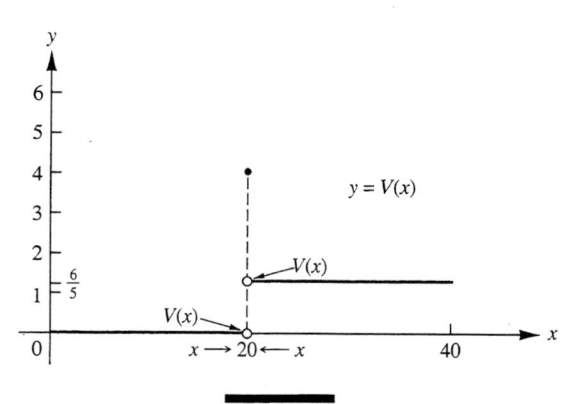

x	$V(x)$
18	0
19.5	0
20	?
20.5	$\frac{6}{5}$
21	$\frac{6}{5}$

Figure 1.19

EXAMPLE 4

Let us change the scene in Example 3 slightly (see Figure 1.20). Our function for vertical speed is now:

$$V(x) = \begin{cases} 4 & x = 20 \\ \frac{6}{5} & x \neq 20 \end{cases}$$

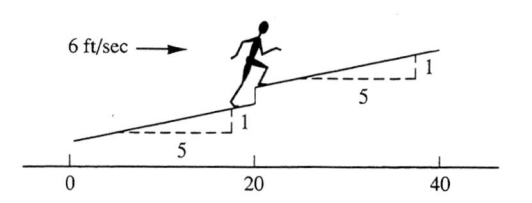

Find $\lim_{x \to 20} V(x)$.

Figure 1.20

Solution Here $\lim_{x \to 20} V(x) = \frac{6}{5}$ (see Figure 1.21). As long as $x \neq 20$, $V(x) = \frac{6}{5}$; thus $V(x)$ is as near (in this case equal to) $\frac{6}{5}$ as we can imagine if x is near, but not equal to, 20.

1.2 The Limit of a Function

x	$V(x)$
18	$\frac{6}{5}$
19.5	$\frac{6}{5}$
20.5	$\frac{6}{5}$
21	$\frac{6}{5}$

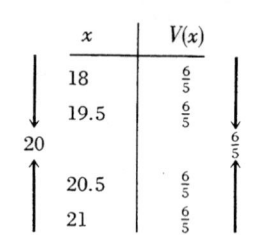

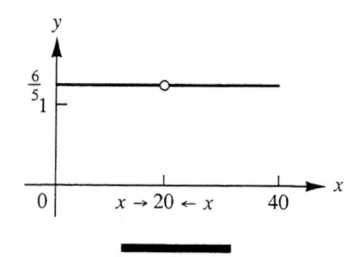

Figure 1.21

You might wonder at this point: If the limits just described ignore the most interesting part of the person's walk up the hill—namely, the process of stepping over the curb—how can limits be useful? This is not easy to answer. The limit concept is one of finding possible continuity at a point of seeming discontinuity. Recall Example 2. There the limit value 6, along with $x = 3$, fills in the hole on the graph.

You may also be wondering now whether a mathematics that depends on so vague an idea as "L is as near to $f(x)$ as we demand" can be relied on to describe, say, the correct rate of treatment of an infection with a potentially hazardous drug. The precise modern definition of limit is as follows: The $\lim_{x \to a} f(x) = L$ if for any number $\varepsilon > 0$ there is a number $\delta > 0$ such that if $0 < |x - a| < \delta$, then $|f(x) - L| < \varepsilon$. Mathematicians have used this definition over the past century to establish precisely and correctly the remaining mathematics in this text. You need not appreciate the full logical complexity and rigor of this definition to use this mathematics dependably.

Let us update our earlier comparison of a function f and its function values with a roll of motion picture film of a changing system. In that analogy we compared a single value $f(t)$ of f with the image on a single frame of the film. Our discussion in this section suggests that we might better compare f with a videotape of the system, where individual function values, or images, do not show as single frames but rather blend smoothly into each other as the system moves through its ever-changing states. In later sections we will compute many limits; the exercises in this section will help you form a geometric notion of this unique idea.

Exercises 1.2

1. Decide whether the $\lim_{x \to b} f(x)$ exists for each of the functions graphed in Figures 1.22–1.27. If the limit does exist, what is its value?

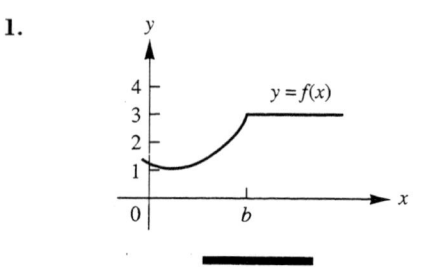

Figure 1.22

72 Chapter 1 Limits and the Derivative

2.

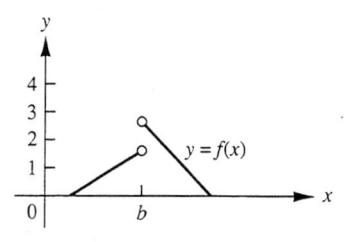

Figure 1.23

3.

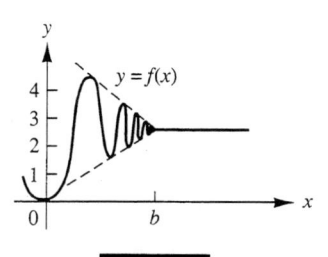

Figure 1.24

4.

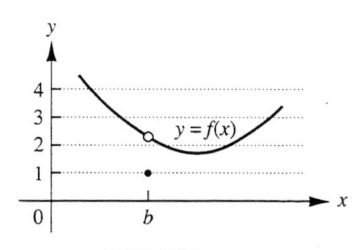

Figure 1.25

5.

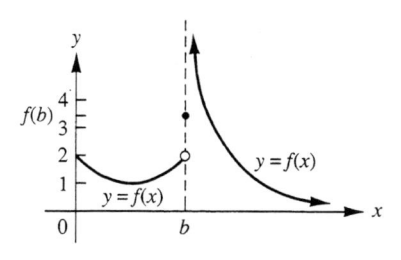

Figure 1.26

6.

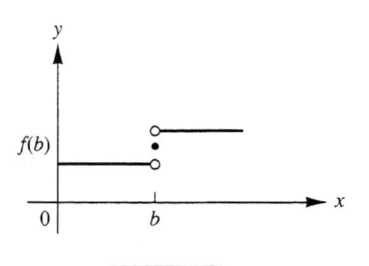

Figure 1.27

7. Write three distinct numbers between 1.0001 and 1.00001 and three more between 0.999 and 0.9999.

8. A subatomic particle is moving toward an atom at 90% of the speed of light, or 0.90(186,000) = 167,400 mi/sec. A physicist is able to measure the time of impact with the atom to an accuracy of 10^{-6} sec. How far could the particle actually be from the target atom at the time the physicist measures the impact?

9. Given the values of a function $y = g(t)$ in Table 1.5, what answer would you like to give to the following questions, based only on the values in the table?

Table 1.5

t	$g(t)$	t	$g(t)$	t	$g(t)$
$\frac{1}{2}$	$\frac{3}{2}$	1.005	1.99	1.7	1
$\frac{3}{4}$	1.90	1.01	1.95	1.95	0.77
$\frac{7}{8}$	1.97	1.1	1.9	1.999	0.755
0.9999	1.99988	1.5	1.6	2	3
				2.0001	0.749

What is

a. $\lim_{t \to 1} g(t)$ **b.** $\lim_{t \to 2} g(t)$ **c.** $\lim_{t \to 1.5} g(t)$

Find the limit and graph each function in Exercises 10–13.

10. $\lim_{x \to 1} 2x + 5$ [*Hint:* See Example 1.]

11. $\lim_{x \to -3} \frac{x^2 - 9}{x + 3}$ [*Hint:* See Example 2.]

12. $\lim_{x \to 2} \frac{x^2 - 4}{x - 2}$ [*Hint:* See Example 2.]

13. $\lim_{x \to 1} \frac{x^2 + x - 2}{x - 1}$ [*Hint:* Factor $x^2 + x - 2$.]

14. Drop a tennis ball to the court from shoulder height. Let $h(t)$ equal its height at time t.

    a. Draw what you believe the graph of $h(t)$ ought to look like, in general form.

    b. Let t_0 be the time when the ball stops bouncing. What is $\lim_{t \to t_0} h(t)$?

15. Estimate the value of $\lim_{x \to 0} (2^x - 1)/x$ using a calculator. [*Hint:* Try $x = \frac{1}{2}, \frac{1}{4}, \frac{1}{6}, \frac{1}{15}, \frac{1}{50}, \frac{1}{100}$, and $\frac{1}{1000}$.]

16. In finding $\lim_{x \to b} f(x) = L$, we do not allow x to equal b. While it is not necessary that $f(b)$ equal L, this is allowed. Convince yourself, based on the limit definition, that if $f(x) = 3$ for every number x, then $\lim_{x \to 5} f(x) = 3$. Thus the limit value of a constant function is the function value itself.

1.2 The Limit of a Function   73

Computer Application Problems

Use the BASIC program LIMVAL to estimate the limit value in Exercises 17–20.

17. $\lim\limits_{x \to 0} 2^{-x}$

18. $\lim\limits_{x \to 0} \dfrac{1}{x(2)^{1/x}}$

19. $\lim\limits_{x \to 0} (1 + x)^{1/x}$

20. $\lim\limits_{x \to 2} \dfrac{x^4 - 16}{x - 2}$

1.3 Continuous Functions and the Limit Properties

The purpose of this section is to learn to calculate limits and recognize mathematical continuity, using general properties of the limit.

If you are walking late one evening and suddenly step off a curb that you did not know was there, you physically experience a *discontinuity!* A discontinuity in the relationship between the variables in a functional correspondence can be just as upsetting mathematically. This section will help you learn to recognize *continuity* in its mathematical form. In so doing, you will also learn how to compute a variety of limits using a few basic rules.

What does continuity mean to you? If the problems at the end of this section turn out to be considerably different from the examples, you will feel a sense of discontinuity. The word *continuous* means that behavior *up to* a point is a reliable predictor of behavior *at* the point. This is unlike the discontinuity of stepping off an unexpected curb on a darkened sidewalk. The words we have just used to describe continuity—*up to* versus *at*—may suggest to you that the limit concept has a role in determining mathematical continuity. Let us see how it does, beginning with a look at the four graphs in Figure 1.28.

In Figure 1.28(a), (b), and (c), the graph of the function is not a continuous, unbroken drawing. In part (a), it is not continuous because there is no function value $f(a)$. In part (b), the function value $f(a)$ does not fill in the hole. In part (c), there is a value $f(a)$, but it is predicted (approximated) only by function values on one side of a. In part (d), we see a continuous graph, and none of the three defects illustrated in (a), (b), and (c) occurs.

Definition

A function f is **continuous at the number** a if *all three* of the following hold true:

1. The number a is in the domain of f. That is, there is a function value $f(a)$ at a.
2. $\lim\limits_{x \to a} f(x)$ exists.
3. $\lim\limits_{x \to a} f(x) = f(a)$.

If any one of these three conditions is not met, then f is **discontinuous** at a. A function is said to be **continuous** if it is continuous at every point a in its domain.

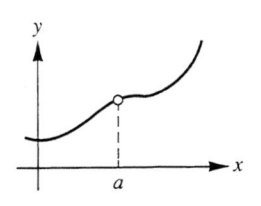

Note that condition 3 says that what happens *near a* [expressed by the limit value $\lim_{x \to a} f(x)$] is what happens *at a* [expressed by the function value $f(a)$]. For this reason, limits of continuous functions are easy limits to evaluate.

The limit of a continuous function at a point is simply the value of the function at the point.

EXAMPLE 1

Let $f(x) = (x - 2)^2 + 1$. Find $\lim_{x \to 3} f(x)$.

Solution The graph in Figure 1.29 tells us that f is continuous at $a = 3$ and thus that $\lim_{x \to 3}(x - 2)^2 + 1 = f(3) = (3 - 2)^2 + 1 = 2$.

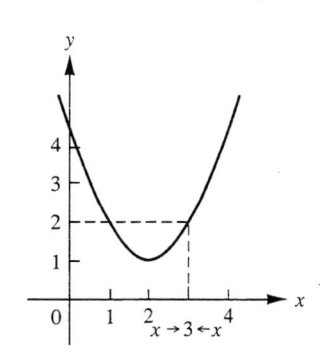

(a) There is no $f(a)$.

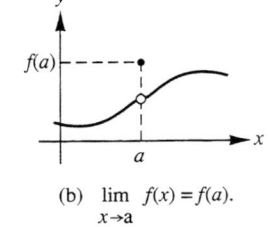

(b) $\lim_{x \to a} f(x) = f(a)$.

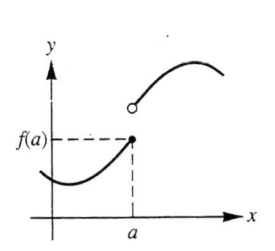

(c) $\lim_{x \to a} f(x)$ does not exist.

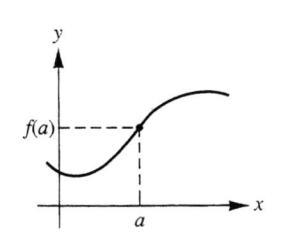

(d) $\lim_{x \to a} f(x) = f(a)$.

Figure 1.28

Figure 1.29

EXAMPLE 2

Let

$$f(x) = \begin{cases} \dfrac{x^2 - 1}{x + 1} & x \neq -1 \\ -2 & x = -1 \end{cases}$$

Explain why f is continuous at $x = -1$.

Solution This is only a bit more involved, despite its appearance. For $x \neq -1$,

$$f(x) = \frac{(x - 1)(x + 1)}{x + 1} = x - 1$$

Thus the graph of f, for $x \neq -1$, looks like the graph in Figure 1.30(a).

Now, in addition, $f(-1) = -2$, so that the *entire* graph of f looks like Figure 1.30(b). The graph is visually continuous. But not all functions are so easy to graph, so let us tie this example to the definition of continuity at $a = -1$.

1.3 Continuous Functions and the Limit Properties 75

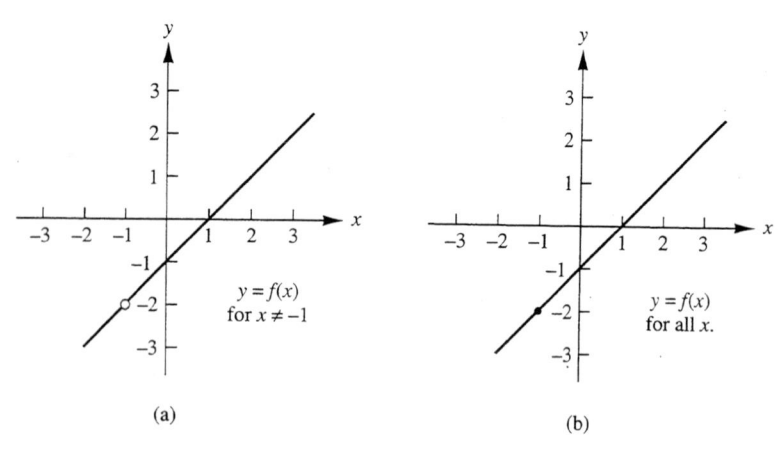

Figure 1.30

(a) $y = f(x)$ for $x \neq -1$

(b) $y = f(x)$ for all x.

To test for continuity of f at $a = -1$, we proceed step-by-step. We must verify that all three conditions of the definition of continuity at $a = -1$ are true:

Condition 1 is true because $f(-1)$ is defined, and $f(-1) = -2$.

Condition 2 is true because

$$\lim_{x \to -1} f(x) = \lim_{x \to -1} \frac{x^2 - 1}{x + 1} = \lim_{x \to -1} (x - 1) = -2$$

Condition 3 is true because the answers in conditions 1 and 2 coincide. That is,

$$\lim_{x \to -1} f(x) = -2 = f(-1)$$

Hence, because conditions 1, 2, and 3 all hold, f is continuous at $a = -1$. These steps show that it is not necessary to graph a function to determine continuity. ∎

Here is a second such example.

EXAMPLE 3

Let

$$g(x) = \begin{cases} \dfrac{x^2 - 2x - 3}{x - 3} & x \neq 3 \\ 5 & x = 3 \end{cases}$$

Determine whether the function g is continuous at $a = 3$.

Solution To try to establish continuity at $a = 3$, we again proceed step-by-step according to the definition.

Step 1. $g(3)$ exists, and $g(3) = 5$.

Step 2.

$$\lim_{x \to 3} \frac{x^2 - 2x - 3}{x - 3} = \lim_{x \to 3} \frac{(x - 3)(x + 1)}{x - 3} = \lim_{x \to 3} (x + 1) = 4$$

76 Chapter 1 Limits and the Derivative

Step 3. The answers in steps 1 and 2 are unequal. Therefore, the function g is discontinuous at $x = 3$. The graph of g (Figure 1.31) illustrates this.

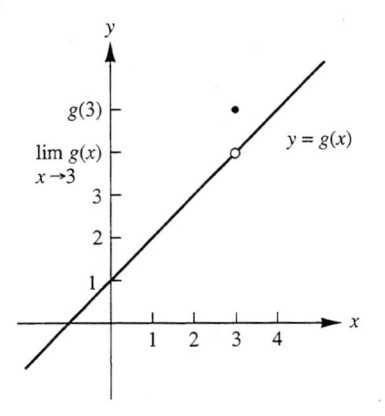

Figure 1.31

Remark. The functions in Examples 2 and 3 can be shown to be continuous at all other points in their respective domains, although we will not do this.

General Limit Properties

These examples illustrate only the more mathematically interesting cases of continuity and discontinuity. Most questions of continuity, and indeed many questions involving limits in general, are answered by using a short list of limit properties that allow us to find a limit value according to general rules. These properties are given here, without proof, and are then followed by further examples. The proofs of these properties were provided years ago by mathematicians using the abstract $\varepsilon - \delta$ definition of limit noted at the close of Section 1.2.

The Limit Properties

Suppose that f and g are two functions and that $\lim_{x \to a} f(x)$ and $\lim_{x \to a} g(x)$ exist. Then

1. $\lim_{x \to a}(f(x) \pm g(x)) = \lim_{x \to a} f(x) \pm \lim_{x \to a} g(x)$.
2. $\lim_{x \to a}(f(x) \cdot g(x)) = \lim_{x \to a} f(x) \cdot \lim_{x \to a} g(x)$.
3. If $\lim_{x \to a} g(x) \neq 0$, then

$$\lim_{x \to a} \frac{f(x)}{g(x)} = \frac{\lim_{x \to a} f(x)}{\lim_{x \to a} g(x)}$$

4. If k is a fixed number, then $\lim_{x \to a} k \cdot f(x) = k \cdot \lim_{x \to a} f(x)$.
5. If $f(x) = c$ is a constant function, then

$$\lim_{x \to a} f(x) = c$$

(since no matter how close x is to a, the value of $f(x)$ is always c).

6. If $f(x) = x^m$, then $\lim_{x \to a} f(x) = \lim_{x \to a} x^m = a^m$, where m and a are positive real numbers.

7. If f is continuous, then $\lim_{x \to a} f(g(x)) = f(\lim_{x \to a} g(x))$.

Property 6 says that all power functions

$$y = x^m$$

are continuous. Hence, property 6, combined with properties 1, 4, and 5, says that all *polynomial functions* (see Background Review 1.1) are continuous. Let us apply these properties to further examples.

EXAMPLE 4

Let $f(x) = 1 - x + 3x^4$. Show that f is continuous at $x = 2$.

Solution

Step 1. $f(2)$ is defined and $f(2) = 1 - 2 + 3(2)^4 = 47$.

Step 2.

$$\lim_{x \to 2}(1 - x + 3x^4) = \lim_{x \to 2} 1 - \lim_{x \to 2} x + 3 \cdot \lim_{x \to 2} x^4 \quad \text{by properties 1 and 4}$$
$$= 1 - 2 + 3(2)^4 \quad \text{by properties 5 and 6}$$
$$= 47$$

Step 3. The answers in steps 1 and 2 coincide. Therefore f is continuous at $x = 2$. ∎

Note that the graph of this function would not be very easy to draw! This illustrates a utility of the properties listed earlier.

Evaluation of Limits

The limit properties can be used to evaluate limits without regard to questions of continuity.

EXAMPLE 5

Evaluate

$$\lim_{x \to 1} \frac{x^2 - x^3 + 2x}{x^4 + 1}$$

Solution Since $\lim_{x \to 1}(x^4 + 1) = 2 \neq 0$, property 3 tells us that

$$\lim_{x \to 1} \frac{x^2 - x^3 + 2x}{x^4 + 1} = \frac{\lim_{x \to 1}(x^2 - x^3 + 2x)}{\lim_{x \to 1}(x^4 + 1)} = \frac{1 - 1 + 2}{2} = 1$$

by properties 1, 4, 5, and 6. ∎

Property 7 is especially useful. If we graph $f(x) = \sqrt{x}$ for $x > 0$, we can see that it is continuous. Let us apply property 7 to the problem of finding $\lim_{x \to 2} \sqrt{x^2 + x - 2}$. Notice that $\sqrt{x^2 + x - 2} = f(g(x))$, where $g(x) = x^2 + x - 2$. Therefore, by property 7, $\lim_{x \to 2} f(g(x)) = f(\lim_{x \to 2} g(x))$ and we have by substitution

$$\lim_{x \to 2} \sqrt{x^2 + x - 2} = \sqrt{\lim_{x \to 2}(x^2 + x - 2)} = f(\lim_{x \to 2} g(x)) = \sqrt{4 + 2 - 2} = 2$$

The next example combines this technique with earlier methods.

EXAMPLE 6

Find

$$\lim_{x \to 2} \sqrt{\frac{x^3 - x^2 + 2x}{(x - 1)(x - 2)}}$$

Solution

$$\lim_{x \to 2} \sqrt{\frac{x^3 - x^2 - 2x}{(x - 1)(x - 2)}} = \sqrt{\lim_{x \to 2} \frac{x^3 - x^2 - 2x}{(x - 1)(x - 2)}}$$

$$= \sqrt{\lim_{x \to 2} \frac{x(x^2 - x - 2)}{(x - 1)(x - 2)}}$$

$$= \sqrt{\lim_{x \to 2} \frac{x(x + 1)(x - 2)}{(x - 1)(x - 2)}}$$

$$= \sqrt{\lim_{x \to 2} \frac{x(x + 1)}{x - 1}} \quad \text{since } x - 2 \neq 0$$

$$= \sqrt{\frac{\lim_{x \to 2} x(x + 1)}{\lim_{x \to 2} (x - 1)}}$$

$$= \sqrt{\frac{2 \cdot 3}{2 - 1}} = \sqrt{6} \quad \blacksquare$$

In this example we could not use property 3 because

$$\lim_{x \to 2}(x - 1)(x - 2) = 0$$

Instead we applied the factorization/cancellation technique seen in Examples 2 and 3. This example also illustrates that we cannot simply "look at" a limit problem and "see" its answer. We must methodically use the limit properties 1 to 7 until we have a final answer.

Vertical Asymptotes

The fact that we cannot divide by zero is closely related, via limit property 3, to the concept of a vertical asymptote, first discussed in Example 4 of Section

0.4. Consider the graph of $f(x) = 1/(x - 3)^2$ shown in Figure 1.32(a). The vertical line $x = 3$ is a **vertical asymptote,** and the graph is said to be **asymptotic** to this line. By this we mean that the curve of the graph becomes ever closer to, but never actually touches, the line $x = 3$. Suppose that we try to compute $\lim_{x \to 3} 1/(x - 3)^2$. We quickly realize that as x approaches 3, the numbers $(x - 3)^2$ approach 0. Consequently, their *reciprocals,* $1/(x - 3)^2$, can be made as large as we demand. In such a case we write

$$\lim_{x \to 3} \frac{1}{(x - 3)^2} = +\infty$$

The symbol "$+\infty$," read "plus infinity," is not a number but rather a notation for our observation that the numbers $1/(x - 3)^2$ become larger and larger as x approaches 3.

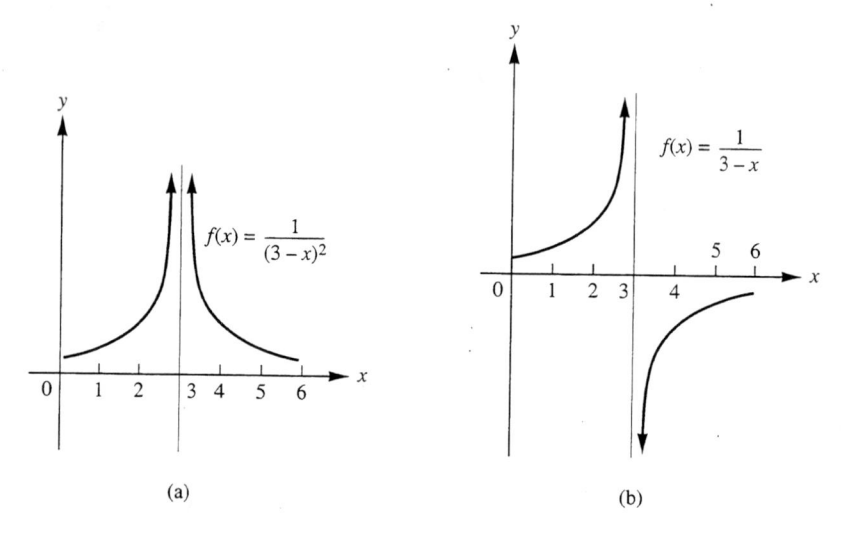

Figure 1.32

(a)

(b)

Figure 1.32(b) illustrates another kind of vertical asymptote. In this case $\lim_{x \to 3} 1/(3 - x)$ cannot exist in any sense, since if $x < 3$ and x approaches 3, then $1/(3 - x)$ again approaches $+\infty$. However, if $x > 3$, then $3 - x < 0$ and the reciprocals $1/(3 - x)$ become ever larger negative numbers and are said to approach "$-\infty$" or "minus infinity." Thus a common direction for the values of $f(x) = 1/(3 - x)$ cannot be found as x approaches 3, and the limit cannot exist. Further study of such examples becomes increasingly technical and is beyond the intent of this text. Instead you should remember the following general alternative to limit property 3.

If $\lim_{x \to a} f(x) \neq 0$ and $\lim_{x \to a} g(x) = 0$, then $\lim_{x \to a} [f(x)/g(x)]$ *cannot exist* as a real number. In such a case the graph of $y = f(x)/g(x)$ should be checked for a possible vertical asymptote at $x = a$.

Horizontal Asymptotes and Limits at Infinity

Consider the graph of $y = f(x)$ shown in Figure 1.33.

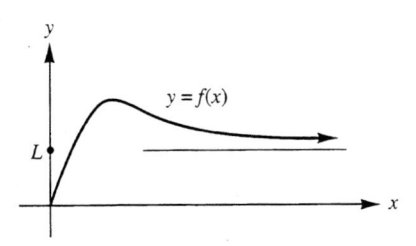

Figure 1.33

The horizontal line at height L, $y = L$, is called a **horizontal asymptote** of the graph of f: The graph moves closer and closer to this horizontal line as x becomes larger and larger. We write this as

$$\lim_{x \to +\infty} f(x) = L$$

since in this case we want to know what number the values of $f(x)$ become closer and closer to, not when x approaches a number, but rather when x becomes larger and larger—that is, when x increases without bound.

Definition

We say that
$$\lim_{x \to \infty} f(x) = L$$

if the following is true: The function values $f(x)$ are as close to L as we demand, if x is a large enough number.

That is, L is the single number that best approximates $f(x)$ when x itself is a very large number. The line $y = L$ is then called a horizontal asymptote for the graph of f, and L is said to be the "limit at infinity of f."

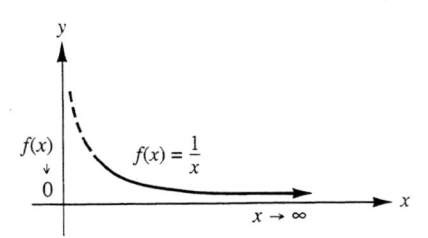

Figure 1.34

The familiar graph of $f(x) = 1/x$ illustrates this definition and the concept of a horizontal asymptote. We see from Figure 1.34 that

$$\lim_{x \to \infty} \frac{1}{x} = 0$$

and thus that $y = 0$ is a horizontal asymptote. Indeed, more generally, $\lim_{x \to \infty} 1/x^m = 0$ for any exponent $m > 0$. This property, combined with the limit properties 1 to 7 (all of which apply to limits at infinity as well), allows us to find many limits at infinity by a single technique.

EXAMPLE 7

Find

$$\lim_{x \to \infty} \frac{3x^2}{x^2 + 4}$$

Solution The single technique we use is to divide both numerator and denominator by the highest power of x that appears in the given formula. Thus

$$\lim_{x \to \infty} \frac{3x^2}{x^2 + 4} = \lim_{x \to \infty} \frac{3x^2/x^2}{(x^2 + 4)/x^2} = \lim_{x \to \infty} \frac{3}{1 + (4/x^2)} = \frac{3}{1} = 3$$

since

$$\lim_{x \to \infty} \frac{4}{x^2} = 4 \cdot \lim_{x \to \infty} \frac{1}{x^2} = 0$$

We illustrate the graph of this function in Figure 1.35.

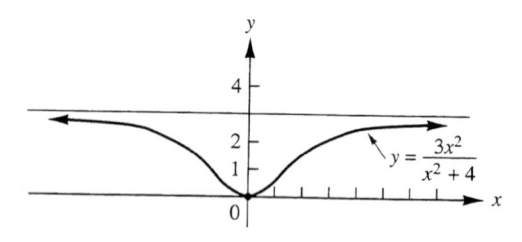

Figure 1.35

Notice in Figure 1.35 that as x moves farther and farther in the *negative* direction, the graph again approaches the line $y = 3$; we summarize this by writing

$$\lim_{x \to -\infty} \frac{3x^2}{x^2 + 4} = 3$$

This section concludes our brief study of the limit value of a function. We study limits because they are needed for computing the exact rate of change in a functional relationship, as we will see in the next section.

Background Review 1.1

Limit problems repeatedly involve factorization and cancellation, as in Examples 2, 3, and 6.

If $f(x) = a_m x^m + a_{m-1} x^{m-1} + \cdots + a_1 x + a_0$ is a polynomial function and the number λ is a root of this polynomial (that is, $f(\lambda) = 0$), then $x - \lambda$ must be a factor of this polynomial. You can find this factor by dividing $a_m x^m + a_{m-1} x^{m-1} + \cdots + a_1 x + a_0$ by $x - \lambda$.

Thus, since $\lambda = 2$ is a root of $x^4 - 4x^2 - x + 2 = 0$, then $x - 2$ must divide this polynomial. By long division

$$\frac{x^4 - 4x^2 - x + 2}{x - 2} = x^3 + 2x^2 - 1$$

Thus $\lim_{x \to 2} \frac{x^4 - 4x^2 - x + 2}{x - 2} = \lim_{x \to 2} \frac{(x-2)(x^3 + 2x^2 - 1)}{x - 2} = \lim_{x \to 2} x^3 + 2x^2 - 1 = 15$

Some useful factorizations (and formulas) are the following:

1. $a^2 - b^2 = (a - b)(a + b)$
2. $a^3 - b^3 = (a - b)(a^2 + 2ab + b^2)$
3. $a - b = (\sqrt{a} - \sqrt{b})(\sqrt{a} + \sqrt{b}),\ a, b \geq 0$
4. $\sqrt{a} - b = (\sqrt{a} - b)\frac{\sqrt{a} + b}{\sqrt{a} + b} = \frac{a - b^2}{\sqrt{a} + b},\ \sqrt{a} + b \neq 0,\ a \geq 0$
5. $(x + a)(x + b) = x^2 + (a + b)x + ab$
6. $\frac{a + b}{c} = \frac{a}{c} + \frac{b}{c},\ c \neq 0$
7. $\frac{\sqrt[m]{a}}{x} = \sqrt[m]{\frac{a}{x^m}},\ x > 0$

Exercises 1.3

Evaluate the limits in Exercises 1–28, if they exist.

1. $\lim_{x \to 25} x^2 - 2x$
2. $\lim_{x \to -3} \sqrt{x^2 - 2}$
3. $\lim_{x \to 1} \frac{1}{2 - x^3}$
4. $\lim_{x \to 1} \frac{2x}{x + 1}$
5. $\lim_{x \to 2} \frac{x^2 - 4}{x + 2}$
6. $\lim_{x \to 3}(x^3 + 1)\sqrt{1 + x}$
7. $\lim_{x \to 15}(\sqrt{x + 1})(x^2 - 1)$
8. $\lim_{x \to 2} x^3 - x^2 + x - 2$
9. $\lim_{x \to 5} \frac{x^2 - 5x + 1}{x - 5}$
10. $\lim_{x \to \sqrt{2}} \frac{x - \sqrt{2}}{x^2 - 2}$
11. $\lim_{x \to 7} \frac{x - 7}{x^2 - x + 42}$
12. $\lim_{x \to 0} \frac{x^2 - x}{x}$
13. $\lim_{h \to 0} \frac{h}{h^2 - h}$
14. $\lim_{x \to 26}(x + 1)^{1/3}$
15. $\lim_{x \to 5} \frac{x - 5}{x^2 - 5x + 5}$
16. $\lim_{x \to 2} \sqrt[3]{\frac{x - 2}{x^2 - x - 6}}$
17. $\lim_{t \to 2}(t^2 - t)^7$
18. $\lim_{x \to 0} \sqrt{\frac{x^2 + x}{x}}$
19. $\lim_{x \to 3} \frac{2 + (x - 1)^5}{13 + \sqrt{x^2 + 7}}$
20. $\lim_{x \to 2} \frac{x^2 - 2x}{x^6 - 2x^5 + x - 2}$
21. $\lim_{x \to 1} \frac{x^4 - x^3 + x^2 - x}{x(x - 1)(x - 2)}$
22. $\lim_{x \to 3} \frac{x^4 - 1}{x(x - 1)(x - 2)}$
23. $\lim_{y \to 2} \frac{y + 2}{y - 2}$
24. $\lim_{x \to 17} \frac{\sqrt{(x - 17)^2}}{|x - 17|}$
25. $\lim_{x \to 0} \frac{\sqrt{x + 9} - 3}{x}$ [Hint: Multiply the fraction by $(\sqrt{x + 9} + 3)/(\sqrt{x + 9} + 3)$ and simplify before finding the limit.]
26. $\lim_{x \to 0} 2^{x^2}$ [Hint: Use a calculator or property 7.]
27. $\lim_{x \to 0} \sqrt{\frac{1}{x^2} + 1} - \frac{1}{x}$ [Hint: First find a common denominator.]
28. $\lim_{x \to 0} \frac{\sqrt{x + 4} - 2}{x^2 + 2x}$ [Hint: Apply the idea used in Exercise 25.]

29. Recall Examples 3 and 4 of Section 1.2. In these two problems, someone is stepping over a curb and up a hill, as in Figure 1.36, or stepping over a curb while walking up a hill, as in Figure 1.37.

Figure 1.36

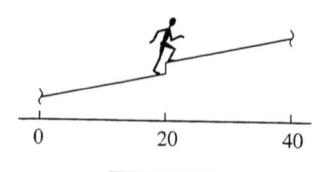

Figure 1.37

Both phenomena appear to be discontinuous at the curb, where $a = 20$. Exactly which of the three properties of continuity are violated in each example?

30. Draw a graph of a function f with the following properties:
   a. The domain of f is all positive numbers between 0 and 4.
   b. f is continuous except at $x = 1$ and $x = 3$.
   c. $f(3) = 4$ and $\lim_{x \to 3} f(x)$ does not exist.
   d. $\lim_{x \to 1} f(x) = 2$.

31. Draw a graph of a function g with the following properties:
   a. The domain of g is all real numbers between 0 and 4, except for $x = 2$.
   b. g is not continuous at $x = 1$, $x = 2$, and $x = 3$.
   c. $\lim_{x \to 2} g(x) = 3.5$.
   d. $\lim_{x \to 1} g(x)$ does not exist and $g(1) = -1$.
   e. $\lim_{x \to 3} g(x) = 2$ and $g(3) = 1$.

32. Suppose that $f(x) = (x^2 - 1)/(x - 1)$. We will define f at $x = 1$ by rolling a die and letting $f(1)$ equal the number 1, 2, 3, 4, 5, or 6 that results. What is the probability that the resulting function will be continuous at $x = 1$? (Use your intuitive idea of probability.)

Determine which of the functions in Exercises 33–36 are continuous.

33. $f(x) = \begin{cases} \dfrac{x^2 - 4x - 5}{x - 5} & x \neq 5 \\ 6 & x = 5 \end{cases}$

34. $g(x) = \begin{cases} x^2 & x < 1 \\ 2x - 1 & x \geq 1 \end{cases}$

35. $h(x) = \begin{cases} \sqrt{x} & 0 \leq x < 4 \\ 3 & x = 4 \\ \dfrac{2}{x - 3} & x > 4 \end{cases}$

36. $k(x) = \begin{cases} \dfrac{x^2 - 36}{x - 6} & x \neq 6 \\ 12.00001 & x = 6 \end{cases}$

37. Show that $x = 3$ is a vertical asymptote of
$$y = \frac{2}{x - 3} + 1$$

38. Show that $x = -2$ is a vertical asymptote of
$$y = \frac{x}{(x + 2)^2} - 1$$

Evaluate the limits in Exercises 39–44.

39. $\lim_{x \to \infty} \dfrac{2x^3 - x}{3x^3 + x^2}$

40. $\lim_{x \to \infty} \dfrac{-x^3 + x^2 - x + 1}{x^3}$

41. $\lim_{x \to \infty} \dfrac{x^2 - (1/x)}{2x^3 + 1}$

42. $\lim_{x \to -\infty} \dfrac{1 - x^2}{x + x^2}$

43. $\lim_{x \to \infty} \dfrac{3x}{\sqrt{1 + x^2}}$ [Hint: Divide by x, not by x^2.]

44. $\lim_{x \to -\infty} \dfrac{\sqrt{x^4 + 2x^2}}{1 - x^2}$

Assume that the graphs of the functions indicated in Figures 1.38–1.41 continue on "out to infinity" in the pattern shown. Decide whether $\lim_{x \to \infty} f(x)$ exists in each case, and determine its value if it does exist.

45.

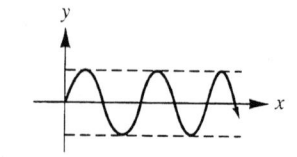

Figure 1.38

84 Chapter 1 Limits and the Derivative

46.

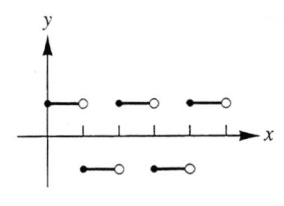

Figure 1.39

47.

$y = f(x)$

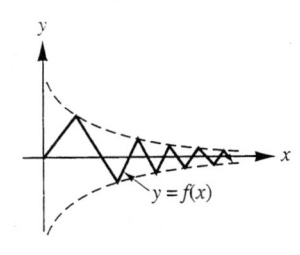

Figure 1.40

48.

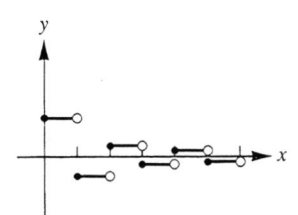

Figure 1.41

49. Show that $y = 3$ is a horizontal asymptote of
$$y = \frac{1}{1-x} + 3$$
Sketch the graph and its asymptote.

50. Show that $y = -1$ is a horizontal asymptote of
$$y = \frac{2}{(x-3)^2} - 1$$
Sketch this graph and its asymptote.

51. Not all limits can be directly evaluated, even when they exist. In Chapter 5, we will begin to see a variety of significant real-world applications for the number that results from computing $\lim_{n \to \infty}[1 + (1/n)]^n$. It is impossible to evaluate this limit by the methods of this section.

  a. Compute the value of $[1 + (1/n)]^n$ for $n = 1, 2, 3, 5, 10, 100, 5{,}000,$ and $100{,}000$ using a calculator.

  b. Compare this problem with that of computing
  $$\lim_{x \to 0}(1 + x)^{1/x}$$
  [*Hint:* What if $x = \frac{1}{500}; \frac{1}{100{,}000}$?]

52. Show that $y = 0$ is a horizontal asymptote of $y = 2^x$ by evaluating $\lim_{x \to -\infty} 2^x$.

53. What is your best guess at the value of
$$\lim_{x \to \infty} \frac{x}{2^x}$$
[*Hint:* Try a few values on your calculator.]

54. Evaluate $\lim_{x \to \infty}(\sqrt{x^2 + 1} - x)$ by multiplying and dividing this expression by $\sqrt{x^2 + 1} + x$.

55. Let $f(x) = 1/x$ for $x > 0$. Graph f. Let $g(h) = [f(1 + h) - f(1)]/h$, $h \neq 0$. The function g measures the slope of the secant line between the two points $(1, 1)$ and $[1 + h, 1/(1 + h)]$.

  a. Draw the best tangent line you can to the graph of f at the point $(1, 1)$, and measure its slope.

  b. Compute $\lim_{h \to 0} g(h)$. (First derive a formula for g, using algebra.)

  c. How can we define $g(0)$ so that g will be continuous at 0?

  d. Interpret your answer to (c) geometrically.

Challenge Problems

56. Compute $f\left(\frac{1}{10}\right), f\left(\frac{1}{100}\right),$ and $f\left(\frac{1}{1{,}000}\right)$ for $f(x) = (\sqrt{x^2 + 0.01})/x$ using a calculator. What is $\lim_{x \to 0} f(x)$?

57. Let $L(x) =$ the sum of the horizontal and vertical steps and rises of the staircase, whose step and rise are both x, in Figure 1.42. What does $L(x)$ equal? What is $\lim_{x \to 0} L(x)$? Is there something strange here? [*Hint:* The first two questions are very simple.]

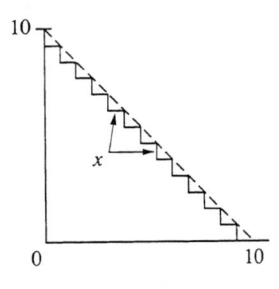

Figure 1.42

1.3 Continuous Functions and the Limit Properties

1.4 The Derivative of a Function

The purpose of this section is to bring together all the ideas of previous sections into the computational tool of calculus: the derivative of a function.

The **derivative of a function** is the measure of the *rate of change* between the variables in a functional correspondence. Introduced by Newton and Leibniz in the seventeenth century, it is arguably the single most important mathematical invention found between the introduction of a letter to represent a variable in the fifteenth century and the advent of the computer in the twentieth. At the same time, the derivative of a function is a geometrically appealing, and seemingly simple, idea.

The Slope of a Curve at a Point and the Derivative of a Function

If you are walking straight up the side of a hill, your legs and lungs will tell you that the curved hillside has a varying slope. We informally define the *slope of a curve* at a point P to be the slope of the straight line *tangent* to the curve at P, as indicated in Figure 1.43. This is reasonable enough, but how can we measure the exact slope of this tangent? To begin, we do only what is actually possible. We approximate the complex idea *tangent slope* by the simpler idea *secant slope* [recall that a secant line is a line that intercepts the graph at two points, say, $(a, f(a))$ and $(a + h, f(a + h))$]. We then take the limit as one point $a + h$, defining the secant line, approaches the other point a.

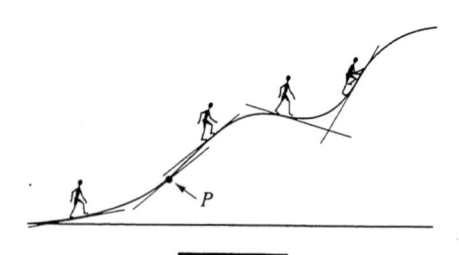

Figure 1.43

Let us make this idea computational. Consider Figure 1.44(a). Imagine a fixed value $x = a$. We want to define the slope of the line T tangent to the graph of $y = f(x)$ at $P = (a, f(a))$. To do this we shift some distance h from the point a to a second point $a + h$. We then compute

$h = (a + h) - a =$ change in variable x from a to $a + h$*

$f(a + h) - f(a)$ = change in function value between a and $a + h$

$$\frac{f(a + h) - f(a)}{h} = \text{average change in } f \text{ over } a \text{ to } a + h$$

---

*Note that if we shift to the left, then $h < 0$, which must be allowed. For purposes of understanding the main ideas, it suffices to think of h as a positive number.

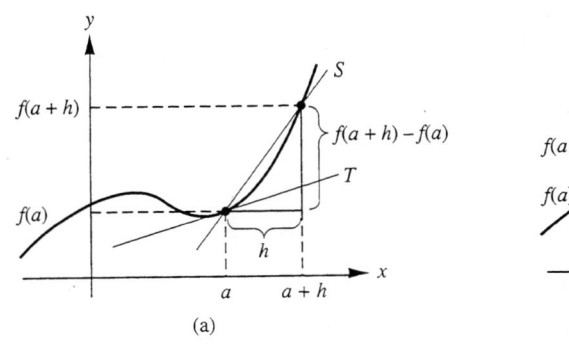

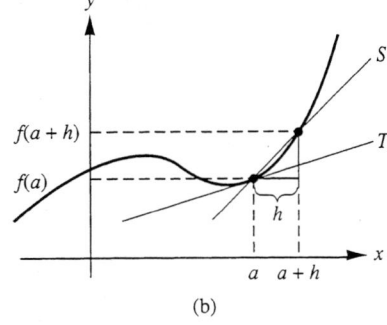

Figure 1.44

This quotient is the slope of the secant line S. We then take the limit as h approaches zero so as to obtain the slope of the tangent line T, as was done in Section 1.1.

Definition

The slope of the curve $y = f(x)$ at the point $(a, f(a))$ is defined to be the slope of the tangent line to this curve at the point $(a, f(a))$ and is the number

$$\lim_{h \to 0} \frac{f(a+h) - f(a)}{h}$$

if this limit exists. When this limit exists, the function f is said to be **differentiable** at a. The numerical value of the limit is denoted by $f'(a)$. This number is called the **derivative of f at a.**

The number $f'(a)$ has two interpretations. One is geometric, as embodied in the definition. In Figure 1.44(a) and then (b), a smaller value of h gives a secant line and its slope

$$\frac{f(a+h) - f(a)}{h}$$

closer to the tangent line. Thus $f'(a)$, being the limit value as $h \to 0$, is the exact slope of the tangent line.

The second interpretation of $f'(a)$ is conceptual, and out of this concept comes a vast array of applications of calculus. Because

$$\frac{f(a+h) - f(a)}{h}$$

measures the *average* rate of change in the function as x changes from a to $a + h$, the number $f'(a)$ is thought of as the *exact* rate of change of f at $x = a$. We make this interpretation because in the limit process the change h in x approaches zero.

1.4 The Derivative of a Function

This second interpretation will be progressively developed and exploited throughout the text. The first interpretation may be explicitly summarized in a single important formula. Because $f'(a)$ is the slope of the line tangent to the curve $y = f(x)$ at the point $(a, f(a))$, we can use the point–slope formula for the equation of a straight line (see Section 0.3) to arrive at the following conclusion, illustrated in Figure 1.45:

The equation of the straight line tangent to the curve of $y = f(x)$ at the point $(a, f(a))$ is given by

$$y - f(a) = f'(a)(x - a) \tag{1}$$

whenever the derivative $f'(a)$ exists.

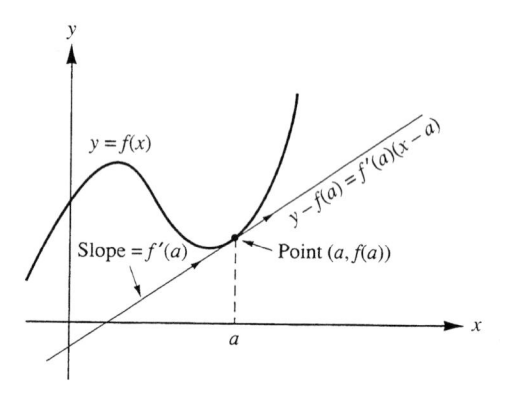

Figure 1.45

The concept of the derivative was the great idea that Newton and Leibniz crystallized from the earlier work of others. Newton used it to describe planetary motion and to study gravity and mechanics. Today it is used, for example, to design a better stereo receiver. A wildlife conservationist can use it to study predator–prey relationships, and an econometrician can apply it to model a changing economy. We will use it throughout this course to understand both abstract and natural phenomena.

Why did it take one of the world's true geniuses to invent the derivative? Perhaps because it seemed impossible to compute the slope of the exact tangent—the exact rate of change—of a curve. Obviously what we want to do is to average the change in f over a "0 instant" in time. This amounts to computing the meaningless quotient

$$\frac{f(a + 0) - f(a)}{0} = \frac{f(a) - f(a)}{0} = \frac{0}{0}$$

Let us see how the limit process accomplishes the needed computation.

Initial Computation of the Derivative

In the next section we will begin to see some significant applications of the derivative. In this section we compute only a few derivatives and explicitly link the derivative with the tangent line to a curve.

EXAMPLE 1

Compute the derivative $f'(2)$ of the function

$$f(x) = \frac{1}{x} \quad \text{at the point } a = 2$$

and illustrate the answer on a graph.

Solution According to the definition,

$$f'(a) = \lim_{h \to 0} \frac{f(a+h) - f(a)}{h}$$

We have, with $a = 2$,

$$f'(2) = \lim_{h \to 0} \frac{f(2+h) - f(2)}{h} = \lim_{h \to 0} \frac{\frac{1}{2+h} - \frac{1}{2}}{h}$$

$$= \lim_{h \to 0} \frac{\frac{2 - (2+h)}{2(2+h)}}{h} \qquad \text{Using } 2(2+h) \text{ as a common denominator in the numerator}$$

$$= \lim_{h \to 0} \frac{-h}{2(2+h)} \cdot \frac{1}{h} \qquad \text{Using } \frac{\frac{a}{b}}{c} = \frac{a}{b} \cdot \frac{1}{c}$$

$$= \lim_{h \to 0} \frac{-1}{2(2+h)} \qquad \text{Canceling } h$$

$$= -\frac{1}{4}$$

We illustrate this answer in Figure 1.46 as the slope of the line T tangent to the graph of $f(x) = 1/x$ at the point $(2, f(2)) = \left(2, \frac{1}{2}\right)$.

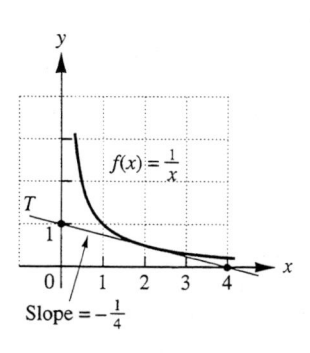

Figure 1.46

Slope $= -\frac{1}{4}$

EXAMPLE 2

Write the equation of the line tangent to the graph of $f(x) = 1/x$ at the point $\left(2, \frac{1}{2}\right)$.

Solution The point $\left(2, \frac{1}{2}\right)$ is both on the curve and on the tangent line. The slope of the tangent is $f'(2) = -\frac{1}{4}$ from Example 1. Thus we know the slope, $-\frac{1}{4}$, and a point, $\left(2, \frac{1}{2}\right)$, on this tangent line. According to Equation 1 earlier, the desired equation must be

$$y - \frac{1}{2} = \left(-\frac{1}{4}\right)(x - 2)$$

or

$$y = -\frac{1}{4}x + 1 \quad \blacksquare$$

Compare the y-intercept of this equation with that indicated on the graph in Figure 1.46.

The next example illustrates that it is often no more difficult to calculate the derivative at an arbitrary point a than to calculate it at the particular number $a = 2$.

EXAMPLE 3

Find a formula for the derivative $f'(a)$ of

$$f(x) = x^3$$

at any point a.

Solution According to the definition, we have

$$f'(a) = \lim_{h \to 0} \frac{f(a + h) - f(a)}{h}$$

$$= \lim_{h \to 0} \frac{(a + h)^3 - a^3}{h} = \lim_{h \to 0} \frac{(a + h)(a^2 + 2ah + h^2) - a^3}{h}$$

$$= \lim_{h \to 0} \frac{(a^3 + 3a^2h + 3h^2a + h^3) - a^3}{h}$$

$$= \lim_{h \to 0} \frac{3a^2h + 3h^2a + h^3}{h}$$

$$= \lim_{h \to 0}(3a^2 + 3ha + h^2) \quad \text{Canceling } h \neq 0$$

$$= 3a^2$$

Therefore the formula for $f'(a)$ at any point a is $f'(a) = 3a^2$. $\blacksquare$

EXAMPLE 4

Use the formula found in Example 3 to compute $f'(1)$, $f'(0)$, and $f'\left(-\frac{1}{2}\right)$ for $f(x) = x^3$. Illustrate these answers on a graph.

Solution From Example 3, $f'(a) = 3a^2$. Therefore
$$f'(1) = 3(1)^2 = 3$$
$$f'(0) = 3(0)^2 = 0$$
and
$$f'\left(-\tfrac{1}{2}\right) = 3\left(-\tfrac{1}{2}\right)^2 = \tfrac{3}{4}$$

These answers are the slopes of the tangents to the graph drawn in Figure 1.47. ∎

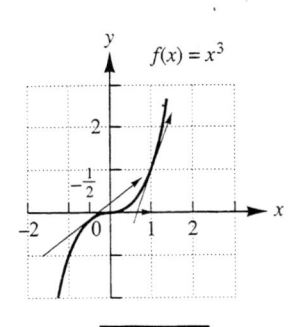

Figure 1.47

Before leaving this example, let us note an important point that it illustrates:

The derivative of a function is itself a function.

That is, we have seen that for the function
$$f(x) = x^3$$
the derivative, found in Example 3, must be
$$f'(x) = 3x^2$$

We can regard this expression as a new function that describes the correspondence between a point x and the slope of the tangent to the graph at the point $(x, f(x))$.

The process of computing the function f' from the function f is called **differentiation**. In Chapter 2, we will learn a few general rules that will allow us to differentiate a function without going through the limit process. Our understanding of how the derivative is defined by a limit will then enable us to understand what the derivative means when we do find it.

A First Applied Interpretation of the Derivative

At some time you have no doubt had the thought, "If things can just continue in this direction for a while, then . . . ." It may be helpful to you to interpret our first application of the derivative in such everyday terms, because we can easily relate this thought to the idea that the slope of the tangent line to a curve indicates the direction of the curve at the point where the tangent is found.

EXAMPLE 5

Shown in Figure 1.48 is a (hypothetical) graph of the sales of a struggling new manufacturer of tennis rackets, the Strung Manufacturing Company. The Strung Company is in its first year of production. In this figure, the unit interval $0 \leq t \leq 1$ represents 1 year's time; for example, $t = \tfrac{1}{2}$ corresponds to 6 months, or midyear. The management of the Strung Company is having a stockholders' meeting at midyear and has gathered sales data showing that by time t in

1.4 The Derivative of a Function

the year, the company has sold $R(t) = t^3$ million tennis rackets. For example, by $t = \frac{1}{4}$ (3 months into the year) the company had sold $R(\frac{1}{4}) = (\frac{1}{4})^3 = 0.015625$ million rackets—that is, 15,625 rackets.

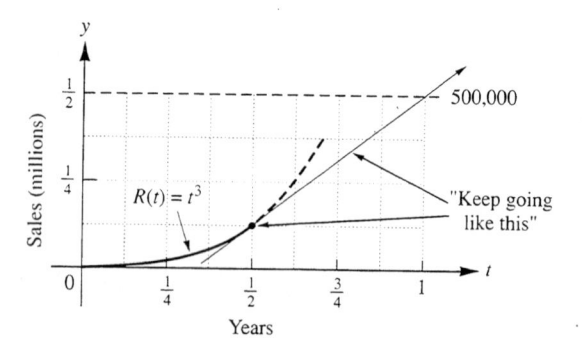

Figure 1.48

At the midyear meeting management announces, "If sales just keep going like this—*increasing at their present rate*—then by year's end we will have sold at least 500,000 rackets." Use the derivative and the equation of the tangent line to the curve of the sales function $R(t) = t^3$ at $t = \frac{1}{2}$ to explain this assertion.

Solution The derivative

$$R'(t) = 3t^2$$

of the sales function was found in Example 3. Management realizes that the slope of the tangent to the sales curve indicates the direction of the sales curve and is the rate of growth of sales.

At midyear, when $t = \frac{1}{2}$, this slope is

$$R'\left(\frac{1}{2}\right) = 3\left(\frac{1}{2}\right)^2 = \frac{3}{4}$$

Therefore, from Equation 1 on page 88, the equation of the tangent line to the graph of the sales curve at midyear is

$$y - R\left(\frac{1}{2}\right) = R'\left(\frac{1}{2}\right)\left(t - \frac{1}{2}\right)$$

or

$$y - \left(\frac{1}{2}\right)^3 = \frac{3}{4}\left(t - \frac{1}{2}\right)$$

$$y = \frac{3}{4}t - \frac{1}{4} = \frac{3t - 1}{4}$$

If sales "keep going like this"—that is, in the direction they are moving at midyear, or *along the tangent line*—then at the end of the year $t = 1$, sales will be

$$y = \frac{3(1) - 1}{4} = \frac{1}{2} \text{ million}$$

or 500,000 rackets. Figure 1.48 illustrates this conclusion geometrically. ∎

The tangent line T, representing the direction of the curve at the moment $t = \frac{1}{2}$, represents the thought, "If things keep going like this . . . ." Of course if the sales curve continues to follow the true graph of $y = t^3$ after midyear, then year-end sales will be even greater; the stated prediction is based only on information known near the time it is made.

Continuity and Differentiation

How are the concepts of this and the previous section related? Briefly, a function f that has a derivative $f'(a)$ at a point a must necessarily be continuous at a. However, a function may be continuous and yet fail to be differentiable. An example of this phenomenon is given in Section A.1 along with a brief indication of why differentiability implies continuity. Thus, of these two conditions, differentiability is the stronger.

Exercises 1.4

1. Let $f(x) = x^2$. In (a), (b), and (c), compute, as in Example 3.

   a. $f'(2)$ **b.** $f'(a)$ for any a

   c. $f'(-1)$, $f'(\frac{1}{2})$, $f'(\sqrt{2})$, $f'(1.021)$

   d. Graph the equations $y = f(x)$ and $y = f'(x)$ on the same coordinate system. In what way are the positive and negative values of f' in accord with the shape of the graph of f?

2. Let $f(x) = \sqrt{x}$, $x \geq 0$. Compute

   a. $f'(3)$ **b.** $f'(a)$

   c. $f'(4)$ and $f'(225)$

   [Hint: $(\sqrt{a} - \sqrt{b})(\sqrt{a} + \sqrt{b}) = a - b$.]

3. Let $f(x) = 1/x$, $x \neq 0$. Compute $f'(a)$ for any point $a \neq 0$. Graph f and its tangent lines at $a = -1$ and $a = 1$, using the formula you derive for $f'(a)$.

4. Let $f(x) = 3x + 1$. Compute $f'(a)$ for any a. Interpret the derivative as the direction of the graph, and explain in words why your answer for $f'(a)$ makes sense.

5. Let $f(x) = \sqrt{7}$ for all x. Compute $f'(a)$ for any a. Interpret the derivative as the direction of the graph, and explain in words why your answer makes sense. Can you make a general statement about the derivative of any constant function?

In Exercises 6–9, use information from earlier exercises to find the equation of the tangent line to the graph of $y = f(x)$ at the given point. Draw the graph of the function and its tangent line at this point.

6. $f(x) = 1/x$; $(1, f(1))$ (Exercise 3)
7. $f(x) = x^2$; $(1, f(1))$ (Exercise 1)
8. $f(x) = 3x + 1$; $(4, f(4))$ (Exercise 4)
9. $f(x) = \sqrt{7}$; $(1, f(1))$ (Exercise 5)
10. If two straight lines L and M are *perpendicular* with slopes S and T, respectively, then the product of their slopes is -1; that is, $S \cdot T = -1$. Find the equation of the straight line perpendicular to the graph of the functions in Exercises 6–9 at the indicated point. [Hint: Let S be the slope of the tangent line that you have already found, and then solve for T in the equation $S \cdot T = -1$.]
11. Suppose that you are driving your car in Glacier National Park on a dark night and around a curve that follows the shape of the graph of $y = 1/x$, as indicated in Figure 1.49. As you round the curve, with only your headlights to see by (and ignoring the width of the beam), do you even see the shining eyes of the mountain lion watching you from the point $P = (1.6, 0.2)$ when your car is at the point $(1, 1)$?*

---

*My thanks to Peter Casazza for this example.

[Hint: What does this have to do with derivatives and tangent lines? If you can answer this, then you may then want to use Exercise 6 to complete the answer.]

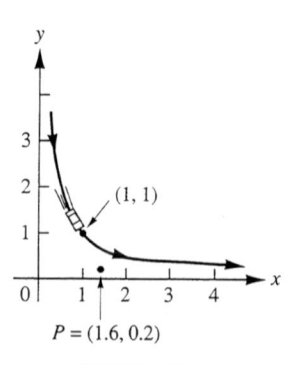

Figure 1.49

12. Mathematicians like to use mathematical concepts informally to make a point, because the concepts of mathematics closely conform to everyday ideas. For example, a friend of the author is fond of saying, "It's not what you have that makes you happy, it's your derivative that really counts!" We take our happiness or wealth for granted and think in terms of being happ*ier* or wealth*ier*. While a function value $f(x)$ measures the *state* of a system, the value of the derivative $f'(x)$ measures its *change-of-state*. Thus if $f(t)$ is your wealth at time t, then $f'(t)$ measures the rate at which you are becoming wealthier. Who would you guess is happier at midyear, $t = \frac{1}{2}$:

a. The owner of a stock investment whose value at time t is
$$S(t) = \frac{1}{t} \text{ thousands of dollars}$$

b. The owner of a new business whose profits by time t are
$$P(t) = t^3 \text{ thousands of dollars}$$

[Hint: Compute $S'(\frac{1}{2})$ and $P'(\frac{1}{2})$, using Exercise 6 and Example 3. How do these compare with $S(\frac{1}{2})$ and $P(\frac{1}{2})$? A graph may also help.]

13. Represent the length of the semester by a unit interval $[0, 1] = \{t: 0 \leq t \leq 1\}$. Thus $t = \frac{1}{2}$ is mid-semester. Represent your possible grade scale by a unit interval with $0.8 < y \leq 1$ representing a grade of A, $0.6 < y \leq 0.8$ representing a B, ... , and $y \leq 0.2$ representing an F. Suppose that your grade at time t in the semester is $g(t) = \sqrt{t}$. Thus at midterm you are earning a C, since $g(\frac{1}{2}) = 0.707$. But your grade is improving all the time, and you think to yourself: "If things keep going like this . . . ." Why is it not even theoretically possible that your grade could continue to improve at the *rate* it is improving at mid-semester? Use Exercise 2(b) to help answer this. **Remark.** The grade scale here is chosen *only* to make the arithmetic easier!

14. Analyze a newspaper article about the economy with respect to the article's attempts to contrast the state of the economic matter it discusses with its change-of-state.

Computer Application Problems

Use the BASIC program DERIV to estimate the value of the derivative of the functions given in Exercises 15–18 at the indicated point. Indicate your conclusions on a graph.

15. $f(x) = 2^x$ at $x = 0$
16. $f(x) = x^4$ at $x = 1$
17. $f(x) = 2^{-x^2}$ at $x = 0$
18. $f(x) = \sqrt{1 - x^2}$ at $x = \frac{1}{2}$

1.5 The Derivative as Rate of Change

The purpose of this section is to help you understand that if $f(x)$ represents *how much*, then $f'(x)$ represents *how fast*.

As useful as it is to understand that the derivative of a function represents the slope of a line tangent to its graph, this view is not enough. To apply the derivative, you must understand that it measures the exact rate of change of values at a point. In this section we reinforce this understanding before we present formal mechanical rules for calculating the derivative in Chapter 2.

94 Chapter 1 Limits and the Derivative

Difference Quotients and Units of Measurement for the Derivative

Let us review the definition of the derivative and introduce some new terminology. Think of $y = f(x)$ as an amount, corresponding to x. The quotient defining the derivative

$$\frac{f(x + h) - f(x)}{h}$$

is called a **difference quotient,** because it is a quotient of differences:

$$\frac{f(x + h) - f(x)}{h} = \frac{\text{difference in function values}}{\text{difference in underlying variable values}}$$

since $h = (x + h) - x$. This quotient measures the change in the amount $f(x)$ at x in proportion to the change in the underlying independent variable. It is the *average rate of change*, an approximate measure of the rate of change at this amount.

When x represents time, we can see relatively easily that the difference quotient measures an average rate as we commonly think of it, since

$$\frac{f(t + h) - f(t)}{h} = \frac{\text{change in amount}}{\text{change in time}}$$

and h measures the change in time between time t and time $t + h$. The derivative

$$f'(t) = \lim_{h \to 0} \frac{f(t + h) - f(t)}{h}$$

is then thought of as the *exact rate* of change at the instant t, since the time difference over which the average is found shrinks to zero.

Even when the variable x does not represent time, the same kind of meaning for $f'(x)$ holds. We can often use the units of measurement of the variables in a given functional relationship to help determine this meaning.

EXAMPLE 1

Let $A(x)$ be the number of customers attracted to a product when x dollars are spent on advertisement. What does $A'(x)$ represent?

Solution The difference quotient

$$\frac{A(x + h) - A(x)}{h}$$

is the ratio

$$\frac{(\text{customers when } x + h \text{ dollars are spent}) \text{ minus } (\text{customers when } x \text{ dollars are spent})}{\text{change } h \text{ in dollars spent}}$$

and is measured in *customers per dollar*. It represents the average change in the number of customers attracted to the product per dollar spent on advertising. The derivative $A'(x)$, being the limit of this ratio as the change in dollars spent on advertising shrinks to zero, is measured in the same units of customers per dollar. Therefore, $A'(x)$ is the *rate* measured in customers per dollar at which customers are attracted to the product, at an advertising level of x dollars. ∎

Remark. In applications of calculus to business or other fields that deal only in whole, discrete units of measure, we often face a theoretical difficulty. In reality our function models can only have whole number values (a business does not have $\sqrt{2}$ customers!) and therefore cannot be continuous where the value changes. We noted at the close of Section 1.4 that a function that is not continuous cannot have a derivative, and thus we cannot easily apply calculus to it. We handle this difficulty by treating a function like that found in Example 1 as though it had continuously changing values between its values that could actually occur in reality. As you will see in more detail in later chapters, this treatment is both practical and effective.

EXAMPLE 2

Let $C(x)$ be the total cost for a manufacturer to produce x units of a given product. What does $C'(x)$ represent?

Solution The difference quotient

$$\frac{C(x+h) - C(x)}{h} = \frac{[\text{cost to produce } (x+h) \text{ units}] \text{ minus } [\text{cost to produce } x \text{ units}]}{\text{change in number of units produced}}$$

$$= \frac{\text{change in cost}}{\text{change in number of units produced}} \text{ dollars per unit}$$

is the average rate of change in costs between production level x and production level $x + h$.

Therefore

$$C'(x) = \lim_{h \to 0} \frac{C(x+h) - C(x)}{h} \text{ dollars per unit}$$

is the rate of change in production *costs per unit*, at a production level of x units. ∎

The number $C'(x)$ is called the **marginal cost,** at a production level of x units. Marginal cost is regarded as the additional cost to the manufacturer to produce one additional unit. Similarly, if $P(x)$ represents the profit to this manufacturer for the production of x units, then $P'(x)$ is the rate of increase in profits at production level x and is called the **marginal profit** at x. The marginal profit $P'(x)$ is regarded as the additional profit resulting from the production of one additional unit, at production level x.

The preceding two instances exemplify the general point that the units of measurement in which the difference quotient

$$\frac{f(x+h) - f(x)}{h} = \text{change in amount per unit}$$

is measured indicate to us what $f'(x)$ measures:

$f'(x)$ = rate of change in this amount per unit, at the level of x units

That is, if $y = f(x)$ is *how much* we have at x, then $f'(x)$ is *how fast* this amount is changing *at x*.

Some Particular Derivatives and Their Measurement and Meaning

We now consider a series of basic examples, chosen for their everyday nature, to illustrate what the derivative means and measures. In the first instance, the derivative gives an answer in agreement with what we already intuitively know.

EXAMPLE 3

Throw a rock up into the air. It rises and then falls back to the ground. A graph of its height (in feet) versus time (in seconds) could look like the graph in Figure 1.50 and be given by the function

$$h(t) = -(t - 2)^2 + 4 = -t^2 + 4t$$

How fast is the rock moving at time $t = 2$?

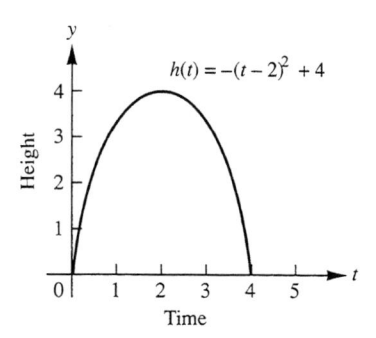

Figure 1.50

Solution The graph in Figure 1.50 suggests that at time $t = 2$, at the peak of its flight, the rock is not moving. That is, its state is fixed for an instant, and its rate of change must be zero. Let us see that the derivative $h'(2)$ confirms this. We have

$$h'(2) = \lim_{s \to 0} \frac{h(2 + s) - h(2)}{s} \quad \frac{\text{change in height}}{\text{change in time}} \quad \frac{\text{ft}}{\text{sec}}$$

$$= \lim_{s \to 0} \frac{[-(2 + s)^2 + 4(2 + s)] - [-(2)^2 + 4(2)]}{s}$$

$$= \lim_{s \to 0} \frac{-4 - 4s - s^2 + 8 + 4s + 4 - 8}{s}$$

$$= \lim_{s \to 0} \frac{-s^2}{s}$$

1.5 The Derivative as Rate of Change

$$= \lim_{s \to 0}(-s)$$

$$= 0 \, \frac{\text{ft}}{\text{sec}} \quad \blacksquare$$

Of course at times other than $t = 2$, the rock has some speed, positive (when it is rising) or negative (when it is falling), which can be computed in the same way. (See Exercise 15.)

Remark. Note how we used s here, rather than h, to represent the imagined time shift from 2 to the later time $2 + s$. The letter "h" in the definition of a derivative, or the letter "s" used in this example, is called a **dummy variable:** It is just a mark—a letter—that we use to carry the idea and compute with.

The next discussion interprets the derivative as it relates to a common concern in any business enterprise, a need for information about the cash flow to a company as a result of its operations. (**Cash flow** is the *rate* at which money is returning to the company via sales of its products.)

Let us return to the hypothetical curve of the Strung Manufacturing Company of Example 5, Section 1.4. As Figure 1.51 shows, the growth in sales is slow early in the year. This confronts management with a cash-flow problem. Unless money from sales is flowing into the company at an adequate rate, management will have to go into the short-term financial market and borrow funds to purchase raw materials to maintain production. In order to estimate its borrowing needs, management needs to know the rate at which money is flowing back into the company from sales. This rate of return will be determined by the rate of sales. The derivative may be used to determine when the *rate* of sales is at a required level.

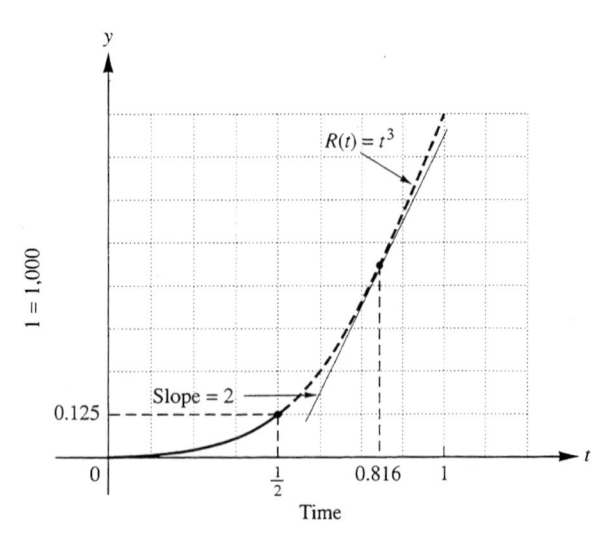

Figure 1.51

EXAMPLE 4

a. Determine when the Strung Manufacturing Company, whose total sales are $R(t) = t^3$ by time t, will be selling its product at the rate of 2 million units per year.

b. If the company earns $20 per racket, what will be its cash flow per day at the time found in (a)?

Solution

a. In Example 3 of Section 1.4 we saw that the derivative of R is

$$R'(t) = 3t^2$$

The derivative represents the rate of sales at time t.* This rate will be 2 million units per year when

$$R'(t) = 2$$

or when

$$3t^2 = 2$$

This equation has the positive solution

$$t = \sqrt{\tfrac{2}{3}} \approx 0.816$$

Therefore the rate of sales will be 2 million units per year at this time, on day $298 = (0.816)(365)$ in the year.

b. On the 298th day of the year, when the sales rate is 2 million units per year, the company will be earning a return at the rate of $2 \times 20 = \$40$ million per year. This means that its cash flow on this day will be

$$\frac{\$40{,}000{,}000}{365} = \$109{,}589 \text{ per day}$$

with which to finance continuing production. ∎

The next example shows how precisely calculus can measure and represent change. It is also a prelude to the Fundamental Theorem of Calculus (to be considered in Chapter 4) and a revealing example of the interplay between calculus and the physical world.

EXAMPLE 5

Toss a pebble into a smooth pool of water. How fast does the *area* enclosed by the expanding circular wavelet of water grow, with respect to its lengthening radius? (See Figure 1.52.)

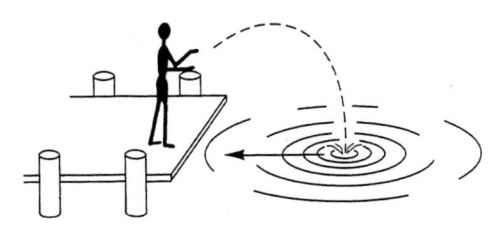

Figure 1.52

---

*The reason that $R'(t)$ represents the rate of sales at any time t is that the difference quotient

$$\frac{R(t+h) - R(t)}{h} = \frac{\text{change in amount sold}}{\text{change in time}}$$

is measured in units sold per year.

Solution When the pebble hits the water, the circular wave begins to move away from the point of impact. As it moves, the circle is expanding, and so its area is growing. We are asked: How fast is it growing? That is, how fast is extra area being added *as the radius lengthens?* Since at radius r the area of the circle is given by the function $A(r) = \pi r^2$, another way of asking this question is: How fast is $A(r)$ changing as r changes? We can answer this question by computing the derivative $A'(r)$. We have

$$A'(r) = \lim_{h \to 0} \frac{A(r+h) - A(r)}{h} \text{ in.}^2/\text{in.}$$

$$= \lim_{h \to 0} \frac{\pi(r+h)^2 - \pi r^2}{h}$$

$$= \lim_{h \to 0} \frac{\pi(r^2 + 2rh + h^2) - \pi r^2}{h}$$

$$= \lim_{h \to 0} \frac{2\pi rh + \pi h^2}{h}$$

$$= \lim_{h \to 0} 2\pi r + \pi h \qquad \text{Canceling } h, \text{ which is not } 0$$

$$= 2\pi r \qquad \text{in.}^2 \text{ of area per in. of radius} \quad \blacksquare$$

The *rate* of area growth coincides with the *circumference* of the expanding circle. That is, the expanding circle increases its area only on its outer edge (rather clearly!) and at precisely the rate $2\pi r$, the length of this edge.

Think of the tracking arm on your stereo turntable as it follows the grooves in a record's surface. These grooves (almost) cover the area of your record. The record revolves at a constant rate of 33 rpm. Hence the tracking arm must move *faster* at the beginning of a song than at the end. How fast does it move? It moves precisely the length of the record groove $2\pi r$ at a distance r from the record's center, per rpm. The groove is analogous to the circular wave, and the circular area bounded by the tracking arm shrinks at a rate equal to the length of the groove.

In Example 5, we computed the rate of change with respect to the lengthening radius. If there were a pier built out into the pond on pilings—say, 5 ft apart—we could observe the expanding wavelet move through the pilings and with a watch measure how fast the radius of the circle was expanding with respect to time. Such a naturally observable measure raises a question: How fast does the area expand over time? The answer must wait for Section 2.2, when we study the *chain rule* for differentiation.

With these examples we can better understand what calculus offers and indeed better understand its essential nature. We have compared a function f to a roll of motion picture film, and more recently to a videotape, of a changing system, where a single function value $f(t)$ shows us the image of the system on "frame t." The preceding examples suggest that we can regard f' as a viewing of the film in motion, where a single value $f'(t)$ tells us the rate at which change is occurring in the image at "frame t." In an application, we may regard the process of differentiation that produces f' from f as the process of finding a formula by which we can view the tape f—the system—in motion. In Chapter 4, we will study the process of antidifferentiation, which will take us from f' back to f, in effect freezing the motion f' and showing us any fixed image $f(t)$ that we wish to see.

Exercises 1.5

Assume that each of the functions in Exercises 1–10 is differentiable. In similar language, state what the derivative $f'(5)$ represents for each of these functions. What does $f'(x)$ represent in each case? [*Note:* The notation $f'(x)$ and $f'(5)$ is used here in the generic sense.]

1. $R(x)$ is the total revenue earned by the Ace Manufacturing Company, from the sale of x items.
2. $R(x)$ is the number of rabbits living in an area populated by x foxes.
3. $R(t)$ is the number of people in a small town who have heard a rumor by day t.
4. $B(t)$ is the total air force budget for fighter aircraft in year t.
5. $V(t)$ is the speed in ft/sec of the car you are driving at time t in seconds.
6. $J(x)$ is the amount of junk food you have eaten by the xth day of this semester.
7. $P(x)$ is the number of bacteria killed by a dosage of x milligrams of penicillin.
8. $C(x)$ is the number of television commercials on a program watched by x millions of viewers.
9. $B(x)$ is the number of people found backpacking into the Grand Canyon in the first x miles from the trailhead.
10. $L(t)$ is the height of a slipping ladder leaning against a wall at time t (in seconds) as it slides down the wall.

In Exercises 11–22, use the following formulas, which were obtained in earlier exercises and examples, to make the needed computations:

If $f(x) = x^3$, then $f'(x) = 3x^2$.
If $f(x) = x^2$, then $f'(x) = 2x$.
If $f(x) = 1/x$, then $f'(x) = -1/x^2$.
If $f(x) = \sqrt{x}$, then $f'(x) = 1/(2\sqrt{x})$.

11. A manufacturing company has a fixed production cost of $50,000 and a unit cost per item of $10.
    a. Derive a formula for the cost to the company to produce x items.
    b. Compute the marginal cost to the company to produce 15 items and 50 items. (Recall that marginal cost is the derivative of cost; see Example 2.)
    c. Thinking of marginal cost as the "cost to produce an additional item," does your answer to (a) make sense?
    d. Suppose instead that the cost function of the company is
    $$C(x) = \frac{100}{x} + 50{,}000$$
    Compute the marginal cost to produce the 50th item.
    e. Why do the marginal costs differ so significantly in (b) and (d)?

12. Suppose that the demand function (Section 0.4) for the production of x units of a certain item is
    $$D(x) = x \text{ dollars}$$
    The revenue function for this item is therefore
    $$R(x) = x \cdot D(x) = x^2$$
    Suppose further that the cost to produce x units of this item is
    $$C(x) = \sqrt{x}$$
    Compute the marginal profit at a production level of nine units.

13. Compute the speed of the rock in Example 3 at $t = 1$ sec. That is, find $h'(1)$.

14. Let $f(x) = 1/x$. At what value of x does the tangent line to the graph have slope -4? Slope $-\frac{1}{100}$?

1.5 The Derivative as Rate of Change 101

15. A rock is tossed *up* into the air from the top of a tower. Its height above the ground at time t is $H(t) = -16t^2 + 64t + 80$ ft at time t. We will see in the next chapter that $H'(t) = -32t + 64$. Use these facts in answering the following questions:

   a. How high is the tower? [*Hint:* At what time was the rock at the height of the tower?]

   b. When does the rock reach the top of its flight? [*Hint:* How fast is it moving then?]

   c. How high does it rise above the tower? [*Hint:* Use (b).]

   d. When does it hit the ground? [*Hint:* How high (that is, low) is it then?]

   e. How fast was it thrown from the tower; that is, what was its initial velocity?

   f. How fast is it moving when it hits the ground? [*Hint:* Use (d).]

   g. How high was it, and how fast was it going, at the end of 1 sec? At the end of 4 sec?

   [*Hint:* Remember that the function H represents the various states of this system and that the function H' represents the motion, or change-of-state, of the system. Thus questions about position are answered using H while questions about motion are answered using H'.]

16. Consider a triangle of altitude x and base y (Figure 1.53), and let S be the slope of its upper side.

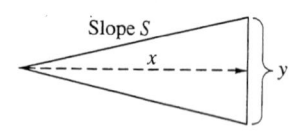

Figure 1.53

Recalling that the slope of a line can be measured anywhere along its length, how fast is y expanding as x lengthens? Before you do this as a differentiation problem, what is your intuitive guess as to the correct answer?

17. a. The volume of a sphere of radius r is $\frac{4}{3}\pi r^3$. How fast is the volume of a balloon growing with respect to its increasing radius as you inflate it?

   b. The surface area of a sphere of radius r is $4\pi r^2$. How fast is the surface area of a balloon increasing with respect to the radius as you inflate it?

   [*Hint:* Consider the properties of limits and the definition of the derivative to convince yourself that the constant factors $\frac{4}{3}\pi$ and 4π do not affect the differentiation of the variable factors, r^3 and r^2, respectively, and use the formulas listed earlier.]

18. Consider a circle of radius $r = 3$ and area A. Then enlarge the radius by an amount $h = 0.05$, to $r = 3.05$, giving a larger circle of area B. Interpret the number $B - A$ geometrically. Then compute the average change in area $(B - A)/h$, and compare this number with the circumference of the smaller circle. Explain the relationship of these calculations to the discussion in Example 5. Do you understand how this problem and Example 5 are related to the rate at which the tape on one of your cassettes moves from reel to reel?

19. A function value $f(x)$ and its derivative $f'(x)$ represent two very different and often contrasting measures of a system. It is advantageous to be able to distinguish between the two in a conceptual way. Figure 1.54 shows the year-long sales curve of two companies, the Topspin Manufacturing Company and the Strung Manufacturing Company of Example 4. Notice that while the sales of Topspin are in a better *state* throughout the year ($S(t) > R(t)$ for $0 < t < 1$), by year's end the sales of both companies are in the same state: $S(1) = R(1)$. This happens because the *change-of-state* of the Strung Company overtakes and surpasses that of Topspin at some point during the year. Using the facts that $S'(t) = 2t$ and $R'(t) = 3t^2$, determine exactly when this will occur. What might reasonably be expected of the future course of their sales in the early part of the next year? (When you are reading an article in economics or related fields, you can enrich your understanding of calculus and of the article if you notice how the analysis presented in the article shifts between an emphasis on the state of the matter at hand and its change-of-state.)

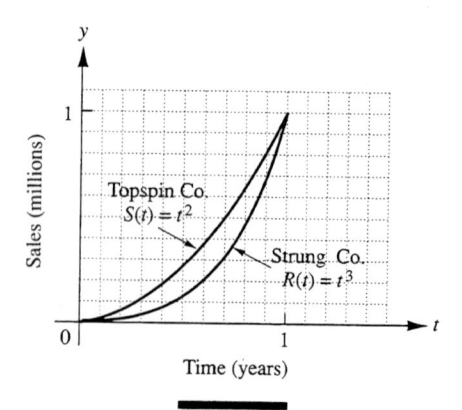

Figure 1.54

102 Chapter 1 Limits and the Derivative

20. Refer to Exercise 19. Is there a point in the year when Topspin's sales are increasing at twice the rate of Strung's? Is there a point where Strung's are increasing at twice the rate of Topspin's?

Challenge Problems

21. "To play good tennis, keep your knees bent and your eye on the ball!" This problem concerns keeping your eye on the ball. Suppose you are playing tennis and your opponent returns the ball and it hits the court 16 ft in front of you. A possible model of the height $H(x)$ of the approaching ball as you prepare to hit it is the graph of $y = \sqrt{x}$ for, say, $0 \le x \le 20$ (see Figure 1.55). That is, $H(x) = \sqrt{x}$ is how high the ball is at distance x from its bounce. How fast is the ball rising per foot of horizontal distance at the point where you will hit it?

22. Referring to Exercise 21, the width of a tennis racket is normally about 8 in. When you hit the ball, your racket maintains contact for a horizontal distance of about 6 in. Find the equation of the tangent line to the graph of $y = \sqrt{x}$ at $x = 16$. Use it to compute approximately how far the ball wants to move vertically against the face of your racket during the 6 in. of horizontal contact. How does physical reality, in the moments after impact, relate to the idea that the tangent line represents the thought, "If things just keep going like this"? (See Figure 1.56.)

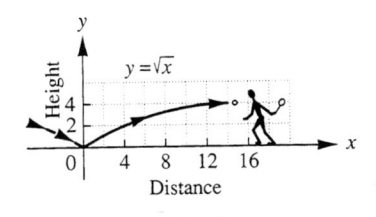

Figure 1.55

Figure 1.56

Chapter 1 Summary

1. The average change in the variable $y = f(x)$, when x changes from t to $t + h$, is the ratio

$$\frac{f(t + h) - f(t)}{h}$$

This computation is also the slope of the straight line through the two points $(t, f(t))$ and $(t + h, f(t + h))$ on the graph of $y = f(x)$.

2. The limit of a function f at a point b is not necessarily related to the function value (if any) at b, but is the single number that best approximates function values $f(x)$ for x as near to b as one cares to imagine. A function is continuous when the limit at b coincides with the function value $f(b)$.

3. The derivative $f'(a)$ of a function f at the point a is the limit of its average rate of change at a. Thus

$$f'(a) = \lim_{h \to 0} \frac{f(a + h) - f(a)}{h}$$

4. The derivative represents both the exact rate of change of $y = f(x)$ with respect to change in x at the point a and the slope of the line tangent to

the graph of f at $(a, f(a))$. The equation of this tangent line is given in point–slope form by

$$y - f(a) = f'(a)(x - a)$$

5. While a function value $f(x)$ represents the state of a changing system at x, the derivative $f'(x)$ represents the rate at which the quantity $f(x)$ is changing with respect to changes in the underlying variable x, and consequently measures the change-of-state at x of the system under consideration.

Chapter 1 Summary Exercises

Evaluate the limits in Exercises 1–10.

1. $\lim\limits_{x \to 11} \sqrt{x - 2}$

2. $\lim\limits_{x \to 2} 3x^3 - 2x^2$

3. $\lim\limits_{x \to \sqrt{7}} \dfrac{x^2 - 7}{x - \sqrt{7}}$

4. $\lim\limits_{x \to 2} \dfrac{x - 2}{x + 4}$

5. $\lim\limits_{x \to 1} \dfrac{x^3 - 1}{x + 1}$

6. $\lim\limits_{x \to -1} (x^2 + 1)^5$

7. $\lim\limits_{x \to -2} \dfrac{x - 2}{x^3 - 8}$

8. $\lim\limits_{x \to 7} (x + 1)^{2/3}$

9. $\lim\limits_{x \to 5} \sqrt{\dfrac{x^2 - 25}{x - 5}}$

10. $\lim\limits_{t \to 0} \dfrac{t^2 - t}{t}$

Which functions in Exercises 11–14 are continuous at $x = 2$?

11. $f(x) = x^3 - x^2$

12. $f(x) = \begin{cases} x^2 & x < 2 \\ 6 - 2x & x \geq 2 \end{cases}$

13. $f(x) = \begin{cases} \dfrac{x^3 - 8}{x - 2} & x \neq 2 \\ 9 & x = 2 \end{cases}$

14. $f(x) = \begin{cases} \dfrac{x^4 - 16}{x^2 - 4} & x \neq 2 \\ 8 & x = 2 \end{cases}$

15. Sketch a graph of a function g such that
    a. $\lim\limits_{x \to 3} g(x) = 5$.
    b. g is not continuous at $x = 3$.

16. Sketch a graph of a function h such that
    a. h is not continuous at $x = 1$ and $x = 2$.

b. h is continuous at $x = 0$.

c. $\lim\limits_{x \to 0} h(x) = 3$.

d. $\lim\limits_{x \to 2} h(x) = 1$.

e. $\lim\limits_{x \to 1} h(x)$ does not exist.

Using Sections 1.4 and 1.5, find the slope and equations of the tangent lines at the indicated point for each of the functions given in Exercises 17–20.

17. $f(x) = x^2$ $(x, y) = \left(\tfrac{1}{2}, \tfrac{1}{4}\right)$

18. $f(x) = x^3$ $(x, y) = \left(-\tfrac{3}{2}, -\tfrac{27}{8}\right)$

19. $g(x) = \dfrac{1}{x}$ $(x, y) = \left(10, \tfrac{1}{10}\right)$

20. $h(t) = \sqrt{t}$ $(9, 3)$

21. Find the y-intercept of the tangent line to the graph of $f(x) = 1/x$ at the point $\left(5, \tfrac{1}{5}\right)$.

22. Let $S(t) = 1/t$ be the total sales (in millions of dollars) of a company in month t, up to $t = 10$ months. If sales decline at the rate $S'(10)$—that is, "if things keep going like this"—thereafter, when will monthly sales be zero?

In the same kind of language with which the functions in Exercises 23–26 are defined, say what the number $f'(15)$ represents.

23. $f(x)$ is the total cost to produce x items.

24. $f(t)$ is the distance you have run by time t seconds.

25. $f(z)$ is the population of a city in year z.

26. $f(s)$ is your income per year, s years after graduation.

27. The total sales (in millions of dollars) of a company is $s(t) = \sqrt{t}$ by time t during the first year of operation. During this period, when were sales growing at the rate of

   a. $2 million per year?

   b. $0.5 million per year?

28. Were sales ever growing at the rate of $120,000 per year in Exercise 27?

29. A rock is tossed into the air and at time t is $h(t) = -16t^2 + 96t + 16$ ft above the ground. Given that $h'(t) = -32t + 96$,

   a. When does the rock reach the top of its flight?

   b. How high did it rise?

   c. Is it rising or falling at $t = 4$? How high is it at $t = 4$?

   d. How fast is this occurring in (c)?

   e. When does it hit the ground?

   f. How fast is it moving at that point?

   It may be helpful in this exercise to carefully distinguish between the state of the rock, its height $h(t)$, and its change-of-state, or rate of movement, at time t, $h'(t)$.

Evaluate the limits in Exercises 30–33.

30. $\lim\limits_{x \to \infty} \dfrac{3x^2 + 2x}{2x^2 + 1}$

31. $\lim\limits_{x \to -\infty} \dfrac{4x^3 - 1}{1 + x^3}$

32. $\lim\limits_{x \to \infty} \dfrac{1 + x}{\sqrt{x^2 + x}}$

33. $\lim\limits_{x \to \infty} \dfrac{(x^2 + 1)^{2/3}}{x}$

Determine the vertical asymptote, if any, of the functions in Exercises 34–37.

34. $f(x) = \dfrac{1}{(x - 2)(x^2 + 1)}$

35. $f(x) = \dfrac{x^2 - 4}{(x + 1)(x - 2)}$

36. $g(x) = \dfrac{1}{(x - 1)^2}$

37. $h(x) = \dfrac{3}{2/(x - 1)}$

38. Show that $y = (1/x) + 2$ is asymptotic to $y = 2$ as $x \to +\infty$.

39. Show that $y = x^2/(4 - x^2)$ is asymptotic to $y = -1$ as $x \to +\infty$.

40. Evaluate $\lim\limits_{x \to \infty} \dfrac{\sqrt{x^2 + 7} - 3}{x}$

Chapter 2

The Derivative

State and Change-of-State

When a function is used as a mathematical model of a changing system, its derivative represents the change-of-state of that system. The derivative is, therefore, an efficient and powerful tool for analyzing a system. We can readily appreciate the derivative's usefulness by a comparison. Let us suppose that you want to analyze your tennis game. If we consider the system to be you, the ball, and your racket, a *state* of the system is shown by a snapshot of you hitting the ball at time t. Thus a state is the system frozen and unmoving; it corresponds to a single function value $f(t)$. More useful to you in improving your game, or in describing the whole system, would be a videotape f of you hitting the ball. Seeing yourself in motion in such a video would show the change in the system. The derivative f' represents the *change-of-state* of a system and shows the tape f in motion. In turn a single value $f'(t)$ is the rate at which change is occurring at time t; it gives us information that even a video could not provide. Thus the derivative holds an advantage in analyzing a changing system more than comparable to a videotape in picturing motion.

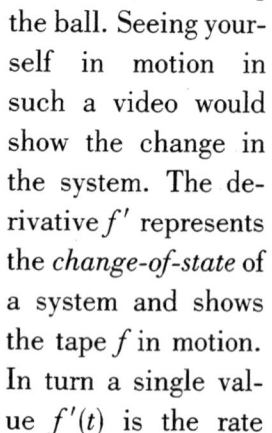

2.1 The Derivative of a Function: Basic Operations

In this section we learn the mechanical rules for differentiating power functions x^n, constant functions, and sums and differences of functions.

The first three sections of this chapter explain the six rules we use to find the derivative of a sum or difference, composition of functions, and the product or quotient of two functions. These six rules give differentiation a procedural form based on an algebraic formula. Our previous concentration on the derivative as a limit will not be lost, however, since it is from this experience that we understand what the derivative means and measures. The last three sections then use the derivative in applications.

The Derivative of $f(x) = x^n$

Let $f(x) = x^n$, where n is any fixed positive whole number: 1, 2, 3, ... We wish to find a general formula for the derivative $f'(x)$. Let us begin by recalling Example 3 in Section 1.4, where we found that when $f(x) = x^3$, then $f'(x) = 3x^2$. This suggests the possibility that, in general, when $f(x) = x^n$, then $f'(x) = nx^{n-1}$. Let us see why this is so.

Consider the expression

$$(x+h)^n = \overbrace{(x+h)(x+h)\ldots(x+h)}^{n \text{ factors}}$$

Expansion of this product yields a sum of the following *form*:

$$x^n + nx^{n-1} \cdot h + \frac{n(n-1)}{2}x^{n-2} \cdot h^2 + \begin{pmatrix}\text{terms having a factor of } h \text{ to the}\\ \text{power of 3 or more, if } n \geq 3\end{pmatrix} \quad (1)$$

The first term, x^n, arises by multiplying one x from each of the n factors $(x+h)$. The second term in Equation 1, $nx^{n-1} \cdot h$, results from choosing an x from *all but one* factor and an h from that single remaining factor. Since one h must be chosen from each of n factors, n copies of the form $x^{n-1}h$ result. The remaining terms result by multiplying h's from at least two factors and x's otherwise. We will see that these further terms are unimportant.

Let us now compute the derivative of $f(x) = x^n$. By definition,

$$\begin{aligned}f'(x) &= \lim_{h \to 0} \frac{f(x+h) - f(x)}{h} = \lim_{h \to 0} \frac{(x+h)^n - x^n}{h} \\ &= \lim_{h \to 0} \frac{x^n + nx^{n-1}h + (\text{terms including } h^2) - x^n}{h} \\ &= \lim_{h \to 0} [nx^{n-1} + (\text{terms including } h)] \quad \begin{array}{l}\text{By eliminating } x^n \\ \text{and by canceling } h\end{array} \quad (2)\end{aligned}$$

If you have difficulty seeing this last step, you may wish to refer to the details of Example 3 in Section 1.4, where the algebraic simplification just shown is done in detail (in the case of $n = 3$).

Let us now compute the limit as $h \to 0$ in Equation 2. Because all terms including an h in this equation then vanish, we have the conclusion that

$$f'(x) = \lim_{h \to 0} \frac{(x+h)^n - x^n}{h} = nx^{n-1}$$

since the term nx^{n-1} is the only term that remains. This reduces differentiation of x^n to an algebraic formula:

Theorem 2.1

The Power Rule

If $f(x) = x^n$, then $f'(x) = nx^{n-1}$.

The **power rule** effectively reduces differentiation of x^n to a bit of arithmetic: Reduce the exponent by 1 and multiply the resulting expression by n. Our partial proof of the power rule applies only to whole numbers $n = 1, 2, \ldots$. The rule, however, is valid for any real number exponent; we will not prove this but will use it from now on, beginning with the first example.

EXAMPLE 1

For each of the functions $f(x) = x^{15}$, $x^{1/2}$, $x^{\sqrt{3}}$, x^{-1}, and x, find $f'(x)$.

Solution If $f(x) = x^{15}$ then $f'(x) = 15x^{14}$

If $f(x) = x^{1/2}$ then $f'(x) = \frac{1}{2}x^{-1/2}$ for $x > 0$

If $f(x) = x^{\sqrt{3}}$ then $f'(x) = \sqrt{3}\,x^{\sqrt{3}-1}$

If $f(x) = x^{-2}$ then $f'(x) = -2x^{-3}$ for $x \ne 0$

If $f(x) = x = x^1$ then $f'(x) = 1 \cdot x^0 = 1$ for $x \ne 0$

(The conclusion remains true for $x = 0$ using the limit definition) ∎

Alternate Notation for the Derivative

The preceding list is a tedious way to write the relationship between f and f'. Another notation for the derivative saves much writing and includes both the function and its derivative in one expression. We now write

$$D_x f(x) \quad \text{or simply} \quad D_x f$$

for the derivative $f'(x)$ of the function $y = f(x)$. In this new notation the power rule becomes

$$D_x x^n = nx^{n-1}$$

and is read, "The derivative of x^n (with respect to x) is nx^{n-1}." Applications similar to Example 1 are then written, for instance, as

$$D_x x^{1/2} = \frac{1}{2}x^{-1/2} \qquad D_x x^{-7} = -7x^{-8} \qquad D_x x^{5/2} = \frac{5}{2}x^{3/2}$$

for appropriate values of x.

Several alternative ways of representing the derivative are widely used. These include

$$\frac{df}{dx} \qquad \frac{dy}{dx} \qquad y' \qquad D_x y \qquad \frac{d}{dx}f(x)$$

The notation dy/dx is due to Leibniz and is a reminder that the derivative arises from a quotient of differences. It is a single, notational form taken as a whole, not a fraction of one number dy divided by another dx. The simplest of these alternatives, $D_x y$, is the notation that we will most often use.

Let us reinforce an earlier point here. There is a significant difference between $D_x f = f'$ and $f'(a)$. Both f' and $D_x f$ represent new *functions* of x. On the other hand, $f'(a)$ is a *number*, equal to the slope of the tangent line to the graph of f at $x = a$. Both the function $f' = D_x f$ and the number $f'(a)$ are important, but they are different.

The Differentiation of Sums and Constant Multiples of Functions

Knowing that $D_x x^3 = 3x^2$ and $D_x x^7 = 7x^6$, how do we compute $D_x(x^3 + x^7)$ or $D_x(9x^4)$? The **sum** and **constant multiple rules** solve these differentiation problems.

Theorem 2.2

The Sum and Constant Multiple Rules

If f and g are differentiable functions, then

1. $D_x[f(x) \pm g(x)] = D_x f(x) \pm D_x g(x)$.
2. $D_x k f(x) = k D_x f(x)$, where k is a constant.

EXAMPLE 2

Find the derivative of (a) $x^3 + x^7$ and (b) $9x^4$ using the sum and constant multiple rules.

Solution

a. From the sum rule, $D_x(x^3 + x^7) = D_x x^3 + D_x x^7 = 3x^2 + 7x^6$.
b. From the constant multiple rule, where $k = 9$ is the constant multiple, $D_x 9x^4 = 9 D_x x^4 = 9(4x^3) = 36x^3$. ∎

EXAMPLE 3

Find $D_x(3x^2 - 5x^{20} + 4x^{-2})$.

Solution Using the sum and constant multiple rules together,
$$D_x(3x^2 - 5x^{20} + 4x^{-2}) = 3D_x x^2 - 5D_x x^{20} + 4D_x x^{-2}$$
$$= 3(2x) - 5(20x^{19}) + 4(-2x^{-3})$$
$$= 6x - 100x^{19} - 8x^{-3} \blacksquare$$

The sum rule tells us something that is not geometrically obvious: The slopes of the tangent lines to the graphs of f and g respectively can be added to obtain the slope of the tangent to the graph of their sum $f + g$. At the same time, the sum rule is an immediate consequence of the limit properties (Section 1.3) applied to a sum of functions. The constant multiple rule, in turn, is an immediate consequence of the limit property $\lim_{x \to a} kg(x) = k \lim_{x \to a} g(x)$.

Higher-Order Derivatives

If, say, $f(x) = x^5$, then $f'(x) = 5x^4$ is itself a function. It too has a derivative, then; this derivative is given by
$$D_x 5x^4 = 5(4x^3) = 20x^3$$
This function is called the **second derivative** of f and we write $f''(x) = 20x^3$, or, alternatively, $D_x^2 x^5 = 20x^3$. This is an example of a **higher-order derivative**: a derivative of a derivative.

Such higher-order derivatives have numerous applications, but here we mention only one. If $y = f(t)$ is the height of a falling rock at time t, the velocity of the falling rock is $D_t y = f'(t)$. But the rate of change of velocity over time is acceleration. Thus $D_t^2 y = f''(t)$ is the *acceleration* of the rock at time t.

This can go on—to third, fourth, . . . , nth derivatives. Thus
$$D_x^3 f(x) \quad \text{or} \quad f'''(x) \quad \text{means} \quad D_x(D_x^2 f(x))$$
That is, the third derivative of f is the derivative of the second derivative. In general, then,
$$D_x^n f(x) \quad \text{means} \quad D_x(D_x^{n-1} f(x))$$
In Leibniz's notation, we write
$$\frac{d^n y}{dx^n} \quad \text{or} \quad \frac{d^n f}{dx^n}$$
for this nth derivative.

EXAMPLE 4
If $f(x) = x^5$, then
$$D_x f = 5x^4$$
$$D_x^2 f = D_x(5x^4) = 5(4x^3) = 20x^3$$
$$D_x^3 f = D_x(20x^3) = 20(3x^2) = 60x^2$$

$$D_x^4 f = D_x(60x^2) = 60(2x) = 120x$$
$$D_x^5 f = D_x(120x) = 120(1) = 120$$
$$D_x^6 f = D_x(120) = 0 \quad \text{(see below)}$$

and thus $D_x^n f = 0$ for any $n \geq 6$. ■

In this example we explicitly encounter the derivative of a constant function for the first time. Why is $D_x(120) = 0$? Intuitively, if $f(x) = C$ is a constant function, the values of f *never change*. Thus the rates of change of f, $f'(x)$, must be 0. This is an important rule, which may be proven rigorously using the definition of the derivative, although we will not do so here.

Theorem 2.3   **The Constant Rule**

If $f(x) = C$, a constant, then $f'(x) = 0$.

In the following exercises you may wish to consult Background Review 5.1, which covers the laws of exponents.

Exercises 2.1

Compute Exercises 1–22.

1. $D_x x^3$
2. $D_x x^{-5}$
3. $D_x x^{1/3}$
4. $D_x x^{3/2}$
5. $D_t(-t^{-1/2})$
6. $D_x x^{1.9}$
7. $D_x(x^2 - x^3)$
8. $D_x(3x^{1/3})$
9. $D_u\left(\dfrac{u^3}{6} + u^{3/2}\right)$
10. $D_x(x^{1/2} + x^{-1/2})$
11. $D_x(1 + x + x^2)$
12. $D_t(3 - t^{4/3})$
13. $D_x(4x^{-3} + 6x^2 - 2x + 29)$
14. $D_x(1 - x + x^2 - x^3)$
15. $D_z(1 + z^3)^2$
16. $D_s(1 + s)^3$
17. $D_v[2 + (v - 3)^2]$
18. $D_x[(x + 1)^2 - x^2]$
19. $D_x[(x + 1)^2 - (x + 1)]$
20. $D_x\left(\dfrac{x^3 + x^2}{6}\right)$
21. $D_x\left(\dfrac{x^4}{4} - \dfrac{x^{-3}}{3}\right)$
22. $D_x(2x^{-1/2} - 3x^{1/3})$

23. Find the 1st, 2nd, 3rd, 5th, and nth derivatives of

    a. $f(x) = x^4$ b. $f(x) = \dfrac{1}{x}$

Compute Exercises 24–28.

24. $D_x^2(1 - x + x^2 - x^3)$
25. $D_x^3(x^3 + x^{-2})$
26. $D_x^{22} x^{23}$
27. $D_x^2(x^{1/2} + 3x^{1/3})$
28. $D_x^3(x^{5/2} - 1)$

29. Find $f'(0)$ and $f''(0)$ for

    a. $f(x) = x^{4/3}$ b. $f(x) = x^{2/3}$

if possible and graph these two functions.

30. A train is leaving the station and by time t has traveled $A(t) = 5t^2$ ft. At the same time, someone has climbed on top of the train and begins running along the roof of a boxcar in the same direction that the train is moving. By time t the person has traveled $B(t) = 8t^2 + 2t$ ft along the roof. How fast does the person appear to be moving to an observer standing

beside the train at the end of 2 sec? When is the person traveling 80 ft/sec?

31. Jim and Bob are waiting in the starting blocks for the 100-yd dash. At time $t = 0$ the starter sounds; in the initial moments Jim has covered a distance of $J(t) = 10t^2$ ft, whereas Bob has traveled $B(t) = 64t^9$ ft, where t is measured in seconds. At the end of $\frac{1}{2}$ sec, who is moving fastest? At the end of $\frac{3}{4}$ sec? Is this a reasonable model for the distance Bob has traveled at the end of $\frac{3}{2}$ sec? How about Jim?

32. There is a mathematical difficulty that we have avoided in this section. You now know how to find

$$D_x x^{\sqrt{3}} = \sqrt{3} x^{\sqrt{3}-1}$$

but if asked to evaluate this at, say, $x = 4$, you might not know how. Your instructor may wish to explain the meaning of expressions such as $4^{\sqrt{3}}$. If you have a calculator with an $\boxed{x^y}$ button, use it to evaluate (approximately) this derivative at $x = 1, 2, 4,$ and $\sqrt{2}$.

2.2 The Chain Rule

The purpose of this section is to help you learn to use the chain rule for differentiation and to understand what this rule tells us in an application.

The chain rule is perhaps the most frequently used rule for differentiation in calculus because of its mathematical form and its meaning in applications.

Imagine a process that passes through an intervening stage (see Figure 2.1). For example, as we saw in Section 0.5, the process could begin with the interest rate on home mortgages, go through the intervening stage of home construction, and end with the price of lumber. Of equal significance are the many food chains found in nature, where a population of large organisms depends on the health of a population of smaller organisms, which in turn depends on the health of a population of even smaller, perhaps microscopic, organisms.

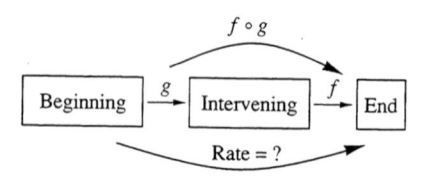

Figure 2.1

Such chain processes are represented by a composition of functions $(f \circ g)(x) = f(g(x))$, where f, g, and $f \circ g$ are as indicated in Figure 2.1. In this section we wish to compute the *rate of change* between the beginning and end of such a chain. We will do so using a formula for the derivative

$$(f \circ g)'(x) = D_x f(g(x))$$

known as the chain rule. The **chain rule** tells us how to calculate the rate of change over a chain relationship in terms of the rate of change over each intervening part. Though the proper formula is easy enough to write, let us first think about what is being asked for.

Since the derivative represents rate of change, we are trying to determine the rate of change of the overall process—from the beginning to the end of the chain. Surely this rate of change must involve the intermediate stage as well.

That is, it must involve the rate of change g' between the beginning and intermediate stage *and* the rate f' between the intermediate stage and the end of the chain. The following mathematical theorem describes the overall rate of change for all single-variable chain processes wherever they occur:

Theorem 2.4

The Chain Rule

If f and g are differentiable functions, then

$$D_x f(g(x)) = f'(g(x))g'(x)$$

The chain rule formula is valid only when both $g'(x)$ and $f'(g(x))$ are defined at x. A proof of the chain rule is beyond the intent of this text. We will only consider its rationale and interpret some of its details.

The chain rule tells us exactly how an overall rate of change is determined by the rates of change in intervening stages. The *overall rate* $(f \circ g)'$ is a *product*

$$f'(g(x)) \text{ times } g'(x)$$

of the rates f' and g' that hold between stages. But there is an additional subtlety in this formula. The factor

$$f'(g(x))$$

is a key part of the chain rule: It tells us that the rate of change f' must be computed *at the amount* $g(x)$ in the intervening stage. We will look at the meaning of this in an application after an initial example of the use of the chain rule.

EXAMPLE 1

Use the chain rule to find $D_x(x^2 + 1)^7$.

Solution One way to do this is to multiply $x^2 + 1$ by itself seven times to obtain a polynomial and then to use the rules of Section 2.1. However, with the chain rule this is unnecessary. We regard $(x^2 + 1)^7$ as $f(g(x))$, where $f(x) = x^7$ and $g(x) = x^2 + 1$. The chain rule tells us that

$$D_x(x^2 + 1)^7 = f'(g(x))g'(x)$$

From Section 2.1, $f'(x) = 7x^6$. Hence, $f'(g(x)) = 7[g(x)]^6 = 7(x^2 + 1)^6$. In addition, $g'(x) = 2x$. Putting these two together, we have

$$D_x(x^2 + 1)^7 = 7(x^2 + 1)^6 \cdot 2x \quad (f' \text{ at } g(x) \text{ times } g'(x))$$

$$= 14x(x^2 + 1)^6 \quad \blacksquare$$

Let us see how the chain rule can be used in an applied setting, before looking at further computational examples.

EXAMPLE 2

Suppose that sample data from the home construction industry yields the following correspondences among interest charged by lending institutions, home construction, and lumber prices: When mortgage money is available at $r\%$, the number of homes constructed is $g(r) = (-2r + 50)/15$ million homes, and when x million homes are constructed, the price per (board) foot of lumber is $f(x) = (x^2 + 5)/12$ dollars per board foot. How fast are lumber prices changing when interest on home mortgages is 10%?

Solution Since $f(g(r))$ is the cost of lumber when interest is $r\%$, we must find $(f \circ g)'(10)$, the value of $D_r f(g(r))$ at $r = 10\%$. By the chain rule this is

$$f'(g(10)) \cdot g'(10)$$

From Section 2.1, $f'(x) = x/6$. Since $g(10) = [(-2)(10) + 50]/15 = 2$, then $f'(g(10)) = \frac{2}{6} = \frac{1}{3}$. Because $g'(r) = -\frac{2}{15}$, then in particular for $r = 10$, $g'(10) = -\frac{2}{15}$. Therefore the overall rate of change in this system is

$$f'(g(10)) \cdot g'(10) = \left(\frac{1}{3}\right)\left(-\frac{2}{15}\right) = -\frac{2}{45}$$

dollars per board foot per unit of interest. ∎

Note how the *level* of home construction $g(10)$ induced by the given interest charge $r = 10$ directly enters the calculation. The chain rule formula involves two factors:

1. The rate of change of home construction at an interest charge of $r = 10\%$, $g'(10)$.
2. The rate of change of lumber prices f' when the number of homes constructed is $g(10) = 2$ million homes, given by $f'(g(10))$.

It is only reasonable that the rate of change between construction levels and lumber prices should be computed at the level of construction $g(10)$ induced by the given interest charge $n = 10$. The chain rule tells us exactly how this consideration must enter the calculation, via the factor $f'(g(10))$. As in Example 1, the derivative of f must be evaluated *at* the amount $g(x)$ of the intervening stage, and this must then be multiplied by $g'(x)$.

When using the chain rule, perhaps more than at any other point in calculus, we must be alert to the matter of pure *form*. We illustrate this by repeating Example 1 but without the bother of isolating and naming functions f and g. Instead we want to think about problems calling for the chain rule in terms of an **inner part** and an **outer form**.

Consider the problem of finding $D_x(x^2 + 1)^7$ in the following way. The expression $(x^2 + 1)^7$ involves an inner part and outer form:

1. The inner part is $x^2 + 1$.
2. The outer form is a 7th-power function: ( )^7.

That is, an expression $f(g(x))$ has the form of an inner part $g(x)$ inside an outer form f. The chain rule formula $D_x f(g(x)) = f'(g(x))g'(x)$ can be thought of as saying that

$D_x f(g(x)) = $ [derivative of the outer *form* f (evaluated at the inner part)] *times* [derivative of the inner part g]

Let us use this literal form of the chain rule to repeat Example 1. We can then compare the two methods and see that, while the result is the same, the procedure is simplified.

EXAMPLE 3

Compute $D_x(x^2 + 1)^7$.

Solution The outer form is a 7th-power function, with derivative $7(\circ)^6$. The inner part is $x^2 + 1$.* We have

$$D_x(x^2 + 1)^7 = 7(\text{inner part})^6 D_x(\text{inner part})$$
$$= 7(x^2 + 1)^6 D_x(x^2 + 1)$$
$$= 7(x^2 + 1)^6(2x) = 14x(x^2 + 1)^6 \quad \blacksquare$$

EXAMPLE 4

Compute $D_x(2 - x^3)^{1/2}$.

Solution Here the inner part is $(2 - x^3)$ and the outer form is $(\quad)^{1/2}$. The outer form has $\frac{1}{2}(\quad)^{-1/2}$ as its derivative. Therefore

$$D_x(1 - x^3)^{1/2} = \frac{1}{2}(\text{inner part})^{-1/2} D_x(\text{inner part})$$
$$= \frac{1}{2}(2 - x^3)^{-1/2} D_x(2 - x^3)$$
$$= \frac{1}{2}(2 - x^3)^{-1/2}(0 - 3x^2) = \frac{-3}{2}x^2(2 - x^3)^{-1/2}$$
$$= \frac{-3x^2}{2\sqrt{2 - x^3}} \quad \blacksquare$$

In the next example we apply the chain rule *twice* to complete the computation, since the given function consists of one composition (cubing) followed by a second (5th power).

EXAMPLE 5

Compute $D_x[(x + 1)^3 + 2]^5$.

---

*Many students have asked, "How can I find the inner part and outer form?" It's easy. Just try to evaluate the formula for some particular value, say, $x = 2$. The *first* "natural" calculation you make uses the inner part. The *last* calculation you make uses the outer form. When evaluating $(x^2 + 1)^7$ at $x = 2$, the first thing you write is $2^2 + 1 = 5$; the last is $5^7 = 78,125$. Since the last operation we perform is the raising of a number to the 7th power, this is the outer form, or function.

Solution

$$D_x[(x+1)^3 + 2]^5 = 5\overbrace{((x+1)^3 + 2)}^{\text{inner part}}{}^4 \cdot D_x\overbrace{((x+1)^3 + 2)}^{\text{inner part}}$$

$$= 5((x+1)^3 + 2)^4[3\overbrace{(x+1)}^{\text{2nd inner part}}{}^2 \cdot D_x\overbrace{(x+1)}^{\text{2nd inner part}}] \quad \text{since } D_x 2 = 0$$

$$= 5((x+1)^3 + 2)^4[3(x+1)^2 \cdot 1]$$

$$= 15(x+1)^2((x+1)^3 + 2)^4 \quad \blacksquare$$

The chain rule is also valid for negative exponents.

EXAMPLE 6
Find $D_x[1/(4-x)]$.

Solution To treat this as a composition of functions, we write $1/(4-x) = (4-x)^{-1}$. Since this outer form has derivative $-1()^{-2}$, then

$$D_x\left(\frac{1}{4-x}\right) = D_x(4-x)^{-1}$$

$$= (-1)(4-x)^{-2}D_x(4-x)$$

$$= (-1)(4-x)^{-2}(-1) = (4-x)^{-2} = \frac{1}{(4-x)^2} \quad \blacksquare$$

The chain rule can be reinforced via the Leibniz notation dy/dx. In the expression $y = f(g(x))$, we can replace the inner part $g(x)$ with the letter u; that is, let $u = g(x)$. We can then write du/dx for $g'(x)$ and dy/du for $f'(u)$. Since $f'(u) = f'(g(x))$, the chain rule can be written as

$$\frac{dy}{dx} = \frac{dy}{du}\frac{du}{dx}$$

While this is easy to remember, to apply this equation correctly, we must also remember that dy/du means f' evaluated at $g(x)$.

We conclude with an example of the chain rule from common experience.

EXAMPLE 7
Recall Example 5 of Section 1.5, in which a person tosses a pebble into a smooth pool of water (see Figure 2.2). How fast is the area of the expanding circle increasing *with respect to time* (rather than with respect to the lengthening radius, as before), given that the radius of the circular wavelet is $g(t) = (3t/2) - 1$ ft at time $t > 1$ sec? How fast is the area expanding at the end of $t = 4$ sec?

Solution It is natural to regard the expanding circle as a function of time, via a chain process and a composition $f \circ g$ (see Figure 2.3).

Figure 2.2

Figure 2.3

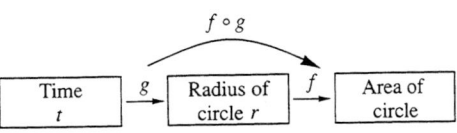

We know one function from geometry: $f(r) = \pi r^2$ when the radius is r. The energy dissipated by the falling pebble creates a circular wave of radius $g(t)$, where we are given that this radius is $g(t) = (3t/2) - 1$ at time t. For example, at time $t = 1$ sec, the radius is $\frac{1}{2} = g(1)$ ft long; at $t = 6$, the radius is $g(6) = 8$ ft long.

Since $(f \circ g)(t)$ represents the area of the circle as a function of time, the *rate of growth* of the area of the expanding circle as a function of time is given by the derivative $D_t f(g(t)) = f'(g(t))g'(t)$. Since $f'(r) = 2\pi r$ and $g'(t) = \frac{3}{2}$, we have

$$D_t f(g(t)) = (2\pi g(t))g'(t) = 2\pi\left(\frac{3}{2}t - 1\right)\left(\frac{3}{2}\right)$$
$$= 3\pi\left(\frac{3}{2}t - 1\right)$$

At the end of $t = 4$ sec, the circular area is expanding at the rate of $(f \circ g)'(4) = 3\pi\left(\frac{3}{2}(4) - 1\right) = 15\pi$ ft^2/sec. ∎

This example again reinforces the meaning of the chain rule. The formula $D_t f(g(t)) = f'(g(t))g'(t)$ tells us that the rate of area expansion is dependent on two distinct factors:

1. The rate of expansion f', which is purely geometric, and the length of the radius $g(t)$.
2. The rate g' at which the radius lengthens over time.

Exercises 2.2

Identify an inner part and an outer form for each of the functions in Exercises 1–6. Your answer need not be unique (but there is a natural answer). Then differentiate *only* the outer form.

1. $(x^3 + x)^5$
2. $\sqrt{1 + x^2}$
3. $\sqrt[3]{1 - x^4}$
4. $\left(\dfrac{1}{x^2 + 1}\right)^3$
5. $(\sqrt{x + 1} + 2)^{-3}$
6. $\left(\dfrac{x + 1}{x - 1}\right)^{5/3}$

Compute the derivative on the appropriate domain in Exercises 7–28 using the chain rule.

7. $D_x(1-x)^8$

8. $D_x\left(3+\dfrac{x}{2}\right)^5$

9. $D_t(1+2t)^9$

10. $D_t(1+\sqrt{t})^7$

11. $D_z\left(\dfrac{1+z}{4}\right)^4$

12. $D_u(4-u+u^2)^3$

13. $D_x(x^2+1)^{-3}$

14. $D_v(1+\sqrt{v})^5$

15. $D_x(x-x^2)^3$

16. $D_x(2-x)^{3/2}$

17. $D_t\sqrt{t^5+4}$

18. $D_x\sqrt[3]{1-x^4}$

19. $D_x(x^3+x^2)^{7/6}$

20. $D_x(x^2+1)^{-1/2}$

21. $D_r(\sqrt{1-r}+2)^3$

22. $D_x\dfrac{1}{1+x^2}$

23. $D_x\dfrac{1}{(1+2x)^5}$

24. $D_v(1+(1+v^2)^3)^5$

25. $D_u\left(\dfrac{4}{2+u^3}\right)$

26. $D_x\left(\dfrac{1}{1-x^2}\right)^3$

27. $D_s\sqrt{1+\sqrt{s+1}}$

28. $D_z[(1+z)(1-z)]^3$

In Exercises 29–31, compute $D_x f(g(x))$, without knowing $f(x)$, but given that $f'(x)=1+x^2$. For example, if $g(x)=x^3+2$, $D_x f(g(x))=[1+(x^3+2)^2](3x^2)$.

29. $g(x)=x^3$

30. $g(x)=1+\sqrt{x}$

31. $g(x)=1+x$

In Exercises 32–34, given that $f'(x)=1/x$, find $D_x f(g(x))$.

32. $g(x)=x^2$

33. $g(x)=1-x$

34. $g(x)=x^2-x$

35. Evaluate $(f\circ g)'(1)$ in Exercises 29–31.

36. Evaluate $(f\circ g)'(2)$ in Exercises 32–34.

37. Find the slope of the tangent line to the graph of $y=f(g(x))$ at $x=4$ in Exercises 32–34.

38. Find the equation of the tangent line to the graph of
   a. $g(x)=x^2$ at $x=2$
   b. $f(x)=\sqrt{12+x}$ at $x=4$
   c. $f\circ g$ at $x=2$

39. Refer to Example 7. Compute how fast the area of the circle is expanding over time if at time $t\geq 1$ the radius is
   a. $r(t)=5t$
   b. $r(t)=5\sqrt{t}$
   c. $r(t)=5[(t-1)^3+1]$

   In each case compute how fast the area is growing at the instant $t=2$ sec. Exactly where does the size of the circle at this particular time enter into the calculation?

40. Suppose that there are $B(x)=100(1+x)^{1/3}$ bass in a farm pond when the pond contains x tons of algae and that there are $A(n)=\sqrt{n}/20$ tons of algae when there are n tons of inorganic nutrients in the pond. How fast is the bass population changing when there are 2 tons of nutrients in the pond?

41. The I. Sputter investment firm says, "When I. Sputter speaks, people listen." Suppose that on day t of a certain week, I. Sputter invests $S(t)=500\sqrt{t}+1{,}000$ dollars in a certain stock. Suppose that when I. Sputter invests x dollars in a stock, their "listeners" invest $D(x)=x^3+100$ dollars in this stock.
   a. What is the rate of investment by these listeners in this particular stock on day $t=3$?
   b. What is the rate of investment by both I. Sputter *and* their listeners?

42. A successful brewery wishes to expand its share of the market. The company decides to spend less on ingredients and more on advertising, emulating its competitors. Marketing surveys show that if the brewery spends x dollars on advertising, then its sales will increase at the rate of $B'(x)=50x^{1/2}$ cans per dollar of advertising. On the tth day of its advertising campaign, the brewery spends $g(t)=100+11t$ dollars on advertising. At what rate will sales be growing on day $t=4$ of the advertising campaign?

2.3 Differentiation of Products and Quotients

The purpose of this section is to help you learn to differentiate products $f(x)g(x)$ and quotients $f(x)/g(x)$ of two functions f and g.

The differentiation of a product, or a quotient, of two functions takes a form we do not naturally expect and must be learned as a rote procedure. Before stating the appropriate formulas and learning how they are used, let us consider an example that illustrates that what we might naturally want to do in finding the derivative of a quotient cannot be correct.

In Example 3 of Section 0.1 we modeled the amount of time $T(x)$ that it takes a grad student to select a pair of nonrecessive and recessive *Drosophila* as

$$T(x) = \frac{20}{7}\left(\frac{75-x}{25-x}\right)$$

after x pairs have been selected from an initial population of 75 nonrecessive and 25 recessive individuals. This function tells us that as the ratio of nonrecessives to recessives grows, the amount of time it takes to select a pair made up of one of each also grows. If we imagine the grad student physically making the selection, we can easily envision her growing frustration at finding a recessive among the ever more comparatively numerous nonrecessives. What of the *rate* at which this frustration grows, or equivalently, the rate at which the time demand function T grows? That is, what is $T'(x)$?

Since

$$T(x) = \frac{20}{7}\left(\frac{75-x}{25-x}\right)$$

and the rates of decline of nonrecessives and recessives are both $-1 = D_x(75-x) = D_x(25-x)$, at first we might want to say that

$$T'(x) = \frac{20}{7}\frac{(-1)}{(-1)} = \frac{20}{7}$$

But this conclusion of a *constant* rate does not fit our intuitive understanding of what is happening.

Nor is this conclusion the correct mathematics, since the derivative of the quotient defining the function T cannot be found so directly but must be done according to the following rule for differentiating a quotient (or a product) of functions f and g:

Theorem 2.5

If f and g are differentiable functions, we have

1. the *product rule*: $D_x[f(x)g(x)] = f(x)\,D_x g(x) + g(x)\,D_x f(x)$ and
2. the *quotient rule*:

$$D_x\left[\frac{f(x)}{g(x)}\right] = \frac{g(x)\,D_x f(x) - f(x)\,D_x g(x)}{g(x)^2} \qquad g(x) \neq 0$$

for differentiating the product and quotient of f and g, respectively.

The **product** and **quotient rules** are often more briefly written as:

$$(fg)' = fg' + gf' \qquad \text{and} \qquad \left(\frac{f}{g}\right)' = \frac{gf' - fg'}{g^2}$$

As an example of the product rule and a quick check against earlier results, consider finding the derivative of, say, $h(x) = x^5$ (where we already know that $h'(x) = 5x^4$) by writing h as f times g with $f(x) = x^2$ and $g(x) = x^3$. According to the product rule, we must have

$$D_x(x^2 \cdot x^3) = x^2 \cdot D_x x^3 + x^3 \cdot D_x x^2 = x^2(3x^2) + x^3(2x) = 5x^4$$

Since this does equal $5x^4$, we are assured that the product rule is consistent with earlier experience. We would, of course, find the derivative of h directly since we can.

Let us now illustrate the use of the quotient rule and find the rate at which the time-demand function T discussed earlier is changing. In this case we must use the quotient rule.

EXAMPLE 1

Find $T'(x)$ for

$$T(x) = \frac{20}{7}\left(\frac{75-x}{25-x}\right)$$

Solution Using first the constant multiple rule and then the quotient rule, with $f(x) = 75 - x$ and $g(x) = 25 - x$, we have

$$T'(x) = \frac{20}{7}\left[\frac{(25-x)D_x(75-x) - (75-x)D_x(25-x)}{(25-x)^2}\right]$$

$$= \frac{20}{7}\left[\frac{(25-x)(-1) - (75-x)(-1)}{(25-x)^2}\right] = \frac{20}{7} \cdot \frac{50}{(25-x)^2} \quad \blacksquare$$

This result is consistent with our intuition about the actual physical process modeled by the function T. Not only is $T' > 0$ and therefore the selection time is increasing, but as x gets closer to 25, T' becomes ever larger and the selection time increases at an ever larger rate. After $x = 10$ selections, we have $T'(10) \approx 0.635$, while after $x = 20$ selections, we have $T'(20) \approx 5.71$ minutes per selection.

Here is a second example.

EXAMPLE 2

Compute

$$D_x \frac{x^2+1}{x^3+2}$$

Solution According to the quotient rule, we have

$$D_x \frac{x^2+1}{x^3+2} = \frac{(x^3+2)\boxed{D_x(x^2+1)} - (x^2+1)\boxed{D_x(x^3+2)}}{(x^3+2)^2} \quad (1)$$

$$= \frac{(x^3+2)(2x) - (x^2+1)(3x^2)}{(x^3+2)^2} \quad (2)$$

$$= \frac{2x^4 + 4x - 3x^4 - 3x^2}{(x^3+2)^2}$$

$$= \frac{-x^4 - 3x^2 + 4x}{(x^3+2)^2} \quad \blacksquare$$

This looks complicated, but think of it this way: The quotient rule gives a first step (*only*) in computing the derivative of the quotient. It yields line (1) and then we are done with it. What remains are *two* differentiation problems that occur on the right-hand side of (1): $\boxed{D_x(x^2 + 1)}$ and $\boxed{D_x(x^3 + 2)}$. We then do these two problems, using earlier rules, in the transition from line (1) to line (2). After that, only algebra remains. To summarize, the point of the quotient rule is to *reduce a quotient-differentiation problem to two easier subproblems;* the product rule has the same function.

EXAMPLE 3

Compute $D_x(x^2 + 1)(x^{3/2} + x)$ using the product rule.

Solution The product rule gives us

$$D_x[(x^2 + 1)(x^{3/2} + x)] = (x^2 + 1)\boxed{D_x(x^{3/2} + x)} + (x^{3/2} + x)\boxed{D_x(x^2 + 1)} \quad (3)$$

$$= (x^2 + 1)\left(\frac{3}{2}x^{1/2} + 1\right) + (x^{3/2} + x)(2x)$$

The differentiation is complete. The product rule leads us in one step to the two easier problems: $D_x(x^{3/2} + x)$ and $D_x(x^2 + 1)$. After differentiation, only algebra remains. The expression in line (3) becomes

$$\frac{3}{2}x^{5/2} + \frac{3}{2}x^{1/2} + x^2 + 1 + 2x^{5/2} + 2x^2 = \frac{7}{2}x^{5/2} + 3x^2 + \frac{3}{2}x^{1/2} + 1 \quad \blacksquare$$

You may wish to check this result against earlier experience by first multiplying the two functions and then applying the power rule.

Before considering more examples of these rules, let us see why the product rule is true. By definition

$$D_x[f(x)g(x)] = \lim_{h \to 0} \frac{f(x + h)g(x + h) - f(x)g(x)}{h}$$

Years ago a mathematician had the following idea: Write

$$\frac{f(x + h)g(x + h) - f(x)g(x)}{h}$$

as

$$= \frac{f(x + h)g(x + h) \overbrace{- f(x + h)g(x) + f(x + h)g(x)}^{= 0} - f(x)g(x)}{h}$$

$$= \frac{f(x + h)g(x + h) - f(x + h)g(x)}{h} + \frac{f(x + h)g(x) - f(x)g(x)}{h}$$

$$= f(x + h)\left[\frac{g(x + h) - g(x)}{h}\right] + \left[\frac{f(x + h) - f(x)}{h}\right]g(x)$$

The factors in brackets are difference quotients for g and f, respectively. Their limits, as $h \to 0$, yield $f(x + h)g'(x) + f'(x)g(x)$. The factor $f(x + h)$ has limit

$f(x)$ as $h \to 0$, since f is continuous because it is differentiable (Section A.1). The final limit must then be

$$f(x)g'(x) + f'(x)g(x)$$

That is, $\quad D_x[f(x)g(x)] = f(x)g'(x) + g(x)f'(x)$

A proof of the quotient rule is more involved, yet similar, and we will omit it.

Mixing the Chain Rule with Products and Quotients

Here we encounter problems involving the use of two or more differentiation rules, applied in sequence—that is, one after the other. A key to solving such problems is to look at the problem calmly and then try to recognize its *form* before performing any differentiation.

EXAMPLE 4

Compute

$$D_x \sqrt{\frac{1+x}{2-x}}$$

Solution The form of this problem is that of a composition of functions—that is, it has an inner–outer form. An inner part is $(1+x)/(2-x)$ within the outer form $\sqrt{} = ()^{1/2}$. The problem first calls for the chain rule:

$$D_x \sqrt{\frac{1+x}{2-x}} = D_x \left(\frac{1+x}{2-x}\right)^{1/2} = \frac{1}{2}\left(\frac{1+x}{2-x}\right)^{-1/2} D_x\left(\frac{1+x}{2-x}\right)$$

To finish the computation, we must compute

$$D_x \left(\frac{1+x}{2-x}\right)$$

using the quotient rule. Continuing, we have

$$D_x \left(\frac{1+x}{2-x}\right)^{1/2} = \frac{1}{2}\left(\frac{1+x}{2-x}\right)^{-1/2} \frac{(2-x)D_x(1+x) - (1+x)D_x(2-x)}{(2-x)^2}$$

$$= \frac{1}{2}\left(\frac{1+x}{2-x}\right)^{-1/2} \frac{(2-x)(1) - (1+x)(-1)}{(2-x)^2}$$

$$= \frac{1}{2}\frac{1}{[(1+x)/(2-x)]^{1/2}} \frac{2-x+1+x}{(2-x)^2}$$

$$= \frac{1}{2}\frac{(2-x)^{1/2}}{(1+x)^{1/2}} \frac{3}{(2-x)^2}$$

$$= \frac{3}{2(1+x)^{1/2}(2-x)^{3/2}}$$

$$= \frac{3}{2\sqrt{(1+x)(2-x)^3}} \blacksquare$$

122 Chapter 2 The Derivative

EXAMPLE 5

Compute $D_x(x + x^3)(x^2 + x^4)^7$.

Solution This problem first of all has the *form* of a product. We first use the product rule to obtain

$$D_x(x + x^3)(x^2 + x^4)^7 = (x + x^3)\boxed{D_x(x^2 + x^4)^7} + (x^2 + x^4)^7 D_x(x + x^3)$$

One of the remaining differentiation problems involves a chain rule, and we continue

$$D_x(x + x^3)(x^2 + x^4)^7 = (x + x^3)[7(x^2 + x^4)^6 D_x(x^2 + x^4)] + (x^2 + x^4)^7(1 + 3x^2)$$
$$= (x + x^3)[7(x^2 + x^4)^6(2x + 4x^3)] + (x^2 + x^4)^7(1 + 3x^2)$$

At this point all differentiation is finished and only algebra remains:

$$= (x^2 + x^4)^6[7(x + x^3)(2x + 4x^3) + (x^2 + x^4)^1(1 + 3x^2)]$$
$$= (x^2 + x^4)^6[(14x^2 + 14x^4 + 28x^4 + 28x^6) + (x^2 + x^4 + 3x^4 + 3x^6)]$$
$$= (x^2 + x^4)^6[15x^2 + 46x^4 + 31x^6] \blacksquare$$

EXAMPLE 6

Compute

$$D_x \frac{(1 + x^2)^3}{\sqrt{1 - x^5}}$$

Solution The expression

$$\frac{(1 + x^2)^3}{\sqrt{1 - x^5}}$$

is, above all else, in the *form* of a quotient. To begin differentiation, we have

$$D_x \frac{(1 + x^2)^3}{\sqrt{1 - x^5}} = \frac{\sqrt{1 - x^5}\,\boxed{D_x(1 + x^2)^3} - (1 + x^2)^3\,\boxed{D_x\sqrt{1 - x^5}}}{(\sqrt{1 - x^5})^2}$$

Each of the two remaining differentiation problems has the form of a composition of functions and calls for the chain rule. We continue:

$$D_x \frac{(1 + x^2)^3}{\sqrt{1 - x^5}} = \frac{\sqrt{1 - x^5}\,\boxed{D_x(1 + x^2)^3} - (1 + x^2)^3\,\boxed{D_x\sqrt{1 - x^5}}}{(\sqrt{1 - x^5})^2}$$

$$= \frac{\sqrt{1 - x^5}[3(1 + x^2)^2 D_x(1 + x^2)] - (1 + x^2)^3[\frac{1}{2}(1 - x^5)^{-1/2} D_x(1 - x^5)]}{1 - x^5}$$

$$= \frac{\sqrt{1 - x^5}[3(1 + x^2)^2(2x)] - (1 + x^2)^3[\frac{1}{2}(1 - x^5)^{-1/2}(-5x^4)]}{1 - x^5}$$

$$= \frac{6x(1 + x^2)^2\sqrt{1 - x^5} + \dfrac{5x^4(1 + x^2)^3}{2\sqrt{1 - x^5}}}{1 - x^5}$$

$$= \frac{\dfrac{12x(1-x^5)(1+x^2)^2 + 5x^4(1+x^2)^3}{2(1-x^5)^{1/2}}}{1-x^5}$$

$$= \frac{(1+x^2)^2(12x + 5x^4 - 7x^6)}{2(1-x^5)^{3/2}} \quad \blacksquare$$

We could write further algebraic variations of this answer, but having reduced it to a single fraction, we might as well stop unless there is some reason to simplify further. You will no doubt find the algebraic calculations in this section challenging—more so than in sections that follow. This is a good section in which to learn to pay close attention to notational detail.

Background Review 2.1

Confronted with a differentiation problem of the form

$$\frac{f(x)^r}{g(x)^r}$$

you can regard it as a quotient problem, or you could rewrite it in the form

$$\left(\frac{f(x)}{g(x)}\right)^r$$

and regard it as a chain rule problem. Depending on the particular functions involved, one or another of these alternate forms *may* be easier to differentiate.

Similarly, a product

$$f(x)^r g(x)^r$$

is alternatively a composition

$$[f(x)g(x)]^r$$

Because differentiation produces negative exponents, we frequently encounter, as the result of differentiation, expressions like

$$g(x)f(x)^{1/3} + h(x)f(x)^{-2/3}$$

This equals

$$g(x)f(x)^{1/3} + \frac{h(x)}{f(x)^{2/3}}$$

which equals

$$\frac{g(x)f(x) + h(x)}{f(x)^{2/3}} = f(x)^{-2/3}[g(x)f(x) + h(x)]$$

because

$$a + \frac{b}{c} = \frac{ac+b}{c}$$

124   Chapter 2   The Derivative

For the same reason, we encounter expressions of the form

$$\left(\frac{a}{b}\right)^{-r}$$

This equals

$$\frac{1}{\left(\frac{a}{b}\right)^r} = \frac{1}{\frac{a^r}{b^r}} = \frac{b^r}{a^r}$$

Exercises 2.3

Compute the derivatives in Exercises 1–14.

1. $D_x(x^2 + 1)(1 - x)$
2. $D_t(1 - t^7)(t + t^3)$
3. $D_x \frac{\pi}{\sqrt{7}}$
4. $D_x(1 + \sqrt{x})(1 + x^{-1})$
5. $D_z \frac{z+1}{z^2}$
6. $D_x(1 - x^{1/2})(1 + x^{1/2})$
7. $D_x \frac{x+1}{x-1}$
8. $D_x \frac{1-x^2}{x^2+x}$
9. $D_t \frac{t}{t^3 + 2t}$
10. $D_x \frac{x-2}{x^2 + x + 1}$
11. $D_x \left(\frac{x^4}{4} - \frac{x^2}{2}\right)\left(\frac{x^3}{3} + x\right)$
12. $D_v \left(\frac{1}{v+1}\right)\left(\frac{1}{v} + \frac{1}{2}\right)$
13. $D_x(2\sqrt{x} + 3\sqrt[3]{x})(\frac{2}{3}x^{3/2} + x)$
14. $D_x(\sqrt{x+1})(x-1)$

23. $D_x \left(\frac{x+1}{x-1}\right)^2$
24. $D_x \left(\frac{x^4}{4} - \frac{x^5}{5}\right)^7$
25. $D_v(v\sqrt{v^2 + 1})$
26. $D_y \sqrt{\frac{y}{y+1}}$
27. $D_x(x^3 - x)^4 \left(\frac{x^2}{2} + 1\right)^{-5}$
28. $D_x \frac{1}{(x+1)^9}$
29. $D_x \frac{2x^3 - x}{x^2 + 1}$
30. $D_x[6(x+1)^{1/2}(x-1)^{1/3}]$
31. $D_t \left(\frac{1 + \sqrt{t}}{1 - \sqrt{t}}\right)$
32. $D_x \frac{(x^2-1)^3}{(x^2-2)^5}$
33. $D_x \left(\frac{\sqrt{x^2 + 1}}{x^3 + 1}\right)^5$
34. $D_x x^2 \sqrt{(x+1)^2 - 9}$
35. $D_x \frac{x^4 - x^3 + x^2 - 1}{1 - x + x^3 - x^5}$
36. $D_x(x^2 + 1)^3(x^{1/3} + x^5)$
37. $D_u \left[(u+1)\frac{u^2+1}{u^3-1}\right]$
38. $D_z \left(z \sqrt{\frac{z^2+1}{z^2-1}}\right)$
39. $D_x[x(\sqrt{x^2 + 1})(x+2)^3]$
40. $D_x \sqrt{x(x^{1/2} + 1)\left(\frac{x^4}{4} + x\right)}$

Using as much ordinary language and as little notation as you can, describe the overall form of Exercises 15–22. What kind of differentiation (quotient, product, chain rule) could be done first and last in these problems? *Do not actually differentiate.* Your answer need not be unique.

15. $\sqrt{\frac{x+1}{x-1}}$
16. $\frac{(x^2+1)^3}{x-1}$
17. $\left(\frac{x+1}{x^2+1}\right)^5$
18. $(2x^3 + 1)\sqrt{x-1}$
19. $(z+1)^{1/2}(z+2)^{1/3}$
20. $[(2x+1)\sqrt{x^2-7}]^3$
21. $\sqrt{x + x\sqrt{x+1}}$
22. $\left[\frac{(3x-2)^3(x+1)}{x-1}\right]^2$

41. Compare answers to the following pair of problems.

$$D_x \frac{x+1}{x} \quad \text{and} \quad \frac{D_x(x+1)}{D_x x}$$

42. In Exercise 41, $(x+1)/x = 1 + (1/x)$. Graph this function. Why does the graph tell you that $[D_x(x+1)]/D_x x$ cannot be the correct derivative of $(x+1)/x$?

Compute the derivatives in Exercises 23–40 after you have first described to yourself the overall form of the problem.

43. Compute the derivative of the function
$$f(x) = \frac{1}{(x^2+1)^3}$$
using the quotient rule. Then compute it again using the chain rule by rewriting it as
$$f(x) = (x^2+1)^{-3}$$
Compare your answers.

44. Suppose that the demand function for an item is
$$D(x) = \frac{3}{x^2+1}$$
dollars when x items are produced. At what rate is total revenue changing when $x = 10$ items are produced?

45. Suppose that the selling price (demand function) and cost functions for the Ace Manufacturing Company to produce and sell x units are $S(x) = 1/(x+1)$ and $C(x) = \sqrt{x^2+21}$, respectively.

   a. Write a formula for the total sales revenue earned from the sale of x items.

   b. Write a formula for net profit (revenue less cost) for the sale of x items.

   c. How fast are net profits changing at a sales level of 10 units?

46. In Einstein's theory of relativity, if your mass sitting still is m_0 and you then become an astronaut working in a spaceship moving at speed v, your moving mass is
$$m = \frac{m_0}{\sqrt{1-(v/c)^2}}$$
where c is the speed of light. How fast is your mass changing if your spaceship is leaving earth's gravitational field at a speed of $v = c/2$, half the speed of light? Is your mass increasing or decreasing? [Hint: Compute $D_v m$.]

47. An assembly-line worker has a box consisting of 30 nuts, 45 bolts, and 70 washers. Every 30 sec he reaches in for 1 nut, 1 bolt, and 2 washers. Let $f(x)$ be his "frustration function": the ratio of nuts, bolts, and washers to nuts remaining x min after this process has begun.

   a. Write a formula for $f(x)$.

   b. At what rate is the worker's frustration increasing at the end of 10 min?

Challenge Problem

48. At one-minute intervals, a robot reaches into a box that initially contains 50 washers and 90 rivets and removes one of each. When the ratio of washers to rivets is r, the robot needs $f(r) = 1/r^2$ sec to find the next washer and rivet.

   a. Describe the ratio of washers to rivets at time t as a function $g(t)$.

   b. Model the overall process as a composition.

   c. At what rate is the ratio of washers to rivets changing at time t? At $t = 30$ min?

   d. At what rate is the search time $f(r)$ changing at ratio r? At ratio $r = \frac{1}{3}$? $r = \frac{1}{5}$?

   e. At what rate is the search time changing at time t? At $t = 30$ min? At $t = 40$ min? At $t = 45$ min?

2.4 Applications to Graphing Functions

In this section we begin learning how to obtain the graph of a function indirectly, via its tangent lines. That is, we will be given information about $f'(x)$ and asked to draw a graph of $y = f(x)$.

Long after you have forgotten its formulas and details you should remember this about calculus: In calculus we learn about something by seeing how it changes. More technically, we learn about the function f via information about the derivative f'. A simple analogy is that we usually learn more from a motion picture than from a single photograph, though each has its own interest, as do both $f'(x)$ and $f(x)$. You may find this attitude of learning from change a familiar one, since it is something you commonly do. Here it is put into a mathematical form.

Consider Figure 2.4. Given a function $y = f(x)$, with unknown graph, each drawing is made by five evaluations at the points $x = -3, 0, 2, 5,$ and 8. In Figure 2.4(a), we evaluate $f'(x)$; in Figure 2.4(b), we evaluate $f(x)$. Figure 2.4(b) suggests that more than one graph can fit these points. Figure 2.4(a) gives a sense of direction to the graph and appears more informative; it indicates how knowledge about $f'(x)$ (often easier to obtain) tells us about $f(x)$. Here is a first example.

EXAMPLE 1

Sketch a graph of a function $y = f(x)$ with the following properties: $f(0) = 2$, $f'(-1) = 2, f'(0) = 1, f'(1) = 0,$ and $f'(2) = -1$.

Solution We first fix a point at $(0, 2)$ because $f(0) = 2$. Next we sketch appropriate tangent lines above the points $0, -1, 1,$ and 2, and then sketch in a curve that conforms to these tangent lines and also goes through $(0, 2)$, as in Figure 2.5. That is, we first indicate that $f'(0) = 1$ by drawing a line with slope 1 through the point $(0, 2)$. Because this slope is positive, we move up and over to $x = +1$, where we indicate that $f'(1) = 0$ by a line with slope 0. Then, moving over to $x = 2$, we indicate that $f'(2) = -1$ by a line with slope -1. We finish with $f'(-1) = 2$.

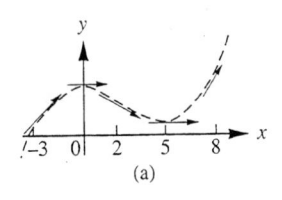

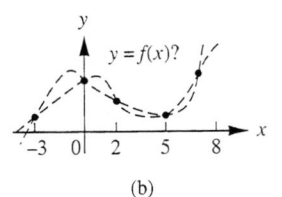

Figure 2.4

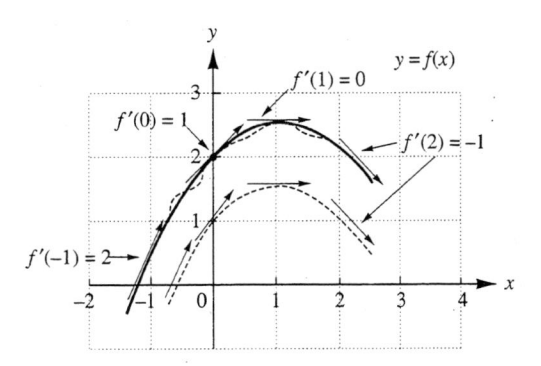

Figure 2.5

Figure 2.5 illustrates two important points. First, since we know the tangent slope only at the points $0, -1, 1,$ and 2, it is possible that between these points the graph could be like the partially "dotted" curve. Second, notice the lower curve "parallel" to $y = f(x)$. While it cannot be the graph of $y = f(x)$, because $(0, 1)$, rather than $(0, 2)$, is on this curve, the slopes of its tangent lines are *exactly the same* as those of $y = f(x)$. This curve satisfies all the conditions of the example except $f(0) = 2$. This means that knowledge about $f'(x)$ alone cannot specify $f(x)$ uniquely; knowledge of f' tells us how the graph is changing, what its *shape* is, not how high or low it is on the coordinate axes.

EXAMPLE 2

Sketch the graph of a function $y = f(x)$ whose derivative $f'(x)$ has the following values: $f'(0) = 0, f'(1) = 2, f'(-1) = -\frac{1}{2}, f'(2) = 0,$ and $f'(3) = 2$.

2.4 Applications to Graphing Functions 127

Solution Since $f'(x)$ measures the slope of the graph of $y = f(x)$, we obtain a sketch by first drawing straight lines with appropriate slope above the points $x = -1, 0, 1, 2$, and 3 (moving from left to right). We then sketch in a curve that conforms to these tangents. That is, we first draw $f'(-1) = -1/2$ at some height, and then, because this slope is negative, draw $f'(0) = 0$ at a lower point; we continue with $f'(1) = 2$ pointing upward, and so on, finishing with $f'(3) = 2$.

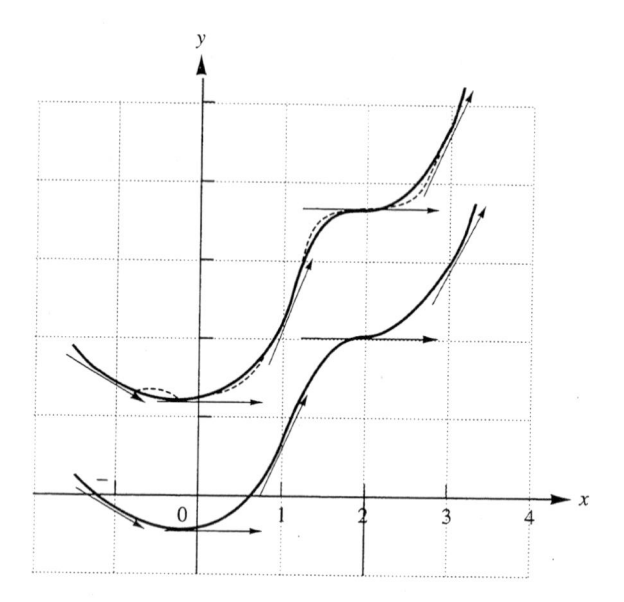

Figure 2.6

The "parallel" graphs in Figure 2.6 indicate again that $f'(x)$ does not specify $f(x)$ uniquely, since both graphs satisfy all the conditions asked for in the example. The discussion following Example 4 indicates why this is not a serious difficulty; notice how in both Figures 2.5 and 2.6 the lower graph apparently may be moved into the upper by simply raising it a *constant* distance.

Increasing and Decreasing Functions

The preceding examples show that a few values of $f'(x)$ suggest a good graph of $y = f(x)$. What if we knew more? What if we knew $f'(x)$ for *all* x? This brings us to two important concepts.

Definition

A function f is said to be **increasing (decreasing) on an interval** (a, b) if the graph of f rises (falls) as we cross the interval from left to right. That is, f is increasing on (a, b) if $f(x_1) < f(x_2)$ whenever $x_1 < x_2$ in the interval (a, b); f is decreasing if instead $f(x_1) > f(x_2)$.

We illustrate these concepts in Figure 2.7(a) and (b).

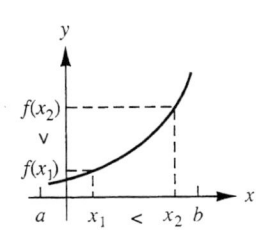

(a) f is increasing on (a, b).

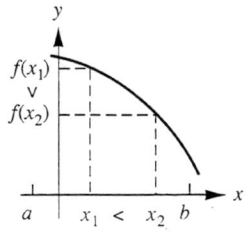

(b) f is decreasing on (a, b).

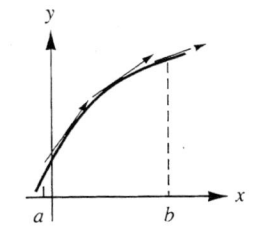

(c) $f'(x) > 0$ for each x.

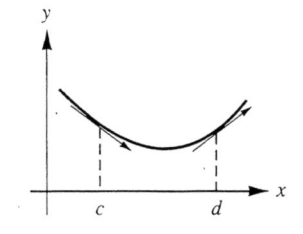

(d) $f'(c) < 0$; f is decreasing at c. $f'(d) > 0$; f is decreasing at d.

Figure 2.7

In practice we cannot easily determine whether a function f is increasing or decreasing using this definition. For this reason the key fact of this section is the following theorem, which tells us how ready information about f' at each x tells us about f itself.

Theorem 2.6

Let f be a function defined on (a, b).

1. If $f'(x) > 0$ at each x in (a, b), then f is increasing on (a, b).
2. If $f'(x) < 0$ at each x in (a, b), then f is decreasing on (a, b).

Theorem 2.6 is illustrated in Figure 2.7(c); it gives rise to a useful concept available to us via the tool of the derivative and is illustrated in Figure 2.7(d).

Definition

A function f is said to be **increasing (decreasing) at a point** c if $f'(c) > 0$ (if $f'(c) < 0$).

Notice the distinction we are making in our two definitions. The derivative f' allows us to speak of a function that is increasing *at a point*. This is distinct from increasing *on an interval*. Theorem 2.6 tells us how these two concepts are initially related: A function that increases at each point of an interval necessarily increases on the whole interval. There is a subtlety here, however, since the converse of Theorem 2.6 is not true. The function whose graph is shown in Figure 2.6 illustrates this point: This function is increasing *on the interval* $(0, 3)$ but is not increasing *at* $c = 2$. Exercise 9 explores this subtlety further. Finally, note that the numerical value of $f'(c)$ tells us more: the rate at which f increases at the point c.

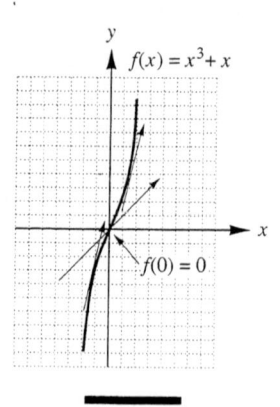

Figure 2.8

EXAMPLE 3

Determine whether the function $f(x) = x^3 + x$ is increasing or decreasing at each point in the interval $(-2, 2)$. Is f increasing on this interval? Graph this function.

Solution We first find that $f'(x) = 3x^2 + 1$. Since 1 and 3 are positive numbers and x^2 is never negative, we must have that $f'(x) = 3x^2 + 1 > 0$ for any x. Thus f is increasing at every point x in the interval $(-2, 2)$, and by Theorem 2.6(1) is increasing on $(-2, 2)$.

Going further, we can evaluate this derivative at a few points—say, $x = -1, 0,$ and 1—and obtain $f'(-1) = 4, f'(0) = 1,$ and $f'(1) = 4$, along with $f(0) = 0$. This allows us quickly to make a sketch of the graph of $f(x) = x^3 + x$ shown in Figure 2.8.

When f is a mathematical model for a system outside mathematics, Theorem 2.6 illustrates the principle that knowledge about how the system is changing, represented by f', informs us about the system itself, represented by f. Example 3 shows how this is done in practice when f is given. The next three examples show how the general shape of the graph of an unknown function f, for which we are given no formula for f itself, can be obtained from a formula for $f'(x)$.

EXAMPLE 4

Suppose that $f'(x) = x$. Sketch a graph of $y = f(x)$.

Solution In this example we know all values of $f'(x)$ rather than just a few. In particular, since $f'(x) = x$, we see that

1. $f'(x) > 0$ if $x > 0$, so f is increasing at every $x > 0$.
2. $f'(x) < 0$ if $x < 0$, so f is decreasing at every $x < 0$.

(Note also that $f'(0) = 0$; f is neither increasing nor decreasing at 0.) Moreover, since $f'(1) = 1, f'(2) = 2, f'(-1) = -1,$ and $f'(-2) = -2$, we can sketch appropriate tangent lines and a graph as indicated in Figure 2.9 (curve A).

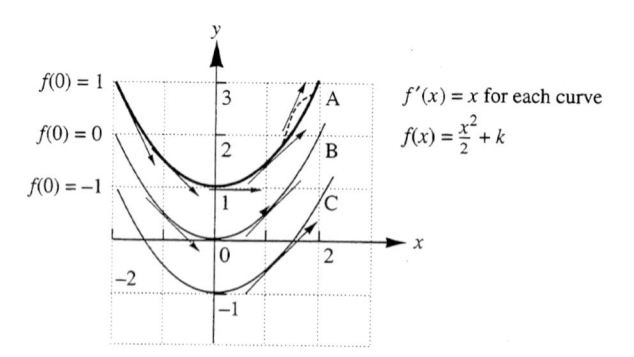

Figure 2.9

Notice that in contrast to the situation in Example 1, knowing that $f'(x) = x$ for all x tells us that the partially dotted graph (curve A) cannot possibly be a part of the solution. [For example, the apparent slope of the

130   Chapter 2   The Derivative

dotted graph at $x = \frac{3}{2}$ is too large (≈ 3).] Notice also that in the absence of information about $f(0)$, curve B or curve C can equally well be a solution. Let us clarify this point in analytic terms.

While the emphasis in this section is on graphing f given a formula for f', we digress for a moment to discuss Example 4 in terms of an algebraic formula for f. Since $D_x x^2 = 2x$ and $D_x k = 0$, where k is a constant, we see that if $f(x) = (x^2/2) + k$, then $f'(x) = D_x[(x^2/2) + k] = (2x/2) + 0 = x$. Thus there are *many* (for any constant k) functions whose derivative is $f'(x) = x$, but these all appear as a constant plus a single function $x^2/2$, whose graph (curve B, Figure 2.9) has the same shape as curve A (and curve C). The collection of functions $(x^2/2) + k$, for k any constant, is called a **family of functions**; all members of this family have derivative $f'(x) = x$. In Figure 2.9, if $k = 1$, we obtain curve A; $k = 0$ gives curve B; $k = -1$ gives curve C. This indicates in analytic terms why f' does not specify f uniquely and why, in each of the preceding examples, "parallel" graphs, obtained by adding a constant to a single graph, have the same tangent slope. In Section A.2, we discuss and illustrate the general mathematical reasons why functions with equal derivatives can differ only by a constant.

The Intermediate Value Theorem

The task of using f' to graph f is made much easier if we use a bit of mathematical theory called the **intermediate value theorem** for *continuous* functions.

Theorem 2.7

The Intermediate Value Theorem

If $y = g(x)$ is a continuous function on an interval $[a, b]$ and if $g(a)$ and $g(b)$ have opposite signs, then $g(x) = 0$ at *some* x between a and b.

Alternatively, if $g(x) \neq 0$ for *any* x in $[a, b]$, then all values of $g(x)$ have the same *sign* (that is, *all* are positive or *all* are negative).

The validity of Theorem 2.7 is suggested by the graphs in Figure 2.10, where the theorem seems intuitively obvious. Actually this "obvious" result is surprisingly deep, and its proof is well beyond the intent of this text. We only want to use the theorem to organize and simplify our task.

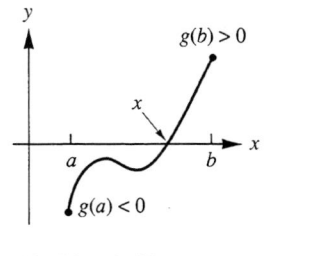

Figure 2.10

(a) $g(a)$ and $g(b)$ have different signs.

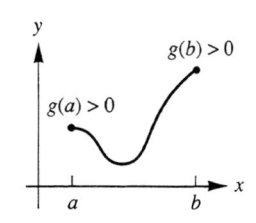

(b) $g(x) \neq 0$ for all x; $g(x)$ is always positive.

Graphing f Given Only f'

Let us now combine Theorems 2.6 and 2.7 in a useful and not entirely obvious way.

EXAMPLE 5

Sketch a graph of a function $y = f(x)$ whose derivative is

$$f'(x) = x(x - 2)$$

by determining all points where this function is increasing and decreasing. On which *intervals* is f increasing and decreasing?

Solution Notice first that $f'(x) = x(x - 2)$ is a continuous function of x and has only two zero values. These zero values are at $x = 0$ and $x = 2$, since $f'(0) = 0$ and $f'(2) = 0$. We first mark these on the number line in Figure 2.11.

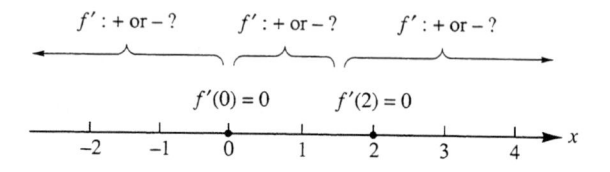

Figure 2.11

Consider first the interval $(0, 2)$ between these two zero values. It follows from the intermediate value theorem that $f'(x)$ has the same sign at any x between 0 and 2. Let us determine this sign by using a "test point," say, $x = 1$, between 0 and 2. Since $f'(1) = 1 \cdot (1 - 2) = -1 < 0$, it follows that $f'(x) < 0$ for all x between 0 and 2 and f is decreasing at each point in $(0, 2)$. We indicate this in Figure 2.12.

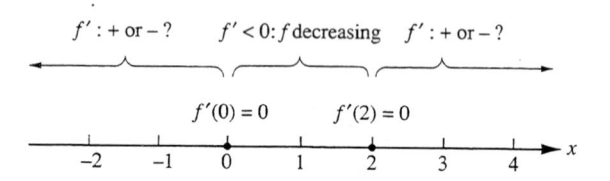

Figure 2.12

The same reasoning may be applied to numbers $x < 0$ and numbers $x > 2$. Because f' is continuous and has no zero values to the left of $x = 0$ or to the right of $x = 2$, Theorem 2.7 again tells us that f' has constant sign on each of these intervals. Since

$$f'(-2) = -2(-2 - 2) = 8 > 0 \quad \text{and} \quad f'(3) = 3(3 - 2) = 3 > 0$$

it follows that $f'(x)$ is positive for all $x < 0$ and all $x > 2$. Our choice of $x = -2$ and $x = 3$ as test points for this sign check is completely arbitrary; any *one* $x < 0$ and any *one* $x > 2$ will do, so we choose values that are easy to use. We summarize these conclusions in Figure 2.13.

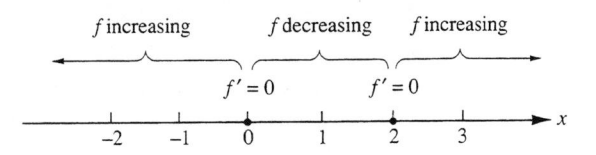

Figure 2.13

Figure 2.13 and Theorem 2.6 tell us that the function f must be decreasing on the interval (0, 2) and increasing on the intervals $(-\infty, 0)$ and $(2, \infty)$. Knowing only this, we would sketch a graph of the general shape shown in Figure 2.14, but we can make that shape more precise by using information obtained earlier:

$$f'(-2) = 8 \quad f'(0) = 0 \quad f'(1) = -1 \quad f'(2) = 0 \quad f'(3) = 3$$

Since we do not have any value of f itself, we can (as in previous examples) draw parallel graphs of the same shape as equally valid solutions.

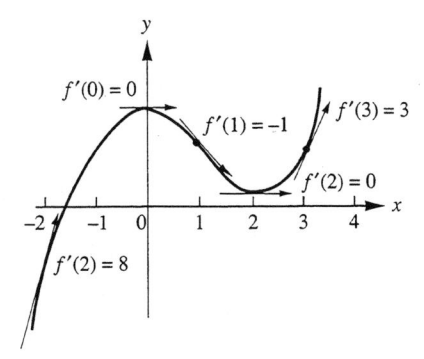

Figure 2.14

An Investment Return Curve

In our concluding example, we have neither a formula for $f'(x)$ nor a formula for $f(x)$. Instead, we have a relationship between f and f' given by an equation relating $f(x)$ to $f'(x)$. Such an equation is called a **differential equation**, which we will study in detail in Chapter 8.

EXAMPLE 6

Suppose that you invest $500 in a mutual fund firm with instructions to reinvest your earnings as they occur. The fund has a record of earning money at the rate of 20% of its invested funds. Graph the amount of your holdings in the fund if this record is maintained.

Solution Let $f(t)$ = the amount of your investment at time t (in years). Thus $f(0) = \$500$, your initial investment. At any time t, $f(t)$ represents the state of your investment—how much is in your account. From Section 1.5, $f'(t)$ is the rate at which your investment is growing. Because earnings occur at the rate of 20% of the amount invested at any time, we have

$$f'(t) = 0.20\, f(t)$$

2.4 Applications to Graphing Functions

This equation is an example of a differential equation, an equation relating an unknown function and its unknown derivative. A graph of the function f will indicate the course and amount of your investment. We begin by marking a point at time 0 representing your initial $500 investment.

Next, since $f'(0) = 0.20 f(0) = 0.20(500) = 100 > 0$, we also draw a tangent to the unknown graph at $t = 0$, indicating a slope of 100. Thus your investment is rising at the rate of $100 per year *at* time 0. This implies a key point: At a later time t_1, $f(t_1)$ *is greater than* $f(0)$. This, in turn, means that

$$f'(t_1) = 0.20 \, f(t_1) > 0.20 \, f(0) = 100$$

so that a second, steeper tangent line should be drawn above point t_1.

Applying the same reasoning at yet a later time t_2, we must have $f(t_2) > f(t_1)$ and

$$f'(t_2) = 0.20 \, f(t_2) > 0.20 \, f(t_1) = f'(t_1)$$

Thus at t_2 we must draw a tangent that is even steeper than that drawn at t_1. Continuing this thought process, we see that the graph of this unknown function $f(t)$, representing your total investment at time t, must look like the graph in Figure 2.15. ∎

Figure 2.15 helps us visualize the effect of continuous reinvestment of earnings (with a 20% rate of return). This is an example of **exponential growth**—growth that "feeds" on itself. While we used an unusually high rate of return to illustrate the solution easily, this model of "continuous" earnings on an investment with continuing reinvestment is quite useful; it will be studied in numerical detail in Chapter 5.

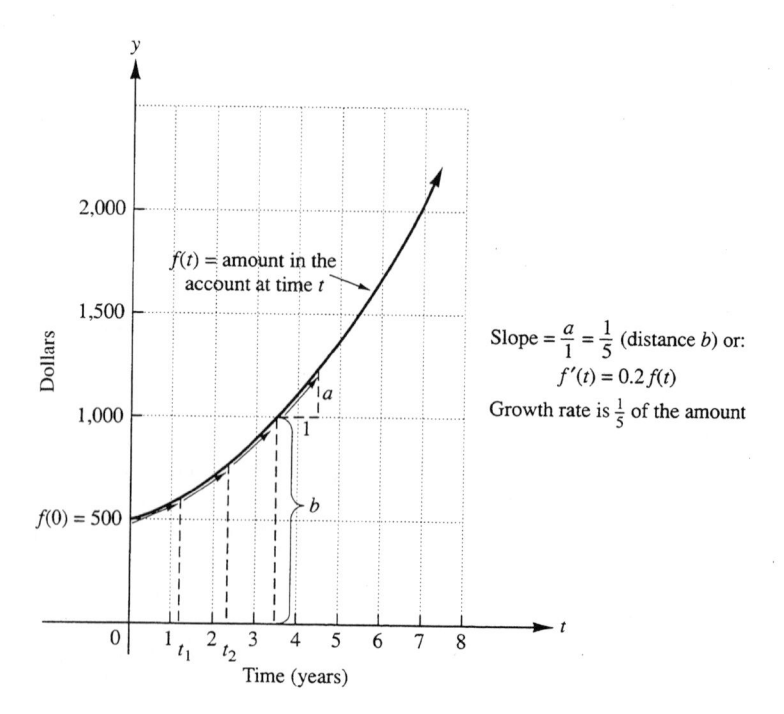

Figure 2.15

Graph showing $f(t)$ = amount in the account at time t, with $f(0) = 500$, values on vertical axis 1,000, 1,500, 2,000 (Dollars), horizontal axis Time (years) from 0 to 8, with points t_1, t_2 marked. Slope $= \frac{a}{1} = \frac{1}{5}$ (distance b) or: $f'(t) = 0.2 f(t)$. Growth rate is $\frac{1}{5}$ of the amount.

Exercises 2.4

Sketch a graph of a continuous function $y = f(x)$ that satisfies the conditions in Exercises 1–6. Remember that $f(x)$ tells you how high, or how low, the graph is at x and that $f'(x)$ tells you how the graph is changing at x.

1. **a.** $f(0) = 0$, $f'(0) = 0$, $f'(-1) = -1$, $f'(1) = +1$.
   b. In addition, $f'(2) = 0$, $f'(3) = -1$, and alternatively, $f(0) = 1$ (instead of $f(0) = 0$).

2. **a.** $f'(1) = f'(-1) = 0$ and $f(1) = f(-1) = 2$.
   b. In addition, $f'\left(\frac{1}{2}\right) = -1$ and $f'\left(-\frac{1}{2}\right) = +1$.
   c. Instead of (b), $f'\left(\frac{1}{2}\right) = \frac{1}{2}$ and $f'\left(-\frac{1}{2}\right) = -\frac{1}{2}$.

3. **a.** $f(0) = 0$, $f'(0) = 0$, $f'(-1) = +1$, $f'(1) = +1$.
   b. In addition, $f'(3) = -1$.
   c. Instead of (b), $f'(2) = 0$, and alternatively, $f(0) = -1$ (instead of $f(0) = 0$).

4. **a.** $f'(1) = 0$, $f'(2) < 0$, and $f(0) > f(2)$.
   b. In addition, $f'(0) = 0$ and $f'(-1) < 0$.
   c. Instead of (b), $f'(0) = 0$ and $f'(-1) > 0$.

5. **a.** $f(0) = 0$, $f'(x) > 0$ if $x > 0$.
   b. In addition, $f'(a) < f'(b)$ if $0 < a < b$.

6. **a.** $f'(x) = 0$ for all x.
   b. In addition, $f(1) = 2$.

7. Let $f(x) = x^2$ on the interval $(-2, 2)$. At which points is f increasing; decreasing? On which *intervals* contained in the interval $(-2, 2)$ is f increasing; decreasing?

8. Let $f(x) = 4$ for all x. Is f increasing on any interval? Is f decreasing at any point? What property of the derivative distinguishes a constant function?

9. Let $f(x) = x^3$. Show that f is increasing on the interval $(-1, 2)$ but that f is not increasing at the point $x = 0$. How does this function show that the converse to Theorem 2.6(1) is not true?

10. If $f(x) = x^3$, then $f'(x) = 3x^2$. Give an example of a function g so that $g(x) \neq f(x)$ and yet $g'(x) = 3x^2$.

For each of the following exercises, you are given three graphs (see Figures 2.16–2.19). The first is the graph of the derivative f' of a function f. Which one of the remaining pair of graphs could be the graph of the function f? Can the other be as well? [*Hint:* Concentrate on the positive/negative properties of f'.]

11.

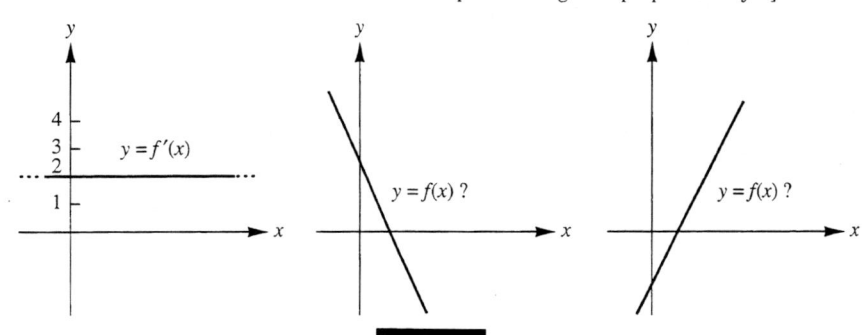

Figure 2.16

12.

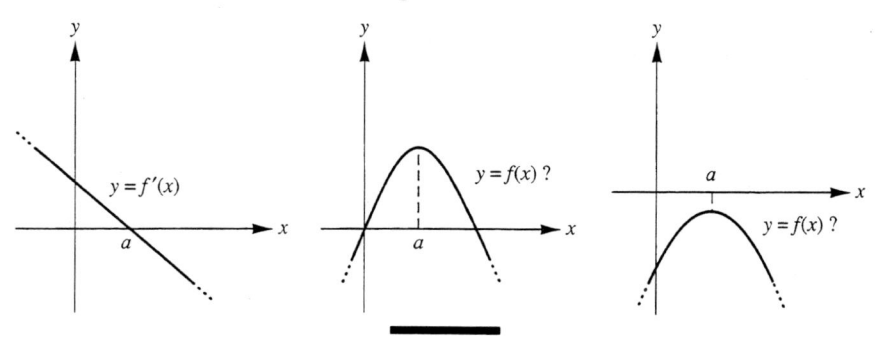

Figure 2.17

13.

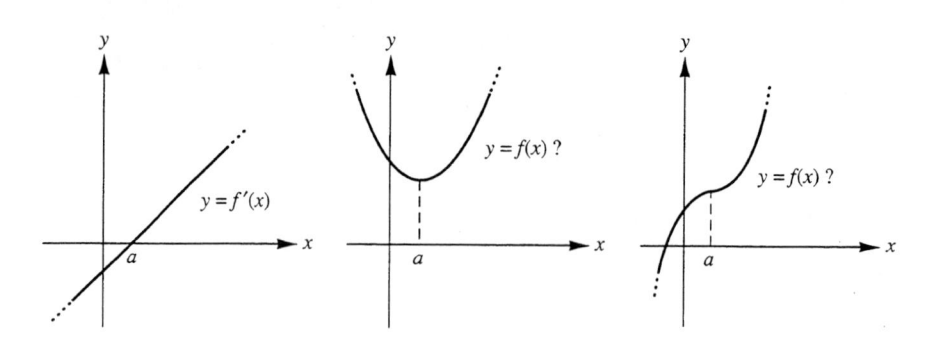

Figure 2.18

14.

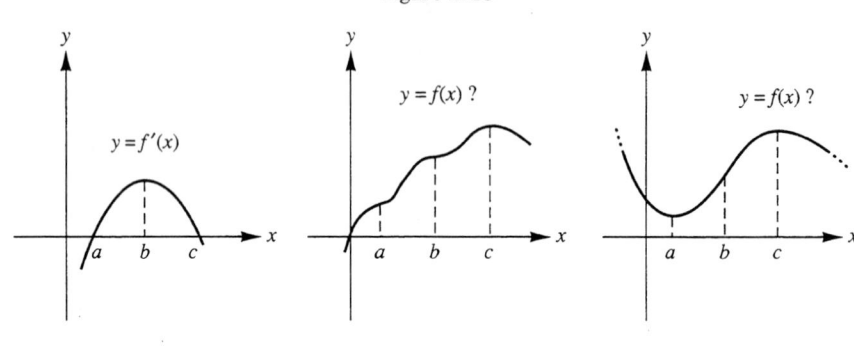

Figure 2.19

Draw the graph of a function $y = f(x)$, on the indicated interval, whose *derivative* is given by Exercises 15–19.

15. $f'(x) = 2x + 1$ $\quad -2 \leq x \leq 2$

16. $f'(x) = x^2$ $\quad -2 \leq x \leq 2$

17. $f'(x) = \dfrac{1}{x + 1}$ $\quad 0 \leq x \leq 4$

18. $f'(x) = 2$ $\quad -1 \leq x \leq 3$

19. $f'(x) = x - 1$ $\quad 0 \leq x \leq 3$

20. In Exercises 15–19, how would you draw each graph if you knew in addition that $f(0) = 1$?

Use Example 5 as a guide to drawing the graph of an unknown function $y = f(x)$ whose derivative is given by Exercises 21–25. In addition, specify each interval of numbers on which the function is increasing or decreasing *at each point* in the interval. On which *intervals* is f increasing/decreasing?

21. $f'(x) = x(2 - x)$ on $[-1, 3]$

22. $f'(x) = x(x + 1)$ on $[-2, 2]$

23. $f'(x) = x^2(x + 1)$ on $[-2, 2]$

24. $f'(x) = x(x + 1)(x - 2)$ on $[-2, 4]$

25. $f'(x) = x(x - 1)(2x - 4)$ on $[-1, 3]$

26. In which of Exercises 15–19 can you write a formula for $y = f(x)$ itself? What is such a formula? What is such a formula if, in addition, $f(0) = 1$? [*Hint:* Think of the formula

$$D_x x^n = n x^{n-1}$$

in "reverse" and consider the discussion following Example 4.]

27. Graph a function $y = f(x)$ so that $f'(x) = 1/x^{1/3}$ if $x \neq 0$ and $f(0) = 0$, on the interval $(-8, 8)$.

How would the derivatives of the functions represented by the pairs of graphs in Figures 2.20 and 2.21 differ? Think in terms of slopes of tangent lines and in terms of increasing/decreasing slopes themselves if necessary.

28.

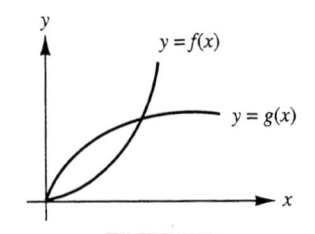

Figure 2.20

136   Chapter 2   The Derivative

29.

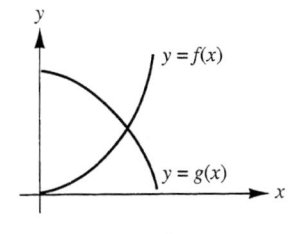

Figure 2.21

30. Suppose that in Example 6 the return on investment is 10% rather than 20%. How would the graph of the unknown function $f(x)$ differ? What if the return were 50%?

31. Sketch the graph of a function $y = f(x)$ satisfying

$$f'(x) = f(x) \quad \text{and} \quad f(0) = 1$$

using Example 6. Compare the shape of this graph with the graph of the function

$$g(x) = 2^x$$

Challenge Problems

32. Radioactive material decays at a rate proportional to the amount of radioactive material present. Suppose that

$f(t)$ = grams of cesium 90 stored in an abandoned Kansas salt mine in year t in the future

with $t = 0$ denoting the present. Suppose further that this radioactive material decays at the rate of 50% of its mass at any time. This physical property can be restated as a differential equation

$$f'(t) = -0.50 f(t)$$

since $f'(t)$ = rate of change of the amount at time t and the material is decaying at the rate of 50% of the amount $f(t)$ present at time t. Suppose that $f(0) = 1{,}000$ g. Use Example 6 to graph this unknown function $f(t)$ representing the amount of cesium 90 remaining in storage at time t.

33. An investment company adopts the following investment strategy for its assets: It wishes to achieve an investment level of $1 million and at any time will invest its assets at the rate of 10% of the difference between this $1 million and the amount already invested. Let $f(t)$ = amount invested at time t. Suppose that $f(0) = \$100{,}000$. Write this investment strategy as a differential equation $f'(t) = 0.10(1{,}000{,}000 - f(t))$, and graph the unknown function representing the course of investments.

2.5 Maximum and Minimum Values of a Function

The purpose of this section is to help you learn to locate the peaks and valleys on a continuous curve $y = f(x)$ using the first derivative $f'(x)$ and the second derivative $f''(x)$.

Consider the graph in Figure 2.22, indicating the total profit due to all sales of a certain item at varying price levels. Three price levels for this item are of obvious interest. A peak profit level is first reached when the item sells for $a per item. If the price is raised, total profits fall and a "local minimum" profit is reached at $b. If, however, the price is raised further, total profits begin to rise again (the higher price is being received on all sales of the item, and perhaps some new customers are found because the item now costs enough to attract them) and a second peak profit level is reached at a price of $c per item.

Figure 2.22

In this section we will study two test procedures that use the derivative of a function to locate such maximum and minimum points on its graph. We first need a key definition.

Definition

Let $y = f(x)$ be a function. A single particular value $f(\bar{x})$ of $f(x)$ is said to be

1. A **local maximum value** (abbreviated *local max*) of $y = f(x)$ if

$$f(\bar{x}) \geq f(x)$$

for all x in some open interval I containing $\bar{x}$.

2. A **local minimum value** (abbreviated *local min*) of $y = f(x)$ if

$$f(\bar{x}) \leq f(x)$$

for all x in some open interval I containing $\bar{x}$.*

---

*Often in mathematics special cases are either interesting or dull. Here dullness prevails: Note that in this definition *every* value of a *constant* function is both a local max and a local min.

Remark. Throughout this section we will use the notation $\bar{x}$ whenever we want to remind ourselves that we are dealing with a particular, single value of the variable x.

In Figure 2.23, with $\bar{x} = a$, b, or c, the indicated values $f(a)$, $f(b)$, and $f(c)$ are, respectively, a local max, a local min, and a local max value. Related intervals I_a, I_b, and I_c are indicated as well. The points on a graph at which either a local max or a local min occurs [such as the points $(a, f(a))$, $(b, f(b))$, and $(c, f(c))$ in Figure 2.23] are called **local extrema** or, simply, **extreme points.**

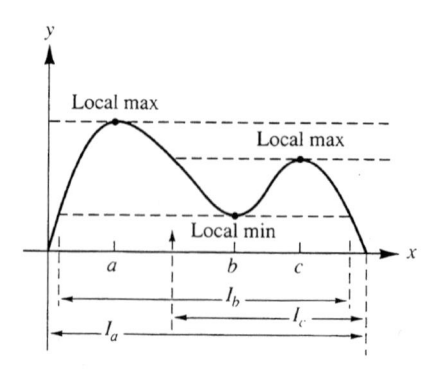

Figure 2.23

It is important that you distinguish between a max or min value $f(\bar{x})$ and the point $x = \bar{x}$ upon which this value is computed. Either $\bar{x}$ or $f(\bar{x})$ may be of interest in a given example; the general procedure is first to locate $\bar{x}$, determine whether a max or a min occurs at $\bar{x}$, and finally compute the max or min value $f(\bar{x})$.

We may outline the development of this entire section via Figure 2.24. In general, we wish to find the extreme points of an unknown graph and identify each as either a local max or a local min; these occur at $(a, f(a))$, $(b, f(b))$, and $(c, f(c))$ in Figure 2.24. Notice that tangent lines to the graph at these points have either zero slope or, as at $(c, f(c))$, there is *no* tangent line to the graph and $f'(c)$ does not exist. This section is founded on the following theorem, whose proof will not be given:

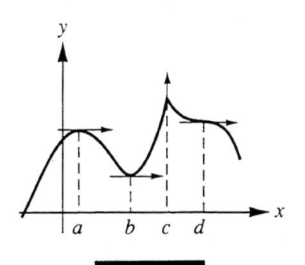

Figure 2.24

Theorem 2.8

The Main Theorem on the Location of Local Extrema

If $f(\bar{x})$ is a local maximum or a local minimum value of a continuous function $y = f(x)$ on an open interval, then necessarily either

1. $f'(\bar{x}) = 0$

or

2. $f'(\bar{x})$ does not exist.

A point $\bar{x}$ at which either $f'(\bar{x}) = 0$ or $f'(\bar{x})$ does not exist is called a **critical point**. Each point $\bar{x} = a, b, c,$ or d in Figure 2.24 is a critical point. However, $(d, f(d))$ is *not* an extreme point. That is, Theorem 2.8 does not guarantee that when $f'(\bar{x}) = 0$ (or fails to exist), then $f(\bar{x})$ is *necessarily* a local max or a local min. Rather, Theorem 2.8 tells us that among the infinitely many values of a function, its extreme points will be found among its typically few critical points. The distinction between critical points and extreme points will call for much of our effort and attention in this section. We will learn to complete the following procedure:

Given $y = f(x)$,

1. Find all critical points.
2. Among these identify all extreme points.
3. Determine whether each extreme point is a local max or a local min.

In our first example we find only the few critical points of a function.

EXAMPLE 1

Find all critical points $\bar{x}$ of $f(x) = x^4 - 2x^3 + 1$.

Solution We are to find all points $\bar{x}$ for which the derivative $f'(x) = 4x^3 - 6x^2 = 2x^2(2x - 3)$ is zero; note that in this example $f'(x)$ exists for all values of x. Since $2x^2(2x - 3) = 0$ is equivalent to the two equations

2.5 Maximum and Minimum Values of a Function

$$2x^2 = 0 \quad \text{or} \quad 2x - 3 = 0$$

there are only two such critical points, $\bar{x} = 0$ and $\bar{x} = \frac{3}{2}$. ∎

We will complete the analysis of this function and its graph in Example 2.

The remainder of this section is divided into three subsections. First we study a test for maxima and minima based on the first derivative of a function. We then study an alternate test based on the second derivative; this second test, while easier to use, is not always applicable, however. We then learn how to find the largest and smallest value of a function on a closed interval.

Subsection A: The First Derivative Test for Extrema

The **first derivative test** for extrema is no more than the formal mathematical statement of the geometrical configurations that can exist near a point $\bar{x}$ on the graph of a function when $f'(\bar{x}) = 0$ (or does not exist). The principal configurations when $f'(\bar{x}) = 0$ are illustrated in Figure 2.25.

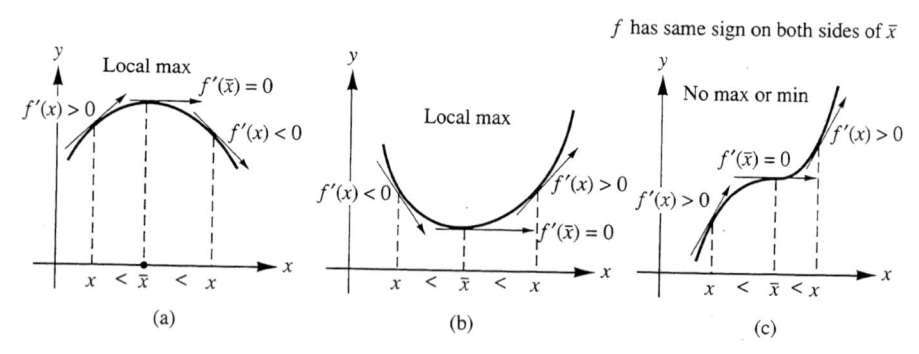

Figure 2.25

(a) Local max: $f'(x) > 0$ for $x < \bar{x}$, $f'(\bar{x}) = 0$, $f'(x) < 0$ for $x > \bar{x}$

(b) Local min: $f'(x) < 0$ for $x < \bar{x}$, $f'(\bar{x}) = 0$, $f'(x) > 0$ for $x > \bar{x}$

(c) No max or min: f has same sign on both sides of $\bar{x}$; $f'(x) > 0$, $f'(\bar{x}) = 0$, $f'(x) > 0$

Figure 2.25 indicates, for example, that a function must *increase to*, and *decrease from*, a point of maximum value. That is, the graph must rise to a peak—this occurs when $f'(x) > 0$ for x less than $\bar{x}$—and then fall from this peak—this occurs when $f'(x) < 0$ for x greater than $\bar{x}$.* When both of these conditions are reversed [Figure 2.25(b)], we obtain instead a point of minimum value. You should try to understand the following formal statement of the first derivative test for extrema in terms of Figure 2.25.

Theorem 2.9

The First Derivative Test for Extrema

Let $y = f(x)$ be a continuous function, and let $\bar{x}$ be a point at which either

$$f'(\bar{x}) = 0 \quad \text{or} \quad f'(\bar{x}) \text{ does not exist}$$

---

*Indeed, exactly these conditions are used in Section 2.4 to give a technical definition of the term *increasing (decreasing) function*. An intuitive understanding of these terms will suffice in this section.

and, additionally, $f'(x)$ exists and is nonzero in an open interval I containing $\bar{x}$ (except possibly for $\bar{x}$ itself). Then

1. If $f'(x) > 0$ for $x < \bar{x}$ and $f'(x) < 0$ for $x > \bar{x}$, for x in I, then $f(\bar{x})$ is a local maximum.
2. If $f'(x) < 0$ for $x < \bar{x}$ and $f'(x) > 0$ for $x > \bar{x}$, for x in I, then $f(\bar{x})$ is a local minimum.
3. If $f'(x)$ has the *same sign* for x on *both sides* of $\bar{x}$ in the interval I, then $f(\bar{x})$ is neither a local max nor a local min.

Notice that Figure 2.25(c) illustrates (one case of) part 3 of the first derivative test. We see there that the graph of f may rise (or fall) to a "momentary" leveling point and then continue to rise (fall) beyond that point; in such a case no maximum or minimum occurs. In practice, testing the sign of $f'(\bar{x})$ on *both* sides of $\bar{x}$ will immediately single out which of the possibilities illustrated in Figure 2.25 actually holds.

Let us use the first derivative test to complete the analysis of the function of Example 1.

EXAMPLE 2

Locate all extrema and sketch the graph of $f(x) = x^4 - 2x^3 + 1$.

Solution In Example 1 we found that $f'(x) = 4x^3 - 6x^2 = 2x^2(2x - 3)$ and that f has two critical points $\bar{x} = 0$ and $\bar{x} = \frac{3}{2}$.

How can we next determine the *sign* of f' at points other than $\bar{x} = 0$ or $\bar{x} = \frac{3}{2}$, as we must in order to apply the first derivative test, Theorem 2.9? This is not difficult. It is visually clear that if a continuous function changes sign on an interval, it must have a value of zero at some point between these two sign changes.* Since f' itself is a continuous function, f' must have *constant sign* on any interval that excludes the two critical points $\bar{x} = 0$ and $\bar{x} = \frac{3}{2}$, where f' has its *only* zero values. Thus we may determine the sign of f' on such intervals by choosing one test point in each such interval and finding the sign of f' at each chosen test point. Let us do this.

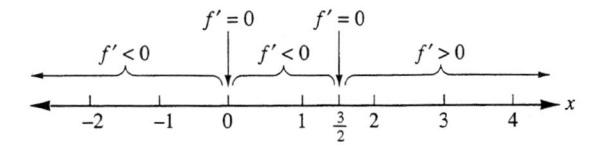

Figure 2.26

---

*This is discussed in more detail in Section 2.4, where Figure 2.10 illustrates this statement and Theorem 2.7 (Section 2.4) is the detailed formulation that we apply here, in exactly the manner of Example 5 of Section 2.4.

In Figure 2.26 we first mark both critical points $\bar{x} = 0$ and $\bar{x} = \frac{3}{2}$ on the number line. We then choose easy-to-evaluate "test points" on *both* sides of each critical point. Let us use $x = -1, 1,$ and 2 as test points. Since

$$f'(-1) = -10 \quad f'(1) = -2 < 0 \quad \text{and} \quad f'(2) = 8 > 0$$

the sign of f' has the pattern illustrated in Figure 2.26. According to Theorem 2.9(2), then, f has a local min at $\bar{x} = \frac{3}{2}$; according to Theorem 2.9(3), f has neither a local max nor a local min at the critical point $\bar{x} = 0$.

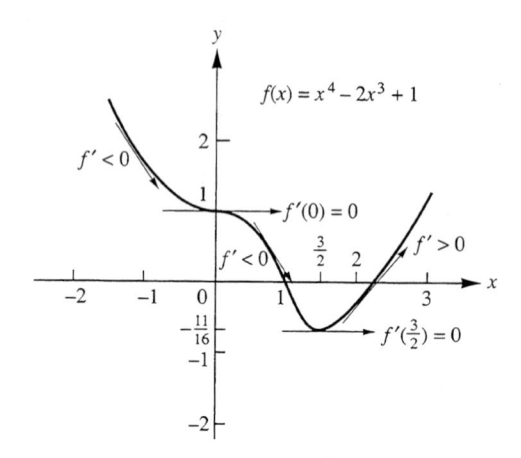

Figure 2.27

We thus sketch the graph of f in Figure 2.27. In doing so we also evaluate $f(0) = 1$ and $f(\frac{3}{2}) = -\frac{11}{16}$ to place these points at the proper height. ∎

Notice that although $f'(0) = 0$, f does not have a local extremum at 0. The graph of f only levels off "for an instant" at $x = 0$ and continues its overall downward trend because f' is negative on both sides of zero. Thus $f(\frac{3}{2}) = -\frac{11}{16}$ is the only extremum of this function and is a local minimum. Notice again that among the infinitely many different values of a function, we need only consider those few occurring at critical points to find all local extrema.

Subsection B: The Second Derivative Test for Extrema and Inflection Points

When both the first and second derivatives exist at a critical point, the task of establishing the nature of a critical point can be made much simpler through the use of the second derivative in many, but not all, cases. To employ the second derivative we need to introduce a new idea.

Consider the graphs presented in Figure 2.28(a) and (b). Both are graphs of increasing functions, and the derivatives $f'(x)$ and $g'(x)$ are both positive. Yet the graphs do differ. Let us see how the *second* derivatives f'' and g'' reflect this difference and allow it to be tested. The graph $y = f(x)$ in Figure 2.28(a) is said to be *concave up:* The graph of f lies *above* the tangent line at $(x, f(x))$

on an interval about x. The graph $y = g(x)$ in Figure 2.28(b) is said to be *concave down:* The graph lies *below* the tangent line at $(x, g(x))$ on an interval about x.

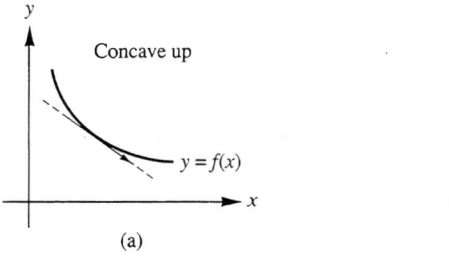

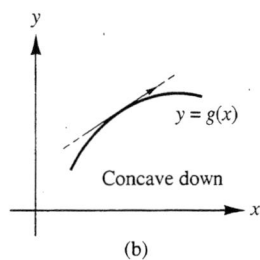

Figure 2.28

(a) Concave up, $y = f(x)$
(b) Concave down, $y = g(x)$

Forget the function $y = f(x)$ for a moment and instead try to imagine the derived function $y = f'(x)$ for a graph that is concave up, as in Figure 2.29(a). The graph of the derivative $f'(x)$, itself thought of as a function of x, is an increasing function, as Figure 2.29(b) indicates. This suggests that the derivative of f'—that is, f''—must itself be positive for a graph to be concave up.

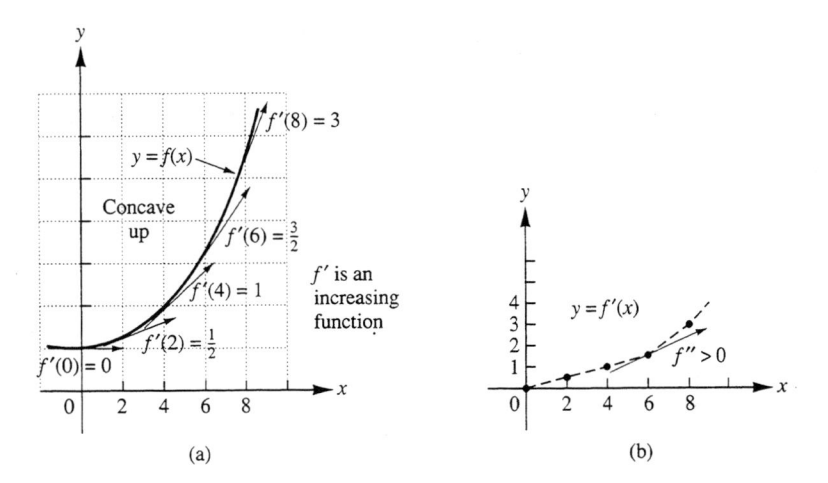

Figure 2.29

(a) $y = f(x)$, Concave up; $f'(0) = 0$, $f'(2) = \frac{1}{2}$, $f'(4) = 1$, $f'(6) = \frac{3}{2}$, $f'(8) = 3$; f' is an increasing function
(b) $y = f'(x)$, $f'' > 0$

In an analogous way, we conclude that for the function $y = g(x)$ of Figure 2.28(b), we must have $g''(x) < 0$. We summarize these ideas in the *test for concavity*, as follows:

Theorem 2.10

The Second Derivative Test for Concavity

Let $y = f(x)$ be a function having a second derivative $f''(x)$.

1. If $f''(x) > 0$, then the graph of $y = f(x)$ is concave up at x.
2. If $f''(x) < 0$, then the graph of $y = f(x)$ is concave down at x.

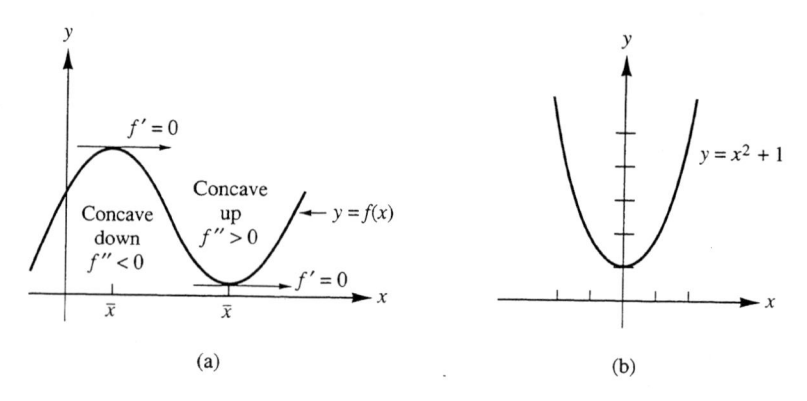

Figure 2.30

(a)

(b)

Let us think about the meaning of Theorem 2.10 applied to a critical point $\bar{x}$, where $f'(\bar{x}) = 0$, as in Figure 2.30(a). We realize immediately that the graph of f is concave down ($f'' < 0$) at a max value and concave up ($f'' > 0$) at a min value. This may be proven in general and gives us parts 1 and 2 of the **second derivative test for extrema.**

Theorem 2.11

The Second Derivative Test for Extrema

Let $y = f(x)$ be a function having a second derivative $f''(\bar{x})$ at a point $\bar{x}$. Then

1. If $f'(x) = 0$ and $\boldsymbol{f''(\bar{x}) < 0}$, then $f(\bar{x})$ is a **local maximum** value of $y = f(x)$.
2. If $f'(\bar{x}) = 0$ and $\boldsymbol{f''(\bar{x}) > 0}$, then $f(\bar{x})$ is a **local minimum** value of $y = f(x)$.
3. If $f'(\bar{x}) = 0$ and $f''(\bar{x}) = 0$, then no conclusion is possible using $f''(\bar{x})$ alone.

We will consider the inconclusive case 3 shortly.

EXAMPLE 3

The familiar graph of $f(x) = x^2 + 1$ is shown in Figure 2.30(b). The graph is concave up with a local min at $\bar{x} = 0$. This is consistent with both Theorems 2.10 and 2.11 in that $f'(x) = 2x$, $f'(0) = 0$, and $f''(x) = 2$ is positive for *any* x (in particular, $f''(0) = 2$). ∎

In the next example we use the second derivative test for extrema as our primary tool and return to the first derivative test only when this primary tool is inconclusive. This is a general rule of procedure.

EXAMPLE 4

Find the local extrema and sketch the graph of

$$f(x) = \frac{x^5}{5} - \frac{x^4}{2} + 2$$

Solution We proceed step-by-step but try to use the simpler second derivative test rather than the first derivative test whenever we can. The second derivative test requires testing only *at* the critical point $\bar{x}$, while the first derivative test asks for information *near* $\bar{x}$ as well.

Step 1. First compute f' and f''. We have
$$f'(x) = x^4 - 2x^3 \quad \text{and} \quad f''(x) = 4x^3 - 6x^2$$
These both exist for all x.

Step 2. Local extrema can occur only at critical points where
$$f'(x) = x^4 - 2x^3 = x^3(x - 2) = 0$$
These points are $\bar{x} = 0$ and $\bar{x} = 2$.

Step 3. By the second derivative test, $f''(2) = 4(2)^3 - 6(2)^2 = 8 > 0$, so that $f(2) = \frac{2}{5}$ is a local minimum.

If we similarly test at $\bar{x} = 0$, we have $f''(0) = 4(0)^3 - 6(0)^2 = 0$. Therefore the second derivative test is inconclusive, and we must use the first derivative test.

Step 4. To use the first derivative test we must check the sign of $f'(x)$ on both sides of $\bar{x} = 0$, as in Example 2. Let us choose test points $x = -1$ and $x = +1$ on both sides of $\bar{x} = 0$ (see Figure 2.31). Since $f'(x) = x^3(x - 2)$, we see that $f'(-1) = 3 > 0$ and $f'(1) = -1 < 0$. Hence, $f'(x) > 0$ if $x < \bar{x}$ and $f'(x) > 0$ if $x > \bar{x}$ (and $x < 2$). Therefore $f(0) = 2$ is a local maximum value in accordance with the first derivative test.

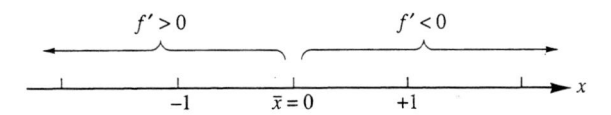

Figure 2.31

Step 5. Putting all these steps together, including the computations
$$f(0) = 2 \quad \text{and} \quad f(2) = \frac{2}{5}$$
we sketch the graph shown in Figure 2.32.

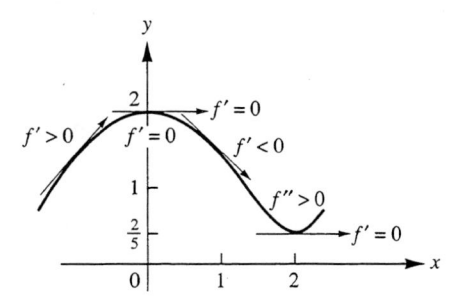

Figure 2.32

In the next example, f' does not exist and so of course f'' cannot be used.

2.5 Maximum and Minimum Values of a Function 145

EXAMPLE 5

Find local extrema and graph $f(x) = x^{2/3} + 1$.

Solution

Step 1. $f'(x) = \left(\frac{2}{3}\right)x^{-1/3} = 2/(3x^{1/3})$.

Step 2. f' is never zero. However, since $f'(x)$ does not exist at 0, $\bar{x} = 0$ is a critical point of possible extrema. Since $f''(0)$ also cannot exist, we *must* use the first derivative test.

Step 3. Thus we test the sign of f' on both sides of $\bar{x} = 0$. Since $f'(x) = 2/(3x^{1/3})$

$$f'(x) > 0 \quad \text{for} \quad x > 0$$
and
$$f'(x) < 0 \quad \text{for} \quad x < 0$$

(for example, evaluate f' at $x = \pm 1$). Therefore, by the first derivative test,

$$f(0) = 0^{2/3} + 1 = 1$$

is a local minimum.

Step 4. Since $f'(0)$ does not exist, the graph of $y = f(x) = x^{2/3} + 1$ cannot turn smoothly at $x = 0$. For example, the slope is quite different at, say, the two points $x = \pm 1/125$; for $f'(\pm 1/125) = \pm 10/3$. The graph of f is as indicated in Figure 2.33.

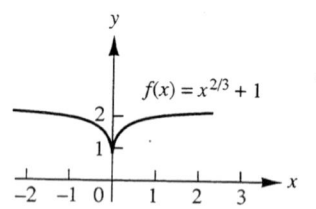

Figure 2.33

We conclude this subsection with a discussion of inflection points, because these are closely related to the inconclusive case 3 of the second derivative test. A point $(\bar{x}, f(\bar{x}))$ on the graph of a continuous function is called an **inflection point** if the *concavity* of the graph *changes* at this point. The point $(\bar{x}, f(\bar{x}))$ in Figure 2.34 is an inflection point, since the graph changes from concave down to concave up at this point. This graph also suggests that inflection points are likely to occur between successive extrema ($\bar{x} = 0$ and $\bar{x} = 2$ in Figure 2.34), because the concavity of f must change between these points.

Since concavity is measured by the sign of the second derivative, it follows that the concavity of f can change at $\bar{x}$ if and only if the sign of f'' is different on alternate sides of the point $\bar{x}$. Necessarily, then, if $(\bar{x}, f(\bar{x}))$ is a point of inflection and f'' is continuous, then $f''(\bar{x}) = 0$. We summarize these remarks as the **second derivative test for inflection points**.

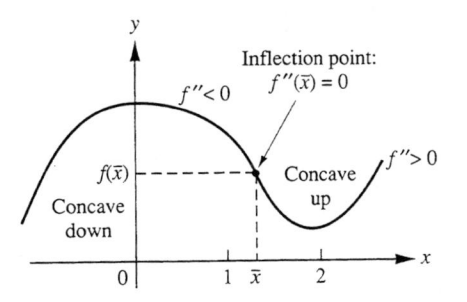

Figure 2.34

Theorem 2.12

The Second Derivative Test for Inflection Points

Let f'' exist in an open interval containing a point $\bar{x}$.

1. The point $(\bar{x}, f(\bar{x}))$ is a point of inflection if and only if the sign of f'' changes when f'' is evaluated at points x on alternate sides of $\bar{x}$.
2. If $(\bar{x}, f(\bar{x}))$ is a point of inflection on the graph of $y = f(x)$ and f'' is continuous at $\bar{x}$, then $f''(\bar{x}) = 0$.

Remark. In Theorem 2.12(1) we make no supposition about the value of $f'(x)$, which can be nonzero (or even undefined) at an inflection point $(x, f(\bar{x}))$.

EXAMPLE 6

Find all inflection points of $f(x) = (x^4/4) + x^3 - 2$, and sketch the graph of f.

Solution Since $f'(x) = x^3 + 3x^2$ and $f''(x) = 3x^2 + 6x = 3x(x + 2)$ is continuous, part 2 of the second derivative test for inflection points tells us that such points can occur only when $3x(x + 2) = 0$ or when

$$\bar{x} = 0 \quad \text{or} \quad \bar{x} = -2$$

If we check the sign of f'' on alternate sides of each of these points, we have the result shown in Figure 2.35.

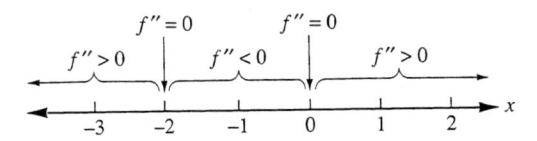

Figure 2.35

Therefore both $\bar{x} = 0$ and $\bar{x} = 2$ are points of inflection. This does not, however, tell us much about the graph of this function. To graph $y = f(x)$, we need to locate extrema as well.

Step 1. Since $f'(x) = x^3 + 3x^2$ exist for all x, the only possible extrema are where $0 = x^3 + 3x^2 = x^2(x + 3)$. The solutions are $\bar{x} = 0$ and $\bar{x} = -3$; hence these are our only critical points.

Step 2. Since $f''(-3) = 9 > 0$, then $f(-3) = -\frac{35}{4}$ is a local minimum. Our work already shows that f has an inflection point at $\bar{x} = 0$, at height $f(0) = -2$.

Step 3. Putting all this together, we conclude that the graph of this function must be as indicated in Figure 2.36.

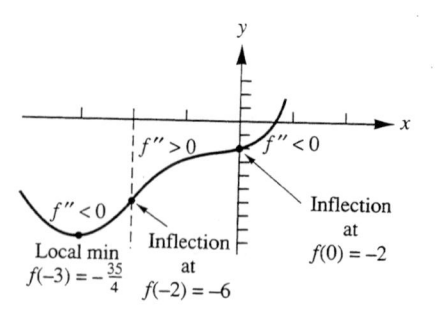

Figure 2.36

Remarks

1. In general it is not true that if $f''(\bar{x}) = 0$, then $(\bar{x}, f(\bar{x}))$ is a point of inflection; see Example 4, steps 3 and 4. Furthermore, if f'' is not continuous at x, then it is not necessary that $f''(\bar{x}) = 0$. Exercises 23, 41, and 42 provide basic examples of these phenomena.
2. Nothing in our analysis tells us where (if at all) a graph crosses the x-axis; this is another question that we are not particular about when finding the general shape of such a graph and its extrema and inflection points. (We need such information only when we want to know where a graph is, not what its shape is.)

Subsection C: The Maximum and Minimum Values of a Continuous Function on an Interval

Consider the problem of a box manufacturing company that wants to make the largest box (in volume) possible from a flat piece of cardboard measuring 20 by 20 in. by cutting a square off each corner and folding the edges up, as shown in Figure 2.37. The problem is to decide how large to make the cut x so as to obtain the largest box (in volume). Note that x can be any size from 0 to 10 in.

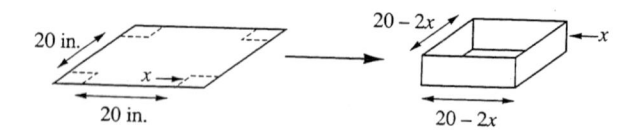

Figure 2.37

Rather than thinking at first in terms of changing sizes of the cut x and varying box sizes, think in terms of what can happen—that is, think in terms of Figure 2.37 itself. With a cut of size x, the sides of the resulting box must have length $20 - 2x$, width $20 - 2x$, and height x. Since volume is the product of these three dimensions, the volume of the resulting box is

$$V(x) = x(20 - 2x)(20 - 2x) = 400x - 80x^2 + 4x^3$$

The company's problem is now a purely mathematical one: Which number x yields the maximum value of this function *on the interval* $0 \leq x \leq 10$ representing the domain of available choice? It is this kind of mathematical problem, the optimization of a function on an interval, that we will now learn to solve.

Could it be that there is no largest box but only a variety of box sizes? We next state the general, and not easily proven, purely mathematical result that ensures that seekers of maximum efficiency need not search in vain. This theorem states that continuous phenomena always have a "best" and a "worst" case.

Theorem 2.13

The Existence of Maxima and Minima on an Interval

If $y = f(x)$ is a *continuous* function, defined on a *closed* interval $[a, b]$, then there must be

1. a point $x_{\max}$ between a and b, and
2. a point $x_{\min}$ between a and b

such that
$$f(x_{\min}) \leq f(x) \leq f(x_{\max})$$

for *all other* points x between a and b.

The values $f(x_{\max})$ and $f(x_{\min})$ are the largest and smallest values, respectively, of $y = f(x)$ on the interval $a \leq x \leq b$. The theorem promises only that such points exist—that is, that the box manufacturer can make a largest box if $x_{\max}$ can be found. The theorem does not tell us how to find such points.

The procedure for finding such max and min points uses much of what you have already learned, together with one additional detail. Consider the graph in Figure 2.38. This graph shows a continuous function defined on an interval $[a, b]$. The function does indeed have a maximum and minimum value as indicated and as Theorem 2.13 asserts that it must. The min value could be discovered by the derivative tests studied earlier. But the max value could not, because $f'(x_{\max}) \neq 0$. The first derivative test for local extrema would not reveal that the max of such a function on an interval occurs at an endpoint. This is the one extra detail that we have to consider in such problems. Figure 2.38

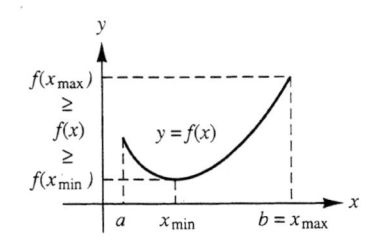

Figure 2.38

presents an example of **endpoint extrema**. Since any one interval has only two endpoints, endpoint extrema can be handled easily.

To manufacture the largest box, and in general to apply the existence Theorem 2.13, we use the following procedure:

How to Find the Largest and Smallest Values of a Continuous Function $y = f(x)$ on an Interval $a \leq x \leq b$

Step 1. Find all points $\bar{x}$, $a < \bar{x} < b$, such that

$$f'(\bar{x}) = 0$$

or $\qquad f'(\bar{x})$ does not exist

Step 2. Evaluate $y = f(x)$ at $x = a$, $x = b$ and at $x = \bar{x}$ for all points $\bar{x}$ found in step 1.

Step 3. The largest value obtained in step 2 is

$$f(x_{max})$$

The smallest value is $\qquad f(x_{min})$

In short, the first derivative is used to find local extrema inside the interval, and the resulting function values are compared with values at the endpoints to see which of (all of) these are the largest and smallest.

EXAMPLE 7

How should you cut a flat piece of cardboard measuring 20 by 20 in. to make the largest possible box?

Solution The function

$$V(x) = 4x^3 - 80x^2 + 400x \qquad 0 \leq x \leq 10$$

represents all possible volumes for all possible cuts $0 \leq x \leq 10$. Following the procedure just outlined, we first have

Step 1. $V'(x) = 12x^2 - 160x + 400 = 4(3x^2 - 40x + 100)$. We can use the quadratic formula to find the roots of the equation $3x^2 - 40x + 100 = 0$. These roots are

$$x = \frac{-(-40) \pm \sqrt{(-40)^2 - 4(3)(100)}}{2 \cdot 3} = \frac{40 \pm \sqrt{400}}{6}$$

$$= \frac{40 \pm 20}{6} = 10, \frac{10}{3}$$

Therefore $\bar{x} = \frac{10}{3}$ is the only possible extreme value in the interval $0 < x < 10$.

150 Chapter 2 The Derivative

Step 2. Evaluate the volume function V at $\bar{x} = \frac{10}{3}$ and at *both* endpoints $a = 0$ and $b = 10$. We have

$$V(10) = 0 \qquad V\left(\frac{10}{3}\right) = \frac{16{,}000}{27} \qquad \text{and} \qquad V(0) = 0$$

Step 3. The largest of these measurements

$$V\left(\frac{10}{3}\right) = \frac{16{,}000}{27} \approx 592.59 \text{ in.}^3$$

gives the box with the largest volume. We must cut a distance of $x = \frac{10}{3} = 3\frac{1}{3}$ in. at each corner. The remaining cuts, $x = 0$ or $x = 10$, obviously give a "box" of minimum volume: zero! ∎

Although it is not necessary to do so, we can conclude from Example 7 that a graph of the volume function $y = V(x)$, representing all possible volumes of resulting boxes, must look like the graph in Figure 2.39. We will solve many problems of this kind in the next section. In this section you will be asked to find the maxima and minima of a function on a closed interval without regard to particular application, using the preceding steps 1, 2, and 3.

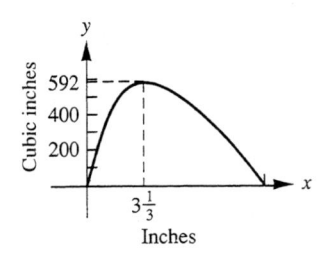

Figure 2.39

Finally, you may wish to use Figure 2.67 of the Chapter 2 Summary as a guide and outline of this section.

Background Review 2.2

In this section you will need to find the roots of many equations.

1. The roots of a quadratic equation $\alpha x^2 + \beta x + \gamma = 0 \qquad \alpha \neq 0$

   can be found either by factoring the equation in the form

   $$(ax + b)(cx + d) = 0$$

   or by the quadratic formula $\qquad \bar{x} = \dfrac{-\beta \pm \sqrt{\beta^2 - 4\alpha\gamma}}{2\alpha}$

2.5 Maximum and Minimum Values of a Function

For example, we may factor
$$2x^2 - 5x - 3 = 0$$
as
$$(2x + 1)(x - 3) = 0$$
with roots
$$\bar{x} = -\frac{1}{2}, 3$$

2. The roots of a polynomial equation of degree greater than two are difficult to find by general methods. A computer or powerful calculator is useful. In this text most higher-degree equations can be solved by factoring a power of x. For example, the equation
$$x^4 - 2x^3 - 3x^2 = 0$$
may be written as
$$x^2(x^2 - 2x - 3) = 0$$
or
$$x^2(x - 3)(x + 1) = 0$$
with roots
$$\bar{x} = 0, 3, -1$$

3. Equations involving a single fractional power such as
$$4x^{1/3} - 5 = 0$$
should be written as
$$x^{1/3} = \frac{5}{4}$$
Since $(x^{1/3})^3 = x$, we then cube both sides to obtain
$$x = \left(\frac{5}{4}\right)^3 = \frac{125}{64}$$

4. If the unknown x appears in a denominator such as
$$2 - \frac{1}{x^2} = -\frac{3}{x}$$
it is best to multiply both sides of the equation by the highest power of x that appears. Thus
$$x^2\left(2 - \frac{1}{x^2}\right) = x^2\left(\frac{-3}{x}\right)$$
or
$$2x^2 - 1 = -3x$$
or
$$2x^2 + 3x - 1 = 0$$
which can then be solved by other methods, but remember that if $x = 0$ arises as solution, it cannot be used in the original equation.

Exercises 2.5

Subsection A: The First Derivative Test for Extrema

In Exercises 1–4, depicted in Figures 2.40–2.43, determine where f' is positive, negative, or zero. Identify local max and local min points if they exist.

1.

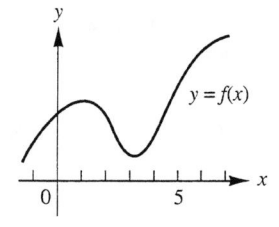

Figure 2.40

2.

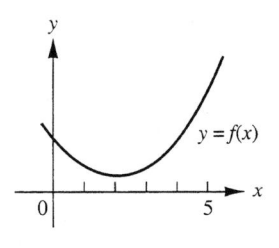

Figure 2.41

3.

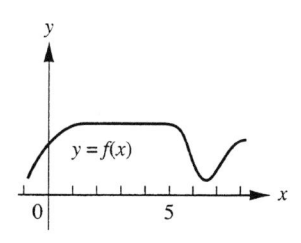

Figure 2.42

4.

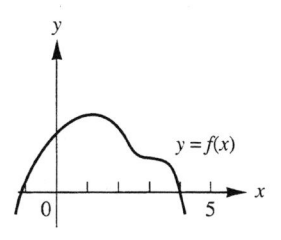

Figure 2.43

5. Suppose that the function $y = f(x)$ has a local max $f(0) = 4$, a local min $f(2) = 1$, and a local max $f(3) = 2$, and no further extrema. Draw the "nicest" graph you can imagine of a continuous function with these properties. Then draw the worst continuous graph you can imagine of a function with these properties. (There is no single correct answer to either of these questions, which draw on your overall sense of the concepts in this section; you will find that the theorems discussed do limit how unshapely even the "worst" graph can be.)

6. Suppose that $y = f(x)$ has a local max at $f(0) = 4$, that $f'(2) = 0$, and that $f(3) = 1$ is a local min. Repeat Exercise 5 for this function.

Subsection B: The Second Derivative Test for Extrema and Inflection Points

7. For Exercises 1–4, determine where $f''(x)$ is positive and negative. Must $f''(x)$ also be zero at certain points? Mark these as closely as you can.

8. Draw the graph of a function $y = f(x)$ such that $f'(1) > 0$ and $f''(1) < 0$.

9. Draw a graph of a function $y = f(x)$ such that $f(0) = 1$ and $f'(0) = 0$; that $f''(0) = 2$; that $f(2) = 0$; and that $f''(2) = -3$. What is the worst continuous graph you can imagine with these properties? Use your overall sense of the subject to draw such a graph, which is not of course unique.

10. Draw a graph of a function $y = f(x)$ such that $f''(-1) > 0$ and $f''(1) < 0$ and so that, alternatively, one of the following is true:

    a. $f'(0) = 0$ **b.** $f'(0) > 0$

In Exercises 11–22, use the second derivative test to find all local max and local min values and draw the graph of each function.

11. $f(x) = x^2 + 2x + 1$ 12. $g(x) = \dfrac{x^2}{2} - 3x + 2$

13. $h(x) = 3x^3 - 6x - 1$

14. $g(x) = 4x^3 - 9x^2 - 12x + 2$

15. $f(x) = x^4 - 2x^2 + 2$

16. $f(x) = \dfrac{x^3}{3} - 2x^2 + 3x - 1$

17. $f(x) = 2x^3 + 3x^2 - 12x + 1$

18. $h(x) = 4x^3 - 3x^2 + 6x - 7$

19. $f(x) = x + \dfrac{1}{x}$ $x > 0$

20. $g(x) = x + \dfrac{1}{x}$ $x < 0$

21. $f(x) = \dfrac{x}{x^2 + 1}$

22. $f(x) = x^{3/2} - \dfrac{8}{3}x,\ x > 0$

23. Consider the three functions $f(x) = x^2 + 1$, $g(x) = x^3 + 1$, and $h(x) = x^4 + 1$.

    a. First show that these functions each have only one critical point $\bar{x} = 0$.

2.5 Maximum and Minimum Values of a Function   153

b. Apply the second derivative test to each of these functions. For which of these is the test conclusive, allowing you to sketch the graph?

c. Apply the first derivative test to the remaining functions. Can you now sketch the graph of each of these?

d. Which of these functions has an inflection point at $\bar{x} = 0$?

In Exercises 24–34, use the second or first derivative test to determine all local max and local min values and draw the graph of each function.

24. $f(x) = 4x^5 - 5x^4 - 1$ 25. $g(x) = \dfrac{x^4}{4} - 8x + 1$

26. $h(x) = \dfrac{x^3}{3} - x^2 - 2x + 1$

27. $f(x) = 5x^6 - 6x^5$

28. $f(x) = -|x - 2| + 1$ 29. $f(x) = \dfrac{x^4}{4} - 2x^2 + 2$

30. $f(x) = 3x^{1/3} + x$ 31. $f(x) = x^{4/3}$

32. $g(x) = \dfrac{3}{2}x^{2/3} - x$ 33. $f(x) = x^4 - 4x^3 - 4$

34. $h(x) = x^2(x^2 - 1)^2$

Determine where the graph of each function in Exercises 35–42 has inflection points and where it is concave up or down. Graph the function.

35. $f(x) = (x - 2)^3$ 36. $f(x) = (x + 1)^5$

37. $f(x) = x^3 + \dfrac{x}{2}$ 38. $f(x) = (x^2 - 3)^2$

39. $f(x) = x^3 + x^2 + x - 4$

40. $h(x) = \left(x^3 - \dfrac{5}{2}\right)^2$ 41. $f(x) = x^{1/3}$

42. $g(x) = x^{6/5}$

Subsection C: The Maximum and Minimum Values of a Continuous Function on an Interval

In Exercises 43–52, determine the maximum and minimum values of each function on the indicated interval and draw its graph.

43. $f(x) = x^2 + 2x + 1$ on $[0, 3]$
44. $g(x) = x^2 - 4x + 2$ on $[0, 3]$
45. $g(x) = 3 - x - x^2$ on $[-1, 1]$
46. $f(x) = x^3 + 1$ on $[-2, 0]$
47. $f(x) = x^4 - 2x^2 + 1$ on $[-1, 2]$
48. $f(x) = x^5 - 32$ on $[-1, 2]$
49. $g(x) = 2x - 3$ on $[-1, 2]$
50. $h(x) = x^2 + \dfrac{1}{x^2}$ on $\left[\dfrac{1}{2}, 2\right]$
51. $f(x) = x^{2/3} + 1$ on $[-1, 2]$
52. $f(x) = x^3 - 3x^2 - 3x + 1$ on $[-1, 3]$
53. The function $f(x) = 2x + 1$ has no maximum value on the (half-open) interval $[0, 3)$. Why? Does this example contradict Theorem 2.13?

Computer Application Problems

54. Use the BASIC program FUNVAL1 to graph
$$f(x) = 20x^4 - 70x^3 + 175x^2 - 45x + 8$$
on the interval from $x = 0$ to $x = 2$. Identify extrema and inflection points as closely as you can.

55. Repeat Exercise 54 for the function
$$f(x) = \dfrac{4x^5}{5} - \dfrac{13x^4}{4} + \dfrac{14x^3}{3} - \dfrac{5x^2}{2} + 3$$
on the interval from $x = -1$ to $x = 2$.

2.6 Optimization

The purpose of this section is to help you learn to mathematically reformulate a problem given in descriptive language and to use this formulation to determine the optimal solution to the problem.

If "efficiency" is a goal and can be given a mathematical description, calculus can often be used to identify the *most* efficient way to accomplish a task. **Optimization** is the name given to mathematical methods used to identify such choices.

You have already learned the technical tools and methods needed for applied optimization. Your new task will be to construct, from the descriptive language in which a problem is given, an **objective function.** This objective function serves as a mathematical model of the problem and implicitly incorporates all the information given in the descriptive language of the problem into the mathematical form of a function of a single variable. Constructing an objective function will require you to create, by your own choice, one or more variables representing unknowns in the problem. If you introduce more than one variable, as is usually necessary, you also need to derive, from the language of the problem, a **constraint equation.** A constraint equation enables a step-by-step reduction of the number of variables in the problem to only *one*. Let us solve a simple problem that illustrates these ideas.

EXAMPLE 1

Among all the rectangles having a perimeter of 36 ft, find the one with the largest area.

Solution Begin by imagining a rectangle of perimeter 36 ft (Figure 2.44).

Our task is to determine the length of each side of this rectangle, which can be of any shape (see Figure 2.45), so as to enclose the largest area.

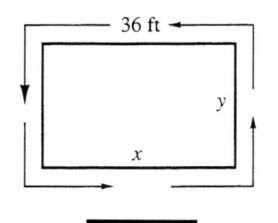

Figure 2.44

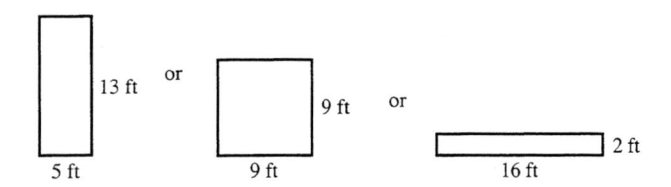

Figure 2.45

The quantity to be optimized—in this case, maximized—is therefore area. We must construct an objective function that models the areas of all possible choices of rectangles with a 36-ft perimeter. Since a function needs a variable, we have to create one or more of these. We do so by deciding exactly where our freedom of choice lies in constructing the rectangle and then assigning letters to these variables. Here we are free to choose the length and width of the rectangle. Therefore let $x =$ (length of the rectangle) and $y =$ (width of the rectangle). Then, as a first attempt at constructing an objective function, let $A =$ area of the rectangle, so that $A = xy$.

Though this last equation does define area as a function of the variables x and y, it does not define area as a function of a *single* variable and thus cannot be optimized by the methods of previous sections. This is why a constraint equation is needed.

In this problem we are dealing not with just any rectangle but with one constrained to have a 36-ft perimeter. Therefore

$$36 = x + y + x + y = 2x + 2y$$

This equation is an example of a constraint equation. We use it to eliminate one variable, say, y, as follows: Since $2x + 2y = 36$, then $y = 18 - x$. The area A can now be written as a function $A(x) = 18x - x^2$ of a single variable, for $A = xy = x(18 - x)$. The function A is the objective function for this application. It and its domain ($0 \le x \le 18$) incorporate *all* the information given in the problem, and its function values represent all the possible areas that a rectangle with a 36-ft perimeter can have. To find the largest area, we

2.6 Optimization   155

must find the maximum value of this function, using the derivative and the methods of the previous sections. We have

$$A'(x) = 18 - 2x$$

Therefore $A' = 0$ when $\bar{x} = 9$. Because $A(0) = A(18) = 0$ and $A(9) = 81$, the number $\bar{x} = 9$ is a critical point that gives the maximum value of the objective function. Additionally, since $y = 18 - x$, then $y = 18 - 9 = 9$ is the length of the remaining side of the rectangle. Thus the largest rectangle is in fact a square measuring 9 by 9 ft and having an area of 81 ft^2. ∎

This example, with its intuitively apparent solution, has introduced you to the ideas of an objective function and a constraint equation in a simple case so you could see how these concepts lead step-by-step to a solution. The solutions of the remaining examples and exercises follow the same general procedure. They will be more or less difficult depending on the difficulty in assigning variables, the determination of a constraint equation (if needed), and the finding of a suitable initial equation to which the constraint equation can be applied so as to obtain an objective function of a single variable. We summarize the procedure as follows:

General Procedure for the Solution of an Optimization Problem

1. Read the problem fully.
2. Draw a picture if it seems appropriate.
3. Read the problem again and assign letters to quantities that may vary.
4. Decide what quantity is to be optimized, and try to express it as some kind of equation in terms of the variables determined in step 3. (This is sometimes a two-step process in which the equation is expressed first in written English and then in mathematical terms.)
5. If there is more than one variable involved, find a constraint equation that relates the variables.
6. Use the constraint equation to eliminate all but one variable in the equation obtained in step 4 so as to arrive at the objective function and its domain.
7. Find the maxima or minima of the objective function on its domain.
8. Relate these maxima or minima to the original problem and reach a conclusion.

Remark. Try not to avoid using extra variables or a constraint equation. It is much easier to introduce more variables than you think you need and then to eliminate these by mechanical algebra than to puzzle over the problem so as to avoid variables. In short, avoid "rushing" the problem, introduce variables freely, and let the mechanics of algebra save you much unnecessary thought.

The remainder of this section is divided into subsections; each subsection addresses a different problem, but all use the procedure just outlined.

Subsection A: Optimum Numerical States

These examples, which do not require that a picture be drawn, are among the simplest optimization problems.

EXAMPLE 2

Find two nonnegative numbers whose sum is 100 and whose product is a maximum.

Solution

Step 1. There are two unknown numbers; let us call them x and y.
Step 2. We are to maximize their product, which we will call P. We then have the equation $P = xy$.
Step 3. There is more than one variable, and so we need a constraint equation. The statement of the problem requires that our two numbers add up to 100. Therefore $x + y = 100$ is the constraint equation and $y = 100 - x$.
Step 4. The function to be optimized is $P(x) = xy = x(100 - x) = 100x - x^2$ with domain $0 \leq x \leq 100$ (since $x \geq 0$ and $0 \leq y = 100 - x$, then $x \leq 100$).
Step 5. We need to find the maximum product—that is, the maximum value of the function $P(x)$. Since $P'(x) = 100 - 2x$, then $P'(x) = 0$ when $\bar{x} = 50$. In addition, $P''(x) = -2 < 0$, so that $\bar{x} = 50$ and $y = 100 - 50 = 50$ give a maximum product $P(50) = 50 \cdot 50 = 2{,}500$ [since at the endpoints $x = 0$ and $x = 100$, $P(x) = 0$]. ■

EXAMPLE 3

Find two positive numbers whose product is 25 and whose sum is a minimum.

Solution

1. Let x and y be the two unknown numbers.
2. We are to minimize $S = x + y$.
3. Needing a constraint equation, we impose the remaining condition that $xy = 25$ or $y = 25/x$, for $x > 0$.
4. Therefore we must minimize $S(x) = x + y = x + (25/x)$ on the open interval $(0, \infty)$.
5. We have $S'(x) = 1 - (25/x^2)$ for all $x > 0$. Thus $S'(x) = 0$ when $1 - (25/x^2) = 0$ or when

$$1 = \frac{25}{x^2}$$

or

$$x^2 = 25$$

or

$$\bar{x} = \pm 5$$

We need only consider the positive solution $\bar{x} = 5$.

2.6 Optimization   157

Now $S''(x) = 50/x^3$ is positive at $\bar{x} = 5$ and indeed for any $x > 0$. Therefore the graph of S is concave up for all $x > 0$, and the two numbers $x = 5$ and $y = \frac{25}{5} = 5$ yield a minimum sum $S(5) = 10$. ∎

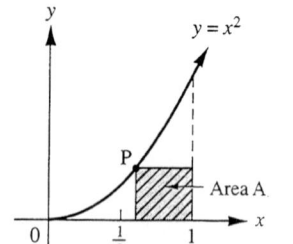

Subsection B: Optimum Geometric States

The examples in this subsection all require the optimization of some quantity that is naturally related to the dimensions of some geometric figure.

EXAMPLE 4

Consider Figure 2.46. Determine the location of the point P on the graph of $y = x^2$ so that the area of the indicated rectangle is a maximum, where $0 \leq x \leq 1$.

Figure 2.46

Solution

1. Try to imagine the nature of the problem by drawing a few rectangles of the kind that result from different choices of the point P (see Figure 2.47).
2. The location of P can be specified by its x and y coordinates, and so we can use these as variables.
3. We are to maximize the area A of the rectangle in Figure 2.48. This area is

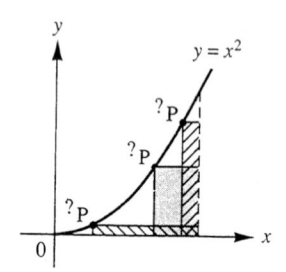

$$A = (\text{length})(\text{height})$$
$$= (1 - x)y$$

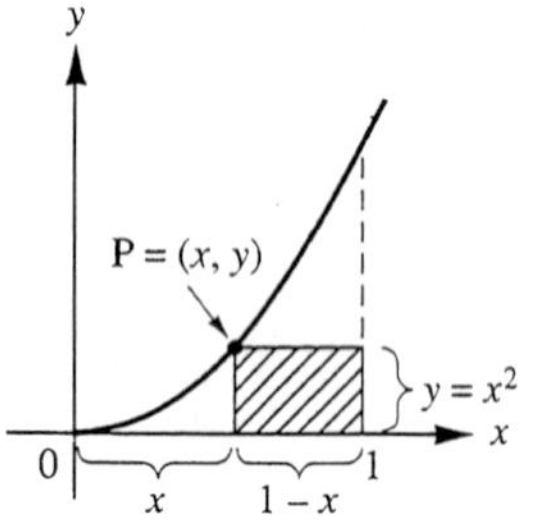

Figure 2.47

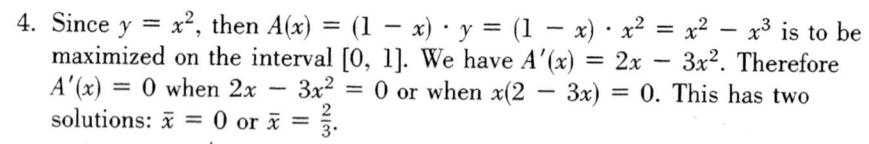

4. Since $y = x^2$, then $A(x) = (1 - x) \cdot y = (1 - x) \cdot x^2 = x^2 - x^3$ is to be maximized on the interval $[0, 1]$. We have $A'(x) = 2x - 3x^2$. Therefore $A'(x) = 0$ when $2x - 3x^2 = 0$ or when $x(2 - 3x) = 0$. This has two solutions: $\bar{x} = 0$ or $\bar{x} = \frac{2}{3}$.

We next have $A(0) = A(1) = 0$ and $A(\frac{2}{3}) = \frac{4}{27}$. Therefore the rectangle of maximum area is obtained by locating the point P at $(\frac{2}{3}, \frac{4}{9})$. ∎

Figure 2.48

EXAMPLE 5

Postal regulations limit the size of a rectangular package to a sum of 84 in. for its length and its girth (see Figure 2.49). What are the volume and the dimensions of the largest package you can mail if it is square on each end?

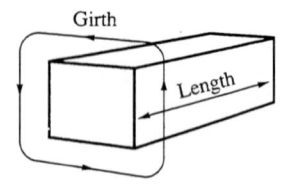

Figure 2.49

Solution

1. The variables in this problem are the length l of the package and the width w of the (square) end of the package.
2. We are to maximize the volume $V = w \cdot w \cdot l = w^2 l$ of the box.
3. Having two variables, we need a constraint equation. This constraint, which is supplied by the post office, is that the length plus the girth is 84 in. Since the length is l and the girth is $w + w + w + w = 4w$, we have the constraint equation $4w + l = 84$ or $l = 84 - 4w$ (with $0 \leq w \leq 21$, since $l = 84 - 4w \geq 0$).

158   Chapter 2   The Derivative

4. Therefore the function (of one variable) to be maximized is

$$V(w) = w^2 \cdot l = w^2(84 - 4w) = 84w^2 - 4w^3$$

Since $V'(w) = 168w - 12w^2 = 12w(14 - w)$, then $\bar{w} = 14$ and $\bar{w} = 0$ are solutions.

5. Evaluating V at the endpoints $w = 0, 21$ and at $\bar{w} = 14$, we find that $V(0) = V(21) = 0$ and that $V(14) = 84(14)^2 - 4(14)^3 = 5,488$ in.^{3} is the maximum volume possible.

The dimensions of this package are $w = 14$ in. and $l = 84 - 4(14) = 28$ in. ∎

EXAMPLE 6

Suppose that you are building a home, and as a step toward energy efficiency, you want to install an attic fan that will exhaust air through the attic. This demands a rectangular opening at the end of the attic. To minimize the back pressure on the fan when it is in operation, as well as to provide maximum ventilation of the attic at all times, you must tell your building contractor the dimensions of the opening so as to maximize its area. If the dimensions of the attic are as indicated in Figure 2.50, determine the largest opening that can be made.

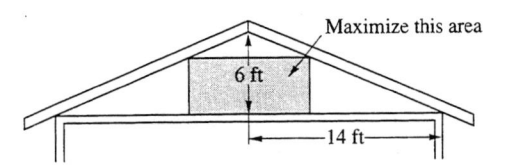

Figure 2.50

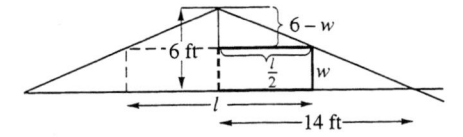

Figure 2.51

Solution

1. As Figure 2.50 indicates, the variables under your control are the length and width of the rectangular opening. Call these l and w, as in Figure 2.51.

2. You wish to maximize the area $A = lw$ of the opening.
3. The needed construction must somehow depend on the rise (6 ft) and run (14 ft) of the roof. One approach that uses all the available information is to think in terms of similar triangles. From Figure 2.51, we see that $(6 - w)/(l/2) = \frac{6}{14}$, because the ratios of the corresponding sides of the triangle formed by the roof, $\frac{6}{14}$, and the triangle formed above the opening, $(6 - w)/(l/2)$, must be the same. Solving for l, we have $l = [14(6 - w)]/3$.
4. Using this constraint equation in the formula for area in step 2, we then have $A(w) = l \cdot w = \frac{14}{3}(6 - w)w$ or $A(w) = \frac{14}{3}(6w - w^2)$.

5. Therefore, $A'(w) = \frac{14}{3}(6 - 2w)$ equals zero when $\bar{w} = 3$. Because $A''(w) = \frac{14}{3}(-2) < 0$, this solution yields the maximum area since the endpoints $w = 0$ and $w = 14$ yield zero area.

Returning to the constraint equation in step 3, we see then that $l = [14(6 - 3)]/3 = 14$. The opening should be 3 ft high and 14 ft wide. It has an area of $A(3) = 42$ ft². ∎

Subsection C: Optimum Allocation of Resources

In the previous subsection we saw how volume could be maximized subject to a geometric constraint (Example 5). Now let us maximize volume subject to an economic constraint.

EXAMPLE 7

A building contractor wishes to build a portable on-site shed in which to store tools and materials overnight at a construction site. The shed will be twice as long as it is wide and have a flat roof. Materials will cost $5 per ft² for the sides, $8 per ft² for the roof, and $10 per ft² for the base. The builder wishes to spend a total of $3,000 to build the shed. What dimensions should be used so as to build a shed of largest possible volume?

Solution

1. We first imagine the shed as in Figure 2.52.
2. What are the underlying variables? They are the length l, height h, and width w of the shed. We label Figure 2.52 accordingly.
3. The builder wishes to maximize the volume $V = lwh$ of the shed.
4. Since the building must be twice as long as it is wide, we have the constraint $l = 2w$. From step 3, $V = 2w \cdot w \cdot h = 2w^2 h$. This does not eliminate enough variables, however. Another constraint equation must be found. Reading the problem again, and taking into account the cost of materials and the fact that opposite sides of the shed have the same area, we see that

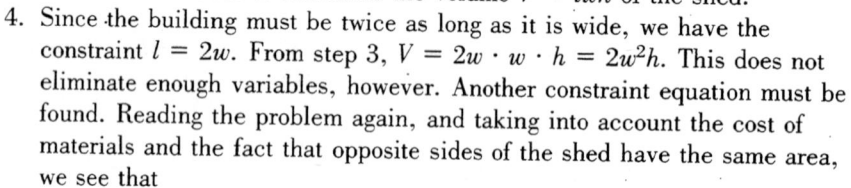

$$5(2lh) + 5(2wh) + 10(wl) + 8(wl) = 3{,}000 \tag{1}$$

since the left side of this equation represents the total cost of materials, which must, on the other hand, equal 3,000. Since $l = 2w$, we can rewrite Equation 1 as

$$20wh + 10wh + 20w^2 + 16w^2 = 3{,}000$$
$$2wh + wh + 2w^2 + 1.6w^2 = 300$$
$$3wh + 3.6w^2 = 300$$
$$h = \frac{300 - 3.6w^2}{3w} = \frac{100}{w} - 1.2w$$

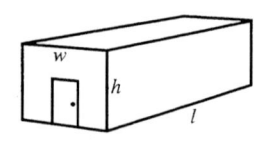

Figure 2.52

Having solved the equation for one of the variables, h, in terms of the other, w, we can say that

$$V(w) = 2w^2 h = 2w^2 \left(\frac{100}{w} - 1.2w\right) = 200w - 2.4w^3$$

as a function of w alone.

5. We have $V'(w) = 200 - 7.2w^2$ and $V''(w) = -14.4w$, which is negative for positive width w. Solving $V'(w) = 0$, we obtain $\bar{w} = \sqrt{200/7.2} \simeq 5.27$ ft as the width yielding a maximum volume, since $V(0) = 0$ at our only endpoint $w = 0$.

What of the length and height? Since $l = 2w$, the optimum length is $l = 10.54$ ft. Since $h = (100/w) - 1.2w$, the optimum height is $h = (100/5.27) - 1.2(5.27) = 12.65$ ft. The (maximum) volume of the resulting shed is $V(5.27) = 702.65$ ft^3 at a cost of \$3,000, or \$4.27 per ft^3. ∎

Since building materials are sold in standard dimensions, the contractor would probably adapt these optimum dimensions to more standard dimensions of $w = 6$, $h = 8$, and $l = 12$, obtaining a volume of $V(6) = 576$ ft^3. Some volume would be lost, but the cost would only be \$2,736 for a better-proportioned toolshed averaging \$4.75 per ft^3.

A word about the use of mathematical models and solutions in business and economics is appropriate here. The ordinary units of business are discrete, whole units. A company does not produce 150 and *one-half* washing machines or earn $\sqrt{200}$ dollars profit. At the same time we cannot apply calculus without using fractions and irrational numbers. We can deal with this matter in two ways. In a very-large-scale enterprise dealing in millions of units, it is perfectly reasonable to speak of 1.76254 million units. On a small scale we pretend to deal in fractional units and then adapt any optimum solution to whole units. While subtleties could arise in doing so indiscriminately, if the scale is in fact small, consequential errors will not practically arise. This view is implicit in the next example, where whole units of goods are the reality, but the mathematical model for the solution "pretends" that fractional units exist.

An everyday question faced by any merchandiser, large or small, is this: "When I reorder goods to replace those that are being sold, how large an order should I make?" Let us analyze this problem for the sale of a single product over a one-year period, where the total number of units of the item to be sold is known from experience. This is called an inventory problem, since it is used to determine the most efficient supply (inventory) of goods to be kept on hand to meet continuing sales.

EXAMPLE 8

Cecilia has been operating the Cheese Villa, specializing in quality European cheeses, for some years now. She knows that she can expect to sell 18,000 lb of cheese in the next year. Thus she could simply order 1,500 lb of cheese each month; however, having to store and maintain the cheese at a cost of

$0.50 per pound per year, she could cut costs by keeping a smaller supply on hand. Still, she would then have to order cheese more frequently and this too has a cost, since (besides what she must pay for the cheese) she must also pay $50 in handling and bookkeeping charges for each order. How large an order should she make to minimize her total costs of ordering and maintaining her inventory of cheese?

Solution

1. The key variable in this problem is

$$x = \text{size (in pounds) of each order}$$

   Because order size is the only variable, we should not expect to need a constraint equation. Note that $0 \leq x \leq 18{,}000$.

2. The quantity to be minimized is

   $T(x) =$ total inventory costs for the year if orders of size x are placed

   We can initially analyze this function in a "literary" form as

$$T(x) = (\text{handling cost per order})(\text{number of orders}) \qquad (2)$$
$$+ (\text{storage costs per pound})(\text{number of pounds in storage})$$

3. Let us put this equation in mathematical form. Since 18,000 lb of cheese are bought in a year, then if each order is of size x lb, Cecilia will place $18{,}000/x$ orders in 1 year. Her total ordering cost will be

$$50\left(\frac{18{,}000}{x}\right) = (\text{handling cost per order})(\text{number of orders}) \qquad (3)$$

   since each order has a handling charge of $50.

   Going further, we already know that it takes $0.50 to store 1 lb of cheese for 1 year. Since the number of pounds in storage at any time will vary, we *make the assumption* that over the course of the year there will always be an *average* of $x/2$ lb in storage at any time (one-half the size of each order). Storage costs for the year will then be

$$0.5\left(\frac{x}{2}\right) = \frac{x}{4} \qquad (4)$$

   Using the algebraic statements in Equations 3 and 4 to replace the "literary" statements in Equation 2, we obtain a formula for the function T that measures inventory costs for orders of size x,

$$T(x) = 50\left(\frac{18{,}000}{x}\right) + 0.5\left(\frac{x}{2}\right) = \frac{900{,}000}{x} + \frac{x}{4}$$

4. To minimize this function, we compute

$$T'(x) = \frac{-900{,}000}{x^2} + \frac{1}{4}$$

   and

$$T''(x) = \frac{1{,}800{,}000}{x^3}$$

Note that $T'' > 0$ for $x > 0$, so that any extreme point we find will yield a minimum inventory cost. Now $T'(x) = 0$ when

$$\frac{-900{,}000}{x^2} + \frac{1}{4} = 0$$

Writing this equation as $4(900{,}000) = x^2$, we have $\bar{x} \approx 1{,}897.366$ lb as the optimum order size. Cecilia's yearly inventory cost will be $T(1{,}897) = \$948$ per year, and she will place approximately $18{,}000/1{,}897.366 \approx 9.49$ orders over the year.

Without this analysis, Cecilia might be tempted to act more simply but at higher cost. She might, as some businesses do, reason as follows: "Since I am selling 18,000 lb a year, or 1,500 lb a month, I will reorder 50% of my monthly sales when my stock shrinks to 50% of my monthly sales volume." This means that Cecilia would place 24 orders of 750 lb each at a cost of $50 per order and, in addition, have an average inventory of $750 + \frac{1}{2}(750)$ lb in storage throughout the year. Cecilia's inventory costs would then be $50(24) + 0.5(1{,}125) = \$1{,}762.50$, a considerably higher cost than the minimum obtainable.

Finally, we leave it to the reader to see, using Figure 2.53, that faced with a choice between placing nine orders or ten, Cecilia should place nine.

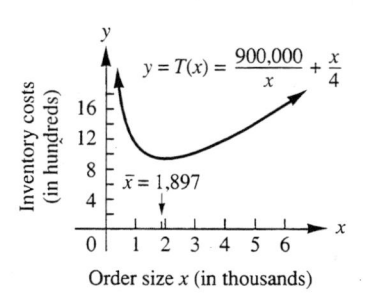

Figure 2.53

Our concluding optimization application in this subsection deals with a common difficulty. A business has a number of production units and each produces a certain return. Management naturally thinks in terms of increasing the number of productive units so as to increase the total return. However, one is then faced with a decline in efficiency of each production unit due to unavoidable causes—for example, a simple overcrowding of the units. Therefore management must seek that number of production units that will *maximize* total return, taking into account declining efficiency as the number of units increases.

EXAMPLE 9

A small New York State winery is planning to plant an additional acre of grapevines. From past experience the vineyard owner knows that if 1,500 vines are planted per acre, each vine will yield 8 liters of wine per vine at maturity in a normal year. The owner also knows that for each 100 additional vines planted on the same acre of land, overcrowding of the vines will reduce the yield per vine by 0.3 liter. How many vines should be planted on this acre so as to maximize yield?

Solution

1. The underlying variable in this problem is the number of vines to be planted on the acre. Let x be this number of vines. Since there appears to be only one variable in this problem, we should not need a constraint equation.
2. The quantity to be optimized is the volume of wine to be produced. Let W = yield (in liters) of the acre of vines. Our task is to write W as a function of x while taking account of all the given information.
3. We reason as follows. The total yield W depends directly on the yield per vine and the number of vines. Therefore

$$W = \text{(number of vines)(yield per vine)}$$
$$= x \text{ (yield per vine)}$$

   since x vines are to be planted. We still need to determine the yield per vine in terms of x, the number of vines planted.

   We are given that each vine yields 8 liters of wine, if the vines are not crowded, but that this yield is reduced by 0.3 liter for each 100 vines planted *above* 1,500 vines. Therefore the yield per vine is $8 - 0.3[(x - 1,500)/100]$ liters, because $x - 1,500$ is the number of vines exceeding 1,500, and division by 100 reduces this figure to the stated terms to which the factor 0.3 applies. Consequently,

4. $$W(x) = x\left[8 - 0.3\left(\frac{x - 1,500}{100}\right)\right], \quad x \geq 1500$$

   is the function expressing total yield as a function of the number of vines planted.
5. A bit of algebra shows that $W(x) = 12.5x - 0.003x^2$, so that $W'(x) = 12.5 - 0.006x$ and $W''(x) = -0.006 < 0$. The graph of W is thus a parabola opening down; hence $W(x)$ is a maximum when $\bar{x} = 12.5/0.006 \simeq 2,083$ vines, yielding $W(2,083) \simeq 13,020$ liters of wine.

The vintner should plant this number of vines on the acre of land so as to maximize yield. Since an acre has 43,560 ft^2, this means that 20.9 ft^2 of space should be allocated for each vine (with the vines being spaced roughly 4.5 ft apart in each direction). ∎

Subsection D: Optimum Profit and Revenue

In this subsection we discuss a few elements of what economists call the **theory of the firm**. These elements deal with the costs incurred and the revenues gained in the production of some item offered for sale and with the search for maximum profit. In the course of this discussion we will derive the general principle that maximum profit occurs at that point where marginal revenue equals marginal cost.

The sale of an item for profit centers around its price p. Let us write $p = P(x)$, where x is the level of production of the item. This is a sensible model because in an unregulated and competitive economic system, the price of an item is determined by the number of units x of the item available. Such a function is called a **demand function.**

A manufacturer's concern is not so much the price of an item but rather the company's total revenue—the dollars coming back to the company from the sale of x items and ultimately its profit. Let

$$R(x) = \text{total revenue from the sale of } x \text{ items}$$

Let us derive a basic equation. Since

$$\text{Total revenue} = (\text{price of each item})(\text{number of items sold})$$

then

$$R(x) = px$$

when x units are produced and sold. In terms of a demand function $p = P(x)$, this means that

$$\boldsymbol{R(x) = xP(x)}$$

Three basic examples of revenue and demand functions are indicated in Figure 2.54.

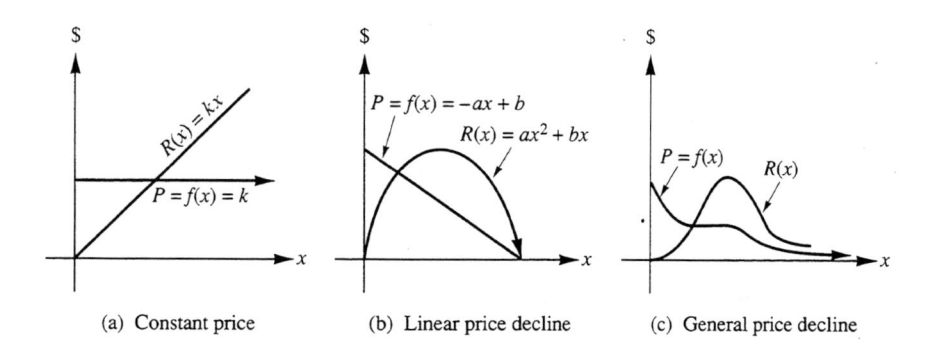

Figure 2.54

(a) Constant price
(b) Linear price decline
(c) General price decline

Before going further to considerations of profit, let us analyze a particular example.

EXAMPLE 10

The Topspin Manufacturing Company has introduced a new racket into the competitive market for tennis rackets. In one city the racket is priced at $50 and sells at the rate of 125 rackets a week. In a second city the racket is priced at $45 and sells at the rate of 140 per week. Assume that the demand function for this racket is linear as in Figure 2.54(b) and that these two trial markets are reliable indicators of price and demand.

a. Write the demand function for this product as a linear function.
b. Write a formula for the revenue function.
c. At what price will revenue be maximized?

Solution

a. While the data are given in terms of demand as a function of price, remember that a demand function represents the reverse correspondence.

Imagine, then, a production/sales level of x items at price p. We are to write $p = f(x) = ax + b$ as a linear function based on the given data.

Since $50 = f(125)$ and $45 = f(140)$ define points on the graph of a straight line, this line has slope

$$\frac{50 - 45}{125 - 140} = -\frac{1}{3}$$

and equation $y - 50 = -\frac{1}{3}(x - 125)$. Therefore

$$p = f(x) = -\frac{1}{3}x + \frac{125}{3} + 50$$

or

$$p = \frac{-x + 275}{3}$$

b. Consequently, the total revenue function is

$$R(x) = xp = \frac{-x^2 + 275x}{3}$$

c. Therefore, optimum revenue is obtained when

$$R'(x) = \frac{-2x + 275}{3} = 0$$

or when $\bar{x} = 137.50$.

Since $R''(x) = -\frac{2}{3} < 0$, a production level of $\bar{x} = 137.5$ rackets yields maximum revenue. Since $p = (-x + 275)/3$, the price per racket that yields maximum revenue is $p = (-137.5 + 275)/3 \approx \$45.83$. At this price, the Topspin Company will earn a total revenue of

$$(\text{price})(\text{demand}) = (45.83)(137.5) \approx \$6,302 \quad \blacksquare$$

A company that sought only to maximize its total revenue (return on investment) would be considered unrealistic, since it must also balance these revenues against the costs of production. Such production costs will have two important components: (1) fixed costs, such as land, buildings, and equipment, and (2) variable costs, such as labor, materials, and research and development. Let us analyze the relationship between costs, revenue, and profit. It doing so, we will uncover a general principle of economic behavior using calculus. One consequence of this principle is that the production level at which maximum profit occurs is independent of fixed costs.

To begin, we will set

$C(x)$ = cost to produce x items
$R(x)$ = revenue from the production and sale of x items
$P(x)$ = profit from the production (and sale) of x items

Since profit is revenue less costs, we have $P(x) = R(x) - C(x)$. We know that maximum profit can occur only at a critical point $\bar{x}$ of the function P. It is in the nature of the revenue and cost functions we encounter that $P''(\bar{x}) < 0$ (precise and general conditions on the cost and revenue functions that ensure this can, with some complexity, be stated), and therefore a critical point $\bar{x}$ of P will typically yield maximum profit. More importantly, however, we wish to observe the following general law of economic behavior:

At a production level $\bar{x}$ yielding maximum profit,

Marginal revenue = Marginal cost

This is because at the critical point $\bar{x}$ we will have $P'(\bar{x}) = R'(\bar{x}) - C'(\bar{x}) = 0$, or

$$R'(\bar{x}) = C'(\bar{x})$$

and these derivatives measure marginal revenue and marginal costs.

EXAMPLE 11

In reckoning its profits the Topspin Company of Example 10 must also consider its costs. Each week it plans to produce at least 100 rackets at a fixed cost of $1,400 and at a variable cost of $(x^2 - 100x)/12$ for the production of x rackets, with $x \geq 100$.

a. Determine how many rackets should be produced in order to maximize its profits.
b. What will its maximum profit be?
c. What is its marginal cost to produce this profit? Its marginal profit?

Solution

a. In Example 10, we determined that the Topspin Company can expect a total revenue of

$$R(x) = \frac{-x^2 + 275x}{3}$$

for the production of x rackets. According to the preceding information, the cost to produce x rackets is

$$C(x) = \text{fixed costs} + \text{variable costs}$$
$$= 1{,}400 + \frac{x^2 - 100x}{12}$$

Therefore, profit is

$$P(x) = R(x) - C(x)$$
$$= \frac{-x^2 + 275x}{3} - \left(\frac{x^2 - 100x}{12} + 1{,}400\right)$$
$$= \frac{-5x^2}{12} + 100x - 1{,}400$$

We could optimize this profit function directly. Instead let us use the general principle just obtained and determine when marginal costs equal marginal revenue. That is, let us solve

$$R'(x) = C'(x)$$

or equivalently

$$-\frac{2}{3}x + \frac{275}{3} = \frac{2x - 100}{12}$$

We reduce this equation to

$$-\frac{5x}{6} = -100$$

with solution $\bar{x} = 120$ rackets.

b. Note that $R''(x) = -\frac{2}{3} < 0$ and $C''(x) = \frac{1}{6} > 0$, and thus $P'' = R'' - C'' < 0$. We conclude that a production level of $x \simeq 120$ rackets will yield a maximum profit of

$$P(120) = -\frac{5}{12}(120)^2 + 100(120) - 1{,}400 = \$4{,}600$$

Returning to the demand function for this problem (Example 10), we see that at a production level of $x = 120$ rackets each racket will sell for $f(120) = (-120 + 275)/3 = \$51.67$. Therefore, given this cost of production, the Topspin Company should raise its price to $51.67. At this price its total revenue, $R(120) = \$6{,}200$, will be smaller than its maximum revenue, $R(137.5) = \$6{,}302$, obtained in Example 10, but its profit ($4,600) will be greater [since $P(137.5) \simeq \$4{,}472.39$].

c. The marginal cost (= marginal revenue) to produce each racket at this production level will be $C'(120) = [2(120) - 100]/12 = \$11.67$. Its *marginal profit* will then be $51.66 - $11.66 = $40 on the 120th racket sold. ∎

Notice from this example that marginal costs are independent of fixed costs since the derivative of a constant is zero. Hence a point of maximum profit is independent of fixed costs.

What if R' never equals C'? If $R' > C'$, profits go on increasing forever once revenue exceeds costs. For example, if it costs $8 to produce an item that sells for $10, with a fixed production cost of $500, then $R(x) = 10x$ and $C(x) = 8x + 500$. After 250 items are produced, profits continue to accumulate at the rate of $R' - C' = \$2$ per unit. In the opposite case, if marginal revenue never exceeds marginal cost, a profit will never be made.

Exercises 2.6

Subsection A: Optimum Numerical States

1. Find two nonnegative numbers whose sum is 35 and whose product is a maximum.

2. Find two positive numbers whose product is 36 and whose sum is a minimum.

3. Find out if there exist two positive numbers whose quotient is 10 and whose product is a maximum.

4. Do there exist two positive numbers whose product is 20 and whose difference is a minimum?

5. Find three nonnegative numbers whose sum is 15 and whose product is a maximum when one of these three is twice the sum of the remaining two.

6. Find two negative numbers whose product is 25 and the sum of whose squares is a minimum.

Subsection B: Optimum Geometric States

7. Determine the rectangle of
   a. Least perimeter having an area of 169 in.^{2}.
   b. Largest area having a perimeter of 50 cm.

8. Cut a piece of wire 30 in. long into two pieces and bend each piece into a square. How should you cut the wire so that the sum of the areas of the two resulting squares is a minimum? A maximum? (See Figure 2.55.)

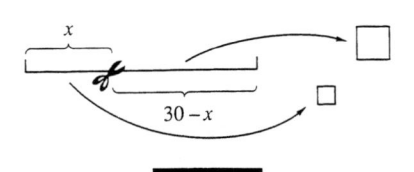

Figure 2.55

9. Cut a piece of wire 20 in. long into two pieces. Bend one piece into a square and the other into a circle. How should you cut the wire so as to
   a. Obtain the largest total area from the square and circle?
   b. Obtain the smallest total area from the square and circle?

   [*Hint:* If x is the length of wire bent into a circle and r is the radius of this circle, then $2\pi r = x$.]

10. Consider a square of length 1 on each side in the Cartesian plane (see Figure 2.56). Where would you place the point P so that
    a. The sum of the areas of the rectangles R_1 and R_2 is a maximum?
    b. The product of these two areas is a maximum?

    [*Hint:* The point P lies on the graph of $y = x$.]

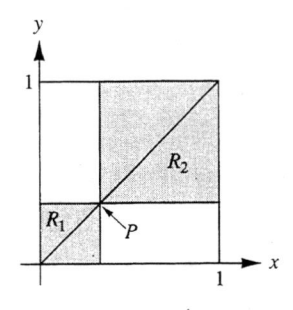

Figure 2.56

11. Consider the drawing shown in Figure 2.57. Where would you place the point x so that the product of the area of the square formed by joining the two triangles R_1 and R_2, and the area of the remaining region R_3, is a maximum?

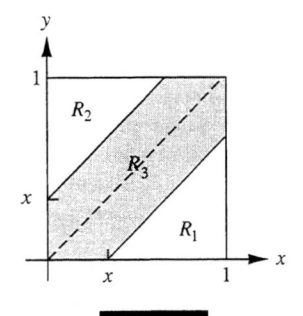

Figure 2.57

12. A box is to have a square bottom, to be open at the top, and to have a volume of 125 in.^{3}. How high should the box be to use the smallest amount of material in making it? [*Hint:* Minimize the total surface area of the box.]

13. What is the length of the shortest ladder that will span a 5-ft-high wall (if the wall is 4 ft from the side of a building) while still resting on the ground and against the building, as Figure 2.58 shows? [*Hint:* Use the idea of similar triangles from Example 6 and the Pythagorean theorem.]

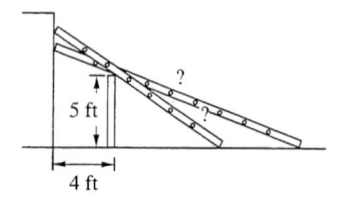

Figure 2.58

14. The local zoo wants to build an enclosure for three zebras that it will acquire next spring (see Figure 2.59). On one side of the enclosure it will build a barn that the zebras can enter whenever they wish. The barn will be 20 yd long. The remainder of the enclosure will consist of 150 yd of fencing. What should the dimensions of the enclosure be so as to enclose a maximum area?

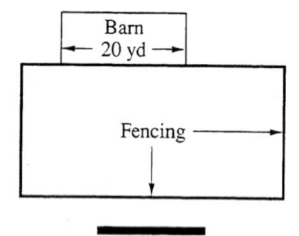

Figure 2.59

15. A canning factory would like to determine the dimensions of a cylindrical can of volume 35 in.^{3} that will use the least amount of material in its construction. Keeping in mind that the circular top and bottom of the can will have to be cut from a square piece of metal, find these dimensions (see Figure 2.60).

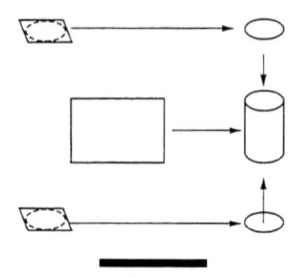

Figure 2.60

16. The roof of a building is to be 30 ft long, as indicated in Figure 2.61. What should the height of the attic and width of the building be so as to obtain an attic of maximum cross-sectional area? [*Hint:* The algebra will be simpler if you maximize the *square* of the area of the attic.]

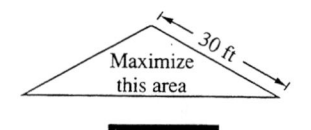

Figure 2.61

17. Gerald has just bought a home in a new subdivision and wishes to have a rectangular swimming pool built in the backyard. Since the builder destroyed all the trees and other natural vegetation, it doesn't matter to Gerald where the pool is put, but he does not want to use more than a third of the total yard area of 1,560 ft^2 for the pool site. The pool is to have concrete walkways along each side that are 3 ft wide and walkways on the two ends that are 4 and 7 ft wide, respectively. What should the dimensions of the pool be so as to have maximum area and so that the area of the pool and its walkways equals one-third of the yard? (See Figure 2.62.)

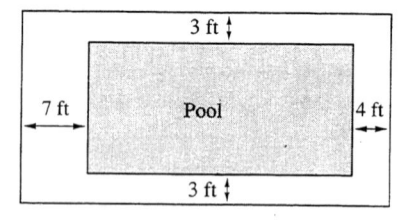

Figure 2.62

Challenge Problem

18. A bridge is to be built to span a waterway 300 ft wide and to be supported on each side of the waterway by a 40-ft-high pier, as indicated in Figure 2.63. Where should the foot (point *A*) of the bridge roadway be placed so that the total length of the bridge is a minimum?

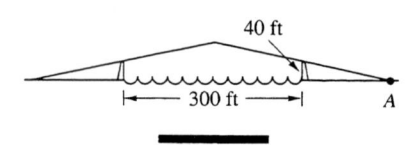

Figure 2.63

Subsection C: Optimum Allocation of Resources

19. A manufacturer wishes to specify the dimensions of a box that will be used to package a particular product. The box is to have a volume of 125 in.^{3}, a square bottom, and a closed top.

  a. If the cost of materials is $0.08 per ft^2 for the sides and top and $0.16 for the bottom, what dimensions should the box have so as to minimize its cost?

  b. If the cost of materials for the bottom is known to be twice that of the sides and top, but the cost of materials is uncertain and may vary from week to week, can the manufacturer still specify the dimensions of the least costly box, no matter what the cost of the basic materials will be? [*Hint:* Let $k =$ cost per ft^2 of the top and side material; the bottom will then cost $2k$.]

20. An oil storage tank is to be built so as to store $250{,}000\pi$ ft^3 of oil. The steel used to construct the sides and top of the tank will cost $50 per ft^2. However, the base of the tank is of a more involved construction. Utilizing steel, concrete, and other materials, it will cost $300 per ft^2 to construct. What dimensions should the tank have so as to minimize cost? (See Figure 2.64.)

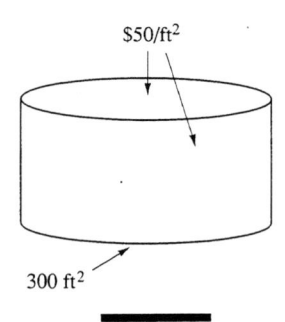

$50/ft^2

300 ft^2

Figure 2.64

21. A publishing company expects to sell 240,000 copies of a new novel during the next year at a fairly even rate of sales. It costs the company $0.80 per year to store one book in its warehouse and $1,400 for the start-up of each printing run. How many copies of the book should be printed in each printing run so as to minimize the year-long inventory costs? How frequently will the company be making a printing?

22. Referring to Exercise 21, and assuming that an equal number of books are removed from the publisher's warehouse each day,

  a. How many books will be on hand when the next printing run is completed and the books are stored in the warehouse together with those already there?

  b. Suppose that the publishing company instead does a printing run of 50% of its monthly sales each time its stock of books shrinks to 50% of its monthly sales volume. What will its inventory costs be? What if it orders 75% of its monthly volume when its stock shrinks to 25% of its monthly volume?

23. A farmer must apply 400,000 lb of herbicide on his soybean crop during its four-month growing season. The herbicide will be applied in a roughly equal amount, moving from one acre to the next each day. In addition to the cost of the herbicide, the local co-op charges a $60 fee each time it sends herbicide out to his farm. Until it is used, the herbicide must be stored away from the weather, so that the farmer must figure in a storage cost of $20 per ton over the four-month season. How large an order of herbicide should he place so as to minimize his inventory and delivery costs?

24. A pharmacist has a large and varied inventory. One of her consistent needs is gel capsules, of which she uses 150,000 per year at a uniform rate. Each capsule costs $0.01, and it costs $0.05 to store a capsule for one year. Reorder costs are $12 per order. What is her optimum reorder size?

25. A tennis club offers memberships at $325 and is able to enroll 85 members. It decides to lower the membership fee for all members by $10 for each additional 5 members who join the club and to restrict the total membership to 140 members. What level of membership fee maximizes its total revenue?

26. A local bowling alley decides to lower its prices to attract more business. At $3 per hour it attracts 75 customers who bowl 2 hours each. For each $0.25 decline in price, it attracts an additional 5 customers and all customers will bowl an extra quarter hour. At what level should the bowling alley set its price so as to maximize its total revenue? [*Hint:* Let x equal the number of $0.25 price declines. Each customer will bowl $2 + (x/4)$ hours at $3 - 0.25x$ dollars per hour, and there will be $75 + 5x$ customers.]

27. A fast-food franchiser plans to open several outlets in a city of 750,000 people. Marketing surveys predict that with 20 franchise outlets, each will attract $80,000 in business each month, but that each additional outlet will reduce this amount by $3,200 per month. How many outlets should be built so as to maximize the total revenue from all outlets?

28. A bacteriologist is using recombinant DNA techniques to produce a strain of bacteria that in turn produces a useful by-product. A culture of 5,000 bacteria will produce 10^{-3} g *per* bacterium of the by-product per month, but this will be reduced by 2% (for each bacterium) for each additional 100 bacteria in the culture. What is the optimum size for the bacteria culture?

29. A roadside campground rents 65 campsites at $8 per night each for overnight camping. The owner guesses that it will be necessary to reduce the charge by $0.50 per campsite for every 5 additional campsites that are put into the already crowded facility. What is the optimum number of campsites that the campground should have?

30. A company employs 25 salespeople to promote its new product in the southwestern United States. Each salesperson generates $160,000 in sales per month at a salary of $2,000 a month and travel costs of $500 per month. The company believes that for each additional salesperson it hires, the sales of existing salespeople will drop by $3,000 per month. What is the optimum number of salespeople that it should employ? [*Hint:* Do not forget to consider the salesperson's salary and travel costs in your calculations.]

Challenge Problems

31. An oil rig 50 mi off the coast of Louisiana is to deliver oil to the strategic petroleum reserve, a vacant underground salt dome (see Figure 2.65). A pipeline runs from the salt dome to the coast with a terminal 80 mi west of the oil rig. It is proposed to build a pipeline east along the coast with a terminal closer to the rig. It is estimated that over the lifetime of this project the oil can be moved from the rig to the coast at a cost of $60 per ton and moved thereafter via the pipeline at a cost of $20 per ton. Where should the terminal of the coastal pipeline be located?

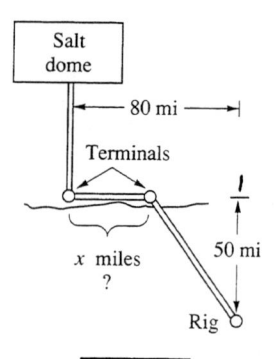

Figure 2.65

32. The human body provides myriad examples of optimal design. In its evolution over several million years, nature long ago solved the following exercise: Where should a branch artery serving point A in Figure 2.66 connect to a main artery between points B and C, where C is 10 mm away from A?

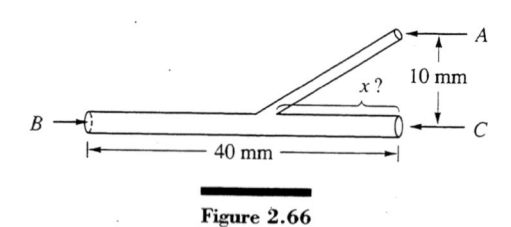

Figure 2.66

The measure of efficiency here is not dollars but energy, supplied by the body's heartbeat. If the radius of the branch artery is half that of the main artery, the cost in energy used to force the blood through the more restricted branch artery is known to be 16 times that of the main artery. At what point should the branch artery meet the main artery so as to minimize the total energy used in pumping blood from point B to point A?

Subsection D: Optimum Profit and Revenue

33. Suppose that a producer can sell 2,100 copies of a product at a cost of $10 and 3,000 copies at a cost of $8. What is the demand function for this product, if this demand function is assumed to be linear?

34. A company has spent $125,000 on land, buildings, and machinery to produce a particular product. The firm finds that it costs $4 in materials and labor to build each of the first 900 items, but thereafter that this cost is reduced by $0.20 for each additional 1,000 items produced. What is the company's total cost function for the production of x items, where $x \geq 900$? What is its marginal cost of production when it produces 5,400 items?

35. Tony's Pizza has determined that it can sell 150 "House Special" pizzas at a price of $9 each and 120 at a price of $10 each (per week). Tony determines that it costs him $(-x/150) + 7$ dollars per pizza to make x pizzas, where $100 \leq x \leq 200$. Assuming that the demand function for pizzas is linear, how should Tony price his pizza so as to maximize his profit? How many will he sell?

36. Marsha wants to produce pastry as good as that found in the neighborhood bakeries of her native Austria, from which she recently emigrated. But quality pastry demands quality ingredients at much higher costs since they are difficult to obtain in a market dominated by processed food. She devises a quality index x for ingredients, ranging from 0 for the quality of ingredients ordinarily available to 1 for the highest-quality ingredients she can find. From this Marsha determines that her cost per ounce of finished pastry is $25x^2 + 35$ cents per ounce. Because quality products do attract buyers despite the high price, Marsha finds that her sales *revenue* per ounce is $-64(x - 0.8)^2 + 40$ cents per ounce.

   a. What is Marsha's profit function at quality index x?

   b. What quality index maximizes her profits per ounce?

37. The Brown Bean is a coffee house that offers a variety of house-roasted bean coffees. The owner has to decide how many different varieties of coffee to stock. If $x \geq 1$ is the number of different kinds of coffee stocked, the owner estimates total revenue to be $10(x + 8)^{3/2}$ dollars per week. However, if coffee beans are not used to make coffee within a week or so, the flavor loss is so great that the owner must discard the beans at a total loss. Thus the coffee house's costs are quite high if there are too many varieties of coffee on hand. The owner estimates the costs due to such losses to be $2x^2$ dollars per week. How many varieties of coffee should be stocked so as to maximize revenue?

38. The demand function for telephone service in a certain locale is
$$P = 28 - \frac{x}{5} \text{ cents per minute}$$
for x minutes of service. It costs the telephone company 17 cents to provide 1 minute of phone service.

   a. What is the profit function for this telephone service?

   b. What price maximizes revenue?

   c. If the government adds a tax of 1.3 cents per minute of phone service, what price for service maximizes revenue to the phone company (which must pay the tax out of its profits)?

   d. Should it pass on the entire tax to its customers?

Challenge Problems

39. An electrical utility has a monthly demand function of
$$P = 5.5 - \frac{x}{10,000} \text{ cents per kilowatt-hour (kwh)}$$
per customer. The utility company also has a fixed cost of $15 per customer and variable costs of 4.2¢ per kwh.

   a. Find its profit function per customer.

   b. What price per kwh will maximize its revenue per customer?

   c. If an increase in coal costs to the utility raises its production costs to 4.35¢ per kwh, should it pass all of the increase in costs on to its customers? Why or why not? [*Hint:* Find the demand level at the higher price.]

40. Let $C(x)$ denote the cost to a manufacturer in order to produce x items. Then the function
$$A(x) = \frac{C(x)}{x} = \frac{\text{total costs}}{\text{total number of items}}$$
is the *average cost per item* to the manufacturer. Determine the level of production yielding the lowest average cost (in terms of marginal costs), assuming that $C''(x) > 0$. [*Hint:* After calculating $A'(x)$, replace the term $C'(x)$ that appears in the calculation by its equal (in terms of $C(x)$ and x) at the optimum level to verify that the optimum value is indeed a *minimum*.]

Chapter 2 Summary

The power rule: $D_x x^n = nx^{n-1}$

The chain rule: $D_x f(g(x)) = f'(g(x))g'(x)$

The product rule: $D_x[f(x) \cdot g(x)] = f(x)D_x g(x) + g(x)D_x f(x)$

The quotient rule: $D_x\left(\dfrac{f(x)}{g(x)}\right) = \dfrac{g(x)D_x f(x) - f(x)D_x g(x)}{g(x)^2} \qquad g(x) \neq 0$

The sum and constant multiple rules: $D_x(af(x) \pm bg(x)) = aD_x f(x) \pm bD_x g(x)$

1. Review the meaning of the following phrases: (a) f is increasing (decreasing) *at a point* x; (b) f is increasing (decreasing) *on an interval*. Review how to graph f knowing only f'.

2. If $f'(\bar{x}_1) = 0$ and $f'(\bar{x}_2) = 0$ and $f' \neq 0$ for no other x between x_1 and x_2 and f' is continuous, then the *sign* of f' is constant between x_1 and x_2 and may be found by evaluating f' at any one point between x_1 and x_2.

3. A function may have local maxima and local minima (peaks and valleys) on its graph. These are called *extrema* and can occur only at *critical points* $\bar{x}$: where $f'(\bar{x}) = 0$ or $f'(\bar{x})$ does not exist.

4. The *first* and *second derivative tests* determine which critical points of a function are actually extrema. These may be remembered and applied by following the options on the scheme in Figure 2.67, where, for example, $f'\ +/-$ means that the sign of f' changes from $+$ to $-$ as $x < \bar{x}$ moves to $x > \bar{x}$.

Find CP: $\bar{x}$

$f'(\bar{x}) = 0$ No $f'(\bar{x})$

Second Derivative Test $f''(\bar{x}) = 0$ First Derivative Test

$f''(\bar{x}) < 0$ $f''(\bar{x}) > 0$ $f'\ +/-$ $f'\ -/+$

Max   Min   Max   Min

$f'\ +/+$ or $-/-$

$f''\ +/-$ or $-/+$ $f''\ +/+$ or $-/-$

Inflection   Nothing special

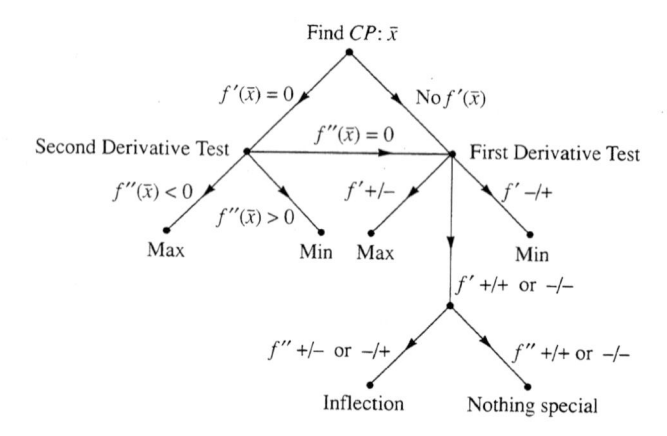

Figure 2.67

5. An *inflection point* is a point on a graph where its concavity changes and is suggested, but not guaranteed, whenever $f'' = 0$ or f'' is undefined.

6. A continuous function on an interval $[a, b]$ has both a largest and a smallest value. These are found by evaluating f at all critical points and at the endpoints a and b and then by choosing the largest and smallest values from among these.

7. In applied optimization we model a system by an *objective function* and determine its extrema so as to find the optimum solution. This is done by introducing letters $x, y, z, \ldots$ to represent those elements that vary in the system. *Constraint equations* can be used to eliminate all but one variable in the solution procedure. It is best to readily introduce notation and then methodically eliminate variables by means of one or more constraint equations.

Chapter 2 Summary Exercises

Find the derivative of the given function in Exercises 1–10.

1. $x - x^2 + 1$
2. π^2
3. $1 - \dfrac{1}{x}$
4. $\sqrt[3]{x}$
5. $(2x + 1)(x^2 + x)$
6. $(x^3 + x)^5$
7. $\dfrac{2x + 1}{2x + 3}$
8. $\left(\dfrac{x + 1}{x + 2}\right)^3$
9. $\dfrac{x^2}{\sqrt{x^2 + 1}}$
10. $\dfrac{x^2 - 1}{x + 1}$

Find the indicated derivative in Exercises 11–22.

11. $D_t(1 + t^2)^3$
12. $D_z z^{0.99}$
13. $f'(x)$ for $f(x) = \dfrac{x^4}{4} + x$
14. $\dfrac{dy}{du}$ for $y = \dfrac{u + 1}{u^2}$
15. y' for $y = \dfrac{w + 1}{w - 1}$
16. $D_x^3\left(x + \dfrac{1}{x} - 1\right)$
17. $f''(3)$ for $f(t) = t^3 + t^2 + t + 1$
18. y''' for $y = \dfrac{x - 1}{x + 1}$
19. $\dfrac{d^2y}{dx^2}$ for $y = (x^2 + 1)(x + 1)^5$
20. $D_x\sqrt{x}(x + 1)$
21. $f'(t)$ for $f(t) = \left(\dfrac{t^2 + 1}{t - 1}\right)^4$
22. $D_x y$ for $y = (1 + 3x^2)^{1/3}$
23. Let $f(x) = \sqrt{x}$ and $g(t) = 1 - t^2$. Find the slope and equation of the tangent line to the graph of $f \circ g$ at $t = \tfrac{1}{2}$.
24. Explain geometrically why the derivative of $x/(x + 1)$ cannot possibly be $(D_x x)/[D_x(x + 1)]$.

25. Let p be the population of catfish in a commercial farming pond that can support a maximum of M catfish. It is known that the rate of growth $G(p)$ of the catfish population at population level p is proportional to p and to $M - p$. Thus $G(p) = kp(M - p)$. At what population level p is the growth G itself most rapid?

26. The cost C and revenue R to a firm that produces and sells x thousand units of its product are given by $C(x) = 10 + 2.5x$ and $R(x) = 3x - 0.2x^2$ in thousands of dollars. Find the marginal cost, marginal revenue, and marginal profit accruing to the firm for the production and sale of 4,000 items.

27. The number of units sold of an item selling at price p is $U(p) = 500 - 0.3p$. Letting the unit interval $0 \le t \le 1$ represent one year, the price p at time t is $p(t) = 4 - 8(t - \tfrac{1}{2})^2$. At what rate are revenues from the sale of this item growing at the end of the first quarter, when $t = \tfrac{1}{4}$?

28. Sketch the graph of a continuous function satisfying the conditions $f(0) = 1$, $f'(1) = 0$, $f'(-1) > 0$, $f(2) = 1$, $f'(0) = f''(0) = 0$, and $f(1) = 2$.

29. Sketch the graph of a continuous function that is increasing at every point on the interval $(0, 1)$ and decreasing on the interval $[1, 2)$.

30. Sketch the graph of a continuous function that is increasing on the interval $[0, 1]$ so that the rate of increase is itself increasing at every point on this interval.

31. Sketch the graphs of two functions, both increasing at every point of an interval, but whose *second* derivatives have opposite signs at each point of the interval.

32. Sketch the graph of a continuous function that has a maximum value at $x = 0$ but $f'(0) \neq 0$ (and where 0 is not an endpoint).

33. Sketch the graph of a continuous function such that $f'(1) = 0$ and $f'(x) < 0$ for all other x.

34. Sketch the graph of a continuous function that has local maxima at the points $x = -1$, 2, and 4.

35. Specify all of the properties of the sign and equality with zero of f, f', and f'' for the graph of $y = f(x)$ shown in Figure 2.68 over the entire real line.

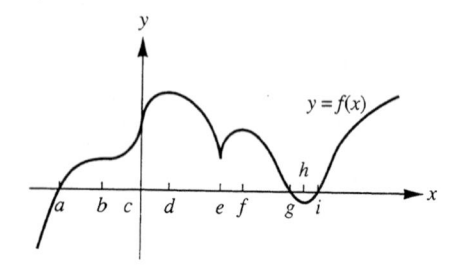

Figure 2.68

Draw the graph of a function whose derivative is given in Exercises 36–41.

36. $f'(x) = x$ on $[-1, 1]$
37. $f'(x) = x^2$ on $[-1, 1]$
38. $g'(x) = (x - 1)x$ on $[-1, 2]$
39. $h'(t) = (t - 1)^2 t$ on $[-1, 2]$
40. $f'(z) = z(z + 1)(z - 1)$ on $[-2, 2]$
41. $f'(x) = \frac{1}{3}$ on $(-\infty, \infty)$

42. Sketch the graph of a function y that satisfies the differential equation $y' = 0.10y$, where $y(0) = 1$ (Section 2.4).

43. Draw the "best," and then the "worst," graph you can imagine of a continuous function f that meets the following conditions: f has a local max at $f(0) = -1$, $f'(1) = 0$, $f'(2)$ is $+\infty$, and $f(3)$ is a local min.

44. Draw the graph of a continuous function such that $f'(0) < 0$, $f''(0) > 0$, and $f'(1) = f''(1) = 0$.

Find all local max and local min values of the functions given in Exercises 45–50. Sketch the graph of each function, and indicate where it is concave up and concave down.

45. $f(x) = x^3 - x - \frac{1}{2}$
46. $g(x) = \frac{x^4}{4} - 4x^2 + 1$
47. $h(t) = t^3 - 6t^2 + 9t - 3$
48. $f(x) = x^2 - 80x + 2$
49. $f(x) = x - \frac{4}{x^2}$, $x \neq 0$
50. $g(t) = \frac{t^4}{4} + \frac{17t^3}{3} - 30t^2$

Find all local max and local min values of the functions in Exercises 51–54.

51. $f(x) = x^{2/3} + 1$
52. $h(u) = \frac{u^4}{4} - u^3 + 1$
53. $g(t) = 1 - t^{1/3}$
54. $f(z) = 10z^9 - 9z^{10}$

In Exercises 55–58, find the maximum and minimum value of the function on the indicated interval.

55. $f(x) = \frac{x^4}{2} + 16x$ on $[-1, 3]$
56. $g(x) = x + \frac{1}{x}$ on $\left[\frac{1}{2}, 2\right]$
57. $g(t) = t^2 + 3t + 1$ on $[-2, 1]$
58. $f(u) = u^3 - \frac{3}{2}u^2 - 6u + 1$ on $[-3, 2]$

59. A foreman wishes to mark off a corner of his shop floor with a 20-ft-long rope behind which scrap will be stored until disposed of. Where should he tie the two ends of the rope so as to enclose the largest possible floor area? (See Figure 2.69.)

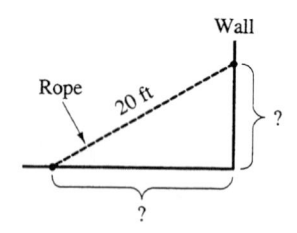

Figure 2.69

60. A brewery has determined that its total sales per month are

$$S = 10{,}000(x + 1.1y - 0.1xy)$$

cases of beer when its cost per case is split between x dollars for advertising and y dollars for added quality of ingredients. It will spend $2 per case on advertising and added quality of ingredients. Which allocation of these two will yield maximum sales; minimum sales? [Hint: Since $x + y = 2$, restrict x to the interval $[0, 2]$.]

61. A company has a contract to produce 8,000 washing machines per year at a uniform monthly rate. The annual storage cost per machine is $75 and the setup cost for a production run is $300. Determine the number of machines that must be produced in each production run so as to minimize total annual costs.

62. The total cost to produce x thousand units is $C(x) = 3x^2 + 5x + 18$ thousand dollars. What level of production will minimize the *average cost* of each thousand units?

63. The demand function for an item is $P = 100 - 5x$ when x items are produced and the average cost to the manufacturer per item is $2.

  a. Write the revenue function for the sale of x items.

  b. Write the profit function using the average cost figure.

  c. At what production level x is profit maximized?

64. Refer to Exercise 63. Suppose that the government imposes a value-added tax of t dollars per item, raising the cost to produce x items to $2x + tx$.

  a. At what production level (this level depends on t) is profit maximized in this case?

  b. Note that total tax revenue to the government accruing from the sale of x items is tx. Use your answer in (a) to determine the tax rate t that will yield maximum tax revenue to the government.

  c. Use (b) to determine the level of maximum profit for the manufacturer and maximum tax revenue for the government.

65. An indoor tennis club charges a $35-per-month membership fee and $5 per hour for court playing time.

  a. The club estimates that it will gain 60 additional members for each $5 decrease in its monthly membership rate. It also expects that each pair of members will use 6 hours of court time per month. At $35 per month the club has 200 members. What monthly membership cost will maximize revenue (each month) and how many members will the club have?

  b. For the optimum membership in (a), the club additionally estimates that for each $0.50 decrease in court fees per hour, each pair of members will play an additional 2 hours per month. What court fee will maximize revenue?

66. Repeat Exercise 65 under the additional conditions that the tennis club has a fixed mortgage cost for its facility of $3,000 per month, a monthly expenditure per member of $4, and a $1-per-hour cost of court maintenance and lighting for its indoor facility.

67. Two power plants emitting particulate matter are located 10 mi apart. One plant emits 5 times the particulate matter of the other, and in both cases the concentration of particulate matter falling to earth is inversely proportional to the square of the distance from the plant, up to $\frac{1}{4}$ mi from either plant. Thus

$$C(x) = \frac{5k}{x^2} + \frac{k}{(10-x)^2} \qquad \frac{1}{4} \le x \le 9\frac{3}{4}$$

is the total concentration at a point x between the two plants. Where will this concentration be least?

68. A vineyard owner must decide when to harvest her crop for sale. Grapes picked later in the season will bring a better price per liter, but the overall yield from the vineyard will be less. The owner will earn $0.50 per liter at the start of the season, with each vine yielding 6 liters. Each week thereafter the price increases by $0.03 per liter, but the yield per vine decreases by 0.25 liter. When should she harvest so as to maximize her return?

Chapter 3

Related Rates and the Differential

Further Applications of the Derivative

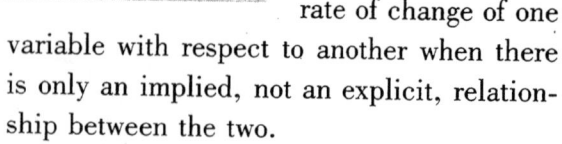

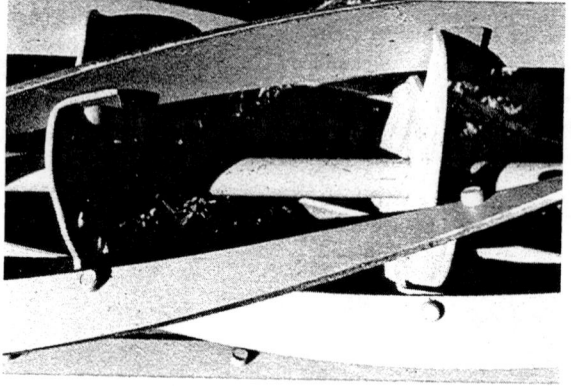

You have now learned how to differentiate functions, and you have seen the most immediate applications of the derivative. Whenever change is part of a problem, you should try to use the derivative as part of your approach to the problem. Indeed, you can expect to find applications of the derivative in the literature of your own field. In this brief chapter we present two additional applications of the derivative. First, we look at problems involving the *related rates* of two variables in a system when these two variables have a fixed relationship over time. Then we consider applications of the *differential estimate* of the change between two values of a function, which allows us to estimate unavoidable errors in measurement quickly. We need to begin, however, with a technique for finding the rate of change of one variable with respect to another when there is only an implied, not an explicit, relationship between the two.

3.1 Implicit Differentiation

In this section we learn to compute the rate of change of one variable with respect to another when the variables are related by an equation in which the relationship between the variables is not known explicitly as a function.

An equation $y = f(x)$, such as $y = x^3 - x + 1$, is said to define y **explicitly** as a function of x. An equation $g(x, y) = 0$, such as

$$y^2 - xy + 1 = 0 \qquad (1)$$

is said to define y **implicitly** as a function of x. For example, if we set $x = 2$, then Equation 1 becomes

$$y^2 - 2y + 1 = 0$$

which is equivalent to $(y - 1)^2 = 0$, or $y = 1$. Thus a given value of x determines a value of y.

Indeed, a given x may determine more than one value of y. This can be seen directly in this example because it is possible to find y *explicitly* as a function of x. If we regard

$$y^2 - xy + 1 = 0$$

as a quadratic equation in y and use the quadratic formula

$$y = \frac{-b \pm \sqrt{b^2 - 4ac}}{2a}$$

with $a = 1$, $b = -x$, and $c = 1$, we then obtain

$$y = \frac{x \pm \sqrt{x^2 - 4}}{2}, \text{ with domain } |x| \geq 2$$

For example, if $x = 3$, then $y = (3 \pm \sqrt{5})/2$ in Equation 1, and more generally, Equation 1 defines *two* functions, $y = (x + \sqrt{x^2 - 4})/2$ and $y = (x - \sqrt{x^2 - 4})/2$, implicitly. Notice that y is defined, then, by two different functions of x and that the procedure of using the quadratic formula is a bit involved. Furthermore, if the equation were, say, $y^5 - xy + 1 = 0$, we would not even have a method for explicitly finding the function(s) $y = f(x)$ implicitly defined by this equation. As we will see, this section is not directly concerned with the problem of multiple formulas defining y, nor with the problem of explicitly finding a formula for y, nor even with the possibility that an equation in x and y may not define y as a function of x in any form.

In calculus we are instead interested in the derivative

$$D_x y$$

of y with respect to x. In this section you will learn a fairly simple method, called **implicit differentiation**, to find this derivative without bothering to solve for y explicitly in terms of x. This method depends on understanding one idea.

Again consider the equation

$$y^2 - xy + 1 = 0 \qquad (2)$$

We can differentiate Equation 2 formally by writing

$$D_x(y^2 - xy + 1) = D_x 0 = 0$$

or

$$\boxed{D_x y^2} - D_x xy + D_x 1 = 0 \tag{3}$$

It would *not* be correct to continue and write

$$\boxed{2y} - D_x xy + D_x 1 = 0$$

Instead we have to recall that Equation 2 *implicitly* defines y by some unspecified function of x—say, $g(x)$. [But we have no formula for $g(x)$!] This means that the computation $D_x y^2$ is really a computation $D_x y^2 = D_x g(x)^2$. Since $D_x g(x)^2 = 2g(x)g'(x)$ by the chain rule, we must, when computing $D_x y^2$ in Equation 3, write

$$D_x y^2 = 2y \cdot D_x y$$

That is, because y is a function of x, the rate of change $D_x y$ *must appear* in computing $D_x y^2$. This computation exemplifies the one new idea you must learn in this section. The chain rule, as we have just applied it, tells us how to take the implicit dependence of y on x into account when we compute the derivative of y with respect to x. We now move from Equation 3 to

$$2y\, D_x y - \boxed{D_x xy} + 0 = 0 \tag{4}$$

since $D_x 1 = 0$.

The remaining computation $\boxed{D_x xy}$ is done with a similar idea in mind. That is, the expression xy must be thought of as a *product* of two functions of x—one of them, y, being implicit, existing only "behind the scenes" and the other being x itself. Thus

$$D_x xy = x\, D_x y + y\, D_x x = x\, D_x y + y \cdot 1 = x\, D_x y + y$$

by the product rule. Equation 4 now becomes

$$2y\, D_x y - (x\, D_x y + y) = 0$$

or

$$2y\, D_x y - x\, D_x y - y = 0 \tag{5}$$

Differentiation can proceed no further. After all, $D_x y$ is the quantity we are trying to compute and is an *unknown* in the equation. Therefore we treat Equation 5 as a (linear) *algebraic equation in the unknown* $D_x y$ as follows:

$$2y\, D_x y - x\, D_x y - y = 0$$
$$(2y - x)D_x y - y = 0 \quad \text{(Factor } D_x y\text{)}$$
$$(2y - x)D_x y = y$$

or

$$D_x y = \frac{y}{2y - x}$$

180   Chapter 3   Related Rates and the Differential

by isolating D_xy on one side of the equation. We have finally computed D_xy. Note also that the final answer involves both x and y.

This procedure is a very detailed example of implicit differentiation. We ordinarily proceed more briefly.

EXAMPLE 1

Compute D_xy where $y^3 + y + x = 2$.

Solution We differentiate both sides of the equation to obtain

$$D_x(y^3 + y + x) = D_x(2) \quad \text{or} \quad D_xy^3 + D_xy + D_xx = 0$$

Applying the chain rule to D_xy^3, we have $3y^2D_xy + D_xy + 1 = 0$. Factoring D_xy gives $(3y^2 + 1)D_xy + 1 = 0$. Isolating D_xy yields $(3y^2 + 1)D_xy = -1$ or

$$D_xy = \frac{-1}{3y^2 + 1} \quad \blacksquare \tag{6}$$

These two examples cause us to summarize the key formula for this section, obtained via the chain rule, as follows:

If y is implicitly a function of x, then

$$\boldsymbol{D_xy^n = ny^{n-1}D_xy}$$

for any exponent n.

Once found, the derivative D_xy may be used as we have learned to use the explicit derivative $f'(x)$ in Chapters 1 and 2.

EXAMPLE 2

Find the slope and equation of the straight line tangent to the graph of the equation $y^3 + y + x = 2$ at the point $(2, 0)$ on the graph.

Solution First, note that if $x = 2$, the equation becomes

$$y^3 + y + 2 = 2$$

Thus $0 = y^3 + y = y(y^2 + 1)$. Consequently, $y = 0$ when $x = 2$ (since $y^2 + 1$ is never 0). The point $(2, 0)$ is indeed on the graph of the given equation.

In Example 1, we found that $D_xy = -1/(3y^2 + 1)$. When $x = 2$ and (necessarily) $y = 0$, we see that

$$D_xy = \frac{-1}{3(0)^2 + 1} = -1$$

is the slope of the tangent line to the unknown graph at $(2, 0)$. The point–slope equation of this tangent line is $y - 0 = (-1)(x - 2)$ or $y = -x + 2$. $\blacksquare$

This example shows that once we have obtained $D_x y$, we proceed as usual to answer related questions. We graph the equation and solution of Example 2 in Figure 3.1; such a graph is not needed to use implicit differentiation.

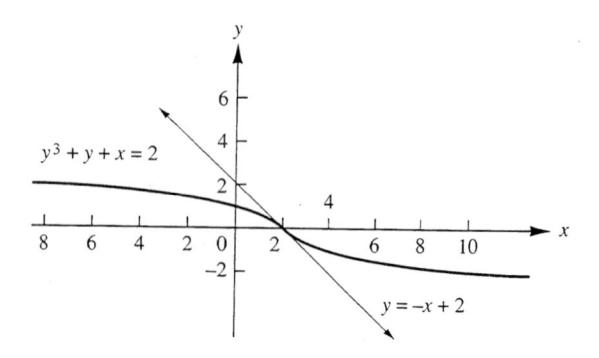

Figure 3.1

The next example is more involved algebraically.

EXAMPLE 3
Find $D_x y$ if $y^4 + (x/y) - x = x^2$.

Solution
$$D_x y^4 + D_x \frac{x}{y} - D_x x = D_x x^2$$

$$4y^3 D_x y + D_x \frac{x}{y} - 1 = 2x$$

We must treat the problem $D_x(x/y)$ as a *quotient differentiation* problem. We have

$$4y^3 D_x y + \frac{y D_x x - x D_x y}{y^2} - 1 = 2x$$

or
$$4y^3 D_x y + \frac{y - x D_x y}{y^2} - 1 = 2x$$

since $D_x x = 1$. The remaining details are all algebraic. We have

$$4y^3 D_x y + \frac{1}{y} - \frac{x}{y^2} D_x y = 2x + 1$$

$$\left(4y^3 - \frac{x}{y^2}\right) D_x y = 2x + 1 - \frac{1}{y} \qquad \text{Gather like terms and factor } D_x y$$

$$\frac{4y^5 - x}{y^2} D_x y = \frac{2xy + y - 1}{y} \qquad \text{Find common denominators}$$

or $\quad D_x y = \dfrac{(2xy + y - 1)y^2}{(4y^5 - x)y} = \dfrac{2xy^2 + y^2 - y}{4y^5 - x}$ ∎

182 Chapter 3 Related Rates and the Differential

The next example illustrates a repeated application of the chain rule in the process of implicit differentiation.

EXAMPLE 4

Find $D_x y$ when $\sqrt{1 - y^3} + x = 0$.

Solution $D_x(1 - y^3)^{1/2} + D_x x = 0$

$$\frac{1}{2}(1 - y^3)^{-1/2} D_x(1 - y^3) + 1 = 0$$

$$\frac{1}{2(1 - y^3)^{1/2}}(0 - D_x y^3) + 1 = 0$$

$$\frac{-3y^2 D_x y}{2\sqrt{1 - y^3}} = -1 \quad \text{or} \quad D_x y = \frac{2\sqrt{1 - y^3}}{3y^2} \quad \blacksquare$$

Implicit differentiation may also be used to find the second derivative (or any higher-order derivative) of y with respect to x.

EXAMPLE 5

a. Find $D_{xy}^2 y$ from the equation $y^3 + y + x = 2$ of Example 1.
b. Recalling that the sign of the second derivative tells us when a graph is concave up (or down), explain why the formula obtained for $D_{xy}^2 y$ is consistent with Figure 3.1.

Solution

a. From Equation 6, we have

$$D_x y = \frac{-1}{3y^2 + 1} = -(3y^2 + 1)^{-1}$$

Therefore
$$D_{xy}^2 y = D_x(-(3y^2 + 1)^{-1})$$
$$= +(3y^2 + 1)^{-2} D_x(3y^2 + 1)$$
$$= \frac{1}{(3y^2 + 1)^2} \cdot 6y \cdot D_x y$$

Now, using Equation 6 again,

$$D_{xy}^2 y = \frac{6y}{(3y^2 + 1)^2} \cdot D_x y = \frac{6y}{(3y^2 + 1)^2} \cdot \frac{-1}{3y^2 + 1} = \frac{-6y}{(3y^2 + 1)^3}$$

b. Since $y^2 \geq 0$ for any value of y, the denominator is always positive and the sign of $D_{xy}^2 y$ is determined by the sign of $-6y$. Thus $D_{xy}^2 y > 0$, and the graph is concave up if $y < 0$ and $D_{xy}^2 y < 0$, and concave down if $y > 0$. These conclusions are seen to be consistent with the graph drawn in Figure 3.1. ∎

Background Review 3.1

Much of the effort in this section involves algebraically solving for $D_x y$. To make this process easier, you may rewrite an equation such as

$$x D_x y + y^2 D_x y = 3 + D_x y$$

as

$$x D + y^2 D = 3 + D$$

by letting $D = D_x y$. Then solve for D as follows:

$$x D + y^2 D = 3 + D$$
$$x D + y^2 D - D = 3$$
$$(x + y^2 - 1)D = 3$$
$$D_x y = D = \frac{3}{x + y^2 - 1}$$

One nice thing about the algebraic solution of equations in this section is that all equations are *linear in the unknown* variable $D_x y$. These can all be solved by

1. Gathering all terms involving $D_x y$ on one side of the equation.
2. Gathering all other terms on the other side of the equation.
3. Factoring $D_x y$ on one side.
4. Dividing the other side by the factor of $D_x y$, as the example illustrates.

Exercises 3.1

Find $D_x y$ for each of the equations in Exercises 1–14.

1. $x^2 + y^2 = 4$
2. $xy = 3$
3. $x^2 y = 1$
4. $x^2 - 3y^2 = x$
5. $2xy + y^2 = x^3$
6. $y^3 - y^2 - x = 1$
7. $x^2 + 3xy^2 - y^3 = 0$
8. $y - \frac{y}{x} = x$
9. $\sqrt{xy} - y = x$
10. $x^3 - y^3 = xy$
11. $\sqrt{y^2 - 1} + x = 0$
12. $y^{1/3} - x = y$
13. $\frac{x - y}{x + y} = 1$
14. $\frac{x}{y} = \frac{1}{x + y}$

In Exercises 15–24, find the equation of the tangent line to the graph of each equation at the indicated point. When only x (or y) is given, use the equation to find the corresponding y (or x) coordinate of the point (x, y) at which the tangent is to be found.

15. $xy - x^2 - 1 = 0$ at $(1, 2)$
16. $\sqrt{xy} - x = 0$ at $(2, 2)$
17. $y^3 - 2xy^2 + xy = 1$ at $(0, 1)$
18. $y^3 + y^2 = x$ at $(12, 2)$
19. $\frac{1}{y} - (x + 1) = 2$ at $\left(-1, \frac{1}{2}\right)$
20. $\frac{1}{y} - \frac{1}{x} = 0$ at $y = 1$ [*Hint:* First replace y by 1 in the equation and solve for x.]
21. $\sqrt{y} - x = 4$ at $y = 16$
22. $(y - 1)^{1/3} - x = 0$ at $x = 2$
23. $x^2 + y^2 = 25$ at $y = 3$ [*Hint:* This is the equation of a circle of radius 5 centered at $(0, 0)$.]
24. $y^2 + xy + 4 = 0$ at $x = 6$

184   Chapter 3   Related Rates and the Differential

Find $D_x^2 y$ in Exercises 25 and 26.

25. $y + 4y^3 = 5x$

26. $xy - \dfrac{1}{y} = 3$

27. Suppose that two natural species X and Y of population sizes x and y, respectively, are competing for the same food supply in such a way that the product of their populations is always constant. This means that $xy = C$ where C is constant. Compute how fast the population of species Y is changing with respect to that of X when $x = 200$ individuals.

28. Suppose that studies show that the number of tennis balls x and the number of tennis rackets y sold by a merchant on a typical day are related by the equation $11x = 50y + 0.001xy^3$. Find $D_y x$, the rate at which sales of tennis balls change with increasing sales of rackets, when $y = 10$ rackets sold. [*Hint:* You will need to find the number x of balls sold when 10 rackets are sold.]

3.2 Related Rates

In this section we learn to model systems where time is a background variable that appears only *implicitly*.

Imagine a system undergoing change *over time*. Suppose that the system contains two variables—say, x and y. Both x and y are changing over time and so should be written as functions

$$x = x(t) \quad \text{and} \quad y = y(t)$$

of an **implicit background variable** t representing time. The typical question we address in this section is the following: If we have information about

$$D_t x(t) \quad \text{the rate at which } x \text{ is changing over time}$$

can we draw conclusions about

$$D_t y(t) \quad \text{the rate at which } y \text{ is changing over time?}$$

If x and y have no relationship to each other, then the answer is no. But if there is a relationship between x and y, *particularly one that is fixed over time*, we can usually compute the rate of change of one variable over time given the rate of the other. The variables $D_t x(t)$ and $D_t y(t)$ are then called **related rates**. Let us consider a system we can easily imagine and use it to introduce the concepts and procedures we will use to solve all related rate problems.

Suppose that a 20-ft-long ladder is leaning against a wall and that the foot of the ladder begins to slide away from the wall at the rate of 3 ft/sec. How fast is the top of the ladder sliding down the wall?

Figure 3.2 illustrates this system. The two variables are

$$x = x(t) = \text{position of the foot of the ladder at time } t$$
(measured in feet from the wall)

and

$$y = y(t) = \text{position of the top of the ladder at time } t$$
(measured in feet up the wall)

Although we are given no explicit information about these variables as functions of time, we are given that

$$D_t x(t) = 3 \text{ ft/sec}$$

since $\quad D_t x(t) = $ *rate* at which the distance of the foot of the ladder from the wall is changing

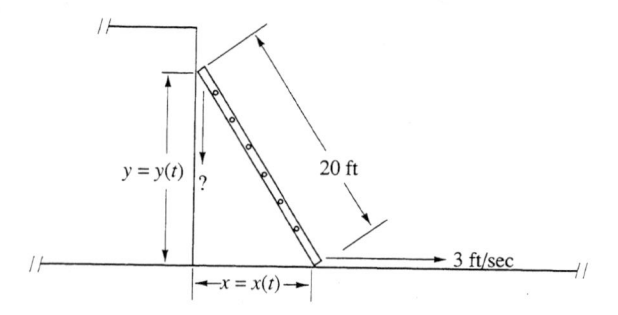

Figure 3.2

We would like to compute

$$D_t y(t) = \text{rate at which the top of the ladder is sliding down the wall}$$

We will discuss how this can be done in a series of three steps.

Step 1. If the two variables x and y had no relationship, we could do nothing. But in fact they have a very specific relationship *fixed over time:* The ladder always completes a *right triangle* with the wall and floor. Therefore, by the Pythagorean theorem,

$$x(t)^2 + y(t)^2 = (20)^2 = 400 \qquad (1)$$

As written here, this equation reminds us of the background time variable t. Normally we will not include the variable t explicitly and will write such an equation instead as $x^2 + y^2 = 400$.

Because this example illustrates a general method for solving related-rate problems, let us pause and consider the role of Equation 1 in a larger context. An equation of this sort will be used in *every* related-rate problem. Equation 1 is called the *state equation* for this system. We regard the *state* of a system as the system frozen, unmoving, as though in a photograph. If you were to take a single photo of the ladder sliding down and away from the wall, the only information you could get from this photo (about this problem) would be Equation 1. The solutions to all exercises in this section, all of which deal with change over time, ironically begin with a state equation. *The* **state equation** *describes the relationship, fixed over time, among the variables as though the system were frozen.* Therefore, do not think of everything moving about, but rather freeze the motion and observe how the variables are related, as though they did not vary!

This ironic viewpoint is effective because differentiation allows us to move routinely from a state equation to a **change-of-state equation,** where we incorporate information about *rates* of change into the solution. In effect then, think of the state equation as a mathematical photograph of the system and the change-of-state equation as a mathematical motion picture of the system.

Let us make the calculations that yield the change-of-state equation in this example.

Step 2. To put into motion our fixed photo of the ladder system, represented by the state equation

$$x^2 + y^2 = 400$$

—that is, to let the ladder slide down the wall—we differentiate the state equation with respect to time to obtain the change-of-state equation for this system

$$D_t(x^2 + y^2) = D_t 400$$

or

$$D_t x^2 + D_t y^2 = 0$$

Now, since the distances x and y are functions of time, this becomes, by implicit differentiation,

$$2x\, D_t x + 2y\, D_t y = 0$$

or

$$x\, D_t x + y\, D_t y = 0 \qquad (2)$$

Equation 2 is the change-of-state equation for a ladder sliding down a wall. It tells exactly how position (x and y) and motion ($D_t x$ and $D_t y$) are related. We can think of this equation as a mathematical motion picture of the ladder sliding down the wall.

Step 3. Continuing with the problem of the falling ladder, we are also given that

$$D_t x(t) = 3 \text{ ft/sec}$$

Therefore, in this particular system we obtain by substitution

$$3x + y\, D_t y = 0$$

or

$$D_t y = \frac{-3x}{y} = -3\left(\frac{x}{y}\right) \qquad (3)$$

This last equation says that the top of the ladder is sliding down (note the *minus* sign) the wall at time t at a rate of three times the ratio of the position of the foot of the ladder to the top of the ladder at time t.

Once we have derived the state and change-of-state equations for such a system we can answer further questions about the system. Suppose that we are asked: How fast is the ladder falling down the wall when the foot of the ladder is 16 ft from the wall? We use the additional information given in this question via an additional step.

Step 4. We are now given further *state* information: At some unspecified time t, $x(t) = 16$. We employ this state information *in the state equation* $x^2 + y^2 = 400$ by substitution to obtain $16^2 + y^2 = 400$ or $y^2 = 144$ with solution $y = 12$ (the second solution $y = -12$ has no meaning here). That is, the top of the ladder is 12 ft above the ground when the foot is $x = 16$ ft from the wall. Therefore, from Equation 3

$$D_t y(t) = -3\left(\tfrac{16}{12}\right) = -4 \text{ ft/sec}$$

This means that when the base is sliding away from the wall at 3 ft/sec, the top of the ladder is falling at the rate of 4 ft/sec, when the base of the ladder is 16 ft from the wall.

This is a typical related-rate problem. All the problems in this section can be solved by following this procedure.

Solution Method for Related-Rate Problems

In a system undergoing change over time,

1. Clearly identify the quantities that are varying over time and represent each by a letter: $x = x(t)$, $y = y(t)$, etc.
2. Imagine the system frozen. Write down the *state equation* that relates these variables in that fixed position.
3. Differentiate the state equation with respect to time to obtain the *change-of-state* equation for the system.
4. Substitute given observations about the *rate* of change of one or more variable(s) in the change-of-state equation to obtain conclusions about the rate of change of the remaining variable.
5. If further information is given, or needed, about the state of the system (as in step 4 in the example) use the given state information *in the state equation* so as to derive further information about the state(s) of the system.

When using this procedure, you must clearly distinguish between state information and rate information. As we have seen, state information is about what the system *is*. Thus, in the case just described, we learn where the top of the ladder *is* when the bottom *is* 16 ft from the wall from the state equation. Rate information is about how the system *changes*.

EXAMPLE 1

Water is pumped into a rectangular swimming pool at the rate of 50 ft^3/min. The pool measures 12 ft by 25 ft. How fast is the water level rising in the pool? (See Figure 3.3.)

Figure 3.3

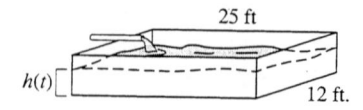

Solution We follow the procedure just outlined.

Step 1. The variables in the system are

$$V(t) = \text{amount (volume) of water in the pool}$$

and

$$h(t) = \text{depth of the water in the pool}$$

Step 2. What is the state equation for the system? Take a photo (or shut off the water for an instant)—and what do we have? The pool has area $12 \cdot 25 = 300$ ft², and so

$$V(t) = 300h(t)$$

since volume is area times height.

Step 3. What is the change-of-state equation? Differentiate the state equation:

$$D_t V = 300 D_t h \qquad (4)$$

Step 4. What else are we given? Answer:

$$D_t V(t) = 50 \text{ ft}^3/\text{min} = \text{rate at which the volume of water is changing}$$

What are we to calculate? Answer:

$$D_t h(t) = \text{rate at which the water is rising}$$

Therefore, substituting in the change-of-state equation (4), we have $50 = 300 D_t h$ or

$$D_t h = \frac{50 \text{ ft}^3/\text{min}}{300 \text{ ft}^2} = \frac{1}{6} \text{ ft/min}$$

That is, the water is rising at the rate of 2 in./min. ∎

In Example 1, no further information about the state of the system at a particular time was needed to solve the problem. This is not always the case.

EXAMPLE 2

In the still waters of a Louisiana swamp an oil rig has begun leaking oil. The oil floats on the surface and starts to form a circular oil slick whose radius is growing at the rate of 2 ft/min. (See Figure 3.4.)

a. How fast is the area of the oil slick growing when the slick is 100 ft in diameter?

b. How fast is the area growing when it covers 5,000 ft² of swamp?

Figure 3.4

Solution

Step 1. The variables in this system are

$$A(t) = \text{area of the oil slick at time } t$$

and

$$r(t) = \text{radius of the oil slick at time } t$$

Step 2. What is the state equation?

An oil company plane flying in to observe the oil slick takes a photo. This photo shows an (approximately) circular oil slick. Since the area of a circle is given by $A = \pi r^2$, we regard this formula as the appropriate description of the state—the amount of area covered by oil—of this system. That is, omitting the variable t,

$$A = \pi r^2 \qquad (5)$$

is the state equation.

Step 3. How is the system changing? (What is the change-of-state equation?)
We have

$$D_tA = \pi D_t r^2 \quad \text{or} \quad D_tA = (2\pi r)D_t r \tag{6}$$

a. We are given $D_t r = 2$ ft/min and $r(t) = 50$ ft at a particular time t. We are to calculate D_tA, the rate at which the area of the slick is growing. Using Equation 6, we have $D_tA(t) = 2 \cdot \pi \cdot 50 \cdot 2 = 200\pi$ ft²/min ≈ 628 ft²/min.

b. We are given $D_t r = 2$ ft/min and $A(t) = 5{,}000$ ft² at a particular time t. We are to again calculate D_tA. As before, we have

$$D_tA = 2r(t) \cdot \pi \cdot 2 = 4\pi r(t)$$

However, we do not yet know what $r(t)$ is. We know only that the area of the slick *is* 5,000 ft². According to step 5, we must use this information about the state of the system in the state equation (5). Doing so, we find that $5{,}000 = \pi r^2$ or that $r = \sqrt{5{,}000/\pi} \approx 39.89$ ft when $A = 5{,}000$. Returning to the change-of-state equation (6), we then find that at this moment $D_tA = 4\pi r(t) = 4\pi(39.89) \approx 501.33$ ft²/min. ∎

The next example is more abstract in that no particular "photo" relates the variables. In this example the state equation is given to us.

EXAMPLE 3

A manufacturing firm can spend resources on either labor or equipment. If x dollars are spent on labor and y dollars on materials, statistical analysis of the company's records shows that the firm is producing

$$P = \frac{x^2 + xy + 3y}{4{,}000}$$

units of a certain item. At what rate is production increasing when the firm is spending \$5,000 on labor and \$8,000 on materials and is increasing its expenditures on labor and materials at the rates of \$300 and \$600 per day, respectively?

Solution

Steps 1 and 2. The variables x and y are already given, and the state equation is

$$P = \frac{x^2 + xy + 3y}{4{,}000}$$

representing the number of units produced *when* labor and equipment expenditures are at levels x and y, respectively.

Step 3. What is the change-of-state equation? Thinking of P, x, and y as functions of a background variable time, we have

$$D_t P = \frac{D_t x^2 + D_t xy + D_t 3y}{4{,}000}$$

$$D_t P = \frac{2x\, D_t x + x\, D_t y + y\, D_t x + 3 D_t y}{4{,}000}$$

Step 4. We are to calculate $D_t P$ = rate of increase in production when $x = 5{,}000$, $y = 8{,}000$, and $D_t x = 300$, $D_t y = 600$. We have

$$D_t P = \frac{2(5{,}000)(300) + 5{,}000(600) + 8{,}000(300) + 3(600)}{4{,}000}$$

$$= \frac{3{,}000{,}000 + 3{,}000{,}000 + 2{,}400{,}000 + 1{,}800}{4{,}000}$$

$$= \frac{8{,}401{,}800}{4{,}000} \text{ units/day}$$

$$\approx 2{,}100 \text{ units/day}$$

That is, at the given rates and expenditure levels, production is increasing at the rate of approximately 2,100 units per day. ∎

Background Review 3.2

The area of a triangle is $bh/2$.

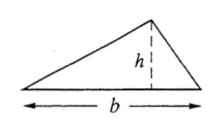

The Pythagorean theorem for a right triangle is

$$a^2 + b^2 = c^2$$

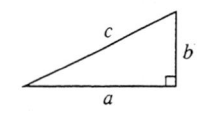

The circumference C and area A of a circle of radius r are

$$C = 2\pi r \quad \text{and} \quad A = \pi r^2$$

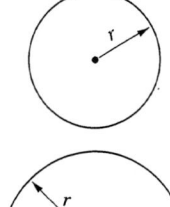

The volume V and surface area A of a sphere of radius r are

$$V = \tfrac{4}{3}\pi r^3 \quad \text{and} \quad A = 4\pi r^2$$

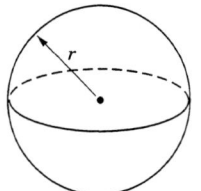

The volume V and surface area A of a cylinder are

$$V = \pi r^2 h \quad \text{and} \quad A = 2\pi rh + 2\pi r^2$$

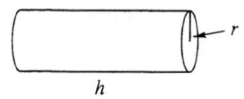

The volume V and surface area A of a cone are

$$V = \frac{\pi r^2 h}{3} \quad \text{and} \quad A = \pi r \sqrt{r^2 + h^2} + \pi r^2$$

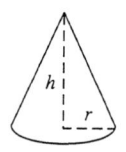

Remember to distinguish between state information and rate information: State information is about what *is*; rate information is about how it *changes*. Use the state equation to determine further needed information about the state of the system, and use rate (and if necessary, state) information in the change-of-state equation to derive conclusions about the rate at which the system is changing when it is in a particular state.

Exercises 3.2

Regard the equation given in Exercises 1–4 as a state equation. Find the change-of-state equation by differentiation with respect to t, and solve the equation for $D_t y$ in terms of x, y, and $D_t x$.

1. $x + y^2 = 200$
2. $x^2 - y^3 = \sqrt{7}$
3. $2x + 3y = xy$
4. $x = \frac{4}{3}\pi y^2$

5. Given a state equation $3y + x^2 = 100$,
   a. Find y when x is 5.
   b. Find $D_t y$ when $D_t x = 4$.
   c. Find $D_t y$ when x is 5 and $D_t x = 4$.

6. Given a state equation $y - (x/y) = 1$,
   a. Find x when y is -2.
   b. Find $D_t y$ when $D_t x = 3$.
   c. Find $D_t y$ when y is -2 and $D_t x = 3$.

   [*Hint*: Remember to use the quotient rule in finding $D_t(x/y)$.]

7. This exercise refers to the first example described on page 185.
   a. How far is the foot of the ladder from the wall when the top of the ladder is 8 ft high along the wall?
   b. If the top of the ladder is instead falling at the rate of 1.5 ft/sec when it is 8 ft high, how fast is the foot of the ladder moving away from the wall?
   c. How fast is the foot of the ladder moving away from the wall at time t, if the top of the ladder is instead falling at the rate of 2 ft/sec? (Your answer must necessarily be in terms of the ratio of base and height of the ladder.)

8. If a tennis ball hits the court 16 ft in front of you, travels a path conforming to the graph $y = \sqrt{x}$, and is moving horizontally at the speed of 35 ft/sec, how fast is the ball moving vertically when you hit it?

9. Suppose that you are riding your bicycle up a hill whose shape conforms to the graph of $y = 0.3(x/10)^2$, for $0 \leq x \leq 100$ yd from the base of the hill (see Figure 3.5).

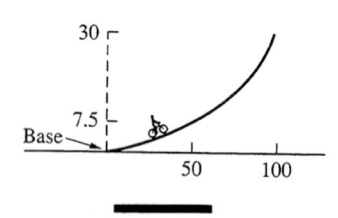

Figure 3.5

a. If you are moving horizontally at the rate of 5 yd/sec, how fast are you moving vertically when you are 30 yd from the base of the hill?

b. If at 80 yd from the base of the hill, in the horizontal direction, you are moving vertically at a rate of 1.5 yd/sec, how fast are you moving horizontally?

c. At a distance of 50 yd from the base of the hill, how fast must you move horizontally in order to maintain a vertical speed of 1 yd/sec? At a distance of 10 yd? 90 yd?

10. The volume of a *torus* (that is, something shaped like a doughnut or an inflated automobile tube) is given by

$$V = (2\pi s)(\pi r^2) = 2\pi^2 s r^2$$

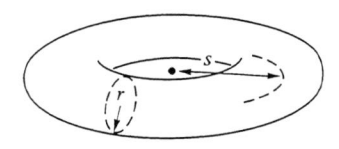

Figure 3.6

where s and r are the dimensions indicated in Figure 3.6. If an automobile tube is inflated in such a way that $D_t r = 1$ cm/min and $D_t s = 0.2$ cm/min, at what rate is air being pumped into the tube when $s = 50$ cm and $r = 10$ cm? [*Hint:* All three variables V, s, and r must be treated as implicit functions of time.]

11. In the making of Vermont maple syrup, foreign matter must be filtered out of the sap collected from maple trees. Suppose that maple sap is being poured from a cylindrical tank with radius 1.5 ft and length 8 ft into a filtering tank that has a conical bottom of depth 1.5 ft joined to a cylinder of radius 1 ft. (See Figure 3.7.)

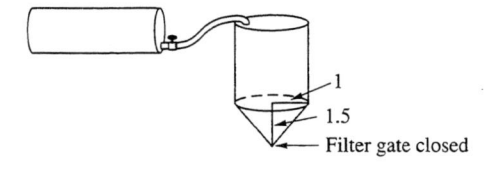

Figure 3.7

If the maple sap is flowing out of the cylindrical tank and into the filtering tank at the rate of 6 ft³/min, how fast is the level of sap in the filtering tank rising when the sap is at a level of

a. 1 ft deep in the tank?

b. 2 ft deep in the tank?

c. 1.5 ft deep in the tank?

12. A spherical balloon of radius 8 in. has a small puncture and is losing air at the rate of 0.1 ft³/sec. How fast is the radius of the balloon shrinking when the diameter of the balloon is 8 in.?

13. A fisherman has cast a floating fishing lure into the center of a 40-ft-wide stream. The current in the stream takes the floating lure downstream, pulling string off the fisherman's reel at the rate of 1.5 ft/sec. How fast is the lure moving downstream when the fishing line is 50 ft long? (See Figure 3.8.)

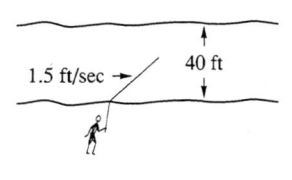

Figure 3.8

14. Imagine the pickup arm on a 33 rpm record player. If each groove in the record is 0.2 mm wide, how fast is the tone arm moving toward the center of the record? How fast is the area bounded by the groove of the record in which the pickup arm is riding shrinking when the arm is 12 cm from the center of the record? How fast is the pickup arm moving in the record groove when it is 20 cm from the center of the record? (Treat each groove as a circle for the purpose of solving this problem.)

15. A California vineyard is beginning to receive acclaim for its Chardonnay. It has to choose whether to dilute its Chardonnay grape with cheap, perfumy Thompson's Seedless grape, up to the legal maximum of 49%, or to maintain the quality of its wine but at a lower volume of production. The vineyard conducts a market survey and determines that if x and y represent the percentage of Chardonnay and Thompson's Seedless blended in its casks, then the vineyard's profit will be

$$P = \frac{8x^2}{y+1} \text{ dollars per case}$$

What will happen to its profits when the percentage of Thompson's Seedless reaches 30% and is increasing at the rate of 3% per year? (Note that $x + y = 100$ since the wine will be made up of only these two grape varieties.)

16. A manufacturing company has a production budget of $200,000, all of which will be spent on x units of labor, costing $10,000 per unit, and y units of capital, costing $30,000 per unit. It can produce

$$P = 50xy$$

items when using x units of labor and y units of capital. At what rate is production changing when it employs 8 units of labor and is decreasing capital expenditures by 0.5 unit per month? [*Hint:* $10,000x + 30,000y = 200,000$.]

17. Suppose that a manufacturer of tennis balls and tennis rackets has found that from the sale of x tennis balls and y tennis rackets, sales revenues R are

$$R = 35y + 0.65x + 0.002xy^2 \text{ dollars}$$

a. What is the revenue from the sale of 10 rackets and 50 tennis balls?

b. At what rate is revenue changing when 3 rackets and 18 tennis balls are sold per day?

18. Suppose that a manufacturer's profit P from the sale of x items is

$$P = 40x - 3{,}000 \text{ dollars}$$

At what rate must sales increase so that profits increase at the rate of $600 per day?

19. In the late nineteenth century the British economist Francis Edgeworth attempted to describe economic systems in mathematical terms. One of Edgeworth's concepts was that of a *consumption-indifference* curve. For reasonable economic assumptions such curves typically have the shape shown in Figure 3.9.

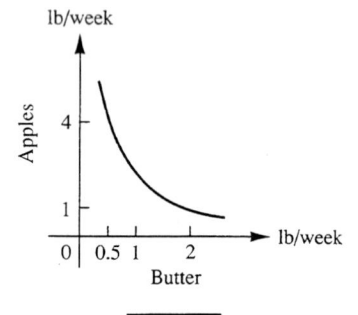

Figure 3.9

Edgeworth's idea was that each point on the curve represents a consumer choice that is equally satisfactory (thus the name *indifference* curve). For example, this indifference curve indicates that consumers are just as happy to have $\frac{1}{2}$ lb of butter and 4 lb of apples, or to have 2 lb of butter and 1 lb of apples, in their food supply for a given week.

If the consumption-indifference curve for two items x and y is given by the equation

$$xy^2 = 3$$

and if the desire for item x is increasing at the rate of 0.25 unit/week, how must the desire for item y change when consumption of y is 1.5 units? 3 units? As part of your solution, graph the equation $xy^2 = 3$ for $x, y > 0$.

20. Two variable resistors are connected in a parallel circuit, as illustrated in Figure 3.10. The resistance R of the overall circuit is given by the electrical law

$$\frac{1}{R} = \frac{1}{x} + \frac{1}{y}$$

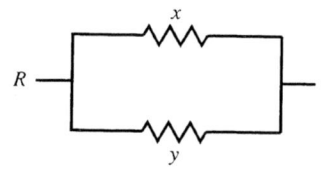

Figure 3.10

where x and y are the resistances in ohms (Ω) of the two variable resistors. If the resistance of one resistor is increasing at the rate of $0.5 \, \Omega/\text{sec}$, while the other is decreasing at the rate of $-0.3 \, \Omega/\text{sec}$, at what rate is the resistance R of the overall circuit changing when $x = 8 \, \Omega$ and $y = 12 \, \Omega$?

21. Suppose that in Exercise 20, the resistor x is controlled by a feedback-sensing device and is increasing at the rate of $0.2 \, \Omega/\text{sec}$. The second resistor is controlled by a computer that must alter the resistance so as to ensure that the overall circuit resistance R is changing at the rate of $0.3 \, \Omega/\text{sec}$. Tell the computer how fast to change the resistance of the other resistor so as to achieve this when $x = 4 \, \Omega$ and $y = 2 \, \Omega$.

22. An observer with a 2-ft-long telescope is watching an approaching airplane. The plane is traveling at a height of 2 mi at a speed of 450 mph. The observer rotates the telescope so as to keep the plane within the scope's field of vision. How fast is the far end of the telescope moving in the horizontal direction when the plane *is directly overhead*? [*Hint:* In Figure 3.11,

$$\frac{T}{L} = \frac{x}{r}$$

where L is the length of the telescope and x is the indicated horizontal measurement.]

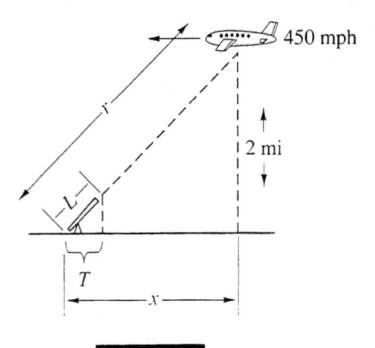

Figure 3.11

23. Do Exercise 22 with a 1-m telescope when the airplane (at an altitude of 2 km) is moving at a speed of 750 kmph and is 6 km from the observer in the direction of the telescope; when it is 2 km from the observer in the horizontal direction.

24. When two species such as the rabbit and the fox have a predator–prey relationship, their rising and falling populations, resulting from predation on one species supplying sustenance for the other, can be represented by an equation that represents their joint population interaction. For example, the equation of the ellipse indicated in Figure 3.12

$$\frac{(x-5)^2}{9} + \frac{(y-2)^2}{1} = 1$$

provides a rough model of rabbit–fox interaction. Notice that as the rabbit population x rises from 500 to 800, the fox population y rises from a low of 100 to a level of 200, whereupon the large fox population begins to cut the rabbit population down to the 500 level again. At the same time, with all these rabbits to eat, the fox population builds up to the 300 level and the cycle continues.

a. How fast is the rabbit population declining when there are 240 rabbits, and the fox population stands at 250 and is declining at the rate of 30 individuals per year?

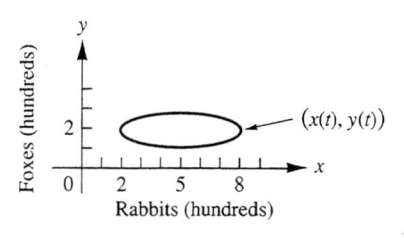

Figure 3.12

b. How fast is the rabbit population increasing when there are 761 rabbits, and the fox population stands at 150 and is increasing at the rate of 30 individuals per year?

25. A building is being demolished. A heavy weight is dropped on the peak of the roof, causing it to fall at the rate of 2 ft/sec. The rafters attached to the center beam are thus forced apart and push the outer walls of the building apart. How fast are the two walls separating when the peak of the roof is 3 ft high, if the original dimensions of the building are as indicated in Figure 3.13? How fast are the walls separating when the roof is flat?

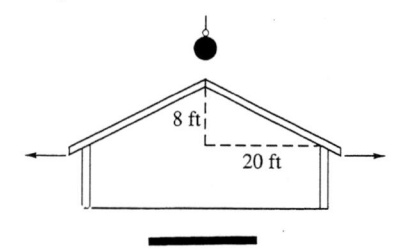

Figure 3.13

26. Two aircraft are approaching the same airport, one from the south and the other from the east. The northbound plane is traveling at 250 mph and the westbound plane at 300 mph. How fast are they approaching each other when they are both 2 mi from the airport? Is the distance between them closing at an increasing or a decreasing rate? [*Hint:* You will need to use a second derivative.]

3.3 The Differential of a Function

The purpose of this section is to compare two function values $f(x + h)$ and $f(x)$ by means of the computation $f'(x)h$.

The human eye can see material objects only as small as $\frac{1}{200}$ in., roughly the thickness of this page. Thus, given a ruler and asked to measure an object, you will *not* be able to measure it exactly. You will instead report an *observed* measurement rather than an *actual* measurement. Such difficulties occur in more complicated systems as well, and measurement error is often unavoidable.

For example, a manufacturer cannot know exactly how much raw material will be used in the manufacture of a particular product, nor can a physician know a patient's precise reaction to a potentially hazardous drug. The difficulties posed by such inherent error are made worse when the needed conclusion is a *function of* the quantity subject to inherent error in measurement. As we will see later, to estimate the inherent error across a functional relationship, we can use the *differential* of a function. Consider Figure 3.14.

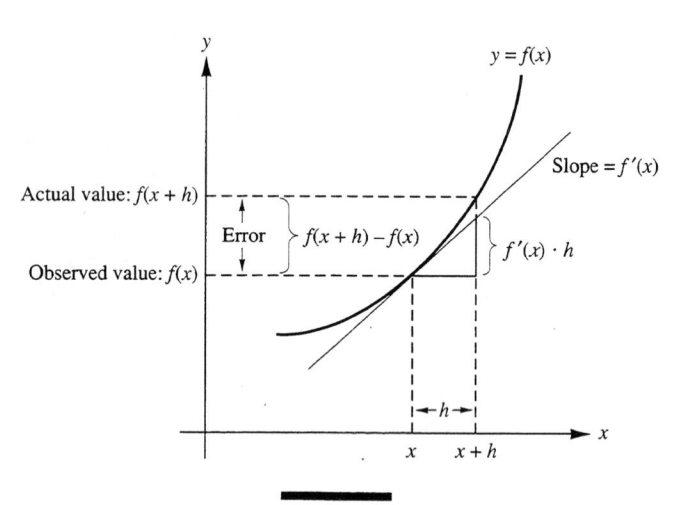

Figure 3.14

We see from Figure 3.14 that if h is a *short* distance, then

$$f(x + h) - f(x) \approx f'(x) \cdot h \qquad (1)$$

For $f'(x)$ is the slope of the tangent line, and this slope, multiplied by the horizontal "run" h, yields the vertical "rise" along the tangent, which *approximates* the rise along the curve of f itself. The rise of the curve is in turn the difference $f(x + h) - f(x)$ on the left in Equation 1.

Definition

The **differential** of a function $y = f(x)$ at a point x, relative to a change in x of amount h, is the number

$$df(x;h) = f'(x)h$$

We simply write **df** for $df(x;h)$ when x and h are understood. We use the notation $df(x;h)$ to remind us that the value of the differential depends on both the point x and the change h in x.

With this definition the approximate equality (1) is written as

$$f(x + h) - f(x) \approx df \qquad (2)$$

EXAMPLE 1

Let $f(x) = 1/x$.
a. Find $df = df(2;0.05)$, the differential of f at $x = 2$ over a change of $h = 0.05$ in the value of x.
b. Compare $f(2 + 0.05) - f(2)$ with the value df found in (a).

Solution

a. We have $f'(x) = D_x x^{-1} = -1/x^2$. Thus $f'(2) = -1/4$. Since $h = 0.05$ we have, from the definition $df = f'(x) \cdot h$, that

$$df = f'(2) \cdot (0.05) = -\frac{1}{4}(0.05) = -0.0125$$

b. We have $f(2 + 0.05) = f(2.05) = 1/2.05 \approx 0.4878$ and $f(2) = 1/2 = 0.5$. Thus

$$f(2 + 0.05) - f(2) \approx 0.4878 - 0.5 = -0.012195$$

(accurate to five decimal places).
Comparing this with $df = -0.0125$, we see that in Equation 2

$$f(2 + 0.05) - f(2) \approx df(2;0.05)$$

with accuracy to three decimal places. ∎

Estimation of Inherent Error

The term on the left in Equation 2 is a useful way of expressing and measuring an error that may arise in the observation of a system. Imagine that we are dealing with some system in which $y = f(x)$ relates two variables x and y in the system. We interpret the problem of *actual* versus *observed* measurement in this system in the following way: An observer wishes to measure the condition $f(x)$ of the system at x. The person makes an appropriate measurement of x and obtains a value

$$f(x) = \text{observed value}$$

At the same time, because of inherent difficulties in making the measurement, the actual value of the measured quantity may well be different—say, instead, a value $x + h$, where h can be negative or positive. This means that

$$f(x + h) = \text{actual value}$$

is the true state of the system. The error, then, between actual and observed value is the difference

$$\text{Error} = E(h) = f(x + h) - f(x) = \text{actual value} - \text{observed value} \quad (3)$$

Combining Equations 2 and 3, we reach the basic conclusion (Figure 3.14)

$$\boldsymbol{E(h) \approx df = f'(x) \cdot h} \quad (4)$$

We regard the error $E(h)$ as a function of h because h is a quantity about which we may have some knowledge in applications. In the next example, $h = \pm\frac{1}{200}$, the limit of accuracy of the human eye.

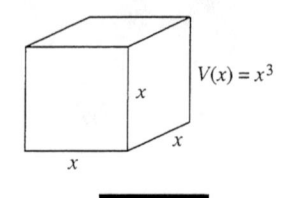

Figure 3.15 $V(x) = x^3$

EXAMPLE 2

Given that your eye can discern differences in measurement no greater than $\frac{1}{200}$ in., estimate the inherent error in measuring the volume of a cube when the observed length of each side is **a.** 1 in.; **b.** 2 in. (See Figure 3.15.)

Solution

a. When $x = 1$, the *observed* volume is

$$V(1) = 1 \text{ in.}^3$$

At the same time, we know that the side x of the cube could actually have length $x = 1 + h$, where $|h| \leq \frac{1}{200}$. (We must use the absolute value of h because our observation could err by as much as $\pm\frac{1}{200}$ in.) Therefore the *actual* volume of the cube could instead be

$$V(1 + h) = (1 + h)^3 \text{ in.}^3$$

where $|h| \leq \frac{1}{200}$. The error in measurement is then

$$E(h) = V(1 + h) - V(1)$$

According to Equation 4, this quantity is approximated by the differential $dV = dV(1;h) = V'(1) \cdot h$. Since $V'(x) = 3x^2$, we have $dV = 3 \cdot (1)^2 \cdot h = 3h$ in.^{3}. Hence, using Equation 4, the inherent error is

$$E(h) \simeq dV(1;h) = 3h$$

Now since $|h| \leq \frac{1}{200}$, the (absolute value of the) error between observed and actual volume is approximately

$$|E(h)| \lesssim 3 \cdot \frac{1}{200} = \frac{3}{200} = 0.015 \text{ in.}^3$$

where the approximate inequality arises because $|h| \leq \frac{1}{200}$.

b. If instead the length of the cube is observed to be $x = 2$ in., we have an observed volume of $V(2) = 2^3 = 8$ in.^{3}. However, the actual volume could be $V(2 + h) = (2 + h)^3$ in.^{3} with $|h| \leq \frac{1}{200}$. (Notice that an error h in measuring the length of a 2-in. side occurs only once, not in each inch of the 2-in. measurement, assuming a normal ruler.) Therefore the error in measurement is

$$E(h) = V(2 + h) - V(h) \simeq dV = dV(2;h)$$

Since $dV = V'(x)h = 3x^2 h$, then $dV = 3 \cdot (2)^2 \cdot h = 12h$ in.^{3}, and with $|h| \leq \frac{1}{200}$, the error is approximately

$$|E(h)| \lesssim 12 \cdot \frac{1}{200} = \frac{12}{200} = \frac{6}{100} = 0.06 \text{ in.}^3 \blacksquare$$

We would think that the inherent error in observed volume would be smaller in Example 2(b) than in (a), since the visual error $\pm\frac{1}{200}$ in. occurs only once in measuring the larger cube. This is true if we consider the error 0.06 in.^{3} relative to the volume 8 in.^{3} of the entire cube. That is, we need to think in terms of the **percentage error** in observation, which is defined as the ratio E/W of the possible error E to the full (observed) size W of the object we are measuring.

198   Chapter 3   Related Rates and the Differential

EXAMPLE 3

What is the percentage error in volume measurement in Example 2(a) and (b)?

Solution

a. In Example 2(a), the observed volume is $V(1) = 1$ in.^{3} with an error of $\frac{3}{200}$ in.^{3}. Thus the percentage error (relative to the volume of the cube) is

$$\frac{\frac{3}{200}}{1} = 0.015 \quad \text{or} \quad 1.5\%$$

b. In Example 2(b), the observed volume is $V(2) = 8$ in.^{3} with an error of $\frac{6}{100}$ in.^{3}. Therefore the percentage error is

$$\frac{\frac{6}{100}}{8} = 0.0075 \quad \text{or} \quad 0.75\% \quad \blacksquare$$

We can formulate the following general conclusion to this discussion:

The Differential Estimate of Error in a Functional Relationship

If $f(x)$ is a given (observed) function value and h is a sufficiently small number, then

$$E(h) = f(x + h) - f(x) \simeq df(x;h) = f'(x) \cdot h$$

where $E(h)$ is regarded as the error between the given value $f(x)$ and a second (unknown or actual) value $f(x + h)$.

The percentage error, relative to the given value $f(x)$, is $E(h)/f(x)$ times 100%.

That is, the differential of f approximates the difference between two values $f(x + h)$ and $f(x)$ of the function f if h is "sufficiently small." The degree of approximation and just how sufficiently small h must be are not specified in this formulation and depend on the size of f'', a topic discussed in more advanced texts.

Differential Approximation of an Unknown Function Value by a Known Function Value

The differential of a function may be used in another way. Given that we know a particular function value $f(x)$, how can we quickly or easily estimate another function value $f(x + h)$, if h is not very large? For example, given that $\sqrt[3]{27} = 3$, what is $\sqrt[3]{27.5}$ approximately equal to? This last problem is easy enough with an electronic calculator, but in fact the calculator itself mimics a procedure like the one we will outline. Recall the approximate equality (2):

$$f(x + h) - f(x) \simeq df$$

Alternatively, we can write this as

$$f(x + h) \simeq f(x) + df(x;h)$$

yielding the following concept, illustrated in Figure 3.16:

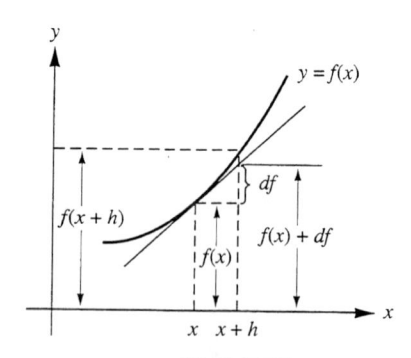

Figure 3.16

The Differential Estimate of an Unknown Function Value

Let $y = f(x)$ be given. Suppose that $f(x)$ is a known value of this function. If h is a sufficiently small number, then

$$f(x + h) \simeq f(x) + df(x;h) = f(x) + f'(x)h \tag{5}$$

This principle may remind you of the discussion in Section 1.4 in which the derivative was related to the idea "if things just keep going like this." The differential $df(x;h) = f'(x)h$ measures how much "things keep going like this" over the distance h.

EXAMPLE 4

Use the differential to estimate the value of $\sqrt[3]{27.5}$.

Solution Let $f(x) = \sqrt[3]{x} = x^{1/3}$. Then $f(27) = 3$ and

$$\sqrt[3]{27.5} = f(27 + 0.5)$$

Our problem is to estimate $f(27 + 0.5)$. Therefore, let $h = 0.5$. By the preceding differential estimate, we have

$$\sqrt[3]{27.5} = f(27 + 0.5) \simeq f(27) + df(27;0.5) = 3 + f'(27)(0.5)$$

Since

$$f'(x) = \frac{1}{3}x^{-2/3} = \frac{1}{3(\sqrt[3]{x})^2}$$

we see that

$$f'(27) = \frac{1}{3(\sqrt[3]{27})^2} = \frac{1}{3(3)^2} = \frac{1}{27}$$

Therefore

$$\sqrt[3]{27.5} \simeq 3 + \frac{1}{27}(0.5) = 3 + \frac{1}{27} \cdot \frac{1}{2} = 3 + \frac{1}{54} = 3.0185 \quad \blacksquare$$

Computation by an electronic calculator, accurate to five digits, gives an answer of 3.018405; thus the differential estimate is quite close.

EXAMPLE 5

The Fast-Bilt Manufacturing Co. finds that its total revenue from the sale of x thousand units of its product is

$$R(x) = 3x - \frac{10}{x} \text{ thousand dollars}$$

Approximate its total revenue from the sale of 105,000 units.

Solution If 100,000 units are sold, then $x = 100$, and total revenue is

$$R(100) = 3(100) - \frac{10}{100} = 300 - 0.10 = 299.9 \text{ thousand dollars}$$

But we wish to know instead $R(105) = R(100 + 5)$. According to the differential-estimate formula, with $x = 100$ and $h = 5$, the total revenue $R(105)$ is approximately

$$R(100 + 5) \simeq R(100) + dR(100;5) = R(100) + R'(100) \cdot 5$$

or

$$R(105) \simeq R(100) + R'(100) \cdot 5$$

Since $R'(x) = 3 + (10/x^2)$, we have

$$R(105) \simeq 299.9 + \left[3 + \frac{10}{(100)^2}\right](5) \simeq 299.9 + 3.001(5)$$

$$= 314.905 \text{ thousand dollars} \quad \blacksquare$$

The advantage to using the differential in an estimate like this is that the additional quantity of five extra (thousands of) units enters the computation as a *linear* factor. This added efficiency is slight in this example but is significant when large or fractional exponents act on the variable x. You may compare the preceding estimate, $314,905, with the *actual* revenue by computing $R(105) = 3(105) - (10/105)$.

Exercises 3.3

1. Let $f(x) = x^2 + 1$. Find $f(x)$, $f(x + h)$, $f(x + h) - f(x)$, and $df(x;h)$ for $x = 2$ and $h = \frac{1}{2}$. Compare your last two answers and indicate these on a graph.

2. Repeat Exercise 1 for $f(x) = 1/(x^2 + 1)$, $x = 2$ and $h = -1$.

In Exercises 3–8, you are given a function and an observed value of this function at a point x, along with a known limit on the error in determining x. Estimate the possible error in the observed function value.

Function	Observed Value	Maximum Error in x
3. $f(x) = x^2$	$f(3) = 9$	± 0.02
4. $f(x) = x^2$	$f(2) = 4$	± 0.05
5. $f(x) = x^3 + \dfrac{1}{x^2}$	$f(1) = 2$	± 0.02
6. $f(x) = x + \sqrt{x} - \dfrac{x^3}{4}$	$f(4) = -10$	± 0.2
7. $f(x) = x(4 - x^2)$	$f(2) = 0$	± 0.1
8. $f(x) = x^{10}$	$f(1) = 1$	± 0.05

9. Suppose that you are given a foot-long ruler and asked to find the area of a square and that you observe the length of each side of the square to be 4 in. with a possible error of $\frac{1}{200}$ in. What is the possible error, then, in your measurement of the area of the square? What are the percentage errors in measuring the length and the area of the square, respectively?

10. Suppose that you are given a foot-long ruler and asked to determine the area and the circumference of a circle. Suppose that you measure the radius of the circle to be 5 in. with a possible error of $\frac{1}{200}$ in. Using $\pi \approx 3.1416$, what are the area and the circumference that you obtain, and how large are the percentage errors in your measurements of the radius, the circumference, and the area of this circle, respectively?

11. The radius of the earth is approximately 4,000 mi. Imagine that you have tied a string tightly around the circumference of the earth at the equator. This string will be $2\pi(4,000) = 25,134$ mi or $132,707,520$ ft long, since there are 5,280 ft in a mile. Now, cut the string momentarily and add to it an additional piece of string 1 ft long. How far above the surface of the earth (at the equator) will you be able to raise this new piece of string, uniformly, all around the earth? [Let $f(x) = x/2\pi$ be the radius of a circle of circumference x. Compute $df(4,000;1/5,280)$.]

In Exercises 12–17, utilize the differential to estimate the indicated value using a nearby function value. In each case, specify x and h and determine $df(x;h)$ first.

12. $f(3.12)$; $f(x) = \dfrac{3}{x}$.   13. $f(4.1)$; $f(x) = \sqrt{x}$.

14. $f(0.12)$; $f(x) = \dfrac{1}{x+1}$.

15. $f(2.8)$; $f(x) = \sqrt{3x}$.

16. Estimate $\sqrt[5]{31}$.

17. Estimate $1.9/2.9$. [Hint: Let $f(x) = x/(x+1)$.]

18. Suppose that $y = \sqrt{x}$ is a model for the height of the bounce of a tennis ball that hits the court 16 ft in front of you. Assume that your racket remains in contact with the ball for a total of 6 in. (horizontally) from the point of initial impact. Use the differential to compute approximately how far the tennis ball wants to move *vertically* along the face of your racket during the 6 in. of horizontal contact.

19. Suppose that you are driving along an interstate highway; your speedometer cable breaks, so you are no longer able to use the speedometer to check your speed. You do have a watch, though, and you realize that markers are placed along the highway at 1-mi intervals. You decide to determine your speed by seeing how long it takes you to drive 1 mi. However, you also realize that with looking from the mile marker down to your watch, you cannot observe the amount of time exactly, and so you guess that your observation of elapsed time between mile markers has an error of 1.5 sec. Recall that

$$d = rt$$

(distance equals rate times time), so that over a distance of 1 mi the function describing your speed (or rate) is

$$r(t) = \dfrac{d}{t} = \dfrac{1}{t}$$

a. Suppose that you observe that it takes you 64 sec to travel 1 mi, resulting in a calculated rate of 56.25 mph. How much error is inherent in this calculation?

b. Suppose instead that you measure your time over a 5-mi interval, resulting in a time of 5.4 min with the same inherent error of 1.5 sec. Calculate your resulting observed speed and the inherent error in this observation.

20. Calculate the *percentage* errors in your time and speed measurements in Exercise 19(a) and (b).

21. The speedometer on your car works roughly like this. A cable connects the speedometer mechanism to the drive shaft, which rotates as your tires rotate along the highway. The mechanism in the speedometer counts the number of revolutions the tires make and converts this to a distance measurement (shown on

the odometer) and a speed measurement (shown on the speedometer). However, the counting mechanism is scaled to the radius of a *new* tire on your car, and after a time your tires wear down to a smaller radius. An old tire must therefore revolve more often to cover the same distance. As a result, your speedometer will show a higher speed than you are actually moving when your tires are worn—a built-in safety edge! There are 5,280 ft/mi, and a common tire radius is 1.25 ft (new). How many revolutions does this tire make per mile? Approximately how many revolutions does it make per mile when the tread is worn by 0.5 in.? [*Hint:* If r is the radius of the tire, then $f(r) = 5{,}280/2\pi r$ is the number of revolutions per mile.]

Challenge Problem

22. The Hard-Gear Manufacturing Co. manufactures a steel rod that is cone shaped on one end (like a sharpened pencil; see Figure 3.17). These rods are used as part of a gearing mechanism, and the cone must be plated with tungsten for durability. This is done by dipping the cone in a plating solution. The surface area of the cone is

$$A(l) = \frac{\pi}{3} l^2$$

from its vertex to a point l in. above the vertex (measured along the axis of the cone). The cone is $4\frac{1}{8}$ in. long and is dipped in the plating solution for a distance of 4 in. with a possible error of $\frac{1}{8}$ in. The plating solution leaves a coating $\frac{1}{64}$ in. thick on the cone, and it can be shown that the volume of plating on the cone is

$$\frac{\frac{\pi}{3} l^2}{64} \text{ in.}^3$$

when it is dipped l in. deep into the solution. What is the volume error in dipping each cone? If the plating solution costs $16 per cubic inch and 100,000 such rods are to be produced, what is the approximate cost of the $\frac{1}{8}$-in. error in dipping to the manufacturer? [*Hint:* Use the differential to estimate how much extra solution might be plated on each cone.]

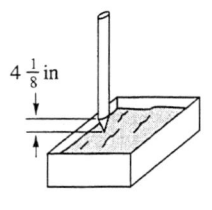

$4\frac{1}{8}$ in

Figure 3.17

Chapter 3 Summary

1. $D_x y^n = n\, y^{n-1} D_x y$ when y is given implicitly as a function of x. All differentiation rules otherwise apply in implicit differentiation.

2. To solve a related-rate problem, first freeze the motion and obtain the state equation for the problem. Put rates of change into the problem by differentiating the state equation to obtain the change-of-state equation.

3. By definition the differential of f at x, for a change of h in x, is $df = f'(x)h$ and is an estimation of the change in function values $f(x + h) - f(x)$. If $f(x)$ is an observed value and $f(x + h)$ is an actual value, the error $E(h) = f(x + h) - f(x)$ between observation and reality is approximately df. Additionally, given $f(x)$ we can estimate $f(x + h)$ by

$$f(x + h) \approx f(x) + df$$

Chapter 3 Summary Exercises

Find $D_x y$ in Exercises 1–4.

1. $x^2 - y^3 = 2x$
2. $x^2 y^3 = 1$
3. $x - xy + y = \sqrt{x}$
4. $\dfrac{x}{y} = x + y$

In Exercises 5–10, find the equation of the tangent line to the graph of the equation at the indicated point.

5. $y + y^2 = x$ at $(2, 1)$
6. $y + y^2 = x$ at $(2, -2)$
7. $\dfrac{1}{y} - x = x^2$ at $x = 1$
8. $y^3 - x^2 = 4$ at $y = 2$
9. $xy = 3$ at $x = 12$
10. $x^2 y^3 = -1$ at $x = 1$

11. Imagine crushed coal being taken out of a hopper car and deposited into a conical pile in the yard of the local power plant (see Figure 3.18). Suppose that the dimensions of the conical pile are such that the radius of the pile equals its height.

Figure 3.18

  a. If the radius is increasing at the rate of 0.5 ft/min, how fast is coal being taken from the hopper car when the radius is 10 ft? 20 ft?

  b. If coal is being deposited on the pile at the rate of 100 ft³/min, how fast is the height of the pile increasing when the radius is 15 ft?

12. A boat is being pulled into a dock by a rope (see Figure 3.19). The dock is 2 ft above the water. If the rope is being hauled in at the rate of 3 ft/sec, how fast is the boat moving toward the dock?

Figure 3.19

13. A company that manufactures power tools finds that when it is selling x units per day, its profit is

$$P = \dfrac{x}{20} - \sqrt{x} \text{ thousands of dollars}$$

How fast are profits increasing when sales are 400 units per day and increasing at the rate of 16 units per day?

14. One aircraft leaves the eastbound runway of an airport at noon and travels east at 300 mph. A second aircraft leaves the northbound runway at 1 P.M. and travels north at 400 mph. How fast are they separating at 1:30 P.M.? How fast is their rate of separation changing at that time?

15. Suppose that a snowball is melting at a rate proportional to its surface area. Show that its radius is decreasing at a constant rate.

In Exercises 16–19, estimate the indicated value of f by using a nearby value and the differential of f.

16. $f(7)$ where $f(x) = \sqrt[3]{x}$
17. $f(3.14)$ where $f(x) = \dfrac{\pi}{x}$
18. $f(8)$ where $f(x) = \dfrac{81}{x^2}$
19. $f(0.15)$ where $f(x) = \sqrt{x}$

20. Suppose that you are measuring a square playing field of dimensions 50 by 50 ft with a 25-ft-long tape measure. You estimate that you will make as much as a $\tfrac{3}{4}$-in. error in measuring the length of each side. How large is the resulting error in the total area of the square playing field?

21. A construction firm is pouring concrete footings of the shape shown in Figure 3.20. It is able to control the top and bottom dimensions at 8 by 8 in. and 16 by 16 in., respectively, by using wooden forms. But it is impossible to exactly control the height x of the footing. The company will try to build these concrete footings 18 in. deep with an error of ± 1 in. in height. If 6 of these footings are built, how much error should be allowed for in the amount of concrete to be used? The volume of a footing of height x is

$$V(x) = \dfrac{16}{243} x^3 - \dfrac{64}{9} x^2 + 256 x$$

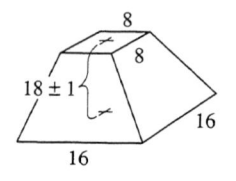

Figure 3.20

204 Chapter 3 Related Rates and the Differential

Chapter 4

Integration

The Whole as the Sum of Continually Varying Parts

In this chapter we begin the study of the other side of calculus: integration and the integral of a function. *Integration is a formal mathematical method for computing the total of a continually varying contribution to a whole.* For example, imagine a stream that runs through farmland of varying slope and contour and then into a city reservoir. The amount of fertilizer and pesticide runoff from the land into the stream varies continually with the land contour along the entire length of the stream. The total of these varying contributions of fertilizer and pesticide ends up in the reservoir. The method of mathematical integration allows us to calculate the resulting total pollution using information about the varying rate of runoff along the stream's length. As a mathematical concept, integration has vastly wider application than this, however, since much of what we experience is the accumulation of many small, diverse contributions. For example, temperature is a measure of the sum of the tiny molecular motions in a heated body.

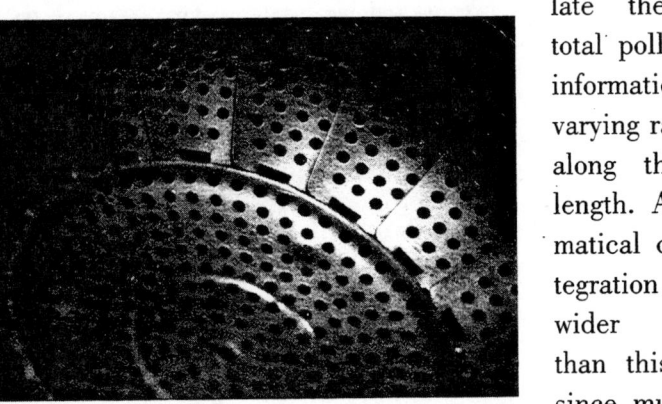

205

4.1 Integration and the Area under a Curve

In this section we define the definite integral of a function and study its approximation via the concept of a Riemann sum.

The process of differentiation finds the rate at which change is occurring given information about how much is available at any point. Suppose that we have the opposite situation. Suppose that we are given information about the rate at which change is occurring. Can we then compute how much is present at any point?

Integration is the name given to the formal mathematical method used to compute the total amount resulting from a *continually varying* contribution to a whole. In the chapter introduction we envision a varying rate of runoff into a stream along its entire edge. When such information is available in numerical or functional form, we can use integration to compute the total amount of runoff into the stream along its entire length.

Differentiation and integration are thus inverses of each other. The Fundamental Theorem of Calculus, which we will study in Section 4.2, is the precise statement of their relationship. The Fundamental Theorem shows that differentiation and integration play a role in calculus analogous to that of addition and subtraction in arithmetic in that they are indeed computational opposites of one another.

We now wish to define the second tool of calculus, the *definite integral* of a function. We will see that the integral of a nonnegative function is the area of the region defined by the varying height of its graph above the axis. We will use a hypothetical but familiar example to motivate the definition of the integral and to see how the area of such a region also coincides with the measurement of a quantity having physical meaning. The integral is a more technically complicated concept than the derivative. Its definition, however, is based on the same theme: Approximate the complex by something simpler.

Suppose you are pedaling your bicycle at a steady rate of 2 yd/sec along a flat, open roadway. Suddenly a gust of wind pushes you from behind for a few seconds and momentarily increases your speed. A graph of your speed at various times could look like Figure 4.1.

This graph represents how fast you are moving at each instant. These sometimes constant, sometimes varying speeds—rates of change of distance over

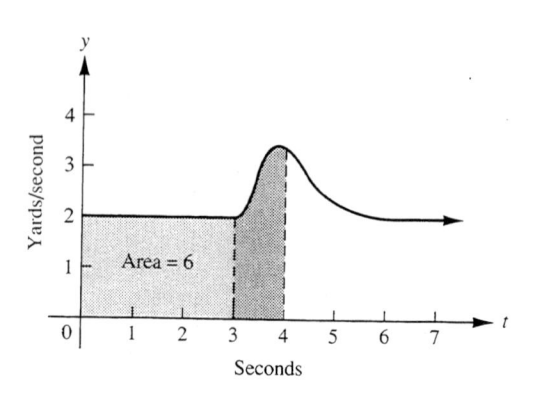

Figure 4.1

time—contribute by accumulation to the total distance you actually travel. How can we calculate this distance using these varying rates?

In the first 3 sec of your journey, before the wind pushes you a little faster, this calculation is not difficult. Since

$$r = \frac{d}{t} \quad \left(\text{rate} = \frac{\text{distance}}{\text{time}}\right)$$

then $d = rt$. So, during the first 3 sec, when your speed r is a constant 2 yd/sec, you will travel a total of

$$2(\text{yd/sec}) \times 3(\text{sec}) = 6(\text{yd})$$

Notice that the number 6 is *also the area* of the region beneath the graph in Figure 4.1 between $t = 0$ and $t = 3$. As we will see shortly, this equivalence is significant and not just a coincidence.

Now let us make a more difficult calculation. How far do you travel during the next second, from $t = 3$ to $t = 4$? This answer is not obvious, because your speed is continually varying—from a low of 2 yd/sec at $t = 3$ to a high of 3.4 yd/sec near $t = 3.75$ and then down to a speed of about 3 yd/sec at $t = 4.25$. How can we compute how far you have moved?

We can take a simple approach to this problem by using the graph to estimate an average, or typical, speed for the time interval from $t = 3$ to $t = 4$ and by using this average speed to compute total distance. Before continuing, consider Figure 4.1 again. Beginning at the point on the graph above $t = 3$, move your pencil along the graph until you reach a point for t somewhere between $t = 3$ and $t = 4$ at which your eye tells you an average, or typical, speed occurs. (Notice that since your speed is near its peak for almost half the time, this average cannot simply be the average of your high and low speeds.)

Now compare your estimate of average speed with the author's choice of about 2.9 yd/sec at about $t = 3.45$ sec (see Figure 4.2). The distance you travel should then be about

$$2.9(\text{yd/sec}) \times (4 - 3)(\text{sec}) = 2.9(\text{yd})$$

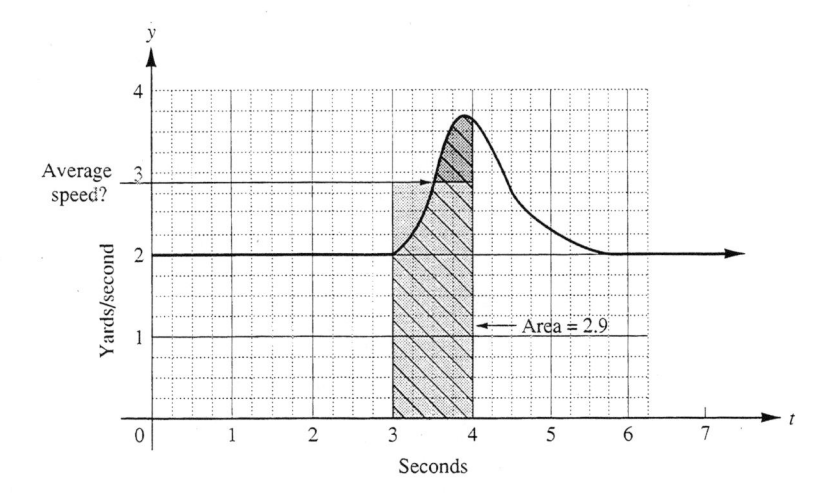

Figure 4.2

4.1 Integration and the Area under a Curve

over the time interval from $t = 3$ to $t = 4$. Again, this product equals the area of a rectangular region (shaded light and medium grey in Figure 4.2). Note that the area of this rectangle appears to equal the cross-hatched area of the region under the curve from $t = 3$ to $t = 4$. The dark grey region below the curve and above the horizontal line (that indicates average speed) seems to balance the light grey region outside the curve. This is again the equivalence noted earlier: The area of the region beneath the rate curve appears to equal the total distance traveled.

What we have uncovered here is a hint of a principle:

If $y = f(t)$ expresses the rate at which a quantity changes (in our example, this quantity is distance), then the area of the region under the graph of $y = f(t)$ represents the total amount (total distance) that this quantity changes.

Mathematically such a relationship is the perfect circumstance, since it tells us that the problem of computing the physical quantity *distance* can be replaced by the conceptually simpler and purely mathematical problem of computing *area*. This brings us to the definition of the definite integral, the third basic concept of calculus, joining the limit and the derivative.

Definition

Let f be a continuous function whose domain includes the interval $[a, b]$, and suppose that $f \geq 0$ on this interval. We define the **definite integral** of f from a to b, denoted by

$$\int_a^b f(x)\, dx$$

to be the area of the region beneath the graph of f above the x-axis that lies between the vertical lines drawn at a and at b. The symbol $\int$ is called an **integral sign**; the function f is called the **integrand** (see Figure 4.3).

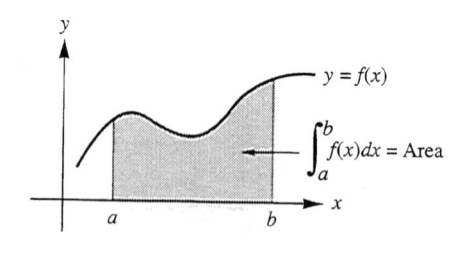

Figure 4.3

Despite the unusual notation, the symbol

$$\int_a^b f(x)\,dx$$

is only a number—an area measurement. Our task here and in Section 4.2 is to learn how to compute this number. We will not discuss the technical reasons why such a number should exist but will only remark that for a continuous function f there are no difficulties: The definite integral $\int_a^b f(x)\,dx$ always exists.

In this and the next section we refer to the definite integral as simply "the integral." We define the integral first for a function with nonnegative values. After discussing the computation of the integral by approximation by Riemann sums, we will extend the definition to functions that have negative values.

Riemann Sums

There are two principal ways to compute the integral of f from a to b. The first is by approximation. Although this is tedious without the aid of a computer or calculator, understanding how the integral is approximated is the first step in learning to apply the integral to practical problems. The second method of computing the integral, which is much more brief, will be given by the Fundamental Theorem of Calculus in Section 4.2.

Recalling our initial example, we could try to estimate the number $\int_a^b f(x)\,dx$ by visually finding an average height on the graph of $y = f(x)$, indicated at Y_{avg} in Figure 4.4, and then multiplying this height by $b - a$, the length of the interval. Thus

$$\int_a^b f(x)\,dx \simeq (b - a)Y_{\text{avg}}$$

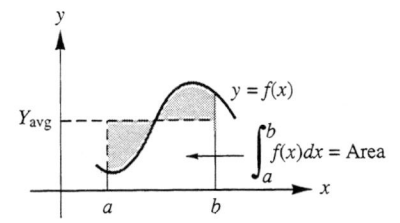

Figure 4.4

We write this as an approximate equality because we cannot easily find the exact average height Y_{avg}. Perhaps we could make a better estimate if the interval were not so long. Let us divide the interval $[a, b]$ into, say, four equal parts, each of length $(b - a)/4$, and try to pick an average height in each interval (see Figure 4.5).

The number $f(x_k)$, for $k = 1, 2, 3, 4$, represents our best estimate of the average height of $y = f(x)$ over the kth interval, for $k = 1, 2, 3, 4$. Since the length of each interval is $(b - a)/4$, we see that

$$\int_a^b f(x)\,dx \simeq f(x_1)\left(\frac{b-a}{4}\right) + f(x_2)\left(\frac{b-a}{4}\right) + f(x_3)\left(\frac{b-a}{4}\right) + f(x_4)\left(\frac{b-a}{4}\right)$$

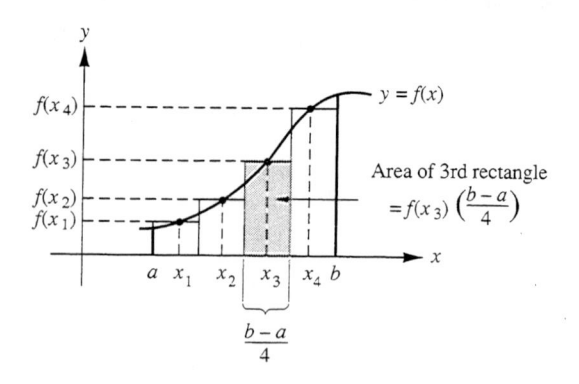

Figure 4.5

Area of 3rd rectangle
$= f(x_3)\left(\frac{b-a}{4}\right)$

Each product in this sum measures a rectangular area of height $f(x_k)$ and width $(b-a)/4$.

This computation may give us a better approximation, but it is still "eye-dependent." In the late nineteenth century the German mathematician Georg Friedrich Bernhard Riemann had a better idea and removed this "eye dependency." Riemann reasoned that if we use a very large number of very short intervals, we need not be concerned what height we choose as an average height. For example, if we were to use 100 intervals, each of length $(b-a)/100$, any value of the function over such a short interval could not differ very much from the average value (height) of the function on this short interval (see Figure 4.6). Riemann's idea is formalized in the following definition.

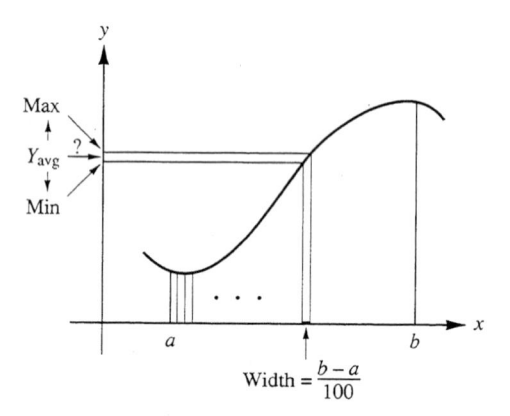

Figure 4.6

Width $= \frac{b-a}{100}$

Definition

Let $y = f(x)$ be a continuous function on the interval $[a, b]$. Let n be a positive whole number, and let $\Delta x = (b-a)/n$. Divide the interval $[a, b]$ into n equal subintervals of length Δx. Choose *any* point x_k in the kth of these intervals, for each $k = 1, 2, \ldots, n$. The number

$$f(x_1)\Delta x + f(x_2)\Delta x + \cdots + f(x_n)\Delta x$$

is called a **Riemann sum**.

210   Chapter 4   Integration

A Riemann sum for a function with values $f(x) \geq 0$ is a sum of areas of rectangles, each rectangle being of height $f(x_k)$ and width Δx, and thus having area

$$f(x_k)\Delta x$$

We show such rectangles in Figure 4.7. The total of these rectangular areas—that is, the numerical value of the Riemann sum—*approximates* the area beneath the graph of $y = f(x)$ from $x = a$ to $x = b$.

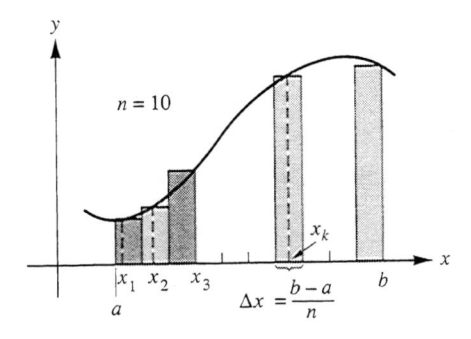

Figure 4.7

This last point is Riemann's theorem, which says that, if we use a large number n of subintervals, we can be as careless as we like in choosing a rectangular height from each subinterval. This theorem will be much used in coming applications of the integral.

Theorem 4.1

Riemann's Theorem

If $y = f(x)$ is continuous on $[a, b]$ and $f(x) \geq 0$ for all x in $[a, b]$, then the area $\int_a^b f(x)\,dx$ beneath the graph of f from $x = a$ to $x = b$ can be approximated to any desired degree of accuracy by a sum of the *form*

$$f(x_1)\Delta x + f(x_2)\Delta x + \cdots + f(x_n)\Delta x$$

where $\Delta x = (b - a)/n$ and x_k is any point in the kth interval of length Δx, if n is a large enough whole number.

Applications of Riemann Sums

EXAMPLE 1

Use Riemann's theorem with $n = 10$ subintervals to approximate the area beneath the curve of $y = f(x) = x^2$ on the interval $[0, 1]$. (Before going on to the solution, consider the graph of $y = x^2$ shown in Figure 4.8. Visually estimate its average height and, using that length, estimate the area beneath the graph.)

4.1 Integration and the Area under a Curve

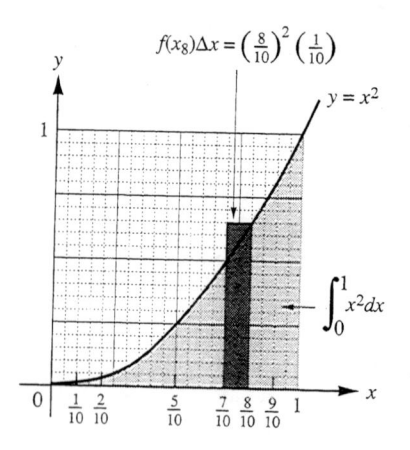

Figure 4.8

$$f(x_8)\Delta x = \left(\frac{8}{10}\right)^2\left(\frac{1}{10}\right)$$

$$y = x^2$$

$$\int_0^1 x^2\, dx$$

Solution With $n = 10$,

$$\Delta x = \frac{1-0}{10} = \frac{1}{10}$$

The intervals we will use begin at 0 and progress to 1 in tenth-of-a-unit steps. Since Riemann's theorem tells us that we may use any height within each interval, let us use the height $f(x_k)$ at $x_k = $ the right endpoint of each interval. These right endpoints are at $\frac{1}{10}, \frac{2}{10}, \ldots, \frac{9}{10}$, and $\frac{10}{10}$, so $x_k = k/10$. The resulting Riemann sum is then

$$f\left(\frac{1}{10}\right)\Delta x + f\left(\frac{2}{10}\right)\Delta x + f\left(\frac{3}{10}\right)\Delta x + \cdots + f\left(\frac{9}{10}\right)\Delta x + f\left(\frac{10}{10}\right)\Delta x$$

$$= \left(\frac{1}{10}\right)^2\left(\frac{1}{10}\right) + \left(\frac{2}{10}\right)^2\left(\frac{1}{10}\right) + \left(\frac{3}{10}\right)^2\left(\frac{1}{10}\right) + \cdots + \left(\frac{9}{10}\right)^2\left(\frac{1}{10}\right) + \left(\frac{10}{10}\right)^2\left(\frac{1}{10}\right)$$

$$= [1^2 + 2^2 + 3^2 + \cdots + 9^2 + 10^2]\left(\frac{1}{10^2}\right)\left(\frac{1}{10}\right)$$

$$= [1 + 4 + 9 + 16 + 25 + 36 + 49 + 64 + 81 + 100]\frac{1}{1{,}000}$$

$$= \frac{385}{1{,}000} = 0.385$$

Thus $\int_0^1 f(x)\, dx = \int_0^1 x^2\, dx \approx 0.385$. There is approximately 0.385 unit of area beneath the graph. As we will see in Section 4.2, the exact answer is $0.3333\ldots$. If we had used $n = 100$ subintervals, our answer would have been more accurate, but it also would have taken quite a bit of work. (How does the answer 0.385 compare with the visual estimate we asked you to obtain earlier?) ∎

As Example 1 indicates, a Riemann sum is not a quick and easy way to estimate the number $\int_a^b f(x)\, dx$. Nor does Riemann's theorem tell us how accurate an estimate a Riemann sum gives. The significance of Riemann sums is conceptual, and its importance lies in applications of the integral.

EXAMPLE 2

Recall the example at the beginning of this chapter. A stream is running through farmland of varying slope and contour and into a city reservoir downstream. Rainfall running off the surrounding fields carries fertilizer and pesticide residues into the stream and so into the city's water supply. The contours of the land cause a varying rate of runoff. Let $y = f(x)$ represent the number of tons of residue per mile running into the stream at a distance of x miles from the reservoir, represented by $x = 0$. Explain why the number $\int_0^9 f(x)\, dx$ represents the total number of tons of residue that go into the stream in the first 9 miles of its length above the reservoir (see Figure 4.9).

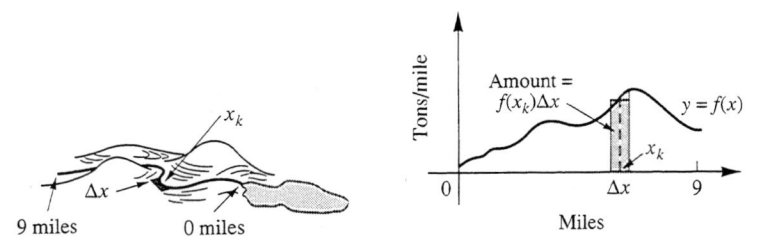

Figure 4.9

Solution By Riemann's theorem the number $\int_0^9 f(x)\, dx$ is approximated by a sum of the form

$$f(x_1)\Delta x + f(x_2)\Delta x + \cdots + f(x_n)\Delta x$$

Consider a single term $f(x_k)\Delta x$. The number $f(x_k)$ is the number of tons per mile running off into the stream at a distance of x_k miles above the reservoir. The number Δx is a length of stream Δx miles long occurring about the point x_k—a subinterval of the stream. Thus the units of $f(x_k)\Delta x$ are (tons/mile) × miles = tons, and we see that $f(x_k)\Delta x$ counts approximately how many tons of residue run off into the stream along a piece of stream Δx miles long containing the point x_k above the reservoir. Therefore the Riemann sum

$$f(x_1)\Delta x + f(x_2)\Delta x + \cdots + f(x_n)\Delta x$$

approximates the contribution of residue along short pieces (each Δx long) of the stream along its entire length from the reservoir to a point 9 miles upstream and so approximates the total contribution. Since this sum *also* approximates $\int_0^9 f(x)\, dx$, this number *is* the total contribution. That is, the integral represents the ultimate runoff into the reservoir. ∎

This example indicates why we must understand the complex notion of Riemann sums. The notation

$$\int_a^b f(x)\, dx$$

represents a purely geometric calculation—the area under the curve $y = f(x)$. Riemann sums enable us to understand what this number measures when $f(x)$ measures a quantity in the external world.

Riemann sum approximation also answers a question that may have occurred to you: Why the peculiar notation $\int_a^b f(x)\,dx$ for the old idea of area? The elongated "S," $\int$, reminds us of the *sum* aspect of Riemann sums, and the dx corresponds to Δx. Each term in a Riemann sum is of the form $f(x)\Delta x$, and all terms of this form, beginning at a and ending at b, are to be added. Thus $\int_a^b f(x)\,dx$ recalls the Riemann method of computing the area by summing products $f(x)\Delta x$ from a to b; the numbers a and b are called the lower and upper *limits of integration*, respectively.

The Integral of a Function with Varying Sign

The integral of a function having both positive and negative function values is *not* defined as the area bounded by its graph. Instead, and consistent with Riemann's theorem, for any continuous function f on $[a, b]$ the definite integral of f from a to b is defined by

$$\int_a^b f(x)\,dx = \lim_{\Delta x \to 0} (f(x_1)\Delta x + f(x_2)\Delta x + \cdots + f(x_n)\Delta x)$$

where $\Delta x = (b - a)/n$ is the length of each subinterval defining the Riemann sum and x_k is any point chosen from the kth subinterval.

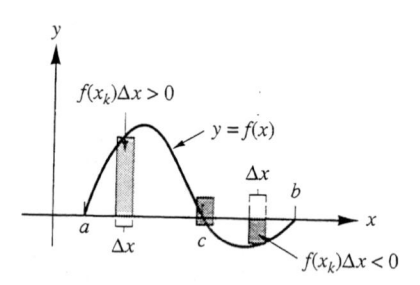

Figure 4.10

This definition gives rise to a number of technical matters that are beyond the intent of this text. For our purposes, we need only realize the following: Figure 4.10 illustrates typical rectangles whose areas arise in the definition. When $f(x_k) < 0$, $f(x_k)\Delta x$ is the negative value of the area of the indicated rectangle. Consequently, the integral of f from a to b is numerically equal to the difference between the (positive) area of the region below its graph and above the x-axis (where $f \geq 0$) less the (positive) area of the region above its graph and below the x-axis (where $f \leq 0$). That is, in the notation of Figure 4.11,

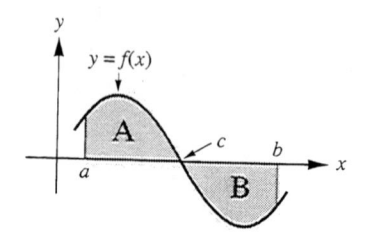

Figure 4.11

$$\int_a^b f(x)\,dx = \text{Area }(A) - \text{Area }(B)$$

The point $x = c$ is found by solving the equation $f(x) = 0$; we can then find $\int_a^b f(x)\,dx$ by finding each area separately and then subtracting. The Fundamental Theorem of Calculus of Section 4.2 will make this detail of calculation obsolete.

Exercises 4.1

Estimate the area $\int_0^2 f(x)\,dx$ for each of the functions whose graph is shown in Figures 4.12–4.15. Make one estimate by estimating the average height of the graph of f and multiplying by the width of the interval. Then make a second estimate by counting squares in the figures.

1.

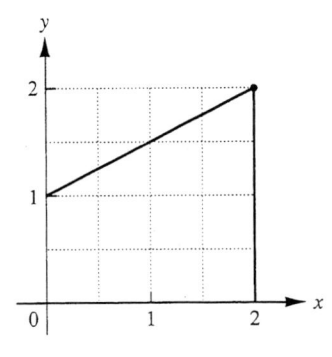

Figure 4.12

2.

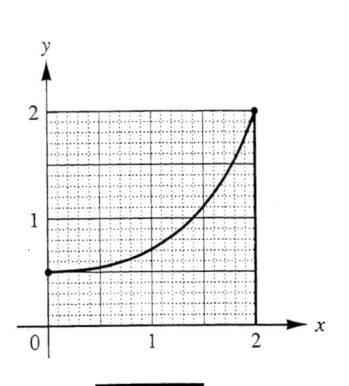

Figure 4.13

3.

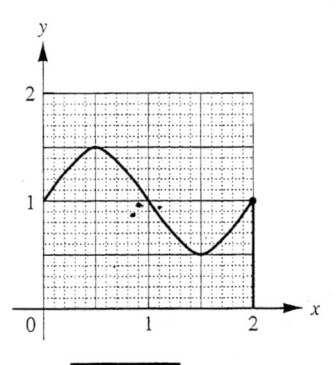

Figure 4.14

4.

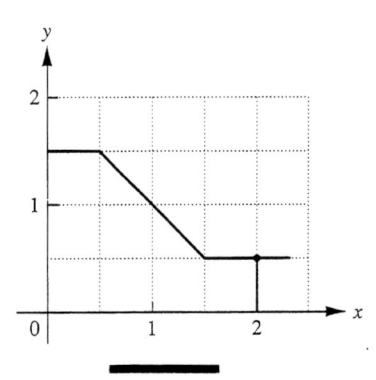

Figure 4.15

Estimate the integral $\int_0^2 f(x)\,dx$ for each function whose graph is shown in Figures 4.16 and 4.17 by counting squares in each figure. Remember that since these functions have negative values, your answer cannot be a simple area measurement.

4.1 Integration and the Area under a Curve   215

5.

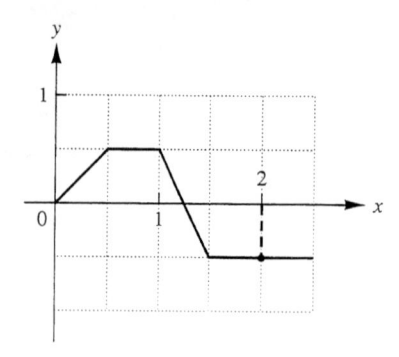

Figure 4.16

6.

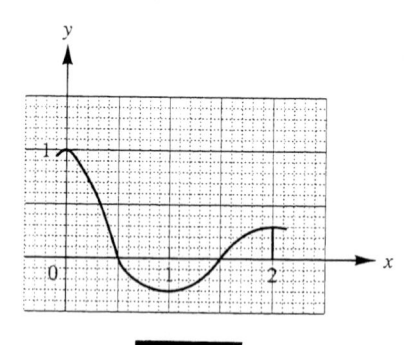

Figure 4.17

7. Let $f(x) = 2x$. The graph of f is shown in Figure 4.18.

   a. Compute $\int_0^1 f(x)\, dx$ as the area of a triangle.

   b. Compute $\int_0^1 f(x)\, dx$ by estimating an average height for f.

   c. Estimate $\int_0^1 f(x)\, dx$ by a Riemann sum with $n = 10$, as in Example 1.

8. Refer to Example 1 and estimate $\int_0^1 3x^2\, dx$ using a Riemann sum with $n = 10$ and with each x_k a right endpoint.

9. Refer to Example 1 and estimate $\int_0^1 x^2\, dx$ using a Riemann sum with $n = 10$, but unlike Example 1, evaluate the term $f(x_k)$ at the left endpoint of the kth interval. In doing this exercise also:

   a. Draw a graph and the rectangular regions whose areas are involved in the computation.

   b. Compare your estimate with the estimate in Example 1. How do you explain the difference?

   c. Make a new estimate, again with $n = 10$, but this time evaluate $f(x_k)$ at the *midpoint* of the kth interval. (Thus you will be using $f\left(\frac{1}{20}\right)$, $f\left(\frac{3}{20}\right), \ldots, f\left(\frac{19}{20}\right)$ as factors of $\Delta x = \frac{1}{10}$ in the Riemann sum.)

In Exercises 10–16, decide if the given sum is a Riemann sum defined over [0, 1]. If it is, specify an $f(x)$ and Δx, and illustrate these. If it is not, explain why.

10. $(0)^3\left(\frac{1}{4}\right) + \left(\frac{1}{4}\right)^3\left(\frac{2}{4}\right) + \left(\frac{1}{2}\right)^3\left(\frac{1}{4}\right) + \left(\frac{3}{4}\right)^3\left(\frac{1}{4}\right)$

11. $\left(\frac{1}{4}\right)^2\left(\frac{1}{4}\right) + \left(\frac{1}{2}\right)^2\left(\frac{1}{4}\right) + \left(\frac{7}{8}\right)^2\left(\frac{1}{4}\right) + (1)^2\left(\frac{1}{4}\right)$

12. $(4)\left(\frac{1}{4}\right) + (2)\left(\frac{1}{4}\right) + \left(\frac{4}{3}\right)\left(\frac{1}{4}\right) + (1)\left(\frac{1}{4}\right)$

13. $\left(\frac{1}{4}\right)^2\left(\frac{1}{4}\right) + \left(\frac{1}{2}\right)^2\left(\frac{1}{4}\right) + \left(\frac{3}{4}\right)^2\left(\frac{1}{4}\right) + \left(\frac{1}{1}\right)^2\left(\frac{1}{4}\right)$

14. $2\pi(0)^2\left(\frac{1}{4}\right) + 2\pi\left(\frac{1}{4}\right)^2\left(\frac{1}{4}\right) + 2\pi\left(\frac{1}{2}\right)^2\left(\frac{1}{4}\right) + 2\pi\left(\frac{3}{4}\right)^2\left(\frac{1}{4}\right)$

15. $(1 - 0)(0)\left(\frac{1}{4}\right) + \left(1 - \frac{1}{4}\right)\left(\frac{1}{4}\right)\left(\frac{1}{4}\right) + \left(1 - \frac{1}{2}\right)\left(\frac{1}{2}\right)\left(\frac{1}{4}\right) + \left(1 - \frac{3}{4}\right)\left(\frac{3}{4}\right)\left(\frac{1}{4}\right)$

16. $\left(\frac{1}{5}\right)^2\left(\frac{1}{5}\right) + \left(\frac{2}{5}\right)^2\left(\frac{1}{5}\right) + \left(\frac{3}{5}\right)^2\left(\frac{1}{5}\right) + \left(\frac{4}{5}\right)^2\left(\frac{1}{5}\right) + \left(\frac{4}{5}\right)^2\left(\frac{1}{5}\right)$

17. Imagine a large city where Main Street runs directly east from the city center to the surrounding countryside, a distance of 15 miles from the city center. The graph of $y = f(x)$ in Figure 4.19 is the graph of the population density along the street.

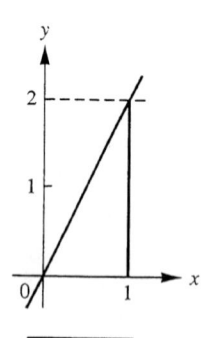

Figure 4.18

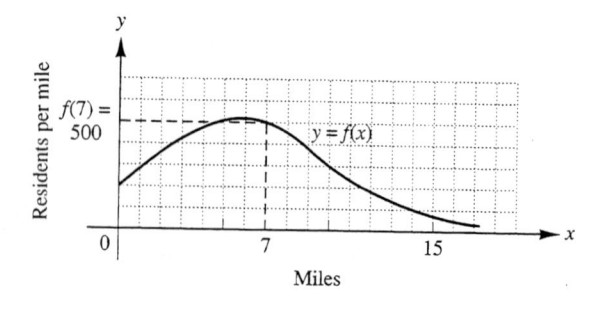

Figure 4.19

216   Chapter 4   Integration

For example, the graph indicates that 7 miles from the city center, there are $f(7) = 500$ residents per mile living on Main Street. Thus one block of Main Street at this 7-mile distance should have approximately $f(7)\left(\frac{300}{5,280}\right) = 28.4$ residents, assuming that a city block is 300 ft long. Use Riemann's theorem to determine what $\int_0^{15} f(x)\, dx$ represents in terms of population.

Challenge Problems

18. A large corporation maintains a bank account in a local bank and daily credits its earnings to the account and debits its costs. Thus money is continually flowing into and out of the account. Figure 4.20 shows the *rate* of income/outgo in the account for 1 year. For example, at $t = \frac{1}{2}$, midyear, the account is receiving money at the rate of $I(t) = \$150$ million per year. Thus, for example, in a 2-week period about midyear, it deposits approximately $\$150(\text{mil/yr}) \times \frac{2}{52}(\text{yr}) = \$5.8(\text{million})$ in its account. Notice that during the last 3 months of the year, the corporation is spending faster than it is earning and the account is losing money at the indicated rate. Use the idea of Riemann sums to explain what the number

$$\int_0^1 I(t)\, dt$$

represents, as area and as dollars. Is there a good correlation here between positive and "negative" area and earnings versus losses?

19. For any whole number n,

$$1^2 + 2^2 + 3^2 + \cdots + (n-1)^2 + n^2 = \frac{n(n+1)(2n+1)}{6}$$

We may replace the limit as $\Delta x \to 0$ in the definition of the integral by the limit as $n \to \infty$ where $\Delta x = (b-a)/n$. With $a = 0$, $b = 1$, and $f(x) = x^2$, let $x_k = k/n$. We may then say, using the general definition of the integral, that

$$\int_0^1 f(x)\, dx = \lim_{n \to \infty}\left[\left(\frac{1}{n}\right)^2\left(\frac{1}{n}\right) + \left(\frac{2}{n}\right)^2\left(\frac{1}{n}\right) + \cdots + \left(\frac{k}{n}\right)^2\left(\frac{1}{n}\right) + \cdots + \left(\frac{n}{n}\right)^2\left(\frac{1}{n}\right)\right]$$

Using the preceding information, simplify the expression inside the limit to equal

$$\frac{2}{6} + \frac{3}{6n} + \frac{1}{6n^2}$$

and then find this limit, thus finding the exact value of $\int_0^1 x^2\, dx$.

Computer Application Problems

20. Use the BASIC program RSUM to see what happens in a Riemann sum approximation of the integral of $f(x) = x^2$ on the interval $[0, 1]$ if one evaluates f at (a) the left endpoint of each subinterval, (b) the midpoint of each subinterval, and (c) the right endpoint of each subinterval using $n = 10$, then $n = 20$ subintervals. Indicate the various results you obtain on a "rough" graph.

21. Use the BASIC program RSUM to approximate the value of the integral of $g(x) = x^3$ on the intervals $[-1, 1]$, $[-1, 2]$, and $[-2, 1]$. Use $n = 25$ subintervals in each case.

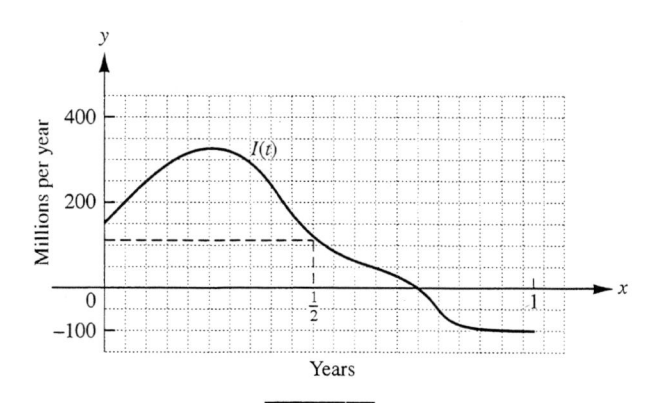

Figure 4.20

4.2 The Fundamental Theorem of Calculus

The Fundamental Theorem of Calculus provides an often easy way to find the exact numerical value of

$$\int_a^b f(x)\, dx$$

The Fundamental Theorem is one of those insightful discoveries that continues to have a hidden but everyday influence on the design and function of technology in the modern world. Surprisingly, the essence of its proof is suggested by a familiar experience.

Recall Example 5 in Section 1.5, where we imagined tossing a pebble into a pool of water and asked, "At what rate is the area enclosed by the expanding circular wave increasing as the radius r lengthens?" We found that this rate is exactly the *length* $2\pi r$ of the moving, *leading edge* of the expanding circle.

In Figure 4.21(b), imagine the region enclosed by the (fixed) vertical lines at $x = a$ and $x = b$, and the (fixed) horizontal axis to be a swimming pool whose remaining boundary is the curved graph of $y = f(x)$. Next think of the line of *length* $f(x)$ at the point x as a straight, leading wave of water moving across the pool from left to right. Let $A(x)$ be the area of the region to the left of this (leading) wave of water when it is at point x. At what rate will this area grow as x moves from a to b? Reasoning by analogy with Figure 4.21(a), the rate of increase $A'(x)$ at x should be $f(x)$, the length of the "leading edge." That is, $A'(x) = f(x)$. This is the essence of the Fundamental Theorem.

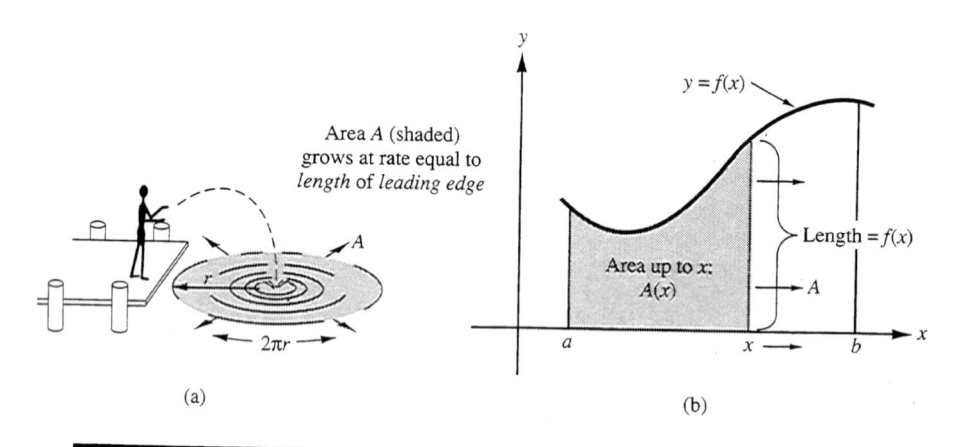

Figure 4.21

(a) Area A (shaded) grows at rate equal to length of leading edge

(b) Area up to x: $A(x)$; Length = $f(x)$

Theorem 4.2

The Fundamental Theorem of Calculus

Suppose that f is a continuous function on the interval $[a, b]$. Then two things are true:

1. $$D_x\left[\int_a^x f(t)\, dt\right] = f(x) \qquad \text{for any } x,\ a < x < b \qquad (1)$$

2. If F is any function whose derivative is f—that is, such that $F'(x) = f(x)$ for all x in $[a, b]$—then

    $$\int_a^b f(x)\, dx = F(b) - F(a) \qquad (2)$$

Equation 1 is the technical statement of the key idea elaborated just above it. The area $A(x)$ beneath the graph of f from a to x varies as x changes. At the same time, the integral $\int_a^x f(t)\, dt$ also represents this same area from a to x; thus

$$A(x) = \int_a^x f(t)\, dt$$

Our central observation, that this area increases at rate $f(x)$ as x moves to the right in Figure 4.21(b), thus has the notational form

$$A'(x) = D_x\left[\int_a^x f(t)\, dt\right] = f(x)$$

We will most often use Equation 2 of the Fundamental Theorem, and we will use it to evaluate a definite integral directly. We will prove Equation 2 after showing how this evaluation is performed.

EXAMPLE 1

Compute $\int_0^1 x^2\, dx$ using the Fundamental Theorem.

Solution To use Equation 2 you must find a new function F such that $F'(x) = f(x) = x^2$. How about $F(x) = x^3/3$? Indeed $F'(x) = 3x^2/3 = x^2$. Therefore, by Equation 2,

$$\int_0^1 x^2\, dx = F(1) - F(0) = \frac{1^3}{3} - \frac{0^3}{3} = \frac{1}{3}$$

Thus the area of the region in Figure 4.22 is $\frac{1}{3}$.

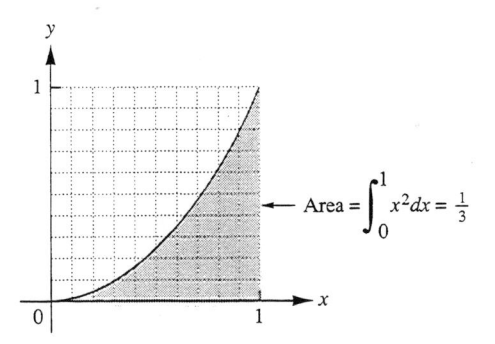

Figure 4.22

This method is considerably easier and more accurate than using a Riemann sum. The only difficulty lies in finding the function F. We will learn systematic ways of finding F for many functions f in later sections.

Assuming that Equation 1 of the Fundamental Theorem is true (a suggestive argument follows), why is Equation 2 correct? Suppose that F is any function whose derivative is f. Then $F'(x) = f(x) = A'(x)$. Since functions with equal derivatives can differ only by a constant (see Theorem A.2, Section A.2), $F(x) = A(x) + C$ where C is a constant. Since $A(a) = \int_a^a f(t)\, dt = 0$ and $A(b) = \int_a^b f(t)\, dt$, then

$$F(b) - F(a) = (A(b) + C) - (A(a) + C) = A(b) = \int_a^b f(t)\, dt$$

Notation. It is convenient to write this last equation using the notation $\int_a^b f(x)\,dx = F(x)]_a^b$ where $F' = f$. That is, we write $F(x)]_a^b$ for $F(b) - F(a)$.

Figure 4.23 suggests an analytical argument for Equation 1 of the Fundamental Theorem. Since $A(x) = \int_a^x f(t)\,dt$ is the area beneath the graph from a to x, then

$$A(x + h) - A(x) = (\text{Area up to } x + h) \text{ minus } (\text{Area up to } x)$$
$$= \text{Area between } x \text{ and } x + h$$
$$= \int_x^{x+h} f(t)\,dt$$

Consequently,

$$A'(x) = \lim_{h \to 0} \frac{A(x+h) - A(x)}{h} = \lim_{h \to 0} \frac{1}{h} \int_x^{x+h} f(t)\,dt$$

Figure 4.23(a) suggests that

$$\frac{1}{h} \int_x^{x+h} f(t)\,dt = \frac{\text{Area of } R}{\text{Bottom width } h} \simeq \text{Average height}$$

As $h \to 0$, the region R begins to look like a tall, narrow rectangle of height approximately $f(x)$. Thus it seems reasonable from Figure 4.23(b) that

$$A'(x) = \lim_{h \to 0} \frac{1}{h} \int_x^{x+h} f(t)\,dt = f(x)$$

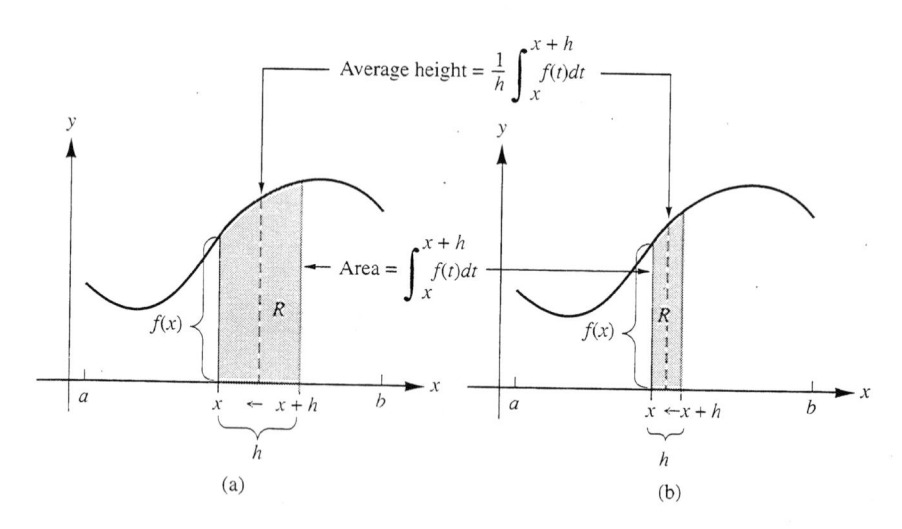

Figure 4.23

(a) (b)

Computation of an Integral Using the Fundamental Theorem

Let us now look at several examples that apply the Fundamental Theorem. We first need the following properties of the integral.

Theorem 4.3 For continuous functions f and g on $[a, b]$ and any (constant) number k, it is always true that

$$1. \quad \int_a^b k f(x) \, dx = k \int_a^b f(x) \, dx$$

and

$$2. \quad \int_a^b [f(x) + g(x)] \, dx = \int_a^b f(x) \, dx + \int_a^b g(x) \, dx$$

For example, we can combine properties 1 and 2 to say that

$$\int_1^3 (x + 5x^3) \, dx = \int_1^3 x \, dx + \int_1^3 5x^3 \, dx \qquad \text{Using property 2}$$

$$= \int_1^3 x \, dx + 5 \int_1^3 x^3 \, dx \qquad \text{Using property 1}$$

We then make a final computation using the Fundamental Theorem. Since

$$D_x\left(\frac{x^2}{2}\right) = x \quad \text{and} \quad D_x\left(\frac{x^4}{4}\right) = x^3$$

we have, using Equation 2 of the Fundamental Theorem,

$$\int_1^3 x \, dx = \left.\frac{x^2}{2}\right]_1^3 = \frac{(3)^2}{2} - \frac{(1)^2}{2} = 4$$

and

$$\int_1^3 x^3 \, dx = \left.\frac{x^4}{4}\right]_1^3 = \frac{(3)^4}{4} - \frac{(1)^4}{4} = \frac{80}{4} = 20$$

Therefore

$$\int_1^3 [x + 5x^3] \, dx = 4 + 5(20) = 104$$

Theorem 4.3 may be quickly proven using the Fundamental Theorem; we will not do so here.

EXAMPLE 2

Compute $\int_1^3 \frac{1}{x^2} \, dx$.

Solution To do this we must find a function F such that $F'(x) = 1/x^2$. Rewrite $f(x) = 1/x^2$ as

$$f(x) = x^{-2}$$

Then $F(x) = -x^{-1}$ because $D_x(-x^{-1}) = -((-1)x^{-2}) = x^{-2}$. Therefore, by Equation 2 of the Fundamental Theorem,

$$\int_1^3 \frac{1}{x^2} \, dx = \left.-x^{-1}\right]_1^3 = \left.\frac{-1}{x}\right]_1^3 = \left(\frac{-1}{3}\right) - \left(\frac{-1}{1}\right) = \frac{2}{3} \quad \blacksquare$$

Compare this answer to the area estimate you can make by counting squares in Figure 4.24, where 1 square unit consists of 100 squares in the plane.

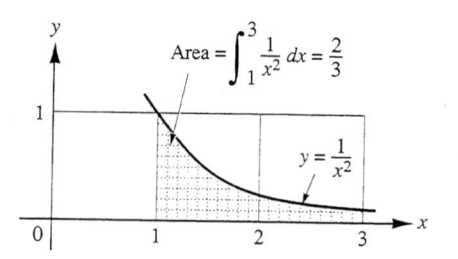

Figure 4.24

In the next example, we use the fact that $(x^{n+1})/(n+1)$ is a function whose derivative is x^n three times, for $n = 1, 2,$ and 3.

EXAMPLE 3

Compute $\int_{-1}^{2} x(1+x)^2 \, dx$.

Solution

$$\int_{-1}^{2} x(1+x)^2 \, dx = \int_{-1}^{2} x(1 + 2x + x^2) \, dx$$

$$= \int_{-1}^{2} (x + 2x^2 + x^3) \, dx$$

$$= \int_{-1}^{2} x \, dx + 2\int_{-1}^{2} x^2 \, dx + \int_{-1}^{2} x^3 \, dx$$

$$= \left.\frac{x^2}{2}\right]_{-1}^{2} + 2\left(\frac{x^3}{3}\right]_{-1}^{2} + \left.\frac{x^4}{4}\right]_{-1}^{2}$$

$$= \left[\frac{2^2}{2} - \frac{(-1)^2}{2}\right] + 2\left[\frac{2^3}{3} - \frac{(-1)^3}{3}\right] + \left[\frac{2^4}{4} - \frac{(-1)^4}{4}\right]$$

$$= \frac{3}{2} + 6 + \frac{15}{4} = \frac{45}{4} \quad \blacksquare$$

A General Rule for Computing $\int_a^b x^n \, dx$

We complete this section with a general rule for the computation of integrals of the form

$$\int_a^b x^n \, dx$$

222 Chapter 4 Integration

In Examples 2 and 3, we found that if $f(x) = x^n$ and $n \neq -1$ then, since

$$D_x\left(\frac{x^{n+1}}{n+1}\right) = (n+1)\frac{x^{(n+1)-1}}{n+1} = x^n$$

we can apply the Fundamental Theorem with

$$F(x) = \frac{x^{n+1}}{n+1}$$

since $F'(x) = f(x)$.

Consequently, we have the repeatedly used formula

$$\int_a^b x^n \, dx = \left.\frac{x^{n+1}}{n+1}\right]_a^b \quad \text{for any } n \neq -1$$

You will use this formula throughout this course. Notice that the formula is true for any real number n, except $n = -1$, and, in particular, is valid for fractional exponents. However, if $n \leq 0$, we cannot apply it when 0 is in $[a, b]$ for then the function $f(x) = x^n$ is not continuous on $[a, b]$.

EXAMPLE 4

Compute $\int_1^4 (x^{-1/2} + 2x^\pi) \, dx$.

Solution
$$\int_1^4 (x^{-1/2} + 2x^\pi) \, dx = \left.\left(\frac{x^{1/2}}{\frac{1}{2}}\right)\right]_1^4 + 2\left.\left(\frac{x^{\pi+1}}{\pi+1}\right)\right]_1^4$$

$$= 2(x^{1/2})]_1^4 + \frac{2}{\pi+1}(x^{\pi+1})]_1^4$$

$$= 2(\sqrt{4} - \sqrt{1}) + \frac{2}{\pi+1}(4^{\pi+1} - 1^{\pi+1})$$

$$= 2(1) + \frac{2}{\pi+1}(4^{\pi+1} - 1)$$

$$= 2\left(1 + \frac{4^{\pi+1} - 1}{\pi+1}\right) \approx 151.95$$

using a calculator to compute the number $4^{\pi+1}$. ∎

A special case of this formula that you might overlook is the next example.

EXAMPLE 5

Compute $\int_1^5 1 \, dx$.

Solution
$$\int_1^5 1 \, dx = \int_1^5 x^0 \, dx = \left.\frac{x^{0+1}}{0+1}\right]_1^5 = x\Big]_1^5 = 5 - 1 = 4 \quad \blacksquare$$

One way to think of this problem is that since $D_x x = 1$ then $F(x) = x$ is the proper function to use in computing the integral. Even better, ask yourself,

"How much area is there between a graph one unit high and the x-axis, along a distance of four units (from $x = 1$ to $x = 5$)?"

We are not able at this time to give a function $F(x)$ such that $F'(x) = 1/x = x^{-1}$. This means that we cannot, without further effort, compute the area

$$\int_a^b \frac{1}{x} \, dx = \int_a^b x^{-1} \, dx \qquad \text{(The case } n = -1 \text{ noted earlier)}$$

by the Fundamental Theorem. We will find this integral in Section 5.3.

Exercises 4.2

1. Let $f(x) = 3$, a constant function.
   a. Compute $\int_1^4 f(x) \, dx$ by drawing a graph of $y = f(x)$ and calculating the appropriate area.
   b. Compute $\int_1^4 f(x) \, dx$ using the Fundamental Theorem.

2. Let $f(x) = x$. Compute $\int_1^3 f(x) \, dx$ by
   a. Drawing a graph and computing the area geometrically.
   b. Using the Fundamental Theorem.

Compute the integrals in Exercises 3–15.

3. $\int_0^4 3x \, dx$

4. $\int_{-1}^2 x^5 \, dx$

5. $\int_{-3}^{-1} x^2 \, dx$

6. $\int_{-3}^{-1} \frac{x^3}{5} \, dx$

7. $\int_0^4 \sqrt{x} \, dx$

8. $\int_1^2 \frac{1}{x^3} \, dx$

9. $\int_0^1 (x^3 - x^2) \, dx$

10. $\int_{-1}^2 (3x + 1) \, dx$

11. $\int_1^2 \left(3 + \frac{2}{\sqrt{x}}\right) dx$

12. $\int_{100}^{400} \left(\sqrt{x} + \frac{1}{\sqrt{x}}\right) dx$

13. $\int_{-1}^1 (2x - 3x^2 + 4x^3) \, dx$

14. $\int_{16}^{64} (4x^{1/4} - 1) \, dx$

15. $\int_0^1 \left(\pi x^{\pi - 1} - 2x^{99} - \frac{1}{3}\sqrt[3]{x}\right) dx$

16. a. Show that the number $\int_0^1 x(1 + x) \, dx$ is not the product of the numbers $\int_0^1 x \, dx$ and $\int_0^1 (1 + x) \, dx$. What does this example tell you about the relationship between $\int_a^b f(x)g(x) \, dx$ and $(\int_a^b f(x) \, dx)(\int_a^b g(x) \, dx)$?

b. Refer to Example 3. Show that, in the second step of the solution,

$$\int_0^2 x(1 + 2x + x^2) \, dx \neq \frac{x^2}{2}\left(x + x^2 + \frac{x^3}{3}\right)\Big]_0^2$$

17. Show that

$$\int_0^1 \frac{x^2 + x}{x} \, dx \neq \frac{\int_0^1 (x^2 + x) \, dx}{\int_0^1 x \, dx}$$

18. Use a graph to determine the answer to

$$\lim_{h \to 0} \left(\int_2^{2+h} \frac{1}{x} \, dx \right)$$

19. Obtain a sheet of graph paper with a 10 by 10 mesh per unit. On this sheet draw a good graph of $f(x) = 1/x$ from $x = 1$ to $x = 6$. Let

$$A(t) = \int_1^t \frac{1}{x} \, dx, \quad t \geq 1$$

Show that

$$A(6) \simeq A(2) + A(3)$$

by counting squares to estimate the appropriate area. In Chapter 5 we will see that exact equality holds in this equation.

20. Let $f(x) = 2x + 1$. Without using the Fundamental Theorem, but using graphs and formulas for the areas of a square and triangle,
    a. Find a formula for the function

$$A(t) = \int_0^t (2x + 1) \, dx$$

b. Graph f and A on the interval 0 to 1.

c. Differentiate the function A using the formula obtained in (a).

d. Discuss this exercise as it relates to Equation 1 of the Fundamental Theorem.

Challenge Problem

21. Investment analysts often use the concept of a *30-day moving average* of a stock's price in order to estimate its merit as an investment. If $S(t) = 5[1 - (1/t^2)]$ is the price of a certain stock on day t, its 30-day moving average price between day x and day $x + 30$ is

$$A_m(x) = \frac{1}{30} \int_x^{x+30} S(t)\, dt$$

Find   **a.** $A_m(x)$ **b.** $A_m(15)$ **c.** $A_m'(x)$

Generally, the moving average price A_m will be much less volatile than the price S itself over an interval of time.

4.3 The Indefinite Integral and Integration by Substitution

In this section we introduce the indefinite integral and the antiderivative and learn the method of integration by substitution.

The principal difficulty in computing the integral

$$\int_a^b f(x)\, dx$$

lies in finding a function F such that $F'(x) = f(x)$. Once F is found, only the arithmetic of computing $F(b) - F(a)$ remains.

Antiderivatives and the Indefinite Integral

Given a function f, a second function F such that $F' = f$ is called an **antiderivative** of f. A single function f has many antiderivatives, but they are of a single form. For example, if $f(x) = x^3$, then $F(x) = x^4/4$ is an antiderivative of F. But so is $F(x) = (x^4/4) + \sqrt{97}$. In fact so is

$$F(x) = \frac{x^4}{4} + C$$

where C is *any* constant. It can be shown that if F is one antiderivative of a given function f, then all antiderivatives of f are of the form $F(x) + C$ where C is an arbitrary constant (see Section A.2). It is common to say then that this last expression represents the **family of antiderivatives** of f; the word *family* is used to emphasize that while an antiderivative is not unique, all antiderivatives $F(x) + C$ have the generic trait $F(x)$.

Because of the Fundamental Theorem, the universally accepted notation for the family of antiderivatives of f is

$$\int f(x)\, dx$$

This expression is called the **indefinite integral** of f. In practice the indefinite integral is thought of as a function of x and is understood to mean any one antiderivative of f. The process of finding this function is referred to as "integrating f," or "finding the indefinite integral of f," or "finding the antiderivative of f." Once we have found the indefinite integral (a function), we use it according to Equation 2 of the Fundamental Theorem to find the definite integral $\int_a^b f(x)\, dx$ (a number). (See the discussion following Example 2 and also Example 7.) We refer to f as the *integrand* in either the definite or indefinite integral, and in both cases speak of "integration" or "integrating f" if the distinction between definite and indefinite is understood.

EXAMPLE 1

Find the indefinite integral

$$\int x^3\, dx$$

Solution Since $D_x(x^4/4) = x^3$,

$$\int x^3\, dx = \frac{x^4}{4} + C$$

because this expression represents the general function in the family of all antiderivatives of $f(x) = x^3$. ∎

The sum and constant-multiple rules for differentiation along with the formula $D_x(x^{n+1})/(n+1) = x^n$ yield the following properties of antiderivatives and indefinite integrals.

Basic Properties of the Indefinite Integral

1. $\displaystyle \int k f(x)\, dx = k \int f(x)\, dx$ (Homogeneity)

2. $\displaystyle \int [f(x) + g(x)]\, dx = \int f(x)\, dx + \int g(x)\, dx$ (Linearity)

3. $\displaystyle \int x^n\, dx = \frac{x^{n+1}}{n+1} + C$, if $n \neq -1$

These properties are used as follows.

EXAMPLE 2

Find $\int (x + \frac{1}{2}x^2 - x^{1/3})\, dx$.

Solution
$$\int \left(x + \frac{1}{2}x^2 - x^{1/3}\right) dx = \int x\, dx + \frac{1}{2}\int x^2\, dx - \int x^{1/3}\, dx$$

$$= \frac{x^2}{2} + \frac{1}{2}\left(\frac{x^3}{3}\right) - \frac{x^{4/3}}{\frac{4}{3}} + C$$

$$= \frac{x^2}{2} + \frac{x^3}{6} - \frac{3}{4}x^{4/3} + C \quad \blacksquare$$

Observe that we need to include only one constant in the general functional expression for an antiderivative. Notice too that to find the definite integral $\int_2^8 (x + \frac{1}{2}x^2 - x^{1/3})\, dx$, we would use the antiderivative

$$F(x) = \frac{x^2}{2} + \frac{x^3}{6} - \frac{3}{4}x^{4/3}$$

and evaluate $F(8) - F(2)$. (Why do we not include the constant C in this formula for F? [*Hint:* Include it in the formula for F and see what happens when you evaluate $F(8) - F(2)$.])

We close this part of the discussion with some important remarks about notation. Just as $f(x) = x^2 - 2x$ and $g(t) = t^2 - 2t$ represent the same function, so also do

$$\int (x^2 - 2x)\, dx \quad \text{and} \quad \int (t^2 - 2t)\, dt$$

represent the same indefinite integral. These all have the same algebraic form, and their solution is the same:

$$\frac{x^3}{3} - x^2 + C \quad \text{and} \quad \frac{t^3}{3} - t^2 + C$$

The letters x and t are called "dummy variables." Some letter must be used to state such a problem, but the letter itself is of no significance.

Integration by Substitution

While you can now easily find $\int x^2\, dx$ or $\int_2^{17} x^2\, dx$, you are probably not able to find

$$\int_0^1 2x\sqrt{x^2 - 1}\, dx$$

We now introduce the first of two principal integration methods, **integration by substitution,** which will solve the preceding problem. (The second method, *integration by parts*, will be studied in Section 7.1.) The substitution method depends on the chain rule for differentiation and, in effect, runs the chain rule in reverse (see Section 2.2).

Suppose that you happen to be given an integral in the *form*

$$\int h(g(x))g'(x)\, dx \tag{1}$$

and *suppose* that you are able to find the simpler integral

$$\int h(u)\, du = f(u) \tag{2}$$

where we (initially) use a different "dummy" variable "u" to distinguish this second, and distinct, integration problem. The very meaning of Equation 2 is that $f'(u) = h(u)$. Consequently, we can say that

$$\int h(g(x))g'(x)\, dx = \int f'(g(x))g'(x)\, dx = f(g(x)) \tag{3}$$

by the chain rule, since $D_x f(g(x)) = f'(g(x))g'(x)$.

What does all this mean? In summary, Equation 3 tells us that an integration problem given in the *form* of Equation 1 *can always be solved* if the apparently simpler integral (Equation 2) can instead be solved. We can turn this observation into a symbol-driven procedure, called the *method of integration by substitution*, if we next *invent* a new notation. Let us think of the letter "u" as a substitute for the function $g(x)$—that is, let $u = g(x)$. At the same time, we invent a new symbol "du" for the expression $g'(x)\, dx$—that is, we agree to also write $du = g'(x)\, dx$. We can then summarize Equations 1, 2, and 3 by straightforward substitution of these new symbols as a single equation

$$\int h(g(x))g'(x)\, dx = \int h(u)\, du = f(u) = f(g(x)) \tag{4}$$

where the substitutions are made in moving from the first to the second integral. Equation 4 indeed gives us a method for integration that we now illustrate.

EXAMPLE 3

Find $\int 2x\sqrt{x^2 - 1}\, dx$.

Solution To use the method of substitution here, we must try to see how the integrand $2x\sqrt{x^2 - 1}$ fits the form $h(g(x))g'(x)$, which specifically involves a function $g(x)$ and its derivative $g'(x)$ as a factor in the integrand.

Since $2x$ is the derivative of $x^2 - 1$, this suggests that we make the substitution

$$u = g(x) = x^2 - 1$$

Since $g'(x) = 2x$, we then write

$$du = 2x\, dx$$

Now, by exact substitution as in Equation 4,

$$\int 2x\sqrt{x^2 - 1}\, dx = \int (\sqrt{x^2 - 1})(2x)\, dx = \int \sqrt{u}\, du$$

$$= \int u^{1/2}\, du = \frac{u^{3/2}}{\frac{3}{2}} + C = \frac{2}{3}u^{3/2} + C$$

To complete the solution we must, as a final step, substitute $g(x) = x^2 - 1$ for u in the resulting antiderivative, obtaining

$$\int 2x\sqrt{x^2 - 1}\, dx = \frac{2u^{3/2}}{3} + C = \frac{2(x^2 - 1)^{3/2}}{3} + C \quad \blacksquare$$

Remarks.

1. Notice that we did not bother to specify the function h of the discussion preceding Example 3. This is the normal procedure. The function h

"comes out in the wash"; in this example, $h(u) = u^{1/2}$ appears automatically as the substitutions are made.

2. Observe that we can always check the correctness of the solution to an indefinite integral by differentiation. We have

$$D_x\left(\frac{2}{3}(x^2 - 1)^{3/2} + C\right) = \frac{2}{3}\left(\frac{3}{2}(x^2 - 1)^{1/2}D_x(x^2 - 1)\right) + 0$$

$$= (x^2 - 1)^{1/2}(2x) = 2x\sqrt{x^2 - 1}$$

which is the original function (integrand) of this problem.

We summarize this discussion and its illustration in Example 3 as the following procedure.

The Method of Integration by Substitution

In order to reduce an integral in the form

$$\int h(g(x))g'(x)\, dx$$

to an integral involving only one function, by the method of substitution,

1. Let $u = g(x)$.
2. Write $du = g'(x)\, dx$ by differentiating $g(x)$.
3. Replace "equals by equals" to obtain

$$\int h(g(x))g'(x)\, dx = \int h(u)\, du$$

4. Find $\int h(u)\, du$ and replace u by $g(x)$ in the answer to complete the solution.

Remark. An important guidepost to you in using this method is this: Since the only function for which we can actually write an antiderivative is a function of the form $h(u) = u^n$, if the method of substitution is being followed correctly, the integral $\int h(u)\, du$ of step 4 will always be of this form (with perhaps a constant factor included).

EXAMPLE 4

Find $\int 3/(\sqrt{1 + 3x})\, dx$.

Solution The derivative of $1 + 3x$ is 3. If we let $u = g(x) = 1 + 3x$, then $du = 3\, dx$. We then have

$$\int \frac{3}{\sqrt{1 + 3x}}\, dx = \int \frac{3\, dx}{\sqrt{1 + 3x}} = \int \frac{du}{\sqrt{u}}$$

$$= \int u^{-1/2}\, du = \frac{u^{1/2}}{\frac{1}{2}} + C = 2u^{1/2} + C$$

$$= 2(1 + 3x)^{1/2} + C$$

or $2\sqrt{1 + 3x} + C$ as a final answer. ∎

In the preceding two examples, we saw that the first step in the solution is to identify one part, $g(x)$, of the given integrand whose derivative $g'(x)$ is another part. In fact it is not necessary that $g'(x)$ be exactly equal to another part, but only "almost" equal to another part, where "almost" means up to a *constant factor*. The next example illustrates this.

EXAMPLE 5

Compute $\int t^2(t^3 - 1)^{99}\, dt$ by substitution.

Solution We let $u = t^3 - 1$, because its derivative $3t^2$ is "almost" the other factor in this problem. To complete the solution we have

$$du = 3t^2\, dt$$

Since only $t^2\, dt$ is part of the integrand, we then write

$$\frac{1}{3} du = t^2\, dt$$

and, by exact substitution,

$$\int t^2(t^3 - 1)^{99}\, dt = \int (t^3 - 1)^{99} t^2\, dt = \int u^{99}\left(\frac{1}{3}\right) du$$

$$= \frac{1}{3} \int u^{99}\, du = \frac{u^{100}}{300} + C = \frac{(t^3 - 1)^{100}}{300} + C$$

You can check this answer by differentiation:

$$D_t\left[\frac{(t^3 - 1)^{100}}{300} + C\right] = \frac{100(t^3 - 1)^{99}}{300}(3t^2) + 0 = t^2(t^3 - 1)^{99} \ \blacksquare$$

We conclude with two special examples.

EXAMPLE 6

Find $\int x(x + 1)^7\, dx$.

Solution If we let $u = x + 1$, from past experience, we see that $du = 1 \cdot dx = dx$, which is *not* the remaining part $(x\, dx)$ of the problem. Nonetheless, a solution can be found. Again let $u = x + 1$. Then $x = u - 1$, and since $du = dx$, we can then say

$$\int x(x + 1)^7\, dx = \int (u - 1)u^7\, du = \int (u^8 - u^7)\, du$$

$$= \frac{u^9}{9} - \frac{u^8}{8} + C = \frac{(x + 1)^9}{9} - \frac{(x + 1)^8}{8} + C \ \blacksquare$$

The next example illustrates how a definite integral can be evaluated via the substitution method.

EXAMPLE 7

Compute $\int_1^2 x\sqrt{4-x^2}\,dx$.

Solution We first find the antiderivative $\int x\sqrt{4-x^2}\,dx$ by substitution. Let $u = 4 - x^2$. Then $du = -2x\,dx$ or $-\frac{1}{2}du = x\,dx$. Therefore

$$\int x\sqrt{4-x^2}\,dx = \int -\frac{1}{2}\sqrt{u}\,du = -\frac{1}{2}\int u^{1/2}\,du$$

$$= -\frac{1}{2}\left(\frac{u^{3/2}}{\frac{3}{2}}\right) + C = \frac{-(4-x^2)^{3/2}}{3} + C$$

Consequently,
$$\int_1^2 x\sqrt{4-x^2}\,dx = \frac{-(4-x^2)^{3/2}}{3}\Big]_1^2$$

$$= \frac{-(4-2^2)^{3/2}}{3} - \left(\frac{-(4-1^2)^{3/2}}{3}\right)$$

$$= \frac{3^{3/2}}{3} = 3^{1/2} = \sqrt{3} \blacksquare$$

The evaluation of the definite integral can be shortened in Example 7 using the equation $\int_a^b h(g(x))g'(x)\,dx = \int_{g(a)}^{g(b)} h(u)\,du$, where $u = g(x)$. In Example 7, since $g(x) = 4 - x^2$, $g(1) = 3$ and $g(2) = 0$. We can then write

$$\int_1^2 x\sqrt{4-x^2}\,dx = -\frac{1}{2}\int_3^0 u^{1/2}\,du = -\frac{1}{2}\left(\frac{2}{3}u^{3/2}\right)\Big]_3^0$$

$$= -\frac{1}{2}\left(\frac{2}{3}\cdot 0 - \frac{2}{3}(3)^{3/2}\right) = \sqrt{3}$$

as before.

We conclude this section with some remarks on misleading procedures when using the substitution method. You can easily fall into such procedures if you mistakenly regard the symbols "du" and "dx" as numbers; these are not numbers but "placeholders" in the mechanical procedures of substitution. For example, consider the integral

$$\int \frac{1}{2x(x^2+1)^3}\,dx$$

It is natural to try to find this integral by letting $u = x^2 + 1$ to obtain $du = 2x\,dx$. We may then be tempted to regard this as an algebraic equation and write $(1/2x)\,du = dx$. Exact substitution then leads to the integral

$$\int \frac{1}{4x^2}\cdot\frac{1}{u^3}\,du$$

While this is not incorrect, this integral cannot be evaluated because du asks for a function of u whose derivative with respect to u contains a different variable x. It is not possible to find such a function. In fact, the given integration problem cannot be solved by any substitution.

The method of substitution is more exact in its procedures than it may appear. Of the three integrals

$$\int \sqrt{1 + x^2}\, dx, \qquad \int x\sqrt{1 + x^2}\, dx, \qquad \text{and} \qquad \int x^2\sqrt{1 + x^2}\, dx$$

only the middle one can be found by methods now available to us. You can avoid misleading and even incorrect steps in using substitution if you permit yourself to move only *constant multiples* from one side of the equation $du = g'(x)\, dx$ to the other. In particular, do not regard du or dx as a number and write things like $1/du$ or $1/dx$. The expression $1/du$ has no meaning: We may as well have written that "stones have blue eyes."

Background Review 4.1

The method of substitution is especially needed in the exercises of this section because in general

$$(a + b)^n \text{ does not equal } a^n + b^n$$

Similarly (for $a, b \neq 0$), $\sqrt{a + b}$ never equals $\sqrt{a} + \sqrt{b}$ and

$$\frac{1}{a + b} \text{ never equals } \frac{1}{a} + \frac{1}{b}$$

Exercises 4.3

In Exercises 1–6, isolate one part—one function—in the integrand whose derivative is "almost"—up to a constant multiple—another part of the integrand. Let u equal this part and write out du. Do not complete the solution.

1. $\displaystyle\int \frac{x}{3}(1 + x^2)^7\, dx$

2. $\displaystyle\int x\, 3^{x^2}\, dx$

3. $\displaystyle\int \frac{z}{z^2 + 3}\, dz$

4. $\displaystyle\int (x^5 - 2x)^3 (10x^4 - 4)\, dx$

5. $\displaystyle\int 3t^2 \sqrt{t^3 - 1}\, dt$

6. $\displaystyle\int 3\sqrt{x + 1}\, dx$

Find the indefinite integrals in Exercises 7–28. In each case, check your answer by differentiation.

7. $\displaystyle\int 2x\sqrt{1 + x^2}\, dx$

8. $\displaystyle\int \frac{1}{\sqrt{x + 1}}\, dx$

9. $\displaystyle\int 3x^2 \sqrt{x^3 - 4}\, dx$

10. $\displaystyle\int \frac{2t}{(1 + t^2)^3}\, dt$

11. $\displaystyle\int t\sqrt{3t^2 + 1}\, dt$

12. $\displaystyle\int 3x(1 + x^2)^5\, dx$

13. $\displaystyle\int x(1 - x^2)^3\, dx$

14. $\displaystyle\int 7(x + 1)^{11}\, dx$

15. $\displaystyle\int 8x^3(2x^4 - 1)^5\, dx$

16. $\displaystyle\int \frac{x}{\sqrt{1 - x^2}}\, dx$

232 Chapter 4 Integration

17. $\displaystyle\int \frac{2x}{(x^2+1)^3}\,dx$

18. $\displaystyle\int \frac{z}{2(z^2+1)^3}\,dz$

19. $\displaystyle\int \frac{1}{\sqrt{3y+2}}\,dy$

20. $\displaystyle\int \frac{1-x}{(x^2-2x-1)^4}\,dx$

21. $\displaystyle\int \frac{2x-1-3x^2}{(x^3-x^2+x)^7}\,dx$

22. $\displaystyle\int v^2(2v^3-7)^{1/3}\,dv$

23. $\displaystyle\int (u+1)(u^2+2u)^3\,du$

24. $\displaystyle\int \frac{x}{\sqrt{x^2+1}}(1+\sqrt{x^2+1})\,dx$

25. $\displaystyle\int x(1-x)^{99}\,dx$

26. $\displaystyle\int x^2(1+x)^7\,dx$

27. $\displaystyle\int x^2\sqrt{x+1}\,dx$

28. $\displaystyle\int \frac{x}{(x+1)^3}\,dx$

29. Convince yourself that the substitution $u=x^2+1$ cannot solve

$$\int (2x+1)(x^2+1)^2\,dx$$

Then solve this problem by writing

$$(2x+1)(x^2+1)^2 = 2x(x^2+1)^2 + (x^2+1)^2$$

and integrating each summand.

Compute the definite integrals in Exercises 30–33.

30. $\displaystyle\int_{-1}^{2} x(x^2+1)^3\,dx$

31. $\displaystyle\int_{0}^{1} \frac{z}{\sqrt{1+z^2}}\,dz$

32. $\displaystyle\int_{0}^{1} x(x+1)^5\,dx$

33. $\displaystyle\int_{-1}^{1} \frac{1}{(2t+3)^3}\,dt$

In Exercises 34–39, find the general antiderivative F of each function via indefinite integration.

34. $f(x) = x^2 + \dfrac{x}{3}$

35. $g(x) = x(x^2-5)^3$

36. $h(t) = 1 - t - 4t^3$

37. $p(v) = \dfrac{v}{\sqrt{v^2+1}}$

38. $f(u) = \dfrac{3u^2}{(u^3-1)^2}$

39. $k(y) = y(1+y)^{1/2}$

4.4 Applications of the Indefinite Integral

In this section we learn to use the indefinite integral to determine the state or condition of a system if we are given information about how the system is changing.

In this and the next section we begin a study of applications of integration. In subsequent chapters we will reapply the underlying themes of these applications to progressively more sophisticated situations.

In earlier applications of calculus we have seen that when a function f is used to model a system under study, its function values $f(x)$ represent the state or condition of the system at x, and the derivative $f'(x)$ represents the rate at which the system is changing at x. Thus differentiation converts state information to rate information.

The reverse process, in which we compute f given f', is called **antidifferentiation**. Thus antidifferentiation converts rate information to state information. Since the indefinite integral represents an antiderivative, the operative form of antidifferentiation is the process of integration via the indefinite integral. We can briefly, but loosely, summarize these interpretations in applications by the mnemonic "equations"

$$D_x(\text{State}) = \text{Rate} \qquad \text{and} \qquad \int (\text{Rate})\,dx = \text{State}$$

In this section we will be concerned with the latter equation. We will be given *rate-of-change* information, and our problem will be to describe the *amount* of change. That is, we will know how a system is changing, and we will determine the state to which it is moving. In terms of our ongoing analogy, we will be shown a motion picture or videotape f' of a system in motion, and we will be asked to stop the film or tape and describe a fixed image of the system $f(t)$ at time t.

The simplest problems of the kind we study in this section are geometrical and were encountered informally in Section 2.4.

EXAMPLE 1

Find a formula $y = f(x)$ for a curve where you are given a formula

$$g(x) = 2x - 2$$

for the slopes of tangent lines to this curve.

Solution The function $g(x)$ tells you, at the point $(x, f(x))$, how the curve is changing. This means then that $f'(x) = g(x)$. Hence

$$f(x) = \int g(x)\, dx = \int (2x - 2)\, dx = x^2 - 2x + C \; \blacksquare$$

As we saw in Section 2.4, without knowing the value of the (unavoidable) constant C, we are unable to describe the function $y = f(x)$ uniquely. But since

$$f(x) = x^2 - 2x + C = (x^2 - 2x + 1) - 1 + C = (x - 1)^2 - 1 + C$$

then f is one of a *family* of curves of the shape indicated in Figure 4.25.

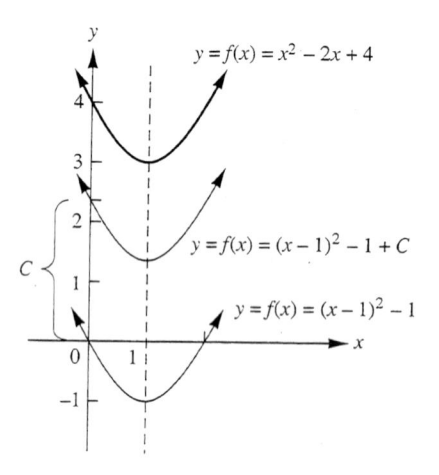

Figure 4.25

EXAMPLE 2

Suppose that in Example 1 we are also given that $f(1) = 3$. Show how this additional fact determines $y = f(x)$ uniquely, and find $f(5)$.

Solution We already know that $f(x) = x^2 - 2x + C$ where C can be any number. Now we are asking, "Which one of these functions (values of C) has a graph that passes through the point $(1, 3)$?" If we impose the additional condition that $f(1) = 3$, we have that $3 = f(1) = 1^2 - 2(1) + C = -1 + C$. Therefore $C = 4$, and $y = f(x) = x^2 - 2x + 4$ is the *unique* solution to this problem. Consequently, $f(5) = 19$. The graph of this function is that *one* member of the family of curves indicated by the heavier line in Figure 4.25. $\blacksquare$

Examples 1 and 2 show that if we know the rate at which change is occurring, $f' = g$, and if we know one value of the unknown function f [here, $f(1) = 3$], then we can find $f(x)$ in general and other values [here, $f(5)$] in particular. The next example puts this method to practical use.

EXAMPLE 3

Suppose that a manufacturing company knows that its marginal cost when producing x items is

$$M_C(x) = 1{,}200/(x + 1)^3 \text{ dollars/unit}$$

Before it begins production, the company's fixed costs for buildings, machines, and so on are \$15,000. Find a formula for $C(x)$, the total cost to produce x items. Additionally find how much it costs the company to produce 10 items.

Solution Recall that marginal cost is, by definition, the derivative of the cost function. Therefore

$$C'(x) = M_C(x) = \frac{1{,}200}{(x+1)^3}$$

Hence $\quad C(x) = \int M_C(x)\, dx = \int \frac{1{,}200}{(x+1)^3}\, dx = 1{,}200 \int (x+1)^{-3}\, dx$

By the substitution $u = x + 1$, we have

$$C(x) = \frac{1{,}200}{-2}(x+1)^{-2} + K = \frac{-600}{(x+1)^2} + K$$

We use "K" here to represent the unknown constant so as to avoid confusion with the letter "C" of the cost function.

Now, given the initial condition that $C(0) = 15{,}000$, we also have

$$15{,}000 = \frac{-600}{(0+1)^2} + K = -600 + K$$

Hence $K = 15{,}600$ and

$$C(x) = \frac{-600}{(x+1)^2} + 15{,}600$$

The cost to produce 10 items is then

$$C(10) = \frac{-600}{(10+1)^2} + 15{,}600 \approx 15{,}595 \text{ dollars} \quad \blacksquare$$

EXAMPLE 4

A 1-acre farm pond (4,840 yd²) is observed to have no algae on its surface on the first day of May. With warming temperatures, algae begin to form and by t days after May 1, algae are covering the surface of the pond at the rate of $\sqrt{100t + 2{,}500}$ yd²/day. When will the pond be half-covered with algae?

Solution Let $A(t)$ denote the number of square yards of pond surface that are covered by algae on day t, where $t = 0$ is May 1. We are given that $A'(t) = \sqrt{100t + 2{,}500}$. Hence $A(t) = \int \sqrt{100t + 2{,}500}\, dt$. Letting $u = 100t + 2{,}500$, we have $du = 100\, dt$ and

$$A(t) = \frac{1}{100}\int \sqrt{u}\, du = \frac{1}{100}\frac{u^{3/2}}{\left(\frac{3}{2}\right)} + C = \frac{1}{150}(100t + 2{,}500)^{3/2} + C$$

Now, since $A(0) = 0$,

$$0 = \frac{(100(0) + 2{,}500)^{3/2}}{150} + C = \frac{(50)^3}{150} + C = \frac{2{,}500}{3} + C$$

Hence $C = \dfrac{-2{,}500}{3}$ and $A(t) = \dfrac{(100t + 2{,}500)^{3/2}}{150} - \dfrac{2{,}500}{3}$

square yards on day t.

When will the pond be half-covered with algae? On that day t when

$$A(t) = \tfrac{1}{2}(4{,}840) = 2{,}420 \text{ yd}^2$$

or when

$$\frac{(100t + 2{,}500)^{3/2}}{150} - \frac{2{,}500}{3} = 2{,}420$$

Simplifying: $(100t + 2{,}500)^{3/2} = 150\left(2{,}420 - \dfrac{2{,}500}{3}\right) = 238{,}000$

Equivalently, $100t + 2{,}500 = (238{,}000)^{2/3}$

or $100t \approx 3840.5 - 2{,}500.0$

or $t \approx 13.4$ days

Therefore the pond will be nearly half-covered on approximately May 13. ∎

An early use of calculus was to describe the motion of a moving body. In the next examples, we solve a basic problem of motion—motion determined only by initial position, initial velocity, and known acceleration.

Imagine a moving object such as a speeding car or a falling rock. We will neglect all forces such as air resistance and friction and consider only idealized motion along a straight line. Let

$s(t)$ denote the position of the object—where it *is*—at time t,
$v(t)$ denote its velocity—how fast its position is changing—at time t, and
$a(t)$ denote its acceleration—how fast its velocity is changing—at time t.

Since the derivative measures rate of change, we immediately know that

$$s'(t) = v(t) \qquad \text{and} \qquad v'(t) = a(t)$$

or, equivalently,

$$s(t) = \int v(t)\, dt \qquad \text{and} \qquad v(t) = \int a(t)\, dt$$

EXAMPLE 5

An automobile leaves an intersection at time $t = 0$, with speed 0, and accelerates at the rate of $a(t) = 4t$ ft/sec. How far does it go in 5 sec? How fast will it be traveling at that moment? If the next block is 500 ft long, how long will it take the car to reach the next intersection? (See Figure 4.26.)

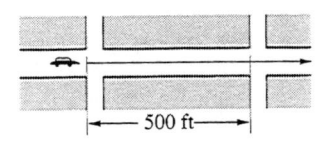

Figure 4.26

Solution We have $v(t) = \int a(t)\, dt = \int 4t\, dt = 2t^2 + C$. Since initial velocity is $0 = v(0) = 2 \cdot 0^2 + C$, then $C = 0$. Therefore $v(t) = 2t^2$. Since $s(t) = \int v(t)\, dt$, we also have

$$s(t) = \int 2t^2\, dt = \tfrac{2}{3}t^3 + C$$

Since $s(0) = 0$, we can additionally say that $0 = s(0) = \tfrac{2}{3}(0)^3 + C = C$. Therefore

$$s(t) = \tfrac{2}{3}t^3$$

By $t = 5$ the car will have traveled a distance of $s(5) = \tfrac{2}{3}(5)^3 = \tfrac{2}{3}(125) \simeq 83.33$ ft. At that moment, $t = 5$, the car will be traveling at a speed of

$$v(5) = 2(5)^2 = 50 \text{ ft/sec}$$

using the formula $v(t) = 2t^2$ we initially obtained.

Since the city block is 500 ft long, the car will reach the next intersection when $s(t) = 500$. This is equivalent to $\tfrac{2}{3}t^3 = 500$, or $t^3 = 750$, with the solution $t = \sqrt[3]{750} = 9.08$ sec. ∎

In the previous example, both constants C were found to be zero. This does *not* happen in general.

EXAMPLE 6

Suppose that you are standing on a cliff 116 ft high and you pick up a rock and toss it up into the air at a *vertical* velocity of 28 ft/sec (see Figure 4.27). Suppose the rock leaves your hand at a height of 4 ft above the ground you are standing on. When will the rock land in the stream at the foot of the cliff (neglecting air resistance)?

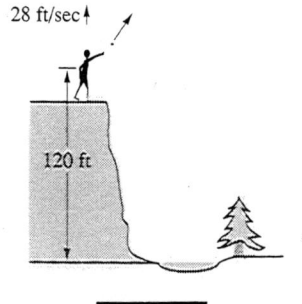

Figure 4.27

Solution Let $s(t)$ denote the height of the rock at time t *above* the stream bed, and $v(t)$ and $a(t)$ its *vertical* velocity and acceleration. Notice that as the solution develops, $v(t)$ will be positive when the rock is rising and negative when the rock is falling; that is, the positive direction of motion is upward.

Once the rock leaves your hand, we know from physics that the only vertical force acting upon it is gravity. Gravity gives the rock an acceleration *back* to earth of $a(t) = -32$ ft/sec^2, where we use a minus sign because the rock is decelerating.

The vertical velocity of the rock at time t is therefore

$$v(t) = \int a(t)\, dt = \int -32\, dt = -32t + C$$

Since you threw the rock upward (against gravity) at the rate of 28 ft/sec, we must impose this condition. In so doing, we will determine C. We have $28 = v(0) = -32 \cdot 0 + C = C$. Hence the velocity of the rock at time t is

$$v(t) = -32t + 28 \text{ ft/sec}$$

By time t the rock has traveled a distance of

$$s(t) = \int v(t)\, dt = \int -32t + 28\, dt = -16t^2 + 28t + K \text{ ft}$$

Since you released the rock at $t = 0$ at a distance of $116 + 4 = 120$ ft above the stream bed, $120 = s(0) = -16 \cdot 0^2 + 28 \cdot 0 + K = K$. Hence, at time t, the height of the rock above the stream bed is

$$s(t) = -16t^2 + 28t + 120 \text{ ft}$$

To answer the original question, note that the rock will hit the stream when $s(t) = 0$, that is, when

$$-16t^2 + 28t + 120 = 0$$

$$4t^2 - 7t - 30 = 0$$

or

$$t = \frac{7 \pm \sqrt{(-7)^2 - 4(4)(-30)}}{2(4)}$$

$$= \frac{7 \pm 23}{8} \text{ sec}$$

(One can factor this equation rather than use the quadratic formula.)
The only *positive* time solution of this equation is

$$t = \tfrac{30}{8} = 3.75 \text{ sec}$$

Thus the rock hits the stream 3.75 sec after leaving your hand. Note the graph of $s(t)$ in Figure 4.28. ∎

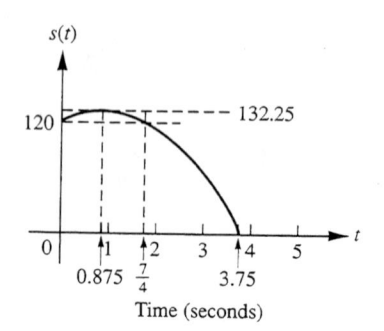

Figure 4.28

Before concluding this section, we take this same example further by answering a series of questions. Some of these answers are indicated in Figure 4.28.

EXAMPLE 7

a. How fast is the rock moving when it hits the stream?

Solution The rock hits the stream at $t = 3.75$ sec. Its velocity at that time is

$$v(3.75) = -32(3.75) + 28 = -92 \text{ ft/sec}$$

The answer here is negative because the rock is falling.

b. When does the rock reach the top of its flight, before it begins to fall back to the stream?

Solution At the instant of the peak of its flight, the *vertical* velocity of the rock must be zero. This instant occurs when $-32t + 28 = v(t) = 0$ or when

$$t = \tfrac{28}{32} = 0.875 \text{ sec}$$

c. How high did the rock rise before beginning its fall?

Solution Using (b), the rock reached a height of

$$s(0.875) = -16(0.875)^2 + 28(0.875) + 120 = 132.25 \text{ ft}$$

above the stream bed at the peak of its flight.

d. When was the falling rock back at the same height from which you threw it, and how fast was it moving at that time?

Solution The rock is at its original height when $s(t) = 120$. Solving for t,

$$-16t^2 + 28t + 120 = 120$$
$$-4t(4t - 7) = 0$$
$$t = \tfrac{7}{4} \text{ sec}$$

Note: The other solution, $t = 0$, represents when the rock left your hand. At $t = \tfrac{7}{4}$ the velocity is $v\left(\tfrac{7}{4}\right) = -32\left(\tfrac{7}{4}\right) + 28 = -28$ ft/sec, exactly the same speed at which it was tossed upward. ∎

Notice that we consider only vertically directed components of the flight of the rock—gravity, vertical velocity, the height of the cliff—relying on physics for the independence of the vertical and horizontal components of motion. None of our computations use, or give, any information about the horizontal displacement of the rock outward from the point at which you threw it.

The kinds of problems we have raised and solved in this section lie in the category of **deterministic models** of natural phenomena. In each problem we use the indefinite integral to determine the future course of an experiment based on knowledge of its initial state and the rate at which change is occurring. The initial state is represented by a single known function value. The rate of change is represented by knowledge about the derivative of the unknown function. Deterministic models ignore random effects. In Example 6, a sudden gust of wind up the cliff would slow the fall of the rock. Deterministic models also assume a stable state of surrounding conditions. In Example 4, a heavy rain shower would dissipate the algae in the pond and could be accounted for only by changing the rate function $A'(t)$ in that model. Nonetheless, to the extent that random effects can be safely ignored and surrounding conditions assumed to be stable, deterministic models derived from rate-of-change analysis can enable us to predict the future states of a system.

Background Review 4.2

Some of the following exercises call for the use of the quadratic formula,

$$x = \frac{-b \pm \sqrt{b^2 - 4ac}}{2a} \qquad a \neq 0$$

for the two roots of the quadratic equation $ax^2 + bx + c = 0$.

An equation in the form

$$ax^3 + bx^2 + cx = 0 \qquad a \neq 0$$

can be factored as

$$x(ax^2 + bx + c) = 0$$

yielding $x = 0$ as one root, with the two remaining roots being solutions to $ax^2 + bx + c = 0$.

Exercises 4.4

In Exercises 1–4, you are given a formula $g(x)$ for the slope of the tangent to the graph of an unknown function $y = f(x)$ at the point x. You are also given one value of the function $y = f(x)$. Find (a) the general formula for $f(x)$ (with an unknown constant); (b) graphs of the family of curves described by this formula; and (c) the particular function having the given value. (See Examples 1 and 2.)

1. $g(x) = x + 1$; $f(1) = 3$
2. $g(x) = 1 - x^2$; $f(2) = 1$
3. $g(x) = x\sqrt{1 - x^2}$, $|x| \leq 1$; $f(0) = 2$
4. $g(x) = \dfrac{1}{(x + 1)^2}$, $x \neq -1$; $f(1) = \dfrac{1}{2}$

Find the unknown function y satisfying the conditions in Exercises 5–9. In each exercise, $y'(t) = g(t)$ so that $y(t) = \int g(t)\, dt$. Each of Exercises 5–9 is called a *differential equation with initial condition*. The function y found in each solution is called a solution of the differential equation.

5. $y'(t) = t + (1 - t)^2$, $y(0) = 1$
6. $y'(t) = 3t^2 - 2t$, $y(0) = 4$
7. $y'(t) = \sqrt{1 - t}$, $y(0) = 3$
8. $y'(t) = t^2/\sqrt{1 + t^3}$, $y(0) = 4$
9. $(t^2 + 1)^2 y'(t) = 2t$, $y(0) = 1$

240   Chapter 4   Integration

In Exercises 10–24, keep in mind that the given rate functions describe the rate of change of the system and that the antiderivative describes the state, or condition, of the system. Therefore questions about how the system is *changing* are answered via the rate function, while questions about when, or what, the system *is* are answered via the antiderivative.

10. At 11 A.M. when you open your picnic basket, there are no ants to be seen near your picnic site. But then ants begin to arrive from three different directions at the combined rate of $10{,}000t^3$ ants/hour, t hours later. At what time will there be 40,000 ants at your picnic?

11. An advertising agency wishes to determine its costs to reach x thousand viewers. Its fixed cost is $15,000, and at a level of x thousand viewers, its marginal cost is

$$M_C(x) = \frac{300x}{\sqrt{x^2 + 100}}$$

How much does it cost the agency to reach 20,000 viewers?

12. A rumor originated by a local gossip at 3 P.M. is spreading through a small town of 127 people at the rate of $81t^2$ people per hour. When will the whole town, with the exception of the one person who is the subject of the rumor, know of it?

13. A company that began its operations with a fixed cost of $8,000 finds that its marginal revenue and marginal cost (in dollars per item) are given by

$$R'(x) = 4 \quad \text{and} \quad C'(x) = \frac{500}{(x+1)^2}$$

at a production level of x items. How many items must it produce to break even? [*Hint:* Compute total revenue $R(x)$, assuming $R(0) = 0$.]

14. For an exam to be taken during your 11 A.M. class, you must memorize a list of formulas for 100 organic molecules. You have learned 95 of these by the time you attend your 10 A.M. class, which precedes the exam period. During this interesting class, you begin to forget the formulas you have memorized at the rate of $0.002t$ formula per minute, t minutes after class has begun. How many formulas will you still recall by exam time?

15. After road racing for 2 hr, Joe Piston is leading Jim Ring by 100 yd and Piston is 300 yd from the finish line. At that moment both Piston and Ring are traveling at a speed of 100 yd/sec, but then Piston's fuel system begins to malfunction and t sec later his speed has slowed to $(-12.35t + 100)$yd/sec while Ring continues on at 100 yd/sec. Who will win the race? By how many yards? At what speed will the winner cross the finish line?

16. After a rainstorm (ending at $t = 0$), the river running through a small community begins rising at the rate of $0.5(4 - t^2)$ ft/hr, t hr later. How many feet will it rise above its normal level? At what time will this occur? When will the river subside to its normal level? [*Hint:* Let N denote its normal level, and let $h(t)$ denote the depth of water at time t. Thus $h(0) = N$ and $h'(t) = 0.5(4 - t^2)$. Use h and h' to answer the first two questions, and find $t > 0$ for which $h(t) = N$ to answer the third. You do not need a value for N to do this exercise.]

17. Driven to the deep forehand corner by your opponent's approach shot, you return with a soft lob. The tennis ball leaves your racket with a vertical speed of 48 ft/sec. Ignoring air resistance, when will the ball return to the same height from which it left your racket (but on your opponent's side of the court)? How high will the ball rise before it begins to fall back down?

18. Referring to Exercise 17, suppose you contact the ball at the baseline on one end of the court. The court is 78 ft long. If the ball has a *horizontal* speed of 22.6 ft/sec throughout its flight, will your shot land in the court?

19. Customers begin returning a defective product to the Reliable Manufacturing Company at the rate of $2x + 3$ products per day, x days after its release for sale. Each return costs the company $5. When will the company's total cost on these returns reach $2,000?

20. Standing at the top of a 150-ft-high cliff, you pick up a rock and throw it *down* toward the creek below at a vertical speed of 20 ft/sec. When will the rock land in the creek? (Assume the rock leaves your hand at ground level, 150 ft above the stream.)

21. Suppose that you shoot an arrow into the air at a speed of 200 ft/sec, while standing at the foot of a hill 100 ft high, and that the arrow lands at the top of the hill. When will it land there? How high did it fly? how fast was it moving when it hit the hilltop? (Ignore wind resistance, and so on, and consider only the effect of gravity on the arrow's flight.)

22. Standing on the flat Kansas plain in the year 1855, a rifleman and a Cheyenne Indian fired their weapons in the horizontal direction toward a cottonwood tree a half-mile away. The bullet left the rifleman's rifle at a *horizontal* speed of 2,000 ft/sec, and the warrior's arrow left his bow at a *horizontal* speed of 500 ft/sec. Assuming both projectiles were fired horizontally at a height of 4 ft, when did each hit the ground? How far did each travel? [*Hint:* Be careful to distinguish vertical speed from horizontal speed and ignore air resistance.]

23. The mosquito population on a Minnesota lake is rising at the rate of $20,000t^3 + 20,000t$ mosquitoes per day on day t, from an initial population of 5,000 mosquitoes. The lake has an area of 40,000 sq yd. When will the mosquito population reach a level of 1,000 mosquitoes per square yard?

24. Mathematics, being abstract, is sometimes easier to remember and understand when it is put in verbal terms, even though such terms are imprecise and cannot be used in the technical solution of a problem. The derivative and integral are closely linked to two of the more basic, powerful, and frequently used verbs in any language: *becoming* and *is*. We can convey this linkage in two "equations" that cross the disciplines of language and mathematics:

$$D_x(\text{Is}) = \text{Becoming} \quad \text{and} \quad \int (\text{Becoming})\, dx = \text{Is}$$

a. A town of 3,500 individuals has been struck by a new, highly contagious influenza virus. On day $t = 0$, 22 individuals have the flu. Then it begins to spread through the population at the rate of $3t^2$ individuals per day on day t. The flu is *becoming* a problem, at this rate of spreading infection. It will *be* a problem when the number of infected individuals *is*, say, one-half the town's population. When will this happen?

b. Local doctors begin an inoculation program on day $t = 12$ to give immunity to those residents who have not yet contracted the flu. There *are* $600(t - 11)$ immune residents by day t resulting from this program. Are more people *becoming* immune than are *becoming* infected on day $t = 13$?

4.5 Applications of the Definite Integral

In this section we return to a major theme in the applications of calculus: approximation of the complex by the simpler. Our purpose is to see how a quantity we wish to measure exactly can be easily approximated by a Riemann sum and thereby found exactly using the definite integral.

In Section 4.1, you learned that the definite integral

$$\int_a^b f(x)\, dx$$

of a nonnegative function f is defined as the area beneath its graph. You then learned that consequently this number can be approximated as closely as you wish by a Riemann sum of the *form*

$$f(x_1)\Delta x + f(x_2)\Delta x + \cdots + f(x_n)\Delta x$$

Let us turn this idea around and state, for a continuous function f, the following principle.

The Approximation Principle (AP) for the Definite Integral

If a real number quantity Q can be approximated, as closely as desired, by a sum of the *form*

$$f(x_1)\Delta x + f(x_2)\Delta x + \cdots + f(x_n)\Delta x$$

then the *exact* value of Q must be

$$Q = \int_a^b f(x)\, dx$$

The number Q is then computed using antidifferentiation and the Fundamental Theorem of Calculus. We will consistently refer to this principle as the **approximation principle** and denote it by **AP**. It is the basis of applied definite integration, and we will use it throughout this text. The validity of AP is founded on the idea that since the quantity Q to be measured *and* the area beneath the curve of f, $\int_a^b f(x)\, dx$, are approximated as closely as we can imagine by the same number (this number taking the *form* of a Riemann sum), then the value of Q and of the integral must be the same.

The approximation principle repeats an underlying theme of calculus: Approximate the complex by something simpler. In the examples, the quantity Q will invariably be the total of a continually varying contribution to a whole—a complex idea. The Riemann sum approximating Q will invariably be a much simpler, discrete calculation that is often suggested by a drawing. We illustrate the use of AP via a series of applications.

A. The Area of the Region between Two Curves

This is a relatively simple application of AP. Consider the region lying *between* the graphs of $y = f(x)$ and $y = g(x)$ and vertical lines at $x = a$ and $x = b$ (see Figure 4.29). Notice that $f(x) \geq g(x)$ for all x.

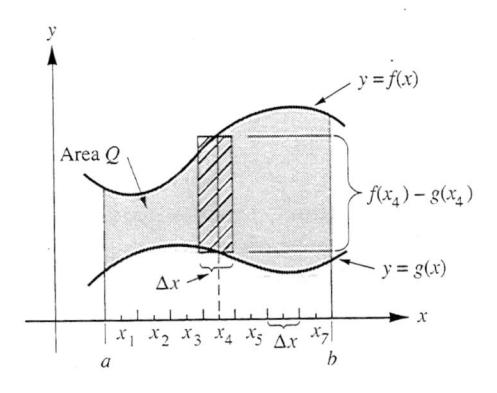

Figure 4.29

Let us use AP to compute the area Q of the region *between* the curves $y = f(x)$ and $y = g(x)$.

As Figure 4.29 illustrates, this number Q can be approximated as closely as we can imagine by a sum of the *form*

$$[f(x_1) - g(x_1)]\Delta x + [f(x_2) - g(x_2)]\Delta x + \cdots + [f(x_n) - g(x_n)]\Delta x$$

4.5 Applications of the Definite Integral

This sum is a Riemann sum for the function $h(x) = f(x) - g(x)$. Therefore, using AP, the exact area Q is

$$Q = \int_a^b [f(x) - g(x)] \, dx \qquad (1)$$

Going further, and using Theorem 4.3 from Section 4.2,

$$Q = \int_a^b f(x) \, dx - \int_a^b g(x) \, dx$$

This says something geometrically clear: The area Q is the difference of the areas of the regions bounded by the x-axis and the graphs of $y = f(x)$ and $y = g(x)$. Notice also that Formula 1 remains valid even if the functions have some negative values [as long as $f(x) \geq g(x)$ for every x] since area is unchanged by moving the horizontal axis about the region of interest.

EXAMPLE 1

Find the area of the region between the graphs of $f(x) = x^2$ and $g(x) = x^3$ on the interval from $a = 0$ to $b = 1$. (Note that $x^2 \geq x^3$ for $0 \leq x \leq 1$.)

Solution We illustrate the graphs of these functions on $[0, 1]$ in Figure 4.30. The area Q of the region between $y = x^2$ and $y = x^3$ lying between 0 and 1 is

$$\int_0^1 [x^2 - x^3] \, dx = \left. \frac{x^3}{3} - \frac{x^4}{4} \right]_0^1 = \frac{1}{3} - \frac{1}{4} = \frac{1}{12}$$

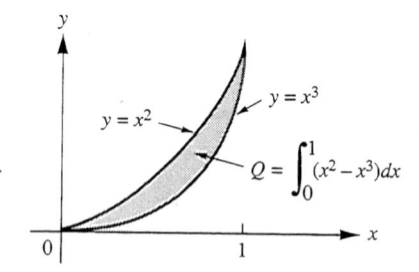

Figure 4.30

Caution. Suppose that in Example 1 we had been asked for the area between the graphs of f and g on the interval from $x = 0$ to $x = 3$. At $x = 1$ the graph of g matches the graph of f and then lies above the graph of f for x between 1 and 3. To find the area between these two graphs on the entire interval we must form two integrals—the first being the integral of $f - g$ on the interval $[0, 1]$ and the second being the integral of $g - f$ on the interval $[1, 3]$, where g is larger than f.

B. The Average Value of a Function

Recall that we used an intuitive idea of average height in our initial discussion of the integral (see Section 4.1). We will now make this intuitive idea an exact calculation by using AP to compute the *average height* Q of the graph of $y = f(x)$ between a and b (see Figure 4.31).

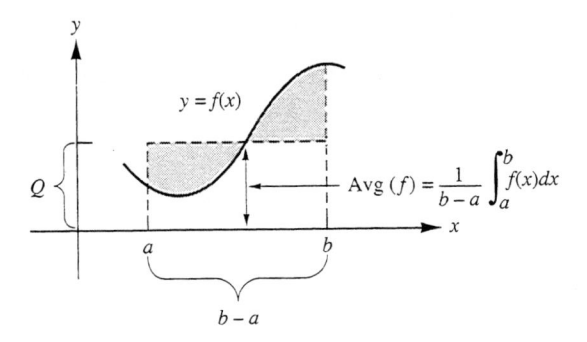

Figure 4.31

Recall that the average of n numbers $a_1, a_2, \ldots, a_n$ is the number

$$\frac{a_1 + a_2 + \cdots + a_n}{n}$$

Let us use this to first approximate Q, the true average height of the function in Figure 4.31, as a statistician would estimate the average of a long list of measurements. (The list, in this case, is the infinitely many values of the function, each value measuring the height of *one* point of the graph.) We *estimate* the average height by first taking a large sample of heights, say,

$$f(x_1), f(x_2), \ldots, f(x_n)$$

where n is very large, and then computing their average:

$$\frac{f(x_1) + f(x_2) + \cdots + f(x_n)}{n}$$

The quantity Q of average height of the entire graph is *approximated* by this number. This expression is not a Riemann sum. However, a little algebraic manipulation makes it so:

$$Q \simeq \frac{f(x_1) + f(x_2) + \cdots + f(x_n)}{n}$$

$$\simeq \left[\frac{f(x_1) + f(x_2) + \cdots + f(x_n)}{n}\right]\left(\frac{b-a}{b-a}\right)$$

$$\simeq \left[f(x_1)\left(\frac{1}{n}\right) + f(x_2)\left(\frac{1}{n}\right) + \cdots + f(x_n)\left(\frac{1}{n}\right)\right]\left(\frac{b-a}{b-a}\right)$$

$$\simeq \frac{1}{b-a}\left[f(x_1)\left(\frac{b-a}{n}\right) + f(x_2)\left(\frac{b-a}{n}\right) + \cdots + f(x_n)\left(\frac{b-a}{n}\right)\right]$$

Now let Δx replace $(b-a)/n$. Then

$$Q \simeq \frac{1}{b-a}[f(x_1)\Delta x + f(x_2)\Delta x + \cdots + f(x_n)\Delta x]$$

The sum on the right is now a Riemann sum if $x_1, x_2, \ldots, x_n$ are chosen so as to lie in distinct intervals of length $\Delta x = (b-a)/n$. Since Q is approximated by $1/(b-a)$ multiplied by this Riemann sum, AP allows us to conclude:

4.5 Applications of the Definite Integral

The average value Q of a function f on an interval $[a, b]$ is

$$\frac{1}{b-a}\int_a^b f(x)\, dx$$

Geometrically, this means that the average value of f on $[a, b]$ is the area beneath its graph divided by the length of the interval. Alternatively, a rectangle $b - a$ units long and Q units high has area $\int_a^b f(x)\, dx$, as indicated in Figure 4.32.

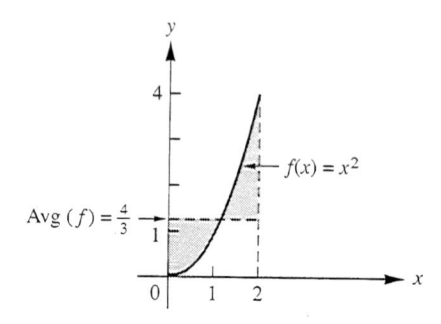

Figure 4.32

Let us find the average value of a familiar function.

EXAMPLE 2

Find the average value of the function $f(x) = x^2$ on the interval $[0, 2]$.

Solution According to our derivation, we use $a = 0$ and $b = 2$ to obtain an average value for f on this interval of

$$\frac{1}{2-0}\int_0^2 x^2\, dx = \frac{1}{2}\left(\frac{x^3}{3}\right)\Big|_0^2 = \left(\frac{1}{2}\right)\left(\frac{8}{3}\right) = \frac{4}{3}$$

We indicate this on the graph in Figure 4.32. ■

C. The Volume of a Solid

A rectangular solid of length l, width w, and height h has the volume, in cubic units (see Figure 4.33):

$$V = lwh$$

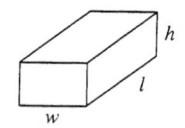

Figure 4.33

Let us reinterpret this formula as $V = (lw)h =$ area $\times$ height $=$ (cross-sectional area) $\times$ (thickness). We then see that a solid of cross-sectional area A and (uniform) thickness h, whose edge is *perpendicular* to the planar surface, has volume $V = Ah$ (see Figure 4.34). Beginning with this formula, we can use AP to derive a general formula for the volume $Q = V$ of a regular solid whose surface (and shape) continually varies about a fixed line.

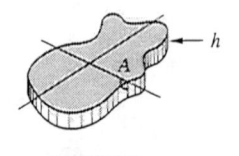

Figure 4.34

Consider Figure 4.35 as a typical example of the kind of solid whose volume we wish to compute. We refer to the line L about which the figure is symmetrically placed as an **axial line**. Notice that the solid figure varies in cross-sectional area at differing heights x along this axial line (with the largest cross-sectional area at height a and the smallest at height b). Let $A(x)$ denote the area of the cross section at height x, perpendicular to L.

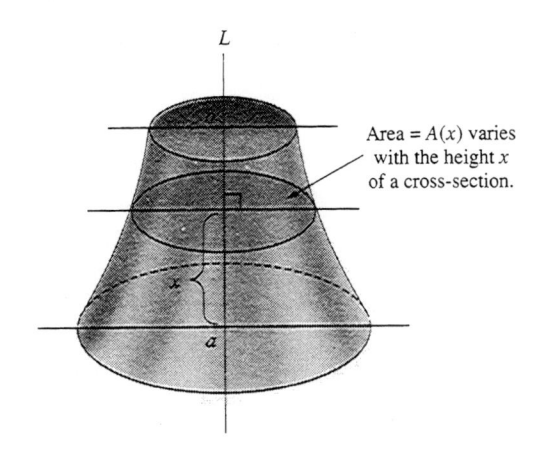

Figure 4.35

We wish to use AP to see that

$$V = \int_a^b A(x)\, dx$$

is the volume of this solid—that is, that *volume is an integral of cross-sectional area*.

Imagine dividing the axial line L in Figure 4.36 into a large number n of short lengths each $\Delta x = (b - a)/n$ long. Within each of these short pieces we imagine a point x_k on the axial line L, and then imagine a thin solid of cross-sectional area $A(x_k)$ and thickness Δx. The volume of this thin solid is

$$V_k = A(x_k)\Delta x$$

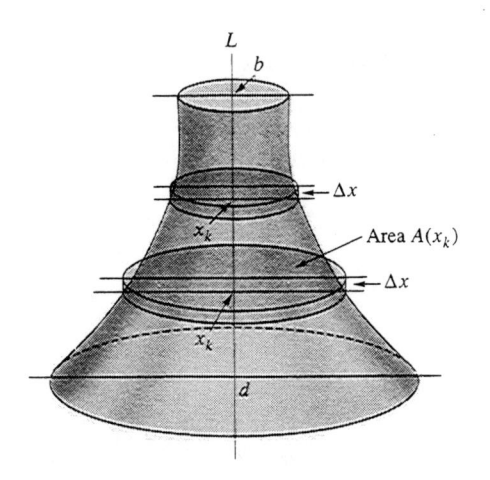

Figure 4.36

4.5 Applications of the Definite Integral   247

The volume of the *entire* solid S is approximated by the *sum* of the volumes of all these thin solids, one for each length Δx. That is,

$$V \simeq A(x_1)\Delta x + A(x_2)\Delta x + \cdots + A(x_n)\Delta x$$

So, applying AP, we have the following:

The volume V of a solid is the number

$$V = \int_a^b A(x)\, dx$$

where $A(x)$ is the cross-sectional area of the region perpendicular to the axial line through the solid, at the point x on this line.

Let us apply this formula to a specific example.

EXAMPLE 3

Find the volume of the cone of length h and radius r in Figure 4.37.

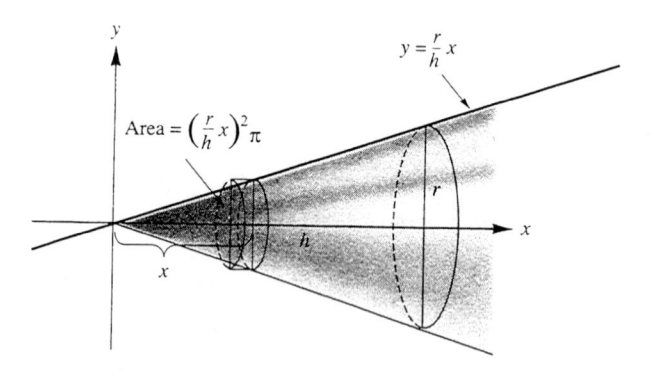

Figure 4.37

Solution Choose as the axial line the x-axis of an xy-coordinate system. We can imagine the cone S as the *solid of revolution* obtained by revolving the line $y = (r/h)x$ about the x-axis. Now imagine a point x on the x-axis between 0 and h. Let us compute $A(x)$, the cross-sectional area of the circle at the point x, perpendicular to the x-axis. This is easy since $y = (r/h)x$ is the *radius* of this circle. Therefore

$$A(x) = \pi\left(\frac{r}{h}x\right)^2 = \frac{\pi r^2 x^2}{h^2}$$

Hence the volume of the cone is

$$V = \int_0^h A(x)\, dx$$

$$= \int_0^h \pi\left(\frac{r}{h}x\right)^2 dx$$

248   Chapter 4   Integration

$$= \frac{\pi r^2}{h^2} \int_0^h x^2 \, dx$$

$$= \frac{\pi r^2}{h^2} \left(\frac{x^3}{3} \right) \Big]_0^h$$

$$= \frac{\pi r^2 h}{3} \quad \blacksquare$$

The idea illustrated in this example may be considerably generalized. Imagine revolving the graph of $y = f(x)$ about the x-axis, as in Figure 4.38, so as to describe the solid S. The volume of S is

$$V = \int_a^b \pi f(x)^2 \, dx$$

since the cross-sectional area at the point x between a and b is

$$A(x) = \pi f(x)^2$$

because $f(x)$ is the radius of the circular cross section, at the point x, perpendicular to the axial line (the x-axis).

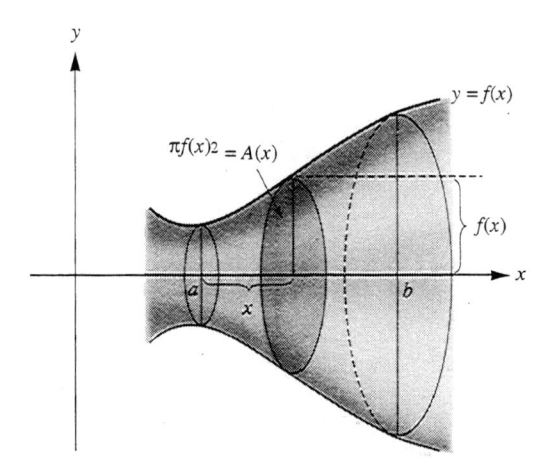

Figure 4.38

The previous applications of the integral deal with purely mathematical models. In the following examples, the integral itself becomes a mathematical model for a physical, biological, or economic process. This kind of application of the integral can be found in your area of study when you want to model, or compute, the sum of a continuously varying contribution to a whole. Example 2 in Section 4.1 was our first encounter with this kind of application.

D. Producers', Consumers', and Polluters' Surplus

Producers' surplus is a measure of the additional profit to producers that arises from the difference between the actual selling price of an item versus the price at which producers would be willing to sell the item (see Figure 4.39).

4.5 Applications of the Definite Integral

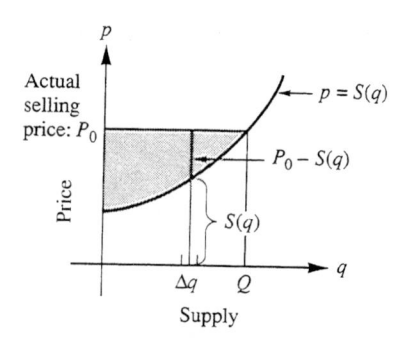

Figure 4.39

The producers' surplus for an item is defined as the area above the supply curve and below the horizontal line at P_0, which represents the price at which the item is actually selling. This definition is understood via a Riemann sum in the following way.

At a lower price p, some producers would be willing to supply a quantity q of this item. Because the item is actually selling at the higher price P_0, their surplus profit is then

$$P_0 - S(q) \text{ dollars/item}$$

This surplus profit is earned on the sale of approximately Δq items. Thus the surplus profit to producers is

$$[(P_0 - S(q)]\Delta q \text{ dollars}$$

at production level q. The total of these surplus profits for the varying quantity levels from 0 to the actual supply available, Q_0, is then approximated by a sum of the *form*

$$[P_0 - S(q_1)]\Delta q + [P_0 - S(q_2)]\Delta q + \cdots + [P_0 - S(q_{n-1})]\Delta q + [P_0 - S(q_n)]\Delta q$$

Therefore, applying AP, the **producers' surplus** for this item is

$$\int_0^{Q_0} [P_0 - S(q)]\, dq$$

A similar concept is that of *consumers' surplus*, defined as the area beneath a demand curve and above the selling price (see Figure 4.40). It represents the total saved by consumers who would be willing to buy the item at a higher price.

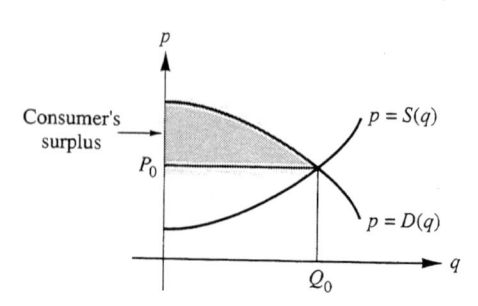

Figure 4.40

250   Chapter 4   Integration

That is, for the supply and demand curves indicated in Figure 4.40, the **consumers' surplus** is

$$\int_0^{Q_0} (D(q) - P_0)\, dq$$

EXAMPLE 4

Find the producers' surplus and the consumers' surplus for an item whose supply and demand curves are given by

$$S(q) = 3q + 1 \quad \text{and} \quad D(q) = 11 - q^2$$

for q thousands of units and prices in dollars per unit.

Solution First we have to find the quantity q at which supply meets demand, so as to determine the selling price. Thus set

$$S(q) = D(q)$$

and solve for q. We have

$$3q + 1 = 11 - q^2$$

or
$$q^2 + 3q - 10 = 0$$

or
$$(q + 5)(q - 2) = 0$$

The positive solution to this equation is $Q_0 = 2$ (thousand units). The selling price P_0 is therefore $S(2) = D(2) = 7$ dollars. Consequently, the producers' surplus is

$$\int_0^2 (7 - (3q + 1))\, dq = \int_0^2 6 - 3q\, dq = 6q - \frac{3q^2}{2}\Big]_0^2 = 6 \text{ thousand dollars}$$

The surplus to consumers, on the other hand, is

$$\int_0^2 ((11 - q^2) - 7)\, dq = \int_0^2 4 - q^2\, dq$$

$$= 4q - \frac{q^3}{3}\Big]_0^2 = \frac{2}{3}(8) = 5{,}333 \text{ dollars} \blacksquare$$

These supply and demand curves and the equilibrium price $P_0 = 7$ at quantity $Q_0 = 2$ are shown in Figure 4.41.

The concept of consumers' surplus has been extended by economists to the problem of how to deal, in economic terms, with the problem of toxic pollutants produced by manufacturers, wherein a manufacturer's desire to maximize profits clashes with society's need to maximize health. A proposed solution is to regard the "right to pollute" as a market item to be purchased from society by business. This leads directly to an example of consumers' (that is, business's) surplus.

Let $p = D(q)$ be the price per ton that business will pay (via governmental licenses) to produce q tons of pollutants. Suppose that society will tolerate the production of Q_0 tons of pollutants. Then $P_0 = D(Q_0)$ is the price per ton

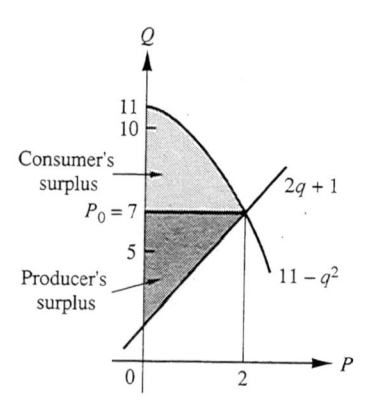

Figure 4.41

[Figure 4.41: Graph showing consumer's surplus and producer's surplus with $P_0 = 7$, curves $2q+1$ and $11-q^2$, intersecting at $q=2$, $Q=5$ to 11.]

polluters will pay to pollute. Consequently, we have a repetition of consumers' surplus (see Figure 4.42), where the area $A = \int_0^{Q_0} (D(g) - P_0)\, dq$ is the surplus (savings) to businesses who pollute.

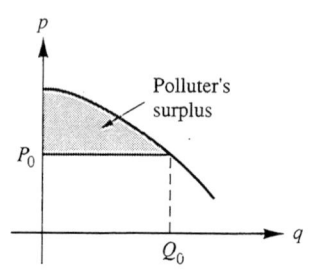

Figure 4.42

[Figure 4.42: Graph showing polluter's surplus above P_0 up to Q_0.]

This amount is important since it is the amount that business, in the aggregate, could then be taxed (via an "equalizing tax") to subsidize general health services, thus neutralizing, at least in economic terms, the impact of pollution on society.

EXAMPLE 5

In a certain locale, the environment can break down 45 tons of an airborne pollutant per year. The manufacturers that produce this pollutant are willing to pay $D(q) = 25 - (q/3)$ dollars per ton to produce the pollutant as a by-product of their production.

a. Find the license fee that will be charged per ton to produce the 45-ton total.
b. Find the surplus to these manufacturers resulting from the ability of this locale to permit this level of pollution.
c. If there are three manufacturers in the area, how much should each be taxed to bring their surplus to zero—that is, what is the "equalizing tax" on their pollution?

Solution

a. At $Q_0 = 45$ tons, manufacturers are willing to pay $D(45) = \$10$/ton to pollute.

b. The surplus to manufacturers is therefore

$$\int_0^{45} \left[\left(25 - \frac{q}{3}\right) - 10 \right] dq = \frac{15}{2}(45) = 337.5 \text{ dollars}$$

c. The three manufacturers should therefore pay an additional pollution tax of $\$337.50/3 = \$112.50$ per year. ∎

E. Accumulation in the Environment

The next example models the accumulation of quantities in the environment.

EXAMPLE 6

Coal is loaded into open-top hopper cars at a coal-mining operation in Montana. As the train hauling the coal pulls out of the mining area, coal dust and small particles of coal fall from each hopper car and settle beside the railroad track. It is found that on each day of operation

$$C(x) = \frac{98}{(x+7)^2} \text{ tons/mile}$$

of coal dust and particles are deposited beside the tracks, at a distance of x miles from the loading site.

a. How much coal is lost to railside deposits each day along the entire length of track from the loading point $x = 0$ to a point 10 miles distant?

b. If the mine operates 365 days a year, how much coal is lost this way in a year? At \$30/ton, how much income is lost per day?

Solution

a. Imagine the 10-mile length of railroad track, broken up into many short pieces each Δx long (see Figure 4.43). (For example, imagine that Δx is the nominal combined width of a railroad tie and the open space between ties.)

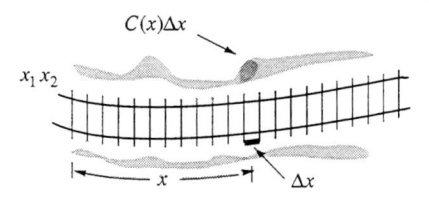

Figure 4.43

Let us first write a Riemann sum that *approximates* the total amount Q of coal dust lost in this 10-mile length of track each day. At a point x miles from the railhead

$$C(x) = \frac{98}{(x+7)^2} \text{ tons/mile}$$

are lost. Thus, along the short length Δx of track at this point x, approximately

$$C(x)\Delta x = \frac{98}{(x+7)^2}\left(\frac{\text{tons}}{\text{mile}}\right) \times \Delta x \text{ (miles)}$$

$$= \frac{98}{(x+7)^2}\Delta x \text{ tons}$$

are lost. Therefore, along the entire 10-mile length of track, a total of approximately

$$C(x_1)\Delta x + C(x_2)\Delta x + \cdots + C(x_n)\Delta x$$

tons are lost, where $x_1, x_2, \ldots, x_n$ is a series of distance markers we imagine along the track from the loading point x_1 to a point 10 miles distant x_M, and $\Delta x = x_{k+1} - x_k$.

Consequently, applying AP, the total amount of coal lost in a single day is

$$Q = \int_0^{10} \frac{98}{(x+7)^2}\,dx = 98\int_0^{10}(x+7)^{-2}\,dx = 98\left[\frac{-1}{(x+7)}\right]_0^{10}$$

(by letting $u = x - 1$; $du = dx$). Hence

$$Q = \frac{98(10)}{7(17)} \approx 8.24 \text{ tons/day}$$

b. Over a year the mining company loses $365 \times 8.24 = 3{,}007.6$ tons of coal. The income lost each day is $\$30 \times 8.24 = \$247.20$. ∎

In Section 4.1 we saw how the accumulation of farmland residues in a downstream reservoir could be modeled by a definite integral. This model has an obvious defect in that it is unreasonable to suppose that *all* of the farming residue carried by rainfall into a stream would actually ever reach a reservoir 10 miles downstream. Similarly, some percentage of the coal dust settling beside the railroad track will be blown away. The calculation of such a residual amount remaining from a continual accumulation requires the study of "Arrival and Accumulation Processes" that we undertake in Chapter 7.

Exercises 4.5

A. The Area of the Region between Two Curves

In Exercises 1–8, compute the area of the region between the graphs of f and g over the indicated interval. Remember to first determine which graph lies above the other over appropriate parts of the interval. In order to determine where one graph crosses the other, you must solve the equation $f(x) = g(x)$.

1. $f(x) = x + 1$, $g(x) = x^2 - 1$, $0 \le x \le 2$
2. $f(x) = x + 1$, $g(x) = x^2 - 1$, $1 \le x \le 3$
3. $f(x) = -(x-2)^2 + 2$, $g(x) = (x-2)^2$, $1 \le x \le 2$
4. $f(x) = -(x-2)^2 + 2$, $g(x) = (x-2)^2$, $0 \le x \le 4$

5. $f(x) = |x - 2|$, $g(x) = -\frac{1}{3}x + 2$, $0 \leq x \leq 3$
   [*Hint:* Use two intervals, [0, 2] and [2, 3].]
6. $f(x) = -x^2 + 3$, $g(x) = x^2 - 1$, on the interval for which $f(x) \geq g(x)$
7. $f(x) = \sqrt{x}$, $g(x) = x - 1$, on the interval for which $f(x) \geq g(x)$
8. $f(x) = (x - 1)^3$, $g(x) = 8$, $0 \leq x \leq 4$

B. The Average Value of a Function

In Exercises 9–12, compute the average value of the given function over the indicated interval.

9. $f(x) = \sqrt{x}$, $0 \leq x \leq 4$
10. $g(z) = z\sqrt{1 - z^2}$, $-1 \leq z \leq 1$
11. $h(t) = \dfrac{t^2}{(1 + t^3)^4}$, $-\dfrac{1}{2} \leq t \leq 2$
12. $l(u) = \dfrac{1}{u^2}$, $1 \leq u \leq 3$

Exercises 13 and 14 are *discrete state problems* in that only finitely many numbers are involved. Their solution using the integral will be "idealized"—not as accurate as, but a lot easier than, a tedious arithmetical calculation. When the scale is very large, solution by the integral is often quite accurate.

13. The owner of an orchard of 40 cherry trees numbers the trees from 1 to 40 in order of estimated productivity and records how many pounds of cherries each tree produces. The owner finds that tree number x produces $25 + (\sqrt{x}/2)$ pounds of cherries in a single season. What is the average production per tree in the orchard?

14. As she gains experience, a new insurance agent increases the number of policies she sells each month. In the tth month of the job, she sells $2t + 3$ policies. What is the average number of policies she sells per month in her first year? First find this number by arithmetic, averaging the numbers of sales 3, 5, 7, ..., 25 in the months $t = 0$ through $t = 11$ over these 12 months. Then find the average by integration, as though this were not a problem of discrete events. How well do the two answers compare? Draw a picture illustrating both answers as the area of some region in each case.

C. The Volume of a Solid

In Exercises 15–21, first sketch the graph of $y = f(x)$ in the usual two-dimensional coordinate system. Then compute the volume of the *solid of revolution* obtained by revolving the graph of $y = f(x)$ about the *x*-axis over the indicated interval. As your first step in the computation, write a formula for the cross-sectional area $A(x)$ of the circle described by revolving the point $(x, f(x))$ about the *x*-axis.

15. $f(x) = \sqrt{x}$, $0 \leq x \leq 2$
16. $f(x) = x^3$, $0 \leq x \leq 1$
17. $f(x) = \sqrt{1 - x^2}$, $-1 \leq x \leq 1$
18. $f(x) = x + 1$, $0 \leq x \leq 2$
19. $f(x) = (x - 1)^2$, $0 \leq x \leq 3$
20. $f(x) = -x^2 + 1$, $0 \leq x \leq 2$
21. $f(x) = 2x^3 - 9x^2 + 12x$, $0 \leq x \leq 2$
22. In the discussion of the formula $V = \int_a^b A(x)\,dx$ for a solid of cross-sectional area $A(x)$ at the point x, $a \leq x \leq b$, it was noted that $A(x)$ must be the area of a cross section *perpendicular* to the axial line L of length $b - a$. This exercise explores the importance of the requirement of perpendicularity. Using ordinary geometry and the formula $V = lwh$ for the volume of a rectangular solid, convince yourself that the volume of the parallelepiped (Figure 4.44) whose sides are of length l, w, h is *not* the number lwh but *is* the number lwk. [*Hint:* The volume of the rectangular solid enclosing the parallelepiped is $(l + x)wk$, and the volume of each of the two triangular solids is $\frac{1}{2}wxk$.]

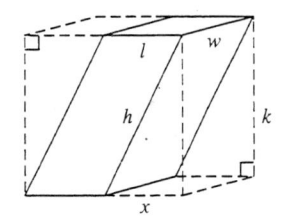

Figure 4.44

23. Use the method of Riemann sums and our discussion of volume as the integral of perpendicular cross-sectional area to convince yourself that the volume of the solid in Figure 4.45 is $(ab)k$ [and not $(ab)s!$].

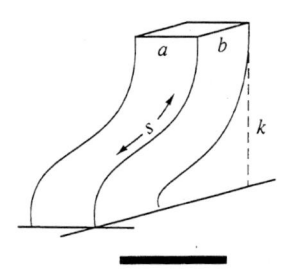

Figure 4.45

24. Compute the volume of the solid obtained by revolving the region *between* the graphs of $f(x) = x$ and $g(x) = x^2$, $0 \le x \le 1$, about the x-axis.

25. Compute the volume of the solid in Figure 4.46. Each cross section of this solid, perpendicular to the axial line L, is a square whose edge is a distance of $f(x) = (x - 10)^2$ in. from L at the point x above the base.

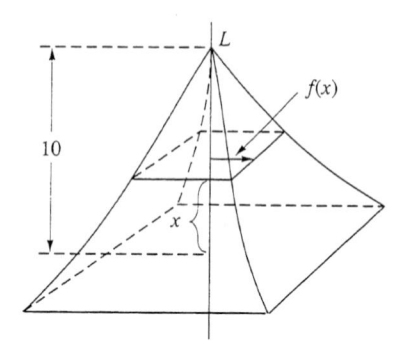

Figure 4.46

26. In adding an outdoor deck to his home, John Handy needed to first pour four concrete footings, of the shape and size in Figure 4.47, on which to set the posts supporting the deck. One bag of ready-mix concrete yields $\frac{2}{3}$ ft^3 of concrete at a cost of $2 per bag. How much will the four concrete footings cost John? [*Hint:* Develop a formula for the (square) cross-sectional area $A(x)$ of the footing at height x along the axial line. Think in terms of the slope of the side of the footing.]

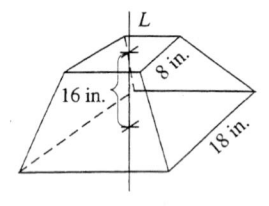

Figure 4.47

27. One pound of dry wood yields 8,600 BTU of heat when burned in a wood stove. A dead hickory tree has a diameter of 20 in. at its base and narrows to a diameter of 10 in. 12 ft above the ground. One cubic foot of dry hickory wood weighs 42 lb. How much heat can be obtained from this 12-ft log, assuming a circular log?

D. Producers', Consumers', and Polluters' Surplus

Compute the producers' surplus and the consumers' surplus for an item whose supply and demand curves are given in Exercises 28–31.

28. $D(q) = \frac{-1}{2}q + 75$ and $S(q) = \frac{q}{4} + 5$

29. $D(q) = 50 - 0.005q^2$ and $S(q) = 0.7q + 3.5$

30. $D(q) = 15 - 0.002q$ and $S(q) = 10$

31. $D(q) = \frac{200 - q}{100}$ and $S(q) = \left(\frac{q}{100}\right)^2$

32. Compute the manufacturers' surplus and equalizing tax for four manufacturers who are licensed to produce 85 tons of pollutants if the demand for pollution rights is $D(q) = 35 - (q/3)$ dollars/ton.

33. One hundred fifty midwestern coal-burning manufacturers are licensed to exhaust 75 million tons of hydrogen sulfide (which in turn produces acid rain in Canada and New England). If the demand for this pollution right is $D(q) = (-q/10^7) + 12$ dollars per ton, compute their surplus and equalizing tax.

E. Accumulation in the Environment

34. One hundred canoe rental operators are licensed to rent canoes to boaters along a 75-mile stretch of the Jack's Fork River. Their customers will leave all manner of trash (pollutants) along the river. The operators are licensed to permit 30 canoes on the river each day, for a 150-day season. This is equivalent to allowing the deposit of 6,750,000 pieces of trash on the river each year (based on an average of 12 items left each day per canoe). If the demand for this level of pollution rights is $D(q) = 0.07 - (q^2/10^{15})$ for q pieces of trash, what should the cost of a canoe rental license be?

The remaining exercises in this section are concrete problems that require you to approximate the quantity Q to be found by a Riemann sum and then apply AP. Make careful use of the *units* of measurement in each problem, as a guide to computation and as a help in combining measurements in the proper way. In each exercise write a formula for the typical term in a Riemann sum that approximates the desired quantity, and show that the units of measurements employed are consistent with this term and with the desired quantity.

35. Along a 6-cm length of artery leading from the heart, the density of cholesterol deposits on the arterial wall is

$$C(x) = \frac{20}{(x+1)^2} \text{ mg/cm}$$

at a point x cm from where the artery enters the heart. Compute the total amount (in mg) of cholesterol deposits along this 6-cm length of artery.

36. Suppose that on the tth day of winter ($t = 0$ is December 21) your home furnace burns $(135 - 2|t - 45|)/9$ gallons of oil. Use an integral to estimate your total oil consumption for the 90 days of winter. [*Hint:* To deal with the quantity $|t - 45|$, consider its value in the first 45 days of winter and then consider its value in the second 45 days of winter.]

37. The velocity at time t sec, $0 \le t \le 10$, of a car traveling between two stoplights is $v(t) = 1.2t(10 - t)$ ft/sec. What is the car's average speed over this time interval? How far did the car travel in these 10 sec?

38. The crawfish is a highly prized edible crustacean of southern Louisiana and thrives in the slowly moving shallow stream called a *bayou*. A certain bayou draining a Louisiana swamp supports $3{,}200 - 50x^2$ pounds per mile of crawfish at a point x miles from the point at which the swamp forms into the bayou. How many pounds of crawfish can be harvested along the bayou's first 7 miles? At a price of $0.90 a pound, what is the monetary value of the harvest?

39. A plot of land 100 ft wide extends 3,000 ft up the side of a mountain (see Figure 4.48). Over the course of 1 hour, $R(x) = 0.002x$ in.³ of rain falls on each square foot of the plot at a distance of x ft up the mountainside. How much rain falls on the entire plot?

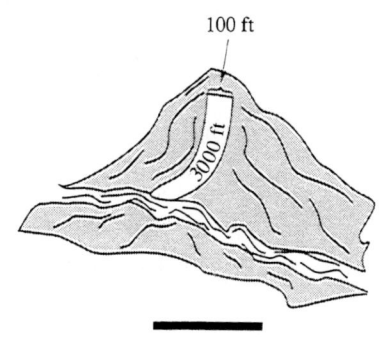

Figure 4.48

40. The fleet of trucks owned by a soft drink distributor consumes gasoline at the rate of $((t - 10)^3/5) + 250$ gallons per day. How much gasoline is consumed from day $t = 0$ to day $t = 20$?

41. The Current River is a popular floating stream, and canoeists leave behind $T(x) = 1/(10 - x)^2$ tons of trash per mile at a distance of $x < 10$ miles down the river from their common starting point on the river. How many tons of trash are left behind in the first 9 miles downriver from their starting point?

42. The heat lost by a closed building is proportional to the difference in temperature between the inside and the outside of the building. (The R-factor of insulation is an expression of this proportion.) Suppose that a building loses 100 BTU of heat per hour for each 1° difference between the inside and outside temperatures. On a given day the temperature outside the building is

$$T(s) = 14 + \frac{144 - (s - 12)^2}{6}$$

at s hours after midnight (when $s = 0$). How many BTUs of heat are lost by the building during a 24-hr period beginning at midnight if the interior temperature of the building is a constant 65°?

Chapter 4 Summary

1. The *definite* integral $\int_a^b f(x)\,dx$ is the numerical measure of the area beneath the graph of $y = f(x)$ from $x = a$ to $x = b$ when $f(x) \ge 0$.
2. The number $\int_a^b f(x)\,dx$ is found by the Fundamental Theorem of Calculus if one can find a function F such that $F'(x) = f(x)$ for each x in $[a, b]$. Then $\int_a^b f(x)\,dx = F(b) - F(a)$.
3. The process of finding F in (2) is called antidifferentiation, and the function F is called an antiderivative of f.

4. The antidifferentiation process is formalized as the problem of finding the *indefinite* integral $\int f(x)\,dx$. The basic formula is

$$\int x^n\,dx = \frac{x^{n+1}}{n+1} + C \qquad n \neq -1$$

and is combined with the method of substitution to find a large class of indefinite integrals.

5. The indefinite integration/antidifferentiation process has direct application to the problem of computing *how much change* has occurred when information is given about *how fast change* is occurring.

6. The definite integral has direct application to computing an unknown quantity Q when the quantity Q may be *approximated* by a (Riemann) sum of the *form*

$$f(x_1)\Delta x + f(x_2)\Delta x + \cdots + f(x_n)\Delta x$$

The approximation principle (AP) is founded on the observation that a sum of this form also approximates the area under the curve $y = f(x)$, and this area is measured exactly by the definite integral $\int_a^b f(x)\,dx$, when $f \geq 0$.

Chapter 4 Summary Exercises

1. Estimate the average height of the graph of $y = x^3$ on the interval [0, 2], and use this number to estimate the value of $\int_0^2 x^3\,dx$. Then use the Fundamental Theorem to find the exact value, and compare your two answers.

2. Let $f(x)$ be the number of residents per mile living x miles from the beginning of a major street that is 10 miles long. Let Δx denote a short length (in miles) of this street. What are the units of measurement of the quantity $f(x)\Delta x$? Explain how your answer is related to what the number $\int_0^5 f(x)\,dx$ measures.

3. Suppose that at time t you are running to class at the rate of $f(t)$ ft/sec. It takes 45 sec for you to run from your dorm to class. Explain what $\int_0^{45} f(t)\,dt$ represents, using an $f(t)\Delta t$ analysis.

4. Explain in complete sentences how the Fundamental Theorem of Calculus tells you how to find the value of an integral.

Use the Fundamental Theorem to find the definite integrals in Exercises 5–14.

5. $\int_0^1 2x\,dx$

6. $\int_{-1}^1 x^3\,dx$

7. $\int_0^2 x(x+1)\,dx$

8. $\int_1^4 \frac{1}{\sqrt{u}}\,du$

9. $\int_1^2 \frac{t+1}{t^3}\,dt$

10. $\int_0^1 (1 - x + x^2 - x^3)\,dx$

11. $\int_{-2}^{-1} \frac{1}{x^3}\,dx$

12. $\int_0^1 \left(4x^3 + \frac{x^2}{3}\right)dx$

13. $\int_0^1 w(1 + w^2)^6\,dw$

14. $\int_{-4}^0 \sqrt{(z+5)}\,dz$

In Exercises 15–16, isolate one part of each integral whose derivative is—up to a constant factor—the remaining part of the integral.

15. $\int x^2 \sqrt{(1-x^3)}\, dx$ 16. $\int \dfrac{5x^6}{x^7+1}\, dx$

Find the indefinite integrals in Exercises 17–24.

17. $\int x\sqrt{(1-x^2)}\, dx$ 18. $\int \dfrac{x}{(\pi+x^2)^3}\, dx$

19. $\int x^3(1-x^4)^{1/2}\, dx$

20. $\int (1-2y)(y-y^2+3)^3\, dy$

21. $\int (t^2+1)^{3/2} t\, dt$ 22. $\int 2(1-u^2)^3 u\, du$

23. $\int x\sqrt{(x^2+a^2)}\, dx$ 24. $\int -3(1-x^3)^{-2} x^2\, dx$

25. Explain why it is necessary to add a constant to each answer in Exercises 17–24 and why you need not do this in Exercises 5–14 and, furthermore, why even if you add a constant to each antiderivative in Exercises 5–14, it has no effect on the final answer.

26. Explain in complete sentences the principal connection between the definite integral and the indefinite integral, concentrating on how one depends on the other.

In Exercises 27–30, solve the differential equation with the given initial condition; that is, find the antiderivative of the given function and use the given initial value to specify the unknown constant that appears in this antiderivative. (See Exercises 5–9 in Section 4.4.)

27. $y'(t) = -t, \quad y(0) = 3$
28. $y'(t) = 3t^2 - 4t^3, \quad y(0) = -2$
29. $y'(t) = t\sqrt{1+t^2}, \quad y(0) = 1$
30. $y'(t) + 2 = 1 - t, \quad y(1) = 0$

Find the integrals in Exercises 31–34.

31. $\int x(1+x)^{13}\, dx$ 32. $\int (1-x^2)(1-x)^5\, dx$

33. $\int_0^3 \dfrac{x^2}{\sqrt{1+x^3}}\, dx$ 34. $\int_0^1 \dfrac{t}{(1+t)^3}\, dt$

35. From a height of 60 ft you toss a rock into the air at the rate of 30 ft/sec. Assume that only gravity acts on the rock thereafter, giving it a deceleration back to earth of -32 ft/sec.

   a. At what time is the rock back at the original height from which it was released?
   b. How fast is the rock moving 1 sec after release?
   c. How fast was the rock moving at the peak of its flight?
   d. When did this occur?
   e. When does the rock hit the ground 60 ft below?
   f. How fast was it moving then?
   g. What was its acceleration at that moment?

36. The Ace Manufacturing Company is selling a new product at the rate of $2x + 7$ thousand units per day, x days after its release to the market. When will sales reach a total of 36,000 units?

37. The marginal revenue to a manufacturer is $M_R(x) = 2 - 0.008x$ thousand dollars for the sale of x thousand items; $500 in display costs are lost before any items are sold. When will its net revenues be $5,500? (**Remark.** Two solutions are possible. Which is more meaningful?)

38. Renée is reeling in the bait of her fishing line at the rate of 9 revolutions per second of the line spool. Each revolution adds $\tfrac{1}{180}$ in. to the radius of the spool, so that $r'(t) = \tfrac{1}{20}$ in./sec is the rate of growth of the radius of the spool. How large is the radius $r(t)$ at the end of t sec, if $r(0) = \tfrac{1}{3}$ in.? If no line is on the spool at $t = 0$, how much line has been wound on the spool 20 sec later? [*Hint:* At any moment t, the circumference of the spool is $C(t) = 2\pi r(t)$ and $C(t)$ is the length of line wound onto the spool in one revolution at time t.]

39. Find the area of the region lying between the graphs of $f(x) = x^2$ and $g(x) = 3x - 2$ on the interval $[0, 6]$.

40. Find the average value of the function $f(x) = \sqrt{x-1}$ on the interval $[1, 4]$.

41. Find the volume of the solid of revolution obtained by rotating the graph of $y = 1/x$ about the x-axis between $x = 1$ and $x = 3$.

42. Find the volume of Figure 4.49, whose base is a square 14 ft long on each side and whose height is 20 ft by (a) finding a formula for the cross-sectional area of the figure x ft above the base and (b) performing the necessary integration.

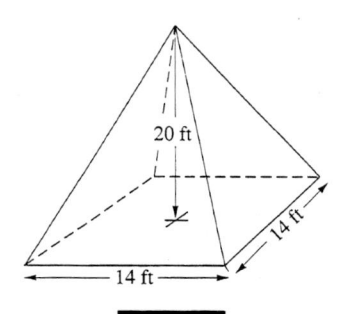

Figure 4.49

43. Find the producers' surplus and the consumers' surplus for an item whose supply and demand curves are given by $S(q) = 10 + 0.6q$ and $D(q) = 50 - 0.2q$.

44. Compute the manufacturers' surplus and equalizing tax for five manufacturers who are licensed to produce 50 tons of toxic waste where the demand for this right to pollute is $D(q) = 100[65 - (q/2)]$ dollars per ton.

45. Let $f(x)$ be the number of pounds of fly ash per square foot falling to the ground x ft to the north of a smokestack that services a local power plant. Explain what the typical term in the Riemann sum approximation to the integral $\int_0^{500} f(x)\,dx$ represents. What does the integral itself represent?

46. An electrical power line is strung in a north–south direction between two supports 300 ft apart. There are $f(x) = 7{,}000/(x + 50)^2$ blackbirds per foot perched on the power line x ft from the north end. How many blackbirds are perched on the entire 300-ft length?

47. Raw materials are arriving at a manufacturing plant at the rate of $f(t) = 5 + 0.07t^2$ tons per day, t days after the start of a production run. How much arrives during the first 15 days?

48. When you slow and stop your car at a stop sign, a small amount of rubber is ground off its tires and settles alongside the roadway. At a distance of x ft from a stop sign, a total of $T(x) = 1{,}250 - \frac{1}{2}(x - 50)^2$ lb/ft (of roadway) of tire rubber are worn off the tires of cars coming to a stop at the sign over the course of a year, out to a distance of 100 ft from the sign. How much tire "dust" (rubber) is deposited alongside the roadway over this 100-ft length over the course of a year?

Chapter 5

The Exponential and Logarithmic Functions

Two Special Functions

A vast array of processes both in nature and in economics may be loosely but usefully recognized by growth that "feeds on itself." Among these are population growth, compound interest, the growth of an investment in which earnings are reinvested, the consumption of a natural resource, and the spread of an epidemic. Calculus offers a unified view of all such processes in the form of a model that goes beyond the familiar algebraic functions and utilizes a special function called the *exponential function*. The inverse of an exponential relationship in turn defines an equally useful nonalgebraic function called the *logarithmic function*. While its applications are perhaps less readily apparent, mathematical psychologists have known for more than a century that the logarithm provides a model of how we perceive such common sensations as taste or sound, or even personal well-being. Both topics of this chapter are essential to your ability to employ the full power of calculus to measure and better understand fundamental processes in the world about you.

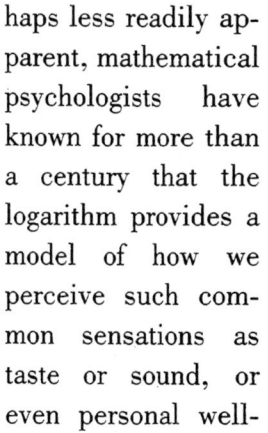

5.1 The Exponential Function and Its Derivative

The purpose of this section is to define and begin the study of the single most important function in the applications of calculus: the exponential function.

Calculus gives us a way to make predictions about the future state of a changing system, but such predictions often call for the use of nonalgebraic functions. Suppose you deposit $100 in a savings account paying annual interest of 8%. A year later you will have $108 in the account; in the next year the amount in the account will grow by $(0.08)(\$108) = \$8.64$. The amount in the account is *growing in direct proportion to the amount present* in the account. In a sense, growth is "feeding on itself." We shall soon see that such a process can best be modeled not by familiar algebraic functions, but by the **exponential function.**

The exponential function plays a much larger and more fundamental role in calculus than this application suggests, however. Indeed the full use of calculus would be impossible without the exponential function, much as arithmetic would be impossible without the numbers 0 and 1 and geometry would be impossible without the number π. The indispensible and shared nature of these elements in arithmetic and geometry is that they act as *invariants* with respect to basic structure in each of these subjects. The numbers 0 and 1 act as invariants in arithmetic in that they leave all other numbers *unchanged under the principal operations* of addition and multiplication: $a + 0 = a$ and $a \cdot 1 = a$, for any number a. In geometry, π is an invariant because it is the unvarying ratio C/D of circumference to diameter in any circle.

An invariant in calculus would have to share analogous properties in this subject. The principal operations in calculus are differentiation and integration. The principal objects are functions. Therefore an invariant for calculus would be a (nonzero) function f that is unchanged under the principal operations

$$D_x f(x) = f(x) \qquad \text{and} \qquad \int f(x)\, dx = f(x) + C$$

Such an invariant is given by the exponential function. This function is the single most important function in calculus. Because of its invariance, it is the easiest function to which we can apply calculus, just as 0 and 1 are the easiest numbers with which we do arithmetic. The *algebra* of the exponential function, on the other hand, will present you with some new challenges that will need your attention.

How Nature Gives Rise to a Differentiation Invariant Function

A differentiation invariant function is an essential link between calculus and the external world. Let us see how a function f such that $D_x f(x) = f(x)$ arises naturally and inevitably in applications.

Single-cell division (mitosis) provides an illustration. Suppose that a single cell divides once each minute to produce two new cells, each of which then divides in the second minute to produce four cells, and so on. Let

$$f(t) = \text{number of cells at time } t$$

A table of values for this function over time, along with the *change in population* per unit of time, appears in Table 5.1.

Table 5.1

Time t (min)	Number of Cells $f(t)$	Population Change $f(t+h) - f(t)$	Elapsed Time h
0	1		
1	2	1	1
2 t	$f(t)$ 4	2	1
3 $t+h$	$f(t+h)$ 8	4	1
4	16	8	1
5	32	16	1
6	64	32	1
7	128	64	1
8	256	128	1

Notice that if we let h denote the change in time between divisions (that is, if $h = 1$), then column 3, compared with column 2, tells us that

$$\frac{\text{Average rate}}{\text{of change}} = \frac{\text{Column 3}}{\text{Column 4}} = \frac{f(t+h) - f(t)}{h} = f(t) = \frac{\text{Amount present}}{\text{at time } t} \qquad (1)$$

For example, $\quad \dfrac{f(2+1) - f(2)}{1} = \dfrac{8-4}{1} = 4 = f(2)$

the number of cells present after 2 minutes. But, of course, the derivative of a function is the limit of its average rate of change. If we replace the average rate by the instantaneous rate $D_t f(t)$ in Equation 1, we find that apparently the law of growth for this cell population is approximately modeled by an equation of the form

$$D_t f(t) = f(t)$$

That is, the invariant of calculus appears to be inevitable in describing natural growth.

This last equation is not correct for single-cell division. It also does not make mathematical sense to speak of differentiation over discrete (1-minute) time intervals. But, in fact, the equation does represent a useful concept here because cells in nature rarely occur individually. Instead, in a body of (perhaps) millions of cells all are dividing at different times.* This division of millions of cells can be accurately modeled by an equation like $D_t f(t) = f(t)$, where $f(t)$ is the total cell population of the body, because the differing division times of the millions of cells allows us to regard the underlying time variable as a continuum.

A function $y = f(x)$ such that

$$D_x f(x) = f(x)$$

certainly *cannot* be a polynomial or a combination of polynomials for, since $D_x x^n = nx^{n-1} \neq x^n$, differentiation cannot leave polynomials invariant (unchanged). Therefore the function we seek—if it exists, and nature suggests that it does—must be fundamentally different from a polynomial function. A close look at the values of the domain and range in Table 5.1 suggests that

$$f(t) = 2^t$$

is a formula for the function f, which models the number of cells at time t. Such a function f is called an "exponential function" because *the variable t occurs as an exponent* in the formula for f. (This is quite different from the familiar function $g(t) = t^2$, where the variable t occurs in the base.) But this function f is not *the* exponential function.

The exponential function—that function which models not the growth of a single cell but the population growth of a body of cells and meets exactly the purely mathematical condition $D_x f(x) = f(x)$—is instead an exponential function of the form $f(x) = e^x$, where the letter e (like the letter π) represents a special number. We will first define the number e and the exponential function and then see why it is invariant under differentiation.

The Definition of the Number e

The definition of e is subtle and not easy to justify. We will simply state it.

The number e is the value of

$$\lim_{n \to \infty} \left(1 + \frac{1}{n}\right)^n \qquad \text{where } n \text{ is a whole number}$$

It is not obvious—indeed it takes a great deal of effort to show—that this limit even exists since, as $n \to \infty$, $1 + (1/n) \to 1$ and $1^n = 1$ no matter how large

---

*See, for example, the concept of *asynchronous* growth in G. G. Meynell and E. Meynell, *Theory and Practice in Experimental Bacteriology* (London: Cambridge University Press, 1970), p. 4.

n is. But in the expression

$$\left(1 + \frac{1}{n}\right)^n$$

the base $1 + (1/n)$ "shrinks to 1" *as* the exponent n "grows to infinity." In the simultaneous shrinkage and growth of base and exponent, respectively, the limit value is unexpected. It can be shown that

$$e = \lim_{n \to \infty} \left(1 + \frac{1}{n}\right)^n = 2.71828182846\ldots$$

This decimal expansion of e is accurate to ten decimal places. The number e is an irrational number, which (like π) has an unending series of digits that repeat in no particular pattern.

The Exponential Function and Its Algebraic Properties

Having the number e, we define the exponential function:

The exponential function is the function given by the formula

$$E(x) = e^x$$

for x any real number.

That is, the value of the exponential function at a number x is the particular number e *raised to the power x*.

EXAMPLE 1

Evaluate the exponential function e^x at

$$x = 2, \; -3, \; \tfrac{1}{2}, \; \tfrac{1}{5}, \; \tfrac{3}{5}, \; -\tfrac{7}{3}, \; \text{and} \; 0$$

Solution Recalling the usual algebraic properties of exponents and using a calculator, we have, without roundoff,

$$e^2 = e \cdot e = (2.718\ldots)(2.718\ldots) = 7.389\ldots$$

$$e^{-3} = \frac{1}{e^3} = \frac{1}{(2.718\ldots)^3} = 0.0497\ldots$$

$$e^{1/2} = \sqrt{e} = \sqrt{2.718\ldots} = 1.648\ldots$$

$$e^{1/5} = \text{fifth root of } 2.718\ldots = 1.2214$$

$$e^{3/5} = (e^{1/5})^3 = (1.2214\ldots)^3 = 1.822\ldots$$

$$e^{-7/3} = \frac{1}{e^{7/3}} = \frac{1}{(e^{1/3})^7} = \frac{1}{(1.3956\ldots)^7} = \frac{1}{10.3112\ldots} = 0.0969\ldots$$

Finally, $e^0 = 1$, since $a^0 = 1$ for any number $a \neq 0$. ∎

Other values of e^x may be found similarly. However, we would normally use a table of values or a calculator with an $\boxed{\text{exp}}$ or $\boxed{e^x}$ or $\boxed{\text{inv}}$ $\boxed{\text{ln}}$ "button" to compute values of e^x. See, for example, Table 5.4 at the close of this section, and compare the values of e^2 and $e^{0.5}$ found in Example 1 with the values given there.

The meaning and calculation of numbers like $e^{\sqrt{2}}$, e^{π}, e^e, and so on, are more complex. The definition of $e^{\sqrt{2}}$ is

$$e^{\sqrt{2}} = \lim_{r \to \sqrt{2}} e^r$$

where r approximates $\sqrt{2}$ and is a rational number. That is, $r = p/q$ when p and q are whole numbers, $q \neq 0$. For example, since

$$\sqrt{2} = 1.414 \ldots \approx \frac{1{,}414}{1{,}000}$$

then

$$e^{\sqrt{2}} \approx e^{1.414/1{,}000} = (e^{1/1{,}000})^{1.414} \approx 4.112 \ldots$$

In practice we would use a calculator, which approximates e^x, using $x = 1.414$. These calculations and Example 1 illustrate that

$$e^x > 0 \qquad \text{for any number } x$$

and that e^x has a definite numerical value for *all* real numbers x. In summary:

The *domain* of the exponential function e^x is

all real numbers

and the *range* of this function consists of only

positive numbers.

The preceding calculations also show that, so far as arithmetic is concerned, we treat e as any other ordinary number.

In calculus we only rarely determine a value of e^x and do so with a calculator or table of values. It is much more important to be familiar with (1) the laws of exponents applied to e^x and (2) the graph of $y = e^x$.

The algebraic laws of exponents (see Background Review 5.1) yield the following algebraic properties of the exponential function.

Algebraic Properties of the Exponential Function

1. $e^0 = 1$
2. $e^{a+b} = e^a \cdot e^b$
3. $e^{ab} = (e^a)^b$
4. $e^{a/b} = (e^{1/b})^a = (e^a)^{1/b}$

5. $e^{-a} = \dfrac{1}{e^a}$

for a and b any real numbers.

In particular, as always, *multiplication* of exponential expressions calls for *addition* of exponents. Notice also from property 3 that, for example,

$$e^{x^2} \neq (e^x)^2$$

for in fact $(e^x)^2 = e^{2x}$ from property 3. For example, with $x = 3$, $e^{3^2} = e^9 \simeq 8{,}103$, whereas $(e^3)^2 \simeq (20.08\ldots)^2 \simeq 403$.

The Graph of $y = e^x$

You must keep in mind the graphs of certain exponential functions, particularly *the* exponential function. Let us begin by graphing

$$y = e^x$$

and the more familiar function

$$y = 2^x$$

First, we make a table of corresponding values using a calculator (exact to three decimal places; see Table 5.2). Then, using Table 5.2, we sketch the two

Table 5.2

x	$y = 2^x$	$y = e^x$
0	1	1
1	2	$e \simeq 2.718$
2	4	$e^2 \simeq 7.389$
3	8	$e^3 \simeq 20.085$
4	16	$e^4 \simeq 54.598$
-1	$2^{-1} = \dfrac{1}{2}$	$e^{-1} = \dfrac{1}{e} \simeq 0.367$
-2	$\dfrac{1}{4}$	$e^{-2} = \dfrac{1}{e^2} \simeq 0.135$
-3	$\dfrac{1}{8}$	$e^{-3} = \dfrac{1}{e^3} \simeq 0.0497$
$\dfrac{1}{2}$	$2^{1/2} = \sqrt{2} \simeq 1.414$	$e^{1/2} = \sqrt{e} \simeq 1.648$
$\dfrac{1}{3}$	$\sqrt[3]{2} \simeq 1.259$	$e^{1/3} = \sqrt[3]{e} \simeq 1.395$
$-\dfrac{1}{2}$	$2^{-1/2} = \dfrac{1}{\sqrt{2}} \simeq 0.707$	$e^{-1/2} = \dfrac{1}{\sqrt{e}} \simeq 0.606$
$\sqrt{2}$	$2^{\sqrt{2}} \simeq 2^{1.414} \simeq 2.664$	$e^{\sqrt{2}} \simeq e^{1.414} \simeq 4.112$
100	A huge number	A huge number
-100	Almost zero!	Almost zero!

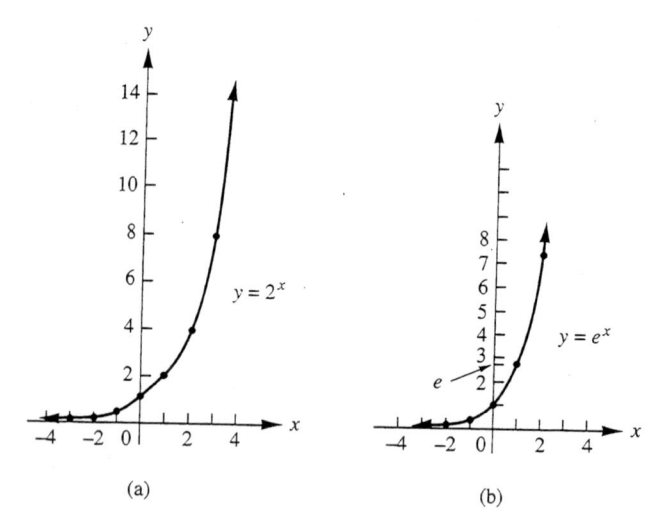

Figure 5.1

(a) (b)

corresponding graphs in Figure 5.1. Both graphs in Figure 5.1, which are similar in form, illustrate exponential growth. This form is quite different from the linear (straight line) growth that we are most used to, and yet it is pervasive in both nature and economics.

Notice that Figure 5.1 also reminds us that e^x is defined and positive for all numbers x. Additionally, $y = e^x$ is a continuous function that is increasing at each point x and indeed seems to increase at an ever-increasing rate.

You should also be able to graph two other (composite) functions closely related to $y = e^x$:

$$y = e^{-x} \quad \text{and} \quad y = e^{-x^2}$$

Again, to draw these graphs, we make a table of function values (see Table 5.3). From Table 5.3, we obtain the graphs shown in Figure 5.2(a) and (b).

Table 5.3

x	$y = e^{-x}$	$y = e^{-x^2}$
0	$e^{-0} = e^0 \approx 1$	$e^{-0^2} = e^0 = 1$
1	$e^{-1} = \dfrac{1}{e} \approx 0.367$	$e^{-1^2} = e^{-1} \approx 0.367$
$\sqrt{2}$	$e^{-\sqrt{2}} = \dfrac{1}{e^{\sqrt{2}}} = 0.243$	$e^{-(\sqrt{2})^2} = e^{-2} \approx 0.135$
2	$e^{-2} = \dfrac{1}{e^2} \approx 0.135$	$e^{-2^2} = e^{-4} \approx 0.018$
-1	$e^{-(-1)} = e^1 \approx 2.718$	$e^{-(-1)^2} = e^{-1} \approx 0.367$
$-\sqrt{2}$	$e^{-(-\sqrt{2})} = e^{\sqrt{2}} = 4.113$	$e^{-(-\sqrt{2})^2} = e^{-2} \approx 0.135$
-2	$e^{-(-2)} = e^2 \approx 7.389$	$e^{-(-2)^2} = e^{-4} \approx 0.018$
$\dfrac{1}{2}$	$e^{-1/2} = \dfrac{1}{\sqrt{e}} \approx 0.606$	$e^{-(1/2)^2} = e^{-1/4} \approx 0.778$
$-\dfrac{1}{2}$	$e^{-(-1/2)} = \sqrt{e} \approx 1.648$	$e^{-(-1/2)^2} = e^{-1/4} \approx 0.778$
100	e^{-100} (almost 0)	$e^{-10,000}$ (almost 0)

Notice that Figure 5.2(a) is just the reflection about the y-axis of the exponential graph $y = e^x$ of Figure 5.1. That this should be so is rather clear, since in replacing x by $-x$ in the exponent, we are in effect reversing the direction of the x-axis. The graph in Figure 5.2(b) may be familiar to you, since it is the famous "bell-shaped curve" that statistically models the (normal) distribution of many phenomena.

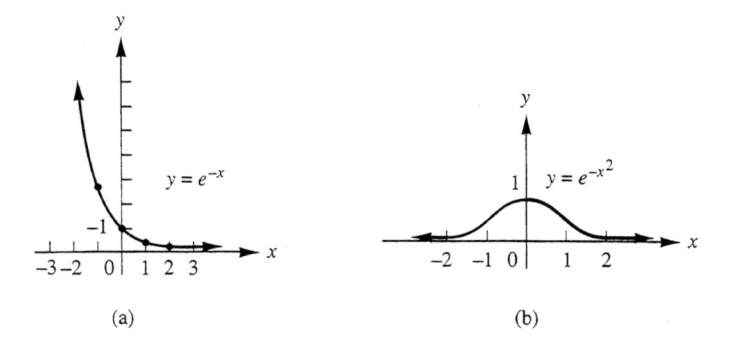

Figure 5.2

(a)   (b)

Differentiation of $y = e^x$: The Invariant of Calculus

Let us now calculate the rate of growth of $y = e^x$, or the slope of tangents to its graph. We will find that

$$D_x e^x = e^x$$

so that *the exponential function is invariant for differentiation.*

Let $f(x) = e^x$. According to the definition of the derivative (Section 1.4), we must find the limit

$$D_x e^x = \lim_{h \to 0} \frac{f(x+h) - f(x)}{h} = \lim_{h \to 0} \frac{e^{x+h} - e^x}{h}$$

First some algebra. By the laws of exponents,

$$\frac{e^{x+h} - e^x}{h} = \frac{e^x e^h - e^x}{h} = \frac{e^x(e^h - 1)}{h} = e^x \left(\frac{e^h - 1}{h} \right)$$

Now, since only h changes in the preceding limit, we can say that

$$D_x e^x = \lim_{h \to 0} e^x \left(\frac{e^h - 1}{h} \right) = e^x \cdot \lim_{h \to 0} \left(\frac{e^h - 1}{h} \right) \qquad (2)$$

We want to see now that $\lim_{h \to 0} (e^h - 1)/h$ in fact equals 1. This is not easy to prove, and we will only try to make a reasonably convincing case. First let us replace h by $1/n$ and $h \to 0$ by $n \to \infty$. That is, we assert that

$$\lim_{h \to 0} \frac{e^h - 1}{h} = \lim_{n \to \infty} \frac{e^{1/n} - 1}{\frac{1}{n}}$$

5.1 The Exponential Function and Its Derivative

Now, by definition, $e = \lim_{n\to\infty}[1 + (1/n)]^n$, so $e \simeq [1 + (1/n)]^n$ for large values of n. Let us then write*

$$\frac{e^{1/n} - 1}{\dfrac{1}{n}} \simeq \frac{\left[\left(1 + \dfrac{1}{n}\right)^n\right]^{1/n} - 1}{\dfrac{1}{n}} = \frac{\left(1 + \dfrac{1}{n}\right)^{n \cdot 1/n} - 1}{\dfrac{1}{n}}$$

(by replacing e on the left by $[1 + (1/n)]^n$ on the right)

$$= \frac{\left(1 + \dfrac{1}{n}\right) - 1}{\dfrac{1}{n}} = \frac{\dfrac{1}{n}}{\dfrac{1}{n}} = 1$$

Therefore we conclude that

$$\lim_{h \to 0} \frac{e^h - 1}{h} = \lim_{n \to \infty} \frac{e^{1/n} - 1}{\dfrac{1}{n}} = 1$$

Combining this with Equation 2, we see that

$$D_x e^x = e^x \cdot \lim_{h \to 0} \frac{e^h - 1}{h} = e^x \cdot 1 = e^x$$

What of the other operation of calculus—integration? Since the indefinite integral is the reverse of differentiation and asks for a function whose derivative is a given function, we can also immediately conclude that

$$\int e^x \, dx = e^x + C$$

These two basic facts should be committed to memory.

The Exponential Function Is Invariant in Calculus

$$D_x e^x = e^x \quad \text{and} \quad \int e^x \, dx = e^x + C$$

The details of algebraic computation with the exponential function are a challenge to most students. But so far as calculus is concerned, this function is the easiest to deal with in calculus because it is invariant under both operations of calculus.

For purely mathematical reasons (the mean value theorem of Section A.2), essentially no other functional form (up to a constant multiple) has these invar-

---

*This step in the "proof" requires much more justification than we give here; complete justification is rather subtle and well beyond the intent of this course.

iance properties. It follows then that if we accept an equation $D_t f(t) = f(t)$ as the appropriate model for growth in some system in the external world, then we must accept the exponential function as the only possible model for that system. There is no other option. This is a significant fact in applications.

Basic Differentiation and Integration of Exponential Functions

This section concludes with six basic examples that show that differentiation and integration of the exponential function are quite straightforward. All computations are based on the invariance rule

$$D_x e^x = e^x$$

and the rules of differentiation already studied.

EXAMPLE 2

Find $D_x(x + e^x)$.

Solution Using the method for differentiating a sum, we have

$$D_x(x + e^x) = D_x x + D_x e^x = 1 + e^x \quad \blacksquare$$

EXAMPLE 3

Find $D_x(xe^x)$.

Solution Using the formula for differentiation of a product, we have

$$D_x(xe^x) = x(D_x e^x) + e^x(D_x x) = x \cdot e^x + e^x \cdot 1 = (x + 1)e^x \quad \blacksquare$$

EXAMPLE 4

Find $D_x(e^x/x)$.

Solution Using the formula for differentiation of a quotient, we have

$$D_x\left(\frac{e^x}{x}\right) = \frac{x(D_x e^x) - e^x \cdot (D_x x)}{x^2} = \frac{xe^x - e^x}{x^2} = \left(\frac{x - 1}{x^2}\right)e^x \quad \blacksquare$$

Let us now do some very basic integration. The next examples all use the formula

$$\int e^x \, dx = e^x + C$$

EXAMPLE 5

Compute $\int 5e^x \, dx$.

Solution

$$\int 5e^x \, dx = 5 \int e^x \, dx = 5e^x + C \quad \blacksquare$$

EXAMPLE 6

Compute $\int (x + e^x) \, dx$.

Solution

$$\int (x + e^x) \, dx = \int x \, dx + \int e^x \, dx = \frac{x^2}{2} + e^x + C \quad \blacksquare$$

EXAMPLE 7

Compute $\int_0^1 (e^x - 1) \, dx$.

Solution Using the Fundamental Theorem of Calculus, we have

$$\int_0^1 (e^x - 1) \, dx = \int_0^1 e^x \, dx - \int_0^1 1 \, dx$$
$$= e^x\Big]_0^1 - x\Big]_0^1$$
$$= (e^1 - e^0) - (1 - 0) = e - 1 - 1 = e - 2$$
$$\approx 2.718 - 2 = 0.718 \quad \blacksquare$$

Calculations such as

$$\int xe^x \, dx \quad \text{or} \quad \int e^{-x^2} \, dx$$

which appear to be only slightly different, turn out to be much more difficult, and we have to defer these.

The Essential Properties of the Exponential Function

The exponential function is defined for all real numbers x by

$$E(x) = e^x \quad \text{where } e = \lim_{n \to \infty} \left(1 + \frac{1}{n}\right)^n \approx 2.71828$$

The exponential function is continuous, always positive, and always increasing (since $D_x e^x = e^x > 0$), and $E(0) = e^0 = 1$. Its graph is shown in Figure 5.3.

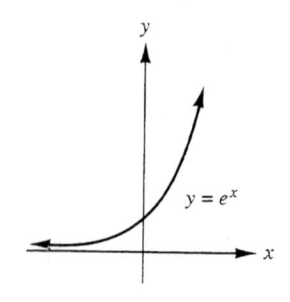

Figure 5.3

272   Chapter 5   The Exponential and Logarithmic Functions

Most important of all:

$$D_x e^x = e^x \quad \text{and} \quad \int e^x \, dx = e^x + C$$

We also remark that it can be shown that the exponential function is asymptotic to the negative x-axis and that it has no vertical asymptote.

Table 5.4
Exponential Function Values

x	e^x	e^{-x}	x	e^x	e^{-x}	x	e^x	e^{-x}
0.00	1.0000	1.0000	0.10	1.1052	0.9048	1.00	2.7183	0.3679
0.01	1.0101	0.9900	0.20	1.2214	0.8187	2.00	7.3891	0.1353
0.02	1.0202	0.9802	0.30	1.3499	0.7408	3.00	20.0855	0.0498
0.03	1.0305	0.9704	0.40	1.4918	0.6703	4.00	54.5981	0.0183
0.04	1.0408	0.9608	0.50	1.6487	0.6065	5.00	148.4132	0.0067
0.05	1.0513	0.9512	0.60	1.8221	0.5488	6.00	403.4288	0.0025
0.06	1.0618	0.9418	0.70	2.0138	0.4966	7.00	1,096.6332	0.0009
0.07	1.0725	0.9324	0.80	2.2255	0.4493	8.00	2,980.9580	0.0003
0.08	1.0833	0.9231	0.90	2.4596	0.4066	9.00	8,103.0839	0.0001
0.09	1.0942	0.9139	1.00	2.7183	0.3679	10.00	22,026.4766	

A complete table of values may be found in a handbook of mathematical functions in the library or via a scientific calculator.

Other values may be found from Table 5.4. For example,

$$e^{1.27} = e^{1+0.2+0.07}$$
$$= (e^1)(e^{0.2})(e^{0.07})$$
$$\approx (2.7183)(1.2214)(1.0725)$$
$$\approx 3.5684$$

Background Review 5.1

The initial definitions are: For b a real number and n a whole positive number

$$b^n = b \cdot b \cdot \cdots \cdot b \qquad \text{to } n \text{ factors}$$

and

$$b^{-n} = \frac{1}{b^n} \qquad \text{if } b \neq 0$$

5.1 The Exponential Function and Its Derivative

For $b \geq 0$, the nth root of b is defined by

$$b^{1/n} = a \quad \text{such that } a^n = b$$

(For $b < 0$ and n odd, the same definition applies.) Thus, for example, $2^5 = 32$ and $32^{1/5} = 2$.
For any rational number p/q, we define

$$b^{p/q} = (b^{1/q})^p = (b^p)^{1/q}$$

Then for any real number x

$$b^x = \lim_{p/q \to x} b^{p/q}$$

since any real number is the limit of a sequence of rational numbers. From these definitions follow the *laws of exponents*:

1. $b^n b^m = b^{n+m}$
2. $\dfrac{b^n}{b^m} = b^{n-m}$, $b \neq 0$
3. $(b^n)^m = b^{nm}$
4. $b^0 = 1$ (if $b \neq 0$; 0^0 is undefined)

where n and m are whole numbers, fractions, or real numbers.

These laws are valid for the same *base* b in the exponential expression b^n. The laws for exponential expressions with a different base are as follows:

1. $a^n \cdot b^n = (ab)^n$
2. $\dfrac{a^n}{b^n} = \left(\dfrac{a}{b}\right)^n$, $b \neq 0$
3. $a^n b^m$ has no (obvious) alternate form.

Finally, since addition/subtraction and multiplication/division cannot be mixed except through the distributive law—$a(b + c) = ab + ac$—the following expressions *do not simplify* in any obvious way:

$$b^n \pm b^m \quad \text{In particular this does not equal } b^{n \pm m}$$
$$(a \pm b)^n \quad \text{In particular this does not equal } a^n \pm b^n$$

Exercises 5.1

The first few exercises ask you to recall algebraic skills dealing with exponents.

1. a. Using (where needed) $\sqrt{2} \simeq 1.4$, compute

$2^7 \quad 2^{-3} \quad 2^{1/2} \quad 2^{-1/2} \quad 2^0$

$2^{3/2} \quad 2^{7/2} \quad 2^{-10}$

b. Given that $2^{1/5} \simeq 1.15$ and $2^{1/10} \simeq 1.07$, compute an approximate value of $2^{\sqrt{2}}$ and 2^π (rounding off $\sqrt{2} \simeq 1.4$ and $\pi = 3.14\ldots \simeq 3.1$).

2. a. Compute

$4^{1/2} \quad 8^{1/3} \quad (-32)^{1/5} \quad 4^{5/2} \quad 8^{-5/3} \quad 32^{1/10}$

b. Approximate

$4^{0.499} \quad 8^{0.3333} \quad 32^{\sqrt{2}}$

3. Given that $e \simeq 2.7$, $e^2 \simeq 7.4$, $e^{1/2} \simeq 1.6$, $e^{1/5} \simeq 1.2$, and $e^{1/10} \simeq 1.1$, use the laws of exponents to

274 Chapter 5 The Exponential and Logarithmic Functions

a. Compute

$$e^{3/2} \qquad e^{5/2} \qquad \frac{e^{5/2}}{e^2} \qquad e^{3/2} \cdot e^{1/2} \qquad e^{5/2} \cdot e^{-3/2}$$

$$(e^8)^{1/4} \qquad (e^{1/3})^6 \qquad \frac{e^{1/2} \cdot e^{7/2}}{e^4} \qquad \left(\frac{e}{2}\right)^2$$

$$e^2 + e \qquad (e^{1/2} + e)^2 \qquad e^{16^{1/4}} \qquad (e^{16})^{1/4}$$

b. Approximate $e^{\sqrt{2}}$ and e^π (rounding off $\sqrt{2} \approx 1.4$ and $\pi \approx 3.1$).

4. Compare the values of 2^x, e^x, and 3^x at $x = 0, \frac{1}{2}, 1$, and -2, and then in general for $x \geq 0$ and for $x < 0$.

On the same coordinate system, graph the pairs of functions in Exercises 5–10.

5. $y = 2(2^x); \quad y = 2e^x$
6. $y = 2^{2x}; \quad y = e^{2x}$
7. $y = 2^{-2x}; \quad y = e^{-2x}$
8. $y = 2^{(-1/4)x}; \quad y = e^{(-1/4)x}$
9. $y = 2^{-(x-1)^2/2}; \quad y = e^{-(x-1)^2/2}$
10. $y = 2^x; \quad y = \left(\frac{1}{2}\right)^x$

In Exercises 11–14, graph each set of functions on the same coordinate system.

11. $y = 2^x; \quad y = e^x; \quad y = 3^x$
12. $y = 2^{3x}; \quad y = 2^{x^3}$
13. $y = \frac{e^x}{e}; \quad y = \frac{e^{x^2}}{e^x}$
14. $y = xe^{-x}; \quad y = xe^{-x^2}$

15. Use a calculator to compute $[1 + (x/n)]^n$ for $x = 3$, $-\frac{1}{2}$, 0.75, and $\sqrt{2}$ and for $n = 100$. Compare each of these values to the value of e^x.

16. Let $f(x) = [1 + (x/n)]^n$. Show that $[1 + (x/n)]f'(x) = f(x)$, and evaluate $f'(2)$ and $f(2)$ when $n = 1{,}000$ using a calculator. Compare these values to e^x with $x = 2$.

Simplify the expressions in Exercises 17–22 algebraically.

17. $\dfrac{e^{2x} - e^x}{e^x}$
18. $(e^x - e^{-x})e^x$
19. $\dfrac{e^x + 1}{e^{2x} - 1}$
20. $\dfrac{(e^{4x})^{1/3}}{(e^{x/27})^9}$
21. $(e^{x^2-1})^{1/(x+1)}$
22. $\dfrac{e^{2x}}{e^2}$

23. Suppose that you deposit $100 in a savings account paying 10% interest per year, which is *compounded yearly*. That is, at the end of year 1, your account will hold $100 + 0.10($100) = $110. At the end of year 2, your account will hold $110 + 0.10($110) = $121. This process, like the process of cell division, "feeds on itself."

a. Compare the amount of money that will be in this account at the end of 1, 2, 3, 5, and 10 years with the values of

$$100e^{0.10x}$$

for $x = 1, 2, 3, 5,$ and 10.

b. Make a graph of the amount of money in this account over time. How does this graph compare with the graph of exponential growth?

c. How does this problem relate to the central ideas of this section and to the discussion of Example 6 in Section 2.4?

Find the derivatives in Exercises 24–31.

24. $D_x\left(\dfrac{1}{x} + e^x\right)$
25. $D_x(x^2 e^x)$
26. $D_x \dfrac{x}{e^x}$
27. $D_x(5e^x)$
28. $D_x(\sqrt{x} \cdot e^x)$
29. $D_t e^{-t}$
30. $D_x \dfrac{x + e^x}{x}$
31. $D_x^5(x^4 + e^x)$

32. Recall that $n! = n(n-1)(n-2) \ldots 3 \cdot 2 \cdot 1$. For example, $5! = 5 \cdot 4 \cdot 3 \cdot 2 \cdot 1 = 120$. Let

$$g(x) = 1 + x + \frac{x^2}{2!} + \frac{x^3}{3!} + \frac{x^4}{4!} + \frac{x^5}{5!}$$

a. Compute $D_x g(x)$.

b. Compare $g(x)$ and $D_x g(x)$. How do they differ?

c. Fill in the blank so that the following is an equation:

$$\underline{} + D_x g(x) = g(x)$$

d. Compare the values of $g(x)$ and e^x for $x = 1$.

Determine the integrals in Exercises 33–37.

33. $\displaystyle\int_0^1 e^x \, dx$
34. $\displaystyle\int (\sqrt{x} + e^x) \, dx$
35. $\displaystyle\int_0^1 (x^2 + e^x) \, dx$
36. $\displaystyle\int (1 - e^x) \, dx$
37. $\displaystyle\int_{-1}^0 e^x \, dx$

38. The number e and the exponential function $y = e^x$ are difficult to deal with, except for the invariance of e^x under differentiation. Could there be some other function, or formula, with the property

$$D_x g(x) = g(x)$$

Suppose there is. Suppose $y = g(x)$ is some *unknown* function that also satisfies

$$D_x g(x) = g(x)$$

5.1 The Exponential Function and Its Derivative

Let us try to compare this supposed unknown function with $y = e^x$. First, recall that e^x never equals zero; thus we can form a new function

$$h(x) = \frac{g(x)}{e^x}$$

We will use the function h to compare e^x with our other supposed differentiation invariant function g, since h measures the ratio of the two functions.

a. Calculate $D_x h(x)$ using the quotient rule.

b. Conclude from (a) that h is a constant function.

c. Therefore $g(x)/e^x = C$, a constant. Hence $g(x) = Ce^x$ is a multiple of e^x.

d. Recall that $e^0 = 1$. Suppose the unknown function $y = g(x)$, besides satisfying $D_x g(x) = g(x)$, also satisfies $g(0) = 1$. What can you conclude? Summarize, in complete English sentences, the meaning of this problem.

Find the maximum and minimum values (if any) of the functions in Exercises 39–41.

39. $f(x) = xe^x$

40. $g(x) = x + e^x$, on the interval $[-1, 1]$

41. $f(x) = \dfrac{x}{e^x}$

42. Give an example (in writing and with as few computations as possible) of some phenomenon from your major area of study that "grows by feeding on itself." Can you graph the course of growth? How does this graph compare to the graph of exponential growth?

43. Show, by differentiation, that the graph of the exponential function is concave up at every point.

44. Let $f(x) = e^x$. Find the first, second, third, and billionth derivatives of $f(x)$.

45. Let $f(x) = xe^x$. Find $f^{(5)}(x)$.

Computer Application Problems

46. Use the BASIC program EXP to estimate the value of $e = \lim_{n \to \infty} [1 + (1/n)]^n$.

47. Compare the estimations obtained by using the BASIC program DXEXP with our discussion of the value of the $\lim_{h \to 0} (e^h - 1)/h$.

5.2 The Calculus of the Exponential Function

In this section we learn to apply the rules and methods we already know to the particular case of the exponential function.

We have just seen that differentiation of $y = e^x$ is the easiest exercise in calculus. But what of differentiation of functions like

$$e^{-x} \qquad e^{3x} \qquad e^{x^2} \qquad e^{\sqrt{x}} \qquad e^{e^x}$$

and so on? The key to such problems is the use of the chain rule for differentiating composite functions. Each of the preceding functions is a composition $f(g(x))$ where, always, $f(x) = e^x$ and in turn

$$g(x) = -x \qquad g(x) = 3x \qquad g(x) = x^2 \qquad g(x) = \sqrt{x} \qquad g(x) = e^x$$

For example, $e^{x^2} = f(g(x))$, where $f(x) = e^x$ and $g(x) = x^2$, since $f(g(x))$ means: Put x^2 in place of x in the formula e^x for $f(x)$.

To differentiate functions in the form $f(g(x)) = e^{g(x)}$, we have to use the chain rule. Since this rule is

$$D_x f(g(x)) = f'(g(x)) g'(x)$$

and since with $f(x) = e^x$ we always have $f'(x) = e^x$, then, with $g(x)$ any other function, we must have

$$D_x e^{g(x)} = D_x f(g(x)) = f'(g(x)) g'(x) = e^{g(x)} g'(x)$$

The last formula is one of the most important in this text, and you should memorize it.

276 Chapter 5 The Exponential and Logarithmic Functions

Differentiation of the General Exponential Function

$$D_x e^{g(x)} = e^{g(x)} g'(x)$$

In terms of the discussion of the chain rule in Section 2.2, the function serving as exponent, $g(x)$, is the "inner function" and the exponential function $f(x)$ is the "outer function."

EXAMPLE 1

a. Differentiate e^{x^2}. Here $g(x) = x^2$ and we have
$$D_x(e^{x^2}) = e^{x^2} \cdot D_x(x^2) = 2xe^{x^2}$$

b. Differentiate e^{-x}. Here $g(x) = -x$ and we have
$$D_x e^{-x} = e^{-x} \cdot D_x(-x) = e^{-x} \cdot (-1) = -e^{-x}$$

c. Differentiate $e^{\sqrt{x}}$.
$$D_x e^{\sqrt{x}} = e^{\sqrt{x}} \cdot D_x x^{1/2} = e^{\sqrt{x}}\left(\frac{1}{2}x^{-1/2}\right) = \frac{e^{\sqrt{x}}}{2\sqrt{x}}$$

As you see, all the work lies in *differentiating the exponent*. The exponential function itself is merely rewritten as part of the computation.

d. Differentiate e^{e^x}. Here we let $g(x) = e^x$ and have
$$D_x e^{e^x} = e^{e^x} \cdot D_x e^x = e^{e^x} \cdot e^x = e^{e^x + x}$$

e. Differentiate e^{kx} where k is constant. Here $g(x) = kx$ and $D_x e^{kx} = e^{kx} \cdot D_x(kx) = ke^{kx}$. ∎

Further Differentiation Involving Exponential Functions

The rule just derived may be used in more involved ways. We illustrate this by a further series of examples.

EXAMPLE 2

Differentiate $x^3 e^{x^2}$.

Solution This function is a product of x^3 and e^{x^2}. Using the product rule, we have

$$\begin{aligned} D_x(x^3 e^{x^2}) &= x^3 D_x e^{x^2} + e^{x^2} \cdot D_x x^3 \\ &= x^3(e^{x^2} \cdot D_x x^2) + e^{x^2}(3x^2) \\ &= x^3 \cdot e^{x^2} \cdot 2x + 3x^2 e^{x^2} \\ &= 2x^4 e^{x^2} + 3x^2 e^{x^2} \\ &= (2x^4 + 3x^2)e^{x^2} \quad \blacksquare \end{aligned}$$

EXAMPLE 3

Differentiate $e^{-x^3}/(x^2 + 1)$.

Solution Because this is a quotient of two functions, we apply the quotient rule.

$$D_x\left(\frac{e^{-x^3}}{x^2 + 1}\right) = \frac{(x^2 + 1)D_x(e^{-x^3}) - e^{-x^3}D_x(x^2 + 1)}{(x^2 + 1)^2}$$

$$= \frac{(x^2 + 1)[e^{-x^3}D_x(-x^3)] - e^{-x^3}(2x)}{(x^2 + 1)^2}$$

$$= \frac{-3x^2(x^2 + 1)e^{-x^3} - 2xe^{-x^3}}{(x^2 + 1)^2}$$

$$= \frac{-3x^4 - 3x^2 - 2x}{(x^2 + 1)^2}e^{-x^3} \quad \blacksquare$$

The chain rule may also be used in a different way in dealing with the exponential function.

EXAMPLE 4

Differentiate $(e^{x^2} + x)^3$.

Solution Note that $e^{x^2} + x$ is "inside" a cubing function. That is,

$$(e^{x^2} + x)^3 = f(g(x))$$

where $f(x) = x^3$ and $g(x) = e^{x^2} + x$.

Using the chain rule, we have

$$D_x(e^{x^2} + x)^3 = f'(g(x))g'(x)$$

$$= 3(e^{x^2} + x)^2 \cdot D_x(e^{x^2} + x)$$

$$= 3(e^{x^2} + x)^2 \cdot [e^{x^2}(D_x x^2) + 1]$$

$$= 3(e^{x^2} + x)^2 \cdot [e^{x^2}(2x) + 1]$$

$$= 3(2xe^{x^2} + 1)(e^{x^2} + x)^2 \quad \blacksquare$$

The key to doing the problems of this section correctly is to determine *at the start* the *form* of the problem and then proceed step-by-step using the appropriate formal rules of differentiation.

We can also differentiate *implicitly* (Section 3.1) with the exponential function.

EXAMPLE 5

Find $D_x y$ if $xe^y = x^2 + y$.

Solution As in Section 3.1,

$$D_x(xe^y) = D_x(x^2 + y)$$

278 Chapter 5 The Exponential and Logarithmic Functions

$$x\,D_x e^y + e^y D_x x = 2x + D_x y$$
$$xe^y D_x y + e^y \cdot 1 = 2x + D_x y$$
$$e^y - 2x = D_x y - xe^y D_x y = (1 - xe^y)D_x y$$

Thus
$$D_x y = \frac{e^y - 2x}{1 - xe^y} \quad \blacksquare$$

Integration of Exponential Functions by the Method of Substitution

Recall how to find $\int x(x^2 + 1)^5\, dx$ by substitution (Section 4.3). First, let $u = x^2 + 1$, find $du = 2x\, dx$, and then replace $x^2 + 1$ by u and $x\, dx$ by $\frac{1}{2}\, du$ in the integrand to obtain

$$\int x(x^2 + 1)^5\, dx = \int \frac{1}{2}u^5\, du = \frac{u^6}{12} + C = \frac{(x^2 + 1)^6}{12} + C$$

The same idea is used in the next example.

EXAMPLE 6
Find $\int xe^{x^2}\, dx$.

Solution Let $u = x^2$. Then $du = 2x\, dx$ or $\frac{1}{2}\, du = x\, dx$. By substitution,

$$\int xe^{x^2}\, dx = \frac{1}{2}\int e^u\, du$$

Now the problem is easy, because $\int e^u\, du = e^u + C$, since the exponential is invariant for integration. Hence

$$\int xe^{x^2}\, dx = \frac{1}{2}\int e^u\, du = \frac{e^u}{2} + C = \frac{e^{x^2}}{2} + C \quad \blacksquare$$

EXAMPLE 7
Find $\int e^{5x}\, dx$.

Solution Recall that in the method of substitution we need to isolate a part of the given integrand whose derivative is "almost" another part, where *almost* means "up to a constant *multiple*." For this example, let $u = 5x$ so $du = 5\, dx$ and $\frac{1}{5}\, du = dx$. Then

$$\int e^{5x}\, dx = \int \frac{1}{5}e^u\, du = \frac{1}{5}\int e^u\, du = \frac{1}{5}e^u + C = \frac{1}{5}e^{5x} + C \quad \blacksquare$$

Example 7 is a special case of a repeatedly useful integral formula

$$\int e^{kx}\, dx = \frac{e^{kx}}{k} + C$$

obtained by letting $u = kx$, k a constant. Note that it is the inverse of the differentiation formula: $D_x e^{kx} = ke^{kx}$.

Amid this forest of calculations are two important trees. Every integration problem in this section reduces by substitution to either

$$\int u^n \, du = \frac{u^{n+1}}{n+1} + C \quad n \ne -1 \quad \text{or} \quad \int e^u \, du = e^u + C$$

There are no other possibilities because these are the only two functions for which we can directly write an antiderivative. The sole point of substitution is to reduce a given problem to one of these two (irreducible) forms.

EXAMPLE 8

Find

a. $\int e^x(e^x + 1)^3 \, dx$ **b.** $\int (2x + 1)e^{x^2+x} \, dx$

Solution In both (a) and (b) we choose u to be a part of the integrand whose derivative is another part.

a. Let $u = e^x + 1$. Then $du = e^x \, dx$. By substitution

$$\int e^x(e^x + 1)^3 \, dx = \boxed{\int u^3 \, du} = \frac{u^4}{4} + C = \frac{(e^x + 1)^4}{4} + C$$

b. Let $u = x^2 + x$. Then $du = (2x + 1) \, dx$. Hence

$$\int (2x + 1)e^{x^2+x} \, dx = \boxed{\int e^u \, du} = e^u + C = e^{x^2+x} + C \quad \blacksquare$$

An Application of the Exponential Function: The Surge–Dissipation Curve

The graph in Figure 5.4 is often used to model the amount of a medicinal drug in a patient's bloodstream or the herd effect of customers surging to a newly available product. Both these phenomena are characterized by an initial surge followed by a slow dissipation from a peak level, indicated at $x = 1/b$ in Figure 5.4. A curve of this kind is known to be given by a function of the form

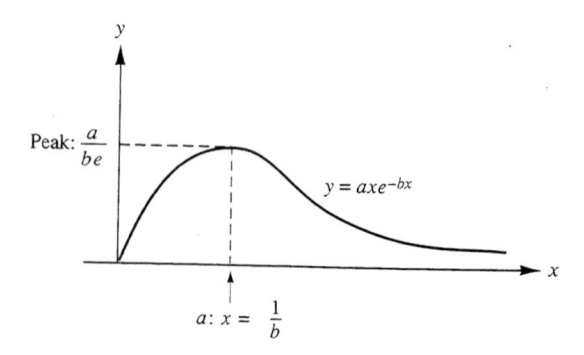

$$f(x) = axe^{-bx} \quad \text{where } a, b > 0$$

Peak: $\dfrac{a}{be}$

$y = axe^{-bx}$

$a: x = \dfrac{1}{b}$

Figure 5.4

280   Chapter 5   The Exponential and Logarithmic Functions

We will establish the location and height of the peak in Figure 5.4 in the next example.

EXAMPLE 9

a. Show that $f(x) = axe^{-bx}$ has one and only one maximum value, that $x_{max} = 1/b$, and that $f(x_{max}) = a/be$.

b. Customers surge to a new product so that after $x = 2$ weeks a peak of 3 million customers is buying the product. How many will be buying it 1 month later ($x = 6$) if the function f is a model for the number of customers in week x?

Solution

a. With $f(x) = axe^{-bx}$ and $a, b > 0$, we have

$$f'(x) = ax\, D_x e^{-bx} + e^{-bx} D_x ax = axe^{-bx}(-b) + ae^{-bx}$$
$$= (-bx + 1)ae^{-bx}$$

Since $ae^{-bx} > 0$, $f'(x) = 0$ when the factor $-bx + 1 = 0$ or when $\bar{x} = 1/b$. Thus $\bar{x} = 1/b$ is the only critical point for f. The factor $-bx + 1$ is positive for $x < 1/b$ and negative for $x > 1/b$. By the first derivative test for extrema, $x_{max} = 1/b$ yields a maximum value for the function f equal to

$$f\left(\frac{1}{b}\right) = a\left(\frac{1}{b}\right)e^{-b(1/b)} = \left(\frac{a}{b}\right)e^{-1} = \frac{a}{be}$$

b. We are given that $x_{max} = 2$ and that $f(x_{max}) = 3$. Consequently, $1/b = 2$ and $a/be = 3$ from part (a). Therefore $b = \frac{1}{2}$ and $a = 3be = 3\left(\frac{1}{2}\right)e = 3e/2$. Thus after 6 weeks there will remain

$$f(6) = \frac{3e}{2}(6)e^{-1/2(6)} = (9e)e^{-3} = 9e^{-2} = 1.218 \text{ million}$$

customers for the product, because $f(x) = (3e/2)xe^{-(1/2)x}$. ∎

What happens to the values $f(x)$ in the "long run"—that is, as $x \to \infty$? It may be shown (using L'Hôpital's rule of Section A.2) that

$$\lim_{x \to \infty} \frac{ax}{e^{bx}} = 0$$

so that the graph of f becomes asymptotic to the x-axis. Thus without additional effort (advertising, and so on), the number of customers will dwindle toward zero as time goes on.

EXAMPLE 10

A small investment firm has $500,000 to invest in stocks, bonds, and other financial instruments. Its investment strategy is to invest $300,000 initially and to *invest the balance* of its funds at the rate of 20% of whatever balance it has.

Thus, for example, if on a certain day t it has \$400,000 invested, then its balance is \$100,000 and it will invest $0.20(\$100,000) = \$20,000$ on that day.

Let $f'(t)$ be the amount in investments on day t. Show that a formula for the amount of the firm's investment on day t is

$$f(t) = 500,000 - 200,000e^{-0.2t}$$

Solution First, at time $t = 0$,

$$f(0) = 500,000 - 200,000e^0 = \$300,000$$

so $f(0)$ does reflect the initial level of investment.

Let us now compute $f'(t)$ to see how the rate of investment predicted by this function compares with the firm's strategy. We have

$$f'(t) = D_t(500,000 - 200,000e^{-0.2t})$$
$$= 0 - 200,000(e^{-0.2t})(-0.2) = 0.2(200,000e^{-0.2t}) \quad (1)$$

Now let us compute the firm's (uninvested) balance on day t. This balance is

$$= 500,000 - \text{(amount in investments on day } t\text{)}$$
$$= 500,000 - f(t)$$
$$= 500,000 - (500,000 - 200,000e^{-0.2t})$$
$$= 200,000e^{-0.2t} \quad (2)$$

If we now compare Equations 1 and 2, we see that

$$f'(t) = 0.2(500,000 - f(t)) \quad \blacksquare$$

This differential equation asserts in ordinary language that "the rate of investment is 20% of the uninvested balance on day t."

Because of the shape of the graph $y = e^{-x}$ (Figure 5.5), functions of the form $f(t) = Ce^{-kt}$ are useful in modeling, for example, the declining effect of a pollutant on the environment at an increasing distance from its source.

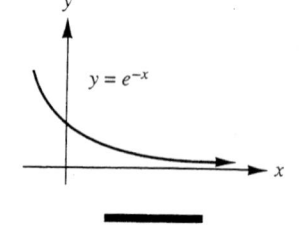

Figure 5.5

EXAMPLE 11

Suppose that an accident has occurred at a paper mill and a chemical used in making paper is leaking into the stream beside the mill. Suppose that at a point x miles downstream the pollutant is killing fish at the rate of $f(x) = 500e^{-0.3x}$ fish per mile (indicating that fewer fish are killed at greater distances from the mill). Compute the total number of fish killed in the first 10 miles downstream from the mill.

Solution We apply the technique of Riemann sums to model this situation.

Imagine the first 10 miles of stream broken into short pieces Δx (of a mile) long (see Figure 5.6). At a point x miles downstream from the mill, fish are

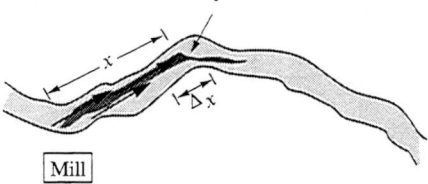

Figure 5.6

Density of fish kill = $500e^{-0.3x}$

Mill

killed at the rate of $500e^{-0.3x}$ fish per mile. Hence, along a distance Δx miles long at this point downstream, we have a total fish kill of approximately

$$(500e^{-0.3x}) \cdot \Delta x \qquad \left(\frac{\text{fish}}{\text{mile}} \times \text{miles} = \text{fish}\right)$$

Reasoning as in Section 4.5D and applying the approximation principle (AP) of Section 4.5, the total fish kill in the first 10 miles is

$$\int_0^{10} 500e^{-0.3x}\, dx$$

We perform the integration using the substitution $u = -0.3x$.

$$\int 500e^{-0.3x}\, dx = 500 \int e^{-0.3x}\, dx = 500\left(\frac{e^{-0.3x}}{-0.3}\right) + C$$

Hence $\int_0^{10} 500e^{-0.3x}\, dx = 500\left[\frac{e^{-0.3x}}{-0.3}\right]_0^{10}$

$$= 500\left(\frac{1}{0.3}\right)\left(1 - e^{-0.3(10)}\right) \qquad \text{Since } e^0 = 1$$

$$\approx 1{,}584$$

Thus 1,584 fish are killed over the 10 miles downstream from the mill. ∎

Exercises 5.2

Write each function in Exercises 1–6 as a composition $f(g(x))$, and specify $f(x)$ and $g(x)$.

1. e^{2x}
2. $e^{\sqrt{x^2-1}}$
3. $\sqrt{e^x + 1}$
4. $2e^x$
5. $e^{(x-1)/x} + 3$
6. $(1 - e^x)^{-1/3} + 2$

Find the derivative of each function in Exercises 7–31.

7. e^{5x}
8. $e^{x/3}$
9. $e^{x^2/2}$
10. $e^{x(x^2+1)^7}$
11. $\sqrt{e^x + 1}$
12. $(x - e^x)^4$
13. $\frac{1}{2}e^{1-t^2}$
14. $x^2 e^x$
15. $e^x - e^{-x}$
16. $e^{x(\sqrt{1-x})}$
17. $e^x \cdot e^{x^2}$
18. $\frac{e^x}{x}$
19. $(e^x)^x$
20. $\frac{x}{e^x}$
21. $e^{1/x}$
22. $\frac{e^x + e^{-x}}{e^x}$ [*Hint:* Simplify first.]

23. $xe^{x^2} + x$
24. $(x^3 - e^x)^{1/2}$
25. $e^{e^{x^2}}$
26. $\sqrt{e^{2x}}$
27. $(e^x + e^{-x})^2$
28. $\dfrac{e^x - 1}{e^x + 1}$
29. $\tfrac{1}{3}e^{3x} + \tfrac{1}{2}e^{2x} + e^x + x$
30. $\dfrac{x(e^{2x} + 1)}{x^2 + 1}$
31. $\dfrac{x^2(1 - e^x)}{e^{x^2}}$

Find the local maxima and local minima (if any) for the functions in Exercises 32–39. Remember that $e^{g(x)} > 0$ for any function g.

32. $f(x) = 2xe^{-3x}$
33. $g(x) = e^{(x-1)^2}$
34. $f(t) = \dfrac{t - 1}{e^t}$
35. $g(t) = e^t - t - 1$
36. $h(u) = (1 + u)^2 e^{-u}$
37. $f(x) = -x^3 e^x$
38. $f(t) = \dfrac{1}{1 + e^{-t}}$
39. $g(x) = 1 - e^{-x}$ for $x \geq 0$

In Exercises 40–45, find $D_x y$ by implicit differentiation.

40. $y^2 e^x = y + 1$
41. $xy = e^x$
42. $xe^y = x + y$
43. $x = e^y$
44. $xy^2 = e^{y^3}$
45. $e^{x-y} = x - y$

46. Show that $e^x > x + 1$ for all $x > 0$ by showing that $f(x) = e^x - (x + 1)$ has a minimum value at $x = 0$, $f(0) = 0$, and $f'(x) > 0$ for $x > 0$.

47. Functions of the form $P(t) = e^{-ae^{-bt}}$, which are known as **Gompertz growth curves**, are used to model certain kinds of population growth in biology. Show that the function
$$P(t) = e^{-0.2e^{-0.03t}}$$
is increasing for all time t.

In Exercises 48–57, find the integrals.

48. $\displaystyle\int e^{2x}\, dx$
49. $\displaystyle\int_0^1 e^{3x}\, dx$
50. $\displaystyle\int_0^1 xe^{x^2/2}\, dx$
51. $\displaystyle\int_1^2 te^{-t^2}\, dt$
52. $\displaystyle\int \dfrac{t}{e^{t^2}}\, dt$
53. $\displaystyle\int e^x \sqrt{e^x + 1}\, dx$
54. $\displaystyle\int \dfrac{e^x + e^{-x}}{(e^x - e^{-x})^3}\, dx$
55. $\displaystyle\int \dfrac{e^{1/x}}{x^2}\, dx$ $\left[\text{Hint: Let } u = \dfrac{1}{x}.\right]$
56. $\displaystyle\int_0^1 x^2 e^{x^3}\, dx$
57. $\displaystyle\int_0^1 x\sqrt{e^{x^2}}\, dx$

Exercises 58–61 use the methods learned in Section 4.5.

58. Find the average value of $y = e^{-x}$ between $x = 0$ and $x = 2$; between $x = -1$ and $x = 3$.

59. Find the volume of the solid of revolution obtained by revolving the graph of $y = e^x$ about the x-axis between $x = 0$ and $x = 1$.

60. Find f'' in Example 9, and show that $f''(1/b) < 0$.

61. In response to a stimulus, the *hypothalamus* (in mammals) secretes a certain hormone, TXY, which flows via a vein to the *adenohypophysis*, causing it to secrete a second hormone enabling the subject to further respond to the stimulus. If the adenohypophysis is located 10mm from the hypothalamus and TXY is secreted and injected into the bloodstream by the hypothalamus at a density of $1.3e^{-0.2x}$ μL/mm (milliliters per millimeter) x mm from the hypothalamus and if the adenohypophysis has to receive 5 μL of the hormone before it will respond at all, will there be sufficient TXY for the adenohypophysis to respond to this stimulus? [*Hint:* Set up a Riemann sum that approximates the total amount of TXY in the bloodstream between the two glands.]

62. Suppose that an investment firm has \$800,000 to invest in stocks and initially invests \$200,000 of this amount. It then adopts the strategy of investing 30% of its uninvested assets each day.

   a. Show that the amount $I(t)$ it will invest on day t is
   $$I(t) = 800{,}000 - 600{,}000 e^{-0.3t}$$
   That is, show that $I'(t) = 0.30\,(800{,}000 - I(t))$.

   b. Ignoring its initial investment of \$200,000, what is the average amount invested by the firm each day over the first 10 days of following this strategy?

63. Find the total number of fish killed in the first 2 miles downstream from the spill in Example 11, and discuss why it should be expected that this number should be as large as it is in comparison with the number found in Example 11.

64. a. Let $f(x) = e^{-x^2/2}$. Show that f has inflection points at $x = \pm 1$.

   b. Show that $g(x) = e^{-(x-3)^2}$ has its maximum value at $x = 3$.

65. Suppose that in Example 9(b) the peak of customers occurs after 1.5 weeks and is 4 million. How many customers remain 1 month later?

5.3 The Natural Logarithm

In this section we introduce the natural logarithmic function as the inverse *of the exponential function.*

Our everyday perception of the world of taste, smell, hearing, economic well-being, and other physiological/psychological functions is innately logarithmic. The same is true of the response of other life forms to their environment. To understand this we first have to define the logarithm and learn its algebraic properties. Consider the graph of the exponential function $x = e^y$ as shown in Figure 5.7.

Notice that we have exchanged the usual notation and labeled the horizontal axis y rather than x, and the vertical x rather than y. After all, these labels are only a matter of choice, and it will make the next step in this discussion easier to follow if we exchange these labels, because we wish to regard a point on the horizontal y-axis as a function of a point x on the vertical x-axis.

Let x mark a point on the vertical axis, as shown in Figure 5.7. The number x is some unique value e^y of the exponential function, since the point (y, x) is on the graph. That is, there is some unique y such that

$$x = e^y$$

We use this equation to define the natural logarithm of x as follows:

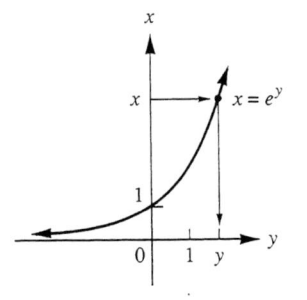

Figure 5.7

Definition

If $x > 0$, we define the **natural logarithm** of x, denoted by ln x, to be the number y such that $x = e^y$. That is,

ln $x = y$ if and only if $e^y = x$

In short, to define the natural logarithmic function

$$f(x) = \ln x$$

we read the exponential graph in the reverse direction—from vertical to horizontal axis. That is all, nothing more.

This definition has two immediate consequences:

1. ln x is defined only for positive x.
2. The graph of $y = \ln x$ is a reflected image of the graph of the exponential function about the line $y = x$: $y = \ln x$ means $x = e^y$.

Statement 1 is true because the range of the exponential function consists of only positive numbers, as illustrated by its graph in Figure 5.7. We illustrate Statement 2 in Figure 5.8, where we return to the usual labeling of the horizontal axis by the letter x and show the graph of $y = \ln x$ as a function of points on the horizontal axis. Notice that if Figure 5.8 is folded along the line $y = x$, the exponential and logarithmic graphs coincide. The graph of $y = \ln x$ shown in Figure 5.8 should be memorized.

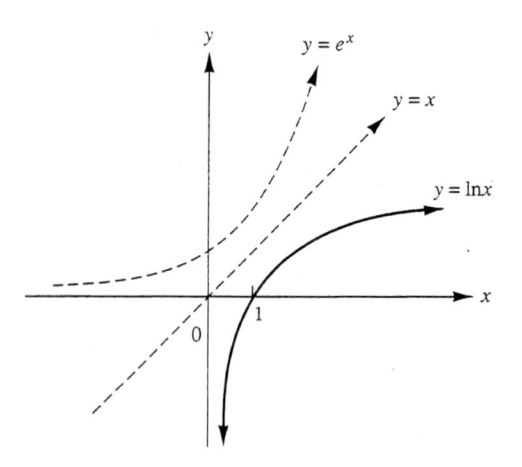

Figure 5.8

Notice also from Figure 5.8 that, while the domain of $y = \ln x$ consists only of positive x, its range does contain all negative y, these arising from numbers x between 0 and 1. We now compute some values of $y = \ln x$, so as to better grasp this *reverse*, or *inverse*, relationship with the exponential function.

EXAMPLE 1

Find the values of

a. $\ln 1$ **b.** $\ln e$ **c.** $\ln e^3$ **d.** $\ln 2$ **e.** $\ln \frac{1}{2}$

Solution   The solutions to this example are indicated in Figure 5.9.

a. By definition, $\ln x = y$ if $e^y = x$. Therefore

$$\ln 1 = y \quad \text{when } e^y = 1$$

Since $e^0 = 1$, then $y = 0$. That is, $\ln 1 = 0$.

b. By definition

$$\ln e = y \quad \text{when } e^y = e$$

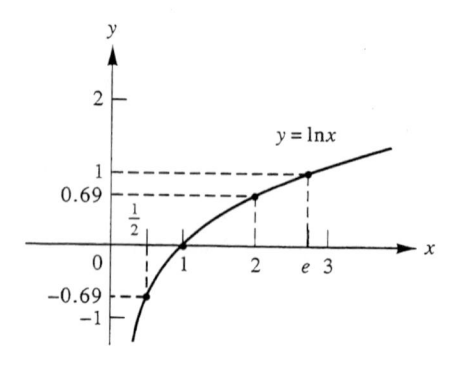

Figure 5.9

286   Chapter 5   The Exponential and Logarithmic Functions

Since $e^1 = e$, we have $y = 1$. That is,
$$\ln e = 1$$

c. By definition
$$\ln e^3 = y \quad \text{when } e^y = e^3$$
Therefore $y = 3$, so $\ln e^3 = 3$.

d. By definition
$$\ln 2 = y \quad \text{when } e^y = 2$$
This is more difficult. It is not obvious what y is. However, as we saw in Section 5.1, *there is some y such that $e^y = 2$*. There is no simple way to calculate this number y. However, if we read Table 5.4 *in reverse*, we see that $e^{0.69} = 2$ (approximately). Therefore $\ln 2 = 0.69$, accurate to *two* decimal places.

e. Let us now try to compute $\ln \frac{1}{2}$. We have
$$\ln \tfrac{1}{2} = y \quad \text{when } e^y = \tfrac{1}{2}$$
Again, we have a difficulty. But, since $e^y = \frac{1}{2}$ is equivalent to
$$e^{-y} = 2$$
(by taking reciprocals of both sides), we see that, by definition,
$$-y = \ln 2 = 0.69$$
This means that $\ln \tfrac{1}{2} = y = -0.69$. ∎

In Table 5.5 at the close of this section, we give further values of the natural logarithm.

Something more significant is happening in Example 1, however. From (d) and (e), we see that $\ln \frac{1}{2} = -\ln 2$. Now, since $\frac{1}{2} = 2^{-1}$,
$$\ln 2^{-1} = -\ln 2$$

This last feature is a principal reason why the logarithmic function is important. *The logarithmic function allows us to convert exponents to linear multiples.* We will shortly see how this property allows us to model common responses to the environment.

Algebraic Properties of the Logarithmic Function

From the definition of the logarithmic function

(a) $\ln x = y$ if and only if (b) $e^y = x$

we can derive the two most useful relationships between the exponential and logarithmic functions. If we take Equation (a) and use it to *substitute* for y in Equation (b), we obtain:
$$e^{\ln x} = x \quad \text{for all } x > 0$$

If, on the other hand, we take Equation (b) and use it to substitute for x in Equation (a), we obtain

$$\ln e^y = y$$

These two formulas should be memorized:

The Inversion Formulas

1. $e^{\ln x} = x$ for all $x > 0$
2. $\ln e^x = x$ for all x

These formulas only state what we have already observed: that the log function reverses—or is the inverse of—the exponential function, and vice versa.

With the inversion formulas we can derive all the principal algebraic rules for logarithms:

The Algebraic Rules of the Logarithm

For real numbers $x, y > 0$, and r,

1. $\ln x^r = r \ln x$
2. $\ln xy = \ln x + \ln y$
3. $\ln \dfrac{1}{x} = -\ln x$
4. $\ln \dfrac{x}{y} = \ln x - \ln y$
5. $\ln 1 = 0$; $\ln e = 1$

Notice that Rule 1 is the promised rule for exponents. It says that the exponent r can be transformed into a *linear factor*. Let us derive this rule. Using the right side of Rule 1 as an exponent for the number e, we have

$$e^{r \ln x} = (e^{\ln x})^r \qquad \text{Since } e^{ab} = (e^a)^b, \text{ applied to } a = \ln x \text{ and } b = r$$

$$= x^r \qquad \text{Since } e^{\ln x} = x \text{ from Formula 1}$$

$$= e^{\ln x^r} \qquad \text{Since } e^{\ln x^r} = x^r \text{ from Formula 1 again}$$

Therefore $e^{r \ln x} = e^{\ln x^r}$ and the two exponents must be equal. That is,

$$r \ln x = \ln x^r$$

The remaining rules can be derived in a similar way. In particular, Rule 5 follows from the equation $\ln e^x = x$ applied to $x = 0$ and $x = 1$, respectively. The remaining rules reflect the fact that a logarithm is an exponent. Since, for example, exponents are added when forming a product, Rule 4 results.

EXAMPLE 2

Given that $\ln 2 = 0.69$ and $\ln 3 = 1.10$ (approximately), calculate

a. $\ln 6$ **b.** $\ln 8$ **c.** $\ln \sqrt{2}$ **d.** $\ln \frac{2}{3}$ **e.** $\ln \frac{1}{9}$ **f.** $\ln \dfrac{\sqrt{2}}{\sqrt[3]{3}}$

Solution

a. From Rule 2,
$$\ln 6 = \ln(2 \cdot 3) = \ln 2 + \ln 3 = 0.69 + 1.10 = 1.79$$

b. From Rule 1,
$$\ln 8 = \ln 2^3 = 3 \ln 2 = 3(0.69) = 2.07$$

c. From Rule 1,
$$\ln \sqrt{2} = \ln 2^{1/2} = \tfrac{1}{2} \ln 2 = 0.35$$

d. From Rule 4,
$$\ln \tfrac{2}{3} = \ln 2 - \ln 3 = -0.4$$

e. From Rules 3 and 1,
$$\ln \tfrac{1}{9} = \ln 9^{-1} = -\ln 9 = -\ln 3^2 = -2 \ln 3 = -2.20$$

f. From Rules 4 and 1,
$$\ln \dfrac{\sqrt{2}}{\sqrt[3]{3}} = \ln 2^{1/2} - \ln 3^{1/3}$$
$$= \tfrac{1}{2} \ln 2 - \tfrac{1}{3} \ln 3$$
$$= 0.35 - 0.37$$
$$= -0.02 \quad \blacksquare$$

Remark. In fact $\ln 6 \simeq 1.79$, from a calculator; each calculation in Example 2 is subject to roundoff error.

Notice that if $0 < x < 1$, then $\ln x < 0$, as already indicated by the graph of $y = \ln x$ (Figure 5.9) and parts (d), (e), and (f) of Example 2.

The preceding algebraic properties of the logarithm can also be combined with the inversion formulas $e^{\ln x} = x$ and $\ln e^x = x$ to solve logarithmic and exponential equations.

EXAMPLE 3

Solve for x in each of the following equations.

a. $2 + \ln x = 5$
b. $4e^x + 2 = 10$
c. $3 \ln(2x) + 1 = 13$
d. $\ln(x^3) + \ln(1/x) = -4$

Solution

a. Since $2 + \ln x = 5$, then $\ln x = 3$. These equal numbers applied as exponents of e yield the equation $e^{\ln x} = e^3$. Using Formula 1, $e^{\ln x} = x$. Therefore $x = e^3 \approx 20.09$.

b. Since $4e^x + 2 = 10$, then $4e^x = 8$ and $e^x = 2$. Thus $\ln e^x = \ln 2$, by applying the logarithm to each side. Using Formula 2, $\ln e^x = x$. Therefore $x = \ln 2 \approx 0.69$.

c. Since $3 \ln(2x) + 1 = 13$, then $3 \ln 2x = 12$ or $\ln 2x = 4$. Thus $e^{\ln 2x} = e^4$ and using Formula 1, $2x = e^4$. Hence $x = e^4/2 \approx 27.30$.

d. Since $\ln(x^3) + \ln(1/x) = -4$, then $3 \ln x - \ln x = -4$. Thus $2 \ln x = -4$ and $\ln x = -2$. Therefore $e^{\ln x} = e^{-2}$ and $x = e^{-2} \approx 0.14$. ∎

Why the Logarithm Is Natural

The function $\ln x$ is called the natural logarithm because it is based on the number e. But the logarithm is natural for another reason that is revealing of human nature and perception and that illustrates why the logarithm must enter applications of calculus.

In 1846, the German physiologist E. H. Weber used the following kind of example to show that human "perception" of very ordinary matters, and deeper ones as well, is naturally logarithmic. Suppose that as a youngster you are paid $0.10 per hour to do a certain task. If your wage is raised to $0.11 per hour, you would *perceive* that you are earning more. But now suppose you are a teenager earning $1.00 an hour. If your wage is raised again by $0.01 to $1.01 per hour, you would hardly notice this. On the other hand, if your wage is raised to $1.10 an hour, you would perceive a "real" increase. Now suppose you are a college senior, working as a tutor in your subject area and earning $10.00 per hour. If you raised your fee to $10.10 per hour, you would hardly notice the raise. On the other hand, a fee of $11.00 per hour would be perceived as a "real" increase.

The "operating rule" in this example is that we perceive a "real" difference when the *increase*—let us call it ΔS—in an existing state S (income level) is a sufficient (constant) *percentage of the state itself*. In this example, we perceive that someone is noticeably earning more when

$$\frac{\Delta S}{S} = 10\%$$

That is, an increase from $1.00 to $1.10 an hour, or from $10.00 to $11.00 an hour, is noticeable, whereas an increase from $10.00 to $10.10 is not. The actual change ΔS in the two cases is $0.10 and $1.00, respectively, but the percentage change (the *ratio* $\Delta S/S$) is in both cases the same: $\frac{1}{10}$, or 10%. This example of a person's *perception* of a noticeable raise versus a raise so small as to be overlooked is an example of a logarithmic relationship. At the same time we should realize that dollars and cents are a directly measurable quantity, whereas a sense of "greater wealth" is a psychological state. Weber's contribution to psychology was to assign a measure to such psychological states using the logarithm.

In his original experiments, Weber exposed each of his subjects to a sequence of stimuli. For example, a subject was given an initial lead ball of weight S_0 and then given several slightly heavier lead balls until he or she picked one, of weight S_1, which the subject felt was just barely heavier than S_0. The subject was then asked to choose, from among several, a second lead ball, of weight S_2, which was perceptively heavier than S_1, and so on. In this way the subject selected a sequence $S_1, S_2, S_3, \ldots$ of stimuli each of which was perceptively greater than the previous one. Weber then noted that the *percentage change* (in weight) between successive stimuli S_n and S_{n+1} was (approximately) constant. For example, if S_0 is a 10-lb weight, the subject might choose S_1 to be an 11-lb weight. Eventually given a 20-lb weight, the subject would choose not a 21-lb weight, but a 22-lb weight as the next noticeably heavier weight, choosing again a 10% heavier weight. Weber then went on to show that this meant that the relationship between perception of a stimulus, recorded by the number n marking the nth level of stimuli, and the actual strength S_n of the stimulus itself is *not a linear* relationship of the form $S_n = an + b$ but rather an exponential relationship of the *form*

$$S_n = ba^n$$

where $b = S_0$ and $a = 1 + (\Delta S/S)$.

This last equation describes the strength of the stimulus S_n in terms of the count n of perception level. It is more natural and useful to describe instead the reverse relationship

$$n = f(S)$$

of a person's perception of a stimulus as a function of the *strength* S of the stimulus. It is this relation that is logarithmic. Let us derive this relationship and see why this is so.

In Weber's model of human perception, we have a sequence of "stimuli"

$$S_0, S_1, S_2, S_3, \ldots$$

where S_0 is an initial stimulus, S_1 is the next highest stimulus one perceives, S_2 the next highest, and so on. In this model,

$$r = \frac{\Delta S}{S}$$

is the (constant percentage) change in stimulus level S_n so as to result in a perceptively higher stimulus S_{n+1}. Consequently, the strength of perceptively higher stimuli is related by the equation

$$S_{n+1} = S_n + r S_n = (1 + r)S_n$$

This means, for example, that

$$S_1 = S_0 + r S_0 = (1 + r)S_0$$
$$S_2 = S_1 + r S_1 = (1 + r)S_1 = (1 + r)[(1 + r)S_0] = S_0(1 + r)^2$$
$$S_3 = S_2 + r S_2 = (1 + r)S_2 = (1 + r)[(1 + r)^2 S_0] = S_0(1 + r)^3$$

In general then
$$S_n = S_0(1 + r)^n$$
where n is the marker that counts, or registers, a person's perception of successively higher stimuli. How does n depend directly on the stimulus itself? Applying the logarithmic function to both sides of this last equation, we have

$$\begin{aligned} \ln S_n &= \ln[S_0(1 + r)^n] \\ &= \ln S_0 + \ln(1 + r)^n \quad \text{Rule 2} \\ &= \ln S_0 + n \ln(1 + r) \quad \text{Rule 1} \end{aligned}$$

Let α represent the *constant* $\ln(1 + r)$. Then
$$\ln S_n - \ln S_0 = n\alpha$$
or
$$n = \frac{1}{\alpha} \ln S_n - \frac{\ln S_0}{\alpha}$$

Now replace $1/\alpha$ by A and $(-\ln S_0)/\alpha$ by B. We then have
$$n = A \ln S_n + B$$

That is, if P is the function that represents a person's perception, or response, to a stimulus of strength S, then $P(S)$ has the *form:*
$$P(S) = A \ln S + B$$

That the perception of change in stimuli is logarithmic has had an influence on measurement in a number of fields. In geology the logarithmic Richter scale is used to measure the power of an earthquake. In soil science (or chemistry) the pH level of soil acidity (a logarithm of the hydrogen ion concentration) is a gauge of the response of plants living off the soil to its minerals. Rather clearly, the small business, or the large corporation, has to deal with its perception of an increasing share of its market. All these are naturally described by logarithmic functions.

The Initial Calculus of the Logarithm: $D_x \ln x = 1/x$

We can obtain the derivative of the logarithm by differentiation of the equation
$$x = e^{\ln x} \tag{1}$$
using the chain rule and the invariance under differentiation of the exponential function. Differentiation of both sides of Equation 1 yields
$$\begin{aligned} D_x(x) &= D_x(e^{\ln x}) \\ 1 &= e^{\ln x} D_x(\ln x) \\ 1 &= x \cdot D_x(\ln x) \end{aligned}$$

Therefore, dividing both sides of this equation by $x \neq 0$, we have
$$\frac{1}{x} = D_x(\ln x)$$

This important formula also furnishes us with an antiderivative for a basic function. Recall that

$$\int x^n \, dx = \frac{x^{n+1}}{n+1} + C \quad \text{if } n \neq -1$$

To this point, we have been unable to write a formula for the antiderivative

$$\int x^{-1} \, dx = \int \frac{1}{x} \, dx$$

We now have it. Since $D_x(\ln x) = 1/x$, we have that for $x > 0$

$$\int \frac{1}{x} \, dx = \ln x + C$$

This formula may be usefully extended to negative values of x. While the function $f(x) = \ln x$ is only defined for $x > 0$, we can consider the function

$$g(x) = \ln|x| \quad \text{for all } x \neq 0$$

When $x < 0$, we have, by the chain rule,

$$D_x \ln|x| = \frac{1}{|x|} \cdot D_x |x|$$

$$= \frac{1}{-x} D_x(-x) \quad \text{Since } x < 0$$

$$= \frac{-1}{-x} = \frac{1}{x}$$

We summarize this discussion in two important formulas that should be memorized:

The Calculus of the Logarithm

For $x \neq 0$, $$D_x \ln|x| = \frac{1}{x}$$

and $$\int \frac{1}{x} \, dx = \ln|x| + C$$

The formula $D_x \ln|x| = 1/x$ says that the graph of $y = \ln x$ grows ever more slowly, always in inverse proportion to the size of x itself. We illustrate this in Figure 5.10 along with a graph of $y = \ln|x|$.

The formula for the derivative of the logarithm may be made more memorable, and perhaps more meaningful, if you realize that it is only the mathematical form of the adage that "the more you have, the more it takes to make you (perceptively) happi*er*." This is true because, according to Weber's model,

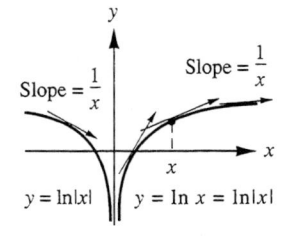

Figure 5.10

5.3 The Natural Logarithm

your perception of happiness is given by $H(S) = A \ln S + B$, where S is the level of stimuli causing this perception. The rate at which happiness increases is then given by the derivative $H'(S) = A/S$. Thus your sense of increasing happiness is *inversely proportional* to S. That is, the larger the stimulus (the more you have) the slower is your perception of improvement.

Exercises 5.3

In Exercises 1–6, find the values of the logarithms and indicate these on the graph of $y = \ln x$. Use Table 5.5 if needed.

1. $\ln 5$
2. $\ln 3.2$
3. $\ln \frac{3}{4}$
4. $\ln e^{1/3}$
5. $\ln|-2|$
6. $\ln 7^{1/5}$

In Exercises 7–18, simplify the expressions using the algebraic rules of the logarithm.

7. $\frac{1}{3} \ln 8$
8. $\ln 6 + \ln \frac{1}{2}$
9. $\ln(x^2 - 1) - \ln(x - 1)$
10. $\ln 25 + 2 \ln \frac{1}{5}$
11. $\ln 3 - \ln 9 + \ln 6$
12. $3 \ln e - \ln e^2$
13. $\frac{3}{2} \ln 4 - 5 \ln 2$
14. $\ln x - \ln y + 2 \ln z$
15. $\ln(a^2 + a) - \ln(a + 1)$
16. $\frac{1}{2} \ln 4 - \frac{1}{3} \ln 8 + \frac{1}{4} \ln 16$
17. $\dfrac{\ln 2 + \ln 3}{\ln 2}$
18. $\dfrac{3 \ln 2 - \ln 4}{5 \ln 2}$

19. Obtain a sheet of graph paper with at least a 10 by 10 mesh per square.

   a. Draw a good graph of $y = 1/x$.

   b. Use the Fundamental Theorem of Calculus and the fact that $D_x \ln x = 1/x$ and that $\ln 1 = 0$ to conclude that $\int_1^a (1/x)\, dx = \ln a$, where $a > 0$.

   c. According to Rule 2 of the algebraic rules of the logarithm,

   $$\ln 6 = \ln 2 + \ln 3$$

   Illustrate this formula by using (a) and estimating the area by counting squares below the graph that you drew in (a). That is, observe directly that the area beneath $y = 1/x$ from 1 to 6 appears to equal the area from 1 to 2 plus the area from 1 to 3:

   $$\int_1^6 \frac{1}{x}\, dx = \int_1^2 \frac{1}{x} + \int_1^3 \frac{1}{x}\, dx$$

20. Show that

    $$1 + \frac{1}{2} + \frac{1}{3} + \frac{1}{4} \geq \int_1^5 \frac{1}{x}\, dx = \ln 5$$

    by representing each summand on the left as the area of a rectangle of base width 1 in the drawing you made in Exercise 19. Can you generalize this observation to $\ln k$ by using a sum with k summands?

The formulas $e^{\ln x} = x$ and $\ln e^x = x$ are especially useful in solving logarithmic and exponential equations, as we saw in Example 3. In Exercises 21–30, solve each equation for the value of x, if possible.

21. $e^{\ln x^2} = 4$
22. $e^{\ln x^2 + 3 \ln x} = 32$
23. $2 \ln\left(\dfrac{1}{x+1}\right) = 8$
24. $2 \ln x + 1 = 3$
25. $e^{-2x} = 5$
26. $e^{x^2 - 1} = 1$
27. $3 = 2e^x - 5$
28. $\ln x^2 \ln x = 1$
29. $-\ln(2 - x) = 3 + \ln(2 + x)$
30. $2 \ln(\ln x) + 1 = 5$

31. In the natural sciences, particularly biology and medicine, an important kind of relationship between variables x and y is an **allometric relationship**, wherein y depends on x in the form

    $$y = Ax^k$$

    (Notice that the stimuli/perception equations derived in this section are of this same form with $x = 1 + r$.) Show that in such an allometric equation the variable $\ln y$ is *linearly* dependent on the variable $\ln x$.

32. The study of the cooling of a warm object placed in a cool container uses the formula

    $$\ln(A - T(t)) = -kt + C$$

    where $T(t)$ is the temperature of the object at time t and A is the temperature of the container. Solve this equation for $T(t)$ to obtain a formula for the temperature of the object at time t.

Table 5.5
Logarithmic Values

x	$\ln x$	x	$\ln x$	x	$\ln x$
0.1000	−2.3026	4.5000	1.5041	44.0000	3.7842
0.2000	−1.6094	4.6000	1.5261	45.0000	3.8067
0.3000	−1.2040	4.7000	1.5476	46.0000	3.8286
0.4000	−0.9163	4.8000	1.5686	47.0000	3.8501
0.5000	−0.6931	4.9000	1.5892	48.0000	3.8712
0.6000	−0.5108	5.0000	1.6094	49.0000	3.8918
0.7000	−0.3567	6.0000	1.7918	50.0000	3.9120
0.8000	−0.2231	7.0000	1.9459		
0.9000	−0.1054	8.0000	2.0794	100.0000	4.6052
		9.0000	2.1972	150.0000	5.0106
1.0000	0.0000	10.0000	2.3026	200.0000	5.2983
1.1000	0.0953				
1.2000	0.1823	11.0000	2.3979	500.0000	6.2146
1.3000	0.2624	12.0000	2.4849	1,000.0000	6.9078
1.4000	0.3365	13.0000	2.5649	1,500.0000	7.3132
1.5000	0.4055	14.0000	2.6391		
1.6000	0.4700	15.0000	2.7081	2,000.0000	7.6009
1.7000	0.5306	16.0000	2.7726	2,500.0000	7.8240
1.8000	0.5878	17.0000	2.8332	3,000.0000	8.0064
1.9000	0.6419	18.0000	2.8904	3,500.0000	8.1605
		19.0000	2.9444	4,000.0000	8.2940
2.0000	0.6931			4,500.0000	8.4118
2.1000	0.7419	20.0000	2.9957	5,000.0000	8.5172
2.2000	0.7885	21.0000	3.0445		
2.3000	0.8329	22.0000	3.0910		
2.4000	0.8755	23.0000	3.1355		
2.5000	0.9163	24.0000	3.1781		
2.6000	0.9555	25.0000	3.2189		
2.7000	0.9933	26.0000	3.2581		
2.8000	1.0296	27.0000	3.2958		
2.9000	1.0647	28.0000	3.3322		
		29.0000	3.3673		
3.0000	1.0986				
3.1000	1.1314	30.0000	3.4012		
3.2000	1.1632	31.0000	3.4340		
3.3000	1.1939	32.0000	3.4657		
3.4000	1.2238	33.0000	3.4965		
3.5000	1.2528	34.0000	3.5264		
3.6000	1.2809	35.0000	3.5553		
3.7000	1.3083	36.0000	3.5835		
3.8000	1.3350	37.0000	3.6109		
3.9000	1.3610	38.0000	3.6376		
		39.0000	3.6636		
4.0000	1.3863				
4.1000	1.4110	40.0000	3.6889		
4.2000	1.4351	41.0000	3.7136		
4.3000	1.4586	42.0000	3.7377		
4.4000	1.4816	43.0000	3.7612		

A more complete table can be found in a handbook of mathematical functions found in the library or via a scientific calculator.

Certain other values may be found using Table 5.5. For example,

$$\ln 60 = \ln 2(30)$$
$$= \ln 2 + \ln 30$$
$$= 0.6931 + 3.4012$$
$$= 4.0943$$

33. An unknown function $S(t)$ describing the sales by time t of a new product in a market of M individuals who are likely to buy the product eventually is known to satisfy the relationship

$$\ln S(t) - \ln(M - S(t)) = kt + C$$

Solve this equation for $S(t)$. How many sales have occurred by day $t = 3$ if $M = 100$, $k = 0.05$, and $C = 2$?

In Exercises 34–39, you are given some information about the function $f(t) = Ae^{kt}$ and then asked to determine other information. [Example 3(b) may help with some of these, along with Formulas 1 and 2, page 288.]

34. $f(0) = 4$. Determine A.
35. $k = 0.05$ and $A = 50$. Determine $f(3)$.
36. $f(1) = 6$ and $A = 3$. Determine k.
37. $f(0) = 100$ and $f(1) = 250$. Determine k and A.
38. $k = -0.02$ and $f(0) = 300$. Determine $f(5)$.
39. $f(0) = 500$ and $f(3) = 250$. Determine $f(6)$.

Graph each equation in Exercises 40–46.

40. $y = \ln \sqrt{x}$, $x > 0$
41. $y = 1 + \ln|x|$
42. $y = x + \ln|x|$
43. $y = \ln(x^2 + 1)$
44. $y = \ln(1 - x)$, $0 < x < 1$
45. $y = \ln\left(\dfrac{1}{1+x}\right)$, $x > -1$
46. $\ln y = x$, $y > 0$

47. Show by differentiation that the graph of $y = \ln x$ is concave down for all x.

48. Compute $f^{(15)}(x)$ for $f(x) = \ln x$, $x > 0$ (that is, the 15th derivative of $\ln x$).

49. Is $f(x) = x + \ln x$ an increasing function for $x > 0$? Is $g(x) = x \ln x$ increasing for $x > 0$?

50. Show that if the rate of change of a function $y = f(x)$ is in constant proportion k to the reciprocal of x, then $f(x) = k \ln x$.

Exercises 51 and 52 illustrate Weber's logarithmic model of human perception.

51. The economist H. F. Clark found that no matter what the level of income, the average person would like to have about 25% more. And when he has *that* income, he would like to have about 25% more, and so on. Each time his income increases by 25% he will, for a short time, perceive himself to be earning enough.

   a. How much additional income would it take for a person earning $15,000 a year to feel that he is earning enough? For a person earning $35,000?
   Suppose a person begins earning $10,000 a year in her first job. Let $S_0 = \$10,000$ and let $S_1, S_2, \ldots$ represent successive income levels at which she believes she is finally earning enough.

   b. Calculate S_1, S_2, S_3, S_5, S_8, and S_{10}.

   c. Let $n =$ the nth level of "perceptibly higher income." Derive a logarithmic formula for n as a function of income level.

   d. Use the formula in (c) to calculate $D_S n$, the rate of change of perception of rising income. This formula measures our feeling, or sense, of having a rising standard of living.

   (i) What is this rate of change at an income level of $10,000? Of $25,000? Of $40,000?

   (ii) Use these calculations to measure and discuss the perception that an individual at each of these income levels would have of a $1,000 raise. [You may wish to discuss this in terms of the differential of a function (Section 3.3).]

52. While Exercise 51 and our discussion "Why the logarithm is natural" make the work of Clark and Weber algebraically explicit, the basic relationship with the logarithm is geometrically transparent. We can use a simplification of the methods of Section 2.4 to see this. Suppose that there is some quantifiable relationship between happiness H as a function of wealth W and that, using Clark's factor of 25%, it always requires a 25% increase in whatever level of wealth we have in order to produce the *same increase* ΔH in happiness. That is, $H(W + 0.25W) - H(W) = \Delta H$ and ΔH is constant. Choose some vertical distance to represent the quantity ΔH (for example, you might try using $\Delta H = \frac{1}{10}$), and try to sketch a graph of H as a function of W such that $H(W + 0.25W) - H(W)$ is always this same distance no matter how large W is, starting with $H(1) = 0$. Does your graph resemble the graph of the logarithm?

Challenge Problems

53. The amount of drug present in a patient's bloodstream t minutes after the drug has been administered can be modeled by a function of the form $f(t) = ate^{-bt}$ studied in Example 9 of Section 5.2. The doctor who injects the drug monitors the patient for 2 minutes and finds that after 1 minute 5 mg of the drug is present and after 2 minutes 3 mg is present; thus $f(1) = 5$ and $f(2) = 3$.

   a. Find the time at which the amount present is a maximum.

b. How much of the drug will be present after 5 minutes?

[*Hint:* Impose the known conditions, and solve for (a) and (b).]

54. Rock is a stimulating musical form that tends to be played loudly. The intensity of sound is measured as energy per unit area: watt/m^2. (When you begin to feel sound, you are feeling this energy.) At a frequency of 1,000 cycles per second, the lowest intensity S_0 that the normal human can hear is 10^{-12} watts/m^2. At a higher intensity S, the *loudness L* of a sound S is measured in the *decibel* (dB) unit and is defined as

$$L = 4.32 \ln(S/S_0) \text{ dB}$$

Because our perception of the loudness of sound is a logarithmic function, we cannot easily relate this perception, measured in decibels, to the amount of energy S used to make the sound. If an ordinary loud band is playing music at a level of 80 dB, how much *more* energy must be used to raise its loudness to 115 dB, near the threshold of pain? [*Hint:* Solve for S when $L = 80$ and when $L = 115$.]

5.4 The Calculus of the Logarithmic Function

In this section we learn to differentiate and integrate various functions of the logarithm.

In the previous section we saw that $D_x \ln|x| = 1/x$ when $x \neq 0$. We can combine this formula with the chain rule to obtain the general formula

$$D_x \ln|g(x)| = \frac{g'(x)}{g(x)} \qquad g(x) \neq 0$$

This formula tells us that the rate of change of $\ln|g(x)|$ is the ratio $g'(x)/g(x)$ of the rate of change of $g(x)$ to $g(x)$ itself. This is a natural ratio encountered in applications, since it expresses the percentage rate of change of the function g.

To derive this formula, think of $\ln|g(x)|$ as a composition, $\ln|g(x)| = f(g(x))$, with $f(x) = \ln|x|$. According to the chain rule, $D_x f(g(x)) = f'(g(x))g'(x)$. With $f(x) = \ln|x|$, we know that $f'(x) = 1/x$ and therefore $f'(g(x)) = 1/g(x)$. Hence

$$D_x \ln|g(x)| = f'(g(x))g'(x)$$
$$= \frac{1}{g(x)} \cdot g'(x)$$
$$= \frac{g'(x)}{g(x)}$$

If $g(x) \geq 0$, we will write $\ln g(x)$ rather than $\ln|g(x)|$. The use of absolute values occurs most often via the solution to an integration problem.

Differentiation and the Logarithm

EXAMPLE 1

Find the derivative of

a. $\ln(x^2 + 1)$ **b.** $\ln(e^x + 1)$ **c.** $\ln|x^3 - 1|$, $x \neq 1$

Solution

a. Let $g(x) = x^2 + 1$. Then $g'(x) = 2x$. Hence

$$D_x \ln(x^2 + 1) = \frac{g'(x)}{g(x)} = \frac{2x}{x^2 + 1}$$

b. Using the same formula in different but more convenient terms, we have

$$D_x \ln(e^x + 1) = \frac{D_x(e^x + 1)}{e^x + 1} = \frac{e^x}{e^x + 1}$$

c.
$$D_x \ln|x^3 - 1| = \frac{D_x(x^3 - 1)}{x^3 - 1} = \frac{3x^2}{x^3 - 1} \quad \blacksquare$$

This then is how to differentiate the *logarithm of a function*. We may also need to differentiate a *function of a logarithm*. For example, consider the problem of finding $D_x(\ln x)^3$.* Think of this function as a composite, $f(g(x)) = (\ln x)^3$, with $f(x) = x^3$ and $g(x) = \ln x$. According to the chain rule, we then have

$$D_x(\ln x)^3 = f'(g(x))g'(x) = 3\, g(x)^2 g'(x)$$

Since $g(x) = \ln x$ and $g'(x) = 1/x$, this becomes

$$D_x(\ln x)^3 = 3(\ln x)^2 \cdot \frac{1}{x} = \frac{3(\ln x)^2}{x}$$

Here are further examples.

EXAMPLE 2

Find the derivative of

a. $(1 + \ln x)^3$ **b.** $\sqrt{\ln x}$ **c.** $(x \ln x)^3$, $x > 0$

Solution

a.
$$D_x(1 + \ln x)^3 = 3(1 + \ln x)^2 \cdot D_x(1 + \ln x)$$
$$= 3(1 + \ln x)^2 \cdot \left(0 + \frac{1}{x}\right) = \frac{3(1 + \ln x)^2}{x}$$

b.
$$D_x \sqrt{\ln x} = D_x(\ln x)^{1/2} = \frac{1}{2}(\ln x)^{-1/2} D_x \ln x$$
$$= \frac{1}{2(\ln x)^{1/2}} \cdot \frac{1}{x} = \frac{1}{2x\sqrt{\ln x}}$$

c.
$$D_x(x \ln x)^3 = 3(x \ln x)^2 D_x(x \ln x)$$
$$= 3(x \ln x)^2[x \cdot D_x \ln x + (\ln x)D_x x]$$
$$= 3(x \ln x)^2 \left(x \cdot \frac{1}{x} + \ln x \cdot 1\right)$$
$$= 3(x \ln x)^2 (1 + \ln x) \quad \blacksquare$$

---

*Note: $(\ln x)^3 \neq \ln x^3$; indeed $\ln x^3 = 3 \ln x$.

We may also combine these differentiation rules with earlier algebraic rules to calculate further derivatives of logarithmic functions.

EXAMPLE 3
Find

a. $D_x \ln\left(\dfrac{x}{x+1}\right)$ **b.** $D_x \ln\left(\dfrac{1}{\sqrt{x}}\right)$, $x > 0$

Solution

a. Since $\ln(a/b) = \ln a - \ln b$, we have

$$D_x \ln\left(\frac{x}{x+1}\right) = D_x[\ln x - \ln(x+1)] = D_x \ln x - D_x \ln(x+1)$$

$$= \frac{1}{x} - \frac{D_x(x+1)}{x+1} = \frac{1}{x} - \frac{1}{x+1} = \frac{1}{x(x+1)}$$

Notice that the "log/quotient" property (step 1) used here allows us to avoid the involved differentiation rule for a quotient.

b. Since $\ln x^r = r \ln x$, we have

$$D_x \ln\left(\frac{1}{\sqrt{x}}\right) = D_x\left[\ln \frac{1}{x^{1/2}}\right] = D_x[\ln x^{-1/2}]$$

$$= D_x\left(-\frac{1}{2} \ln x\right) = -\frac{1}{2} D_x \ln x$$

$$= -\frac{1}{2} \cdot \frac{1}{x} = -\frac{1}{2x} \quad\blacksquare$$

Thus, ordinarily bothersome exponents $\left(\text{here}, -\tfrac{1}{2}\right)$ become simple linear factors that do not enter the differentiation.

Finally, the logarithm may appear as part of a product or quotient.

EXAMPLE 4
Find

a. $D_x\left[\dfrac{\ln(x+1)}{e^x}\right]$ **b.** $D_x[(\ln x)(\ln(x+1))]$, $x > -1$

Solution

a. Using the quotient rule for differentiation, we have

$$D_x\left[\frac{\ln(x+1)}{e^x}\right] = \frac{e^x D_x \ln(x+1) - [\ln(x+1)] \cdot D_x e^x}{(e^x)^2}$$

$$= \frac{e^x\left(\dfrac{1}{x+1}\right) - [\ln(x+1)] \cdot e^x}{e^{2x}}$$

5.4 The Calculus of the Logarithmic Function

$$= \frac{e^x\left[\dfrac{1}{x+1} - \ln(x+1)\right]}{e^{2x}}$$

$$= \frac{e^x[1 - (x+1)\ln(x+1)]}{(x+1)e^{2x}}$$

$$= \frac{1 - (x+1)\ln(x+1)}{(x+1)e^x}$$

b. Using the product rule for differentiation, we have

$$D_x[(\ln x)(\ln(x+1))] = (\ln x)D_x\ln(x+1) + [\ln(x+1)]D_x(\ln x)$$

$$= (\ln x)\left(\frac{1}{x+1}\right) + [\ln(x+1)] \cdot \frac{1}{x}$$

$$= \frac{\ln x}{x+1} + \frac{\ln(x+1)}{x} \quad \blacksquare$$

Integration and the Logarithm

Since
$$D_x\ln|g(x)| = \frac{g'(x)}{g(x)} \quad (g(x) \neq 0)$$

then
$$\int \frac{g'(x)}{g(x)}\, dx = \ln|g(x)| + C$$

This formula may be combined with the method of integration by substitution to do a variety of previously insoluble problems. When an integration problem occurs in "almost" (up to a constant factor) the *form* $g'(x)/g(x)$, it can be solved by the substitution $u = g(x)$.

EXAMPLE 5

Find $\displaystyle\int \frac{x}{x^2 - 1}\, dx$.

Solution Let $u = x^2 - 1$. Then $du = 2x\, dx$, or $\frac{1}{2} du = x\, dx$, and

$$\int \frac{x}{x^2 - 1}\, dx = \int \frac{\frac{1}{2} du}{u} = \frac{1}{2}\int \frac{du}{u} = \frac{1}{2}\ln|u| + C$$

$$= \frac{1}{2}\ln|x^2 - 1| + C = \frac{\ln|x^2 - 1|}{2} + C \quad \blacksquare$$

Notice that $x/(x^2 - 1)$ is "almost" in the form $g'(x)/g(x)$ except for the constant factor 2. Here is a second example.

EXAMPLE 6

Find $\displaystyle\int \frac{e^x}{e^x + 7}\, dx$.

Solution This is in the form $g'(x)/g(x)$ with $g(x) = e^x + 7$. Let $u = e^x + 7$. Then $du = e^x \, dx$ and we have

$$\int \frac{e^x \, dx}{e^x + 7} = \int \frac{du}{u} = \ln|u| + C = \ln(e^x + 7) + C \quad \blacksquare$$

Note: Since the function $e^x + 7$ is positive, we can omit the absolute value sign.

Let us turn now to integration of logarithmic functions. These may *sometimes* be done by substitution.

EXAMPLE 7

Find $\int \frac{\ln x}{x} \, dx$, $x > 0$.

Solution Although it is not immediately clear how to find this integral, notice that if we rewrite the integrand $(\ln x)/x$ as $(\ln x)(1/x)$, we see both the function $\ln x$ and its derivative $1/x$. Thus we let $u = \ln x$, obtain $du = (1/x) \, dx$, and have

$$\int \frac{\ln x}{x} \, dx = \int (\ln x) \cdot \frac{1}{x} \, dx = \int u \, du = \frac{u^2}{2} + C = \frac{(\ln x)^2}{2} + C \quad \blacksquare$$

EXAMPLE 8

Find $\int \frac{dx}{x(1 + \ln x)}$.

Solution Since $D_x(1 + \ln x) = 1/x$, we try the substitution $u = 1 + \ln x$. Then $du = (1/x) \, dx$ and we have

$$\int \frac{dx}{x(1 + \ln x)} = \int \frac{1}{(1 + \ln x)} \cdot \frac{1}{x} \, dx = \int \frac{1}{u} \, du$$

$$= \ln|u| + C = \ln|\ln x + 1| + C \quad \blacksquare$$

The key to the preceding examples is that the derivative of $\ln x$ appears in the formula in just the right way. Were these examples changed slightly, other methods of integration not yet studied might have to be used. The following is an example of an integral that cannot be done by substitution.

EXAMPLE 9

Show that

$$\int \ln x \, dx = x \ln x - x + C \qquad x > 0$$

Solution $\int \ln x \, dx$ is not done by substitution since $D_x \ln x = 1/x$ does not appear in the integrand. Thus we can only differentiate the right-hand side of the given equation and see that it is a correct antiderivative of $\ln x$. We have

$$D_x[x \ln x - x + C] = (x D_x \ln x + \ln x \cdot D_x x) - D_x x + D_x C$$

$$= x \cdot \frac{1}{x} + (\ln x) \cdot 1 - 1 + 0$$

$$= 1 + \ln x - 1 = \ln x \quad \blacksquare$$

Exercises 5.4

Differentiate the functions in Exercises 1–24 on their respective domains.

1. $\ln(x + 3)$
2. $\ln(e^x - x)$
3. $\ln(x^2 + x)$
4. $\ln|x - 1|$
5. $\ln(x^3 + \sqrt{x})$
6. $(1 + \ln x)^3$
7. $(x - \ln x)^2$
8. $(\ln x)^{1/3}$
9. $\dfrac{\ln x}{x}$
10. $\dfrac{(\ln x)^2}{e^x}$
11. $\ln(\ln x)$
12. $\dfrac{1}{\ln x}$
13. $e^{\ln t}$
14. $e^{x \ln x}$
15. $\ln ax$, a constant
16. $\dfrac{x}{\ln x}$
17. $t \ln t - t$
18. $\ln(xe^x)$
19. $e^{-x \ln x}$
20. $e^x \ln x$
21. $\dfrac{\ln x}{\ln(x + 1)}$
22. $\ln\left(\dfrac{x^3}{x^2 + x}\right)$
23. $\sqrt[3]{7 + \ln 3x}$
24. $(e^x - \ln x)^3$

25. Which of the two functions $\ln(x^2)$ or $(\ln x)^2$ is "easier" to differentiate?

26. Apply the formula $a = e^{\ln a}$ to $a = 2^x$ to obtain $2^x = e^{x \ln 2}$. Then find $D_x 2^x$.

27. Suppose that the number of ants arriving at your picnic basket is doubling each minute, beginning with one ant at time $t = 0$. Therefore at time t there are $A(t) = 2^t$ ants searching for food in your picnic basket. How many new ants arrive during the 4th (to 5th) minute? How fast is the ant population of your picnic basket growing at the end of $4\frac{1}{2}$ minutes? [Hint: Use Exercise 26.]

Use Examples 5–9 and the Fundamental Theorem of Calculus to find the integrals in Exercises 28–31.

28. $\displaystyle\int_0^4 \dfrac{x}{x^2 + 1}\, dx$
29. $\displaystyle\int_1^e \dfrac{\ln x}{x}\, dx$
30. $\displaystyle\int_1^{-2} \dfrac{dx}{x(\ln x + 1)}$
31. $\displaystyle\int_1^4 \ln x\, dx \qquad (\ln 2 \approx 0.69)$

In Exercises 32–41, find the indefinite integrals.

32. $\displaystyle\int \dfrac{1}{2x}\, dx$
33. $\displaystyle\int \dfrac{3x^2}{x^3 + 9}\, dx$
34. $\displaystyle\int \dfrac{4x - 2}{x^2 - x}\, dx$
35. $\displaystyle\int \dfrac{\ln(x + 1)}{x + 1}\, dx$
36. $\displaystyle\int \dfrac{e^x + e^{-x}}{e^x - e^{-x}}\, dx$
37. $\displaystyle\int \dfrac{dx}{x \ln x^2}$
38. $\displaystyle\int \dfrac{x\, dx}{1 - x^2}$
39. $\displaystyle\int \ln(2x + 3)\, dx$ [Hint: Let $u = 2x + 3$.]
40. $\displaystyle\int \dfrac{e^{-2x}}{1 - e^{-2x}}\, dx$
41. $\displaystyle\int \dfrac{\ln(\ln x)}{x}\, dx$

42. This exercise shows why the natural logarithm is the preferred logarithm in calculus. The logarithm to a base $b > 0$ other than the base e is defined by $\log_b x = y$ if and only if $b^y = x$.

   a. Explain why $\log_{10} 100 = 2$; $\log_2 8 = 3$; $\log_5 25 = 2$.

   b. Since $\ln b^y = y \ln b$, apply the natural logarithm to both sides of the equation $b^y = x$ and solve for y to conclude that $y = (\ln x)/(\ln b)$.

302   Chapter 5   The Exponential and Logarithmic Functions

c. Use (b) to show that $D_x(\log_b x) = 1/(x \ln b)$, since $\ln b$ is a constant and $\log_b x = y = (\ln x)/(\ln b)$.

d. Since $\ln 2 = 0.69$ (to two decimal places), show that $D_x \log_2 x = 1.449/x$.

43. Find the local maxima and local minima for $f(x) = x \ln x$, and graph this function for $0 < x \leq 2$.

44. Find the equation of the tangent line to the graph of $y = \ln|1 - x^2|$ at $x = \frac{1}{2}$ and at $x = 2$.

45. Does $f(x) = (\ln x)/x$ have a maximum or minimum value at some $x > 1$?

46. Sketch the graph of $g(x) = \ln x - x$. Find the highest point on this graph.

47. Does $f(x) = x/(\ln x + x)$ have a maximum or minimum value for $x > 1$?

48. For which values of x is the graph of $f(x) = x \ln x$, $x > 0$ concave up? Concave down? Graph this function.

49. Which of the two graphs, $y = \ln x$ and $y = \ln(\ln x)$, where $x > 1$, grows more slowly as x increases?

50. Compare the *rates* of growth in sales of two items, one of which has total sales $S_1(t) = \ln t$ on day $t > 1$ and the other of which has sales $S_2(t) = t \ln(\ln t)$, $t > 1$,

a. On day $t = 10$.   **b.** On day $t = 2$.

Challenge Problems

51. For over a century the concept of **entropy** has been crucial for understanding thermodynamics, information theory, and now the processing of signals in computer circuitry. It may be crucial to our future understanding of how life itself is organized via DNA. Roughly speaking, entropy is a measure of the disorder in a system.

Imagine a *binary signal source*, such as a blinking light or a neuron in the brain. That is, imagine a source of energy, or a signal, that is either *on* or *off*. Suppose that this source is either *on* with probability p, $0 \leq p \leq 1$, or *off*, necessarily with probability $1 - p$.

You can imagine the output of such a system as a sequence of 0's and 1's, where we record the output as

0   if the source is *off*
1   if the source is *on*

Thus, over time, the output of such a source might be recorded as

1, 1, 0, 1, 1, 0, 1, 1, 0, . . .

or as

1, 0, 1, 0, 0, 1, 0, 0, 0, 1, 1, 0, 1, 0, 1, 1, 0, . . .

The first sequence has some order, with 1's appearing two-thirds of the time. The second sequence is more disorderly, with 1's appearing roughly half of the time.

The entropy of such a binary source is defined by the function

$$E(p) = -p \ln p - (1 - p)\ln(1 - p)$$

For what probability p is the entropy of the system a maximum? That is, what probability p maximizes the overall disorder of the signals from such a binary source?

52. For $x > 0$, let $f(x) = x^x$. Here both base and exponent are variables. Write $x^x = e^{\ln x^x} = e^{x \ln x}$ and find $f'(x)$, $f'(1)$, and $f'(e)$.

5.5 Exponential Growth and Decay

In this section we learn to model growth that "feeds on itself," using the exponential function.

The same experience, repeated often enough, soon becomes a part of you. You have so often experienced the "rule" that "amount done equals rate times time," whether in a footrace or in a race to complete an exam, that this way of thinking about rate and time is second nature. This section asks you to think about amount, rate, and time quite differently, since our second nature way of thinking about these is limited to situations where amount is a *linear* function of rate and time. It is equally common that *amount is an exponential function* of rate and time. Examples include resource depletion, the growth of investments, environmental degradation, the dating of artifacts, radioactive decay, and population growth.

When you are running a footrace, your speed does not increase the further you run (rather, it tends to slow down!). On the other hand, the speed at which money in your savings account grows does increase with the amount you have saved, because the interest rate is applied directly to this amount. Having run at the rate of 8 yards a second for 3 seconds, you know that you have run 24 yards. Having earned 8% interest on a savings account for 3 years, starting with, say, $100, you might be uncertain as to how much is in the account.

In this second example, growth is "feeding on itself," since the growth of money in the account is determined by how much is in the account (and the interest rate) and this amount itself is growing. We will see that growth that feeds on itself is characterized by exponential change.

The Differential Equation of Exponential Change and Its Solution

Imagine what happens when a new recording by a popular singer is released. At first a few people hear the recording and they tell several friends that they liked it enough to buy a copy. These friends, in turn, trouble themselves to hear the recording, a few buy a copy, and each of them passes on the word to other friends. As this process continues, new sales grow out of existing sales—all at different times in different places.

Let $A(t)$ be the number of people who have bought the recording by day t after its release. In this model, new purchases occur in some proportion k to the number $A(t)$ of existing purchases and in proportion to a given span of time. For example, suppose that on day t, $A(t) = 220$ people have bought the recording and 30% (that is, $k = 0.30$) of the distinct friends they tell this to on day t themselves buy a copy. There will then be $220 \times 0.30 = 66$ new purchases that day. But how many new purchases would there be in the *first half* of day t? In this rough model we would have to say $220 \times 0.30 \times \frac{1}{2} = 33$ new purchases. Thus growth over a period of time is proportional to the period as well.

Thinking this way, we can derive a general model as follows. Imagine a short span of time h between time t and time $t + h$. The increase in the amount A over this span of time is given by

$$A(t + h) - A(t)$$

In this model, the increase is in the proportion k to the amount $A(t)$ at time t and to the span of time h. Thus, just as we saw in the earlier computation, we have, in general,

$$A(t + h) - A(t) = k \cdot A(t) \cdot h$$

Dividing both sides by h yields

$$\frac{A(t+h) - A(t)}{h} = kA(t) \qquad (1)$$

Notice that the right side of Equation 1 is independent of h. Taking now the limit as the time span h shrinks to 0, we obtain

$$A'(t) = \lim_{h \to 0} \frac{A(t+h) - A(t)}{h} = \lim_{h \to 0} kA(t) = kA(t)$$

This equation
$$A'(t) = kA(t) \qquad (2)$$

models the instantaneous rate of growth of the amount A.

Equation 2 is a **differential equation.** In ordinary language, Equation 2 says that the *rate of growth of the amount is proportional to the amount present.* In short, "growth feeds on itself." This one equation has a wide range of application, for it is *the* model for the growth over time of such systems.

The Differential Equation of Exponential Change

If $A(t)$ is the amount present of some quantity that changes over time and if the *rate of growth is proportional to the amount present*, then the function $A(t)$ satisfies the differential equation

$$A'(t) = kA(t)$$

The proportionality constant k is called the **growth constant.**

We call this equation the differential equation of *exponential change* because of its solution. Let us determine this solution.

Recall that the number $x = -2$ is called the solution of the equation $2x + 5 = 1$ because -2 is the *only number* that can be used in place of the "unknown" letter x in the original equation and result in a true statement. The equation

$$A'(t) = kA(t)$$

is called a differential equation because it is an equation involving, not an unknown number, but an *unknown function* and its derivative. To solve such a differential equation we must find not a number, but a *function* that, when substituted in the differential equation, makes it a true statement.

The unknown function satisfying the equation $A'(t) = kA(t)$ is the function $A(t) = Ce^{kt}$, where C can be any number (we will shortly see that C plays a useful role in applications). To see this we substitute in the left side of the differential equation and have

$$A'(t) = D_t(Ce^{kt}) = C\, D_t e^{kt} = Ce^{kt} \cdot k = k \cdot Ce^{kt} \qquad (3)$$

and on the right
$$kA(t) = k \cdot Ce^{kt} \qquad (4)$$

Since the right sides of Equations 3 and 4 are identical, we conclude:

Solution of the Equation of Exponential Change

The differential equation

$$A'(t) = kA(t)$$

is called the differential equation of exponential change because its (only) solution is an exponential function of the form

$$A(t) = Ce^{kt}$$

where C can be any constant.

It is most significant in understanding the world about us that, just as the algebraic equation $2x + 5 = 1$ has *only one* solution, $x = -2$, so too does the differential equation $A'(t) = kA(t)$ have *only one possible* functional form, Ce^{kt}, as its solution. Not only is the exponential function the "easiest" function to deal with in calculus, it is also unavoidable in applications.

A basic example foretells the kinds of applications in this section. Some banks offer continuously compounded interest on money invested in an account. Suppose that the account is advertised as paying 8% interest compounded continuously. This means that over any period of time interest is being credited to the account at the rate of 8% of whatever amount is in the account at that time, *in proportion to that period of time*. Consequently, from our earlier discussion

$$A'(t) = 0.08A(t)$$

That is, continuously compounded interest is precisely modeled by the differential equation of exponential change, in which the growth constant is the rate of interest. This is as it should be since the meaning of "interest rate" is an expression of the proportion of growth in an account.

EXAMPLE 1

Suppose you put $100 in a savings account that pays 8% interest compounded continuously. How much money will be in the account 3 years from now?

Solution Let $A(t)$ be the amount of money in the account at time t. Since $A'(t) = 0.08A(t)$, we have the solution $A(t) = Ce^{0.08t}$. Therefore the amount in the account at the end of 3 years is

$$A(3) = Ce^{0.08(3)} = Ce^{0.24} \approx 1.27125\, C$$

This does not, however, answer the question in terms of dollars. On the other hand, we have not yet used all the facts—in particular, that there was an initial amount of money in the account. We will now see why the number C is useful.

Since
$$A(t) = Ce^{0.08t}$$
at *any* time t, then in particular at time $t = 0$, we have
$$A(0) = Ce^{0.08(0)} = Ce^0 = C \cdot 1 = C$$
That is, C is the initial amount $A(0) = \$100$.
Therefore at time t the account holds
$$A(t) = 100e^{0.08t} \text{ dollars}$$
and so, at the end of 3 years, holds
$$A(3) = 100e^{0.08(3)} = 100e^{0.24} = \$127.125$$
The graph of this solution is shown in Figure 5.11.

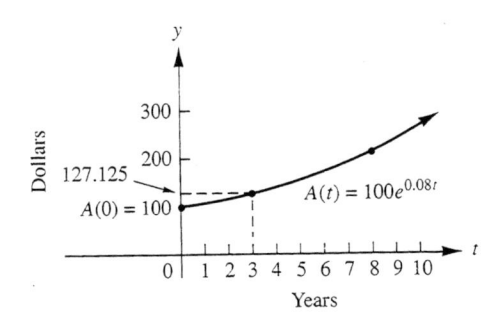

Figure 5.11

The general exponential growth process $A'(t) = kA(t)$ proceeds along the same lines. The process begins in an initial state, or initial condition. The initial amount is always the amount present at time zero, $A(0)$. Reasoning in general as we did in Example 1, we then have

$A(t) = A(0)e^{kt}$, $k > 0$

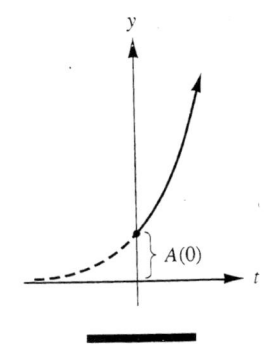

Figure 5.12

The Unique Solution to the Differential Equation of Exponential Change

$$A'(t) = kA(t)$$

beginning with initial amount $A(0)$ is

$$A(t) = A(0)e^{kt}$$

For $k > 0$, this solution is shown in Figure 5.12. This figure reminds us too of an important limitation on the application of such exponential growth models: The model would yield ludicrous results if applied over too long a period of time, because exponential growth is very rapid if k is at all large.

Exponential Decay

Exponential decay is a mirror-image-in-time of exponential growth and differs in only one detail: The "growth" constant is *negative* and is instead a **decay**

constant. For example, we can consider the decline (decay) in the value (purchasing power) of money due to inflation. We commonly hear of monthly or yearly reports of the inflation rate, the rate at which prices are rising. Economists recognize, however, that no one decrees higher prices at such regular intervals and that inflation occurs continuously.* A production firm is charged a higher price for its raw materials; the firm then charges a higher price for its product to a retail store, which then raises the price of its merchandise; this eventually leads to a demand for higher wages by a buyer, who may even be an employee of the company that supplied the original raw materials. In this way inflation feeds on itself over time.

In an inflationary economy, where the rate of inflation is, say, 6%, $100 stashed safely away in a shoebox will lose value, or purchasing power; it will buy less of any particular item as time goes on. Let us analyze this in more detail.

Suppose that D dollars in hand today will buy C copies of a given item. Thus the price per item is D/C dollars, or the *value*—or purchasing power—of one dollar is C/D items.

If the inflation rate is r percent per year, the price per item 1 year later will be $(1 + r)(D/C)$. Consequently, the value of one dollar 1 year later will be $C/[(1 + r)D]$. The percentage change in value—purchasing power of one dollar—is then

$$\frac{C/(1 + r)D - C/D}{C/D} = \frac{1}{1 + r} - 1 = \frac{-r}{1 + r}$$

Thus money is losing value at the rate of $-r/(1 + r)$ percent per year. That is,

---

If the inflation rate is r percent per year, then the rate of decline in purchasing power, or value, of money is $-r/(1 + r)$ percent per year.

Returning to your $100 safely stashed away in a shoebox, while the inflation rate is $r = 0.06$, let

$A(t)$ = purchasing power of this $100 t years in the future

From our earlier analysis,

$$A'(t) = \left(\frac{-r}{1 + r}\right)A(t) = \left(\frac{-0.06}{1.06}\right)A(t) = -0.0566 A(t)$$

---

*Inflation is treated differently by economists and by financiers, who loan money to be repaid at regular intervals. A good discussion can be found in the article by R. C. Thompson, "The True Growth Rate and the Inflation Balancing Principle," *American Mathematics Monthly* 90, no. 3 (1983): 207–10. Our discussion is based on Theorem 3 of that article.

Thus money is losing value at the rate of (approximately) 5.66% per year, and this loss of value is modeled by a differential equation of the form

$$A'(t) = -kA(t)$$

where, in this instance, $k = -0.0566$.

As in our discussion of exponential growth, the solution of the general differentiation equation of exponential decay

$$A'(t) = -kA(t) \quad k > 0$$

is

$$A(t) = Ce^{-kt}$$

where again

$$A(0) = Ce^{k(0)} = Ce^0 = C$$

That is,

$$A(t) = A(0)e^{-kt}$$

furnishes the general model of exponential decay. We illustrate this solution in Figure 5.13. We will apply this model shortly.

Exponential decay

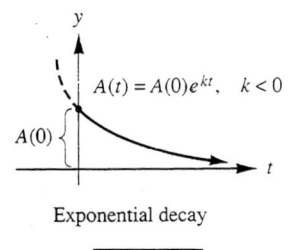

Figure 5.13

Doubling-Time and Half-Life

The number

$$\frac{\ln 2}{k}$$

plays an important and simplifying role in the study of both exponential growth and exponential decay.

In the case of exponential growth, $A(t) = A(0)e^{kt}$, $k > 0$, we have that at time $t_d = (\ln 2)/k$

$$A(t_d) = A(0)e^{kt_d} = A(0)e^{k[(\ln 2)/k]} = A(0)e^{\ln 2} = 2A(0)$$

since $e^{\ln x} = x$. That is, at time t_d the amount present $A(t_d)$ is *twice* the original amount $A(0)$; t_d is called the **doubling-time** for exponential growth.

EXAMPLE 2

If $100 is deposited in a savings account paying 8% annual interest compounded continuously, when will there be $200 in the account?

Solution Let $A(t)$ be the amount of money in the account at time t years in the future. Then from Example 1,

$$A(t) = A(0)e^{0.08t}$$

Since $A(0) = 100$, there will be $200 = 2A(0)$ at time

$$t_d = \frac{\ln 2}{0.08} \approx \frac{0.69}{0.08} \approx 8.63 \text{ years}$$

in the future. ∎

EXAMPLE 3

When will there be $400 in the account discussed in Example 2?

5.5 Exponential Growth and Decay

Solution The account contains $200 after 8.63 years. Since the amount in the account continues to grow from this point in time according to the same interest rate, $200 will double again, to $400, in another 8.63 years. Therefore the original $100 will yield $400 within $8.63 + 8.63 = 17.26$ years. ∎

We illustrate the solutions to Examples 2 and 3 in Figure 5.14.

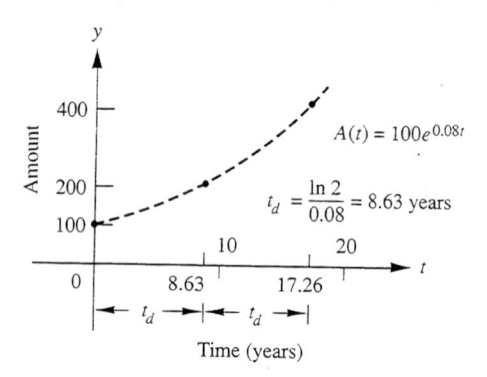

Figure 5.14

In the case of exponential decay, $A'(t) = -kA(t)$, $k > 0$, the number $(\ln 2)/k$ remains a positive number and at time $t_h = (\ln 2)/k$ we have

$$A(t_h) = A(0)e^{-kt_h} = A(0)e^{-k[(\ln 2)/k]} = A(0)e^{-\ln 2} = \frac{A(0)}{e^{\ln 2}} = \frac{A(0)}{2}$$

That is, at time t_h, the amount present $A(t_h)$ is *one-half* the original amount $A(0)$; t_h is then called the **half-life** for exponential decay.

Let us apply this concept to the case of an inflation rate of 6% applied to $100 in dollar bills "safely" stashed away in your shoebox. We can model the purchasing power of this $100 by $A'(t) = -0.0566A(t)$ (as noted in our initial discussion of exponential decay) with solution $A(t) = 100e^{-0.0566t}$. It follows that in $t_h = (\ln 2)/0.0566 \simeq 12.25$ years this $100 will have lost one-half of its purchasing power.

This means that if $100 could buy two pairs of shoes today, then left "safely" stashed away this same $100 will buy only one pair of shoes 12.25 years from now. Such facts are useful guides even in personal finance. In times of inflation, paper money loses value unless it is converted by purchase into a "real" holding, such as land, whose value tends to inflate with the overall economy.

Figure 5.15 illustrates the relationship of the half-life $t_h = (\ln 2)/k$ to the exponential decay curve.

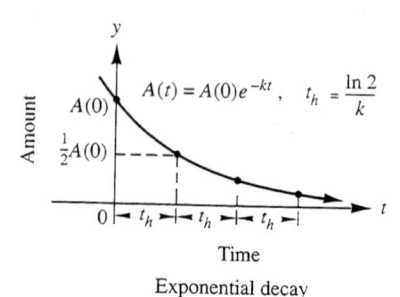

Figure 5.15

Exponential decay

310 Chapter 5 The Exponential and Logarithmic Functions

Applications of the Exponential Change Model

Let us first form an overview of the preceding discussions. In every case we are concerned with an amount—denoted by $A(t)$—that is changing over time t. We have found that if we know $A(0)$, the initial state (of this amount), and if we know the law of change,

$$A'(t) = kA(t)$$

then we must necessarily know the amount present at any later time t:

$$A(t) = A(0)e^{kt}$$

Such a model for growth is called **deterministic**: Future states of the system, $A(t)$, are determined by where things start, $A(0)$, and how things change, $A'(t) = kA(t)$. We will shortly see that the same model can help us look not only into the future but also into the past.

EXAMPLE 4

Environmental Impact For some time, scientists have been concerned about the depletion of ozone in the upper atmosphere. Ozone shields life on earth from certain harmful radiation from the sun. The problem is that fluorocarbons (used in air conditioning systems and aerosol sprays) migrate into the upper atmosphere and convert ozone to another chemical compound, thus depleting it. This example of environmental degradation has the character of exponential growth for the following social reason. Once your neighbor possesses an air-conditioned car or home or uses aerosol sprays, you want to also. Aided by advertising, the growth in the use of fluorocarbons tends to feed on itself.

Suppose that the use of fluorocarbons in industrial societies has been growing at the rate of 6% per year since 1960, when, say, 300 tons were in use, and assume that this rate of growth continues into the future.

a. How many tons of fluorocarbons will be used 15 years later, in 1975? 30 years later, in 1990?

b. Suppose that scientists are able to determine that a certain level of fluorocarbon use will have serious effects on life on earth. Once half this level of use is reached, in how many more years will this level itself be reached?

Solution

a. Let $A(t)$ denote the amount of fluorocarbons in use in year t, where $t = 0$ is 1960. Since the use of fluorocarbons grows at 6% a year, $A'(t) = 0.06A(t)$.

Therefore, as long as this model is valid, in year t, industrial societies worldwide will be using $A(t) = A(0)e^{0.06t}$ tons of fluorocarbons. Since $A(0) = 300$, then

$$A(t) = 300e^{0.06t}$$

In 1975, when $t = 15$, this means that

$$A(15) = 300e^{0.06(15)} = 300e^{0.9} \approx 300(2.4596) = 737.88 \text{ tons}$$

were used. In 1990,

$$A(30) = 300e^{1.8} \approx 1{,}814.9 \text{ tons}$$

will be used.

b. The doubling-time for this exponential growth process is given by

$$t_d = \frac{\ln 2}{0.06} \approx \frac{0.69}{0.06} = 11.5 \text{ years}$$

Therefore, once fluorocarbon use reaches half the dangerous level—whatever that level actually is—the danger will be real in only 11.5 more years, and this conclusion is independent of the initial state $A(0)$. ∎

This example illustrates how matters that feed on themselves can quickly move from a cause for concern to an actual threat. Since much of our quantitative thinking naturally tends to be *linear*—$A(t) = kt$—rather than exponential—$A(t) = e^{kt}$—it is difficult for societies to organize an appropriate response.

EXAMPLE 5

Predicting Crop Losses An agriculture specialist has been asked to observe the ongoing damage to this year's corn crop due to a microorganism that attaches itself to the plant's root system. It is a reasonable assumption that the organism's population is growing exponentially, because population growth in a supportive environment naturally feeds on itself—parents give birth to offspring at a rate proportional to the population of parents.

Because the spreading organism is affecting the productivity of the corn crop, the specialist's task is to predict the size of the harvest 5 weeks from now. The specialist chooses a test plot and by examination determines that of the 300 plants in this test plot, 35 have been attacked by the organism. One week later the specialist examines the plot again and determines that 50 plants are now affected. Assuming that the organism's population is growing exponentially and then spreading exponentially through the corn crop,

a. Predict the number of plants that will be affected by the time of harvest, 5 weeks after the first testing.

b. If previous infestations of the organism are known to decrease a plant's yield by 40%, what level of corn production should the specialist predict to the Department of Agriculture?

Solution

a. Let $A(t)$ equal the number of affected corn plants in the test plot by week t, $0 \leq t \leq 5$.

Since the organism's population and hence its effect are growing exponentially, $A'(t) = kA(t)$ for some unknown constant k. Therefore

$$A(t) = A(0)e^{kt} = 35e^{kt}$$

The second testing by the specialist has determined that $A(1) = 50$. Therefore

$$50 = A(1) = 35e^{k \cdot 1}$$

or
$$e^k = \frac{50}{35} = \frac{10}{7}$$

Hence
$$k = \ln \frac{10}{7} \simeq 0.357$$

It appears that the number of affected plants t weeks later will be
$$A(t) = 35e^{0.357t}$$

Hence, at the time of harvest 5 weeks later, there will be $A(5) = 35e^{0.356(5)} \simeq 35(5.93) = 207.55$ plants whose harvest has been affected.

b. Let N be the normal production of a single cornstalk. According to part (a), at the time of harvest there will be $300 - 208 = 92$ healthy plants producing N units of corn each and 208 affected plants producing only $(1 - 0.40)N = 0.6N$. Therefore, whereas we would normally expect corn production from this test plot to be $300N$, it will instead be $92N + 208(0.6N) \simeq 217N$.

Consequently, the specialist should predict that the corn crop will only be approximately
$$\frac{217N}{300N} \simeq 0.72$$

or 72% of the normal level in the test plot.

Thus, if the test plot is a good representative of the overall problem in the county, the specialist should report to the Department that the corn crop in this county will be approximately 28% below normal due to this microorganism. ∎

Notice that the solution to this problem is a bit different. Rather than being given the growth constant k, we determine k from *two* observations of the state of the system. Our next example shows that there is yet a third possibility: Given the growth constant k and a future state $A(t)$, we can determine the initial state $A(0)$.

Both businesses and individuals routinely face the following kind of problem: What is the real value *now* of an amount A that will be received T years in the future? For example, a business may invest $100,000 in new equipment now and expect a return of $200,000 on this investment 4 years from now. Is this a wise investment? Or an individual may buy a piece of real estate today for $50,000 and expect to sell it in 10 years for $80,000. Is this a good investment? How can $80,000 ten years from now be compared to $50,000 in hand today?

The key concept in understanding such (oversimplified, but basic) transactions is the concept of the **present value of future income.** Suppose that you enter into an agreement with a second party who agrees to pay you A dollars T years from now in return for some action on your part today, time $t = 0$. The question you must ask yourself is: What is A dollars received T years from now worth right now? To analyze this situation, imagine that you were instead paid right now (time $t = 0$) an amount $A(0)$. If you could invest this sum at interest rate k, then the amount $A(t)$ you would have at any time in the future grows according to the differential equation $A'(t) = kA(t)$ with solution $A(t) = A(0)e^{kt}$. Therefore, if instead you are to receive A dollars T years from now, you should expect that

$$A = A(T) = A(0)e^{kT}$$

Hence
$$A(0) = Ae^{-kT}$$

That is, Ae^{-kT} is the amount of money you would need *now* in order to earn A dollars T years from now. In summary,

---

The *present value of a future income* of A dollars received T years in the future, at continuously compounded interest rate k, is

$$Ae^{-kT}$$

EXAMPLE 6: *The Present Value of Future Income*

a. At an available interest rate of 12%, is $200,000 received 4 years from now worth an investment of $100,000 right now?

b. Is the sale at $80,000 today of a real estate investment costing $50,000 ten years earlier a profit or a loss if the available interest rate over this time period is 8.5%?

Solution

a. The present value of $200,000 received $T = 4$ years from now at an interest rate $k = 0.12$ is

$$200{,}000 e^{-0.12(4)} = \$123{,}756.68$$

Since this is more than the $100,000 initially needed to earn the $200,000, this is a wise investment.

b. The present value of $80,000 received $T = 10$ years from now at growth rate $k = 0.085$ is

$$80{,}000 e^{-0.085(10)} = \$34{,}193.19$$

Since this is (far) less than the $50,000 originally invested in this piece of real estate, one has taken a decided loss in a very real sense. ■

There is, of course, another way to look at, say, part b. We could have invested the original $50,000 at 8.5% and in 10 years have a return of

$$50{,}000 e^{0.085(10)} = \$116{,}982.34$$

This sum, being considerably more than $80,000, suggests that it would be better to invest the original $50,000 at 8.5% rather than buy the real estate.

This alternate (time-forward rather than time-backward) view is not used by economists, because it is difficult to use in more involved situations (for example, comparing alternate investments). For such reasons, the (time-backward) concept of *present value of future income* is preferred.

Before reading further, note that the previous three examples taken together illustrate the following:

In any exponential change model with

initial amount $A(0)$,
growth constant k, and
future amount $A(t)$,

knowledge of any two of these three quantities will determine the third.

Perhaps the most natural example of exponential decay is that involving radioactive elements in the environment. Radioactivity is a physical process in which a chemical substance emits, or loses, part of its physical mass through the release of what physicists call "α-particles." Since natural radioactivity is extremely low in the environment, life on earth has never needed to evolve protection against radiation; thus significant amounts of radiation are a hazard.

Let $A(t)$ denote the amount of a radioactive substance at time t. It is known from physical experiments that

$$A'(t) = -kA(t)$$

This means that a radioactive substance decays—loses part of its mass—at a rate proportional to its size. The proportionality constant k is called the **rate of decay**. As before, this differential equation has the solution

$$A(t) = A(0)e^{-kt}$$

It follows, as before, that the half-life of such a radioactive substance is

$$t_h = \frac{\ln 2}{k}$$

This means that at time $t = t_h$, half of the original amount of radioactive substance has been lost through the emission of α-particles. Some half-lives for certain radioactive elements are shown in Table 5.6.

Some forms of radioactivity, such as that due to the strontium 90 produced in nuclear explosions, are extremely harmful. Radioactivity produced by carbon 14 (^{14}C), an unstable isotope of the basic element carbon, occurs naturally; ^{14}C decays very slowly in the environment and is no more hazardous than other normal events in nature because it occurs at such a low level. In fact the radioactivity due to ^{14}C has uses, particularly in archaeology and anthropology, where we need to know the age of an object, such as a handle or bone fragment.

Table 5.6

Substance	Half-life t_h
Radium	1,620 years
Strontium 90	29 years
Barium 140	13 days
Carbon 14	5,760 years

5.5 Exponential Growth and Decay

If the object was once a part of a living organism, establishing its age is possible through the technique known as **carbon-dating.** By simple experiments we can determine the amounts of radioactive ^{14}C and ordinary carbon ^{12}C present in the environment today and thus measure the *ratio R* of ^{14}C to ^{12}C. This ratio is then assumed to be uniform in the environment in place and time and, consequently, present in a constant ratio in all *living* plants and organisms. However, when an organism dies, it ceases to eat. When this happens, the amount of carbon ^{12}C in its once living system remains the same, but the amount of ^{14}C begins to decay. Therefore the ratio R of ^{14}C to ^{12}C in its body begins to change. This observation can be used to date an object.

EXAMPLE 7

Radioactive Decay An archaeological dig in the southwestern United States unearths a stone hammer with a wooden handle. The ratio of ^{14}C to ^{12}C in the handle is found to be one-third of the normal ratio R. How old is the handle?

Solution Let $A(t)$ be the ratio of ^{14}C to ^{12}C in the handle at time t, with $t = 0$ being the time at which the tree used to make the handle was cut.

Since radioactive decay is an exponential process (and ^{12}C does not decay)

$$A'(t) = -kA(t)$$

with solution

$$A(t) = A(0)e^{-kt} = Re^{-kt}$$

since $A(0) = R$.

Since the half-life t_h of ^{14}C is $t_h = 5{,}760$ years and since $t_h = (\ln 2)/k$, we have

$$k = \frac{\ln 2}{5{,}760}$$

Let $t = T$ be the time when the handle was unearthed. Since the observed ^{14}C to ^{12}C ratio is $\frac{1}{3}R$, we know that

$$\frac{R}{3} = A(T) = Re^{-[(\ln 2)/5{,}760]T}$$

or

$$\frac{1}{3} = e^{-[(\ln 2)/5{,}760]T}$$

Taking the logarithm of both sides, we have

$$-\ln 3 = -\frac{\ln 2}{5{,}760}T$$

or

$$T = \frac{\ln 3}{\ln 2} \cdot 5{,}760 \approx 9{,}129$$

That is, the handle is approximately 9,129 years old. ∎

Notice that for the preceding analysis to be correct, we must assume that the naturally occurring ratio of ^{14}C to ^{12}C in the environment, which is R today, was also R in the past.

The examples and exercises of this section are only the first and most basic uses of the exponential function. Many more will be found in the following chapters. This function is one of the most useful in mathematics.

Exercises 5.5

Locate an instance in your field of study that can be modeled by a differential equation of exponential change, and analyze your example using this model. The following exercises may suggest ideas to you.

1. In 1626, Peter Minuit bought Manhattan Island from the Brooklyn Indians for $24. If the Indians had then invested the money in the Bank of England (or any other investment paying a steady return) at 6% interest (compounded continuously), what was the value of their holding in 1988?

2. For centuries people believed that the speed of a falling object was proportional to the distance it had fallen. Show that this cannot be true. How? Imagine a brick dropped from the top of a 500-ft-tall building. Let $A(t)$ equal the distance the brick has fallen by time t. If this old belief is true, then $A'(t) = kA(t)$. What is the consequence? (This problem illustrates why the fact that there is only one possible solution to this differential equation with initial state $A(0)$ is an important mathematical fact, now carried to the physical world.)

3. What is the doubling-time for an investment paying a 12% return? An 8% return?

4. What is the half-life for an uninvested holding of paper money at an inflation rate of 10%? Of 20%?

5. Solve the equation $A(t) = 2A(0)$ for t, using techniques from Section 5.3, where $A(t) = A(0)e^{kt}$.

6. If an individual at age 20 makes a one-time deposit of $2,000 into a retirement account paying 10% continuous interest, how much will be in the account at age 65 upon retirement?

7. The half-life of the popular drug Valium is 12 hours. That is, 12 hours after taking a 10-mg tablet of Valium, 5 mg remains in the bloodstream. If your doctor prescribes a dosage of one 10-mg tablet of Valium every 8 hours, how much Valium is in your bloodstream at the end of 7 hours? 12 hours? 16 hours? (*Note:* 20 mg of Valium is considered to be a dangerous level in that reaction time and involuntary processes in the body are noticeably slowed.)

8. When you buy your first automobile, you will probably face the decision of how large a down payment to make. Suppose you wish to buy an automobile costing $8,000. The dealer will loan you the money for a 4-year period at 12% interest. You initially decide to make a down payment of $2,000 on the car, leaving a loan of $6,000. But at the last moment you decide to buy one more option for the car at an extra cost of $100. Thus you now have a loan of $6,100 to pay. In a crude but realistic sense, this extra $100 will not be paid back until the very last loan payment, 4 years later. Thus you will essentially borrow the $100 for 4 years. How much will this extra $100 *really* cost you? That is, what is the cost of $100 at 12% interest 4 years later?

9. Photographs taken in Germany during the great inflation of the early 1920s show people going to a bakery with a wheelbarrow full of cash to buy a loaf of bread. Suppose that a loaf of bread that cost one-fourth of a deutsche mark (DM) in 1919 cost a wheelbarrow full of deutsche marks in 1922, and suppose that a wheelbarrow holds 2,400,000 DM bills. What was the German rate of inflation during this 3-year period? [*Hint:* You are given $A(0)$ and $A(3)$ and must find the growth constant.]

   Remark. The actual figures are these: On November 20, 1923, when the currency collapsed, it took 10^{11} DM to buy what 1 DM did in 1914.

10. Suppose that the inflation rate from 1970 to 1980 was 9% a year. If you happened to buy an investment in 1970 for $10,000 and you sold it in 1980 for $20,000, did you profit or lose in this transaction?

11. How many dollars should you put into a savings account today so as to have $400 saved 4 years later if the interest rate is 8%?

12. A home bought in 1970 at a price of $30,000 was selling for $90,000 12 years later. If its value continues to appreciate at this rate, when will it sell for $120,000? At an annual inflation rate of 9% during the period 1970 to 1982, did the home really increase in value?

13. Suppose you invest $2,000 in real estate earning a return of 12% a year and $4,000 in stocks earning 8% a year. When will the two investments be equal in value?

14. In exchange for a favor, a friend has agreed to pay you $75 three years from now or $90 five years from now. Which is the better offer if the available interest rate over this period is 8%?

15. Suppose you have a great new idea and in one week convince two friends how absolutely right your idea is. During the next week they each convince two more friends how right you are, and this process continues. In how long a time will the whole world of 4 billion people be convinced of your idea?

16. The rock group "Can-This-Be-Music!" has just released a new recording. In the first week, 15,000 copies are sold. Three weeks after its release the recording has sold a total of 90,000 copies, and further sales are feeding on its now apparent popularity. Assuming that sales are growing exponentially:

    a. Determine the growth constant.
    b. When will sales reach 1,000,000?
    c. In how many weeks will sales double?
    d. If the group is earning $5 per album sold, how much will they have earned at the end of 6 weeks? 9 weeks?
    e. Can this go on forever? What does this suggest about the validity of exponential models?

17. If the discovery of new oil reserves is *falling* at the rate of 5% a year, in how many years will the amount of newly discovered oil be only one-half the amount discovered this year? One-third this amount?

    Remark. It is not necessary to know how much was discovered this year to solve this problem; assume a hypothetical amount $A(0)$.

18. a. Draw a rectangular coordinate system, mark an initial amount $A(0)$ on the y-axis, choose any fixed length of time T on the x-axis, and graph a doubling of the amount $A(0)$ at time T. Graph twice this amount at $2T$, and continue this way. Compare the shape of this graph with an exponential growth curve.

    b. Show that if $A(t) = A(0)e^{kt}$ and $t_d = (\ln 2)/k$, then $A(t + t_d) = 2A(t)$ for any time t.

19. The Rocky Soil Vineyard, which has decided to upgrade the quality of its wines, invests $40,000 in the planting of Merlot grapes. The vineyard expects to earn its first return on the planting 8 years later and estimates this return will be $25,000. If the interest rate during this period was 9%, what proportion of its original investment has been returned in this first year of harvest? [*Hint:* Compare the *present value* of its harvest with the original investment.]

20. It is often said that the total of human knowledge doubles every 10 years. What does this imply about the percentage of new knowledge generated by the world's population each year, if we assume that the growth of knowledge feeds on itself?

21. Suppose that you know 50% of what is known in your field today and that you are continually increasing the amount you know by 20% a year. General knowledge in your field is itself increasing at the rate of 8% a year. In how many years will you know 95% of what there is to know in your field?

22. A colony of *E. coli* bacteria can grow very rapidly under favorable conditions, having a doubling-time of 20 minutes. When will a colony beginning with 1,500 bacteria reach a population of 15,000 bacteria? If this exponential model were valid for as long as 3 days, what would you conclude?

23. A physician, faced with the usual year-end task of deciding where to invest surplus earnings so as to have a "tax loss" for the year, has the choice of investing $15,000 in either (a) a tree farm that will produce Christmas trees, which will be sold 7 years from now for a total return of $50,000, or (b) an oil-drilling operation that will earn the doctor $36,000 in 5 years. If the interest rate is 13%, which of these investments compares best with the original investment? [*Hint:* Compute the *present value* of each investment.] What is the "real-dollar" profit in each of these ventures?

24. If the world population is growing at the rate of 2% a year, in how many years will we be twice as crowded as we are now? Three times as crowded? Four times as crowded?

25. A bacteriologist is studying the spread of a new strain of harmful bacteria in the cherry trees of the Okanagan Valley of British Columbia. Field samples taken 10 days apart show that the percentage of affected plants has grown from 15% to 23%. If the spread of this bacteria is feeding on itself, when will 50% of the cherry trees be affected? If the cherries can be harvested at that time and each affected tree has a 30% loss of normal production, by how much will the normal harvest be reduced?

26. Scientists are studying a new radioactive element. Their initial sample weighs 20 mg. Thirty days later it weighs 15 mg. What is its half-life?

27. The half-life of the radioactive element strontium 90 is 28 years. In a small nuclear war, an acre of Kansas wheatland 400 miles east of Denver, Colorado, receives a dose of strontium 90 from the nuclear explosion that destroyed Denver. If this dose is five times the "safe" level for human habitation, when will those who survive again be able to make bread from this particular acre of land?

28. A bone fragment found in the Olduvai Gorge in Africa has a $^{14}C:^{12}C$ ratio that is 0.01% of the normal ratio. How old is the bone fragment?

29. a. An accident at a nuclear power plant results in the release of the radioactive elements iodine 131 (half-life of 8 days) and strontium 90 (half-life of 28 years) into the dairy farming region downwind from the plant. When will 90% of the iodine 131 be gone due to radioactive decay?

    b. Strontium 90 makes its way into the milk produced by the local dairy industry and is shipped to a nearby city, where it is drunk by the inhabitants. In how many years will 80% of (whatever) the original amount of strontium 90 ingested by those inhabitants still remain in their bodies, assuming it is lost only by decay?

30. Practically anyone alive in the early 1960s ingested some of the strontium 90 produced in the atmospheric testing of nuclear bombs, since strontium 90 was spread worldwide by the winds of the upper atmosphere. What proportion of strontium 90 ingested by your parents in 1960 remains in their bones today?

31. An article by James Fallows in the *Atlantic Monthly* reported that the cost of fighter aircraft purchased by the U.S. Air Force has increased by 36.8% *per year* since the end of World War II in 1945. (For example, in the early 1970s the F-5 fighter cost $4 million apiece, while in 1980 the F-14 cost $30 million.) In 1982, the Air Force bought 230 F-14 fighters. If the budget for fighter aircraft increases by 10% a year for the foreseeable future and if the cost of a single fighter continues to increase at the level of its historical trend, in how many years will the entire budget buy only one aircraft? (See also *Newsweek*, Feb. 22, 1988, p. 63, for a related conclusion.)

32. If an initial amount A_0 is deposited in a savings account paying an annual interest rate r, one year later the amount in the account will be

$$A_1 = A_0 + rA_0 = (1 + r)A_0$$

Two years later the amount will be

$$A_2 = A_1 + rA_1 = (1 + r)A_1 = (1 + r)^2 A_0$$

Continuing this way, n years later, the amount will be

$$A_n = (1 + r)^n A_0$$

Suppose that $A_0 = \$500$ at rate $r = 0.08$ (8% interest).

a. Compute the amount A_3 in the account 3 years later.

b. Suppose that at another bank the interest rate is 8% but is compounded continuously. With $A_0 = \$500$, how much will be in the account 3 years later?

c. Comparing your answers to (a) and (b), what do you conclude?

Challenge Problems

33. Recall from Section 5.1 that

$$e = \lim_{n \to \infty}\left(1 + \frac{1}{n}\right)^n$$

a. Using the substitution $n = m/r$, show that

$$\lim_{m \to \infty}\left(1 + \frac{r}{m}\right)^m = e^r$$

b. If $r = 0.08$ and $m = 4$ in the expression in (a), interpret the quantity

$$500\left(1 + \frac{r}{m}\right)^m$$

as the amount of money in a savings account that was opened with $500 and earned 8% interest compounded quarterly. [*Hint:* Reason as in the opening explanation of Exercise 32.]

c. How would you interpret

$$500\left(1 + \frac{0.08}{365}\right)^{365}$$

d. How do (a), (b), and (c) suggest the term "interest compounded *continuously*"?

34. This exercise compares linear growth to exponential growth in a Malthusian fashion. The English economist Thomas Malthus (1766–1834) hypothesized that food production tended to increase arithmetically while population tended to increase geometrically.

    Suppose that a village in southern Asia has already cleared 30 acres of land and, owing to limited access to arable land, is clearing 2 more acres per year for food production. Suppose that the population of the village is 100 individuals, each needing one-quarter acre of cleared land per year for a food supply. Finally, suppose that the population of the village is increasing at the rate of 3% per year. When will the population outstrip the cleared land available for food production? [*Hint:* To solve this problem, you will need a calculator to help you find an approximate solution to an equation of the form $e^{0.03t} = at + b$.]

Chapter 5 Summary

1. The exponential function $y = e^x$ is invariant under differentiation and integration: $D_x e^x = e^x$ and $\int e^x \, dx = e^x + C$. More generally,
$$D_x e^{g(x)} = e^{g(x)} \cdot g'(x)$$

2. The function $f(x) = \ln x$ is defined for all $x > 0$ by $y = \ln x$ if and only if $x = e^y$. Consequently, $\ln e^x = x$ and $e^{\ln x} = x$.

3. The function $y = \ln x$ grows in inverse proportion to x: $D_x \ln|x| = 1/x$ and $\int (1/x) \, dx = \ln|x| + C$, $x \neq 0$.

4. The graphs of $y = e^x$ and $y = \ln x$ are shown in Figure 5.16.

5. If the rate of change of an amount $A(t)$ at time t is directly proportional to the amount $A(t)$, then $A'(t) = kA(t)$ and, consequently, the amount $A(t)$ changes exponentially:
$$A(t) = A(0)e^{kt} \quad \text{for } t \geq 0$$

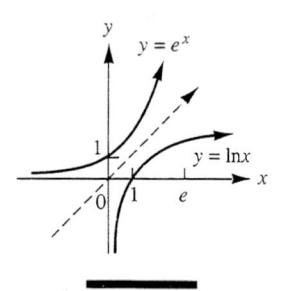

Figure 5.16

6. The number $T = (\ln 2)/k$ is the doubling-time for exponential growth, modeled by $A'(t) = kA(t)$, and the half-life for exponential decay, modeled by $A'(t) = -kA(t)$, where $k > 0$.

Chapter 5 Summary Exercises

Evaluate or simplify each expression in Exercises 1–12.

1. $27^{2/3}$
2. $(3^2)^{-2}$
3. $(8^{-1/2})^2$
4. $e^{0.04(50)}$
5. $\dfrac{e^{2x}}{e^x}$
6. $(e^x - e^{-x})(e^x + e^{-x})$
7. $\dfrac{e^x + e^{3x}}{e^x}$
8. $\ln \sqrt{2}$
9. $\ln 3e$
10. $\ln\left(\dfrac{2}{e}\right)$
11. $\ln e^{15}$
12. $\ln(x^2 - 1) - \ln(x - 1)$

Find the derivative of each function in Exercises 13–29.

13. xe^{-x}
14. e^{kt}
15. $\ln e^t$
16. $e^x + \ln x$
17. $\dfrac{1-x}{e^x}$
18. $\ln(x^2 - 1)$
19. $e^t + e^{-t}$
20. $t^3 e^{-t/3}$
21. $e^x \ln x$
22. $\sqrt{\ln x}$
23. $\dfrac{\ln x}{e^x}$
24. $e^{\ln x^2}$
25. $\ln x^x$
26. $\dfrac{\ln x - 1}{\ln x + 1}$
27. $\ln\left(\dfrac{1 + x^2}{e^x}\right)$
28. $\ln|1 - t^2|$
29. $(\ln x)^{1/2}$

Find the integrals in Exercises 30–45.

30. $\int_0^{20} e^{-0.05t} \, dt$

31. $\int e^x \sqrt{(1 - e^x)} \, dx$

32. $\int u e^{u^2} \, du$

33. $\int_0^1 \frac{2x}{x^2 + 1} \, dx$

34. $\int 6x^4 e^{x^5} \, dx$

35. $\int_0^1 t e^{(1-t^2)/2} \, dt$

36. $\int \frac{e^x}{1 - e^x} \, dx$

37. $\int \frac{1}{1 - x} \, dx$

38. $\int_e^3 \frac{\ln x}{x} \, dx$

39. $\int \frac{2x - 1}{x^2 - x} \, dx$

40. $\int \frac{2x}{9 - x^2} \, dx$

41. $\int e^x (1 + e^x)^2 \, dx$

42. $\int x + e^x - \frac{\ln x}{x} \, dx$

43. $\int \ln e^x \, dx$

44. $\int_{-2}^{-1} \frac{1}{x - 1} \, dx$

45. $\int \frac{1}{x} e^{\ln x} \, dx$

Find the maximum and minimum values, if any, of the functions in Exercises 46–51.

46. $f(x) = e^{-x^2}$
47. $f(t) = 3t e^{-t/2}$
48. $g(x) = x - \ln x$
49. $f(x) = x \ln x^2$
50. $h(x) = x - e^{-x}$
51. $f(x) = \ln|1 - x^2|$

52. Consider the function $A(t) = Be^{kt}$.

   a. Given $B = 150$ and $k = 0.05$, determine $A(7)$.
   b. Given $A(1) = 25$ and $A(3) = 100$, determine k and $A(4)$.
   c. Given $k = 0.09$ and $A(15) = 7{,}000$, determine B and $A(0)$.
   d. Given $A(0) = 50$ and $A(3) = 6.25$, determine $A(1)$.

53. Suppose that the inflation rate is 6% a year for the next 10 years and that you buy an ounce of gold today for $500 and sell it 10 years later for $900. Would you profit in this transaction?

54. Suppose that you know 90% of all that is known about flying a kite and that you are increasing your knowledge of kite flying at the rate of 7% a year and that general knowledge about kite flying is increasing at the rate of 6% a year. When will you know 95% of all there is to know about flying a kite?

55. What is the half-life of a radioactive element that initially weighs 10 mg and 20 days later weighs 8 mg?

56. What is the doubling-time for an investment that was initially worth $1,500 and is worth $2,000 three years later?

57. What is the present value of $5,000 received 5 years from now at an interest rate of 8%?

58. Domestic copper production presently exceeds demand by $\frac{1}{10}$. If production is increasing at the rate of 3% a year and demand is increasing at the rate of 5% a year, when will demand exceed production? [*Hint:* Let D denote present demand. Then present production is $\left(1 + \frac{1}{10}\right)D$. Use these as initial amounts, and equate production and demand curves.]

59. A disease affecting the Colombian coffee crop is spreading according to the differential equation $C'(t) = 0.002C(t)$, where $C(t)$ is the number of affected coffee trees on day t. Of 100 coffee trees, initially 25 are affected. How many will be affected at harvest 180 days later? If affected trees produce only $\frac{1}{3}$ of the normal harvest, by how much will the harvest from these 100 trees be reduced?

60. A bone fragment has a $^{14}C : ^{12}C$ ratio that is $\frac{1}{15}$ of the normal ratio. How old is the bone fragment?

Chapter 6

Functions of Several Variables

Three Dimensions and Beyond

A manufacturing firm's production depends on its level of investment in two distinct entities—labor and equipment. The growth of food crops depends on three distinct chemicals in the soil: nitrogen, phosphate, and potassium. The temperature of a warming object depends on where the temperature is measured on the object and the time since warming has begun. These examples suggest that a more complete and accurate mathematical model of a system in the external world may have to be built upon a function of more than one variable. When this happens, our familiar two-dimensional representation of the graph of a function of one variable $y = f(x)$ no longer suffices, and we must picture a function of two variables $z = f(x, y)$ in a coordinate space of three dimensions. If a system in turn depends on three variables, we quickly realize that we will not be able to picture it at all, for this would require a coordinate space of four dimensions. Fortunately, calculus offers analytical methods for dealing with a function of any finite number of variables whose basic forms are independent of dimension and can largely be learned in the context of a function of only two variables, where understanding through graphical illustration remains possible.

6.1 Functions of Several Variables and Their Graphs

In this section we introduce examples of functions of more than one variable and learn how to graph a function of two variables in a three-dimensional space.

This chapter extends the methods of previous chapters to systems involving more than one variable. While we cannot directly use these methods without developing some new ideas, the guiding concepts remain the same. The value of a function of two or more variables is again a measure of the state of a system. The *partial derivatives* of such a function measure the rate of change in the system but with respect to change in only one variable at a time. The integral of a function again measures accumulation, but in multiple dimensions, and an approximation principle underlies the applications of multiple integration.

The simplest case of a function of more than one variable is, of course, a function of only two variables. We will generally deal only with this case because the basic ideas can often be immediately extended to a function of three or more variables. Even so, the difficulties encountered in studying functions of two variables are more than doubled over those encountered with a single variable. This subject is unavoidably technically demanding, and we have only the space to scratch its surface.

A function f of two variables x and y will be written as $z = f(x, y)$. Such functions will appear, for example, in the form

$$f(x, y) = x + y^2 \qquad f(x, y) = xe^{xy}$$

$$z = \sqrt{x - y} \qquad \text{or} \qquad z = x \ln y$$

We ask the reader to extend earlier concepts of a function as a correspondence and as a dependency relation—$z = f(x, y)$ depends on x and y—to this new setting without further explanation. It is important to realize that having made this step, we are not limited to any particular number of variables. Thus

$$w = x^2 + 2xyz + z + 2xy$$

and

$$f(u, v, z, w) = u - v + z - w$$

are examples of functions of three and four variables, respectively.

The techniques of function evaluation and its meaning in applications are easily extended to this setting.

EXAMPLE 1

A metal rod 20 cm long has a temperature of 80°. We begin heating the rod at one end to a temperature of 150°. The temperature t minutes later at a point x cm from the end of the rod is given by

$$f(x, t) = 70\left(1 - \frac{x}{20}\right)(1 - e^{-0.05t}) + 80$$

What is the temperature of the rod at a point 10 cm from the heated end 7 min after heating has begun? (See Figure 6.1.)

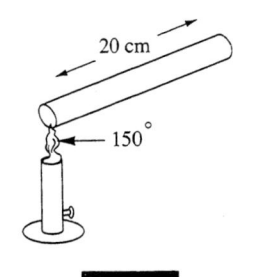

Figure 6.1

Solution We are asked to determine the temperature of the rod 10 cm from its heated end after 7 min of heating. According to the information given, this temperature is

$$f(10, 7) = 70\left(1 - \tfrac{10}{20}\right)\left(1 - e^{-0.05(7)}\right) + 80$$
$$= 70\left(\tfrac{1}{2}\right)(1 - e^{-0.35}) + 80 \approx 90.34° \blacksquare$$

The Graph of a Function $z = f(x, y)$ by the Method of Sections

Consider the function $f(x, t) = xe^{-t}$. For the fixed values $x = \tfrac{1}{2}$, $x = 1$, and $x = 2$, we can obtain three functions of t alone. These are

$$f\left(\tfrac{1}{2}, t\right) = \tfrac{1}{2}e^{-t} \qquad f(1, t) = e^{-t} \qquad f(2, t) = 2e^{-t}$$

Note that all of these are exponential decay functions with initial values $\tfrac{1}{2}$, 1, and 2 at $t = 0$, respectively. We illustrate the graphs of these three functions simultaneously in Figure 6.2.

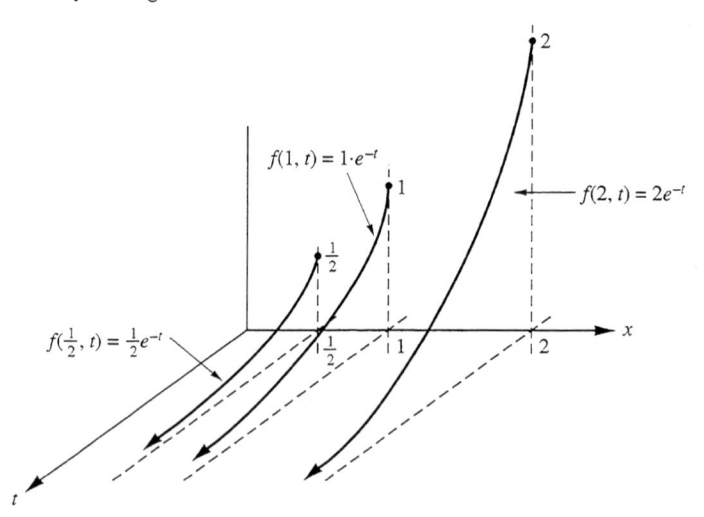

Figure 6.2

Let us do the same kind of thing again, but this time considering fixed values $t = 0$, 1, and 2, and obtaining three functions of x alone. These are

$$f(x, 0) = x \qquad f(x, 1) = xe^{-1} \qquad f(x, 2) = xe^{-2}$$

These three are linear functions of x with slope 1, $e^{-1} \approx 0.37$, and $e^{-2} \approx 0.14$, respectively. We simultaneously illustrate these three functions in Figure 6.3.

If we now superimpose Figure 6.2 upon Figure 6.3, we obtain Figure 6.4. The shaded surface in Figure 6.4 is regarded as the (partial) graph of the function

$$f(t, y) = xe^{-t}$$

considered to be a function of x and t simultaneously, for $x \geq 0$ and $t \geq 0$.

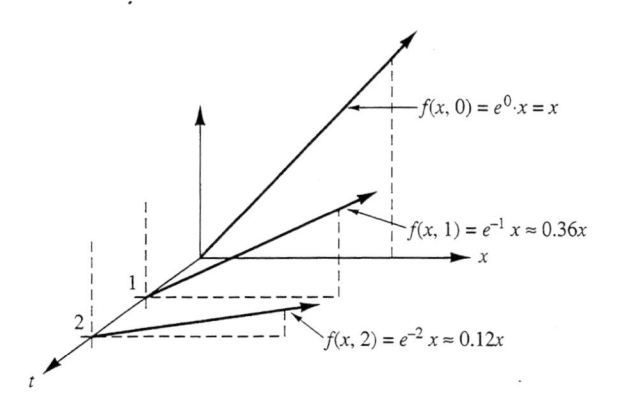

Figure 6.3

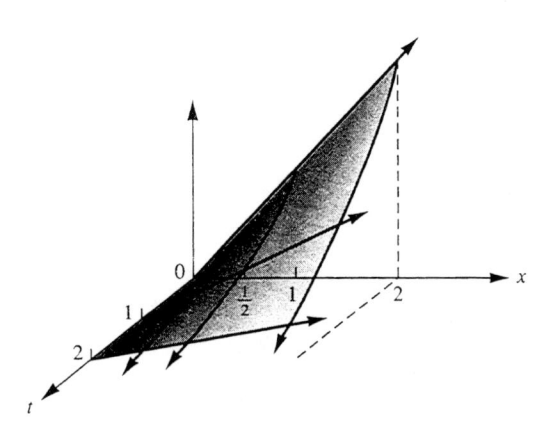

$f(x, 0) = e^0 \cdot x = x$

$f(x, 1) = e^{-1} x \approx 0.36x$

$f(x, 2) = e^{-2} x \approx 0.12x$

Figure 6.4

(The remainder of this graph, for $x < 0$, $t < 0$, and so on, is quite different in appearance.) Notice that the graph must be drawn in a three-dimensional space that utilizes a third (vertical) axis to represent the values of $z = f(x, t)$ above points (x, t) in the horizontal plane formed by the xt-axes.

This example is typical of the graph of a function $z = f(x, y)$ of two variables. The graph will be a *surface* in three-dimensional space. While it can be difficult to obtain an accurate picture of such a graph, the method we have used in obtaining Figure 6.4 is an often-used one and is known as the **method of sections**. In Figure 6.2, we obtained three sections of the graph of f in the t-direction by fixing three values of x (at $\frac{1}{2}$, 1, and 2 on an x-axis) to obtain three functions of the one variable t that we can graph in vertical planes, positioned at the points $x = \frac{1}{2}$, 1, and 2. Each such section is called an **x-section** of the graph of f. A graph in three dimensions has infinitely many x-sections, one for each point x on an x-axis. In practice, graphs of a few of these sections often suggest the entire graph.

We obtain Figure 6.3 in the same manner but instead draw three sections of the graph of f in the x-direction; each of these is called a **t-section** of the graph of f. When we envision Figures 6.2 and 6.3 together, we are able to sketch a full (three-dimensional) picture of the graph of f.

6.1 Functions of Several Variables and Their Graphs

Naturally, then, we realize the need for a *three-dimensional coordinate space* if we are to graph functions of two variables (just as in earlier chapters, we needed a *two*-dimensional space to graph functions of *one* variable). The standard illustration of three-dimensional space is indicated in Figure 6.5, using three axes, two horizontal (*x* and *y*) and one vertical (*z*).

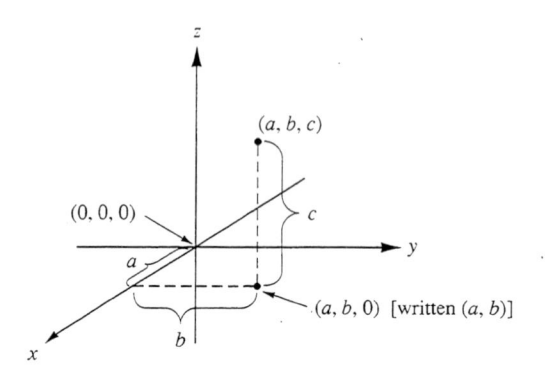

Figure 6.5

Each point in three-dimensional space is fixed by three coordinates (a, b, c) measured in the *xyz*-directions from the origin $(0, 0, 0)$. Conversely, any such measurement parallel to the three axes locates a unique point in three-dimensional space.

In relation to Figure 6.5, a function *f* of the two variables *x* and *y* is traditionally written as $z = f(x, y)$. Thus the first variable, here called *x*, is represented by the axis pointing "outward," and the second variable *y* is represented by the axis pointing to the right. A function value $f(x, y)$, computed at a point (x, y) in the horizontal plane, is then represented by measurement in the vertical direction corresponding to the *z*-axis and above the point (x, y). Finally, the graph of $z = f(x, y)$ is the set of all points $(x, y, f(x, y))$ drawn in such a three-dimensional space.

Three Basic Graphs and Surfaces in Three-Dimensional Space

While it is not easy to sketch the graph of a function of two variables, it is worthwhile to have some familiarity with the three particular shapes that arise from three particular functional forms.

EXAMPLE 2

Sketch the graph of the function $f(x, y) = x^2 + y^2$, and indicate the point on this graph corresponding to $(x, y) = (1, 1)$.

Solution Fixing $y = -1, 0,$ and 1, we obtain three functions

$$f(x, -1) = x^2 + (-1)^2 = x^2 + 1 \qquad f(x, 0) = x^2 + (0)^2 = x^2$$

and
$$f(x, 1) = x^2 + (1)^2 = x^2 + 1$$

of *x* alone. These three each yield a parabolic graph in two dimensions, with vertex at height 1, 0, and 1, respectively, as shown in Figure 6.6.

326 Chapter 6 Functions of Several Variables

If we repeat the same analysis by fixing $x = -1$, 0, and 1, we obtain

$$f(-1, y) = 1 + y^2 \qquad f(0, y) = y^2 \qquad \text{and} \qquad f(1, y) = 1 + y^2$$

whose graphs yield parabolic sections in the alternate direction, as again indicated in Figure 6.6. These sections allow us to conclude that the graph of $f(x, y) = x^2 + y^2$ is as shown in Figure 6.6, where we also indicate that $f(1, 1) = 1^2 + 1^2 = 2$. This surface is called a **paraboloid.**

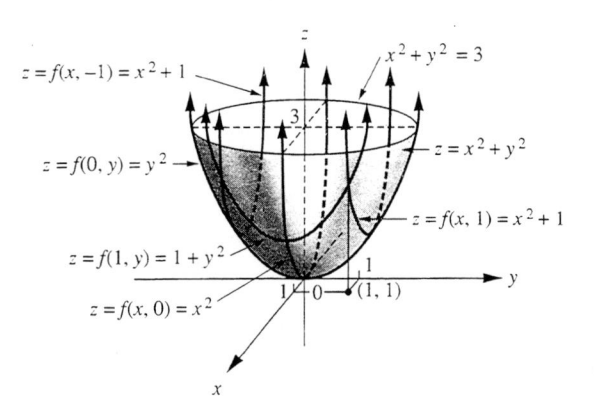

Figure 6.6

Notice in Figure 6.6 that the horizontal cross section of the paraboloid at height $z = 3$ is a circle. This is reflected by the equation

$$f(x, y) = x^2 + y^2 = 3$$

which is the equation of a circle of radius $\sqrt{3}$ centered at $(0, 0)$.

An alternate method of graphing is particularly useful for *linear functions* of x and y. Such a linear function occurs in a form like $f(x, y) = 2x + y + 1$, where x and y appear to a single power. Linear functions of two variables are analogous to linear functions of one variable [for example, $g(x) = 2x + 1$, whose graph is a *line* in an xy-coordinate system]. The graph of the function $f(x, y) = 2x + y + 1$ will be a *plane* in an xyz-coordinate system. Since a plane is determined by any three noncolinear points lying on it, in many cases we need only find the points (x, y, z) on the surface of $z = 2x + y + 1$ where the graph intercepts the x, y, and z axes, respectively, as indicated in Figure 6.7. [An exception, of course, is the case of a plane that is parallel to an axis; in such a case, one (or more) variables may not appear in the equation describing the plane. Alternatively, the plane may intercept all axes at the same point $(0, 0, 0)$.] Notice that such intercepts always have two coordinates equal to zero. We use this observation in the next example.

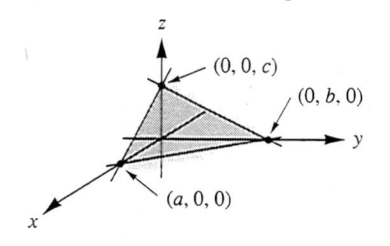

Figure 6.7

6.1 Functions of Several Variables and Their Graphs

EXAMPLE 3

Graph $f(x, y) = 1 - x - \frac{1}{2}y$, and indicate the point on this graph corresponding to $(x, y) = (1, 1)$.

Solution Let us think of this function in the form of an equation:

$$z = 1 - x - \frac{1}{2}y$$

According to Figure 6.7, we can obtain the x-intercept of the graph by setting $z = 0$ and $y = 0$ in this equation. We have

$$0 = 1 - x - \left(\frac{1}{2} \cdot 0\right)$$

or

$$x = 1$$

We indicate this intercept in Figure 6.8 at the point $(x, y, z) = (1, 0, 0)$.

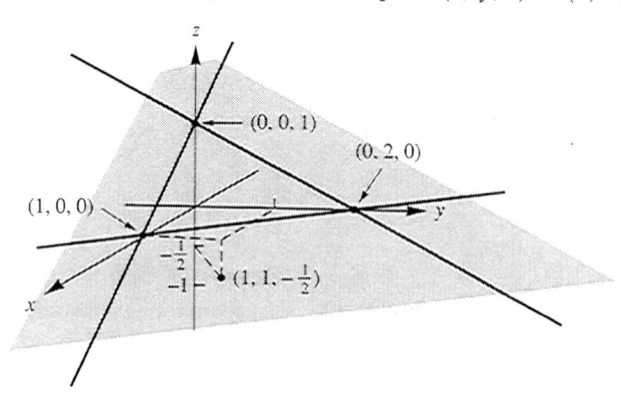

Figure 6.8

The y-intercept is obtained by setting $z = x = 0$ to obtain

$$0 = 1 - 0 - \frac{1}{2}y$$

or

$$y = 2$$

We indicate this solution in Figure 6.8 at the point $(0, 2, 0)$.

Finally, setting $x = y = 0$, we obtain

$$z = 1 - 0 - \left(\frac{1}{2} \cdot 0\right) = 1$$

This intercept, $(x, y, z) = (0, 0, 1)$, is also shown in Figure 6.8. We then sketch straight lines connecting these points so as to indicate the planar shape of the graph of

$$z = 1 - x - \frac{1}{2}y$$

containing these three points.

Now note that $f(1, 1) = -\frac{1}{2}$ and that the location of this point, $(x, y, z) = \left(1, 1, -\frac{1}{2}\right)$, is consistent with the planar shape of the graph. ∎

We conclude this section with the graph of a **hyperbolic paraboloid,** shown in Figure 6.9. This graph is that of the function $f(x, y) = y^2 - x^2$. The existence of such a graph is important because of the peculiar role played by the point $(0, 0, 0)$. Notice that in the vertical plane containing the yz-axes,

328 Chapter 6 Functions of Several Variables

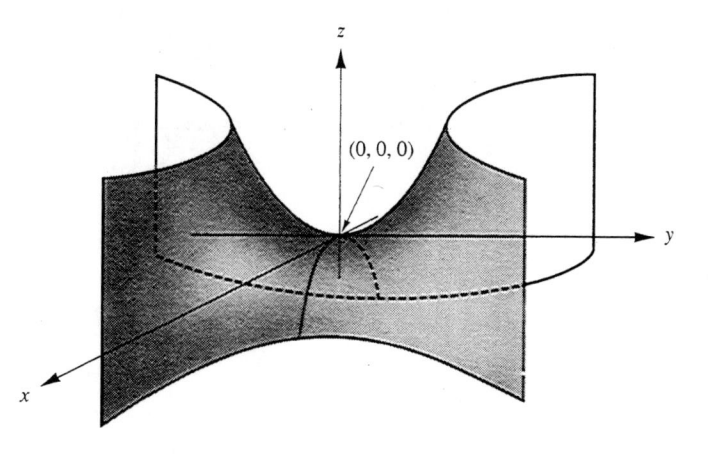

Figure 6.9

this point is a *minimum* value on the parabolic section in that plane. In the vertical plane of the *xz*-axes, the same point is a *maximum* value on the parabolic section in that plane. Such a point is called a *minimax*, or saddle, point and has a significant influence on the techniques available for finding the maximum and minimum values of a function of two variables.

Students frequently ask, "What about the graph of a function $w = f(x, y, z)$ of *three* variables?" Since we are unable to visualize a space of four dimensions, such graphs are unavailable. That is, to picture, say, the relationship $w = x + 2y + 4z$, we would first need three axes to represent the variables x, y, and z and then a fourth to represent a resulting value w.

Using a computer, we can now obtain three-dimensional graphs of complicated functions and shapes rather easily. An example of such a computer-generated graph is shown in Figure 6.10 for the function $f(x, y) = y/(x^2 + y^2)$, x or $y \neq 0$.

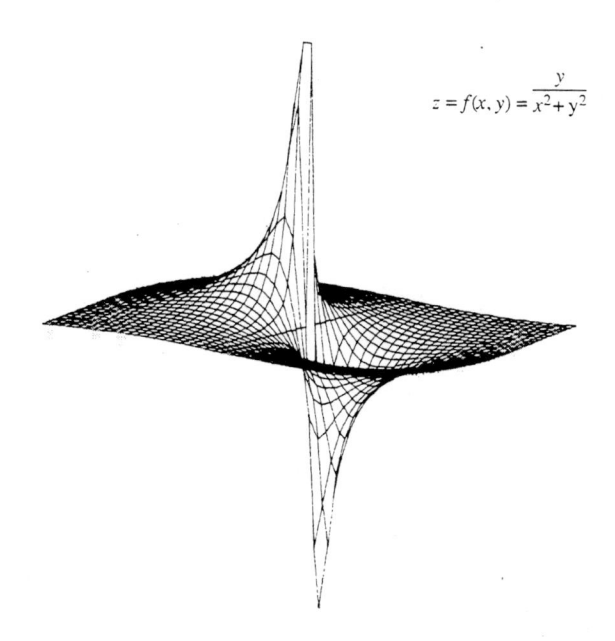

Figure 6.10

6.1 Functions of Several Variables and Their Graphs

Exercises 6.1

Evaluate each of the functions in Exercises 1–6 at the three points $(1, -1)$, $(0, 1)$, and $(4, 0)$, if possible.

1. $f(x, y) = 2x + 3y - 7$
2. $f(x, y) = \dfrac{xy}{1 + y^2}$
3. $g(x, y) = \dfrac{x + y}{x - y}$
4. $g(u, v) = \ln ue^v$
5. $h(s, t) = s^2 - t^2 + e^{st}$
6. $F(w, z) = \sqrt{w + z^2}$

7. Plot each of the given points (x, y, z) in a three-dimensional coordinate system: $(0, 0, 0)$; $(0, 1, 0)$; $(-1, 0, 0)$; $(0, 0, -1)$; $(0, 0, 1)$; $(1, 1, 0)$; $(0, 1, 1)$; $(1, 0, 1)$; $(1, 2, 3)$; $(1, -1, 2)$; $(1, -2, -3)$; $(-2, -2, -2)$.

8. Try to draw a plane in three dimensions that contains the points $(2, 0, 0)$; $(0, 1, 0)$; $(0, 0, 3)$.

Sketch the graph of each plane defined by the functions in Exercises 9–12.

9. $f(x, y) = x + 2y - 1$
10. $g(x, y) = x - y - 2$
11. $h(s, t) = 2 - t + 2s$
12. $R(U, V) = 1 - U - V$

13. Please refer to Figure 6.6. Consider the section drawn there corresponding to $y = 0$ along with the function $f(x, 0) = x^2$. Determine by any method you like the slope of a line tangent to the sectional curve above the point $(1, 0)$ in the xy-plane.

14. Sketch the graph of $f(x, y) = y(1 - e^{-x})$ for $x > 0$ and $y > 0$.

15. Sketch the four sections of $f(x, y) = x^2 + y$ in the x-direction (in vertical planes located) at $y = -1, 0, 1$, and 2. Then attempt to sketch the graph of f itself in three-dimensional space. Is this graph consistent with sections of the graph in the y-direction in the vertical planes located at $x = 1, 0, -1$?

16. How would the graph of $g(x, y) = 2x^2 + \tfrac{1}{2}y$ differ from that obtained in Exercise 15?

17. Graph $f(x, y) = 2x^2 + \tfrac{1}{2}y^2$ using Example 2 as a model.

18. Sketch three sections of $f(x, y) = xy$ in each of the xy-directions. Can you then sketch the graph of f itself? If not, try plotting a few points of the graph, say, for $(x, y) = (1, 1), (1, -1), (-1, 1)$, and $(-1, -1)$.

19. Sketch the graph of $f(x, y) = \ln y$. (This is especially simple since all x-sections look the same.)

20. Graph $f(x, y) = \sqrt{4 - (x^2 + y^2)}$ for $x^2 + y^2 \leq 4$.

21. Graph $f(x, y) = y^2 - x^2 + 3$ using the discussion accompanying Figure 6.9.

22. Graph $g(x, y) = x^2 - y^2$.

23. Graph $f(x, y) = 3 - (x^2 + y^2)$. [Hint: Use Example 2.]

24. Graph $f(x, y) = x^2 + (y - 1)^2$. [Hint: Recall the idea of translation (Section 0.3) and use Example 2.]

25. Graph $f(x, y) = x/y$, for $x \geq 0$, $y > 0$, using x-sections at $x = 0, 1$, and 2 and using y-sections at $y = 1, 2$, and 3.

26. Graph $f(x, y) = x/y$ for $y > 0$ and all possible x.

27. It costs a container manufacturer $2 to produce a bottle and $1 to produce a can, and the company has a fixed cost of production of $100. This yields the cost function

$$C(x, y) = 100 + 2x + y$$

for the production of x bottles and y cans. Shorten the scale on the z-axis so that one unit represents $100, and graph the function $z = C(x, y)$.

28. By selling coffee at x dollars a pound and tea at y dollars a pound, a firm finds that the demand for coffee and tea is $C = 150 - 2x + 6y$ and $T = 200 - y + x$ pounds, respectively. The firm's total revenue is therefore

$$R(x, y) = xC + yT$$
$$= x(150 - 2x + 6y) + y(200 - y + x)$$

Determine total revenue when coffee sells for $2 a pound and tea for $3 a pound; for $1.50 and $2, respectively.

29. Sketch the y-section of the graph of the revenue function in Exercise 28 at $y = $3 per pound of tea. Locate the value of $R(2, 3)$ along the graph of this section. What does this value represent? What do other values along this section represent? What do values of R along the x-section, where $x = 3$, represent?

Computer Application Problems

Use the BASIC program FUNVAL2 to sketch a graph of the functions in Exercises 30–33. This program will print values of $f(x, y)$ along y-sections that you specify. [Enter e^x in BASIC as $\exp(x)$.]

30. $f(x, y) = x^2 + y^2$ for $-1 \leq x \leq 1$ and $-2 \leq y \leq 2$
31. $f(x, y) = yx^2$ for $-1 \leq x \leq 1$ and $0 \leq y \leq 2$
32. $f(x, y) = e^{-(x/y)^2}$ for all $x \geq 0$ and for $0 \leq y \leq 2$
33. $f(x, y) = 1 - (x - y)^2$ for all x and for $0 \leq y \leq 2$

6.2 Partial Differentiation

In this section we study the differentiation of a function of two (or more) variables with respect to each variable separately.

Consider the cost $C(x, y)$ to produce an item that requires x units of labor and y units of materials for its manufacture, where labor costs \$2 per unit and materials cost \$4.50 per unit. The total cost is then $C(x, y) = 2x + 4.5y$ dollars. If one additional unit of labor is used for production and materials remain constant, what effect will this have on the cost of production? Since each unit of labor costs \$2, the answer is obviously \$2. But note that this answer coincides with two particular calculations:

$$\frac{C(x + 1, y) - C(x, y)}{1} = \frac{2(x + 1) + 4.5y - (2x + 4.5y)}{1} = \$2 \quad (1)$$

and
$$D_x(2x + 4.5y) = 2(1) + 4.5(0) = 2 \quad (2)$$

where, in the differentiation in Equation 2, we regard y (the number of units of materials) as a constant.

Notice the close relationship between Equations 1 and 2. In Equation 1, we see a difference quotient that reminds us of the formula defining a derivative, but we also see that *no change* is made in the variable y. In Equation 2, we see that if we regard y as a *constant*, then $D_x(2x + 4.5y) = 2$. This observation underlies the definition of the **partial derivative(s)** of a function of two variables.

Definition

If $z = f(x, y)$ is a function of two variables x and y, we define

1. The *partial derivative* of f with respect to x as

$$\lim_{h \to 0} \frac{f(x + h, y) - f(x, y)}{h}$$

regarding y as a constant.

2. The *partial derivative* of f with respect to y as

$$\lim_{k \to 0} \frac{f(x, y + k) - f(x, y)}{k}$$

regarding x as a constant.

We denote these limits by $\dfrac{\partial f}{\partial x}$ and $\dfrac{\partial f}{\partial y}$ $\left(\text{or alternatively by } \dfrac{\partial z}{\partial x} \text{ and } \dfrac{\partial z}{\partial y}\right)$, respectively.

Notice that in the limit defining $\partial f/\partial x$, the variable y does not change. This allows us to treat y as a constant when we differentiate. The same may be said of the variable x in the limit defining $\partial f/\partial y$. We make use of this in finding $\partial f/\partial x$ and $\partial f/\partial y$ in Example 1. Following that example, we will give geometrical meaning to these definitions.

EXAMPLE 1

Find $\partial f/\partial x$ and $\partial f/\partial y$ for $f(x, y) = x^2 + y^3$.

Solution To find $\partial f/\partial x$, we simply differentiate $x^2 + y^3$ with respect to x in the usual way *and* regard y, and thus y^3, as a constant. We have, in appropriate notation,

$$\frac{\partial f}{\partial x} = \frac{\partial}{\partial x}(x^2 + y^3) = \frac{\partial}{\partial x}x^2 + \frac{\partial}{\partial x}y^3 = 2x + 0 = 2x$$

To find $\partial f/\partial y$, we treat x^2 as a constant and differentiate by y in the usual way, obtaining

$$\frac{\partial f}{\partial y} = \frac{\partial}{\partial y}(x^2 + y^3) = \frac{\partial}{\partial y}x^2 + \frac{\partial}{\partial y}y^3 = 0 + 3y^2 = 3y^2 \quad \blacksquare$$

EXAMPLE 2

Find $\partial f/\partial x$ and $\partial f/\partial y$ for $f(x, y) = ye^{-x}$.

Solution Treating y as a constant, we have

$$\frac{\partial f}{\partial x} = \frac{\partial}{\partial x}(ye^{-x}) = y\frac{\partial}{\partial x}e^{-x} = y(-e^{-x}) = -ye^{-x}$$

Similarly, treating x as a constant and hence also e^{-x} as a constant, we have

$$\frac{\partial f}{\partial y} = \frac{\partial}{\partial y}ye^{-x} = e^{-x}\frac{\partial}{\partial y}y = e^{-x} \cdot 1 = e^{-x} \quad \blacksquare$$

Let us now try to understand what $\partial f/\partial x$ and $\partial f/\partial y$ measure geometrically. Consider the graph of $f(x, y) = ye^{-x}$ for $x, y \geq 0$ in Figure 6.11. In the x-direction the graph is decaying exponentially. Thus its slope is negative and is always changing. But as we saw in Figure 6.3, in the y-direction the graph is growing linearly. At, say, $x = 1$, this linear graph has the equation $z = f(1, y) = ye^{-1} \approx 0.37y$ with slope approximately 0.37.

Now, from Example 2, $\partial f/\partial x = -ye^{-x}$ is defined by

$$\lim_{h \to 0} \frac{f(x + h, y) - f(x, y)}{h}$$

Let us interpret this limit in Figure 6.11. We treat y as a constant by fixing a particular value y on the y-axis. Consider the *vertical plane* at the point y and the section of the graph of f in this plane, as indicated in Figure 6.11. The difference quotient

$$\frac{f(x + h, y) - f(x, y)}{h}$$

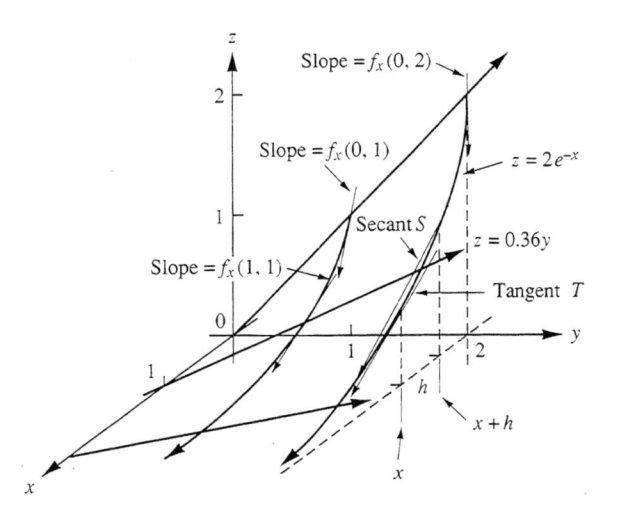

Figure 6.11

is the slope of the secant line S in Figure 6.11 in this vertical plane. If we now imagine letting $h \to 0$ along the horizontal axis, we see that the limit value $\partial f/\partial x$ is the slope of the tangent line T above the point (x, y) in this plane. As found in Example 2, this slope is $-ye^{-x}$ and is both negative (as it should be) and dependent on both x and y. For instance, note that at $x = 0$ this slope is "more negative" than it is, say, at $x = 1$, and that it is more negative at $y = 2$ than at $y = 1$.

The remaining partial $\partial f/\partial y$ must be viewed in the y-direction, *perpendicular* to the view we have just taken. In Figure 6.12, we see that the graph rises (positive slope) linearly (constant slope) in the y-direction. If we fix x, thus establishing the indicated vertical plane, we then note that $\partial f/\partial y$, computed with x fixed, is e^{-x}, the constant slope of the line indicated in Figure 6.12 in this vertical plane.

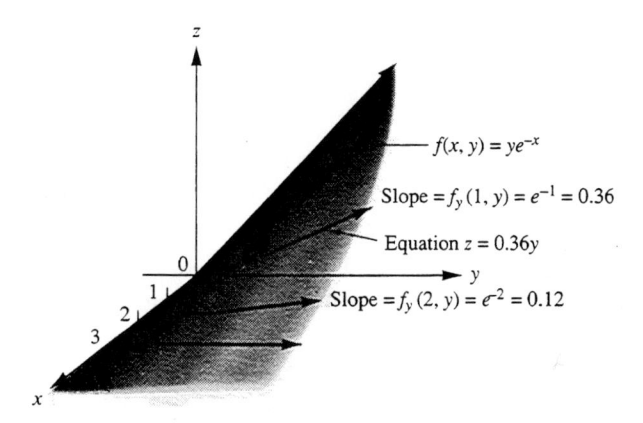

Figure 6.12

In general, if $z = f(x, y)$ has as its graph a three-dimensional surface like that shown in Figure 6.13, then $\partial f/\partial x$ and $\partial f/\partial y$ represent the slope and rate of change of this graph in the entirely *separate, perpendicular* directions of the x and y axes, respectively. For such reasons the adjective *partial* is used.

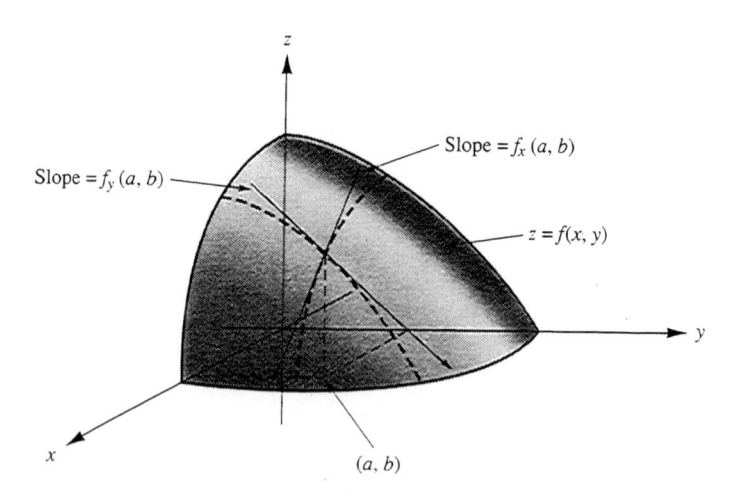

Figure 6.13

When studying functions of one variable, we saw that the derivative of $y = f(x)$ yields the equation of a line tangent to the graph of $y = f(x)$. Partial derivatives of $z = f(x, y)$ yield tangent lines in two-dimensional vertical planes [for the sections of the graph of $z = f(x, y)$ in those planes]. We would like, in a three-dimensional space, to have a single derivative that yields the *tangent plane* to the graph of $z = f(x, y)$. This is not possible, however. On the other hand, it can be shown that the equation of the tangent plane at the point $(a, b, f(a, b))$ can be written as

$$z = f(a, b) + (x - a)\frac{\partial f}{\partial x}(a, b) + (y - b)\frac{\partial f}{\partial y}(a, b)$$

We will not make use of this formula in this section.

Let us return to partial differentiation. Since the notation $\partial z/\partial x$, $\partial f/\partial x$ is so unwieldy, it is common to use an alternate notation for partial derivatives. This notation is

$$f_x = \frac{\partial f}{\partial x} \quad \text{and} \quad f_y = \frac{\partial f}{\partial y}$$

EXAMPLE 3

If $f(x, y) = y^2 - x^2$, find $f_x(1, 1)$ and $f_y(0, -\frac{1}{2})$ and interpret these numbers geometrically.

Solution It is best to find f_x and f_y for general x and y before evaluation at the particular points $(1, 1)$ and $(0, -\frac{1}{2})$. We have

$$f_x = \frac{\partial f}{\partial x} = \frac{\partial}{\partial x}(y^2 - x^2) = 0 - 2x = -2x$$

and

$$f_y = \frac{\partial}{\partial y}(y^2 - x^2) = 2y$$

Therefore $f_x(1, 1) = -2(1) = -2$ and $f_y(0, -\frac{1}{2}) = 2(-\frac{1}{2}) = -1$.

The graph of $z = f(x, y)$ is the hyperbolic paraboloid shown in Figure 6.14. Note how lines drawn tangent to this surface, above $(1, 1)$ and in the x-direction

334 Chapter 6 Functions of Several Variables

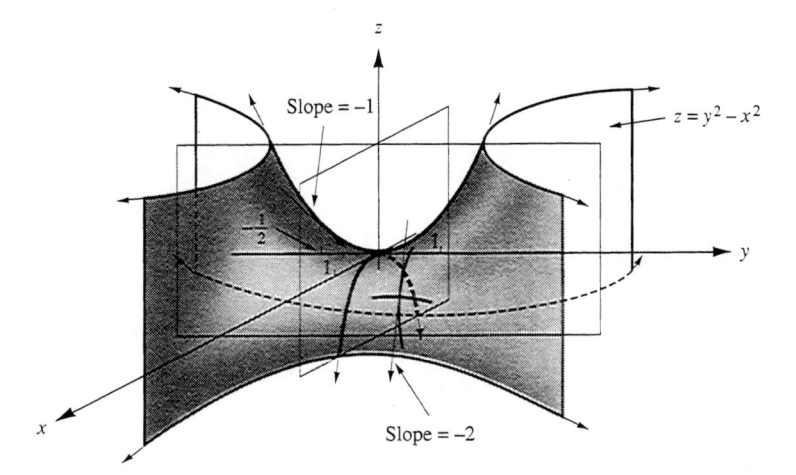

Figure 6.14

and above $\left(0, -\frac{1}{2}\right)$ and in the y-direction, appear to have numerical slope coincident with the values of the partial derivatives evaluated earlier. ∎

The next example applies the chain rule in partial differentiation.

EXAMPLE 4
Find f_x and f_y for $f(x, y) = \ln(x^2 + y)$.

Solution We use the chain rule (regarding $x^2 + y$ as the inner part of the logarithmic function) to obtain

$$f_x = \frac{\partial}{\partial x} \ln(x^2 + y) = \frac{1}{x^2 + y} \frac{\partial}{\partial x}(x^2 + y)$$

$$= \frac{1}{x^2 + y}\left(\frac{\partial}{\partial x}x^2 + \frac{\partial}{\partial x}y\right) = \frac{1}{x^2 + y}(2x) = \frac{2x}{x^2 + y}$$

and

$$f_y = \frac{\partial}{\partial y} \ln(x^2 + y) = \frac{1}{x^2 + y} \frac{\partial}{\partial y}(x^2 + y)$$

$$= \frac{1}{x^2 + y}(0 + 1) = \frac{1}{x^2 + y} \quad \blacksquare$$

EXAMPLE 5
Find f_u and f_v for $f(u, v) = (u^2 + v)^5$.

Solution Again using the chain rule, we have

$$f_u = \frac{\partial}{\partial u}(u^2 + v)^5 = 5(u^2 + v)^4 \frac{\partial}{\partial u}(u^2 + v) = 10u(u^2 + v)^4$$

and

$$f_v = \frac{\partial}{\partial v}(u^2 + v)^5 = 5(u^2 + v)^4 \frac{\partial}{\partial v}(u^2 + v)$$

$$= 5(u^2 + v)^4 \cdot 1 = 5(u^2 + v)^4 \quad \blacksquare$$

Higher-Order Partial Derivatives

Given a function $y = f(x)$ of a single variable and having obtained $f'(x)$, we have a new function of x that may again be differentiated, yielding the second derivative $f''(x)$.

Given $z = f(x, y)$ and having obtained f_x and f_y, we then have *two* new functions of x and y. These may again be differentiated by x, and by y, yielding *four second* (partial) derivatives. These four second partials are defined and denoted by the following:

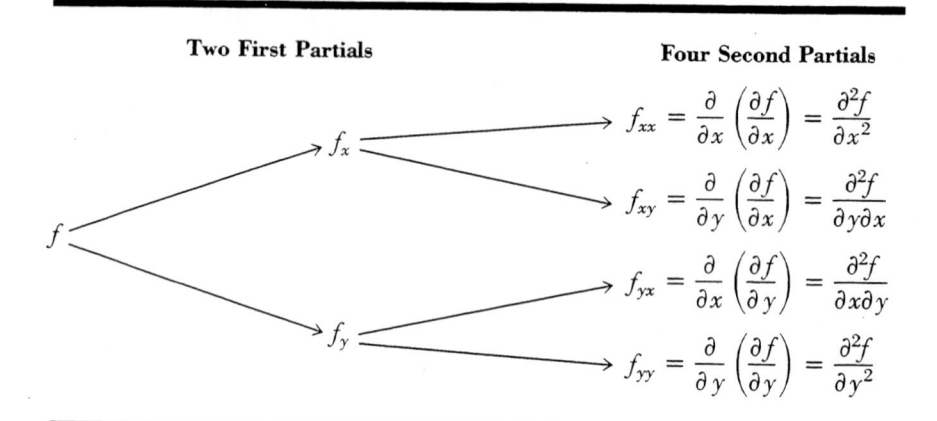

Two First Partials   **Four Second Partials**

$$f \longrightarrow \begin{cases} f_x \longrightarrow \begin{cases} f_{xx} = \dfrac{\partial}{\partial x}\left(\dfrac{\partial f}{\partial x}\right) = \dfrac{\partial^2 f}{\partial x^2} \\[6pt] f_{xy} = \dfrac{\partial}{\partial y}\left(\dfrac{\partial f}{\partial x}\right) = \dfrac{\partial^2 f}{\partial y \partial x} \end{cases} \\[20pt] f_y \longrightarrow \begin{cases} f_{yx} = \dfrac{\partial}{\partial x}\left(\dfrac{\partial f}{\partial y}\right) = \dfrac{\partial^2 f}{\partial x \partial y} \\[6pt] f_{yy} = \dfrac{\partial}{\partial y}\left(\dfrac{\partial f}{\partial y}\right) = \dfrac{\partial^2 f}{\partial y^2} \end{cases} \end{cases}$$

It may be shown that for the functions encountered in this text, $f_{xy} = f_{yx}$. If this were not true, we would need to pay close attention to the order of differentiation: f_{xy} means differentiation by x and then by y; f_{yx} means differentiation by y and then by x.

EXAMPLE 6

Find all four second partials of $f(x, y) = 5x^2y^3 - 2x + 3y - (x/y)$, $y \neq 0$.

Solution We must first find f_x and f_y. We have

$$f_x = 10xy^3 - 2 - \frac{1}{y} \quad \text{and} \quad f_y = 15x^2y^2 + 3 + \frac{x}{y^2}$$

Therefore

$$f_{xx} = \frac{\partial}{\partial x}\left(10xy^3 - 2 - \frac{1}{y}\right) = 10y^3$$

$$f_{xy} = \frac{\partial}{\partial y}\left(10xy^3 - 2 - \frac{1}{y}\right) = 30xy^2 + \frac{1}{y^2}$$

$$f_{yy} = \frac{\partial}{\partial y}\left(15x^2y^2 + 3 + \frac{x}{y^2}\right) = 30x^2y - \frac{2x}{y^3}$$

$$f_{yx} = \frac{\partial}{\partial x}\left(15x^2y^2 + 3 + \frac{x}{y^2}\right) = 30xy^2 + \frac{1}{y^2} \quad \blacksquare$$

Note that indeed $f_{xy} = 30xy^2 + (1/y^2) = f_{yx}$.

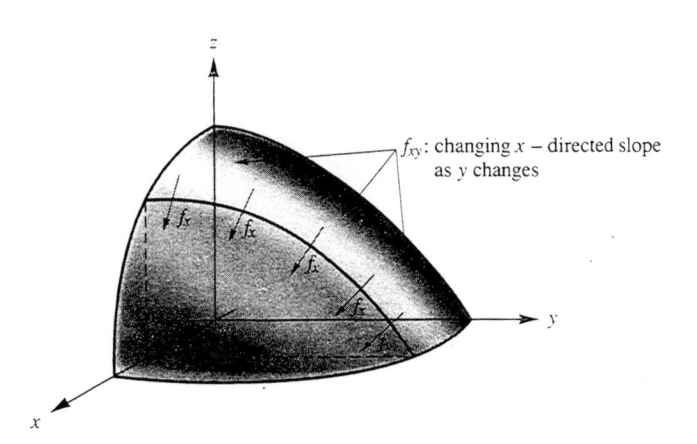

Figure 6.15

f_{xy}: changing x – directed slope as y changes

It is possible to give geometric meaning to the second partial. Very briefly, f_{xx} and f_{yy} (like f'') measure concavity in the x, and separately in the y, direction. The mixed partial f_{xy} measures the rate at which the slopes of sectional tangent lines in the x-direction change as the sectional planes move in the y-direction, as indicated in Figure 6.15.

In the following application, we extend the idea of partial differentiation to a function of three variables.

EXAMPLE 7

Consider a building that faces north–south, of length x, width y, and height z, as shown in Figure 6.16. Due to varying exposure to the warming southern sun and blowing northern wind, the heat loss (measured in BTU's per square foot) varies with each side of the building, as well as with the roof and floor. Appropriate heat-loss factors are shown in the accompanying table.

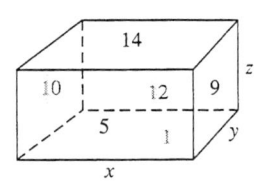

Figure 6.16

Exposure	N	S	E	W	Floor	Roof
Heat loss per ft^2	12	5	9	10	1	14
Area	xz	xz	yz	yz	xy	xy

a. Let $H(x, y, z)$ be the total heat loss of a building of dimensions x, y, and z. Determine a formula for H.

b. Find H_x, H_y, and H_z, and interpret each factor that appears in H_x.

Solution

a. Each of the xz square feet of area with northern exposure loses 12 units of heat. Therefore a total of $12xz$ units is lost on the north side. Similarly, $5xz$ units are lost on the south side. Continuing this reasoning, the appropriate formula for the function H is

$$H(x, y, z) = 12xz + 5xz + 9yz + 10yz + xy + 14xy$$

$$= 17xz + 19yz + 15xy$$

b. Extending the notion of partial differentiation to this function of three variables, we have

$$H_x = \frac{\partial}{\partial x}(17xz + 19yz + 15xy)$$

$$= (17z)\frac{\partial}{\partial x}x + 0 + (15y)\frac{\partial}{\partial x}x$$

$$= 17z + 15y$$

Similarly, $H_y = 19z + 15x$ and $H_z = 17x + 19y$.

These results may be interpreted as follows: The partial derivative H_x measures the rate at which heat loss changes with respect to changes in the length x of the building, keeping height z and width y fixed. Such changes in length affect only the northern and southern exposures, where heat losses of $12 + 5 = 17$ units occur per foot of height z of the building (thus the heat-loss rate of $17z$ appearing in H_x) and affect the roof and floor of the building, where heat losses of $1 + 14 = 15$ units occur per foot of width y of the building (thus leading to additional heat loss at the rate of $15y$). The total $17z + 15y = H_x$ is the heat-loss rate affecting all sides of the building due to a change x in *length* alone. ∎

An Application to Business and Economics: Cobb–Douglas Production Functions

For reasons like those explored in Exercises 28–32, economists have found that a useful mathematical model for the number $N(x, y)$ of units produced using x units of labor and y units of capital is a **Cobb–Douglas production function** of the form

$$N(x, y) = kx^\alpha y^{1-\alpha}$$

where $0 < \alpha < 1$ and k is a constant. The partial derivatives N_x and N_y of N are then called the **marginal rates of change in production** due to a change in labor employed, and capital expended, *separately*.

EXAMPLE 8

If $N(x, y) = 125x^{1/3}y^{2/3}$, then

a. Determine the number of units produced when 8 units of labor and 27 units of capital are employed.

b. At what rate is production increasing with respect to a change in the number of units of labor employed, at the level of labor and capital in (a)?

c. Answer (b) with respect to a change in the number of units of capital employed.

d. Would it make more sense to employ an additional unit of labor or an additional unit of capital at the level of activity in (a)?

Solution

a. We have $N(8, 27) = 125(8^{1/3})(27^{2/3}) = 125(2)(9) = 2{,}250$ units produced.

b. The desired rate is measured by $N_x(8, 27)$. We have

$$N_x = 125 y^{2/3} \frac{\partial}{\partial x} x^{1/3} = 125 y^{2/3} \frac{1}{3} x^{-2/3} = \frac{125}{3}\left(\frac{y}{x}\right)^{2/3}$$

Hence $N_x(8, 27) = \dfrac{125}{3}\left(\dfrac{27}{8}\right)^{2/3} = \dfrac{125(9)}{3(4)} = \dfrac{375}{4} = 93.75$ units

per unit of increase in labor.

c. As in (b), we wish to compute $N_y(8, 27)$. We have

$$N_y = 125 x^{1/3} \frac{\partial}{\partial y} y^{2/3} = 125 \cdot \frac{2}{3} x^{1/3} y^{-1/3} = \frac{250}{3}\left(\frac{x}{y}\right)^{1/3}$$

Therefore

$$N_y(8, 27) = \frac{250}{3}\left(\frac{8}{27}\right)^{1/3} = \frac{250}{3} \cdot \frac{2}{3} = \frac{500}{9} \approx 55.6 \text{ units}$$

per unit of increase in capital.

d. The rates of increase in production per unit of labor and capital found in (b) and (c) are 93.75 and 55.6, respectively. Production will thus increase at a much greater rate due to the use of an additional unit of labor. That is,

$$\frac{N(8 + h, 27) - N(8, 27)}{h} \approx N_x(8, 27) \approx 93.75$$

when h is small. If $h = 1$, corresponding to one additional unit of labor, the *change* in production resulting from the employment of 9 units of labor and (still) 27 units of capital is

$$N(9, 27) - N(8, 27) \approx 93.75(1) = 93.75 \text{ units} \quad \blacksquare$$

Exercises 6.2

Find $f_x(1, -1)$ and $f_y(1, -1)$ for each function in Exercises 1–6.

1. $f(x, y) = 2x + 3y$
2. $f(x, y) = x^2 - y^3$
3. $f(x, y) = x^7 y^8$
4. $f(x, y) = x^2 + 2x + y^3 + xy + y^2 - y$
5. $f(x, y) = \dfrac{x + y}{x - y}$
6. $f(x, y) = e^y \ln x$

Find $\partial f/\partial x$ and $\partial f/\partial y$ for each function in Exercises 7–16.

7. $f(x, y) = (x - y)^3$
8. $f(x, y) = \sqrt{2x + y}$
9. $f(x, y) = xe^{x+y}$
10. $f(x, y) = \ln xye^x$
11. $f(x, y) = (xy + 1)^5$
12. $f(x, y) = xe^{xy}$
13. $f(x, y) = \dfrac{x^2 - y^2}{x^2 + y^2}$
14. $f(x, y) = x^2 y e^y$

15. $f(x, y) = \ln(e^x + e^y)$

16. $f(x, y) = \dfrac{1}{e^x + e^y}$

17. Let $f(s, t) = (s + t)/(s - t)$. Find $f_{ss}(1, -1)$, $f_{tt}(1, -1)$, and $f_{st}(1, -1)$.

18. Let $z = x^2 + y^3$. Find $\partial z/\partial x$, $\partial z/\partial y$, $\partial^2 z/\partial x^2$, $\partial^2 z/\partial y^2$, $\partial^2 z/\partial x \partial y$, and $\partial^2 z/\partial y \partial x$.

19. Let $f(u, v) = uve^{u+v}$. Find $\partial f/\partial u$, $\partial f/\partial v$, $\partial^2 f/\partial u^2$, $\partial^2 f/\partial v^2$, $\partial^2 f/\partial u \partial v$, and $\partial^2 f/\partial v \partial u$.

20. Let $g(w, z) = w^2 e^{wz}$. Evaluate g_w, g_z, g_{ww}, g_{zz}, g_{wz}, and g_{zw} at the point $(1, 2)$.

21. Write $f(x, y) = \ln x^2 y = \ln x^2 + \ln y = 2 \ln x + \ln y$. Then find f_x and f_y.

22. Let $f(x, y) = x^2 y$, and let $x(t) = e^t$ and $y(t) = (t^3 + 1)$.

   a. Find f_x and f_y; $x'(t)$ and $y'(t)$.

   b. Find $f_x(x(t), y(t)) x'(t) + f_y(x(t), y(t)) y'(t)$.

   c. Let $F(t) = f(x(t), y(t))$. Find $F'(t)$.

   d. Compare answers to (b) and (c). What does (b) remind you of?

23. A company is producing bicycles and tricycles at a profit of \$15 and \$7, respectively.

   a. Represent the total profit from the production of x bicycles and y tricycles as a function of two variables.

   b. What is the marginal profit with respect to bicycles and the marginal profit with respect to tricycles, separately, at any level of production?

24. Other things being equal, the distance it takes a moving car to come to a stop when braking will be proportional to its kinetic energy. Consequently, this distance D can be represented by $D(w, v) = kwv^2$ where w is its weight, v its velocity, and k a constant.

   a. Find $\partial D/\partial w$ and $\partial D/\partial v$. What do each of these represent?

   b. Find $(\partial D/\partial w)/D$ and $(\partial D/\partial v)/D$. What do each of these represent?

25. The revenue and cost functions for the sale and manufacture of x tennis rackets and y tennis balls are $R(x, y) = 50x + 0.70y + 0.0005xy$ and $C(x, y) = 3{,}000 + 30x + 0.35y$, respectively. Find the marginal revenue, marginal cost, and marginal profit at the sales level of 50 rackets and 4,000 tennis balls for the sale of tennis balls; of tennis rackets.

26. The mathematician G. Birkhoff defined the "aesthetic measure" A of a work of art (such as a painting, a play, or a piece of music) as the ratio x/y of its "order" x to its "complexity" y. Thus $A(x, y) = x/y$. Find the marginal rates of change of Birkhoff's aesthetic measure due to changes in order and changes in complexity, respectively.

27. At a price of x dollars and y dollars, respectively, the demand for two competing items produced by the same manufacturer is $150 + 2x - 3y$ and $200 - 3x + 5y$. The total revenue to the manufacturer from the sale of both of these items is then

$$R(x, y) = x(150 + 2x - 3y) + y(200 - 3x + 5y)$$

Find the marginal revenue from the sale of each item at a cost of $x = 6$ and $y = 5$ dollars, respectively.

28. We would expect that if one unit of labor produces seven units of goods, then x units of labor will produce $7x$ units. Consider the Cobb–Douglas function $N(x, y) = 7x^\alpha y^{1-\alpha}$, and suppose that labor units x and capital units y happen to cost the same. We may then consider $x = y$ by thinking in terms of like *costs* of production. Show that, in this case, $N(x, y) = 7x$.

29. Show that in the Cobb–Douglas model of production, if labor and capital are both doubled, then production is doubled. In mathematical terms, this means that you must show, using algebra, that

$$N(2x, 2y) = 2N(x, y) \quad \text{when } N(x, y) = kx^\alpha y^{1-\alpha}$$

30. Let $N(x, y) = 500x^{1/2}y^{1/2}$ be a Cobb–Douglas production function.

   a. Find the level of production using $x = 4$ units of labor and $y = 16$ units of capital.

   b. Find the marginal rate of productivity due to labor, and due to capital, at this level of production.

   c. What will have a greater effect on increasing production at this level—one additional unit of labor or one additional unit of capital?

Challenge Problems

31. In a production system, if five units of capital are matched with one unit of labor, the natural tendency is to increase both labor and capital expenditures so as to maintain this relationship. Mathematically, this means that $5x = y$, or that the ratio y/x of labor to capital is maintained at a constant ratio of 5 to 1. [Here, of course, we assume that labor and capital are measured in the like units (dollars) of their respective costs.] In a Cobb–Douglas model

$N(x, y) = kx^\alpha y^{1-\alpha}$ of production, show by substitution that in such a case both

$$N(x, y) = k5^{-\alpha}y \quad \text{and} \quad N(x, y) = k5^{1-\alpha}x$$

32. In Exercise 31, we see that production may be regarded as a function of labor alone, $L(x) = k5^{1-\alpha}x$, or as a function of capital alone, $C(y) = k5^\alpha y$, when capital is matched to labor in the ratio of 5 to 1.

   a. Show that increases in production due to labor or capital are then constant at the rates of $k5^{1-\alpha}$ and $k5^\alpha$, respectively.
   b. Show then that $L' = 5C'$, or that the ratio of the production *rate* with respect to labor to that of the rate with respect to capital is also constant at 5 to 1.

6.3 Optimization

In this section we learn to locate the maximum and minimum values of a function of two variables.

A manufacturing company employing people and equipment is naturally interested in determining an optimal state for its production system: that mix of labor and equipment that will yield maximum profit. A relatively simple test is available for determining the maximum and minimum of such a function of two variables.

Consider Figure 6.17(a, b). The function f in Figure 6.17(a) is said to have a **local maximum value** $f(a, b)$ at the point (a, b) because $f(a, b) \geq f(x, y)$ for all other points (x, y) "near" (a, b). The function f in Figure 6.17(b) is said to have a **local minimum value** $f(a, b)$ at the point (a, b) because $f(a, b) \leq f(x, y)$ for all other points (x, y) near (a, b). We will use the term **local extrema** if f has either a local max or a local min at (a, b). Note also that the continuous function shown in Figure 6.17(c) has *two* local maximum values, one larger than the other, and no local minimum values. This contrasts with a differentiable function of a single variable, which will always have a minimum between two maxima. The function indicated in Figure 6.17(d) has no local maxima or minima.

Critical Points and the Location of Extreme Values

Our method for finding maximum and minimum values for a function of two variables is similar in procedure to the methods learned in Section 2.5 for dealing with a function of one variable.

In Figure 6.17(a, b), we observe that at the point (a, b) of interest, the slope of the tangent to the sectional curves in the direction of both axes is zero. That is, both $f_x(a, b) = 0$ and $f_y(a, b) = 0$.

Definition

A point (a, b) such that both $f_x(a, b) = 0$ and $f_y(a, b) = 0$ is called a **critical point**.

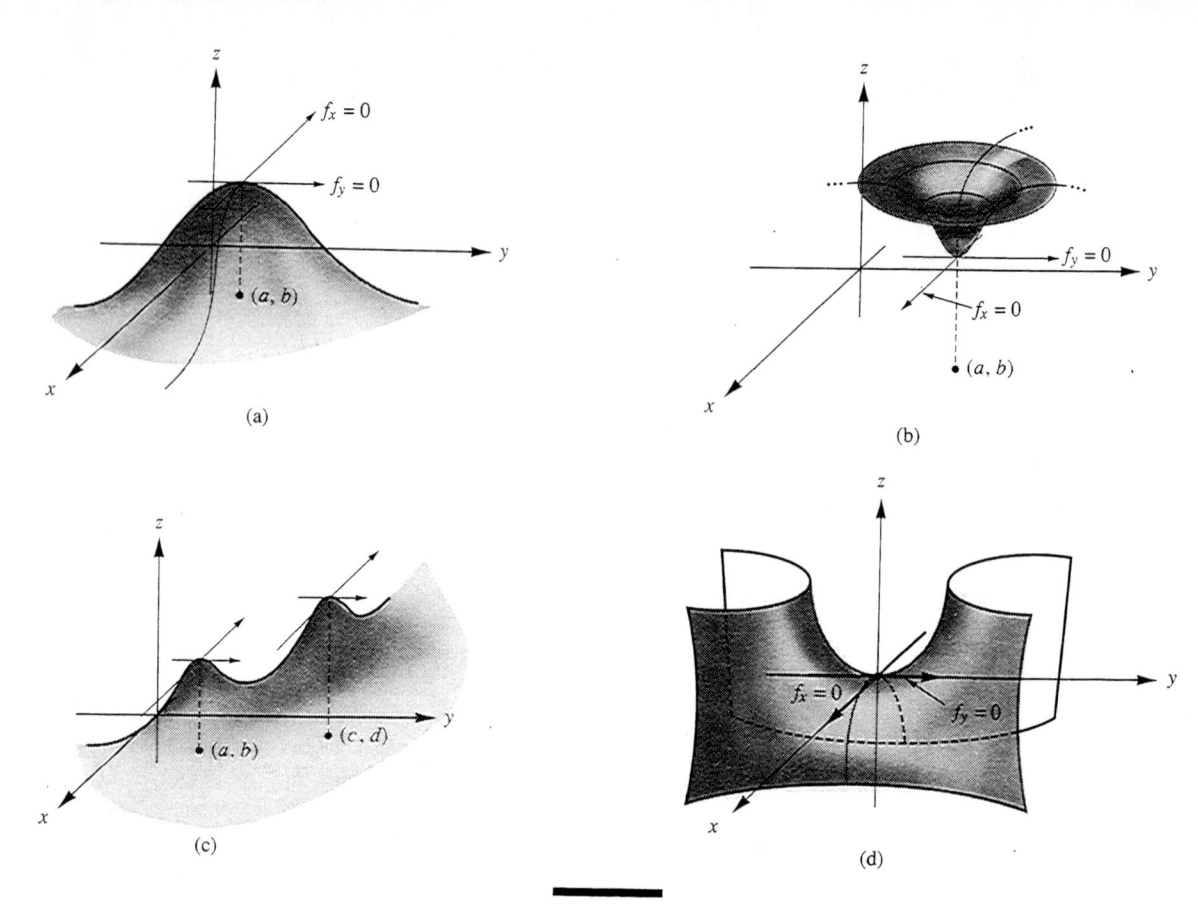

Figure 6.17

In a more general study, we would also consider points (a, b) such that $f_x(a, b)$ and $f_y(a, b)$ fail to exist at critical points. We will not encounter any such functions in this text.

Figure 6.17(a, b) suggests that local extrema can occur only at a critical point. We can prove this in general for the functions we will encounter.

Theorem 6.1 If $f(a, b)$ is a local maximum or a local minimum value of a differentiable function f, then (a, b) is a critical point. That is, both $f_x(a, b) = 0$ and $f_y(a, b) = 0$.

EXAMPLE 1

Find all critical points of

$$f(x, y) = 2x^2 + y^2 - xy + 3y$$

Solution We first find that

$$f_x = 4x - y \quad \text{and} \quad f_y = 2y - x + 3$$

342   Chapter 6 Functions of Several Variables

We must now find all solutions to the system of equations

$$4x - y = 0$$

and

$$2y - x + 3 = 0$$

In algebra you learned several ways of solving such a system; any way that works for you is fine. Here it is easy to see from the first equation that $y = 4x$, and therefore, substituting in the second,

$$2(4x) - x + 3 = 0$$

or

$$7x = -3$$

or

$$x = -\frac{3}{7}$$

Since $y = 4x$, then $y = 4\left(-\frac{3}{7}\right) = -\frac{12}{7}$. Therefore $f_x\left(-\frac{3}{7}, -\frac{12}{7}\right) = 0$ and $f_y\left(-\frac{3}{7}, -\frac{12}{7}\right) = 0$ and the point $(a, b) = \left(-\frac{3}{7}, -\frac{12}{7}\right)$ is a critical point for f. Since this system of equations has no other solution, this is the only critical point for this function. ∎

Theorem 6.1 allows us to narrow the search for local maxima and local minima to only those points (a, b) that are critical points. Example 1 indicates that such points are scarce. Unfortunately the story does not end there, since the class of critical points is larger than the class of maxima or minima [see Figure 6.17(d)]. There (a, b) is a critical point, since both $f_x(a, b) = 0$ and $f_y(a, b) = 0$, but the function has no maxima or minima. This is reminiscent of Section 2.5, where in determining which critical points yield maxima or minima, we used a second derivative test. The procedure in the case of functions of two variables is not so straightforward, however. An additional criterion must be tested in terms of the *discriminant* of a function of two variables.

Definition

The **discriminant** $D(a, b)$ of a function $z = f(x, y)$ at the point (a, b) is the number

$$D(a, b) = f_{xx}(a, b)f_{yy}(a, b) - [f_{xy}(a, b)]^2$$

The number $D(a, b)$ is not difficult to calculate.

EXAMPLE 2

Find $D\left(-\frac{3}{7}, -\frac{12}{7}\right)$ for the function

$$f(x, y) = 2x^2 + y^2 - xy + 3y$$

Solution In Example 1, we found that

$$f_x = 4x - y \quad \text{and} \quad f_y = 2y - x + 3$$

Therefore $f_{xx} = 4 \quad f_{yy} = 2 \quad \text{and} \quad f_{xy} = -1$

6.3 Optimization

Consequently, $f_{xx}f_{yy} - (f_{xy})^2 = 4(2) - (-1)^2 = 7$ for *any* value of x and y. In particular, $D\left(-\frac{3}{7}, -\frac{12}{7}\right) = 7$. ∎

We introduce the discriminant because it gives us a method for determining local extrema for functions of two variables exactly like the method found in Section 2.5 for functions of one variable.

Theorem 6.2

The Second Derivative Test for Extrema for $z = f(x, y)$

Suppose that (a, b) is a critical point of $z = f(x, y)$ and that $D(a, b) > 0$. Then

1. If $f_{xx}(a, b) < 0$, then $f(a, b)$ is a local maximum value.*
2. If $f_{xx}(a, b) > 0$, then $f(a, b)$ is a local minimum value.*

EXAMPLE 3

Determine the local extrema of
$$f(x, y) = 2x^2 + y^2 - xy + 3y$$

Solution In Example 1, we found that $\left(-\frac{3}{7}, -\frac{12}{7}\right)$ is the only critical point for this function. In Example 2, we found that $D\left(-\frac{3}{7}, -\frac{12}{7}\right) = 7 > 0$. Therefore we can apply Theorem 6.2. Since $f_{xx} = 4 > 0$, condition 2 assures us that $f\left(-\frac{3}{7}, -\frac{12}{7}\right) = -\frac{126}{49} \approx -2.57$ is a local minimum value. Thus this function has one minimum and no maximum value. ∎

Examples 1, 2, and 3 show that local extrema can be determined here, as in Chapter 2, *when the discriminant is positive.* The next theorem tells us how to handle the remaining alternatives.

Theorem 6.3

For a critical point of $z = f(x, y)$:

1. If $D(a, b) < 0$, then $f(a, b)$ is *not* an extreme value.
2. If $D(a, b) = 0$, the available information is *inconclusive* and further information about the function must be sought.

Part 1 of Theorem 6.3 is illustrated in the next example.

EXAMPLE 4

Show that $f(x, y) = y^2 - x^2$ has one critical point but no extrema.

---

*We can use f_{yy}, rather than f_{xx}, just as well.

Solution

Step 1. We have $f_x = -2x$ and $f_y = 2y$.

Step 2. Critical points can occur only when $-2x = 0$ and $2y = 0$.
Therefore the only critical point is $(a, b) = (0, 0)$.

Step 3. To compute $D(0, 0)$, we find

$$f_{xx} = -2 \qquad f_{yy} = 2 \qquad \text{and} \qquad f_{xy} = 0$$

Therefore $D = f_{xx}f_{yy} - (f_{xy})^2 = -2 \cdot 2 - (0)^2 < 0$ at any point. In particular, $D(0, 0) < 0$, and applying Theorem 2, $f(0, 0)$ is not an extreme value. ∎

The graph of f, shown in Figure 6.18, is the graph of the hyperbolic paraboloid. The minimax point at $(0, 0)$ appears as a solution, and therefore as a critical point, in steps 1 and 2 in Example 4. But the discriminant rules out $(0, 0)$ as an extreme point in step 3. We will refer to the use of Theorems 6.2 and 6.3 as the **D-test for extrema** of a function of two variables.

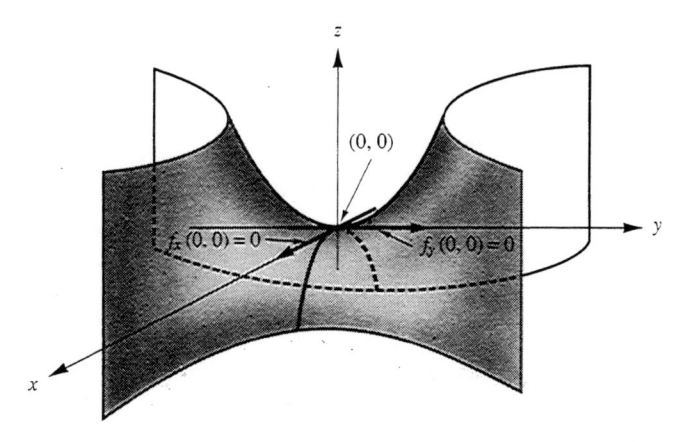

Figure 6.18

Part 2 of Theorem 6.3 is more subtle and will not be dealt with. The growing availability of high-speed computation makes the case in part 2 much easier to handle than just a few years ago. We can ask the computer to compute many values of f near the critical point and judge the result for the likelihood (but not the absolute certainty) of an extremum (see Exercise 30). This is the kind of additional information that must be sought.

Here is a complete example of an application of the D-test for extrema.

EXAMPLE 5

Find all local extrema of

$$f(x, y) = x^2 e^y - xe^y + x$$

Solution

Step 1. We first find that

$$f_x = 2xe^y - e^y + 1 \qquad \text{and} \qquad f_y = x^2 e^y - xe^y$$

6.3 Optimization   **345**

Step 2. We must solve for x and y such that

$$2xe^y - e^y + 1 = 0$$

and

$$x^2 e^y - xe^y = 0$$

We can write the second equation as $(x^2 - x)e^y = 0$, which, because e^y is *never* zero, is equivalent to $x^2 - x = 0$. This equation has two solutions: $x = 0$ and $x = 1$.

Returning now to the first equation, when $x = 0$, we have

$$2 \cdot 0 \cdot e^y - e^y + 1 = 0$$

or $e^y = 1$, and therefore $y = 0$. Thus $(a, b) = (0, 0)$ is one critical point.

The second solution $x = 1$ may yield another. Substituting in the first equation again, we have

$$2 \cdot 1 \cdot e^y - e^y + 1 = 0$$

and therefore $e^y + 1 = 0$. This has no solution since $e^y > 0$ for any value of y.

Step 3. The only critical point is $(0, 0)$. To find $D(0, 0)$, we have

$$f_{xx} = 2e^y \qquad f_{yy} = x^2 e^y - xe^y \qquad \text{and} \qquad f_{xy} = 2xe^y - e^y$$

Therefore

$$\begin{aligned} D(0, 0) &= f_{xx}(0, 0)f_{yy}(0, 0) - f_{xy}(0, 0)^2 \\ &= (2e^0)(0^2 e^0 - 0e^0) - (2 \cdot 0 \cdot e^0 - e^0)^2 \\ &= 0 - (-1)^2 = -1 < 0 \end{aligned}$$

Step 4. Since $D(0, 0) < 0$, we conclude that this function has no extreme values. ∎

Applications

Two-variable optimization allows wider application. The next example may seem familiar, but note that, unlike previous examples of this type, we are no longer confined to a square and can consider an arbitrary rectangular parallelepiped.

EXAMPLE 6

A manufacturer of cereal wishes to sell the cereal in boxes of volume 64 in.³ and minimum surface area. What size box should be used?

Solution The total surface area of an arbitrary rectangular box is seen from Figure 6.19 to be

$$S = 2xz + 2yz + 2xy \tag{1}$$

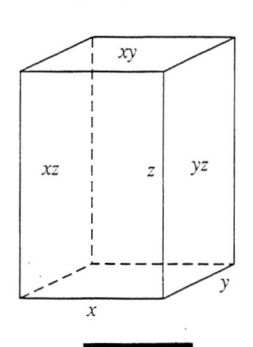

Figure 6.19

This is a function of *three* variables. We must eliminate one variable to apply Theorem 6.2. Since the volume must be 64 in.³, we know that $xyz = 64$; therefore $z = 64/xy$. Substituting this for z in the expression for surface area (1), we obtain a function of two variables

$$f(x, y) = \frac{128}{y} + \frac{128}{x} + 2xy$$

which we are to minimize.

Steps 1 and 2. We must find all solutions (x, y) to

$$f_x = -\frac{128}{x^2} + 2y = 0 \tag{2}$$

and

$$f_y = -\frac{128}{y^2} + 2x = 0 \tag{3}$$

From (2), we have $y = 64/x^2$, $x \neq 0$. If we replace y in (3) by this expression, we obtain

$$-\frac{128}{\left(\frac{64}{x^2}\right)^2} + 2x = 0$$

or

$$-\frac{x^4}{64} + x = 0$$

This is equivalent to $x[1 - (x^3/64)] = 0$, with solutions $x = 0$ and $x = 4$. Since $x = 0$ cannot be used to make a box, we have that $(4, 4)$ is the only critical point [because $y = 64/(4)^2 = 4$].

Step 3. We have $f_{xx} = 256/x^3$, $f_{yy} = 256/y^3$, and $f_{xy} = 2$. Therefore

$$D(4, 4) = \frac{256}{4^3} \cdot \frac{256}{4^3} - 2^2 = 4(4) - 4 > 0$$

and we may apply Theorem 6.2.

Step 4. Since $f_{xx}(4, 4) > 0$, we conclude that $x = 4$, $y = 4$ (and, consequently, $z = 4$ since $xyz = 64$) yields a box of minimum surface area. ∎

Remarks.

1. This box will not be aesthetically pleasing!
2. We see that an assumption of a square-sided box is, in the end, no restriction. Since we are able to consider dimensions of *any* size x, y, and z, the theory of functions of two variables allows us to conclude that the minimum surface area can result only with a square side, and even more, a box in the shape of a cube.
3. It may be shown that this local minimum surface area is *the* minimum surface area that can be used for the desired volume.

EXAMPLE 7

The owner of a catfish farm has the following problem: The farmer owns two lakes L_1 and L_2 to which fingerling catfish are to be transferred (from a controlled hatching pond) and allowed to grow to salable size. Lake L_2 is twice as big as L_1, so that more fingerlings can be put into L_2, where they can be expected to gain more weight in a season as well. Samples from previous years have shown that if x fingerling catfish are put into L_1, then each fish will gain $(7{,}500 - x)/500$ oz in a season. If y fish are put into L_2, then each fish will gain $(15{,}000 - 1.25y)/1{,}000$ oz in a season. Additionally, the fish farmer has a "cost" of $12{,}000 + 5(x + y)$ oz of fish weight gain incurred in raising the fingerlings in the hatching pond before transfer.

How many fish should be put into L_1 and how many into L_2 to obtain a (local) maximum total additional weight gain in a season? (Note: Both profit and cost to the farmer depend directly on weight gain and food weight cost; we could state the problem in terms of dollars for each factor as well.)

Solution Each of x fish in L_1 will gain $(7{,}500 - x)/500$ oz for a total gain in L_1 of $[(7{,}500 - x)/500]x$ oz of fish. By similar reasoning, L_2 will produce $[(15{,}000 - 1.25y)/1{,}000]y$ oz of fish.

The *net* gain in ounces of this operation is

$$f(x, y) = \frac{7{,}500 - x}{500}x + \frac{15{,}000 - 1.25y}{1{,}000}y - [12{,}000 + 5(x + y)]$$

oz of fish weight gain less fish weight cost before transfer. Hence

$$f(x, y) = 15x - \frac{x^2}{500} + 15y - \frac{1.25y^2}{1{,}000} - 12{,}000 - 5x - 5y$$

$$= 10x - \frac{x^2}{500} + 10y - \frac{1.25y^2}{1{,}000} - 12{,}000$$

is to be maximized.

Steps 1 and 2. We must find x and y so that

$$f_x = 10 - \frac{x}{250} = 0$$

and

$$f_y = 10 - \frac{2.5y}{1{,}000} = 0$$

This system has the solution $x = 2{,}500$, $y = 4{,}000$.

Step 3. $f_{xx} = -1/250$, $f_{yy} = -2.5/1{,}000$, and $f_{xy} = 0$. Therefore $D(x, y) = (-1/250)(-2.5/1{,}000) > 0$ for any x and y. Applying Theorem 6.2, we see that $f_{xx}(2{,}500, 4{,}000) = -1/250 < 0$, and therefore $x = 2{,}500$ fingerlings should be put into L_1, $4{,}000$ should go into L_2, and the owner should raise a total of $6{,}500$ fingerlings, yielding a net gain of $f(2{,}500, 4{,}000) = 20{,}500$ oz of fish weight. ∎

Background Review 6.1

The determination of critical points involves the solution of two simultaneous equations

$$P(x, y) = 0$$
$$Q(x, y) = 0$$

in two unknowns x and y.

The technique of solution is *first* to eliminate one variable and then determine the solution for the remaining variable. This elimination may be done by *any* algebraically correct procedure that meets this objective. For example, we may

1. Solve for y in terms of x in $P(x, y) = 0$ and substitute this expression for y in $Q(x, y) = 0$ (this was done in Example 1), or
2. Make some algebraic change in $Q(x, y) = 0$, usually multiplication of both sides by a constant, and subtract the new equation from $P(x, y) = 0$ with the result that one of the variables is eliminated, or
3. Factor $Q(x, y)$ as a product $g(x)h(y)$ and set each factor $g(x) = 0$ and $h(y) = 0$. This was done in Example 5, where $x^2 e^y - xe^y = (x^2 - x)e^y$.

Further elaboration on these methods of elimination is possible.

After eliminating one variable, say, y, we solve for the remaining variable x. Each solution $x = a$ should then be substituted for x in one of the original equations. The result is a second equation in y (alone) that can then be solved for $y = b$. This yields a pair of solutions $x = a$ and $y = b$; these together locate a critical point. If $x = a$ yields two or more solutions for y, then two or more critical points arise with the same x-coordinate $x = a$.

Exercises 6.3

In Exercises 1–20, (a) find all critical points (a, b); (b) evaluate $D(a, b)$ at each point (a, b); and (c) when $D(a, b) > 0$, determine whether $f(a, b)$ is a local maximum or a local minimum value.

1. $f(x, y) = 1 - x^2 - y^2$
2. $f(x, y) = xy$
3. $f(x, y) = x^2 - y^2 + 4y - 16$
4. $f(s, t) = -s^2 - 2s - t^2 - 4t - 2$
5. $f(u, v) = uv^3 - u$
6. $f(x, y) = y^3 x - xy$
7. $f(w, z) = w^2 + z^2 - 2w - 4z + 5$
8. $f(x, y) = 4x - 3y + y^3 - x^2$
9. $f(x, y) = x^3 e^y - xe^y - x$
10. $f(x, y) = xe^{-xy}$
11. $f(x, y) = x^2 - y^2 + xy + y$
12. $f(x, y) = x^3 - y^2 + \dfrac{x^2 y}{2} - y$
13. $f(x, y) = x(\ln y)$ $\quad y > 0$
14. $f(x, y) = xe^{x+y} + y$
15. $f(x, y) = x^3 - y^3 + \dfrac{x^2 y}{2} - y^2$
16. $f(s, t) = e^{-(s^2 + t^2)}$

17. $f(x, y) = xye^{x+y}$

18. $f(x, y) = xe^{-xy} + y$

19. $f(x, y) = e^{-x^2/y}$, $y \neq 0$

20. $f(x, y) = \dfrac{x^2}{1 - y}$, $y \neq 1$

21. a. Graph $f(x, y) = \sqrt{1 - x^2 - y^2}$.
    b. Use this graph to estimate $f_x(0, 0)$ and $f_y(0, 0)$.
    c. Find $f_x(0, 0)$ and $f_y(0, 0)$.
    d. Find $f_x(0.2, y)$ and illustrate this answer on the graph. Using this graph, can you illustrate why $f_{xy}(0.2, 0) = 0$?

22. a. Graph $f(x, y) = \sqrt{4 - (x^2 + y^2)}$.
    b. Use your eye and this graph to estimate $f_x(0, 0)$ and $f_y(0, 0)$.
    c. Find $\partial f/\partial x$ and $\partial f/\partial y$ and evaluate these at $(0, 0)$.

23. Please refer to Example 7. Suppose that the catfish produced in the two lakes can be sold for 10 cents per ounce and that the cost of feeding the fish before they are released into the lakes is 2 cents per ounce. Optimize an appropriate dollar cost function so as to maximize net profit. (*Hint:* The cost of feeding before transfer is $0.02[12,000 + 5(x + y)]$.)

24. Take a piece of wire 20 in. long. Cut it at some point and let y be the length on the right in Figure 6.20. Bend this length into a circle of circumference $y = 2\pi r$ and radius r. Then bend the remaining piece into a rectangle of area xz. How would you choose x, y, and z so as to maximize the total area of the resulting rectangle and circle, using the fact that $20 = y + 2x + 2z$ to reduce the number of variables to two? How would you minimize this area?

25. Please refer to Example 6. Suppose that material for the front, back, and sides of the box costs 0.2 cent per square inch, and for the top and bottom, 0.3 cent per square inch. Optimize the cost of the surface area of the box subject to the same contraint on volume.

26. A manufacturing firm has two production facilities A and B. Facility A is closer to a source of raw materials but less efficient in its use of these. If x units of raw materials are sent to A, the manufacturer will have a shipping cost of $1.3x$ dollars. Of these x units, $0.4x$ finished products will result, each selling for $4[1 - (x/1,000)]$ dollars. If y units are sent to B, shipping costs will be $1.6y$ dollars, but $0.5y$ finished products will result, each selling for $4[1 - (y/1,000)]$. How many units x and y should be sent to A and B so as to maximize net profit (revenue less shipping cost)?

27. A producer sells two products at prices p and q, respectively. Let $C(x, y)$ be the cost to produce x units of the first product and y of the second. Show that maximum profit occurs at the point $x = a$, $y = b$, where

$$\frac{\partial C}{\partial x}(a, b) = p \quad \text{and} \quad \frac{\partial C}{\partial y}(a, b) = q$$

28. An auto manufacturer sells its product in two different countries. The price in one country is $10,000[1 - (x/500,000)]$ dollars and in the second is $9,000[1 - (y/600,000)]$ dollars when x and y cars are sold in each country, respectively. The company's production cost is $100,000,000 + 7,000(x + y)$. How many cars should be sold in each country so as to maximize profit?

29. What is the area of the largest rectangle that can be drawn inside the circle of radius 1 in Figure 6.21 (whose equation is $x^2 + y^2 = 1$)?

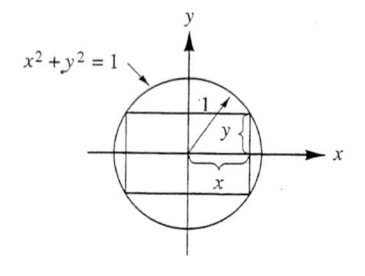

Figure 6.21

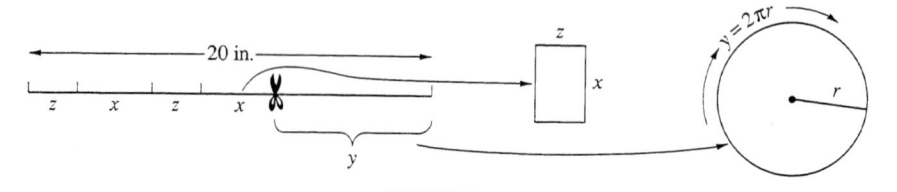

Figure 6.20

Computer Application Problem

30. Consider the function
$$f(x, y) = \sqrt{3 - (0.48x^3 + y^3)}$$

a. Show that $(0, 0)$ is a critical point for f; ignore other critical points.

b. Show that $D(0, 0) = 0$; the D-test fails for f.

c. Use the BASIC program FUNVAL 2 to evaluate f at points near $(0, 0)$. For example, in that program let $A = -\frac{1}{2}$; $B = \frac{1}{2}$; $C = -\frac{1}{2}$; $D = \frac{1}{2}$; $N = 5$; and $K = 5$. Run the program. Does the result suggest that $f(0, 0)$ is a maximum or a minimum?

d. Can you imagine how the graph of f might change in such a way that the conclusion of (c) is false but this was not detected by the program?

6.4 Linear Regression by Least Squares

The purpose of this section is to use linear regression, by the method of least squares, to fit linear and exponential curves to given sets of data.

We have seen two principal ways in which we can obtain a formula for a function $y = f(x)$ relating two variables x and y in a system. In Section 2.6, formulas were derived by analysis of information given about a particular system. In Section 5.5, the formula for an unknown function was found by the solution of a differential equation.

We now learn how to determine functions of the *forms* $y = Ax + B$ or $y = Ae^{kx}$, which only "best fit" *incomplete* information about the relationship of x and y. This information will always consist of several known, related, but particular values of x and y. We will then use the minimization techniques of Section 6.3 to find this "best fit." This technique is known as **linear regression** by the method of **least squares** for reasons that will soon be clear. Because this technique, applied to the case of linear functions $y = Ax + B$, extends immediately to the exponential case $y = Ce^{kx}$ by a logarithmic transformation of the given data, it is the linear case that is essential, and we consider it first.

Best-Fit Linear Approximation

We approach this topic as the following geometrical problem and then turn to applications: Given n points (x_1, y_1), (x_2, y_2), . . . , (x_n, y_n) in the plane, as indicated in Figure 6.22, find the slope A and intercept B of a straight line $y = Ax + B$ that lies "closest" to all n points.

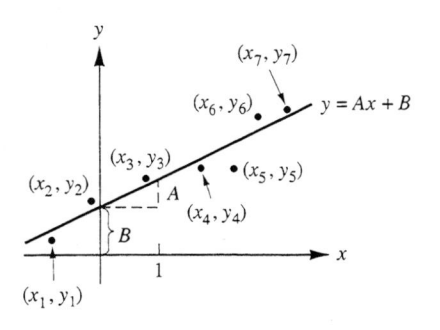

Figure 6.22

Since two points determine a unique straight line, we see that in general it is impossible to find a line that actually goes through all points. Once we have abandoned this hope, we find that a line that is closest to *all* points may very well miss every point!

Imagine that we have such a "closest" line and that this line is given by an equation $y = Ax + B$; what can "closest" mean? First, we should expect that $y_1 \simeq Ax_1 + B$, or that $y_1 - (Ax_1 + B)$ is a minimum. But we should also expect that $y_2 - (Ax_2 + B)$ is a minimum. Continuing this way, we should like it that *all* differences

$$y_k - (Ax_k + B)$$

between the given y-value y_k and the y-value computed using A, B, and x_k are minimized. This suggests that the sum

$$[y_1 - (Ax_1 + B)] + [y_2 - (Ax_2 + B)] + \cdots + [y_n - (Ax_n + B)]$$

ought to be a minimum. However, this is illusory, since points lying above and below the line will contribute addends of varying ($\pm$) sign to this sum, thus ruining it as a measure of the approximation [see the two points $(-1, 0)$ and $(2, 1)$ in Figure 6.23 that are approximately the same distance, but in opposite directions, from the line]. Instead we choose to find A and B so as to minimize the sum of *squares*

$$f(A, B) = (y_1 - (Ax_1 + B))^2 + (y_2 - (Ax_2 + B))^2 + \cdots + (y_n - (Ax_n + B))^2 \tag{1}$$

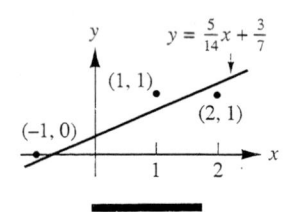

Figure 6.23

Since the points $(x_1, y_1), \ldots, (x_n, y_n)$ are given, and the slope A and intercept B are to be found, the problem of matching the straight line $y = Ax + B$ to the given points becomes one of minimizing this function f of the two variables A and B. This approach to a "closest" or "best" fit is known, then, as the method of least squares.

EXAMPLE 1

Find the slope A and intercept B of the straight line that lies closest (in the sense of least squares) to the points $(-1, 0)$, $(1, 1)$, and $(2, 1)$.

Solution Here we have $n = 3$ points, with $x_1 = -1$, $y_1 = 0$; $x_2 = 1$, $y_2 = 1$; and $x_3 = 2$, $y_3 = 1$, as indicated in Figure 6.23. Accordingly, we form the sum of squares

$$f(A, B) = (0 - (A(-1) + B))^2 + (1 - (A \cdot 1 + B))^2 + (1 - (A \cdot 2 + B))^2$$

or $\quad f(A, B) = (A - B)^2 + (1 - A - B)^2 + (1 - 2A - B)^2$

To minimize this sum, we first find

$$f_A = 2(A - B) + 2(1 - A - B)(-1) + 2(1 - 2A - B)(-2)$$
$$= 2A - 2B + (-2 + 2A + 2B) + (-4 + 8A + 4B)$$
$$= 12A + 4B - 6$$

352   Chapter 6   Functions of Several Variables

and $f_B = 2(A - B)(-1) + 2(1 - A - B)(-1) + 2(1 - 2A - B)(-1)$
$= 2B - 2A + (-2 + 2A + 2B) + (-2 + 4A + 2B)$
$= 4A + 6B - 4$

As in Section 6.3, we then find A and B, so that

$$f_A = 12A + 4B - 6 = 0$$

and
$$f_B = 4A + 6B - 4 = 0$$

The solution to this system of equations is found to be $A = \frac{5}{14}$ and $B = \frac{3}{7}$.

Using the discriminant $D = f_{AA}f_{BB} - (f_{AB})^2$, we may show that these values do minimize the least square sum f. The line

$$y = \frac{5x}{14} + \frac{3}{7}$$

with slope $A = \frac{5}{14}$ and intercept $B = \frac{3}{7}$, is shown in Figure 6.23, indicating how well it "fits" all three points (but matches none!). ∎

With only moderate effort we can determine formulas for the unknown slope A and intercept B of a straight line that best fits an arbitrary number n of points $(x_1, y_1), (x_2, y_2), \ldots, (x_n, y_n)$, by using the techniques of Section 6.3 to minimize the general sum of squares

$$f(A, B) = (y_1 - (Ax_1 + B))^2 + (y_2 - (Ax_2 + B))^2 + \cdots + (y_n - (Ax_n + B))^2$$

The resulting general formulas for A and B are as follows:

General Linear Regression

The straight line $y = Ax + B$ that best fits the n points $(x_1, y_1), \ldots, (x_n, y_n)$ in the sense of least squares has slope

$$A = \frac{n \sum x_k y_k - \left(\sum x_k\right)\left(\sum y_k\right)}{n \sum x_k^2 - \left(\sum x_k\right)^2}$$

and intercept

$$B = \frac{\left(\sum y_k\right)\left(\sum x_k^2\right) - \left(\sum x_k y_k\right)\left(\sum x_k\right)}{n \sum x_k^2 - \left(\sum x_k\right)^2}$$

The Greek letter Σ denotes the word *sum*. For example, the symbol $\sum x_k$ is shorthand for

$$x_1 + x_2 + \cdots + x_n$$

Similarly, $\Sigma\, x_k y_k$ denotes

$$x_1 y_1 + x_2 y_2 + \cdots + x_n y_n$$

and so on. With these formulas we are able to avoid the work of partial differentiation and solution of the equations $f_A = 0$ and $f_B = 0$ seen in Example 1 and can proceed as in the next example. In Example 2, we also adopt the simpler notation $\Sigma\, x$, $\Sigma\, x^2$, etc., for $\Sigma\, x_k$, $\Sigma\, x_k^2$, etc., for the required sums.

EXAMPLE 2

Let us make a prediction about something that has actually happened. Once we have calculated our prediction, we can compare it to the actual result and see the amount of error. Per capita personal income in the United States in the year 1970 (year 0), 1975 (year 5), 1980 (year 10), and 1981 (year 11) is shown in the graph and accompanying chart (Figure 6.24). Imagine you are an economist in 1981. Determine the slope A and intercept B of the straight line that best fits these data, and use these values to predict per capita income for 1982.

Year x	Income (dollars) y
0 (1970)	3,900
5 (1975)	5,800
10 (1980)	9,500
11 (1981)	10,600
12 (1982)	?

$y = 609x + 3{,}491$

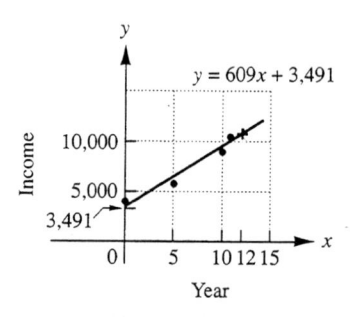

Figure 6.24

Solution We will use the formulas for A and B given for general linear regression. It is convenient to prepare an extended chart of the values of x^2 and xy and the needed sums $\Sigma\, x$, $\Sigma\, y$, $\Sigma\, x^2$, and $\Sigma\, xy$. Note also that we are given four pairs of data consisting of a numbered year and the per capita income in that year; thus $n = 4$. Using a calculator, we have the values given in Table 6.1.

Table 6.1

x	y	x^2	xy
0	3,900	0	0
5	5,800	25	29,000
10	9,500	100	95,000
11	10,600	121	116,600
26 $\Sigma\, x$	29,800 $\Sigma\, y$	246 $\Sigma\, x^2$	240,600 $\Sigma\, xy$

Accordingly, the desired linear regression slope and intercept are

$$A = \frac{n \sum x_k y_k - \left(\sum x_k\right)\left(\sum y_k\right)}{n \sum x_k^2 - \left(\sum x_k\right)^2} = \frac{4(240{,}600) - (26)(29{,}800)}{4(246) - (26)^2}$$

$$= \frac{187{,}600}{308} \approx 609$$

and

$$B = \frac{\left(\sum y_k\right)\left(\sum x_k^2\right) - \left(\sum x_k y_k\right)\left(\sum x_k\right)}{n \sum x_k^2 - \left(\sum x_k\right)^2}$$

$$= \frac{(29{,}800)(246) - (240{,}600)(26)}{308} \approx 3{,}491$$

The regression line in Figure 6.24 is therefore

$$y = 609x + 3{,}491$$

If per capita income increases in 1982 (year 12) in the same pattern as indicated by these data, we would expect it to be

$$y = 609(12) + 3{,}491 = \$10{,}799 \quad \blacksquare$$

The historically correct figure was actually $11,100. The straight line prediction is thus in error by $400, or about 3.6%. With so little data available, as in Example 2, it is questionable to assume that income is growing at a linear (straight line) rate. Yet the predicted income is surprisingly close to being correct. It is equally plain that many data-based relationships cannot be linear. Let us now see how the method of linear regression may be adapted to nonlinear data.

Best-Fit Exponential Curves

If the given data $(x_1, y_1), \ldots, (x_n, y_n)$ are seen to lie approximately in a straight line, it makes sense to attempt to model the data by linear approximation. If instead the data appear as in Figure 6.25, a straight line is not appropriate. From our knowledge of the shape of the exponential curve $y = Ce^{kx}$, we would sensibly prefer to fit such data to an exponential curve. This is quite easy to do, using linear regression, if we apply a logarithmic transformation first to the curve $y = Ce^{kx}$ and then to the data itself.

Consider the functional form $y = Ce^{kx}$ and apply the logarithm to both sides to obtain

$$\ln y = \ln Ce^{kx} = \ln C + \ln e^{kx}$$

or

$$\ln y = \ln C + kx(\ln e) = kx + \ln C \tag{2}$$

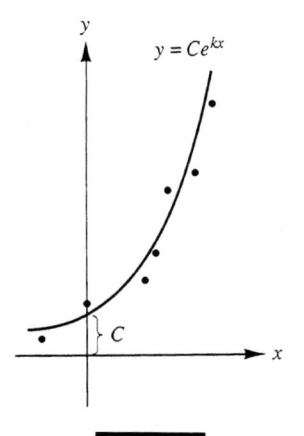

Figure 6.25

Now let $Y = \ln y$, $A = k$, and $B = \ln C$. Equation 2 then becomes $Y = Ax + B$. This conclusion means that when x and y are exponentially related, then x and $Y = \ln y$ are *linearly* related and may be analyzed using linear regression.

EXAMPLE 3

During the 1970s, women marathon runners improved their world-record time at an impressive rate. Figure 6.26 shows the number of minutes by which the record marathon time set previous to 1970 had been *reduced* by the years 1972, 1976, and 1978.

a. Fit an exponential curve $y = Ce^{kx}$ to these data.
b. Imagine you are in 1978. If this trend continues, by how much will a runner have to improve the record time in order to lead the pack in 1979?
c. At what rate is the record time being lowered per year? Can this be sustained?

Year	Minutes
2 (1972)	4
6 (1976)	12
8 (1978)	17

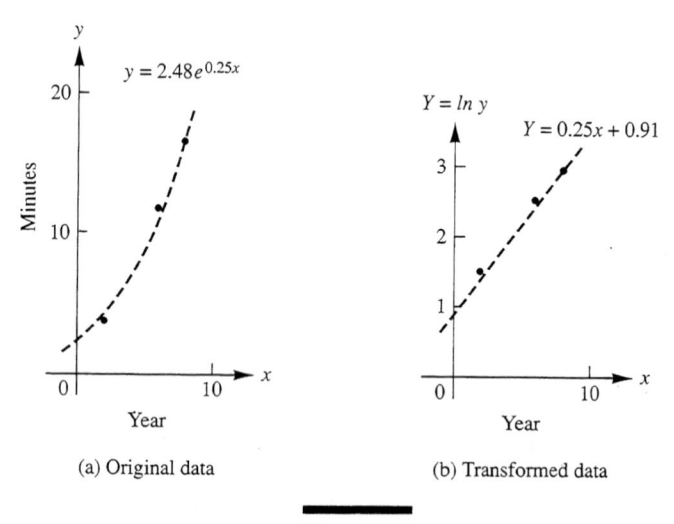

(a) Original data: $y = 2.48e^{0.25x}$

(b) Transformed data: $Y = \ln y$, $Y = 0.25x + 0.91$

Figure 6.26

Solution

a. To fit the exponential curve $y = Ce^{kx}$ to these data, we must fit a linear curve $Y = Ax + B$ to the linearly related data x and $\ln y = Y$. This means no more than that we must include one more column in a chart like that used in Example 2 and we must use this column to fit $Y = Ax + B$ to the data $(x, \ln y)$. Using a calculator to obtain $\ln y$ in each row, we have the figures given in Table 6.2, with $n = 3$. Before going on, note that x and Y appear to be linearly related, as indicated in Figure 6.26(b). Also note that the original data y are *no longer a part* of the calculations.

Table 6.2

	Year x	Minutes y	$Y = \ln y$	xY	x^2
1972	2	4	1.39	2.78	4
1976	6	12	2.48	14.88	36
1978	8	17	2.83	22.64	64
	16		6.70	40.30	104
	Σx		ΣY	ΣxY	Σx^2

Using linear regression, we now have

$$k = A = \frac{3(40.30) - 16(6.70)}{3(104) - (16)^2} = \frac{13.7}{56} \simeq 0.24$$

and

$$\ln C = B = \frac{6.70(104) - 40.30(16)}{56} = \frac{52.0}{56} \simeq 0.93$$

Therefore $Y = 0.24x + 0.93$.

We are not through, however. We must next perform an inverse transform on this straight line, returning it to the desired exponential curve. Since $\ln y = Y$, we have

$$\ln y = 0.24x + 0.93$$

Therefore $y = e^{\ln y} = e^{0.24x + 0.93} = (e^{0.93})(e^{0.24x}) = 2.5e^{0.24x}$. That is, the desired exponential curve is

$$y = 2.5e^{0.24x}$$

This answers part (a).

b. To answer part (b), we evaluate the solution in part (a) at $x = 9$ (corresponding to 1980) to obtain

$$y = 2.5e^{0.24(9)} \simeq 21.7 \text{ min}$$

Thus a world-record time in 1979 would have been about $21.7 - 17 = 4.7$ min better than in 1978. (The actual improvement was about 5.1 min.)

c. The derivative of $y = 2.5e^{0.24x}$ is $y' = 2.5(0.24e^{0.24x}) = 0.24y$. Thus the improvement on world-record time was increasing by 24% a year during the decade.

This pace can hardly be sustained! For example, if things continued this way, by the year 2000 the old world-record time prior to 1970 would have been improved by $2.5e^{0.24(30)} = 3,349$ min in a race that takes only approximately 2 hr! Thus this exponential fit to the data is only valid over a short time span. ∎

It is possible to make a realistic prediction as to what the ultimate world-record time is ever likely to be. In Section A.3, we outline transforms appropriate to other functional forms so as to best-fit data that are neither linear nor exponential, but rather polynomial, hyperbolic, or—like world-record times—sinusoidal, using the procedures learned here. These techniques are closely related to logarithmic scaling, also studied in Section A.3.

Exercises 6.4

Find the best-fit linear regression line for the following data sets, and graph both the data and this line. Identify any points that actually lie on the line.

1. (1, 1), (2, 2), (3, 3), (4, 5)
2. (−2, 0), (0, 1), $\left(1, \frac{3}{2}\right)$
3. (0, 3), (2, 1), (4, 0), (5, −1)
4. (−1, 1), (0, 1.1), (2, 0.9)

Find the best-fit exponential curve $y = Ce^{kx}$ for the following data sets, and graph both the data and this curve.

5. (1, 0.3), (2, 0.7), (3, 2)
6. (1, 1), (2, 4), (3, 8), (4, 16)
7. (1, 11), (2, 8), (3, 6), (4, 4.5)
8. (1, 90,000), (3, 95,600), (4, 97,500), (6, 101,500)
9. The December 6, 1982, issue of *Newsweek* magazine reported on the studies of Dr. George Gerbner of Pennsylvania State University's Annenberg School of Communications on the effect of American television programming on American culture. Gerbner found, for example, that heavy television viewers described their fear of crime as "very serious" at a rate twice that of light viewers and also vastly overestimated the true number of doctors and lawyers in the population. Such results suggested that the number of hours spent watching television affects the accuracy of people's "perception of reality." *Suppose* that Table 6.3 represents the accuracy of perception of reality as a function of the number of hours an individual watches TV. Use linear regression to determine the accuracy of perception due to watching $3\frac{1}{2}$ hours of television per day. First, however, determine which piece of data could be omitted so as to obtain a more accurate conclusion.

Table 6.3

Hours of TV Viewing	Accuracy of Perception (percentage) of Reality
0	85
1	95
2	95
5	75
6	65
8	60

10. Predict U.S. exports of mineral fuels in 1985 based on Table 6.4.

Table 6.4

Year	Dollar Value of Exports (billions)
1980	8
1981	10
1982	12

11. Using linear regression, estimate U.S. domestic oil production in 1990, based on the figures in Table 6.5 (given in billions of 42-gal barrels). Inappropriate data should be omitted (that is, data from earlier years may be misleading).

Table 6.5

Year	Barrels (billions)
1950	1.9
1960	2.4
1967	3.0
1970	3.2
1974	3.0
1977	2.8

12. At about age 28, most people are at the peak of their lifetime physical capacity. Table 6.6 treats 28 as year 0 and peak physical capacity as 1 (unit). The figures corresponding to years beyond age 28 are the percentages of peak capacity available to average individuals at that age. Fit an exponential curve $y = Ce^{kx}$ to these data and determine

   a. The rate per year at which capacity decreases.
   b. The residual capacity available at age 62 (retirement?).
   c. The age at which a person's physical capacity is one-half of its peak.

Table 6.6

Years Beyond Age 28	Proportion of Physical Capacity Remaining
0	1.0
2	0.98
7	0.93
12	0.88
22	0.80

13. Table 6.7 shows the wind-chill index at a temperature of 20° F for various wind speeds. A wind-chill index tells how fast heat is lost; thus a 20° temperature in a 10-mph wind is equivalent to a temperature of 3° in still air. Using Table 6.7, determine the wind chill at a wind speed of 15 mph. You will first have to determine whether to model these data by a straight line with negative slope or by an exponential decay function $y = Ce^{-kx}$. [*Hint:* If an exponential curve is needed, then since ln y is not defined for $y < 0$, first add, say, $D = 20$ to each degree reading and find C and k. Then subtract $D = 20$ from the resulting function.]

Table 6.7

Wind Speed (mph)	Chill Index (°F)
5	19
10	3
20	−10
30	−18

14. As humidity rises, a given temperature becomes more uncomfortable. If you are engaged in a vigorous physical activity, such as playing tennis, the effect can be severe. One measure of this problem is a *heat-stress index*, which measures the combined effects of temperature and humidity. Table 6.8 gives a heat-stress index at an air temperature of 90° at various levels of humidity. For purposes of understanding this index, you can think of a heat stress of 70 as idyllic and of 110 as verging on the miserable.

   a. Model the heat-stress index by an exponential function $y = Ce^{kx}$.
   b. What is the heat stress at a humidity level of 90%?

Table 6.8

Humidity (air temperature: 90°)	Heat Stress
0	83
20	87
40	93
60	100
80	113

Challenge Problem

15. A certain country has experienced the growth of both its imports and its exports (in billions of dollars), as shown by the data given in Table 6.9. Use 1960 as a base year, and choose curves suitable to each data base to model the data. Determine, by visually inspecting the graphs of the resulting functions, when imports will (likely) equal exports.

Table 6.9

	Imports		Exports
Year	Dollar Value (billions)	Year	Dollar Value (billions)
1960	1.0	1960	3.0
1970	2.1	1970	9.5
1975	3.9	1975	12.0
1980	7.9	1980	16.0

6.5 Constrained Optimization and Lagrange Multipliers

In this section we study an efficient technique discovered by the eighteenth-century French mathematician Joseph Lagrange for finding maximum and minimum values of a function subject to a constraint on the underlying variables.

Consider Figure 6.27. In this section we learn to find the maximum and minimum only on that part of the surface defined by $z = f(x, y)$ that is found directly above a curve C lying in the xy-plane. Such problems arise naturally and repeatedly in applications of functions of two (or more) variables but are easiest to introduce first in geometric terms. The surprising feature in such a problem is that it has a simple, straightforward solution using a technique developed by Lagrange called the method of Lagrange multipliers.

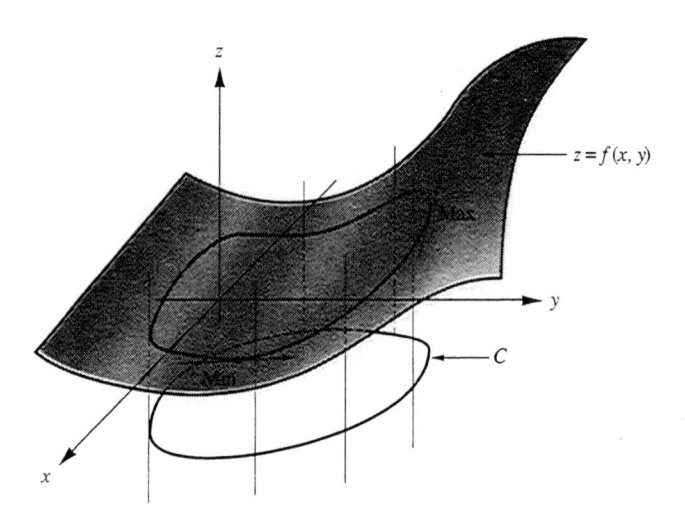

Figure 6.27

To explain and illustrate this method, we have to discuss the idea of a **constraint** and how it is related to the curve C in Figure 6.27. You may recall from Section 2.6 that a constraint equation is one that limits the range of variability of the variables in a problem. In that section a constraint equation was used to eliminate one variable in the problem. Lagrange's method allows us to avoid this often difficult and sometimes impossible step and deal directly with the problem as it arises.

Consider the graph of the hyperbolic paraboloid $f(x, y) = y^2 - x^2$ in Figure 6.28. Suppose that we wish to find the maximum and minimum of the graph *lying above* (and below) *the circle* C of radius 2 and center $(0, 0)$ in the xy-plane. You can visualize this task by imagining a vertical cylinder whose cross section is this circle. The intersection of this cylinder with the hyperbolic paraboloid is the curve whose maximum and minimum are sought. [In Figure 6.28, these appear to be at $(0, \pm 2)$ and $(\pm 2, 0)$, respectively.]

Let us convert this geometric problem into an algebraic one. The equation of a circle of radius 2 and center $(0, 0)$ is

$$x^2 + y^2 = 2^2 = 4$$

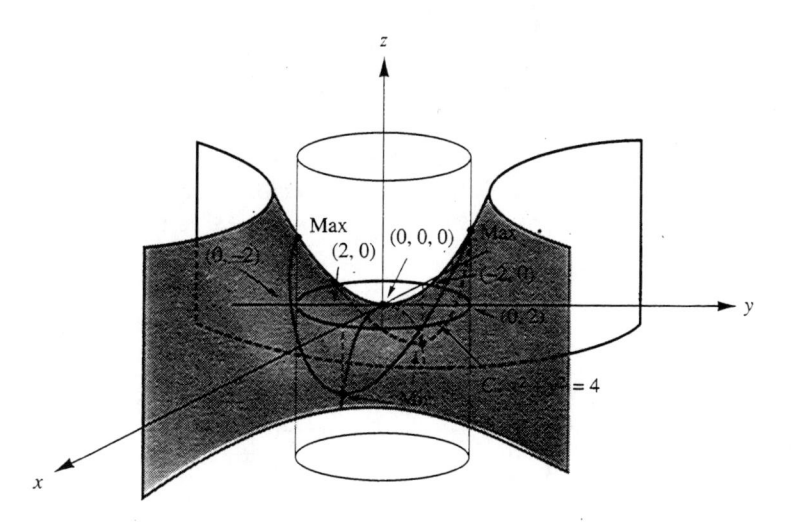

Figure 6.28

by the Pythagorean theorem. Rewrite this equation as

$$x^2 + y^2 - 4 = 0$$

and let

$$g(x, y) = x^2 + y^2 - 4$$

A point (x, y) is on the circle if and only if it satisfies the equation $g(x, y) = 0$.

The equation $g(x, y) = 0$ is called the **constraint equation** for the optimization problem illustrated in Figure 6.28. By considering only points such that $g(x, y) = 0$, we are confined, or constrained, to dealing only with points on the circle. The geometric problem illustrated in Figure 6.28 is described algebraically by the **constrained optimization problem:** Find the maximum and minimum values of

$$f(x, y) = y^2 - x^2$$

subject to the constraint

$$g(x, y) = x^2 + y^2 - 4 = 0$$

Here is Lagrange's method for solving such a problem:

Lagrange's Method of Constrained Optimization

To find *possible* maximum and minimum values of

$$f(x, y)$$

subject to a constraint

$$g(x, y) = 0$$

6.5 Constrained Optimization and Lagrange Multipliers

1. Let
$$L(x, y, \lambda) = f(x, y) + \lambda g(x, y)$$

The function **L** is called the **Lagrangian function,** and the variable λ is called a **Lagrange multiplier.**
2. Determine L_x, L_y, and L_λ.
3. Find all solutions (x, y, λ) to the three simultaneous equations

$$L_x = 0 \qquad L_y = 0 \quad \text{and} \quad L_\lambda = 0$$

Conclusion. Each such solution is called a *critical point* of the Lagrangian system. The desired maximum and minimum values, if any, will be found among these critical points.

EXAMPLE 1
Find the critical points of the Lagrangian system: Optimize $f(x, y) = y^2 - x^2$ subject to $g(x, y) = x^2 + y^2 - 4 = 0$.

Solution

Step 1. Form the Lagrangian function
$$L(x, y, \lambda) = y^2 - x^2 + \lambda(x^2 + y^2 - 4)$$

Step 2. Find
$$L_x = -2x + 2\lambda x$$
$$L_y = 2y + 2\lambda y$$
$$L_\lambda = x^2 + y^2 - 4$$

Step 3. Find all solutions to the simultaneous system of equations

$$-2x + 2\lambda x = 0 \tag{1}$$
$$2y + 2\lambda y = 0 \tag{2}$$
$$x^2 + y^2 - 4 = 0 \tag{3}$$

From Equation 1, canceling the number 2, we have

$$\lambda x - x = 0 \quad \text{or} \quad x(\lambda - 1) = 0 \tag{4}$$

This equation has two solutions: $x = 0$ or $\lambda = 1$.
Using $x = 0$ in (3) gives
$$0^2 + y^2 - 4 = 0 \quad \text{or} \quad y = \pm 2$$

Thus $(0, 2)$ and $(0, -2)$ are two critical points.
Using $\lambda = 1$ in Equation 2, we have
$$0 = 2y + 2(1)y = 4y$$

with solution $y = 0$. Now, putting this value of y into Equation 3 gives
$$0 = x^2 + (0)^2 - 4 = x^2 - 4$$

or $x = \pm 2$. Thus $(2, 0)$ and $(-2, 0)$ are also critical points.

This Lagrangian system has four critical points: (2, 0), (−2, 0), (0, 2), and (0, −2), and the Lagrange multiplier $\lambda = 1$. ∎

Remark. When we are finding the critical points of a Lagrangian system, it can be useful to solve for λ, but it is not necessary to do so. In Section 6.6, we will see that the value of λ has significance in applications.

We have seen twice before (Sections 2.5 and 6.3) that initial derivative tests locate only critical points of *possible* extreme value. In those sections each critical point had to be tested with a specified procedure to determine whether it yields a max or a min or neither. While a complete method is available for functions of a single variable, in the Lagrangian case, as in Section 6.3, only an incomplete method is available. And, like the D-test of Section 6.3, this test makes use of second derivatives.

The H-Test for Lagrangian Optimization

To test a critical point (a, b) for a Lagrangian system, let

$$H = (g_x)^2 f_{yy} - 2g_x g_y f_{xy} + (g_y)^2 f_{xx}$$

and evaluate $H(a, b)$.

1. If $H(a, b) > 0$, then $f(a, b)$ is a local minimum.
2. If $H(a, b) < 0$, then $f(a, b)$ is a local maximum.

If $H(a, b) = 0$, this test is inconclusive.

EXAMPLE 2

Apply the H-test to the critical points (0, 2), (0, −2), (2, 0), and (−2, 0) of Example 1. Classify each as yielding a maximum or a minimum.

Solution With $f(x, y) = y^2 - x^2$ and $g(x, y) = x^2 + y^2 - 4$, we first find that

$$g_x = 2x \qquad g_y = 2y$$

and $\qquad f_{xx} = -2 \qquad f_{yy} = 2 \qquad \text{and} \qquad f_{xy} = 0$

It is often much more practical to form H as a function of x and y before evaluating H at a critical point, particularly when we have a number of critical points to test. We have

$$\begin{aligned} H &= (g_x)^2 f_{yy} - 2g_x g_y f_{xy} + (g_y)^2 f_{xx} \\ &= (2x)^2 \cdot 2 - 2 \cdot 2x \cdot 2y \cdot 0 + (2y)^2(-2) \\ &= 8x^2 - 8y^2 = 8(x^2 - y^2) \end{aligned}$$

We then have

$$H(2, 0) = 8(2^2 - 0^2) = 32 > 0$$

and $f(2, 0) = -4$ is a local minimum value. Next
$$H(0, 2) = 8(0^2 - 2^2) = -32 < 0$$
and $f(0, 2) = 4$ is a local maximum value. The remaining two critical points, where $H(-2, 0) > 0$ and $H(0, -2) < 0$, yield a minimum and a maximum, respectively. Notice how these conclusions are in accord with the graph in Figure 6.28. ∎

When $H(a, b) = 0$ at a critical point (a, b), this procedure is inconclusive. Later, in Example 4, we indicate what may be done in such a case.

Working under a constraint is the natural order of things. One important application of the Lagrangian method is to Cobb–Douglas production functions in the presence of a constraint on available resources.

EXAMPLE 3

A firm utilizing x units of labor and y units of raw materials is able to produce $N(x, y) = 80x^{1/4}y^{3/4}$ units of a certain product. At the same time, each unit of labor costs \$5 and each unit of capital costs \$15. The firm has \$45,000 to expend in production. What level (x, y) of labor and raw materials should be employed to maximize production, subject to these constraints?

Solution We first put the constraints into the form of an algebraic equation. The firm will employ x units of labor at a cost of \$5 per unit, and y units of capital at a cost of \$15 per unit, and will expend \$45,000 between these two. Thus
$$5x + 15y = 45{,}000$$
We are to maximize production
$$N(x, y) = 80x^{1/4}y^{3/4}$$
subject to
$$g(x, y) = 5x + 15y - 45{,}000 = 0$$
This is done via Lagrange's method as follows:

Step 1. Let $L(x, y, \lambda) = 80x^{1/4}y^{3/4} + \lambda(5x + 15y - 45{,}000)$.
Step 2. Solve the three equations
$$L_x = 20x^{-3/4}y^{3/4} + 5\lambda = 0 \qquad (5)$$
$$L_y = 60x^{1/4}y^{-1/4} + 15\lambda = 0 \qquad (6)$$
$$L_\lambda = 5x + 15y - 45{,}000 = 0 \qquad (7)$$

Step 3. From Equation 5, $\lambda = -4x^{-3/4}y^{3/4}$ by solving for λ. From Equation 6, $\lambda = -4x^{1/4}y^{-1/4}$. Therefore
$$-4x^{-3/4}y^{3/4} = -4x^{1/4}y^{-1/4}$$
or
$$\frac{y^{3/4}}{y^{-1/4}} = \frac{x^{1/4}}{x^{-3/4}}$$
or
$$x = y$$

(Note from this preliminary conclusion that $\lambda = -4x^{1/4}x^{-1/4} = -4$.)
Substituting $x = y$ in Equation 7, we obtain

$$5x + 15x - 45,000 = 0$$

or

$$x = \frac{45,000}{20} = 2,250$$

Hence $y = x = 2,250$ also.

(Notice that the Lagrange multiplier λ is useful in solving for x and y in this system of equations.)

Having a solution to the Lagrangian system, we turn to the next step.

Step 4. We have

$$N_{xx} = \frac{-15y^{3/4}}{x^{7/4}} \qquad g_x = 5$$

$$N_{yy} = \frac{-15x^{1/4}}{y^{5/4}}$$

and

$$N_{xy} = \frac{15}{x^{3/4}y^{1/4}} \qquad g_y = 15$$

Because $x = y = 2,250$, the algebra here can be simplified by replacing y by its equal x and writing

$$N_{xx} = \frac{-15x^{3/4}}{x^{7/4}} = \frac{-15}{x}$$

and, similarly,

$$N_{yy} = \frac{-15}{x} \qquad \text{and} \qquad N_{xy} = \frac{15}{x}$$

Then
$$H = (g_y)^2 N_{xx} - 2g_x g_y N_{xy} + (g_x)^2 N_{yy}$$
$$= (15)^2 \left(\frac{-15}{x}\right) - 2(5)(15)\left(\frac{15}{x}\right) + (5)^2 \left(\frac{-15}{x}\right)$$
$$= \frac{-(15)^3 - 10(15)^2 - 15(5)^2}{x}$$

Since it is now clear that $H < 0$ (for any $x > 0$), we need not even bother to simplify further or to evaluate H at $(x, y) = (2{,}250, 2{,}250)$ and can immediately conclude that

$$N(2{,}250, 2{,}250) = 80(2{,}250)^{1/4}(2{,}250)^{3/4} = 80(2{,}250) = 180{,}000$$

is the maximum product possible subject to the given constraint on resources. ∎

Before going on, let us use this example to illustrate the following law of economics: *At the optimum level of production, the ratio of unit costs of labor and materials equals the ratio of their marginal productivities.* That is, note that the ratio of unit costs is

$$\frac{\text{Labor}}{\text{Materials}} = \frac{5}{15} = \frac{1}{3}$$

while the ratio of marginal productivities is

$$\frac{\frac{\partial N}{\partial x}}{\frac{\partial N}{\partial y}} = \frac{20x^{-3/4}y^{3/4}}{60x^{1/4}y^{-1/4}} = \frac{1}{3}\frac{y}{x}$$

At the optimum level in Example 3, $y = x = 2{,}250$ and this last ratio is again $\frac{1}{3}$. Exercise 25 asks you to show that this law of economics holds in general.*

In Section 6.6, we will also see that the Lagrange multiplier $\lambda = -4$ has a "real-world" interpretation as well. The number $4 = -\lambda$ coincides with the *"marginal productivity of investment"* in labor and material." This means that each additional dollar spent on labor and materials will yield four additional units of production.

Our next application shows how the Lagrangian method can sometimes be concluded when $H = 0$. In this example we also impose the additional constraint that x and y be positive.

EXAMPLE 4

Optimize $f(x, y) = x + 2y$ subject to $g(x, y) = 2 - xy = 0$ and x and y positive numbers.

Solution We ignore the added constraint that x and y be positive until step 4.

Step 1. Let $L(x, y, \lambda) = x + 2y + \lambda(2 - xy)$.

Step 2. Find x, y, and λ such that

$$L_x = 1 - \lambda y = 0 \tag{8}$$

$$L_y = 2 - \lambda x = 0 \tag{9}$$

$$L_\lambda = 2 - xy = 0 \tag{10}$$

Step 3. From Equation 8, $\lambda = 1/y$; and from Equation 9, $\lambda = 2/x$. Therefore $2/x = 1/y$ or $x = 2y$. Substituting this in Equation 10, we have

$$2 - (2y)y = 0$$

or

$$y^2 = 1$$

Thus $y = \pm 1$, and, consequently, $x = \pm 2$ and $\lambda = \pm 1$.

---

*A nice discussion of Lagrangian applications to economics can be found in J. Baxely and J. Moorhouse, "Lagrange Multiplier Problems in Economics," *American Mathematics Monthly* 91, no. 7 (1984): 404–12.

Step 4. Since we are only concerned with positive solutions, we discard the negative solutions and consider only $x = 2$ and $y = 1$. Since $f_{xx} = f_{yy} = f_{xy} = 0$, we see that

$$H = (g_x)^2(0) - 2g_x g_y(0) + (g_y)^2(0) = 0$$

and the H-test is inconclusive. However, for $x, y \geq 0$ the graph of f is the plane shown in Figure 6.29, and the constraint is the hyperbola $y = 2/x$ in the xy-plane. It appears from this figure that a minimum (and *no* maximum) value of f occurs above this hyperbola, and we conclude then that $f(2, 1) = 4$ is a minimum value. Since the Lagrangian system of equations has no other solutions (for positive x and y), we conclude that f has no maximum values.

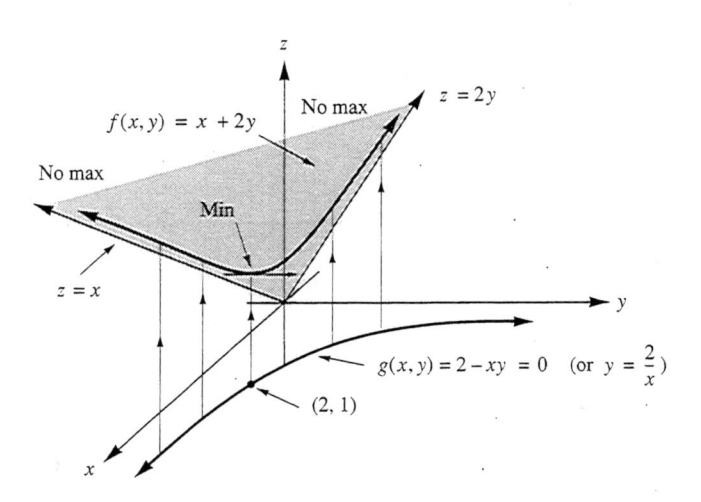

Figure 6.29

Since the graph of a function of two variables can be difficult to obtain, the method of Example 4 may not always suffice. If the graph of f is unavailable, we can form a reasonable conclusion by checking a number of values of f "near" the Lagrangian solution *that also satisfy the constraint*. If the value of f obtained at the solution is larger (smaller) than other values, this value is likely to be a maximum (minimum), but such a conclusion is *not* absolutely certain. In Exercise 15, we give another method of determining maxima and minima that avoids the H-test entirely but can be used only in the case of a "bounded constraint."

The method of Lagrange multipliers may be extended to any finite number of variables; only the algebra of solution becomes more involved. While an H-test is available in the presence of more than two variables, it is beyond the scope of this text.

Background Review 6.2

To solve the three equations in three unknowns resulting from $L_x = 0$, $L_y = 0$, and $L_\lambda = 0$, keep the following guidelines in mind:

1. It is often helpful to solve for λ first, but this was only partially true in Example 3.
2. Normally you must combine at least two equations to make any (and typically, partial) progress toward the solution. You may combine these in any algebraically correct manner that you choose. Your *aim is to eliminate* one (or more) *variables* or to obtain a simpler equation that must then be combined with the remaining equation.
3. If you have obtained a solution for λ (following guideline 1), substitute it in a *different* equation (following guideline 2). Then combine this with a third equation, if need be.
4. Once any one of the variables x, y, or λ is found, this value should be substituted in any remaining equation.
5. Once two variables are found, their substitution in any equation involving all three variables will result in a solution for the third remaining variable.

Remark. If in the course of solution you "cancel" a variable—for example, by writing $2\lambda x^2 = xy$ as $2\lambda x = y$—you *must* then also *include* $x = 0$ as a possible solution, since the cancellation is not valid if x is 0.

Exercises 6.5

Optimize the given function subject to the given constraint in Exercises 1–8.

1. $f(x, y) = x^2 + y^2 \qquad x - 2 = 0$
2. $f(x, y) = 1 - x^2 - y^2 \qquad y - x^2 = 0$
3. $f(x, y) = x^2 + y^2 - 2y + 1 \qquad x - y = 0$
4. $h(x, y) = y^2 - x^2 \qquad y + 2 = 0$
5. $h(x, y) = 1 - x^2 - y^2 \qquad 1 - x - y = 0$
6. $J(s, t) = s^2 - \dfrac{t^2}{2} \qquad s^2 + t - 1 = 0$
7. $f(x, y) = x^2 + xy - y^2 \qquad x^2 + xy - 1 = 0$
8. $P(x, y) = 9x^{1/3} y^{2/3} \qquad x + y - 6 = 0$

In Exercises 9–12 the H-test fails. Sketch the graph of f and the constraint g to determine maxima and minima.

9. $f(x, y) = y^3$
   $g(x, y) = x - 2$
10. $f(x, y) = y - x + 1$
    $g(x, y) = y - x^2 - 1$
11. $f(x, y) = y^2 - x^2$
    $g(x, y) = y + x - 2 = 0$
12. $f(x, y) = xy \qquad x \geq 0$ and $y \geq 0$
    $g(x, y) = x^2 + y^2 - 1$
13. Algebraic cancellation in the course of the Lagrangian solution yields eight critical points for the Lagrangian system: Optimize $f(x, y) = x^2 y^2$ subject to $x^2 + y^2 - 8 = 0$. Find and classify as maxima or minima all eight critical points.
14. Optimize $f(x, y) = e^{-x^2/y}$ subject to $y - x^2 - 1 = 0$, being careful in the course of solving the related algebraic equations not to eliminate solutions lost via algebraic cancellation.
15. The graph, in the xy-plane, of the continuous constraint $g(x, y) = x^2 + y^2 - 20 = 0$ is a circle. That is, the set of all points (x, y) satisfying this equation lies in a "bounded" region in the plane: It does not extend indefinitely in any direction. In such a case the H-test can be omitted entirely in a Lagrangian solution.

Instead we need only evaluate f at each critical point (a, b). The largest value obtained will be the maximum, the smallest will be the minimum, and such points must exist. (This is like the situation in Section 2.5 concerning the maximum and minimum values of a function $y = f(x)$ on a bounded, closed interval $[a, b]$.)

Optimize $f(x, y) = x + 2y + 3$ subject to $x^2 + y^2 - 20 = 0$ by finding all critical points and then evaluating f at each of these.

16. The function $f(x) = x^3$ provides a basic example of the insufficiency of the second derivative test in testing for extreme values in the case of one variable. In what way is the problem of optimizing $f(x, y) = x^3$ subject to $g(x, y) = x - 2 = 0$ similar? [*Hint:* After trying the H-test, graph f and g.]

17. A family wants to add a solar-heated family room—greenhouse to the south side of their home, as indicated in Figure 6.30. The cost of glass and other construction materials convinces them to constrain the length of the new walls enclosing this area to 36 ft. What dimensions should be used so as to maximize the area of the room?

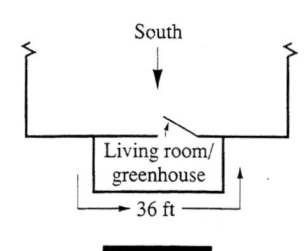

Figure 6.30

18. Please refer to Exercise 26, Section 6.3. Suppose that the manufacturer can afford to spend only $5,200 on shipping costs. How many units x and y should be sent to facilities A and B under this constraint so as to maximize net profit?

19. The number of breakdowns of a piece of production machinery is $B(x, y) = 3 + 0.5(xy)^2 - x - y$, where x and y are the numbers of replacements each month of two related parts P and Q of the machine. If P is replaced $\frac{3}{2}$ times as often as Q, so that $x = \frac{3}{2}y$, what number of replacements each month minimizes the number of breakdowns?

20. Please refer to Exercise 28, Section 6.3. Suppose that import constraints cause the auto manufacturer to set its price $200 higher in the country where x cars are sold. That is,

$$10,000\left(1 - \frac{x}{500,000}\right) = 9,000\left(1 - \frac{y}{600,000}\right) + 200$$

How many cars should be sold in each country so as to maximize profit under this constraint?

21. A company has an advertising budget of $20,000 and will spend x and y thousand dollars, respectively, on two kinds of advertisements for its product. It estimates its total sales to be

$$S(x, y) = \frac{100}{10 + (2/x)} + \frac{150}{15 + (3/y)} \text{ thousand units}$$

What allocation of its advertising budget will yield maximum sales?

22. In the manufacture of a certain product, total production is given by $N(x, y) = 2(xy)^2 - 8x - y$, where x and y are the number of units of labor and machinery used in production, respectively. Each unit of labor costs $2, each unit of machinery $4, and the total budget for labor and machinery is $10,000. How should these two resources be allocated so as to maximize total production?

23. A company has a market for 1,350 units of its product. Its production function is $P(x, y) = 25x^{1/2}y^{1/2}$ for x units of labor and y units of capital. Unit labor costs are $50, and unit capital costs are $150. Minimize the company's total cost $C(x, y) = 50x + 150y$, subject to the constraint $g(x, y) = 25x^{1/2}y^{1/2} - 1,350 = 0$.

24. Show that in Exercise 23 the ratio of marginal productivities P_x/P_y coincides with the ratio of unit costs, $50/150 = 1/3$, at the optimum cost level.

Challenge Problems

25. Let $P(x, y)$ be the production resulting from the employment of x units of labor and y units of capital. Suppose that one unit of labor costs a dollars and one unit of capital costs b dollars and the production budget is c dollars. Thus $g(x, y) = ax + by - c = 0$ is the appropriate constraint. By solving for λ in the resulting Lagrangian equations, show that at the optimum level of production subject to this constraint we have

$$\frac{\text{Marginal productivity of labor}}{\text{Unit labor costs}} = \frac{\text{Marginal productivity of capital}}{\text{Unit capital costs}}$$

That is,

$$\frac{P_x}{a} = \frac{P_y}{b}$$

26. Let $P(x, y) = kx^p y^q$ be the general Cobb–Douglas production function of labor x and machines y. If x costs C dollars per unit and y costs D dollars per unit, show that in order to maximize production

$$\frac{y}{x} = \frac{qC}{pD}$$

If $p = \frac{1}{3}$, $q = \frac{2}{3}$, $C = 3$, and $D = 6$, explain what this formula means, in ordinary English, in terms of the most efficient number of workers assigned to a machine.

27. A firm producing solar-heating equipment wishes to design a passive solar-heat-absorption device consisting of a metal rectangular box to be painted black and filled with water (see Figure 6.31). The box will face south and absorb the sun's heat (which will then be collected in a coil within the box and pumped elsewhere) according to the heat-absorption factors given in the table. So as to control overall weight, the box will be limited to a volume of 20 ft³.

Show that there is *no* set of dimensions for the box that maximizes solar-heat gain. [*Hint:* After finding the critical points, find a value of the function along the constraint that is larger than its value at any of these critical points.]

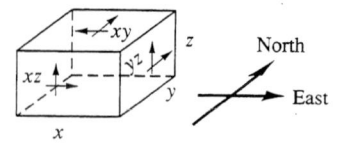

Figure 6.31

Heat-Absorption Factor, per Day (per square foot)

South	East	West	North	Top	Bottom
9	3	5	-4	12	-3

Remark. Since $xyz = 20$, if we set, say, $z = 1$, we obtain $y = 20/x$ as a constraint. By substitution in the objective function, how does this show that the lack of any optimal design is not surprising?

6.6 The Differential

The differential of a function of two variables allows us to estimate a change in function values in any direction based on the rates of change f_x and f_y in only the directions of the coordinate axes.

The graph of $z = f(x, y)$ is typically a curved surface like that in Figure 6.32. The two partial derivatives $f_x(a, b)$ and $f_y(a, b)$ measure the rate of change, or slope of the tangent lines, to this surface in the direction of the coordinate axes above the point (a, b).

Imagine, as in Figure 6.32, a *plane* that contains these two tangent lines to the surface. This plane will be tangent to the surface. If we now imagine moving from (a, b) to a nearby point $(a + h, b + k)$,* the surface will curve away from the tangent plane.

In this section we make use of only one idea: If we do not move very far—that is, if h and k are small—then the tangent plane will remain very close to the surface. This means that

$$f(a + h, b + k) \simeq f(a, b) + \text{(change in height of the tangent plane)} \quad (1)$$

---

*By moving h units in the x-direction and k units in the y-direction, where h and k can be positive or negative. In Figure 6.32, $h > 0$ and $k < 0$.

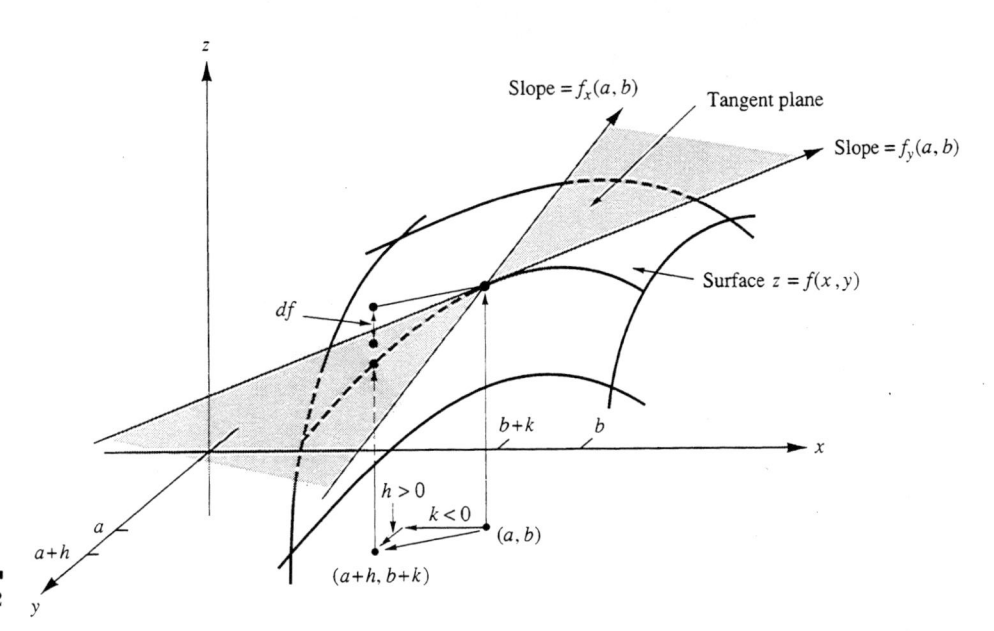

Figure 6.32

Definition

The change in height of the tangent plane to the surface of $z = f(x, y)$ above a point (a, b) when (x, y) moves from (a, b) to a nearby point $(a + h, b + k)$ is called the **differential** of f at (a, b) and is denoted by df.

The numerical value of df is given by

$$df = f_x(a, b) \cdot h + f_y(a, b) \cdot k$$

With this definition we can summarize the basic tool (1) of this section as an approximate equality:

$$f(a + h, b + k) \simeq f(a, b) + df \qquad (2)$$

How good an approximation is the expression (2)? A precise answer involves the second partials of f, as well as the size of h and k, and can be found in more advanced texts. The approximation is quite good in our first example.

EXAMPLE 1

Let $f(x, y) = xy$. (See Figure 6.33.)

a. Find the differential df at $(a, b) = (1, 2)$ for a change of $h = 0.05$ in x and $k = -0.03$ in y.

b. Evaluate $f(a, b) = f(1, 2)$ and

$$f(a + h, b + k) = f(1 + 0.05, 2 + (-0.03)) = f(1.05, 1.97)$$

c. Test the precision of (2) in this case.

6.6 The Differential **371**

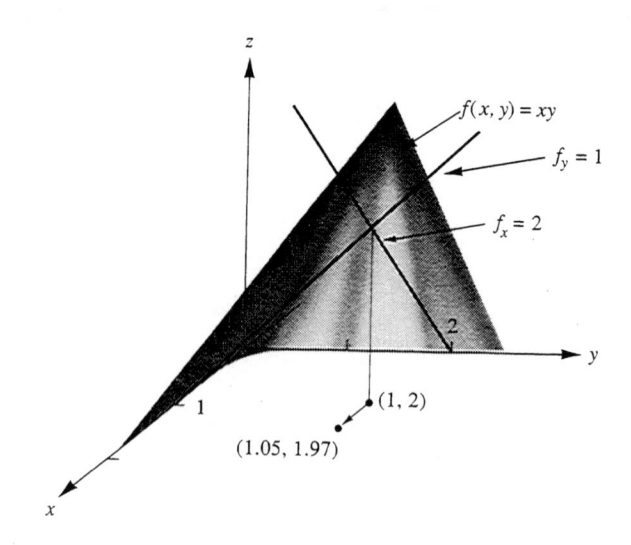

Figure 6.33

(figure shows surface $f(x,y)=xy$ with $f_y=1$, $f_x=2$, and points $(1,2)$ and $(1.05, 1.97)$)

Solution

a. Since $f(x, y) = xy$, then $f_x = y$ and $f_y = x$. Therefore
$$df = f_x \cdot h + f_y \cdot k = y \cdot h + x \cdot k$$
The value of df at $(1, 2)$ is then
$$df = 2(0.05) + 1(-0.03) = 0.07$$

b. We have $f(1, 2) = 1(2) = 2$ and $f(1.05, 1.97) = 1.05(1.97) = 2.0685$.

c. Substituting the numbers obtained in (a) and (b) in
$$f(a + h, b + k) \approx f(a, b) + df$$
we have
$$2.0685 \approx 2 + 0.07 = 2.07$$
a difference of only 0.0015. ∎

The differential of a function of two variables is much like that for a function of one variable (Section 3.3), though it plays a more important role for this reason: The number df is determined by knowledge of f itself in *only two* directions (along the axes, via f_x and f_y) but gives information about the change in f in *all* directions [as (2) tells us] when h and k are chosen appropriately.

Error Estimation Using the Differential

During medical checkups, one's height and weight are routinely measured. However, these cannot be measured *exactly*, since the ruler is accurate to no more than $\pm \frac{1}{8}$ in. and scales measure only pounds and, at best, ounces. This kind of imprecision, which is of no great importance in this case, is inherent in observing many systems and can be significant in some circumstances.

Suppose that in observing a system represented by a function $z = f(x, y)$, we observe (measure) a value $f(a, b)$. Due to unavoidable or inherent error in measuring techniques, however, the actual state of the system may instead be a value $f(a + h, b + k)$. Thus,

$$E = |f(a + h, b + k) - f(a, b)| = |\text{actual value} - \text{observed value}| \quad (3)$$

is the difference between what is actual and what is observed; it is the *error* in measurement. (It is necessary in Equation 3 to write the expression within absolute values because we have no way of knowing whether we err on the high side or on the low side.)

Because of the approximate equality (2), the error E can be estimated using the differential. If we move $f(a, b)$ in (2) to the left side, we obtain

$$f(a + h, b + k) - f(a, b) \simeq df \quad (4)$$

If we now combine Equations 3 and 4, we find that the error

$$E \simeq |df| = |f_x(a, b)h + f_y(a, b)k|$$

Consequently, we conclude that

$$E \lesssim |f_x(a, b)| \cdot |h| + |f_y(a, b)| \cdot |k| \quad (5)$$

where $\lesssim$ denotes an approximate inequality. The error estimate (5) is a practical tool and in some cases indicates the (mathematical) nature of the source of error as well.

EXAMPLE 2

For the purpose of this example, suppose that we are using an inexpensive calculator to multiply the number 8.1234 by 9.5678 and that the calculator discards the fourth digit in both factors before forming the product. Use $f(x, y) = xy$ to estimate the size of calculator error in the resulting product.

Solution The *actual* product is $f(8.1234, 9.5678)$. However, you will observe $f(8.123, 9.567)$ on the calculator screen because of the built-in (inherent) error in this system. Therefore, let $a = 8.123$, $h = 0.0004$, $b = 9.567$, and $k = 0.0008$.

The calculator error is

$$E = |\text{actual} - \text{observed}| = |f(8.1234, 9.5678) - f(8.123, 9.567)|$$

According to (5)

$$E \lesssim |f_x(a, b)| \cdot |h| + |f_y(a, b)| \cdot |k|$$
$$\lesssim |f_x(a, b)| \cdot (0.0004) + |f_y(a, b)| \cdot (0.0008)$$

With $f(x, y) = xy$, we have $f_x = y$ and $f_y = x$. Since $a = 8.123$ and $b = 9.567$, we have

$$f_x(a, b) = 9.567 \quad \text{and} \quad f_y(a, b) = 8.123$$

Consequently,

$$E \leqslant 9.567(0.0004) + 8.123(0.0008)$$
$$\leqslant 0.0038268 + 0.0064984$$
$$\leqslant 0.0103252 \ \blacksquare$$

Direct calculation (for example, using a calculator that can handle up to eight digits) shows that the *exact* error is 0.01032553. Thus the differential estimate of error in this example is a good one.

Example 2 also tells us something important about machine calculation in general. Notice that although the hypothetical calculation ignored the fourth decimal place at the start of its calculation, the resulting error in multiplication "migrated" to the second decimal. Here is where the differential is particularly useful, because we see that use of the differential suggests a *principle* about the accumulation of errors in machine multiplication: Since, with $f(x, y) = xy$, we have $df = yh + xk$, we see that errors in a product accumulate *as a sum* of errors in each factor. Even a very accurate computer, when told to compute a product with many factors, can yield a result whose error is considerably larger than the error arising from each individual factor.

EXAMPLE 3

In Figure 6.34, we illustrate the human lung as a compartment. We wish to estimate the flow rate F of blood through the lung in liters per minute. We will do so indirectly, in keeping with routine hospital procedures, by measuring the amount of oxygen consumed by the lungs in a 1-min period and by measuring the change in the concentration of oxygen (O_2) in the blood as it enters and then leaves the lungs.

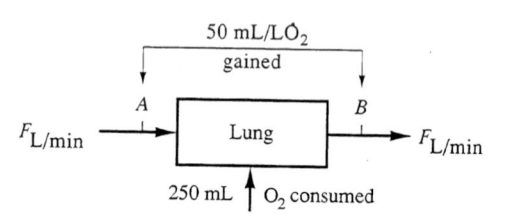

Figure 6.34

Suppose that obervations of a patient at rest for a 1-min period yield the following data:

1. At point A on the vein leading into the lung there are 150 mL of O_2 per liter of blood, and at the artery leaving the lung there are 200 mL of O_2 per liter, a *gain* of 50 mL of O_2 for each liter of blood flowing through the lung. This number may be in error by as much as 2% due to inherent equipment and procedure error.
2. During the 1-min period, the patient inhales 9,250 mL of O_2 and exhales 9,000 mL, thus consuming 250 mL of O_2. There is a known error of at most 1% in this observation.

Find (a) the (observed) amount of blood (in liters) that flows through the lungs during this 1-min period, (b) the inherent error in this measurement, and (c) the percentage error in this measurement of cardiac output.

Solution

a. Suppose that F liters per minute flow through the lung. Let x be the amount of oxygen consumed by the lungs in a 1-min period. This oxygen consumed can be accounted for only by the oxygen gained by blood flowing through the lung. If we let y be the amount of oxygen gained by each liter of blood, then we must conclude that

Oxygen consumed = (oxygen gain per liter) · (number of liters flowing through)

or $\qquad x = y \cdot F$

Consequently, $F = x/y$. The given observations about this patient are that $x = 250$ mL and $y = 50$ mL/L. Therefore we observe a flow rate of $F = \frac{250}{50} = 5$ L through the lung in this 1-min period.

b. However, neither observation x nor y can be made precisely accurate, and we must now deal with the inherent errors of 1% and 2%, respectively, in this system. That is, the actual amount of oxygen consumed could be $x = 250 \pm 0.01(250) = 250 \pm 2.5$, and the gain in concentration could be

$$y = 50 \pm 0.02(50) = 50 \pm 1 \text{ mL/L}$$

To estimate the inherent error in this system, we let $F = f(x, y) = x/y$ and regard the inherent error in measurement as

$$E = |f(250 \pm 2.5, 50 \pm 1) - f(250, 50)|$$

and next form the differential at (250, 50) for a change of $h = \pm 2.5$ and $k = \pm 1$. We first find $f_x = 1/y$ and $f_y = -x/y^2$. Consequently,

$$f_x(250, 50) = \frac{1}{50} = 0.02 \quad \text{and} \quad f_y(250, 50) = \frac{-250}{(50)^2} = -0.1$$

and using the differential estimate of error (5), we have

$$E \leq |f_x| \, |\pm 2.5| + |f_y| \, |\pm 1|$$
$$\leq 0.02(2.5) + 0.1(1) = 0.15 \text{ L/min}$$

c. Since our observation of blood flow was 5 L/min with an error of 0.15 L/min, our percentage error in observation is $(0.15/5)(100) = 3\%$. ∎

Thus the error in measuring the flow rate F is *triple* that of the known error in one of the factors determining this rate. Since the mathematics of this inherent error cannot change, it follows that only better instruments and careful personal attention to detail can improve the accuracy of such procedures.

The differential of a function of more than two variables is defined and used in exactly the same way as that of two variables; one additional partial derivative must be included for each additional variable. For example, the differential of a function $w = f(x, y, z)$ of three variables is defined by

$$df = f_x \cdot h + f_y \cdot k + f_z \cdot l$$

where h, k, and l are changes in the three variables x, y, and z, respectively. Indeed in Example 3, it is appropriate to use the function $f(x, y, z) = x/(z - y)$, where z and y represent the two measurements (of oxygen concentration entering and exiting the lung) that have to be physically made in such a hospital test procedure.

Analysis of, and by, the Lagrange Multiplier

The differential can be used to understand how the Lagrange multiplier λ, which appears in the solution of a constrained optimization problem, can be used to estimate how changes in the constraint affect the system under study.

Suppose that we wish to optimize $f(x, y)$ subject to $g(x, y) = 0$. From Section 6.5, we let $L(x, y, \lambda) = f(x, y) + \lambda g(x, y)$ and then solve for x, y, and λ in the simultaneous equations

$$L_x = f_x + \lambda g_x = 0 \tag{6}$$

$$L_y = f_y + \lambda g_y = 0 \tag{7}$$

and

$$L_\lambda = g = 0 \tag{8}$$

Suppose that a, b, and λ are the solutions. Then, for a change of h and k in the variables x and y at the solution (a, b), we have

$$df = f_x(a, b)h + f_y(a, b)k$$
$$= [-\lambda g_x(a, b)]h + [-\lambda g_y(a, b)]k$$

since from Equations 6 and 7, $f_x = -\lambda g_x$ and $f_y = -\lambda g_y$. Therefore

$$df = (-\lambda)[g_x(a, b)h + g_y(a, b)k]$$

or

$$df = (-\lambda) \, dg \tag{9}$$

Equation 9 gives a useful meaning to the value of the Lagrange multiplier λ, since dg represents a change in the constraint g.

In constrained optimization, the optimal state of the system changes in direct proportion $-\lambda$ to a change in the constraint, where λ is the Lagrange multiplier.

EXAMPLE 4

The firm in Example 3 of Section 6.5 produces $N(x, y) = 80x^{1/4}y^{3/4}$ units using x units of labor costing \$5 per unit and y units of material costing \$15 per unit, and has \$45,000 to expend on labor and material. How many additional units can be produced by the expenditure of one additional unit of labor and/or material at the optimal level of production?

Solution In Example 3 of Section 6.5, we saw that we must optimize $N(x, y) = 80x^{1/4}y^{3/4}$ subject to $g(x, y) = 5x + 15y - 45,000 = 0$. The optimal solution was found to be $x = 2{,}250$, $y = 2{,}250$, and $\lambda = -4$. According to Equation 9 and the preceding discussion,

$$dN = -(-4)\, dg = 4\, dg \qquad (10)$$

Therefore, if $dg = 1$, then $dN = 4$. Notice that $dg = 5h + 15k$ represents an increase in labor and/or material expenditure. Consequently, Equation 10 tells us that one additional unit of either or both expenditures will result in approximately four additional units of production *at the optimal level of production*. ∎

This example and Equation 9 tell us that the value of the Lagrange multiplier allows us to make direct comparisons, at the optimal level, between the level of a constraint g and its effect on an optimal amount f, as though these were directly related.

Exercises 6.6

In Exercises 1–6, use the approximate equality (2) and Example 1 as a model in order to estimate the indicated value of f using an easily calculated "nearby" value.

1. $f(x, y) = 3xy$; estimate $f(1.75, 9.05)$ using $f(2, 9)$ and $h = -0.25$, $k = 0.05$.
2. $f(x, y) = x^2 y$; estimate $f(3.2, 1.9)$ using $f(3, 2)$ and $h = 0.2$, $k = -0.1$.
3. $f(x, y) = \sqrt{x + y}$; estimate $f(0.9, 2.8)$ using $f(1, 3)$.
4. $f(x, y) = x^5 y^4$; estimate $f(0.9, 2.1)$ using $f(1, 2)$.
5. $f(x, y) = \sqrt{xy}$; estimate $f(-2.8, -26.5)$ using $f(?, ?)$.
6. $f(x, y) = x^{1/3} y^{2/3}$; estimate $f(7.9, -8.1)$ using $f(?, ?)$.

In Exercises 7–10, you are given a function and an observed function value at a point (x, y) along with an error in measuring (x, y). Use the differential via the approximate inequality (5) to estimate the possible error in the observed function value.

Function	Observed Value at (x, y)	Error (h, k) in Measuring (x, y)
7. $f(x, y) = \dfrac{x}{y}$	$f(1, 2) = \dfrac{1}{2}$	$(\pm 0.1, \pm 0.1)$
8. $f(x, y) = x^2 + y^2$	$f(1, 1) = 2$	$(\pm 0.02, \pm 0.49)$
9. $H(x, y) = x^9 (3 - y)^7$	$H(1, 4) = -1$	$(\pm 0.2, \pm 0.3)$
10. $g(x, y) = xe^{xy}$	$g(2, 0) = 2$	$(\pm 0.1, \pm 0.25)$

11. What size error is inherent in measuring the area of a tennis court 78 ft long and 36 ft wide with a 50-ft measuring tape that causes you to expect an error of $\frac{1}{4}$ in. in measuring the width and $\frac{3}{4}$ in. in measuring the length?

12. Suppose that you measure a cylinder and observe it to have length 3 ft and radius 6 in., with a possible error of 2% in both dimensions. What is the percentage error in determining the volume of the cylinder?

13. Since it is difficult to measure the radius of a cylinder, you instead decide to measure its length and *circumference*, finding these to be 2.5 ft and 4.2 ft, respectively, with an inherent error of 1% in the length and 2% in the circumference. What is the observed volume of the cylinder? Estimate the size of inherent error in determining this volume. What is the possible percentage error?

14. Suppose that in Example 3 there is a gain of 90 mL of O_2 per liter of blood leaving the lung, that the lungs inhale 300 mL of O_2 more than is exhaled in a 1-min period, and that both measurements are subject to a 0.75% error. Find
   a. The observed amount of blood flowing through the lung.
   b. The inherent error in this observation.
   c. The inherent percentage error.

15. A computer is told to find the product
$$(1.15)(2.14)(3.13)(4.12)(5.11)$$
and for some reason instead finds
$$(1.1)(2.1)(3.1)(4.1)(5.1) = 149.73651$$
Extend your concept of the differential of a function of two variables to a function of five variables, and use it to estimate the error in this calculation.

16. Gold weighs 1,200 lb/ft^3, or 11.8 oz/in.^{3}. You contract to buy a bar of gold measuring 1 by 2 by 6 in. and weighing 141.6 oz at a price of $350 per ounce. You wish to see your bar of gold before locking it away and want to measure it with a ruler. Because of its very slightly rounded edges, you realize it could possibly be 0.05 in. shorter than advertised in each dimension, so far as you can observe. Estimate the monetary value of this possible error in dimensions.

17. In Example 4, approximately how many additional units can be produced through the additional expenditure of $25 on labor and/or capital?

18. Statistical studies determine that the sales $Q(x, y)$ (as a function of unit price x and advertising cost y per unit) of a certain company are
$$Q(x, y) = 250{,}000 + 10{,}000xe^{-x/4}(1 - e^{-y}) \text{ units}$$
Determine

   a. $Q(4, 0.5)$.

   b. The effect of raising the price $0.15 and increasing advertising by $0.10 per unit.

   c. The effect of lowering the price by $0.10 and decreasing advertising by $0.15.

19. Pentagon officials assure Armed Services Committee members that the five critical components of the Hyper-Armed-Remote-Maser-Laser-Effect-Star-Shot orbiting gunner they wish to acquire have individual reliability levels of 0.9, 0.95, 0.93, 0.98, and 0.95, for an overall reliability of
$$(0.9)(0.95)(0.93)(0.98)(0.95) \approx 0.74$$
a 74% reliability level. However, the officials allow that under wartime conditions these factors have a possible error of 8%, 10%, 12%, 10%, and 15%, respectively. Use the differential of $f(x, y, z, w, t) = xyzwt$ to estimate the inherent error in the 74% level of overall reliability. For example, $\Delta x = 0.9(0.08)$.

20. The stopping distance S for a braking car is proportional to its kinetic energy. That is, $S = S(w, v) = kwv^2$, where w is the weight of the car and v is its speed.

   a. Compute the stopping distance of a car weighing 3,000 lb and traveling 55 mph. [*Hint:* k will be a part of your answer.]

   b. Use the differential to approximate this distance if the car is 100 lb heavier and moving 5 mph faster.

   c. Find the ratio of the answer in (b) to that of (a) in order to obtain the additional *percentage increase* in stopping distance.

6.7 The Double Integral

This section introduces the integral of a function f of two variables, over a rectangle, as two partial integrations, one followed by the other, allowing the interpretation of this integral as the volume beneath the surface f and above the rectangle.

Consider a function $z = f(x, y)$ and suppose that its graph in three-dimensional space is as shown in Figure 6.35. Let R be the rectangle lying in the xy-plane bounded by the lines $x = a$, $x = b$, and $y = c$, $y = d$; we will denote this rectangle by $R = [a, b] \times [c, d]$.

We wish to define the double integral $\iint_R f(x, y) \, dxdy$ of f over the region R but will do so in terms of a sequence of two "partial" integrations, one followed by the other, in which we treat one of the variables (x or y) as a constant, much as partial differentiation treats one variable as a constant. Note that in this discussion f must be a continuous function and can have positive and/or negative values; for clarity, our figures will often indicate a nonnegative function.

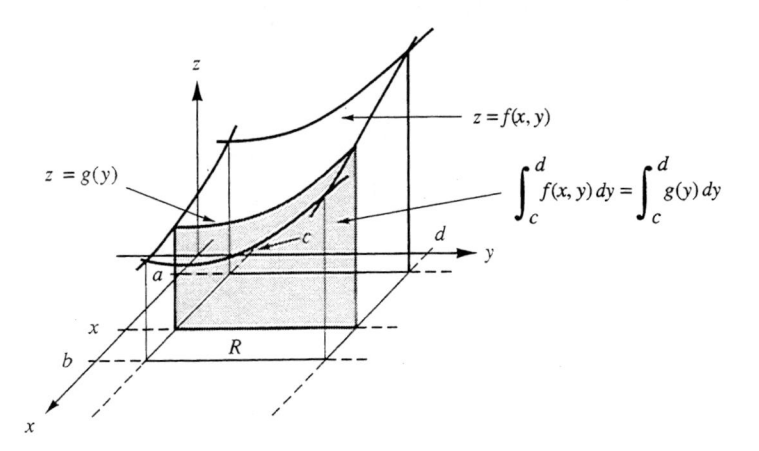

Figure 6.35

Fix a point x between a and b in Figure 6.35. Then consider the function $g(y) = f(x, y)$ of y alone, for $c \leq y \leq d$. The graph of g is, of course, the x-section of the graph of f in the vertical plane at x (parallel to the zy-plane).

We define the integral of f with respect to y (at x) as the definite integral of the function g of y alone:

$$\int_c^d f(x, y)\, dy = \int_c^d g(y)\, dy$$

Notice, then, that in Figure 6.35 $\int_c^d f(x, y)\, dy$ is the area beneath the x-section of the graph of f.

EXAMPLE 1

Let $f(x, y) = xy$, $R = [1, 3] \times [2, 4]$. Find $\int_2^4 f(x, y)\, dy$ as a function A of x and explain what this function represents.

Solution By the definition, with x treated as a constant, and $g(y) = xy$ (a linear function of y, with slope x), we have

$$\int_2^4 f(x, y)\, dy = \int_2^4 g(y)\, dy = \int_2^4 xy\, dy$$

$$= x\left(\int_2^4 y\, dy\right) = x\left(\frac{y^2}{2}\right)_2^4 = 6x$$

Therefore $\int_2^4 f(x, y)\, dy = 6x$ is a function A of x with $A(x) = 6x$. That is, after first treating x as a constant and "integrating y out," we obtain an answer for the integral that depends on (or varies with) x. If we then let x again take on its customary role as a variable, we see that the integral of f by y is a function of x, given by $A(x) = 6x$.

Next, consider what $A(x)$ represents. Figure 6.36 shows the graph of $f(x, y) = xy$. At the point x, $A(x)$ represents the area beneath the x-section of the graph of f. Now, as a function—that is, as x changes—A represents the changing area of the vertical region beneath the graph as x moves from 1 to 3.

6.7 The Double Integral

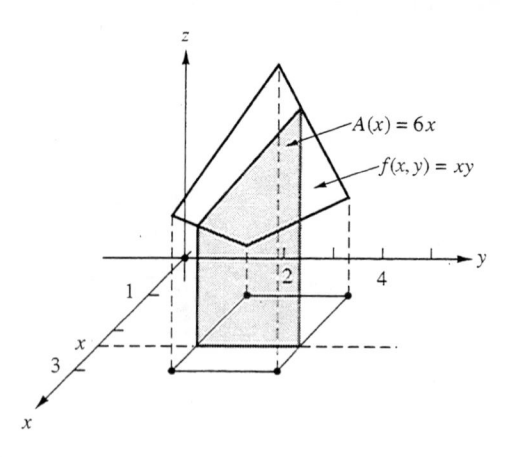

Figure 6.36

With this understanding we are ready to define the double integral of a function $z = f(x, y)$ over a rectangular region $R = [a, b] \times [c, d]$.

Definition

Let $z = f(x, y)$, $R = [a, b] \times [c, d]$, and $A(x) = \int_c^d f(y, x)\, dy$. Define the **double integral** of f over R by $\iint_R f(x, y)\, dydx = \int_a^b A(x)\, dx$. That is,

$$\iint_R f(x, y)\, dydx = \int_a^b \left[\int_c^d f(x, y)\, dy \right] dx$$

If $f \geq 0$, the fact that the function A is the area beneath x-sections of the graph of f helps us understand the following:

The double integral of $f \geq 0$ over R is the *volume* of the solid bounded by the graph of f over the region R.

We use Figure 6.37 to illustrate this, along with our understanding of the integral $\int_a^b A(x)\, dx$ as approximately equal to a Riemann sum of the form

$$A(x_1)\, \Delta x + A(x_2)\, \Delta x + \cdots + A(x_n)\, \Delta x \simeq \int_a^b A(x)\, dx \qquad (1)$$

That is, notice in Figure 6.37 that the typical term $A(x_i)\, \Delta x$ in this sum coincides with the *volume* of a solid of width Δx and planar area $A(x_i)$. The sum of all such volumes, represented by Equation 1, approximates the volume of the solid region in three-dimensional space beneath the graph of f over the region R.

380 Chapter 6 Functions of Several Variables

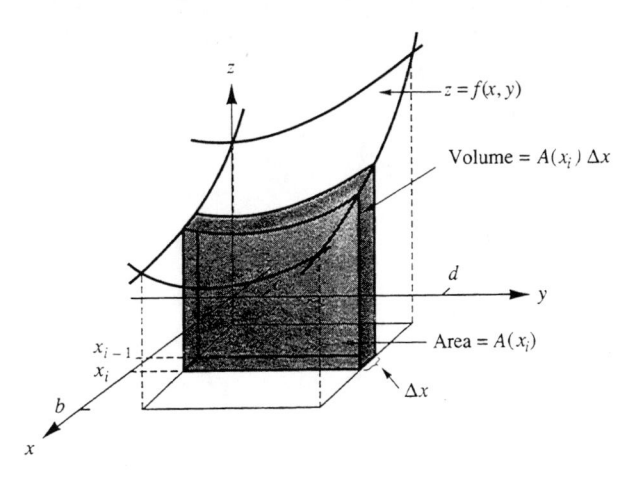

Figure 6.37

Consequently, the double integral just defined is the volume of this solid region. In less precise terms, the second integral (by x) in the defining equation

$$\iint_R f(x, y)\, dy\, dx = \int_a^b \left[\int_c^d f(x, y)\, dy \right] dx \qquad (2)$$

"sums" the cross-sectional areas $A(x)$ of x-sections of the graph of f, from a to b, to obtain the volume of the solid region. Indeed, this conclusion follows directly from our discussion in Section 4.5 of volume as the integral of cross-sectional area along a perpendicular "axial line" if we take the x-axis in Figure 6.37 as an axial line. Equation 2 is only a special case of that discussion.

EXAMPLE 2

Find the volume beneath the graph of $f(x, y) = xy$ above the region $R = [1, 3] \times [2, 4]$.

Solution By definition

$$\iint_R xy\, dy\, dx = \int_1^3 \left[\int_2^4 xy\, dy \right] dx$$

In Example 1, we found that $\int_2^4 xy\, dy = 6x$. Therefore

$$\iint_R xy\, dy\, dx = \int_1^3 6x\, dx = 6\left(\frac{x^2}{2}\right)\Big|_1^3 = 24 \quad \blacksquare$$

Equality of the Iterated Integrals

Let us now indicate why, for continuous functions f, the order of integration (first by y, then by x; or first by x, then by y) does not matter. That is, for a continuous function f,

$$\int_a^b \left[\int_c^d f(x, y)\, dy \right] dx = \int_c^d \left[\int_a^b f(x, y)\, dx \right] dy \qquad (3)$$

Each of these integrals is called an **iterated integral**.

In Figure 6.38, we interpret $\int_a^b f(x, y) \, dx$ (where $f \geq 0$) as the area $B(y)$ beneath a *y-section* of the graph of f for y between c and d.

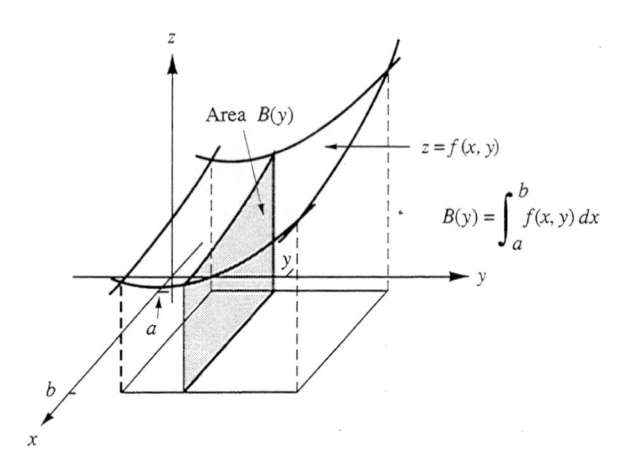

Figure 6.38

As before, the integral

$$\int_c^d B(y) \, dx = \int_c^d \left[\int_a^b f(x, y) \, dx \right] dy$$

"sums" these y-sectional areas from c to d, resulting in an integral that again measures the volume of the solid beneath the graph of f over R. Thus both iterated integrals yield the same volume and so must be equal; this can be rigorously shown as well, using Riemann sum arguments.

EXAMPLE 3

Verify that indeed

$$\int_0^1 \int_0^2 (3xy^2 + x^2) \, dx \, dy = \int_0^2 \int_0^1 (3xy^2 + x^2) \, dy \, dx$$

as an example of Equation 3.

Solution We first find the integral on the left:

$$\int_0^1 \int_0^2 (3xy^2 + x^2) \, dx \, dy = \int_0^1 \left[3y^2 \int_0^2 x \, dx + \int_0^2 x^2 \, dx \right] dy$$

$$= \int_0^1 \left(3y^2 \left[\frac{x^2}{2}\right]_0^2 + \left[\frac{x^3}{3}\right]_0^2 \right) dy = \int_0^1 \left(6y^2 + \frac{8}{3} \right) dy$$

$$= 6\left[\frac{y^3}{3}\right]_0^1 + \frac{8}{3}\left[y\right]_0^1 = 4\frac{2}{3}$$

Now the integral on the right is

$$\int_0^2 \int_0^1 (3xy^2 + x^2) \, dy \, dx = \int_0^2 \left[3x \int_0^1 y^2 \, dy + x^2 \int_0^1 1 \, dy \right] dx$$

$$= \int_0^2 \left[3x \left(\frac{y^3}{3} \right) \Big|_0^1 + x^2 (y) \Big|_0^1 \right] dx = \int_0^2 x + x^2 \, dx$$

$$= \frac{x^2}{2} + \frac{x^3}{3} \Big|_0^2 = 4\tfrac{2}{3}$$

verifying Equation 3 in this example. ∎

The next example indicates that there is sometimes an advantage in choosing to integrate first with respect to one variable rather than with respect to another.

EXAMPLE 4

Find $\int_0^1 \left[\int_{-1}^2 y e^{xy} \, dy \right] dx$.

Solution The function $f(x, y) = y e^{xy}$ is known to be continuous over $R = [0, 1] \times [-1, 2]$. The first integral $\int_{-1}^2 y e^{xy} \, dy$ cannot be done by methods now available to us. However, this can be avoided if we use Equation 3, since

$$\int_0^1 \left[\int_{-1}^2 y e^{xy} \, dy \right] dx = \int_{-1}^2 \left[\int_0^1 y e^{xy} \, dx \right] dy$$

In the integral $\int_0^1 y e^{xy} \, dx$, y is a constant. Therefore

$$\int_0^1 y e^{xy} \, dx = y \int_0^1 e^{xy} \, dx = y \left(\frac{e^{xy}}{y} \right) \Big|_0^1 = e^{xy} \Big|_0^1 = e^y - e^0 = e^y - 1$$

Consequently,

$$\int_0^1 \left[\int_{-1}^2 y e^{xy} \, dy \right] dx = \int_{-1}^2 \left[\int_0^1 y e^{xy} \, dx \right] dy$$

$$= \int_{-1}^2 [e^y - 1] \, dy$$

$$= (e^y - y) \Big|_{-1}^2 = e^2 - 2 - (e^{-1} - (-1))$$

$$= e^2 - \frac{1}{e} - 3 \quad \blacksquare$$

Remark. It is customary to write either iterated integral and the double integral without brackets; thus

$$\iint_R f(x, y) \, dy dx = \iint_R f(x, y) \, dx dy$$

$$= \int_a^b \int_c^d f(x, y) \, dy \, dx$$

$$= \int_c^d \int_a^b f(x, y) \, dx \, dy$$

6.7 The Double Integral

Applications of the Double Integral

Our understanding of the double integral as volume allows an *approximation principle* (AP) for the double integral like that of Section 4.5 for the definite integral of a function of one variable. Through this principle we can recognize how the double integral can be applied.

In Figure 6.39, we illustrate a division of the intervals $[a, b]$ and $[c, d]$ into several shorter intervals of constant length Δx and Δy, respectively. These lengths in turn define a *grid* of small rectangles in the region R, each of area $\Delta x \cdot \Delta y$.

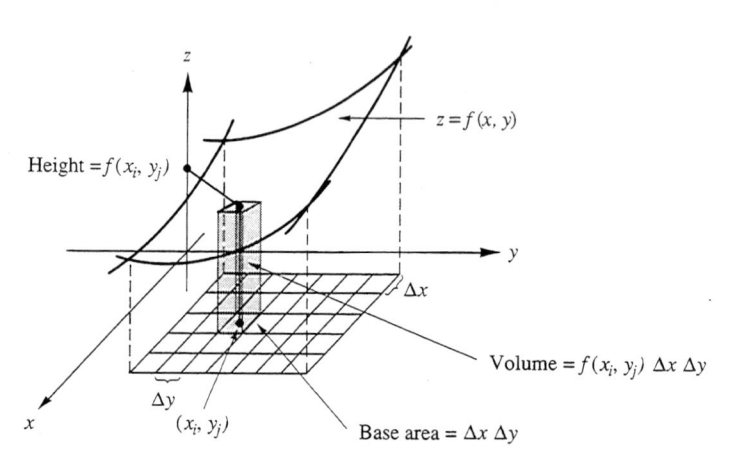

Figure 6.39

If (x_i, y_j) is a point in the indicated rectangle of area $\Delta x \cdot \Delta y$, the product $f(x_i, y_j) \Delta x \cdot \Delta y$ is the *volume* of the rectangular column of height $f(x_i, y_j)$ and base area $\Delta x \cdot \Delta y$. We can conclude, then, that the volume beneath the graph of f over the entire region R, $\iint_R f(x, y)\, dxdy$, can be approximated as closely as we like by a *sum* of the volumes of such rectangular columns of base area $\Delta x \cdot \Delta y$ and varying height $f(x_i, y_j)$, where (x_i, y_j) is chosen to be a point in each rectangular base. This indicates why the following approximation principle is valid for functions of two variables:

If a real number quantity Q can be approximated as closely as we wish by a sum of the form

$$f(x_1, y_1) \Delta x \cdot \Delta y + f(x_1, y_2) \Delta x \cdot \Delta y + \cdots + f(x_1, y_m) \Delta x \cdot \Delta y +$$
$$f(x_2, y_1) \Delta x \cdot \Delta y + \cdots + f(x_m, y_m) \Delta x \cdot \Delta y \quad (4)$$

then
$$Q = \iint_R f(x, y)\, dxdy$$

EXAMPLE 5

Suppose that $p(x, y) = 25x^{1/4}y^{3/4}$ is the production due to x units of labor and y units of materials at a single workstation in which not all workers and all materials are in constant use. Find the average production using up to 16 units of labor and up to 81 units of materials.

Solution As in Section 4.5, we imagine taking a sample of production levels $P(x_i, y_j)$ for $x_1, \ldots, x_n$ units of labor and $y_1, \ldots, y_m$ units of material; this means that a total of mn units of both variables are employed. The average production is then

$$\frac{P(x_1, y_1) + P(x_1, y_2) + \cdots + P(x_1, y_m) + P(x_2, y_1) + \cdots + P(x_2, y_m) + \cdots + P(x_n, y_m)}{mn} \quad (5)$$

Let $[a, b] = [0, 16]$ and $[c, d] = [0, 81]$ denote the ranges of the two variables, respectively. (That is, the interval from 0 to 1 counts as one worker, 1 to 2 counts as a second, and so on.) Now let $\Delta x = (b - a)/n$ and $\Delta y = (d - c)/m$. Then $mn = ((b - a)(d - c))/(\Delta x \, \Delta y)$, and Expression 5 equals

$$\frac{1}{(b - a)(d - c)} [P(x_1, y_1) \, \Delta x \cdot \Delta y + \cdots + P(x_i, y_j) \, \Delta x \cdot \Delta y + \cdots + P(x_n, y_m) \, \Delta x \cdot \Delta y]$$

The sum in this expression also approximates $\iint_R P(x, y) \, dxdy$, where $R = [0, 16] \times [0, 81]$. Therefore, taking the range of variability of x and y into account, the *average* production is from (4)

$$\frac{1}{(16)(81)} \iint_R 25 x^{1/4} y^{3/4} \, dxdy = \frac{1}{(16)(81)} \int_0^{81} \int_0^{16} 25 x^{1/4} y^{3/4} \, dx \, dy$$

$$= \frac{25}{1{,}296} \int_0^{81} y^{3/4} \left(\frac{x^{5/4}}{5/4}\right)_0^{16} dy$$

$$= \frac{25}{1{,}296} \cdot \frac{4}{5} \int_0^{81} y^{3/4} (32) \, dy$$

$$= \frac{5}{324} \cdot 32 \int_0^{81} y^{3/4} \, dy$$

$$= \frac{160}{324} \left(\frac{y^{7/4}}{7/4}\right)_0^{81}$$

$$= \frac{160}{7(81)} (81^{7/4}) = \frac{160}{7} (81)^{3/4} = \frac{160}{7} (27) \approx 617.14 \quad \blacksquare$$

A comparison of the approximation principle (AP) for both the double and single (Section 4.5) integrals shows that they are essentially the same. For this reason, the double integral can be applied to at least as wide a variety of experience as the single integral, based on the same principles. In addition, although we will not do so, the double integral can be defined over a nonrectangular region, and in practice can be computed over quite general regions R in the xy-plane if these are bounded by the graphs of known functions of x or of y. (A rectangle is only the special case in which such functions are constant and define straight sides for the region R.)

Finally, a multiple integral with similar properties may be defined for a function f of three, four, or any finite number of variables. While such integrals no longer represent area or volume, the approximation principle and its spirit of application still hold.

Exercises 6.7

In Exercises 1–6, find each integral as a function of x or y, as the case may be, treating the other letter as a constant.

1. $\int_0^1 x^2 y \, dy$

2. $\int_1^3 (xy + 1) \, dy$

3. $\int_0^2 (y^2 + x) \, dy$

4. $\int_{-1}^1 (3y + 1) \, dy$

5. $\int_{-1}^2 xe^{-y} \, dx$

6. $\int_0^3 \frac{2x}{y^2} \, dx$

Find the iterated integral of each continuous function in Exercises 7–14.

7. $\int_1^2 \int_0^1 (y - x) \, dx \, dy$

8. $\int_{-1}^1 \int_0^1 st \, ds \, dt$

9. $\int_0^1 \int_0^1 (x^2 y + xy) \, dx \, dy$

10. $\int_0^2 \int_{-1}^1 12x^2 y^3 \, dx \, dy$

11. $\int_0^2 \int_0^1 (x - y)^2 \, dx \, dy$

12. $\int_0^1 \int_0^1 e^{x-y} \, dy \, dx$

13. $\int_0^1 \int_0^1 2xye^{-x^2} \, dx \, dy$

14. $\int_2^4 \int_1^2 \frac{1}{xy} \, dy \, dx$

Find the double integral over $R = [0, 2] \times [1, 3]$ of each of the continuous functions f in Exercises 15–18.

15. $\iint_R (x - y) \, dx \, dy$

16. $\iint_R \frac{y}{x + 1} \, dx \, dy$

17. $\iint_R xe^{xy} \, dx \, dy$

18. $\iint_R \ln[(x + 1)y] \, dx \, dy$

19. Find the average production due to the use of 1 to 9 units of labor and 1 to 16 units of material in a workstation whose production function is $P(x, y) = 2x^{1/2} y^{1/2}$.

20. Find the volume of the solid bounded by the graph of $f(x, y) = 2 - x^2 - y^2$ above the square $[-1, 1] \times [-1, 1]$. Graph f and indicate the shape of this solid.

21. A jeweler's worktable measures 50 by 100 cm. The density (mass per unit area) of gold dust on the table is $2e^{-(x+y)} \text{g/cm}^2$ at a point x cm and y cm removed from the center of the table, measured in the direction of each of its two sides. At a price of $12.35/g, what is this gold dust worth?

22. The mouth of a small stream is 10 yd wide. At a point x yd from one side, $f(x, t) = e^{-t}((x - 5)^2 + 1) + 7$ million gal/hr/yd of water flow out of the stream, t hr after a rainstorm that took place a short distance upstream. How many gallons flow out of the stream during the first 5 hr (across the entire width of the steam)? [*Hint:* Over a length of time Δt containing time t, along a short width of stream Δx, and centered at a point x yd from one side, $f(x, t) \Delta x \Delta t$ gal flow out of the stream's mouth.]

Chapter 6 Summary

1. The graph of a function $z = f(x, y)$ of two variables is a surface in three-dimensional space that may be obtained by graphing sections of this surface in planes parallel to the xz-plane and the yz-plane separately. If $f(x, y) = ax + by + c$, then the graph of f is a plane that may often be obtained by finding the (at most three) intercepts of this plane with the three axes.

2. The *partial derivatives* $f_x = \partial f / \partial x$ and $f_y = \partial f / \partial y$ of a function of two variables are obtained by using the ordinary rules for differentiation and treating the remaining variable (and functions of it) as a constant. Each partial derivative f_x and f_y gives rise to two second partials f_{xx} and f_{xy}, and f_{yy} and f_{yx}. In this text $f_{xy} = f_{yx}$. The partial derivatives f_x and f_y represent the rate of change and slope of the tangent to the sectional curves of the graph of f in the x and y directions, respectively.

3. The first step in the *optimization* of a function $z = f(x, y)$ is to find all *critical points* (a, b), where both $f_x(a, b) = 0$ and $f_y(a, b) = 0$. The second step is to determine whether (a, b) yields a local max or local min for f. This is done by first determining whether $D(a, b) > 0$, where the discriminant $D = f_{xx} f_{yy} - (f_{xy})^2$. If $D(a, b) > 0$, then we check the sign of $f_{xx}(a, b)$. If $f_{xx}(a, b) < 0$ (> 0), then $f(a, b)$ is a local max (min). If $D(a, b) < 0$, then $f(a, b)$ is not a local max (min). If $D(a, b) = 0$, further information must be sought.

4. *Linear regression by least squares* is a method for determining the slope A and intercept B of a straight line $y = Ax + B$ that best fits (in the sense of least squares) a given set of data, $(x_1, y_1), \ldots, (x_n, y_n)$. (The relevant formulas for A and B are found in Section 6.4.) Exponential curves $y = Ce^{kx}$ may be linearized by the transformation $Y = \ln y = kx + B$ (where $B = \ln C$) so as to fit the curve to appropriate nonlinear data.

5. In *constrained optimization* we wish to *optimize* $f(x, y)$ subject to $g(x, y) = 0$ by means of the Lagrangian functions $L(x, y, \lambda) = f(x, y) + \lambda g(x, y)$, where $g(x, y) = 0$ is called a *constraint equation*. To do so we find solutions to the three equations

$$L_x = 0 \quad L_y = 0 \quad \text{and} \quad L_\lambda = 0$$

and use the *H*-test to classify each resulting *critical point* as a maximum or minimum, if possible.

6. The *differential* of a function $z = f(x, y)$ is the number

$$df = f_x(x, y)h + f_y(x, y)k$$

The fundamental relationship is that

$$f(x + h, y + k) - f(x, y) \approx df$$

or equivalently,

$$f(x + h, y + k) \approx f(x, y) + df$$

This relationship may be used to (1) estimate $f(x + h, y + k)$, given $f(x, y)$, and (2) estimate (inherent) error in observing a system modeled by f.

7. The double *integral* $\int_R \int f(x, y) \, dx dy$ of a (continuous) function f of two variables over a rectangular region $R = [a, b] \times [c, d]$ in the xy-plane is defined by

$$\iint_R f(x, y) \, dxdy = \int_a^b \left[\int_c^d f(x, y) \, dy \right] dx = \int_c^d \left[\int_a^b f(x, y) \, dx \right] dy \quad (1)$$

and when $f \geq 0$, it can be interpreted as the volume of the solid bounded by R and the graph of f. The integrals on the right in Equation 1 are called *iterated integrals*.

Chapter 6 Summary Exercises

1. Sketch the graph of $f(x, y) = x - y + 3$, and indicate the value of $f(1, 1)$ and $f(1, 2)$ on this graph.
2. Sketch the graph of $f(x, y) = 2x^2 - 3y^2$.
3. Explain in words what an x-section and a y-section of the graph of $z = f(x, y)$ are, and how these sections relate to each other so as to form the surface in three-dimensional space that is the graph of f.
4. Extend your answer to Exercise 3 to a discussion of how the partial derivatives $f_x = \partial f/\partial x$ and $f_y = \partial f/\partial y$ may be given geometrical meaning.

Find $\partial f/\partial x$, f_y, $\partial^2 f/\partial x^2$, f_{yy}, $\partial^2 f/\partial y \partial x$, and f_{yx} in Exercises 5–10.

5. $f(x, y) = x + 2xy + 3y$
6. $f(x, y) = x^2 e^y$
7. $f(x, y) = x^2 y^3$
8. $f(x, y) = e^x - e^{-y}$
9. $f(x, y) = \sqrt{x + y}$; $x, y \geq 0$
10. $f(x, y) = e^{x^2 y^2}$
11. Find the marginal rates of productivity due to labor x and capital y for the Cobb–Douglas production function $N(x, y) = 250 x^{1/5} y^{4/5}$ at an investment level of 32 units of labor and 8 units of capital.
12. Suppose that in Exercise 11, labor costs $7 per unit and capital $150 per unit, and that revenue is $30 for each unit produced. If fixed costs are $10,000, write a function that expresses profit in this enterprise and find the marginal profit separately due to labor, and due to capital, at an investment level of $(x, y) = (32, 8)$.
13. Find all critical points of $f(x, y) = xye^{x-y}$.
14. Find all extreme values of the function f in Exercise 13.
15. Find all critical points and all extreme values of
$$f(x, y) = 2x^2 - 3y^2 + xy + 5y - 5x$$
16. Find all critical points and all extreme values of $f(x, y) = y^3 - x^2$. [*Hint:* You will ultimately find it necessary to sketch a graph of this function.]
17. What is the sign ($\pm$) on f_x and f_y at the point (a, b) if $z = f(x, y)$ has the graph shown in Figure 6.40?

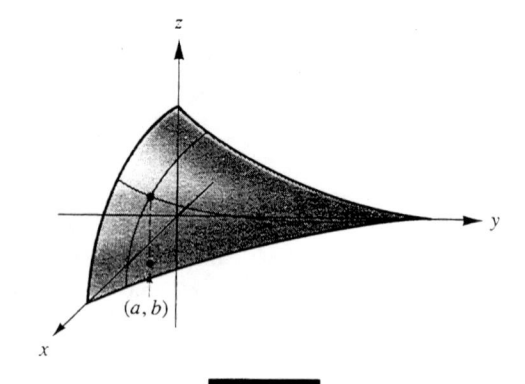

Figure 6.40

18. A wooden beam is to be cut from a log having a radius of 1 ft (see Figure 6.41). The load-bearing strength of the beam is proportional to xy^2. Determine the dimensions of the beam so as to maximize its load-bearing strength. [*Hint:* Let $f(x, y) = Bxy^2$, where B is a constant; note that your final answer is independent of B.]

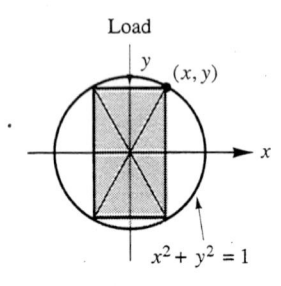

Figure 6.41

19. The profit $P(x, y)$ due to the sale of an item on which x thousand dollars is spent on newspaper advertisements and y thousand dollars on television advertisements is
$$P(x, y) = 30xy - 5xy^2 - 6yx^2$$
How much should be spent in each medium of advertising so as to maximize profit?

20. The variable cost of production of a certain item using x units of labor and y units of material is

$$C(x, y) = \frac{x^2}{8} + \frac{y^2}{50} - \ln xy$$

Find the allocation of labor and material that will minimize cost.

21. Maximize the triangular area enclosed by a 10-yd-long rope that is attached to the walls of an inside corner of a building, as shown in Figure 6.42. Use the Lagrange multiplier λ to estimate how much additional area would be created in this optimum configuration if the rope were $\frac{1}{2}$ yd longer.

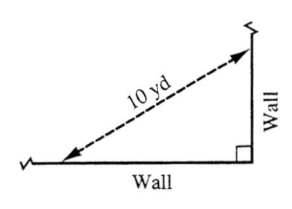

Figure 6.42

22. Apply the Lagrangian method to the constrained optimization problem: Optimize $f(x, y) = -x^2y$ subject to $y - x^2 - 1 = 0$.

23. Minimize the number of returns due to defects in an item that is checked x times for appearance and y times for performance during production when the number of returns per 1,000 units sold is

$$100 - x - 5y - 2xy$$

The cost of each appearance check is $0.01 per item and of each performance check, $0.04 per item; the budget for such checking is $155.

24. Optimize $f(x, y) = 1/(1 + x) + 1/[2(1 + 2y)]$ subject to $x + y = 3$.

25. Optimize $f(x, y) = xy$ subject to the constraint $x^2 + y^2 = 1$.

26. Find the best linear fit $y = Ax + B$ to the given data using the method of least squares.

x	y
-1	-2
0	-1
2	$\frac{1}{2}$
2	1

27. Find the best-fit exponential curve $y = Ce^{kx}$ to the given data using the method of a logarithmic transformation and the method of least squares.

x	y
0	$\frac{3}{2}$
1	1.0
2	0.9
-1	2.0
5	0.3

28. Find df for $f(x, y) = x^2y^3$ where $x = 4$, $y = 2$, $h = \frac{1}{8}$, $k = \frac{1}{6}$.

29. Let $f(x, y) = x/y$. Estimate $7.9/2.2$ using $f(8, 2) = 4$.

30. Estimate the possible error in observation of $f(x, y) = xy^2$ if $f(8, 2)$ is observed with a possible error of $(h, k) = (\pm 0.1, \pm 0.3)$ in measurement of x and y.

31. Give a geometric interpretation of the function values of $A(x) = \int_0^3 f(x, y)\, dy$ where $f(x, y) = xy$, and then of the double integral $\int_1^4 \int_0^3 f(x, y)\, dy\, dx$. Find the integrals in both cases.

32. Find the iterated integral $\int_{-1}^{1} \left(\int_0^2 xe^{-2xy}\, dx \right) dy$.

33. Find $\iint_R x + xy\, dx dy$ where $R = [-1, 2] \times [0, 1]$.

34. Find $\iint_R e^{x-y}\, dx dy$ where $R = [-2, -1] \times [1, 2]$.

35. Find $\int_{-1}^{1} \int_0^1 [4 - (x^2 + y^2)]\, dx\, dy$, and interpret your answer in geometrical terms.

36. Find the average production due to the use of 1 to 4 units of labor applied to 1 to 16 units of material for the Cobb–Douglas production function $P(x, y) = 50x^{1/4}y^{3/4}$.

37. A commercial forest plot has rectangular dimensions 10 mi north by 15 mi east measured from its southwest corner. At a point x miles north and y miles east from this point, there are

$$f(x, y) = 500e^{-0.5x}\left(1 + \frac{(y - 7.5)^2}{60}\right)$$

trees per square mile of trunk diameter ready for harvest. How many trees in the entire plot are ready for harvest?

Chapter 7

Applied Integration

Accumulation and Concentration

Integral models of natural systems have both wide and deep application because they allow us to compute a whole as the accumulation of its parts when these parts continually vary in the size of their contribution to the whole. A particularly broad category of application concerns contributions of varying density, where density expresses our natural concept of an amount per unit of length, area, volume, time, or other such common units of relative measurement. For example, we might wish to know the total population living alongside a length of roadway given information about the number of residents per mile. Alternatively, we can compute the rate of continuous flow of material through a compartment by measuring the density of the material exiting the compartment and combining this measurement with an integral model of the process, yielding applications to both medicine and business. These applications become even more powerful when we recognize that, in the presence of continual accumulation, additional gains and losses in the ongoing process are normal. Such continual arrivals and departures can be naturally modeled by an integral that allows us, for example, to compute the accumulating earnings of an investment, the future population of an urban area, or the present value of a stream of future income. All of these models become especially useful when we realize that we can apply them even when we have only incomplete information in the form of numerical data, using techniques of approximate integration.

7.1 Integration by Parts

In this section we learn the technique of integration by parts and learn to use a table of integrals before deriving further applications of the definite integral.

All formal integration techniques have the single aim of trying to *change the form* of a given integration problem to an "irreducible" integral form for which an antiderivative is known. So far as we are concerned here, there are only two irreducible integral forms: $\int u^n \, du$ and $\int e^u \, du$. Therefore **integration by parts** is a method for reducing a given integral to one of these two.

The Integration by Parts Formula

Integration by parts is based on the formula for differentiation of a product:

$$D_x[f(x)g(x)] = f(x)g'(x) + g(x)f'(x)$$

If we apply the indefinite integral to both sides of this equation, we obtain

$$\int D_x[f(x)g(x)] \, dx = \int f(x)g'(x) \, dx + \int g(x)f'(x) \, dx$$

Since $\int D_x[f(x)g(x)] \, dx = f(x)g(x) + C$, we can write this equation as

$$f(x)g(x) = \int f(x)g'(x) \, dx + \int g(x)f'(x) \, dx$$

Using algebra, we can then rewrite this last equation in the integration-by-parts formula

$$\int f(x)g'(x) \, dx = f(x)g(x) - \int g(x)f'(x) \, dx \qquad (1)$$

It will now take some explanation and several examples before you understand how to make use of this formula. For the majority of applications, we can simplify matters by introducing a special notation.

Definition

Let h and k be two functions. We define the *integral of h by k*, denoted by $\int h(x) \, dk(x)$, by

$$\int h(x) \, dk(x) = \int h(x)k'(x) \, dx$$

While this notation does remind us somewhat of the method of integration by substitution of Section 4.3, the concept here is a much different one. The integral of h by k is called a Riemann-Stieltjes integral. This integral has much application in its own right, but we will use it only as a notational form through which we record our computations in the integration-by-parts process.

EXAMPLE 1

Let $h(x) = e^x/x$ and $k(x) = x^2$. Find

a. $\displaystyle\int h(x)\, dk(x) = \int \frac{e^x}{x}\, d(x^2).$

b. Find $\displaystyle\int 3\, d(e^{2x}).$

Solution

a. From the definition, we begin by differentiating the function $k(x) = x^2$ and have

$$\int \frac{e^x}{x}\, d(x^2) = \int \frac{e^x}{x}(2x)\, dx$$

[for the left integral is $\int h(x)\, dk(x)$ and the right is $\int h(x)k'(x)\, dx$]. Canceling an x and continuing, we have

$$\int \frac{e^x}{x}\, d(x^2) = \int \frac{e^x}{x}(2x)\, dx = \int 2e^x\, dx = 2e^x + C$$

b. Because $D_x\, e^{2x} = 2e^{2x}$ we have, proceeding as in (a),

$$\int 3\, d(e^{2x}) = \int 3(2e^{2x})\, dx = 6 \int e^{2x}\, dx$$
$$= 6\left(\frac{e^{2x}}{2}\right) + C = 3e^{2x} + C\ \blacksquare$$

We now use this new notation (and it is *only* a notation!) to write the integration-by-parts formula (1) in the more efficiently used form

$$\int f(x)g'(x)\, dx = f(x)g(x) - \int g(x)\, df(x) \qquad (2)$$

by replacing the integral $\int g(x)f'(x)\, dx$ on the right side in Formula 1 by $\int g(x)\, df(x)$ in Formula 2.

Inspection of Formula 2 tells us that, given an integration problem involving a product of functions $f(x)g'(x)$, as on the left, we must determine the function $g(x)$ to use the formula for integration by parts. That is, if we regard an integration problem given in the form $\int f(x)g'(x)\, dx$ as made up of two distinct parts, $f(x)$ and $g'(x)$, we must *integrate the part $g'(x)$* [to find $g(x)$].

EXAMPLE 2

Find $\int x \ln x\, dx$ using integration by parts.

Solution The integrand consists of two parts, x and $\ln x$. We must first decide which of these is to play the role of $g'(x)$ in the integration-by-parts formula (2). Since we need to integrate the part $g'(x)$ to find the function $g(x)$, we let $g'(x)$ be the "simpler" of these two functions. In this example we let

$$g'(x) = x \quad \text{and} \quad f(x) = \ln x$$

so that $g(x) = x^2/2$ is an antiderivative of g'.

With $f(x) = \ln x$, $g'(x) = x$, and $g(x) = x^2/2$, we then have, by substitution in (2),

$$\int x \ln x \, dx = \int (\ln x) x \, dx = (\ln x)\left(\frac{x^2}{2}\right) - \int \frac{x^2}{2} \, d(\ln x)$$

To complete the solution, we proceed as in Example 1 and apply the definition of $\int g(x) \, df(x) = \int g(x) f'(x) \, dx$ to obtain

$$\int \frac{x^2}{2} \, d(\ln x) = \int \frac{x^2}{2}\left(\frac{1}{x}\right) dx$$

Therefore

$$\int x \ln x \, dx = \frac{x^2 \ln x}{2} - \int \frac{x^2}{2}\left(\frac{1}{x}\right) dx$$

$$= \frac{x^2 \ln x}{2} - \int \frac{x}{2} \, dx$$

$$= \frac{x^2 \ln x}{2} - \frac{1}{2}\int x \, dx$$

$$= \frac{x^2 \ln x}{2} - \frac{x^2}{4} + C \quad \blacksquare$$

Thus integration by parts changes the form of the given problem ($x \ln x$) to a new (but final) integration problem ($x/2$), which can easily be solved. This transformation of a given form to a simpler form is always our goal.

Let us now see how we may use the notation $\int h \, dk$ introduced earlier to shorten the integration-by-parts process further. Specifically, since

$$\int f(x) g'(x) \, dx = \int f(x) \, dg(x)$$

we may replace the integral on the *left* in the integration-by-parts formula (2) by this expression and obtain a form that may be easier to memorize:

$$\int f(x) \, dg(x) = g(x) f(x) - \int g(x) \, df(x) \qquad (3)$$

Much of what you must learn in this section is to "track" the notation of a given problem carefully through the notational *form* (3).

To use Formula 3, we must first write a given integral in the form on the left in (3). This is essentially the first step in the integration-by-parts process of Example 2.

EXAMPLE 3

Write $\int xe^{-x} \, dx$ in the form $\int f(x) \, dg(x)$.

Solution There are in fact two solutions, but let us first be clear about the meaning of the problem: We must regard $\int xe^{-x} \, dx$ as a problem in the form $\int f(x) g'(x) \, dx$, find $g(x)$, and then write the integral as $\int f(x) \, dg(x)$.

If we let $f(x) = x$ and $g'(x) = e^{-x}$, then $g(x) = -e^{-x}$ and we have $\int xe^{-x} \, dx = \int f(x) \, dg(x) = \int x \, d(-e^{-x})$.

A different integral results if we instead let $f(x) = e^{-x}$ and $g'(x) = x$, so that $g(x) = x^2/2$. Then

$$\int xe^{-x}\,dx = \int e^{-x}x\,dx = \int e^{-x}\,d\!\left(\frac{x^2}{2}\right) \blacksquare$$

Note that in either solution, if we apply the definition $\int h\,dk = \int hk'\,dx$ by differentiating the function that is written after the d, we obtain the original integral.

EXAMPLE 4

Use Formula 3 to find the integral $\int xe^{-x}\,dx$.

Solution Choosing $f(x) = x$ and $g'(x) = e^{-x}$ and using the first solution to Example 3 and then Formula 3, we have

$$\int xe^{-x}\,dx = \int x\,d(-e^{-x})$$

$$= x(-e^{-x}) - \int (-e^{-x})\,d(x)$$

$$= -xe^{-x} + \int e^{-x}\cdot 1\,dx$$

$$= -xe^{-x} + \int e^{-x}\,dx$$

$$= -xe^{-x} - e^{-x} + C \blacksquare$$

Formula 3 is the preferred formula for integration by parts because in many cases the part g' that we wish to integrate is sufficiently simple so that we can perform the integration mentally and write the solution using Formula 3 very directly.

EXAMPLE 5

Find $\int x^5 \ln x\,dx$.

Solution Regarding x^5 as $g'(x)$ and using (3), we have

$$\int x^5 \ln x\,dx = \int (\ln x)x^5\,dx = \int \ln x\,d\!\left(\frac{x^6}{6}\right)$$

$$= \frac{x^6}{6}\ln x - \int \frac{x^6}{6}\,d(\ln x) = \frac{x^6 \ln x}{6} - \frac{1}{6}\int x^6\!\left(\frac{1}{x}\right)dx$$

$$= \frac{x^6 \ln x}{6} - \frac{1}{6}\int x^5\,dx = \frac{x^6 \ln x}{6} - \frac{x^6}{36} + C \blacksquare$$

Remarks

1. In the following chapters, we will use Formula 3 as the preferred integration-by-parts formula.
2. Formula 3 allows us to see quickly when we have made a wrong choice of the part to integrate. If, using Formula 3 and integrating the part x in

Example 4, we write

$$\int xe^{-x}\, dx = \int e^{-x}\, d\left(\frac{x^2}{2}\right) = \frac{x^2 e^{-x}}{2} - \int \frac{x^2}{2}\, d(e^{-x})$$

$$= \frac{x^2 e^{-x}}{2} + \frac{1}{2}\int x^2 e^{-x}\, dx$$

the integration problem has become more, not less, complicated. This second method of first integrating x rather than e^{-x} will not lead to a solution. ∎

As a general rule we choose $g'(x)$ to be the "easier" of the two parts to integrate: e^{ax} is the easiest function to integrate, x^{n+1} is next, and so on. If this rule fails to result in a solution using Formula 3, we begin again and choose a different part to integrate.

The next example yields a formula that will repeatedly be used hereafter.

EXAMPLE 6

Show that

$$\int xe^{kx}\, dx = \frac{e^{kx}}{k^2}(kx - 1) + C \qquad (4)$$

for $k \neq 0$.

Solution Integrating the part e^{kx}, we have

$$\int xe^{kx}\, dx = \int x\, d\left(\frac{e^{kx}}{k}\right) = \frac{xe^{kx}}{k} - \int \frac{e^{kx}}{k}\, d(x)$$

$$= \frac{xe^{kx}}{k} - \frac{1}{k}\int e^{kx} \cdot 1\, dx = \frac{xe^{kx}}{k} - \frac{e^{kx}}{k^2} + C$$

$$= \frac{e^{kx}}{k^2}(kx - 1) + C \quad \blacksquare$$

In more difficult applications of the integration-by-parts formula, we need a separate integration process to determine $g(x)$ once we have chosen that part of the given problem that will play the role of $g'(x)$.

EXAMPLE 7

Find $\int [x/(x + 1)^2]\, dx$.

Solution While x is easy to integrate, this approach will not solve the problem (try it!). Instead let us integrate the remaining part, $1/(x + 1)^2$. We have

$$\int \frac{1}{(x + 1)^2}\, dx = \int (x + 1)^{-2}\, dx$$

$$= \int u^{-2}\, du \qquad \text{Letting } u = x + 1$$

$$= \frac{u^{-1}}{-1} = \frac{-1}{x + 1}$$

7.1 Integration by Parts 395

Therefore $\int \dfrac{x}{(x+1)^2}\,dx = \int x \cdot \dfrac{1}{(x+1)^2}\,dx = \int x\,d\!\left(\dfrac{-1}{x+1}\right)$

$$= x\!\left(\dfrac{-1}{x+1}\right) - \int \dfrac{-1}{x+1}\,d(x)$$

$$= \dfrac{-x}{x+1} + \int \dfrac{1}{x+1}\cdot 1\,dx$$

$$= \dfrac{-x}{x+1} + \ln|x+1| + C$$

where the final integral is found by letting $u = x + 1$. ∎

Finally, this technique can be used as well to find a definite integral.

EXAMPLE 8

Find $\int_1^2 \ln x\,dx$.

Solution Let us first find the indefinite integral $\int \ln x\,dx$. This has a subtle, and unusual, solution. Write $\int \ln x\,dx = \int (\ln x)\cdot 1\,dx = \int \ln x\,d(x)$.

Then
$$\int \ln x\,dx = \int \ln x\,d(x) = x\ln x - \int x\,d(\ln x)$$

$$= x\ln x - \int x\dfrac{1}{x}\,dx$$

$$= x\ln x - \int 1\,dx$$

$$= x\ln x - x + C$$

Hence
$$\int_1^2 \ln x\,dx = (x\ln x - x)\Big]_1^2$$

$$= 2\ln 2 - 2 - (1\cdot \ln 1 - 1)$$

$$= \ln 4 - 1 \qquad \text{Since } \ln 1 = 0$$

$$= \ln 4 - 1 \approx 0.3863 \quad \blacksquare$$

While the method of integration by parts is a good one, in general we use this method only after first inspecting a given problem for a possible solution by the *method of substitution* (Section 4.3) because this method is generally easier to apply.

Using a Table of Integrals

When the integration techniques of integration by substitution or by parts can simplify a problem no further, we may consult a table of integrals for its solution (see Table 7.1). A complete table of integrals, numbering into the hundreds of integrals, can be found in mathematical reference manuals in the library. The following examples indicate how to use Table 7.1.

EXAMPLE 9

Use Table 7.1 to find

$$\int x^{13} \ln x \, dx$$

Solution We begin by looking through the table for a formula having the same *form* for its integrand: $x^{13} \ln x$. Notice Formula 15 in Table 7.1:

$$\int x^n \ln x \, dx = x^{n+1} \left[\frac{\ln x}{n+1} - \frac{1}{(n+1)^2} \right] + C \quad n \neq -1$$

Applying this to $n = 13$, we have

$$\int x^{13} \ln x = x^{14} \left[\frac{\ln x}{14} - \frac{1}{14^2} \right] + C = \frac{x^{14}}{14} \left[\ln x - \frac{1}{14} \right] + C \quad \blacksquare$$

EXAMPLE 10

Find $\int_1^3 [1/x(2x+5)] \, dx$ using Table 7.1.

Solution Formula 13 in Table 7.1 gives

$$\int \frac{1}{x(ax+b)} \, dx = \frac{1}{b} \ln \left| \frac{x}{ax+b} \right| + C$$

Therefore, with $a = 2$ and $b = 5$,

$$\int_1^3 \frac{1}{x(2x+5)} \, dx = \frac{1}{5} \ln \left| \frac{x}{2x+5} \right| \Big]_1^3$$

$$= \frac{1}{5} \left(\ln \left| \frac{3}{11} \right| - \ln \left| \frac{1}{7} \right| \right)$$

$$= \frac{1}{5} \ln \left(\frac{\frac{3}{11}}{\frac{1}{7}} \right)$$

$$= \frac{1}{5} \ln \frac{21}{11} \quad \blacksquare$$

Formula 13, used in this example, is in fact derived by applying the method of integration by parts.

Table 7.1
Table of Integrals

1. $\int x^n \, dx = \frac{x^{n+1}}{n+1} + C, \quad n \neq -1$

2. $\int \frac{1}{x} \, dx = \ln|x| + C$

3. $\int e^x \, dx = e^x + C$

4. $\int \sqrt{x^2 \pm a^2} \, dx = \frac{x}{2} \sqrt{x^2 \pm a^2} + \frac{a^2}{2} \ln|x + \sqrt{x^2 \pm a^2}| + C$

5. $\int \frac{1}{\sqrt{x^2 + a^2}} \, dx = \ln|x + \sqrt{x^2 + a^2}| + C$

6. $\int \frac{1}{\sqrt{x^2 - a^2}} \, dx = \ln|x + \sqrt{x^2 - a^2}| + C$

(*Continued*)

(Table 7.1, continued)

7. $\int \dfrac{1}{a^2 - x^2}\,dx = \dfrac{1}{2a}\ln\left|\dfrac{a+x}{a-x}\right| + C,\quad x^2 < a^2$

8. $\int \dfrac{1}{x^2 - a^2}\,dx = -\dfrac{1}{2a}\ln\left|\dfrac{x+a}{x-a}\right| + C,\quad a^2 < x^2$

9. $\int \dfrac{1}{x\sqrt{a^2 - x^2}}\,dx = -\dfrac{1}{a}\ln\left|\dfrac{a+\sqrt{a^2-x^2}}{x}\right| + C,\quad 0 < x < a$

10. $\int \dfrac{1}{x\sqrt{a^2 + x^2}}\,dx = -\dfrac{1}{a}\ln\left|\dfrac{a+\sqrt{a^2+x^2}}{x}\right| + C$

11. $\int \dfrac{x}{ax+b}\,dx = \dfrac{x}{a} - \dfrac{b}{a^2}\ln|ax+b| + C$

12. $\int \dfrac{x}{(ax+b)^2}\,dx = \dfrac{b}{a^2(ax+b)} + \dfrac{1}{a^2}\ln|ax+b| + C$

13. $\int \dfrac{1}{x(ax+b)}\,dx = \dfrac{1}{b}\ln\left|\dfrac{x}{ax+b}\right| + C$

14. $\int \dfrac{1}{x(ax+b)^2}\,dx = \dfrac{1}{b(ax+b)} + \dfrac{1}{b^2}\ln\left|\dfrac{x}{ax+b}\right| + C$

15. $\int x^n \ln|x|\,dx = x^{n+1}\left[\dfrac{\ln|x|}{n+1} - \dfrac{1}{(n+1)^2}\right] + C,\quad n \neq -1$

Exercises 7.1

Find each integral in Exercises 1–4.

1. $\int \dfrac{1}{x}\,d(x^3)$

2. $\int x\,d\left(\dfrac{1}{x}\right)$

3. $\int x^9\,d(\ln x)$

4. $\int \dfrac{1}{x+1}\,d(xe^x)$

Write each integral in Exercises 5–8 in the form $\int h(x)\,dk(x)$. Give two solutions if you can.

5. $\int xe^{3x}\,dx$

6. $\int 3x^2 \ln x\,dx$

7. $\int 2x(x+1)^5\,dx$

8. $\int x^3 e^{-x/2}\,dx$

Use integration by parts to find each integral in Exercises 9–26.

9. $\int xe^{3x}\,dx$

10. $\int (x+1)^2 e^x\,dx$

11. $\int \dfrac{\ln x}{x^2}\,dx$

12. $\int x(1+x)^7\,dx$

13. $\int \sqrt{x}\ln x\,dx$

14. $\int xe^{-x}\,dx$

15. $\int x^2 e^x\,dx$

16. $\int (x+1)^2 e^{-x}\,dx$

[Hint: In Exercises 15 and 16, use integration by parts twice in succession.]

17. $\int x \ln 3x\,dx$

18. $\int x^{-3}\ln x\,dx$

19. $\int \dfrac{x+1}{e^x}\,dx$

20. $\int \dfrac{1}{\sqrt{x+1}}\,dx$

21. $\int \dfrac{xe^x}{(x+1)^2}\,dx$ [Hint: See Example 7.]

22. $\int x^3 e^{x^2}\,dx$

[Hint: Integrate the "part" xe^{x^2} by substitution.]

23. $\int e^{\sqrt{x}}\,dx$

[Hint: First, let $u = \sqrt{x}$; then find the resulting integral by parts integration.]

24. $\int x \ln x^3\,dx$

25. $\int 3xe^{1-2x}\,dx$

26. $\int (x^2 + x)e^x\,dx$

Find the area beneath the curve between $x = 0$ and $x = 1$ for the functions in Exercises 27 and 28.

27. $h(x) = xe^{-x}$

28. $f(x) = x \ln(x+1)$

Find the integral in Exercises 29–38 using either integration by parts or integration by substitution (Section 4.3).

29. $\int_1^3 \ln x\,dx$

30. $\int_0^1 xe^{x^2}\,dx$

31. $\int_{-1}^1 x^2 e^{x^3}\,dx$

32. $\int_0^1 e^x \ln e^x\,dx$

33. $\int_1^4 \dfrac{\ln(\ln x)}{x}\,dx$

34. $\int_0^1 e^{t+e^t}\,dt$

35. $\int_1^{100} te^{-0.05t}\,dt$

36. $\int_0^{40} (2t + 3)e^{-(1/10)t}\,dt$

37. $\int_0^1 x^2 \ln(x^3 + 1)\,dx$

38. $\int_1^2 x \ln x\,dx$

Find each integral in Exercises 39–47 using Table 7.1, the method of substitution, or integration by parts.

39. $\int \dfrac{1}{\sqrt{x^2 + 9}}\,dx$

40. $\int x^3 e^x\,dx$

41. $\int \dfrac{1}{y^2 - 3}\,dy$

42. $\int x^7 e^{3x}\,dx$

43. $\displaystyle\int_0^2 \frac{z}{3z-7}\,dz$

44. $\displaystyle\int_{-1}^1 3x^2 e^{5x^3}\,dx$

47. $\displaystyle\int \frac{1}{y(1-y)}\,dy$

45. $\displaystyle\int \sqrt{x^2-1}\,dx$

46. $\displaystyle\int_0^1 \frac{1}{4-x^2}\,dx$

7.2 Density and Accumulation

In this section we apply the derivative and integral to the study of density and distribution in order to measure the total amount of continuous accumulation.

Density is an abstract concept that you have used routinely for many years. A density measurement is characterized by the use of the word *per*, as in "people per square mile," "earnings per day," "bacteria per cubic milliliter," and so on. Calculus is particularly useful in describing the relationship between density and accumulation when density continually varies.

Told that an auto dealer uniformly sells 35 cars per month, you would immediately conclude that $35(12) = 420$ cars are sold in a year or, alternatively, that $35\left(\frac{1}{4}\right) = 8.75$ cars are sold in a week. In the first instance, knowledge of the "density" of cars sold *per month* immediately informs you of the **accumulation** of cars sold in a year. In the second instance, knowledge of density tells you, via a **proportion,** the number sold over a smaller unit of time.

On the other hand, the number of cars sold by a dealer is not likely to be uniform over each month but, more realistically, will *vary* with each month. We would still wish to know the total accumulation of sales in a year and, for purposes of scheduling services, we might also wish to know the number sold in a week's time using the relationship of density and proportion. When density continually varies, the Fundamental Theorem of Calculus allows us to compute total accumulation readily.

Density is most often encountered as an expression of amount per unit of length, area, or volume. Suppose that Section G of the football stadium has an area of $30 \times 50 = 1{,}500$ ft^2 and that among all the fans in Section G there are 75 especially dedicated ones wearing hats that resemble the head of the team's mascot. You could then conclude that there is $\frac{75}{1{,}500} = 0.05$ fan *per* square foot who is so decorated and, therefore, that in any 20-ft^2 area of Section G there is likely—on the average—to be at least

$$0.05 \text{ fan/ft}^2 \times 20 \text{ ft}^2 = 1$$

such fan. Of course there is no guarantee, from the available information, that such fans are *uniformly distributed* throughout Section G, and indeed they may instead all be concentrated among their own kind in some small corner of Section G. Thus while density, proportion, distribution, and accumulation are all related, only knowledge of *density at a point* can give you an accurate picture of the fans in Section G. The derivative will allow us to define the concept of density at a point.

Discrete Density and Accumulation

Imagine a 5-ft-long section of a conveyor belt upon which small boxes weighing 1 lb each are stacked at 1-ft intervals, as in Figure 7.1(a). The weight per foot of boxes on the conveyor belt varies along the belt. That is, the boxes are *not uniformly* distributed. Let $\rho(x)$ be the *weight* (of boxes) *per foot* at the xth foot. We see that $\rho(0) = 1$ lb/ft, $\rho(1) = 2$ lb/ft, $\rho(2) = 4$ lb/ft, $\rho(3) = 3$ lb/ft, $\rho(4) = 2$ lb/ft, and $\rho(5) = 1$ lb/ft.

At the same time, the total weight on the belt accumulates with increasing length. Let $F(x)$ be the total weight up to the point x on the belt. Then

$$F(0) = \rho(0) = 1 \text{ lb}$$

$$F(1) = \rho(0) + \rho(1) = 3 \text{ lb}$$

$$F(2) = \rho(0) + \rho(1) + \rho(2) = 1 + 2 + 4 = 7 \text{ lb}$$

and so on.

But perhaps, as in Figure 7.1(b), the boxes are not stacked so neatly. Then we cannot say exactly what the weight of the boxes per foot is at a point x, because this weight seems to vary *continually*. Nonetheless the boxes accumulate to a total weight. How can we describe a measure of density from a knowledge of accumulation? An answer is suggested when we observe in Figure 7.1(a) that, for example, if we divide the *change* in accumulation between $x = 2$ ft and $x = 1$ ft by the distance, 1 ft, between these marks, we obtain

$$\rho(2) = \frac{F(2) - F(1)}{1} \frac{\text{lb}}{\text{ft}} = \frac{7 - 3}{1} = 4 \text{ lb/ft}$$

a quotient that reminds us of the definition of the derivative of a function.

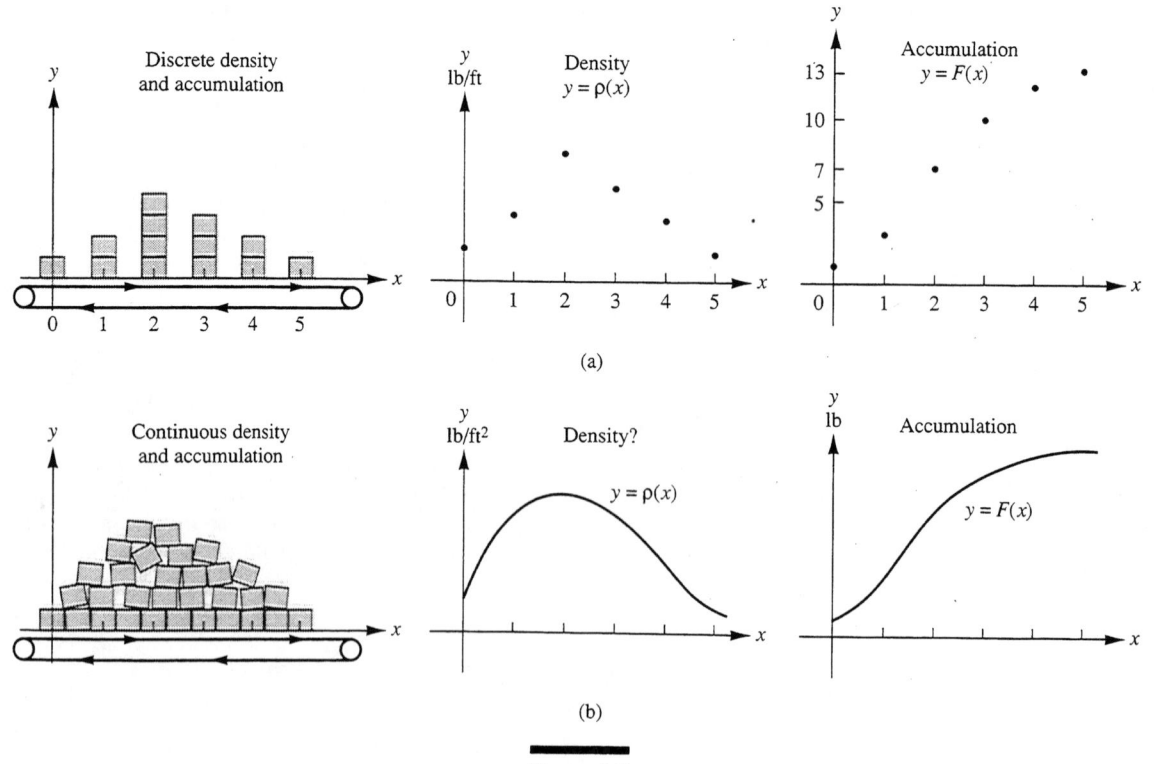

Figure 7.1

400   Chapter 7   Applied Integration

Continuous Density and Accumulation

Although many of our examples involve discrete units—single boxes weighing 1 lb each, individuals living along a street, and the like—we will find it much simpler, having the calculus at hand, to regard discrete items as being *continuously distributed, as though one blends smoothly into the other.* Let us evolve a general concept of density, distribution, and accumulation and their precise relationship under this assumption.

Imagine a major street leading out of the center of a city to its outskirts (see Figure 7.2). Let

$F(x) =$ total number of people living *within* x mi of city center along this street

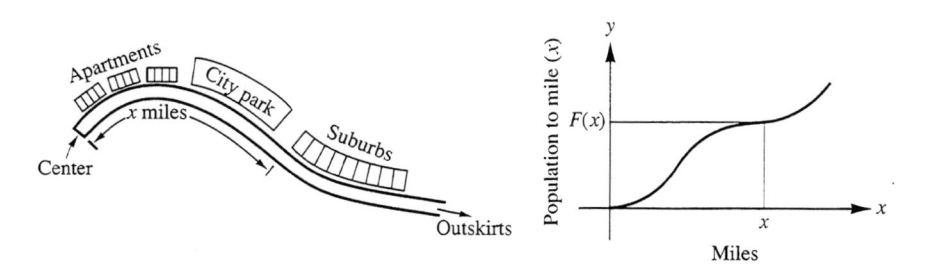

Figure 7.2

The function F is called a **cumulative distribution,** since it measures the distribution of the population as it accumulates with increasing distance from the city center.

In turn, the **difference quotient**

$$\frac{F(x + \Delta x) - F(x)}{\Delta x} : \frac{\text{residents}}{\text{mile}}$$

measures the average number of people per mile living between point x and point $x + \Delta x$ along the street. That is, $F(x)$ is the number of people living at locations up to point x, $F(x + \Delta x)$ is the number living up to point $x + \Delta x$, and the difference, $F(x + \Delta x) - F(x)$, measures the number of people living between points x and $x + \Delta x$. The difference quotient, then, is the average concentration of population between x and $x + \Delta x$.

Let us now take the limit

$$\rho(x) = \lim_{\Delta x \to 0} \frac{F(x + \Delta x) - F(x)}{\Delta x} \text{ residents/mile}$$

when it exists. The function ρ is called the population **density (or concentration) at the point** x.

What does $\rho(x)$ measure? This is difficult to express, because no one actually lives *at* a point x mi from the city center, since a point has no length. Instead $\rho(x)$ gives us a sense of how thickly population is concentrated *near* the point x, for, by definition of the limit defining $\rho(x)$, when Δx is sufficiently short,

$$\rho(x) \simeq \frac{F(x + \Delta x) - F(x)}{\Delta x}.$$

Therefore

$$\rho(x)\Delta x \simeq F(x + \Delta x) - F(x)$$

is approximately how many residents there are along the length Δx between x and $x + \Delta x$; this number of residents is approximately proportional to the length Δx by the proportionality factor $\rho(x)$.

EXAMPLE 1

Subway riders tend to board the central cars of a 200-ft-long subway train. There are $F(x) = 0.0001(100 - x)^3 + 3x - 100$ passengers riding on the first x feet of train between its front end, $x = 0$, and its rear, $x = 200$.
a. How many passengers are riding on the first 150 ft of the train? On the entire train?
b. Find $\rho(x) = F'(x)$, the density of riders per foot of subway, and graph this function.
c. What is the density of riders per foot, 150 ft from the front of the train? Approximately how many riders will be found in a 30-ft-long subway car located 150 ft from the front of the train? 190 ft?

Solution

a. There are $F(150) = 337.5$ passengers riding on the first 150 ft of the train and $F(200) = 400$ on the entire train, an average of $\frac{400}{200} = 2$ passengers per foot of train.
b. Here

$$\rho(x) = F'(x) = D_x(0.0001(100 - x)^3 + 3x - 100) = -0.0003(100 - x)^2 + 3$$

(see Figure 7.3).

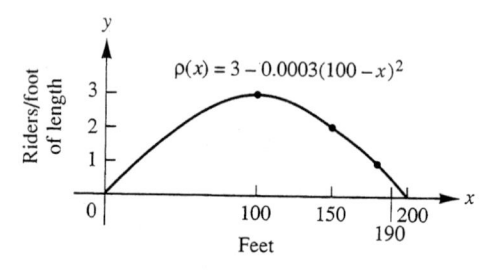

Figure 7.3

[Graph showing $\rho(x) = 3 - 0.0003(100 - x)^2$, with Riders/foot of length on y-axis and Feet on x-axis, marking points at 100, 150, 190, 200]

c. The density of passengers per foot of train length at $x = 150$ is $\rho(150) = 2.25$ people per linear foot. On a 30-ft-long car located at a point $x = 150$ ft from the beginning of the train, there will therefore be *approximately* $2.25(30) = 67.5$ riders since $\Delta x = 30$. At $x = 190$ ft, $\rho(190) = 0.57$, and there will be approximately $0.57(30) = 17.1$ riders on the last car on the train. ∎

Notice that while this discussion began with the idea of a cumulative distribution F, the density ρ seems more familiar, and in fact this is a quantity we can use and measure in practice.

The Relationship Between a Distribution Function F and a Density Function ρ

1. A density function is the derivative of a cumulative distribution function. That is,
$$\rho(x) = F'(x)$$

2. Therefore, from the Fundamental Theorem of Calculus,
$$\int_0^x \rho(t)\, dt = F(x) - F(0)$$
or
$$F(x) = F(0) + \int_0^x \rho(t)\, dt \tag{1}$$

where $F(0)$ is the *amount* at $x = 0$.

The function $f(x) = axe^{-bx}$ is an often-used model of varying density along a line (see Figure 7.4). Recall Formula 4 of Section 7.1, Example 6,

$$\int xe^{kx}\, dx = \frac{e^{kx}}{k^2}(kx - 1) + C \tag{2}$$

for use in the next example.

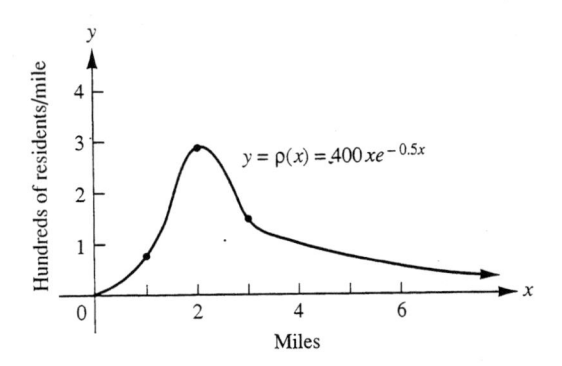

Figure 7.4

[Graph showing $y = \rho(x) = 400xe^{-0.5x}$, with y-axis labeled "Hundreds of residents/mile" and x-axis labeled "Miles"]

EXAMPLE 2

The population density along East Main St. of Hometown, USA, is
$$\rho(x) = 400xe^{-0.5x} \text{ residents/mile}$$

How many residents live along East Main St. within 4 miles of the center of Hometown?

Solution The graph of this density function is shown in Figure 7.4 and indicates that few residents live near the center of Hometown and that the highest concentration (density) of residents is approximately 290 residents per mile and occurs about 2 miles from the city center.

The number of residents at $x = 0$ is $F(0) = 0$ and using Formulas 1 and 2, with $k = -0.5$, the total number of residents living along East Main St. within 4 miles of city center is

$$F(4) = \int_0^4 400xe^{-0.5x}\, dx = 400 \int_0^4 xe^{-0.5x}\, dx$$

$$= 400 \left(\frac{e^{-0.5x}}{(-0.5)^2}((-0.5)x - 1) \right)\bigg|_0^4$$

$$= 400[4 - 12e^{-2}]$$

$$= 400(2.38) \simeq 952 \text{ residents} \quad\blacksquare$$

Up to now we have discussed only one-dimensional *linear density*, amount per unit of length along a line or curve. In the next example we ask you to extend this concept intuitively to two-dimensional density, amount per unit area, where a density function is given in pounds per square foot. In this example the density *changes in only one direction*, and using the approximation principle of Section 4.5, we can measure accumulation in two, and indeed three, dimensions.

EXAMPLE 3

A conical mound of sand of radius 5 ft has a density of $\rho(x) = 500 - 20x^2$ lb/ft², x ft from the center of the mound. How many pounds of sand are in the sandpile?

Solution We first envision the mound of sand and then the density of sand in the pile x ft from the center (see Figure 7.5). The density value $\rho(x)$ of course reflects the weight of the sandpile x ft from the center.

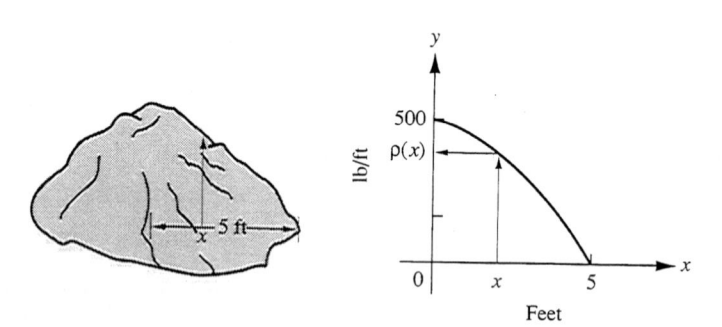

Figure 7.5

Next let us envision the mound as seen from above, as a circle of radius 5 ft (see Figure 7.6). Fix a point P along the radius, r ft from the center, and imagine a short distance Δr about this point. Then consider the ring illustrated in Figure 7.6(a). This ring is approximately $2\pi r$ ft long and Δr ft wide, as indicated in Figure 7.6(b). Its area, then, is approximately $(2\pi r)\Delta r$ ft².

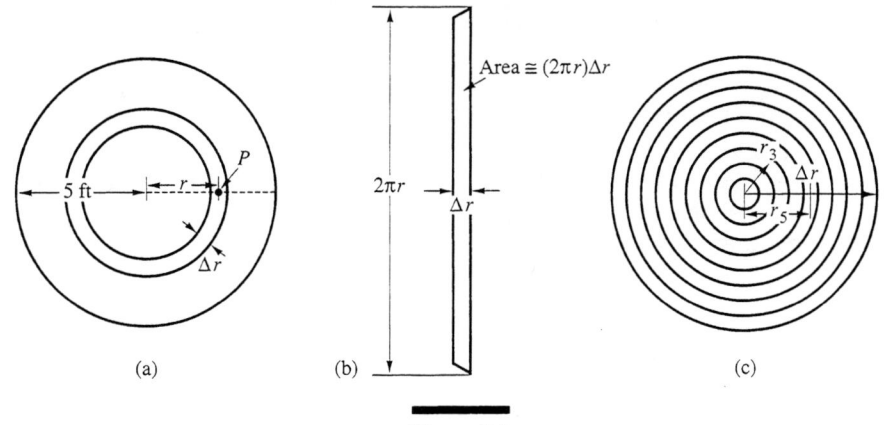

Figure 7.6

At the point r, the density of the sand is $\rho(r) = 500 - 20r^2$ lb/ft². Consequently, the weight of the sand piled only above the ring [in Figure 7.6(a)] is approximately

$$(500 - 20r^2)\,\frac{\text{lb}}{\text{ft}^2}\;\text{times}\;2\pi r \cdot \Delta r\;\text{ft}^2$$

or
$$2\pi r(500 - 20r^2)\Delta r \quad \text{lb}$$

Now imagine the circle in Figure 7.6(a) as the sum of (say, n) concentric rings, each Δr wide, emanating from its center [Figure 7.6(c)]. An approximation of the total weight of the mound of sand is then a Riemann sum of the form

$$2\pi r_1(500 - 20r_1^2)\Delta r + 2\pi r_2(500 - 20r_2^2)\Delta r + \cdots + 2\pi r_n(500 - 20r_n^2)\Delta r \quad \text{lb} \quad (3)$$

The approximation principle (AP) for the definite integral of Section 4.5 tells us that if an unknown quantity Q is *approximately equal* to a sum of the *form*

$$f(x_1)\Delta x + f(x_2)\Delta x + \cdots + f(x_n)\Delta x$$

then
$$Q = \int_a^b f(x)\,dx$$

If we apply AP to Equation 3, we obtain the total weight W of sand in the entire mound as the definite integral

$$W = \int_0^5 2\pi r(500 - 20r^2)\,dr \quad \text{lb}$$

since the form of the sum (3) tells us that it is a Riemann sum for the function $h(r) = 2\pi r(500 - 20r^2)$. Consequently, the weight of the sandpile is

$$W = 1{,}000\pi \int_0^5 r\,dr - 40\pi \int_0^5 r^3\,dr$$

$$= 1{,}000\pi \left[\frac{r^2}{2}\right]_0^5 - 40\pi \left[\frac{r^4}{4}\right]_0^5$$

$$= 6{,}250\pi \approx 19{,}635 \text{ lb} \quad \blacksquare$$

We will make repeated use of AP throughout the remainder of this chapter, in much the same way as was done in Example 3. In each case we will model some accumulating quantity by a Riemann sum approximation to its true value. Its true value will then be exactly equal to an appropriate definite integral.

Geologists and mining engineers often need to know the pressure existing at a given depth below the surface of the earth. This pressure depends on the weight of soil, rock, and liquids as this weight accumulates with increasing depth. A geologist, wishing to know the weight of the earth at a depth D, can obtain samples of materials at various depths by drilling and then construct a density function $y = \rho(x)$ in pounds per *cubic* foot at depth x from this data. The geologist then forms the integral

$$\int_0^D \rho(x) \, dx$$

to obtain the total weight per square foot at depth D. Note that the typical term for a Riemann sum approximating this integral is

$$\rho(x) \, \frac{\text{lb}}{\text{ft}^3} \times \Delta x \text{ ft} = \rho(x) \Delta x \, \frac{\text{lb}}{\text{ft}^2}$$

This integral model becomes an even more practical tool when coupled with the concept of approximate integration of Section 7.5, which enables us to calculate the integral directly from drilling data without the need to construct a formula for the density function ρ.

EXAMPLE 4

Core drilling into the Ogalala Aquifer beneath the Great Plains yields a density of material below the surface of

$$\rho(x) = 50 - \frac{x}{3} \quad \frac{\text{lb}}{\text{ft}^3}$$

x ft below the surface, to a depth of 60 ft (here the density changes only in the direction of changing depth).
a. Find a formula for the cumulative distribution $F(x)$ of the weight of material in a single square-foot-wide column of material above the point x, for $0 \le x \le 60$.
b. Estimate the depth at which the weight of material in this single square-foot column is 1 ton.

Solution The graph of the density function is shown in Figure 7.7(a). The density of materials is 50 lb/ft³ at the surface ($x = 0$) and declines to 30 lb/ft³ at a depth of $x = 60$ ft.
a. The cumulative distribution function F expresses the accumulating weight of materials as we move deeper below the surface. Since $F(0) = 0$, because

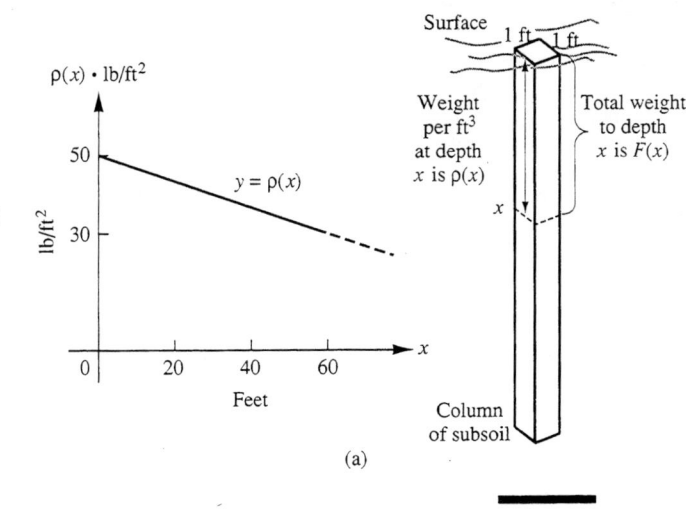

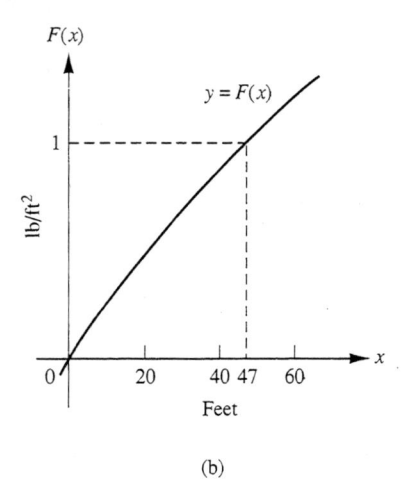

Figure 7.7

there is no material *at* the surface $x = 0$, from Equation 1 (p. 403) we have

$$F(x) = \int_0^x \rho(t)\, dt = \int_0^x \left(50 - \frac{t}{3}\right) dt$$

$$= \left(50t - \frac{t^2}{6}\right)\bigg|_0^x = 50x - \frac{x^2}{6} \text{ lb/ft}^2$$

b. The weight of materials in this square-foot column will reach 1 ton, or 2,000 lb, at that depth at which

$$2{,}000 = F(x) = 50x - \frac{x^2}{6}$$

We solve this quadratic equation by writing it as

$$-\frac{x^2}{6} + 50x - 2{,}000 = 0$$

or

$$x^2 - 300x + 12{,}000 = 0$$

and then applying the quadratic formula, yielding a depth of

$$x = \frac{300 \pm \sqrt{42{,}000}}{2} = 47.53,\ 252.47 \qquad \text{(ft)}$$

We disregard the answer of 252 ft [where $\rho(x)$ no longer applies] and conclude that at a depth of $x = 47.53$ ft the weight of materials is 1 ton per square foot. ■

Although well beyond the scope of this text, the modern medical techniques of tomography, PET scans, and the like are closely related to this discussion of density. All these methods avoid the hazard of x-rays to produce a cross-sectional picture of tissue inside the human body. Each relies on the construction of a density function whose values accurately reflect the type of tissue in the body; the values of this function are transformed by a computer into a picture. These techniques rely on an old theorem of pure mathematics discovered by J. Radon in 1914. The scanner records the value of the integral of an *unknown* density function, representing the accumulation of tissue along a line through the patient's body. Radon's theorem then asserts that if we know the value of this integral along *every* line through the body, then the function representing the density of tissue inside the body is uniquely defined, resulting in a picture of the body's interior organs (see Figure 7.8).

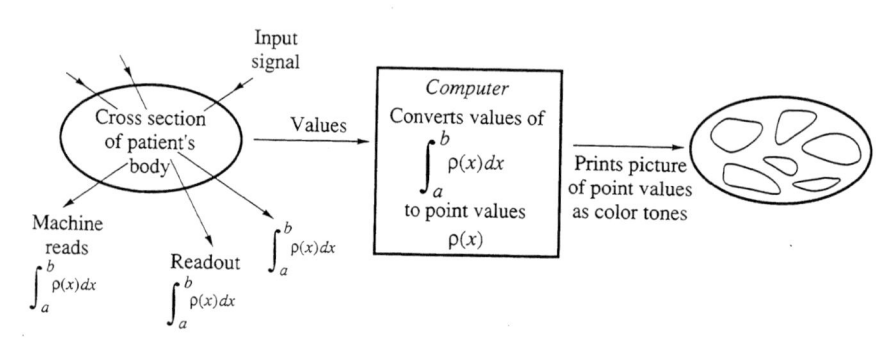

Figure 7.8

Exercises 7.2

Use Equation 2 (p. 403) to evaluate the integrals in Exercises 1 and 2.

1. $\int_0^{10} xe^{-2x}\,dx$

2. $\int_1^3 5xe^{x/4}\,dx$

In Exercises 3–8, find the cumulative distribution function F for the density function ρ. Graph F and ρ.

3. $\rho(x) = x, \quad 0 \le x \le 2$

4. $\rho(x) = 1 - x, \quad 0 \le x \le 1$

5. $\rho(x) = \dfrac{x}{1 + x^2}, \quad 0 \le x \le 5$

6. $\rho(x) = 3\sqrt{x}, \quad 0 \le x \le 16$

7. $\rho(x) = xe^{-x}, \quad 0 \le x < \infty$

8. $\rho(x) = 4xe^{-0.08x}, \quad 0 \le x < \infty$

9. Please refer to Example 1.

   a. How many passengers are riding on the first 50 ft of subway train? 100 ft?

   b. What is the density of riders per foot, $x = 50$ ft from the front of the train?

10. Please refer to Example 1. *Approximately* how many riders will be found on a 30-ft-long subway car centered at $x = 75$ ft from the front of the train?

11. The density of cholesterol deposits in an artery leaving the heart is $\rho(x) = 5e^{-0.02x}$ mg/cm along a 6-cm length of the artery.

    a. Determine the total amount (accumulation) of cholesterol along this 6-cm length.

    b. Determine the distribution $F(x)$ of cholesterol deposits up to a point x cm from the initial measurement $x = 0$.

    c. At what distance x do the deposits accumulate to a total of 20 mg?

    d. Graph the density and distribution functions.

12. The density of a certain species of Gulf Stream shrimp caught for commercial use is $\rho(x) = 75e^{0.15(20-x)}$ shrimp per cubic meter at a depth of x m from the ocean's surface, for $x \le 20$. Sketch the density function $y = \rho(x)$, and sketch the distribution

$y = F(x)$ of shrimp down to a depth of 20 m. How many shrimp will be found in the upper 5 m³ of the ocean at this particular place? [*Hint:* See Example 4.]

13. The western slope of a small mountainside in the Cascades is 15,000 ft long. Rain is falling at a density of $f(x) = 1 + 4x/15,000$ in.³ on each square foot of the mountainside at a distance of x ft from the foot of the mountain. How many cubic feet of rain fall on a 1-ft-wide path up the side of the mountain?

14. At a distance of x cm from the center of the site of a local infection, the density of white corpuscles is $\rho(x) = 20,000/(x^2 + 1)$ per square centimeter. If the infection site has a radius of 0.7 cm, how many white corpuscles are at the infection site? [*Hint:* See Example 3.]

15. At a distance of r mi from the detonation point of a nuclear bomb, the amount of radiation received at ground level is $\rho(r) = (1.875 \times 10^{12})e^{-0.125r}$ rems/mi. Regard the area affected by this radiation as a circle centered at the target, and compute the total rems striking the earth in this circle for a radial distance of 40 mi. (The acronym *rem* stands for *r*oentgen *e*quivalent in *m*an. A 1,000-rem dose of radiation is lethal in about 2 weeks.) [*Hint:* See Example 3.]

16. The centers of many European cities still retain a high population density. If the population of a certain city in France is $\rho(x) = 300e^{-0.08x}$ persons per square kilometer from the center and the city has a radius of 6 km, what is the population of the city?

17. A business is planning an advertising campaign to attract new customers. Because of its location and competition, it estimates that there are $\rho(x) = 45 - x^2$ customers per square mile who would be interested in its product at a distance of $x \le 6$ mi from its location. In how wide a circular area about its location should it advertise so as to reach 3,000 potential customers? [*Hint:* Let R be the unknown radius; solve the equation $\int_0^R 2\pi x(45 - x^2)\, dx = 3{,}000$ for an answer $R \le 6$, eventually using the substitution $z = x^2$ to solve the resulting algebraic equation by means of the quadratic formula.]

Challenge Problem

18. A lead smelter plant in a valley in Idaho is emitting 500 tons of lead particles into the valley area in a given fixed year. Assume that the valley is circular with a radius of 10 mi and that it is centered at the factory site. Because mountain ranges define the valley, assume that all 500 tons of lead settle in the valley. Measurements indicate that 250 tons of the particles settle in the immediate vicinity of the factory. An ecologist is studying this problem and decides to model the distribution of lead particles in the valley using the density function

$$\rho(x) = 250e^{-kx} \text{ tons/mi}^2$$

of lead particles deposited x mi from the factory site.

a. Model the total amount of lead particles deposited in the valley as an integral using Example 3.

b. Set this integral equal to the known total of 500 tons, and estimate the value of k by setting the term e^{-kx} that appears in the resulting equation equal to 0 and then solving for k. [This step is necessary because the equation cannot be solved without special methods (see Section A.5); if k is found to be large, then $e^{-5k} \approx 0$.] Insert this value of k into the original equation and check its accuracy.

c. Graph the density function using this value of k.

d. Using the value of k obtained in (b), calculate approximately how many pounds of lead settle on a 100-by-300-ft suburban lot located 3 mi from the factory site.

7.3 Flow Rate Through a Compartment

In this section we derive a formula for the rate of flow of a substance through a compartment using the concentration of trace materials in this substance.

Imagine a compartment whose interior is inaccessible and through which a substance is flowing at an unknown but constant rate F (measured in liters per second). Assume that thorough mixing occurs within the compartment. We will see how the integral may be used to determine *indirectly* the unknown rate of flow F.

This method joins mathematics and the physical world through the device of adding an amount A of "tracer" substance into the input side of the closed compartment followed by measurement of the **concentration** (*density* in grams per liter) of this tracer substance at the exit side of the compartment, until no further tracer appears at the exit (see Figure 7.9). The mathematical computation explained in this section then yields the flow rate F.

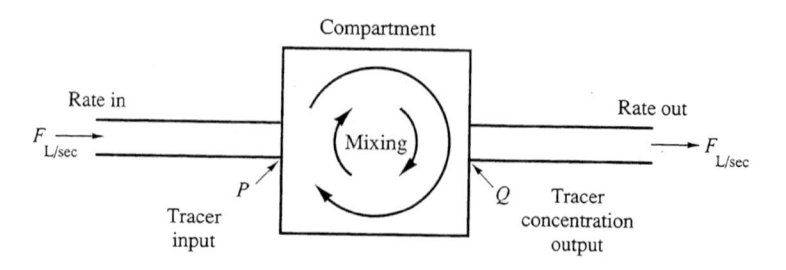

Figure 7.9

A good example of such a **compartment-mixing problem** involves a test procedure on the human heart. With the patient at rest, the rate of flow F of blood through the heart is to be measured. A harmless dye is injected into a major vein entering the heart, wherein it mixes with the blood flowing through the heart. The concentration of this dye is then measured at the aorta leaving the heart. With modern equipment this entire testing procedure can be performed in less than 1 minute.

Consider Figure 7.10 and imagine that, at time $t = 0$, A grams of a tracer substance has been injected into the compartment at point P. Let $C(t)$ denote the concentration, in grams per liter, of the tracer substance as measured at the exit point Q at time t. Suppose that T seconds after injection all of the tracer substance has been mixed in the compartment *and* has passed through the exit point Q.

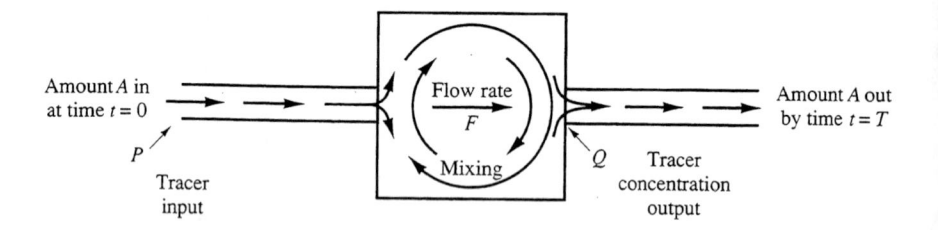

Figure 7.10

The key idea is this. The *same amount* A of substance entering the compartment must exit by time T, and this total amount must be determined by the varying concentration of the tracer as it exits the compartment. Let us see how this total amount is calculated from the concentration $C(t)$.

Consider a short span of time Δt centered about some time t_0 between $t = 0$ and $t = T$ (see Figure 7.11). At time t_0, the concentration of tracer at the exit is $C(t_0)$ grams per liter. Since fluid is exiting the compartment at the rate of F liters per second, then, *over this span* of time Δt, a total of $F \cdot \Delta t$ liters

410   Chapter 7   Applied Integration

exits the compartment Q. Therefore, since the concentration of tracer is $C(t_0)$ grams per liter at time t_0, *approximately* the amount

$$C(t_0) \cdot F \cdot \Delta t \text{ grams}$$

of tracer leaves the compartment over the span of time Δt.

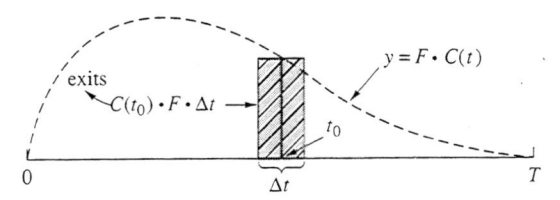

Figure 7.11

If we imagine the time interval from 0 to T to be the sum of many (say, n) short spans of time Δt, we see that the total amount of tracer leaving the compartment is *approximately*

$$F\ C(t_1)\Delta t + F\ C(t_2)\Delta t + \cdots + F\ C(t_n)\Delta t$$

Consequently, applying AP (Section 4.5), we conclude that the total amount A of tracer leaving the compartment is

$$A = \int_0^T F\ C(t)\ dt$$

Since the flow rate F is constant, we may then write

$$A = F \cdot \left(\int_0^T C(t)\ dt \right)$$

or, solving for F, that

$$F = \frac{A}{\int_0^T C(t)\ dt} \tag{1}$$

That is, the flow rate through the compartment can be measured as the amount of tracer injected into the compartment divided by the integral of the concentration of tracer exiting the compartment, over the amount of time it takes to clear the compartment of the tracer.

Note that while the preceding discussion is in terms of liters, grams, and seconds, it could equally well be in terms of liters, milligrams, and minutes, or any combination of similar units.

EXAMPLE 1

A veterinarian wishes to know the rate of flow F of blood through the heart of a sedated dog. At time $t = 0$, 3.5 mg of a harmless dye is injected into a vein entering the heart, and every second thereafter blood is sampled as it leaves the heart. The concentration of dye in the exiting blood is measured using an appropriate device. This process yields the graph of concentration shown in Figure 7.12. The function $C(t) = -0.096(t - 2.5)^2 + 0.6$ closely fits this data, and the veterinarian decides to use it as a model of the concentration of dye in the blood at time t.

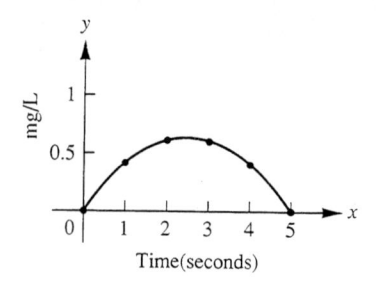

Figure 7.12

[Graph: y-axis labeled mg/L with marks at 0.5 and 1; x-axis labeled Time(seconds) with marks at 0, 1, 2, 3, 4, 5; curve shows an inverted parabola peaking near t = 2.5]

Calculate the rate of flow F of the blood leaving the dog's heart.

Solution According to Formula 1, we have

$$F = \frac{3.5}{\int_0^5 C(t)\, dt}$$

Now
$$\int_0^5 C(t)\, dt = \int_0^5 -0.096(t - 2.5)^2 + 0.6$$

$$= -0.096 \frac{(t - 2.5)^3}{3} + 0.6t \Big]_0^5$$

$$= 3 - \frac{2}{3}(0.096)(15.62) \simeq 2.00\ \frac{\text{mg-sec}}{\text{L}}$$

Therefore the flow rate is

$$F = \frac{3.5}{2.001}\ \frac{\text{mg}}{\frac{\text{mg-sec}}{\text{L}}} \quad \text{or} \quad 1.75\ \text{L/sec} \quad \blacksquare$$

In Example 1, $T = 5$ sec is taken as the *stopping time* for this process. In theory a few molecules of tracer might remain some hours after $T = 5$, but these are so few that no instrument could observe them. That is, $C(t)$ is only *approximately* zero for some time after $t = T$, but in practice we regard $C(t)$ as equal to zero. Exercise 9 at the end of this section explores the effect of changing the designation of a stopping time.

In Section 7.5, we will see how the sometimes impractical step of finding a formula for the concentration function C can be avoided, using the idea of approximate integration, which deals directly with the data as they are naturally obtained.

The next example shows that Formula 1 may also be used to determine the amount A of tracer exiting a compartment over a period of time if instead the flow rate F is known. That is, we are interested in the total amount of "tracer" that has been through the compartment over an interval of time.

EXAMPLE 2

A supermarket chain purchases 3,000 tons of tomatoes per year. The concentration of rotten tomatoes that arrive as part of each shipment varies with the season of the year and is found to be

$$C(t) = 15 + 3te^{-2t}$$

pounds per ton at time t in the year, $0 \leq t \leq 1$. Regard the rotten tomatoes as a tracer substance in the flow of tomatoes through the delivery system.

How many pounds of rotten tomatoes are bought but cannot be sold during the year?

Solution Here $F = 3{,}000$ and we are to determine A, the amount (in pounds) of rotten tomatoes flowing through the transit system that brings tomatoes to the supermarket. Solving for A in Formula 1, we have

$$A = F \cdot \int_0^1 C(x)\, dx = 3{,}000 \int_0^1 (15 + 3xe^{-2x})\, dx$$

$$= 3{,}000 \left(15x + 3\left(\frac{e^{-2x}}{(-2)^2}((-2)x - 1) \right) \right) \Big|_0^1$$

using Formula 2, Section 7.2. Thus,

$$A = 3{,}000(15 + 0.445) = 46{,}335$$

pounds (or 23 tons) of rotten tomatoes, approximately 0.8% of the total yearly shipment. ∎

If we want instead to determine the number of rotten tomatoes over the first 6 months of the year, we would find the integral over $\left[0, \frac{1}{2}\right]$.

There is a second, somewhat hidden, subtlety related to Example 2. If F is given in, say, tons per month while $C(t)$ is given at time t in the year, then Δt in an approximating Riemann sum represents a portion of the year and the units of measure that arise for the product

$$F \cdot C(t) \cdot \Delta t \quad \text{are} \quad \frac{\text{tons}}{\text{mo}} \times \frac{\text{lb}}{\text{ton}} \times \text{yr} = \frac{\text{lb-yr}}{\text{mo}}$$

do not measure an accumulation of tracer. Instead we should adjust F by the number of months in a year to obtain the units

$$12 \cdot F \cdot C(t) \cdot \Delta t : \frac{\text{mo}}{\text{yr}} \times \frac{\text{tons}}{\text{mo}} \times \frac{\text{lb}}{\text{ton}} \times \text{yr} = \text{lb}$$

in order to measure the accumulation of tracer over the year, or any portion of it, using the formula $12 \cdot F \int C(t)\, dt$.

Exercises 7.3

In Exercises 1–4, find the flow rate F (L/sec) accounting for an amount A (g) flowing through a compartment by time T at the given concentration level $C(t)$ (g/sec), for $0 \leq t \leq T$.

1. $C(t) = \dfrac{1}{t+1} - \dfrac{1}{2}$; $A = 10$, $T = 1$
2. $C(t) = 1 - 0.4t$; $A = 8$, $T = 2.5$
3. $C(t) = e^{-2t} - 0.0025$; $A = 3$, $T = 3$
4. $C(t) = 2te^{-3t}$; $A = 1$, $T = 4$ where $C(4) = 0$

In Exercises 5 and 6, determine the amount A that has entered and exited a compartment in the indicated time T, for the given flow rate and concentration $C(t)$.

5. $T = 3$; $F = 25$; $C(t) = 9 - \sqrt{t}$
6. $T = 0.5$; $F = 100$; $C(t) = \dfrac{t^{10}}{25}$

7. Seven mg of a harmless dye has been injected into a vein leading into the heart of a patient at rest. The concentration of dye exiting the patient's heart t sec later is $C(t) = 0.25te^{-0.05t}$ mg/L, and measurement is stopped 2 min after $t = 0$ (where $C(120) = 0.074 \approx 0$ mg/L). Graph this concentration function. At what constant rate F is blood flowing through the patient's heart?

8. The management of an amusement park wishes to know the (flow) rate of customers leaving the park near closing time so as to handle exiting traffic patterns properly. It randomly tags the wrist of 200 entering customers with a fluorescent dye when they enter the park. These customers mix thoroughly with other (untagged) customers during the course of the day. During the final 2 hours of the day, when most customers leave at a fairly constant rate F, a device placed at the exit determines that the proportion of tagged to untagged customers is $C(t) = 0.03(t-1)^2 + 0.01$ for $0 \leq t \leq 2$. Determine the flow rate of customers from the park during this time.

9. A village depends on a lake fed by a spring with a fairly constant flow rate F. The city council wishes to determine the rate of flow of water into the lake. A hole is drilled into the underground stream feeding the spring, and 500 lb of a harmless tracer are poured into the stream. Two hours later ($t = 0$), measurements are begun at the mouth of the spring, and these measurements yield a concentration of $C(t) = 0.002te^{-0.15t}$ lb/gal of tracer at time t after measurement is begun. At time $T = 48$ hr after measurement is begun, $C(t) \approx 0.00007$ lb/gal.

   a. Use $T = 48$ as a stopping time for the measurement process and determine the flow rate F.

   b. What if $T = 24$ hr were used?

10. An unknown amount A of waste sulfuric acid was dumped into a stream at the rear of a factory that makes automobile batteries. The stream flows away from the site at the rate of $F = 350$ ft³/min. Environmental quality technicians measure the concentration of sulfuric acid in the stream water and arrive at a concentration of $C(t) = 0.02e^{-0.3t}$ lb/ft³ at time t (in hours) downstream from the spill site. Determine how much sulfuric acid was dumped into the stream, if measurement is stopped after $T = 12$ hr when the concentration reaches $C(12) = 0.0005$ lb/ft³. (Be careful to use the same units of time in your calculations. In this exercise you are solving for the amount of tracer A *given the flow rate F*.)

11. Management has long known that workers perform best on Tuesday and Wednesday and worst on Fridays. Suppose that 300 cars flow through a single assembly line each day and that the concentration of cars exiting with defects is

$$C(t) = 0.01\left(t + \dfrac{4}{t+1}\right)$$

for $0 \leq t \leq 5$ days. Graph this concentration function, regarding the interval [0, 1] as representing Monday and [4, 5] as representing Friday. How many defective cars are produced during a 5-day week?

12. Graph the concentration function C of Example 2 and also the concentration function $C_1(t) = 15 + 3te^{-5t}$. Determine the amount A of rotten tomatoes if instead C_1 is the appropriate concentration of rotten tomatoes.

7.4 Arrival and Accumulation Processes

In this section we begin to use the versatile tool of mathematical modeling by compartment-mixing analysis to find the total accumulation in a compartment subject to arrivals and departures or further accumulation.

Suppose that at the start of your sophomore year in high school you began to plan for personal expenses at college by setting aside $50 in each of the 36 remaining months before college began. You would then leave home with $50 \cdot 36 = \$1{,}800$.

On the other hand, if you deposited each $50 into a savings account that paid you continuous interest of 1% per month on your accumulating deposits, the amount of your accumulated savings is not so obvious. Since the first $50 deposit would earn interest for 36 months, we know from Section 5.5 that this deposit will grow to $\$50 e^{0.01(36)} = \$71.67$ when you begin college. Your *second* $50 deposit will, however, earn interest for only the *remaining* $35 = 36 - 1$ months before you begin college and will grow to only $\$50 e^{0.01(35)} = \$70.95$. Similarly, the $50 you deposit at the beginning of your junior year of high school will earn interest for only the remaining $24 = 36 - 12$ months in the account before you begin college and grow to only $\$50 e^{0.01(24)} = \$63.56$. Finally, your accumulated savings over this 36-month period is a *sum*

$$50 e^{0.01(36)} + 50 e^{0.01(36-1)} + \cdots + 50 e^{0.01(36-12)} + \cdots + 50 e^{0.01(36-35)}$$

of 36 addends, each addend arising from your $50 deposit and the *remaining time* during which the deposit draws interest. This sum, if we use a calculator, is approximately $2,177.49, which is notably more than the total deposit of $1,800 over these 3 years.

Compartment-Mixing Analysis of Continuous Arrival and Accumulation

This example illustrates the fundamental ideas needed to model the total *accumulation*, by the end of a stated time period T, of a stream of arrivals subject to further growth (or decay) after arrival. Once we couple these notions with a Riemann sum analysis leading to an integral whose value will be the total accumulation, we will have a powerful tool for finding such accumulation in a surprising variety of applications. The first step in such application is to envision a compartment-mixing process as the single model for the variety of applications we intend.

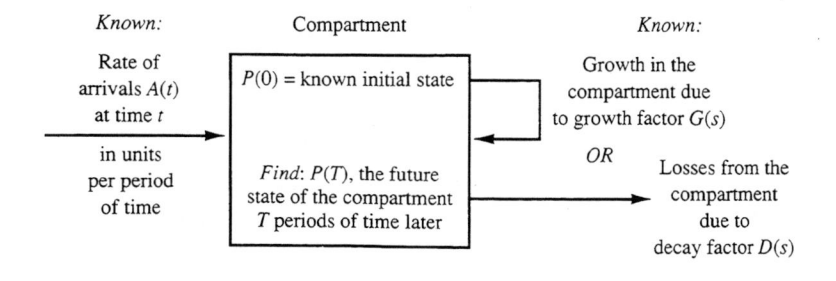

Figure 7.13

Known: Rate of arrivals $A(t)$ at time t in units per period of time

Compartment: $P(0) =$ known initial state; Find: $P(T)$, the future state of the compartment T periods of time later

Known: Growth in the compartment due to growth factor $G(s)$ OR Losses from the compartment due to decay factor $D(s)$

Our preceding discussion is related to Figure 7.13 in the following way. We regard the savings account as a compartment subject to arrivals of $A(t) = 50$ dollars/month in month t for each of $T = 36$ months. Once these dollars arrive in the compartment they will grow according to the given interest rate by the growth *factor* $G(s) = e^{0.01s}$ for s months. If we let $P(t)$ denote the total amount in the compartment at time t, we wish to find $P(36)$, the total amount (accumulation) in the compartment by the end of $T = 36$ months.

Let us derive the following general model for such an arrival and accumulation process.

Given a compartment in a known initial state $P(0)$, subject to arrivals at the rate of $A(t)$ units per period of time at time t, and in which the contents of the compartment grow by the factor $G(s)$, s units of time later, the total contents of the compartment T units of time later will be

$$P(T) = P(0)G(T) + \int_0^T A(t)G(T - t)\,dt \tag{1}$$

The integral in Formula 1 arises from an analysis exactly like that of our initial discussion, coupled with the approximation principle (AP) of Section 4.5. Imagine the interval of time from 0 to T as made up of many short intervals of time Δt, each centered at some intermediate point $t_1, t_2, \ldots, t_n$ between 0 and T (see Figure 7.14).

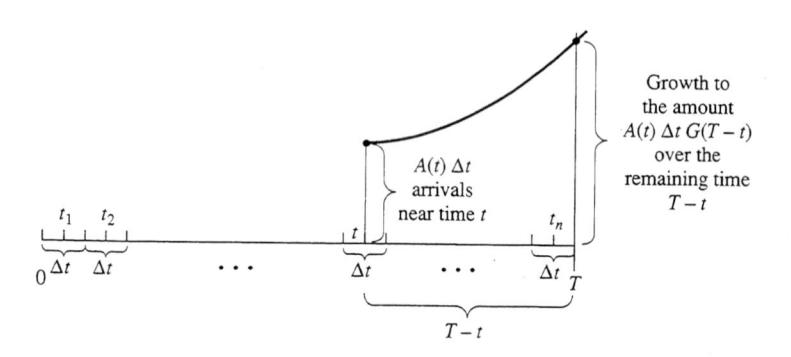

Figure 7.14

Since $A(t)$ units arrive *per* unit of time at time t, then over the span of time Δt centered at time t, *approximately*

$$A(t) \frac{\text{units}}{\text{period of time}} \times \Delta t(\text{periods of time}) = A(t) \cdot \Delta t \quad \text{units}$$

arrive in the compartment. These arrivals then grow for the *remaining time* $T - t$ that they are in the compartment by the growth factor $G(T - t)$. That is, $T - t$ is the amount of time remaining between the time t of arrival and the end of the period T at which we wish to compute total accumulation, and we can apply the growth factor for only that amount of time. Therefore, by the end of the period T the new arrivals during the period t will grow to approximately the amount

$$A(t) \cdot \Delta t \cdot G(T - t)$$

The *total accumulation* in the compartment due to new arrivals over the whole span of time from $t = 0$ to $t = T$ must then be approximated by a sum of the *form*

$$A(t_1)G(T - t_1) \Delta t + A(t_2)G(T - t_2) \Delta t + \cdots + A(t_n)G(T - t_n) \Delta t$$

[This sum is entirely analogous to the sum in our initial discussion.] Since this sum is a Riemann sum for the function $f(t) = A(t)G(T - t)$, we have from AP of Section 4.5 that the total accumulation in the compartment due to new arrivals is the integral

$$\int_0^T A(t)G(T - t) \, dt$$

What of the term $P(0)G(T)$ that appears in Formula 1? Recall that $P(0)$ is the amount initially in the compartment. This initial amount grows by the factor G for the entire period and thus grows to the amount $P(0)G(T)$.

EXAMPLE 1

Suppose that you begin saving at the rate of $A(t) = 50$ dollars per month for $T = 36$ months in a savings account that pays 1% interest continuously compounded per month, and that your parents give you $P(0) = \$200$ as initial encouragement. How much will be in your account at the end of 36 months?

Solution We envision your savings account as a compartment subject to arrivals at the rate of $A(t) = 50$ units per month (see Figure 7.15). Under continuously compounded 1% interest per month, the contents of the compartment are subject to a growth factor of $G(s) = e^{0.01s}$ over s months. Moreover, we are initially given $P(0) = 200$ units in the compartment. According to Formula 1, the accumulated total in the compartment at the end of $T = 36$ months will be

$$P(36) = P(0)G(36) + \int_0^{36} A(t)G(36 - t) \, dt$$

$$\pm\ 200 e^{0.01(36)} + \int_0^{36} 50 e^{0.01(36 - t)} \, dt$$

$$= 200(1.4333) + 50 e^{0.36} \int_0^{36} e^{-0.01t} \, dt$$

$$= 286.66 + 71.67 \left(\frac{e^{-0.01t}}{-0.01} \right)\Big]_0^{36}$$

$$= 286.66 + 71.67 \left(\frac{1 - e^{-0.36}}{0.01} \right)$$

$$= 286.66 + 7{,}167(0.30232)$$

$$= 286.66 + 2{,}166.73 = 2{,}453.39 \text{ dollars}$$

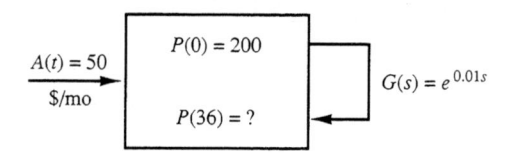

Figure 7.15

[Diagram: $A(t) = 50$ \$/mo arrow into box with $P(0) = 200$, $P(36) = ?$; feedback $G(s) = e^{0.01s}$]

Notice from this calculation that the initial $200 deposit grows to $286.66 and that the arriving deposits of $50/month grow and accumulate to $2,166.73.

As we have seen before, when we apply the continuous mathematics of calculus (we are treating the $50/month deposit as though it is arriving continuously over the days of each month) to a process that is in reality discrete (the $50 deposit is actually made all at once), we do not obtain the exact total that would be found in a "real" savings account. In our earliest discussion, we found the sum to be $2,177.49. This is a small difference, however. Moreover, the gain we make in the way of wider application is considerably more, since with Formula 1 we can treat such a process when the arrival rate itself varies.

EXAMPLE 2

Suppose that in Example 1 you decide to deposit one additional dollar into the account in each successive month, depositing 50, 51, 55, . . . , 85 dollars in the 0th, 1st, 5th, . . . and 35th months. What is the accumulated total in the account after $T = 36$ months?

Solution The arrival rate is now $A(t) = 50 + t$ dollars per month in the tth month, since by month t the deposit has increased by t dollars, for $0 \le t \le 35$. From Formula 1, the monthly deposits will accumulate (by the *end* of 36 months) to

$$\int_0^{36} A(t)G(36-t)\,dt = \int_0^{36} (50+t)e^{0.01(36-t)}\,dt$$

$$= \int_0^{36} 50 e^{0.01(36-t)}\,dt + \int_0^{36} t e^{0.01(36-t)}\,dt$$

$$= 2{,}166.73 + e^{0.36}\int_0^{36} t e^{-0.01t}\,dt \qquad \text{From Example 1}$$

$$= 2{,}166.73 + e^{0.36}\left(\frac{e^{-0.01t}}{(-0.01)^2}(-0.01t - 1)\right)\Big|_0^{36} \qquad \text{From Formula 2, Section 7.2}$$

$$= 2{,}166.73 + e^{0.36}\left(\frac{e^{-0.36}(-1.36) + 1}{0.0001}\right)$$

$$= 2{,}166.73 + 733.29 \text{ dollars}$$

Your increasing rate of deposits yields an additional savings of $733.29. Combined with the initial $200 deposit, the total in your account will be $286.66 + 2{,}166.73 + 733.29 = 3{,}186.68$ dollars. ∎

Compartment-Mixing Analysis of Arrivals and Departures

Compartment-mixing analysis is a versatile mathematical modeling technique since we do not need to be able to envision the contents of the compartment (that is, "see inside" the compartment) or the "mixing" that occurs therein. Rather we need only to be able to observe the (external) arrival rate into the compartment and the growth or decay factors operating on its contents. In the next application, we consider an *arrival-and-departure process* in which we determine the population of an urban area from an analysis of the arrival rate of new inhabitants (due to births and immigration) and the departure of residents (due to deaths and emigration), both of which can be accurately determined from available city records.

Let us envision the urban area itself as a compartment, subject to a known arrival rate $A(t)$ and a departure factor $D(s)$ representing the proportion of existing inhabitants who still remain in the area s years later (see Figure 7.16).

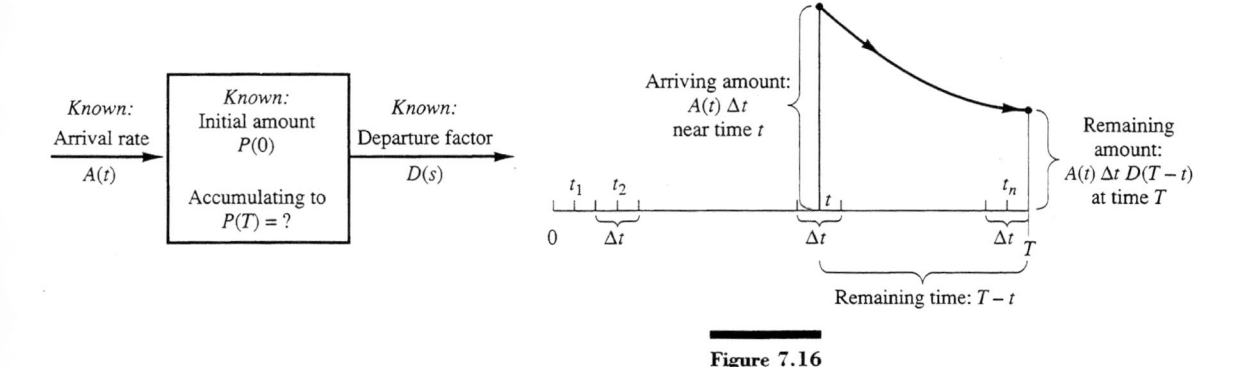

Figure 7.16

Since losses cannot occur until arrivals first occur, the same care must be used in regarding the remaining time $T - t$ as the time period over which the departure factor must be applied to new arrivals arriving at time t. Thus, for an arrival-and-departure process acting on a compartment we have that the accumulation $P(T)$ in the compartment by time T is, as in Formula 1, given by

$$P(T) = P(0)D(T) + \int_0^T A(t)D(T - t)\, dt \qquad (2)$$

It is helpful to remember this formula for the accumulating arrivals and departures in English as the "integral sum" of departures $D(T - t)$ acting on arrivals $A(t)$, for the remaining time $T - t$.

EXAMPLE 3

The initial population of an urban area is 150,000 individuals at time $t = 0$. New residents arrive at the rate of $A(t) = 10,000t$ individuals per year. Existing residents depart by the factor of $D(s) = e^{-0.05s}$ over an s-year span of time.

a. What is the population of this urban area $T = 25$ years later?

b. What percentage of this future population will consist of original residents? New residents?

Solution

a. We regard the urban area as a compartment subject to the given arrival and departure factors (see Figure 7.17).

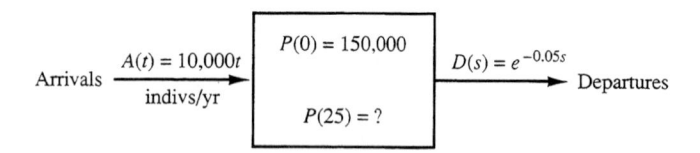

Figure 7.17

Arrivals $A(t) = 10,000t$ indivs/yr → [$P(0) = 150,000$; $P(25) = ?$] → Departures $D(s) = e^{-0.05s}$

According to Formula 2, the population $T = 25$ years later will be

$$P(25) = 150,000 e^{-0.05(25)} + \int_0^{25} A(t)D(25 - t)\, dt$$

$$= 42,976 + \int_0^{25} 10,000 t e^{-0.05(25-t)}\, dt$$

$$= 42,976 + 10,000 e^{-1.25} \int_0^{25} t e^{0.05t}\, dt$$

$$= 42,976 + 10,000(0.287)\left(\frac{e^{0.05t}}{(0.05)^2}(0.05t - 1)\right)\Big|_0^{25} \quad \text{Formula 4, Section 7.1}$$

$$= 42,976 + \frac{2,870}{0.0025}[e^{1.25}(0.25) - (-1)]$$

$$= 42,976 + 2,149,728 = 2,192,704 \text{ inhabitants}$$

b. Original residents make up $42,976/2,192,704 \approx 0.02$, or 2% of the population 25 years later, and new residents therefore make up 98% of the population. ∎

Remark. Because departures from a population tend to occur in proportion to the amount present, a departure factor satisfying the condition $PD' = -kPD$, where P is the population from which departures occur, is appropriate. According to Section 5.5, we may expect a departure factor to have the form $D(s) = e^{-ks}$, as in Example 3. For example, if we are given that a population is declining at, say, 4% a year, then $k = 0.04$ and an appropriate departure factor for this population is $D(s) = e^{-0.04s}$.

The Present Value of a Stream of Future Income

Having learned to think of remaining time $T - t$ *forward*, we now wish to learn to think of *elapsed time* $-t$ backward to compute the value, in the present, of (a stream of) income received in the future.

Recall from Section 5.5 that at an available interest rate r the present value of a single amount A (dollars, perhaps) received t years in the future is Ae^{-rt}. We may think of the present value Ae^{-rt} of A dollars received t years from now as a departure in value of the amount A due to the time that has elapsed before it is actually received (see Figure 7.18). How can we determine the present value of a continuing series of amounts (a stream of future income) received at different times over a span of time from the present, $t = 0$, to some future time T?

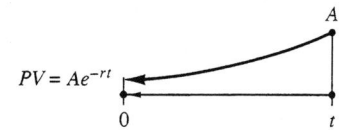

Figure 7.18

We can regard this question as a compartment model in which future amounts arrive at the compartment at the rate $A(t)$, but departures occur from the value of these amounts in the compartment due to present value considerations (see Figure 7.19).

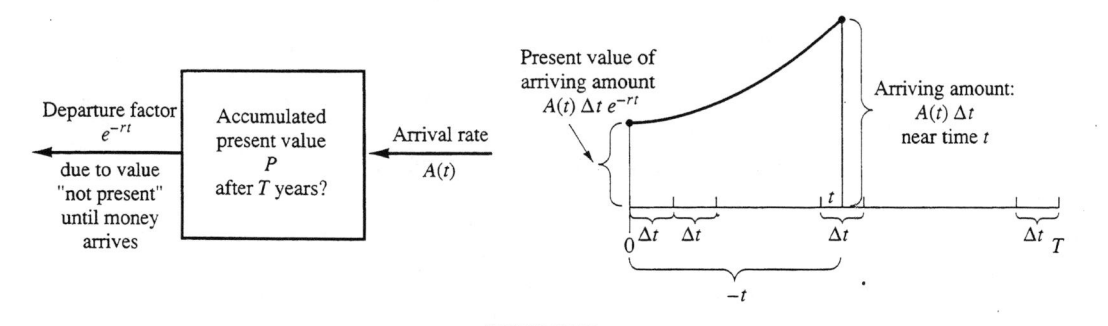

Figure 7.19

Since dollars arrive in the compartment at the rate $A(t)$ over a span of time Δt centered at time t in the future, an amount $A = A(t)\Delta t$ dollars arrive. However, the present value of this amount is

$$[A(t)\Delta t]e^{-rt} \qquad (3)$$

since t years elapse between the present, $t = 0$, and the time t when the amount A is actually received. If we next account for the *stream* of income via a Riemann sum

$$A(t_1)e^{-rt_1}\Delta t + A(t_2)e^{-rt_2}\Delta t + \cdots + A(t_n)e^{-rt_n}\Delta t$$

of terms of the *form* found in Formula 3, we have from the approximation principle (Section 4.5),

At an available interest rate r, the *accumulated* present value P of income received at the rate of $A(t)$ dollars per period at time t in the future, over a span of T periods, is

$$P = \int_0^T A(t)e^{-rt}\, dt \qquad (4)$$

Whereas in our two previous "time-forward" models the remaining time $T - t$ is important, in a present-value model it is the elapsed time, $-t$, with the minus sign measuring this time into the past, that is important. It may be helpful to remember Formula 4 in English: The present value of a stream of income is the (integral) sum of future receipts $A(t)$, received at time t, "discounted" t years into the past (that is, multiplied by e^{-rt}).

EXAMPLE 4

The Ace Manufacturing Company has purchased a new production machine at a cost of $200,000. Ace expects to produce products valued at $50,000 per year over the next 5 years, at which time the machine will be worn out and have no remaining cash value. If the available interest rate is 10%, what is the accumulated present value of the output of this machine?

Solution Let us imagine a compartment into which the return on sales of the machine's output will be placed over the 5-year period. We wish to determine the accumulated present value of these returns over this period (see Figure 7.20).

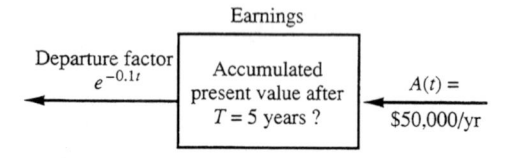

Figure 7.20

Earnings

Departure factor $e^{-0.1t}$ ← Accumulated present value after $T = 5$ years ? ← $A(t) = \$50{,}000/\text{yr}$

Since $A(t) = 50{,}000$ dollars arrive in the compartment per year and the available interest rate is $r = 0.10$ and this rate of income will be received over $T = 5$ years, we have from Formula 4 that the present value of these receipts is

$$P = \int_0^5 A(t)e^{-0.10t}\, dt = \int_0^5 50{,}000 e^{-0.10t}\, dt = \frac{50{,}000}{-0.10}(e^{-0.10t})\Big]_0^5$$

$$= 500{,}000(1 - e^{-0.5}) = \$196{,}735 \text{ dollars} \quad \blacksquare$$

Since Ace paid $200,000 for this machine, it realizes a net *loss* of $3,265 in the purchase and use of the machine in real terms.

The use of a hypothetical "interest rate" r in computing present value is only a convenient conceptual device whereby we recognize that in any decision to use money to make money, alternate investments are possible; in particular we could simply leave the money in an account that paid a return at rate r. It is against this alternate possibility that we must evaluate a decision to use the money in any other way. It should not be surprising that in Example 4 a loss occurs. The total return over 5 years is $250,000, against a cost of $200,000; this 25% return on investment cannot compare favorably with a return of 10% per year over 5 years. The calculations in Example 4, using Formula 4, give us a precise measure of this nonfavorability.

Exercises 7.4

In Exercises 1–6, given the initial amount $P(0)$ in a compartment subject to the arrival rate $A(t)$ and growth (or departure) factor $G(s)$ ($D(s)$), find the amount in the compartment at the indicated time T.

1. $P(0) = 50$; $T = 20$
   $A(t) = 3$; $G(s) = e^{0.05s}$

2. $P(0) = 15$; $T = 150$
   $A(t) = 4t$; $D(s) = e^{-0.002s}$

3. $P(0) = 0$; $T = 5$
   $A(t) = 4e^t$; $D(s) = e^{-(1/2)s}$

4. $P(0) = 3$; $T = 10$
   $A(t) = 5 - \dfrac{t}{2}$; $G(s) = \dfrac{s}{10}$

5. $P(0) = 0$; $T = 5$
   $A(t) = \left(\dfrac{1}{2}\right)t^2$; $G(s) = e^{0.02s}$
   [*Hint:* Use integration by parts (Section 7.1).]

6. $P(0) = 1$; $T = 8$
   $A(t) = 32 - 4t$; $D(s) = e^{-s^2}$
   [*Hint:* Use integration by substitution (Section 4.3).]

7. An account initially contains $500, and annual deposits are made at the rate of $300 per year. How much will be in the account at the end of 5 years at an interest rate of 8%?

8. Repeat Exercise 7 for an interest rate of 6% over a 10-year period.

9. If you deposit $1{,}000 + 150t$ dollars into an account in year t for a period of 10 years and the account draws continuous interest at the rate of 9% a year, how much will be in the account at the end of 10 years? If the account in addition initially contained $3,000, how much will be in the account 10 years later?

10. An investment property earns $400 per month in rental profit. This sum is placed each month in an account bearing 9% continuous interest for a period of 8 years. How much will be in the account at the end of that period? [*Hint:* Treat the monthly deposits as a yearly deposit of $4,800.]

11. In the course of research into his family's roots, John discovers the existence of a parcel of land still in his family's name but on which no one has paid taxes for 40 years. To reclaim the land, John learns that he must pay back taxes at the rate of $250 a year for 40 years plus 8% interest of these taxes for each year since the taxes were due. What is the total payment that John must make to reclaim the land?

12. Please refer to Exercise 10. Suppose that as each year begins, the monthly rent is increased by $15 each year. Thus 5 years later the rent is $475 per month.

    a. At 9% continuous annual interest, how much will be in the account 10 years later?

    b. If in year t the monthly rent is increased by $15e^{0.10t}$ so that the rent is $400 + 15te^{0.10t}$ in year t, how much will be in the account 10 years later?

13. A hospital patient is intravenously treated with a drug at the increasing rate of $A(t) = 5 + 0.7t$ mg/hr. The patient's body eliminates this drug at the rate of 30% per hour, so that of an amount in his system at any time only $e^{-0.3t}$ remains t hours later. How many mg are present 4 hours after medication has begun?

14. On May 1, a farmer notices one lily pad in the stock pond and after a few weeks realizes that new lily pads are forming at the rate of $A(t) = e^{2t}$ lily pads per week. But of the lily pads in the pond at any

time, only $e^{-1.99t}$ remain t weeks later. How many lily pads will be in the pond on September 1?

15. A sample of pond water initially contains 200 *paramecia*, which reproduce at the rate of $150e^{0.12t}$ new paramecia per hour. *Amoebae*, which consume paramecia at the rate of 25% of their population per hour, are introduced into the sample. What will be the population of paramecia at the end of 5 hours? [*Hint:* Use $D(s) = e^{-0.25t}$ as the departure function.]

16. A diabetic's body is producing insulin at the rate of $A(t) = 70$ mg/hr. The amount of insulin in her body must be kept above the level of (approximately) 250 mg. Her body eliminates insulin at the rate of 35% per hour. If at time $t = 0$ the patient has 300 mg in her body, how much additional insulin should she be given at time $t = 0$ so as to ensure a level of 250 mg 8 hours later? [*Hint:* Treat this as a compartment-mixing problem in which $P(0)$ is unknown. Note that $D(t) = e^{-0.35t}$ (why?).]

17. In the book *In Search of Excellence* by Peters and Waterman, the authors seek to identify those traits distinguishing the most successful corporations from their less successful competition. Many of these attributes turn out to be not hard assets like plant and equipment but "soft" values, or company "culture," such as ways of generating and recognizing new ideas.

    The transformation of culture can be regarded as a compartment-mixing process. If at time $t = 0$ a group possesses 150 "traits" that determine its "culture," and of whatever particular traits it holds at any time only $e^{-0.03t}$ of these remain t years later, and if new cultural traits are introduced or originate in the group at the rate of $A(t) = 25te^{-0.5t}$ per year in year t, then:

    a. How many culture-defining traits will be found in this group 40 years later?

    b. Of the original 150 traits, how many remain at that time?

    c. What percentage of traits held at any time T are original traits?

18. What is the present value of an investment that yields a return of $500 per year for 10 years at an interest rate of 9%?

19. What is the present value of an investment that yields a return of $R(t) = 1{,}500te^{-t}$ dollars in year t for a period of 10 years at an interest rate of 10%?

20. Show that if the available interest rate is r and the inflation rate is s, then the present value of A dollars received T years from now is $Ae^{(r-s)T}$. [For a full discussion, see Theorem 3 in the article by R. C. Thompson, "The True Growth Rate and the Inflation Balancing Principle," *American Mathematical Monthly* 90, no. 3 (1983), 207–10.]

21. Suppose that you bought the $20 million ticket in the state lottery and have won the prize. You then learn that you will receive your winnings as follows: In each of the next 10 years you will receive $500,000, and at the end of that time you will receive the remaining $15 million. What is the present value of your winning ticket, if you could obtain an interest rate of 9% for your investments over this 10-year period?

22. Please refer to Exercise 13 of Section 7.2. If, in that exercise, $e^{-0.03x}$ of the rainfall striking any point moves downhill x feet before being absorbed into the ground, how much rainfall will reach the foot of the mountain?

7.5 Approximate Integration

In this section we learn to estimate the value of a definite integral by a procedure called the trapezoidal rule.

While you can easily evaluate the integral

$$\int_1^2 \frac{1}{x^2}\, dx = \left.\frac{-1}{x}\right]_1^2 = \frac{1}{2}$$

you would find it quite impossible to evaluate the integral

$$\int_0^1 \frac{1}{1+x^2}\, dx$$

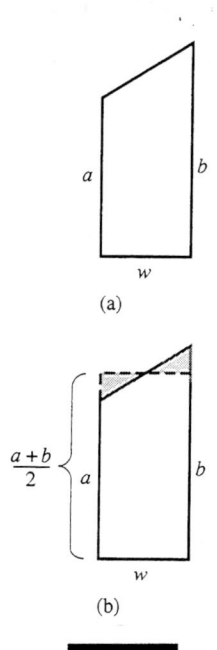

without an extensive amount of trigonometry! Yet it is apparent that the function involved, $1/(1 + x^2)$, is not very complicated and algebraically is hardly different from $1/x^2$. Because such slight changes in an integrand can produce often intractable problems in evaluating a definite integral, we develop in this section a simple and fairly accurate method of approximating the value of a definite integral.

The Trapezoidal Rule

To develop this method we first need to recall the formula for the area of the trapezoid, shown in Figure 7.21(a), of base width w and heights a and b on opposite sides.

Notice that the area of the rectangle indicated in Figure 7.21(b) equals the area of the trapezoid itself. Since this rectangle is w units wide and $(a + b)/2$ units high, we see that the area of the rectangle, and therefore the area of the trapezoid, is

$$\frac{a + b}{2} w$$

That is, this area is the width of the trapezoid multiplied by the *average* of the height of its two sides. Let us use this formula to develop the trapezoidal rule for approximate integration.

Figure 7.22(a) recalls the fact that the definite integral $\int_a^b f(x)\, dx$ is approximately equal to a Riemann sum, $f(x_1)\Delta x + f(x_2)\Delta x + \cdots + f(x_n)\Delta x$, and that this approximation need not be very good if the number n of intervals is small and if the points $x_1, \ldots, x_n$ are not well chosen. In Figure 7.22(a), $n = 5$ and $\Delta x = (b - a)/5$, and the points $x_1, \ldots, x_5$ are all chosen to be the right endpoint of an interval.

Now notice Figure 7.22(b), where we again partition the interval $[a, b]$ into five parts but sketch in a *trapezoid* above each subinterval of width Δx. The *sum* of the areas of these five trapezoids is, by visual inspection, much closer to the area $\int_a^b f(x)\, dx$ beneath the graph of $y = f(x)$. We use this idea to formulate the trapezoidal rule.

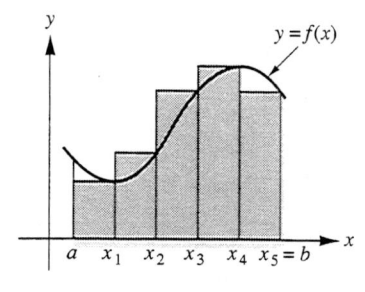

Figure 7.21

Figure 7.22   (a) Rectangular approximation   (b) Trapezoidal approximation

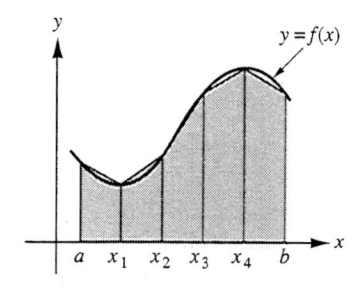

7.5 Approximate Integration

Consider the single trapezoid of base width Δx located at points x_k and x_{k+1} in the interval $[a, b]$ in Figure 7.23. The height of the trapezoid is $f(x_k)$ and $f(x_{k+1})$ on its two vertical sides. Therefore the area of this trapezoid is

$$\frac{f(x_k) + f(x_{k+1})}{2} \Delta x$$

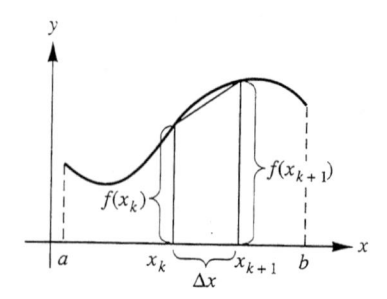

Figure 7.23

Now consider a sum of areas of such trapezoidal regions approximating the area under the curve in Figure 7.24. This sum will have the form

$$\frac{f(x_1) + f(a)}{2}\Delta x + \frac{f(x_2) + f(x_1)}{2}\Delta x + \frac{f(x_3) + f(x_2)}{2}\Delta x + \cdots + \frac{f(x_{n-1}) + f(b)}{2}\Delta x$$

Notice that the number $f(x_k)/2$ occurs *twice* in this sum, for $k = 1, 2, \ldots, n - 1$. Therefore the sum equals

$$\left(\frac{f(a)}{2}\right)\Delta x + f(x_1)\Delta x + f(x_2)\Delta x + \cdots + f(x_{n-1})\Delta x + \left(\frac{f(b)}{2}\right)\Delta x$$

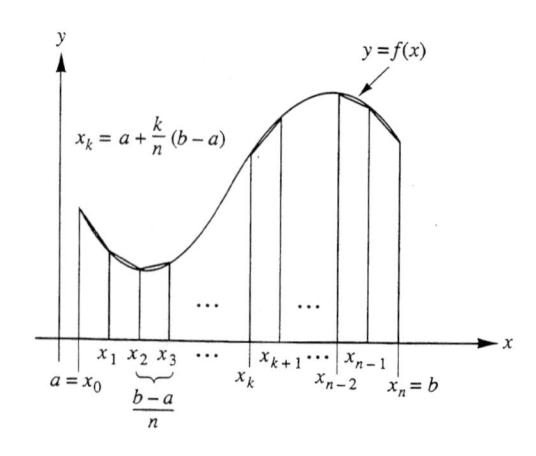

Figure 7.24

426   Chapter 7   Applied Integration

If we now factor the number $\Delta x/2$ from this sum, we obtain

$$[f(a) + 2f(x_1) + 2f(x_2) + \cdots + 2f(x_{n-1}) + f(b)] \cdot \frac{\Delta x}{2}$$

This last sum is the formula we seek.

The Trapezoidal Rule (Refer to Figure 7.24)

Let $y = f(x)$ be a continuous function on the interval $[a, b]$. Let n be a whole positive number. Let $\Delta x = (b - a)/n$ and let $a, x_1, \ldots, x_{n-1}, b$ be $n + 1$ equally spaced (by the distance Δx) points on the interval $[a, b]$, beginning at a and ending at b.
Then

$$\int_a^b f(x)\, dx \simeq [f(a) + 2f(x_1) + 2f(x_2) + \cdots + 2f(x_{n-1}) + f(b)]\frac{\Delta x}{2} \qquad (1)$$

Remark. Under these conditions, $x_k = a + (k/n)(b - a)$.

Notice that all the trapezoidal rule does is estimate the integral from $n + 1$ particular known values of the function by *averaging* successive values [namely, $(f(x_{k+1}) + f(x_k))/2$] and treating this average as typical for the whole interval $[x_k, x_{k+1}]$.

EXAMPLE 1
Approximate $\int_0^1 1/(1 + x^2)\, dx$ by the trapezoidal rule with $n = 4$.

Solution Since $n = 4$, then $\Delta x = (1 - 0)/4 = 1/4$, and the interval $[0, 1]$ is partitioned into four parts with $a = 0$, $x_1 = \frac{1}{4}$, $x_2 = \frac{1}{2}$, $x_3 = \frac{3}{4}$, and $b = 1$ (see Figure 7.25).

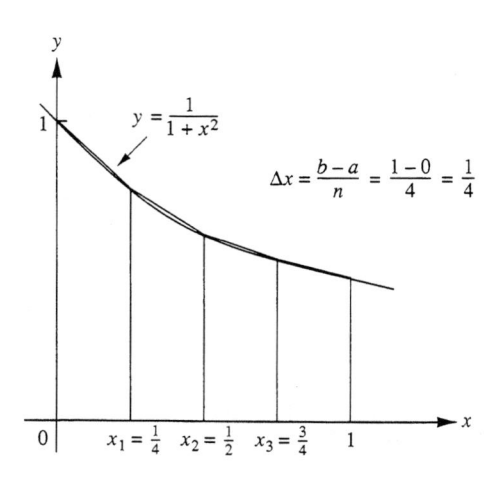

Figure 7.25

7.5 Approximate Integration

By the trapezoidal rule (1),

$$\int_0^1 \frac{1}{1+x^2} dx \approx \left[f(0) + 2f\left(\frac{1}{4}\right) + 2f\left(\frac{1}{2}\right) + 2f\left(\frac{3}{4}\right) + f(1) \right] \frac{\frac{1}{4}}{2}$$

$$\approx \left[\left(\frac{1}{1+0^2}\right) + 2\left(\frac{1}{1+\left(\frac{1}{4}\right)^2}\right) + 2\left(\frac{1}{1+\left(\frac{1}{2}\right)^2}\right) + 2\left(\frac{1}{1+\left(\frac{3}{4}\right)^2}\right) + \left(\frac{1}{1+1^2}\right) \right] \frac{1}{8}$$

$$\approx \left[1 + 2\left(\frac{16}{17}\right) + 2\left(\frac{4}{5}\right) + 2\left(\frac{16}{25}\right) + \frac{1}{2} \right]\left(\frac{1}{8}\right) \approx \left[1 + \frac{32}{17} + \frac{8}{5} + \frac{32}{25} + \frac{1}{2} \right]\frac{1}{8}$$

$$\approx \frac{3}{2} \cdot \frac{1}{8} + \frac{4}{17} + \frac{1}{5} + \frac{4}{25} \qquad \text{Distribute the factor } \tfrac{1}{8}$$

$$\approx \frac{3}{16} + \frac{100 + 85 + 68}{17(25)} = \frac{5{,}323}{6{,}800} \approx 0.7828 \quad \blacksquare$$

The exact value of this integral, obtained by alternate methods, is $\pi/4 \approx 0.7854$. Therefore the trapezoidal rule applied to this example with as few as $n = 4$ points yields an approximation accurate to two decimal places.

Applications to Integral Models of Accumulation

Any of the integral models of previous sections may be combined with the trapezoidal rule to yield an estimate of the quantity in question when only incomplete information (data), rather than a complete function-based model, is given.

EXAMPLE 2

A marathon runner is observed to be running at the times indicated in Figure 7.26, with varying speeds over the course of 1 hour. Approximately how far did the person run?

Time (hr)	mph
0	10.5
$\frac{1}{4}$	10.0
$\frac{1}{2}$	9.4
$\frac{3}{4}$	11.0
1	10.0

Figure 7.26

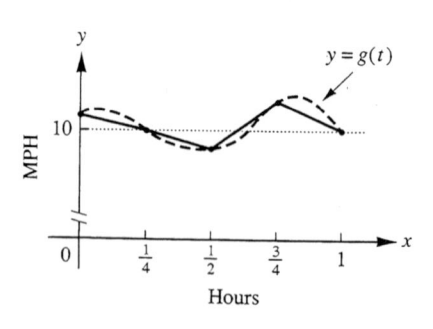

Solution Let $g(t)$ denote the runner's speed at time t. We are given that $g(0) = 10.5$ mph, $g\left(\frac{1}{4}\right) = 10.0$, and so on. This means that we are given a few points on the graph, but not the whole graph of, the runner's speed at time t. With these known values, we can estimate the area beneath this graph, $\int_0^1 g(t)\, dt$, which measures the total distance run in 1 hour.

428   Chapter 7   Applied Integration

Using the given information and the trapezoidal rule, we can approximate the integral

$$\int_0^1 g(t)\, dt \quad \text{by} \quad \left[g(0) + 2g\left(\frac{1}{4}\right) + 2g\left(\frac{1}{2}\right) + 2g\left(\frac{3}{4}\right) + g(1)\right]\frac{\frac{1}{4}}{2}$$

since in this model $\Delta t = \frac{1}{4}$. Therefore

$$\int_0^1 g(t)\, dt \simeq [10.5 + 2(10.0) + 2(9.4) + 2(11.0) + 10.0]\left(\tfrac{1}{8}\right) = 10.16 \text{ mi} \quad \blacksquare$$

EXAMPLE 3

The Topspin Manufacturing Company is producing and selling the seasonal product of tennis balls and must borrow money at 9% per year to capitalize its production. Estimate the present value of its income based on the following projection of sales over a year (see Table 7.2).

Solution Let $r(t)$ denote Topspin's rate of return on sales at time t in the year, for $0 \le t \le 1$.

According to the discussion of the previous section, the present value of the stream of income $r(t)$, arriving at time t, is the present value integral:

$$\int_0^1 r(t) e^{-0.09t}\, dt$$

Let us use the trapezoidal rule with $n = 4$ along with the given data so as to estimate this present value. We have

$$\int_0^1 r(t) e^{-0.09t}\, dt \simeq \left[r(0)e^{-0.09(0)} + 2r\left(\tfrac{1}{4}\right)e^{-0.09(1/4)} + 2r\left(\tfrac{1}{2}\right)e^{-0.09(1/2)} + 2r\left(\tfrac{3}{4}\right)e^{-0.09(3/4)} + r(1)e^{-0.09(1)}\right]\left(\tfrac{\frac{1}{4}}{2}\right)$$

$$\simeq [0 + 2(15)(0.978) + 2(75)(0.956) + 2(50)(0.935) + 20(0.914)]\left(\tfrac{1}{8}\right)$$

$$\simeq 35.565 \text{ thousands of dollars} \quad \blacksquare$$

The application exercises of this section ask you to use the integral-modeling techniques of previous sections *as though a function were present*, as in Example 3, and then to approximate the integral that arises using the data (rather than a function) that are given.

Table 7.2

Time (yr)	Thousands of dollars per year
0	0
$\frac{1}{4}$	15
$\frac{1}{2}$	75
$\frac{3}{4}$	50
1	20

Exercises 7.5

1. You know that $\int_0^1 x^2\, dx = x^3/3\Big|_0^1 = \frac{1}{3}$. How well does the trapezoidal rule with $n = 4$ approximate this answer?

2. Will the trapezoidal rule give an *exact* answer when used to approximate the integral $\int_0^2 x\, dx$? Why or why not?

3. Approximate the value of $\int_0^1 \sqrt{1-x^2}\,dx$ using the trapezoidal rule with $n=4$.

4. Approximate $\ln 2 = \int_1^2 (1/x)\,dx$ using the trapezoidal rule with $n=5$.

5. Repeat Exercise 4 with $n=10$.

6. Approximate $(1/\sqrt{2\pi})\int_0^1 e^{-x^2/2}\,dx$ using the trapezoidal rule with $n=8$.

7. Approximate $\int_0^1 1/(1+x^2)\,dx$ using the trapezoidal rule with $n=8$. (Compare your answer with that in Example 1.)

8. Let f be the function whose graph is shown in Figure 7.27.

   a. Calculate the area $\int_0^2 f(x)\,dx$ beneath the graph of $y=f(x)$ by any method you like.

   b. Approximate this answer using the trapezoidal rule with $n=4$.

   c. Repeat part (b) using $n=8$.

   d. Will the approximation be improved if instead $n=16$?

   e. What do (b) and (c) suggest in general about the nature of trapezoidal approximation?

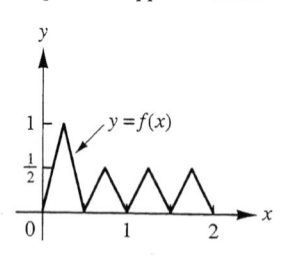

Figure 7.27

9. The speedometer on Joe Piston's Formula 1 racing car shows the following readings at $\tfrac{1}{2}$-sec intervals as he drives through an S-curve on a Grand Prix track (see Figure 7.28 and the table at the top of the next column). It took him 4 sec to negotiate the curve. Approximately how long was the curve (in miles)?

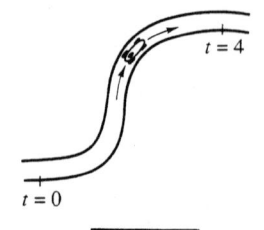

Figure 7.28

Time (sec)	mph
0	100
$\tfrac{1}{2}$	80
1	90
$\tfrac{3}{2}$	110
2	115
$\tfrac{5}{2}$	95
3	90
$\tfrac{7}{2}$	105
4	135

10. Environmental quality specialists are trying to estimate the amount of fertilizer residue running off the farmland bordering a 2-mile length of stream. Rather than attempt to model the runoff by a formula for $f(x)$, the rate of runoff in pounds per mile at *each* point x along the stream, they instead set up monitoring devices at $\tfrac{1}{4}$-mile intervals and measure the runoff rate at those points (see Figure 7.29). Use the following data to estimate the total runoff into this 2-mile length of stream.

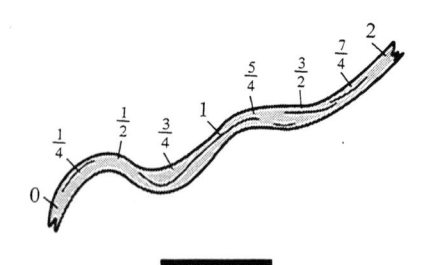

Figure 7.29

Mile Marker	lb/mi
0	100
$\tfrac{1}{4}$	50
$\tfrac{1}{2}$	60
$\tfrac{3}{4}$	80
1	80
$\tfrac{5}{4}$	30
$\tfrac{3}{2}$	20
$\tfrac{7}{4}$	25
2	30

11. Estimate the present value of the stream of income recorded in Table 7.3 if the interest rate is 10% over the indicated 5-year period.

430 Chapter 7 Applied Integration

Table 7.3

Year	Income (dollars/yr)
0	0
1	15,000
2	10,000
3	23,000
4	37,000
5	25,000

12. The Big Pine Timber Company, a large lumber corporation, planted 20,000 acres in a single variety of yellow pine timber 8 years ago. The timber should grow to salable size in 12 more years. This variety is, however, particularly susceptible to a form of leaf rust that attacks the timber, slowing its growth. The entire stand of timber, which is of identical genetic makeup, is seriously threatened. The company starts a control program to contain and diminish the problem. Experts know that, if properly treated, only $e^{-0.10t}$ of the timber presently affected by the leaf rust at any time will still be affected t years later. If the leaf rust spreads at the rates in Table 7.4 for the next 12 years (in 2-year periods), how many acres will be affected in the year of harvest?

Table 7.4

Year t	New Infestations (acres/yr)
8	100
10	200
12	400
14	1,000
16	1,500
18	800
20	600

13. A family plans to save money at the rate of $2,000 per year to partially finance their child's college education, which begins 5 years from now. They expect to earn 10.5% interest per year on these savings. Since their business is seasonal, they will not be able to save at a regular rate over the year but will deposit more during the part of the year when the business is doing well and less at other times—yielding an overall rate of $2,000 per year. Let $r(t)$ be the rate of savings at year t. Model the value of their investment at the end of the 5-year period, incorporating the given interest rate in an integral, and then estimate its value using the trapezoidal rule.

14. Use the BASIC program TRAPGEN to estimate the value of $\int_0^1 1/(1 + x^2)\, dx$ using $n = 50$ subintervals.

The remaining exercises can be done with a calculator. However, if a great deal more data were given, a computer would be more practical. Your instructor may wish you to find the trapezoidal rule estimates in Exercises 15–18 using the BASIC programs TRAPDATA, ARRIDEPT, or PRESVAL.

15. A supermarket chain buys 500 tons of tomatoes a year, and seasonal factors cause the concentration of rotten tomatoes arriving in each shipment to vary over the course of the year. Sampling at 2-month intervals yields the data in Table 7.5 for the number of pounds per ton arriving during the year. Approximately how many pounds of rotten tomatoes did the chain buy during the year?

Table 7.5

Time	lb/ton
0	20
2	20
4	35
6	45
8	40
10	15
12	15

16. Core drillings by geologists in search of oil at a particular site yield the data in Table 7.6 for the density of materials beneath the earth's crust down to a depth of 100 ft. Estimate the weight of these materials at a depth of 100 ft using the trapezoidal rule.

Table 7.6

Depth	lb/ft^3
0	50
20	65
40	55
60	85
80	80
100	70

17. An urban area initially contains 30,000 individuals. Of the population in this area only the proportion $e^{-0.05t}$ remains t years later. Over a 10-year period new individuals arrive at the rate indicated in Table 7.7. Approximate the population of this urban area at time $T = 10$.

Table 7.7

Year	People/yr
0	2,000
2	1,500
4	2,500
6	3,000
8	3,000
10	2,800

18. An urban area is regarded as circular with a 5-mile radius. The population density is measured at 1-mile intervals from the center of the region and is recorded in Table 7.8. Approximately how many people live in this region?

Table 7.8

Mile	People/mi²
0	5,000
1	4,000
2	6,000
3	5,000
4	4,000
5	2,000

Chapter 7 Summary

1. A *density function* ρ is the derivative of a cumulative *distribution function* F, where $F(x)$ is the total accumulation to the point x. Consequently, from the Fundamental Theorem of Calculus,

$$F(x) = F(0) + \int_0^x \rho(t)\, dt$$

2. If F is a flow rate and the *concentration* of an initial amount A (incorporated into the medium of flow) is $C(t)$, t units of time later, then

$$F = \frac{A}{\int_0^T C(t)\, dt}$$

where $C(t) = 0$ for $t \geq T$.

3. Arrivals into, and departures from, a closed compartment may accumulate in, or disperse from, the compartment over time. If $A(t)$ is the rate of arrivals into the compartment at time t, then:

   a. If only the proportion $D(s)$ of the amount in the compartment remains in the compartment s units of time later, then over a span of time from 0 to T, the contents $P(t)$ of the compartment are given by

$$P(T) = P(0)D(T) + \int_0^T A(t)D(T-t)\, df$$

   Population growth and decline is a common example.

   b. If $G(s)$ is the proportion of growth in the amount in a compartment s units of time later, then the total accumulating in the compartment due to growth acting on new arrivals (only) is

$$\int_0^T A(t)G(T-t)\, dt$$

   Accumulating deposits, with interest applied, is a common example.

c. If $A(t)$ is a stream of income going into a compartment, then, at an interest rate r, the present value of the contents of the compartment T years later of this stream of income is

$$\int_0^T A(t)e^{-rt}\, dt$$

4. The trapezoidal rule

$$\int_a^b f(x)\, dx \simeq [f(a) + 2f(x_1) + 2f(x_2) + \cdots + 2f(x_{n-1}) + f(b)]\frac{\Delta x}{2}$$

where $\Delta x = (b - a)/n$ and the points x_i are Δx units apart between a and b, is a generally better approximation to the integral of f than a Riemann sum.

The trapezoidal rule may be incorporated with data and the models given in Summary points 1, 2, and 3 to obtain good approximations in the presence of incomplete information.

Chapter 7 Summary Exercises

1. Find the density function ρ for the distribution function $F(x) = 100(1 - e^{-0.3x})$.

2. Find the distribution function F for a density function $\rho(x) = 10e^{-2x}$ if $F(0) = 0$ and $x \geq 0$.

3. Find the density function ρ of people per foot for the distribution $F(x) = -0.00005(x - 200)^3 + 6x - 400$ of the number of people riding on a 400-ft-long subway train up to a distance of x ft from the front end of the train. Approximately how many individuals would be found in a 40-ft-long subway car 100 ft from the *end* of the train?

4. What is a density function? How is it related to a distribution function?

5. The distribution of a trail of ants leading to a picnic basket is $F(x) = 200(1 - e^{-x})$ ants x ft from the picnic basket. What is the density of ants 18 in. from the basket? What are the units of measurement of your answer? Approximately how many ants will be found along a 1-in. length of this trail of ants 12 in. from the picnic basket?

6. If $\rho(x) = 10xe^{-0.5x}$ is the density of microorganisms, in thousands per cm^3, x cm below the surface of a freshwater pond, how many microorganisms inhabit the first 15 cm below the surface in a vertical rectangular water column with a 1-cm^2 square cross section?

7. Determine the flow rate F through a compartment into which 5 g of a dye has been introduced if the concentration of the dye exiting the compartment is $C(t) = (-t/3) + 3$ g/L t seconds later, $0 \leq t \leq 9$.

8. If we know the flow rate of some medium through a closed compartment and are able to measure the concentration of some foreign substance in this medium, can we then find how much of the foreign substance was originally introduced into the medium? If so, how?

9. How many individuals live in a circular region of radius 10 miles if the population density is $\rho(x) = 500e^{-0.02x}$ individuals per square mile x miles from the center of this region?

10. Discuss the general concept of an arrival-and-departure process, including how the factor of remaining time of the process enters the model of such a process. How does a model of accumulating wealth fit this general model? Does the present value of an income stream fit this model?

11. Determine the amount $P(T)$ present at $T = 10$ if $P(0) = 5$ and arrivals occur at the rate of $A(t) = 8$ units per minute subject to the departure function $D(s) = e^{-0.04s}$ for $0 \leq t, s \leq 10$.

12. If $1,500 per year is deposited into an account paying an interest rate of 8%, how much will be in the account at the end of 10 years?

13. Advertising is drawing new customers to a product at the rate of $A(t) = 5\sqrt{10t}$ customers per day, but only the proportion $D(s) = (10 - s)/10$ remain customers s days after trying the product. How many faithful customers remain 10 days after advertising starts, if there were no customers on day zero?

14. If new arrivals are entering a compartment at the rate of $A(t) = 100e^{-0.40t}$ per second and departures occur continuously at the rate of 30% of the amount in the compartment and the compartment initially contains 150, what does it contain 8 sec later?

15. Compute the present value of a stream of income of $A(t) = 35t + 25$ dollars/year in year t at an interest rate of 10% over a 15-year period.

16. If money is deposited into an account bearing 12% interest for a period of 20 years at a rate of $1,500e^{0.18t}$ dollars/year in year t, how much will be in the account at the end of this period?

17. A lottery winner wins $40 million in the state lottery. However, he then learns that the money will be given to him at the rate of $2 million a year for the next 20 years. At an average interest rate of 10% over this period, what is the present value of his winning ticket?

18. Seven mg of a harmless dye is injected into one side of the heart of a patient at rest, and the concentration of dye in the blood leaving the heart is measured t sec later, yielding Table 7.9. Estimate the flow rate of blood through the heart in L/min.

Table 7.9

Time (sec)	Concentration (mg/L)
0	0
10	8.0
20	6.0
30	0.1

19. Approximately how many defective items are produced over an 8-hr period, at a rate of 50 items per hour, if the ratio of defective items to well-made items, as measured every 2 hr during this period, is as given in Table 7.10?

Table 7.10

Time (hr)	Ratio
0	0
2	0.05
4	0.07
6	0.15
8	0.12

20. Sampling the population of potential customers in a circular area of radius 2 miles centered at a newly opened restaurant yields the data in Table 7.11. Approximately how many customers live in this circular area?

Table 7.11

Distance (mi)	Potential Customers/mi²
0	25
$\frac{1}{2}$	30
1	15
$\frac{3}{2}$	10
2	5

Chapter 8

Differential Equations

Laws of Change in a Dynamical System

Calculus enables us to make remarkably precise descriptions and measurements of the future states of complex systems, even when we start with little information. The information that we do have will always consist of two kinds. If we use the values of a function, say, $y = f(t)$, to represent the states of a system, then one piece of information will be a single, *initial* state of the system. That is, we will usually know $f(0)$. The other piece of information will always be a law, or rule, about how the system *changes*. The form of this law will be a *differential equation*, which models the change-of-state of the system. To describe the states of the system, we solve this differential equation for a formula that yields the values of the up-to-then unknown function $f(t)$. We then have a formula that describes the state of the system at time t. Concurrently—or even alternatively—we may want to have a holistic view of how the system can evolve over time, in a sense to see the process modeled by the differential equation at work. This approach can be especially insightful and is at the same time quite practical, since it is met by having a graph of the family of all possible solutions to the differential equation. In many of the most important cases, these graphs can be found without actually solving the equation.

8.1 Differential Equations and Their Solutions

In this section we learn what it means for a function to be a solution of a differential equation and study a useful extension of the fundamental differential equation $y' = ky$ of exponential growth.

A **differential equation,** which we discussed briefly in Section 5.5, is an equation involving an unknown function—say, $y(t)$—its derivatives, and, explicitly or implicitly, the underlying variable t. The equations

$$y' = 3y \qquad y' = y + t \qquad 2y' + ty = 3 \qquad y'' = -y$$

are all examples of differential equations. A solution to a differential equation is a known formula for $y(t)$ that, when substituted in the equation, satisfies the equation. You may recall from Chapter 5, for example, that $y(t) = e^{3t}$ satisfies the equation $y' = 3y$.

There is a close analogy here with the solution of algebraic equations. For example, $x = 3$ is seen upon substitution to satisfy the algebraic equation $2x + 1 = 7$, and no other value of x will satisfy it. Let us try substitution in a differential equation.

EXAMPLE 1

Show that $y(t) = 3e^t - t - 1$ is a solution of the differential equation $y' = y + t$.

Solution We must substitute the formula $y(t) = 3e^t - t - 1$ in each side of the equation and see if equality results. On the left side we have

$$y'(t) = D_t(3e^t - t - 1) = 3e^t - 1$$

Then on the right,

$$y(t) + t = (3e^t - t - 1) + t = 3e^t - 1$$

The same answer results. Therefore $y'(t) = y(t) + t$ is a true equation when $y(t) = 3e^t - t - 1$. ∎

Notice in this example that although we write a differential equation as $y' = y + t$, we are to understand that y is a function of t. That is, strictly speaking, the equation should be written as $y'(t) = y(t) + t$.

Differential equations differ from algebraic equations in one particular way that is crucial to their application. Differential equations typically have not one, two, or a few solutions but infinitely many. However, these solutions generally (but not always!) have a certain basic form and differ only by a suitably placed constant. For example, consider the trivial differential equation $y'(t) = 2$. Here we are asking for a function whose derivative is the constant value 2. Evidently $y(t) = 2t$ is such a function. But then so is $y(t) = 2t + 1$ or $y(t) = 2t + \pi$. Indeed $y(t) = 2t + C$ is a solution for any constant value C. Thus the differential equation $y'(t) = 2$ has infinitely many different solutions, each one determined by a specific value assigned to C, but all have the same *form* $y(t) = 2t + C$. We may illustrate the solutions to $y'(t) = 2$ as a family of graphs, corresponding to the functional form $2t + C$, of straight lines of constant slope 2, as indicated in Figure 8.1.

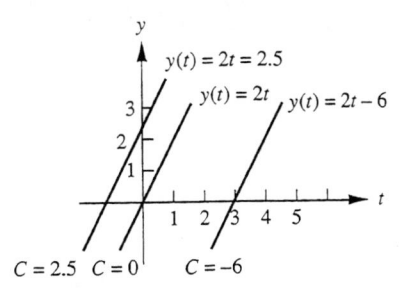

Figure 8.1

Throughout this chapter we will see how constants systematically arise in the solution of an equation and that they often have a critical meaning in applications. The next example is more exemplary of the kinds of equations and solutions we will encounter and shows that such a constant may appear as a factor of the solution as well.

EXAMPLE 2

Show that, no matter what the value of the constant C, the function $y(t) = Ce^{-t^2/2}$ is a solution of the differential equation $y' = -ty$.

Solution Substituting on the left side, we have

$$y'(t) = C \cdot D_t e^{-t^2/2} = C\left(\frac{-2t}{2}e^{-t^2/2}\right) = -Cte^{-t^2/2}$$

And on the right,

$$-ty(t) = -t(Ce^{-t^2/2}) = -Cte^{-t^2/2}$$

Since the same answer results, this formula, for *any* value of C, is a solution. ∎

The solution of a differential equation involves, then, what is called a **family of functions,** each member of this family being determined by the choice of the constant C. A few members of the family of solutions to Example 2 (for $C = -1, 0, 1, 2$) are shown in Figure 8.2.

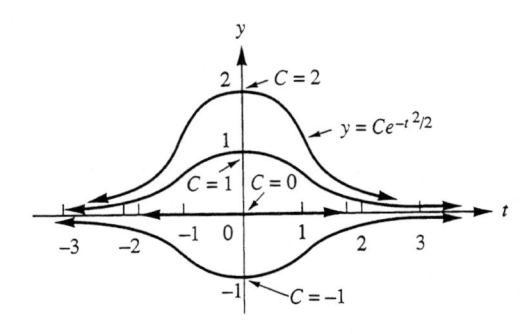

Figure 8.2

An **initial value problem (IVP)** is a differential equation coupled with a single known value of the function called an **initial condition.** An IVP of the kind we will study, unlike a differential equation standing alone, has *only one* solution.

8.1 Differential Equations and Their Solutions   437

EXAMPLE 3

Show that $y(t) = 3e^t - t - 1$ is the solution to the IVP $y' = y + t$ and $y(0) = 2$.

Solution We have already seen in Example 1 that the function y satisfies the equation. Moreover, this function also satisfies the initial condition $y(0) = 2$, since $y(0) = 3e^0 - 0 - 1 = 3 - 1 = 2$. ∎

Our next example indicates why an IVP has only one solution.

EXAMPLE 4

Among all the possible solutions $y(t) = Ce^{-t^2/2}$ of the equation $y' = -ty$ in Example 2, find the one solution that satisfies the initial condition $y(0) = 4$.

Solution We already know that this formula satisfies the differential equation. In order for it to satisfy the initial condition $y(0) = 4$, we must have

$$4 = y(0) = Ce^{-0^2/2} = C \cdot 1 = C$$

Therefore, if we let $C = 4$, then $y(t) = 4e^{-t^2/2}$ is the only solution to the IVP $y' = -ty$ and $y(0) = 4$. ∎

An IVP may involve a function value other than one at $t = 0$.

EXAMPLE 5

Determine C so that $y(t) = Ce^{-t^2/2}$ is a solution to the IVP $y' = -ty$ and $y(2) = -3$.

Solution We insist that $-3 = y(2) = Ce^{-2^2/2} = Ce^{-2}$. Therefore $C = -3e^2$ is the needed value of C. The solution to this IVP is therefore $y(t) = (-3e^2)e^{-t^2/2} = -3e^{2-(t^2/2)} = -3e^{(4-t^2)/2}$. ∎

Let us now turn to some applications of differential equations and IVP's that indicate their use. We will also get a first look at how the solution to a differential equation may be found.

An Application: Time-Dependent Exponential Differential Equations and Relative Rate of Change

In Section 5.5, we saw that the (differential) equation $A'(t) = kA(t)$ [with solutions $A(t) = Ce^{kt}$, where $A(t)$ is the amount present (state) at time t] is a useful model for many natural systems. We also recognize that it has obvious limitations. For example, provided with a stable environment, a colony of the bacterium *E. coli*, which doubles its population every 20 minutes, could easily take over the entire world if an exponential model for its growth actually held for an unlimited amount of time. Since this does not happen in reality, we need a more sophisticated model. In this and the next section, we study three basic extensions of this model that better reflect the external world in a variety of circumstances.

The first of these models asks that we first rewrite the differential equation $A'(t) = kA(t)$ as $A'(t)/A(t) = k$. The quantity on the left

$$\frac{A'(t)}{A(t)} = \frac{\text{Rate of change in amount at time } t}{\text{Amount present at time } t}$$

measures the **percentage** (or **relative**) **rate of change** (see Section 5.4) in the amount present. This is often a more important, and natural, quantity to consider than $A'(t)$ itself. For example, imagine a young mouse, weighing 2 oz and growing at the rate of 0.10 oz per day, and a young elephant, weighing 500 lb and gaining 10 lb per day. Which is growing faster? At first, we might say the young elephant. On the other hand, if we consider the percentage rates of growth (relative to existing size), we have that the percentage rate of growth for the mouse is

$$\frac{0.10 \text{ oz}}{2 \text{ oz}} = \frac{1}{20}$$

or 5%, and the percentage rate of growth for the elephant is

$$\frac{10 \text{ lb}}{500 \text{ lb}} = \frac{1}{50}$$

or 2%. The mouse is growing at a rate $2\frac{1}{2}$ times as fast as the elephant relative to its size! Notice also that a percentage rate of growth is a *dimensionless* quantity.

The concept of percentage rate of growth is particularly useful when this rate is *not constant*. In such a case the percentage rate of growth $A'(t)/A(t)$ is some function $g(t)$. That is, we have a differential equation of the *form*

$$\frac{A'(t)}{A(t)} = g(t)$$

[For example, if $g(t) = 0.02t$, we would have the equation $A'/A = 0.02t$.] Let us apply the calculus to find the solution to such an equation.

Assume that we know that $A(t) > 0$ and recall that

$$D_t(\ln A(t)) = \frac{1}{A(t)} \cdot D_t A(t) = \frac{A'(t)}{A(t)}$$

by the chain rule. Therefore the equation $A'(t)/A(t) = g(t)$ is equivalent to the equation

$$D_t(\ln A(t)) = g(t)$$

If we integrate both sides of this equation, we have

$$\int D_t \ln A(t) \, dt = \int g(t) \, dt$$

or

$$\ln A(t) = \int g(t) \, dt$$

by the Fundamental Theorem of Calculus. Now, since $e^{\ln x} = x$, we have

$$A(t) = e^{\ln A(t)} = e^{\int g(t)\,dt}$$

or

$$A(t) = e^{\int g(t)\,dt}$$

Implicit in this solution is an unknown constant that results when the indefinite integral is found.

In summary:

The general solution of the differential equation

$$\frac{A'(t)}{A(t)} = g(t)$$

where $A(t) > 0$ for all t is

$$A(t) = e^{\int g(t)\,dt} \qquad (1)$$

EXAMPLE 6

A company has been marketing a certain product for some time using a heavy advertising campaign. It then decides to stop advertising and observe what happens to sales. If $A(t)$ is the amount of sales on day t after advertising stops, marketing surveys show that

$$\frac{A'(t)}{A(t)} = -0.01t$$

t days after advertising stops. That is, sales are declining, and the percentage rate of decline is in proportion to the number of days without advertising.

If, on the day advertising stops, $A(0) = 150$ items were sold, how many will be sold 14 days later?

Solution Note that this is an IVP with $A'/A = -0.01t$ and $A(0) = 150$. According to our preceding discussion, with $g(t) = -0.01t$, we have

$$A(t) = e^{\int -0.01t\,dt} = e^{-0.01\int t\,dt} = e^{-0.01t^2/2 + C} = e^{-0.01t^2/2} \cdot e^C$$

Note that since C is an unknown constant, so is e^C; thus we may replace e^C by a "new" unknown constant C_1. This is a typical maneuver in dealing with such constants as they appear. Thus we write this solution as

$$A(t) = e^{-0.005t^2} \cdot e^C = C_1 e^{-0.005t^2}$$

Now, since $A(0) = 150$, we have

$$150 = A(0) = C_1 e^{-0.005(0)^2} = C_1 \cdot 1 = C_1$$

Hence $A(t) = 150 e^{-0.005t^2}$ is the solution to this IVP.

Therefore, on day $t = 14$, $A(14) = 150 e^{-0.005(14)^2} \approx 56.30$ items will be sold. We graph this solution in Figure 8.3.

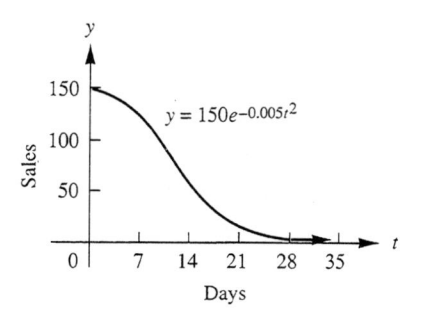

Figure 8.3

(Graph: Sales vs Days, showing $y = 150e^{-0.005t^2}$, with y-axis marked 50, 100, 150 and t-axis marked 7, 14, 21, 28, 35 Days)

Again, although it is convenient to model a system where items are measured in discrete, whole units, answers in fractional units will inevitably appear. Consequently, we interpret 56.30 as 56 items.

Toxicity in the Environment

Expressions of the form

$$\frac{A'(t)}{A(t)}$$

are particularly useful in environmental studies. If a population $A(t)$ is in decline, this quantity, which is often called the **mortality rate** of the population, is a measure of the *percentage rate* of decline in the population.

The next example shows how two simple, verifiable conditions, written as differential equations, allow us to derive a precise model of a realistic system.

EXAMPLE 7

Leakage of a broken container vessel has allowed a lethal chemical to run into a Missouri stream containing a healthy population of smallmouth bass. An environmental team finds that

i. The amount $C(t)$ of the chemical in the stream t days after the spill is declining at the rate of 25% a day (due to the current in the stream). That is,

$$C'(t) = -0.25C(t)$$

Furthermore, $C(0) = 250$ gallons initially leaked into the stream.

ii. The mortality rate $B'(t)/B(t)$ for the population $B(t)$ of smallmouth bass on day t after the chemical spill is directly proportional to the amount of the chemical in the stream. That is,

$$\frac{B'(t)}{B(t)} = -kC(t)$$

Furthermore, sampling by the environmental team in a marked-off test area of the stream shows that $k = 0.002$.

Find the number of smallmouth bass still alive in the stream 6 days after the spill, if the healthy stream initially contained 15,000 smallmouth bass.

Solution

Step 1. Let us first determine how much of the chemical is in the stream on day t after release. Accordingly, we must solve the differential equation $C'(t) = -0.25C(t)$. As we saw in Section 5.5, this equation of exponential decay has the solution

$$C(t) = C(0)e^{-0.25t} = 250e^{-0.25t}.$$

since $C(0) = 250$.

Step 2. Combining this solution with the condition in (ii), we see that the mortality rate is exponentially declining according to the differential equation

$$\frac{B'(t)}{B(t)} = -kC(t) = -0.002(250e^{-0.25t}) = -0.5e^{-0.25t}$$

Consequently, from (1) on p. 440

$$B(t) = e^{\int -0.5e^{-0.25t}\,dt} = e^{-0.5(e^{-0.25t})/(-0.25) + C}$$
$$= e^{2e^{-0.25t}} e^C = C_1 e^{2e^{-0.25t}}$$

[The function $B(t)$ is new to us. It is an exponential of an exponential. Such functions appear so frequently in applications that they are known as **Gompertz growth functions.**]

Now, since $B(0) = 15,000$, we have

$$15,000 = B(0) = C_1 e^{2e^{-0.25(0)}} = C_1 e^{2e^0} = C_1 e^2$$

Therefore $C_1 = 15,000/e^2 \approx 2,030$. Hence we take

$$B(t) = 2,030 e^{2e^{-0.25t}}$$

as the population of smallmouth bass t days after the spill.

Step 3. Consequently, on $t = 6$ days after the spill the smallmouth bass population will be reduced to

$$B(6) = 2,030 e^{2e^{-0.25(6)}} = 2,030 e^{2(0.2231)} \approx 3,172$$

fish, from an original population of 15,000. ∎

Notational Conventions in the Written Form of Differential Equations

We commonly suppress the variable t as a part of the unknown function when writing a differential equation. For example, the equation $y'(t) = k\, y(t)$ is written as $y' = ky$.

If the equation is instead, say, $A'(t)/A(t) = e^t$, the variable t cannot be suppressed entirely, but the equation would commonly be written as

$$\frac{A'}{A} = e^t \quad \text{or} \quad A' = e^t A$$

Finally, the particular letter representing the unknown function is immaterial, and each of the following is considered to be the *same* differential equation:

$$A'(t) = kA(t)[M - A(t)] \quad \text{or} \quad y' = ky[M - y]$$

Similarly, $A'(t) = t^2 A(t)$ and $y' = t^2 y$ are considered to be the same equation.

We will begin in the next section to consistently use the more brief of these notations.

Exercises 8.1

Show that

1. $y(t) = e^{-t^2}$ is a solution of $y' = -2ty$.
2. $y(t) = 7$ is a solution of $y' = 0$.
3. $y(t) = t + 1$ is a solution of $y' = y - t$.
4. $u(t) = \frac{1}{t}$ is a solution of $u' = \frac{-u}{t}$.
5. $u(t) = 1 - 3e^{2t}$ is a solution of $u' = 2(u - 1)$.
6. $x(t) = \frac{e^t}{1 + e^t}$ is a solution of $x' = x(1 - x)$.

Verify that the function $y = f(t)$ is a solution of each IVP in Exercises 7–10.

7. $f(t) = 1 - e^{-t}$ for $y' = 1 - y$; $y(0) = 0$
8. $f(t) = 3$ for $y' = 3 - y$; $y(0) = 3$
9. $f(t) = 2e^{e^t}$ for $y' = e^t y$; $y(0) = 2e$
10. $f(t) = 1 + 2e^{kt}$ for $y' = k(y - 1)$; $y(0) = 3$

Each function f in Exercises 11–14 is a solution of the indicated differential equation for any choice of the constant C. Determine C so that the initial value for f is also satisfied.

11. $f(t) = 1 - Ce^{-t}$ for $y' = 1 - y$; $y(0) = 3$
12. $f(t) = C$ for $y' = 0$; $y(0) = \sqrt{2}$
13. $f(t) = \frac{C}{t^2 + 1}$ for $y' = \left(\frac{-2t}{t^2 + 1}\right)y$; $y(1) = 2$
14. $f(t) = t^3 + C$ for $y' = 3t^2$; $y(2) = 9$
15. Show that, although $y(t) = t + 1$ is a solution of $y' = y - t$, neither $y(t) = t + 1 + C$ nor $y(t) = C(t + 1)$ is a solution. Then show that $y(t) = t + 1 + Ce^t$ is a solution for any value of C.

Remark. While an arbitrary constant C must appear in the general solution to a differential equation, it cannot be haphazardly inserted but must arise through an integration process, as seen in Examples 6 and 7. We will deal with this more fully in Sections 8.3 and beyond.

16. Verify that
    a. $y(t) = Ce^{kt} \pm 1$ is a solution of $y' = k(y \pm 1)$.
    b. $y(t) = 1/(1 + Ce^{-kt})$ is a solution of $y' = ky(1 - y)$.

17. The percentage rate of growth of sales of a new product is

$$\frac{A'(t)}{A(t)} = 0.02t$$

t days after the product is introduced on the market, where $A(t)$ is the total number of sales by day t, with $A(0) = 10$.

a. How many sales will be made in 14 days?

b. How long will it take to sell 4,000 items?

18. Suppose that you invest $S(0) = \$500$ in a certain stock. Let $S(t)$ be the value of this stock on day t. Suppose that the percentage rate of change of the value of this stock is

$$\frac{S'(t)}{S(t)} = 0.05\left(\frac{1}{t+1}\right)$$

What is the value of this stock 14 days later?

19. A woman who is pregnant has one glass of wine containing 18 mL of alcohol. Her body metabolizes (eliminates) the alcohol at the rate of 40% of the amount present. On the other hand, the baby in her womb absorbs alcohol from her body at the rate of 5% of the amount present in her body. How much alcohol will the baby have consumed at the end of 2 hours? [*Hint:* Model this system as $W'(t) = -KW(t)$ and $B'(t) = kW(t)$, where $W(t)$ and $B(t)$ are the amounts of alcohol present in woman and baby at time t, and where $K = 0.4$, $k = 0.05$, $W(0) = 18$, and $B(0) = 0$.]

Challenge Problems

20. This exercise shows that the concept of a percentage rate of growth $A'(t)/A(t)$ can be quite useful since percentages are *dimensionless quantities*. This means that percentages for two very different items can be directly compared.

Let $S(t)$ be the dollar value of the stock of a certain oil company. Let $\theta(t)$ be the known oil reserves (in millions of barrels) of this particular oil company. We cannot easily compare $S(t)$ and $\theta(t)$ since one is measured in dollars and the other in barrels. On the other hand, we can expect that the percentage rate of change $S'(t)/S(t)$ of the value of oil company stock is directly proportional to the percentage rate of change of its known reserves $\theta'(t)/\theta(t)$. That is,

$$\frac{S'(t)}{S(t)} = k\frac{\theta'(t)}{\theta(t)}$$

Integrate both sides of this equation to show that

$$S(t) = C\,\theta(t)^k$$

This equation means that the value of the company's stock is directly dependent on a *power* of its known reserves. Such a relationship is called an **allometric relationship.**

21. Refer to Example 7 and suppose that the initial amount of chemical spilling into the stream was 500 (rather than 250) gallons.

a. How many bass will remain in the stream 6 days later?

b. Answer the same question if instead the mortality constant k were 0.004 and the spill were again 250 gallons.

22. The use of antibiotic tetracyclines in animal feed stocks in the United States has created local biological systems favoring the growth of bacteria resistant to these "wonder drugs." These resistant bacteria show some evidence of having now spread to humans, including the transfer, by mere physical proximity, of immunity-conferring cell constituents called "R-plasmids" from resistant bacteria to ordinary bacteria, thus creating immunities even in humans who have not overused these drugs [*Science '86* 7, no. 4 (May 1986): 40–43].

Consider a model of the human system populated by the common and prolific bacterium *E. coli* (which has, largely harmlessly, been a part of the human system for millions of years) in conjunction with a small initial population of resistant bacteria, capable of transferring their R-plasmids directly to *E. coli*. Let $N(t)$ and $R(t)$ be the numbers of these common and resistant bacteria at time t. Let us make the following assumptions about this system:

a. $N(0) = 10^6$ and $R(0) = 5$.

b. Over a "short" span of time—say, $0 \le t \le 2$ hours—the population of *E. coli* is growing exponentially and $N'(t) = 0.02\,N(t)$.

c. The population of resistant bacteria increases in direct proportion to the total number, $N(t)R(t)$, of possible physical contacts between the number $N(t)$ of *E. coli* and the number $R(t)$ of bacteria carrying R-plasmids. That is, $R'(t) = k\,N(t)R(t)$. Suppose that the proportion k of physical contacts in which resistance is transferred is very small, say, $k = 10^{-7}$.

Under these assumptions determine the number of resistant bacteria present after $t = 6$, 12, and 24 hours. [*Hint:* The solution is entirely like that of Example 7 except that decline there is replaced by growth here.]

8.2 Growth in a Limited Environment

In this section we study Newton's law of heating or cooling as an example of change in proportion to remaining capacity, and then turn to the differential equation of logistic growth, characterizing percentage rate of change in proportion to percentage of remaining capacity.

A second exponential model begins with the everyday experience of heating and cooling. Isaac Newton, an inventor of calculus, was also a great natural scientist. One of the principles he discovered was

---

Newton's Law of Heating or Cooling

The rate of heating (or cooling) of an object is directly proportional to the temperature difference between the object and its surroundings.

Let us convert this English language statement into a mathematical one. Imagine that you have just taken a can of soda from your refrigerator, at a temperature of 45°F, and set it on a table in your room, where the temperature is 68°. Let $T(t)$ be the temperature of the soda at time t. The soda will begin to warm, so $T(t)$ will increase with t, but certainly $T(t) \leq 68$ for all t! Newton's law says that the rate of change of temperature, $T'(t)$, is directly proportional to the temperature difference, $68 - T(t)$. That is, in mathematical language,

$$T'(t) = k(68 - T(t)) \qquad \text{where } k > 0$$

If for some reason you wanted to know a formula for the temperature of your can of soda at time t minutes later, you would need to solve this differential equation. Let us do so but by a substitution.

Let $h(t) = T(t) - 68$ be the temperature difference between the soda and the room. Then

$$h'(t) = T'(t) - 0 = T'(t) = k(68 - T(t)) = k(-h(t))$$

That is, the function $h(t)$ satisfies the equation of exponential decay: $h'(t) = -kh(t)$. Therefore

$$h(t) = Ce^{-kt}$$

Hence, since $h(t) = T(t) - 68$, then $T(t) - 68 = Ce^{-kt}$, or

$$T(t) = 68 + Ce^{-kt}$$

This is the solution function for Newton's equation of heating or cooling.

Since $T(0) = 45$ when you took the can of soda from the "fridge," we also must have that

$$45 = T(0) = 68 + Ce^{-k0} = 68 + C$$

or $C = 45 - 68 = -23$.

Therefore, at time t, your warming can of soda has a temperature of

$$T(t) = 68 - 23e^{-kt} \text{ degrees}$$

We graph this function in Figure 8.4. Notice that the values of this function exponentially "decay upward" from an initial temperature of 45° up to (very near) 68°. Notice too how naturally the given data for this system lead to an IVP and how the constant C, which must arise for purely mathematical reasons, is essential for modeling the physical system.

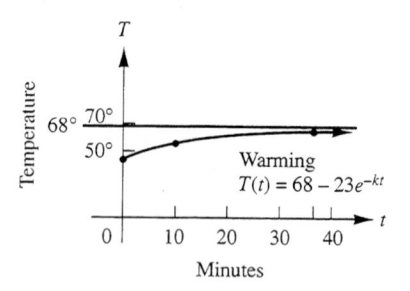

Figure 8.4

Are you wondering about the number k? It is determined by how you treat your can of soda. The *warming constant* k will be different for different materials. For example, if the can of soda were put in a foam hand-holder, k would be a very small number since warming would take place very slowly. Let us see how k can be determined.

EXAMPLE 1

Suppose that you have become so interested in this discussion that you forgot to drink your soda. Using a thermometer, you find that after 10 minutes outside the refrigerator, your can of soda is at a temperature of 55°.
a. How warm will it be 10 minutes later?
b. When will it be 65°?
c. Will it ever be 70°?

Solution

a. We have seen that the equation

$$T'(t) = k(68 - T(t))$$

with initial condition $T(0) = 45°$ models the warming can of soda and has the solution

$$T(t) = 68 - 23e^{-kt}$$

Since $T(10) = 55$, we have $55 = 68 - 23e^{-k(10)}$. Consequently, $23e^{-10k} = 13$ or $e^{-10k} = 0.565$. Therefore, taking the natural log of both sides, $-10k = \ln 0.565$, so that $k = 0.057$. That is, one actual observation, $T(10) = 55°$, allows us to determine the warming constant k and a complete mathematical model of your warming can of soda:

$$T(t) = 68 - 23e^{-0.057t}$$

Therefore, another 10 minutes later, at $t = 20$, the soda will be at a temperature of $T(20) = 68 - 23e^{-0.057(20)} = 60.64°$.

b. The soda will be at 65° when
$$65 = T(t) = 68 - 23e^{-0.057t}$$
or when $-3 = -23e^{-0.057t}$. Solving for t, we have $\ln 0.130 = -0.57t$ or $\ln 0.130 = -0.057t$ or $t = 35.79$ minutes after leaving the fridge.

c. No. Besides there being no physical cause for such an event, notice that since $e^{-0.057t} > 0$ for all t then
$$T(t) = 68 - 23e^{-0.057t} < 68 \qquad \text{for all } t$$
The graph of this solution is shown in Figure 8.4. ∎

Notice, then, that very little information yields a complete mathematical model of a can of soda warming to the temperature of its surroundings.

The Differential Equation of Growth in Proportion to Remaining Capacity

A can of soda at 45° set in a room at 68° has a *remaining capacity* for warming of $68° - 45° = 23°$. Newton's law of heating or cooling is an example of a general concept with wide application: **rate of growth in proportion to remaining capacity,** whose mathematical form is

$$A'(t) = k[M - A(t)] \tag{1}$$

Here $A(t)$ is an amount present at time t, and M is an upper (or lower) limit to which A can grow (decline). The number $M - A(t)$ is the measure of the remaining capacity for change in A. In Example 1, $M = 68°$, the temperature to which warming occurs, and therefore $68 - T(t)$ is the remaining capacity for warming of the can of soda.

Equation 1 will arise repeatedly in various forms in the next four sections. While it may be solved as earlier, its solution is also a special case of the methods of Section 8.4 and will only be stated here.

A differential equation of the form
$$A'(t) = k(M - A(t))$$
where k and M are constants, has the unique solution
$$A(t) = M - Ce^{-kt} \tag{2}$$
With initial condition $A(0)$, the constant C is given by
$$C = M - A(0)$$

In case $A(t) \geq M$ and $k > 0$, we have *decline* of the values of A to the remaining capacity M. For example, a warm meal brought into a cool room will cool to the temperature of the room in the same manner, yielding a solution to Equation 1 like that indicated in Figure 8.5.

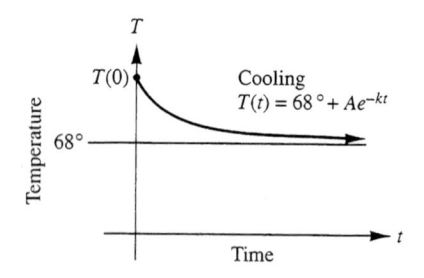

Figure 8.5

The solution (Equation 2) may of course be applied without regard to a physical example such as heating or cooling.

EXAMPLE 2

a. Use Equation 2 to find the solution of
$$y' = 2(30 - y)$$
b. Find the unknown constant C when $y(0) = 10$.
c. Determine that value of t for which $y(t) = 25$.

Solution

a. Comparing $y' = 2(30 - y)$ with Equation 1, we see that $k = 2$ and $M = 30$. Using Equation 2, we have
$$y(t) = 30 - Ce^{-2t}$$

b. We have $C = 30 - y(0) = 30 - 10 = 20$. Thus
$$y(t) = 30 - 20e^{-2t}$$

c. We must determine t so that
$$25 = 30 - 20e^{-2t}$$
$$-5 = -20e^{-2t}$$
$$0.25 = e^{-2t}$$
$$\ln 0.25 = \ln e^{-2t} = -2t$$

or
$$t = \frac{-1}{2} \ln 0.25 \simeq \frac{-1}{2}(-1.386) = 0.693 \blacksquare$$

The Logistic Equation: Percentage Rate of Growth to Maximum Capacity

A final introductory example adds one more step in complexity and yields a model with wide application in business, biology, medicine, economics, and other fields. At the same time, this model incorporates ideas found in all previous models.

The differential equation of Newton's law of heating or cooling has the *form*
$$A' = f(A) \qquad \text{where } f(A) = k(M - A)$$
Notice that the rate of change of A is a function of A alone, with time t only a background variable appearing implicitly in the equation.

Let us combine this observation with a consideration of the percentage rate of change of A. That is, suppose that the percentage rate of change of A is a function of A alone. This means that
$$\frac{A'}{A} = f(A)$$
If now $f(A) = k(M - A)$ is a simple linear function, we obtain the equation
$$\frac{A'}{A} = k(M - A) \qquad \text{or} \qquad A' = kA(M - A)$$
Bringing in the background variable t, we have
$$A'(t) = kA(t)[M - A(t)] \qquad (3)$$

The differential equation (3) is called the **logistic differential equation.** It very naturally describes growth in a limited environment for the following reason. Suppose that a natural environment is capable of supporting at most M individuals; M is called the *carrying capacity* of the environment. Let $A(t)$ be the size of the population at time t. When the population growth in a limited environment is being modeled, it is natural to suppose that the percentage rate of growth in the population, $A'(t)/A(t)$, is *directly proportional to the percentage of unused capacity* in the environment. Since this percentage is $[M - A(t)]/M$, we immediately have the equation
$$\frac{A'(t)}{A(t)} = r\left(\frac{M - A(t)}{M}\right)$$
or
$$\frac{A'}{A} = \frac{r}{M}(M - A)$$
That is, we have the logistic equation
$$A' = kA(M - A) \qquad \text{where } k = \frac{r}{M}$$
Thus, for very general but plausible reasons, the logistic equation readily arises.

The logistic equation can be recalled in these terms: Growth A' (in an environment limited to M individuals) is directly proportional (k) to its presence, A, and its absence, $M - A$. This statement is a convenient guide for recognizing when the logistic equation may be a useful model for a system. This is well shown by the next example, wherein we see the logistic equation arise for specific, but alternate, reasons as we *build a model* of this system.

EXAMPLE 3

A local high school has 900 students. A new variety of flu virus has been contracted by some of the students, who come to class on Monday morning. It is known that:

i. The virus is spread in the incubation stage by proximity and can be passed from an infected person (carrier) to another person before either is aware that one has the virus.
ii. Each carrier is in close enough proximity to pass on the virus to six other students each hour.
iii. In 2% of the cases where a noninfected student is in proximity to an infected student, he or she will contract the illness.

Construct a differential equation that models the spread of this flu epidemic through the high school.

Solution Let $f(t)$ = the number of students carrying the flu virus t school hours after 9 A.M. Monday morning. We wish to determine the state $f(t)$ of the epidemic at time t by analyzing its change-of-state. That is, we will try to express

$$f'(t) = \text{rate at which the flu is spreading}$$

in an analytic way.

To do this, we imagine a fixed time t and a short span of additional time h. We will try to calculate

$$f(t + h) - f(t) = \text{increase in the number of flu victims}$$

between time t and time $t + h$.

Let us first try to understand what *one* flu carrier can accomplish. Over an interval of time h (hours), this carrier will be in proximity to

$$6 \cdot h \text{ individuals}$$

since he or she is in proximity to six individuals *per hour* [from (ii)].

Now, at time t there are

$$f(t) \text{ carriers} \quad \text{and} \quad 900 - f(t) \text{ noncarriers of this virus}$$

Therefore the *percentage* of noninfected students at time t is

$$\frac{900 - f(t)}{900}$$

Consequently, of the $6h$ students contacted,

$$\left[\frac{900 - f(t)}{900}\right] \cdot 6h$$

are not infected. According to (c), 2% of these, an amount

$$0.02 \left[\frac{900 - f(t)}{f(t)}\right] \cdot 6h = \frac{0.12}{900}[900 - f(t)]h$$

will become infected.

This last quantity is the number of new infections due to *one* carrier. But at time t there are $f(t)$ carriers. Therefore the total number of new infections is

$$f(t)\left[\frac{0.12}{900}(900 - f(t))\right]h \qquad (4)$$

At the same time, the number of new infections is also

$$f(t + h) - f(t) \qquad (5)$$

If we set expressions 4 and 5 equal, we obtain

$$f(t + h) - f(t) = f(t)\left[\frac{0.12}{900}(900 - f(t))\right]h$$

or

$$\frac{f(t + h) - f(t)}{h} = \frac{0.12}{900}f(t)[900 - f(t)]$$

The quantity on the left is the familiar quotient used to define the derivative of a function. The quantity on the right is independent of h. If we take the limit as $h \to 0$ of both sides of this equation, we obtain the law of change-of-state for this system:

$$f'(t) = 0.00013 f(t)[900 - f(t)] \qquad (6)$$

This differential equation models the spreading flu epidemic under the given conditions. Its solution can be used to predict (or estimate) the number of students affected by time t, as we will see in the next example. ∎

Note that $M = 900$ is the "carrying capacity" of the environment in Example 3 and that Equation 6 has exactly the logistic *form* $A' = kA[M - A]$ with $A(t) = f(t)$, $k = 0.00013$, and $M = 900$. This is not surprising, again by general reasoning, since an epidemic can reasonably be expected to spread at a rate proportional to both its presence, $f(t)$, and its absence, $900 - f(t)$.

We defer a derivation of the solution of the general **logistic equation** to Section 8.4 and only state the solution here for use in the next example.

The solution of the logistic differential equation

$$A'(t) = kA(t)[M - A(t)]$$

with initial state $A(0)$ is

$$A(t) = \frac{M}{1 + Ce^{-Mkt}} \qquad (7)$$

where

$$C = \frac{M - A(0)}{A(0)}$$

If $A(0) = 0$, the solution is $A(t) = 0$ for all t.

EXAMPLE 4

If $f(0) = 20$ is the initial number of students in Example 3 carrying the flu virus on Monday morning, how many will be carrying it Friday noon, assuming a 6-hour school day?

Solution From Example 3, the logistic equation (6) modeling the flu epidemic is $f'(t) = 0.00013 f(t)[900 - f(t)]$. Accordingly (from Equation 7), this equation has the solution

$$f(t) = \frac{900}{1 + Ce^{-900(0.00013)t}}$$

with $C = (900 - 20)/20 = 44$. That is,

$$f(t) = \frac{900}{1 + 44e^{-0.117t}}$$

Noon Friday is 27 school hours after Monday morning. Therefore the number of students carrying the flu virus then is

$$f(27) = \frac{900}{1 + 44e^{-0.117(27)}} = \frac{900}{1 + 44(0.042)} \approx 316 \ \blacksquare$$

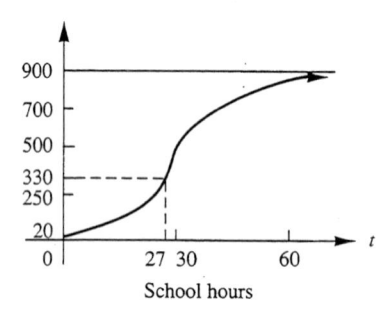

Figure 8.6

The graph of the solution in Examples 3 and 4, shown in Figure 8.6, is typical of the graph of a solution to the logistic equation. We will derive such graphs in Section 8.3. The amount $A(t)$ in question rises ever more rapidly to the level $M/2$ and then continues to rise, but at a declining rate thereafter. This "makes sense" in the spread of an epidemic, where the virus quickly spreads to, say, half the population and then, there being fewer and fewer uninfected victims, spreads at a slower and slower rate. Of course, even in this model, where we realistically assume a maximum capacity, we achieve a still idealized model and solution. Rarely does any infectious agent actually affect an entire susceptible population. This model should also not be viewed as literally true for long measures of time.

The important differential equations of Sections 5.5, 8.1, and 8.2, their solutions, and the intuitive nature of their application are summarized in Table 8.1.

Table 8.1
Differential Equations

Name	Equation	Solution	Application
Exponential growth or decay	$y' = ky$	$y(t) = Ce^{kt}$	Growth in proportion to amount present; growth that "feeds on itself"
Exponential growth to remaining capacity	$y' = k(M - y)$	$y(t) = M - Ce^{-kt}$	Growth in proportion to amount *not* present
Relative growth or mortality	$\dfrac{y'}{y} = g(t)$	$y(t) = e^{\int g(t)\,dt}$ if $y(0) \neq 0$	Percentage rate of growth varying over time
Logistic growth	$\dfrac{y'}{y} = r\left(\dfrac{M-y}{M}\right)$	$y(t) = \dfrac{M}{1 + Ce^{-rt}}$ if $y(0) \neq 0$	Percentage rate of growth in proportion to percent of remaining capacity
	Alternate form: $y' = ky(M - y)$	$y(t) = \dfrac{M}{1 + Ce^{-Mkt}}$ if $y(0) \neq 0$	Growth in proportion to amount present *and* remaining capacity

Exercises 8.2

Use the general solution (Equation 2) to Newton's law of heating or cooling to find the solution to each of the differential equations in Exercises 1–4 and determine the unknown constant C that appears in the solution.

1. $y' = 3(5 - y); \quad y(0) = 1$
2. $y' = \frac{1}{2}(-3 - y); \quad y(0) = 0$
3. $A' = 0.5(2 - A); \quad A(0) = 2$
4. $2y' = 4 - 8y; \quad y(0) = -1$

Use the general solution (Equation 7) to the logistic equation to find the solution to each of the differential equations in Exercises 5–8, and determine the unknown constant C that appears in the solution.

5. $A' = 2A(3 - A); \quad A(0) = 5$
6. $A' = \frac{1}{3}A(6 - A); \quad A(0) = 6$
7. $y' = 2y(y - 5); \quad y(0) = 1$
8. $y' = 0.3y(150 - y); \quad y(0) = 100$
9. Find t so that $y(t) = 2$ in the solution to Exercise 1.
10. Find t so that $A(t) = 1$ in the solution to Exercise 3, if such a number t exists.
11. Find t so that $A(t) = 1.5$ in the solution to Exercise 5.
12. Find t so that $y(t) = 100$ in the solution to Exercise 8, if such a number t exists.
13. Repeat Example 1 if, 5 minutes after removal, the can of soda is at a temperature of 52°.
14. Find a suitable thermometer, obtain a can of soda from a nearby vending machine, and repeat Example 1 for this can of soda. Be sure to first determine the

temperature at your desk using the thermometer and second to determine the initial temperature of the soda when you obtain it.

15. An investment management service has $150,000 of its clients' money to invest. It chooses the following general investment strategy. On each day t it invests 12% of its uninvested balance in money market instruments. Let $A(t)$ denote the amount of money it has invested by day t.

a. Write a differential equation that expresses the service's investment strategy.

b. With $A(0) = 0$, find the solution to this differential equation.

c. How much will be invested by day $t = 5$? $t = 10$?

16. Verify that $A(t) = M - Ce^{-kt}$ is the solution to $A'(t) = k(M - A(t))$ by substitution. Then verify that $C = M - A(0)$.

17. Glucose (blood sugar) is commonly administered to a hospital patient by intravenous injection at a constant rate R. The patient's body metabolizes (uses) the glucose at a rate k proportional to its concentration in his or her bloodstream. If $G(t)$ denotes the concentration of glucose in the patient's system at time t, the process of glucose metabolism can be modeled by a "rate-in less rate-out" equation

$$G'(t) = R - kG(t)$$

a. Write this equation in the form (1) by factoring k and letting $M = R/k$.

b. Write the solution of this equation.

c. Show that $\lim_{t\to\infty} G(t) = R/k$. What does this mean?

18. A rumor started by one individual is spreading through an office of 50 employees at the rate of 20% per hour of the individuals who have not yet heard the rumor. When will 30 people have heard the rumor?

19. Imagine that you have a certain unit of material to learn for the next exam. Let 1 denote this unit amount of material. Let $A(t)$ denote the amount of the material that you know by time t. It is not unreasonable to expect that your rate of learning the new material depends on two things: (1) how much you already know, $A(t)$, and (2) how much you do not yet know, $1 - A(t)$. Let us suppose for the sake of specificity that you initially know $A(0) = 0.15$, or 15% of the material. Suppose that you learn new material at the rate of 25% of the product of what you do know and do not know.

a. Express this last sentence as a differential equation.

b. Write the solution of this differential equation with the given initial value.

c. At what time will you know 50% of the material? 80%? 90%?

d. Graph the function in (b).

20. Refer to Example 4 and evaluate the solution at noon Wednesday of the first week and then at noon Monday, Wednesday, and Friday of the next week. Compare your answers with the level of infection indicated by the graph in Figure 8.6.

21. In a purely rational market the value $S(t)$ of one issue of stock in a company would be A/N, the ratio of the company's total assets to the total number N of shares of stock issued. Suppose instead that the percentage rate of change in $S(t)$ is directly proportional to the percentage of "rational" stock value not yet achieved. Derive a differential equation that models this assumption, and write it in simplest algebraic form.

22. Use the answer found in Exercise 21, along with the formula for the solution of the logistic equation, to determine $\lim_{t\to\infty} S(t)$, the long-run value of the stock.

23. This exercise concerns the production of a product A using a fixed amount M of natural resources (raw materials). Assume that sales $S(t)$ at time t are directly proportional to the amount $A(t)$ produced at time t. Thus $S(t) = \beta A(t)$. Conclude that the growth in sales is governed by an equation of the *form*

$$S'(t) = kS(t)[M\beta - S(t)]$$

if the percentage rate of growth of the amount produced is directly proportional to the percentage $[M - A(t)]/M$ of resources remaining.

24. In chemistry we encounter an autocatalytic reaction wherein one substance A is converted into a new substance B, which then catalyzes ("feeds") its own formation. Let $B(t)$ be the amount of substance B at time t, and let M be the original amount of substance A. A reasonable model for this system is

$$\frac{B'(t)}{B(t)} = k\left[\frac{M - B(t)}{M}\right]$$

since $M - B(t)$ is the amount of substance A left from which to form substance B.

Consider the rate of formation $B'(t)$ as a function $f(B)$ of the amount of B that is present. That is, regard

$$f(B) = \frac{k}{M}B(M - B)$$

454   Chapter 8   Differential Equations

as a function of amount present rather than time.

At what amount B is this rate f a maximum? [*Hint:* Use the first and second derivative tests to determine the maximum value of f.]

25. Imagine a typical manufacturing process beginning with a fixed number R of units of raw material, each of which will be fashioned into a finished product by the manufacturing process. Let P be the unknown number of finished units beginning with R units of raw material. It is reasonable to assume the following:

   i. Some of the initial R units of raw material will no doubt be defective. Let $U = \alpha R$ be the number of usable units; that is, a fixed proportion α of the given units is usable.

   ii. Let $N(t)$ be the number of units, beginning with the number U of usable units, which are still being processed at time t. Thus $N(0) = U$. There is a certain "mortality rate" $N'(t)/N(t)$ for the number of units still in processing at time t, since some will be lost, others will be broken, others will be ruined, and so on. Assume that $N'(t)/N(t) = -k$. This is a reasonable assumption for an experienced manufacturing process.

   iii. Let T denote the amount of time it takes to produce P finished units. We assume that $T = \beta U$. That is, T is proportional to the number of original units to be processed. Notice then that $N(T) = P$.

   a. Show that $P = \alpha R e^{-\alpha \beta k R}$. [*Hint:* Solve the equation in (ii), impose the initial condition, evaluate the solution at $t = T$, and make all the substitutions you can, using (i) and (iii).]

   b. Graph the equation in (a) on an RP-axis system for $\alpha = 0.9$, $k = 0.05$, and $\beta = 1.5$.

   c. Show that a maximum number of finished products results when the initial input of raw materials is $R = 1/\alpha\beta k$ units. [*Hint:* Maximize the function $f(x) = axe^{-bx}$, where $a = \alpha$, $b = -\alpha\beta k$, $x = R$.]

   d. If α, β, and k are as in (b), what are R and P equal to in (c)? If α, β, and k are unknown, but R is as in (c), what is P equal to?

   e. If you were a manager in charge of this manufacturing process, how would the equation in (c) advise you to go about improving the efficiency of your production?

8.3 Evolution of Time-Autonomous Systems

The method of this section will allow us to see, via a graph, how a system evolves over time if the system can be described by a time-autonomous differential equation, without actually solving this equation.

The general form of a differential equation in previous sections was

$$y' = g(t, y)$$

For example, we can think of the equation $y' = t^2 + y$ as $y' = g(t, y)$, where $g(t, y) = t^2 + y$. A differential equation is called **autonomous** if the variable t (usually representing time) need not appear *explicitly* in the equation. That is, an autonomous differential equation is one in which

$$y' = g(y)$$

is a function of y alone, as in Newton's law of heating or cooling, where $g(y) = k(M - y)$. Understand, however, that even in an autonomous differential equation y itself still represents a function of the background (implicit) variable t, which only happens not to be needed in expressing the equation.

We can sketch the graphs of solutions to the autonomous differential equation $y' = g(y)$ without solving the equation. To do this we first have to understand the role of **constant solutions** to a differential equation.

Newton's law of heating or cooling offers a ready example of the existence and meaning of constant solutions. Suppose that a cool object has been brought into a room in which the temperature is 68°F (and remains so) and that the temperature of this object at time t is $y(t)$. According to Newton's law, the object will warm according to the differential equation $y' = k(68 - y)$. But what if the object *itself* is 68° when it is brought into the room? That is, what if $y(0) = 68°$? Then nothing will happen, since the object can neither warm nor cool and therefore at all times t, $y(t) = 68°$. This is physically clear, of course. But notice that this *constant function* is also a mathematical solution of the differential equation, since (1) $y'(t) = 0$, and (2) $k(68 - y(t)) = k(68 - 68) = 0$. Recalling then that a differential equation has many solutions, we see that among these there may be constant solutions. In Figure 8.7(a), we graph the constant solution to the equation $y' = k(68 - y)$ with initial value $y(0) = 68$.

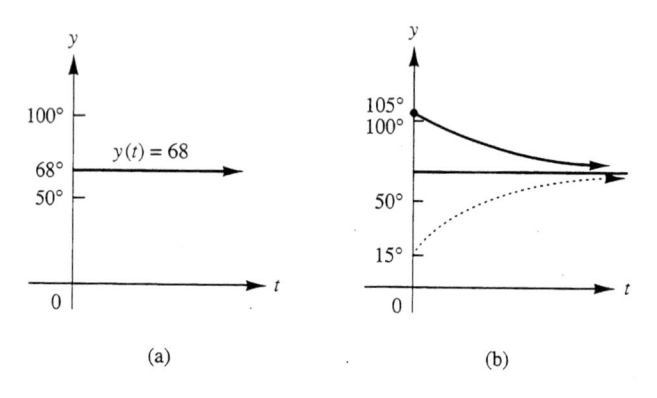

Figure 8.7

(a) (b)

Let us next ask, What if the object brought into the room was a warm meal, with $y(0) = 105°$? We should then expect that the solution to the IVP $y' = k(68 - y)$ and $y(0) = 105$ would have a graph like that shown in Figure 8.7(b). [Notice also that a cooler object would warm to the room temperature, as indicated by the dotted line graph in Figure 8.7(b).]

This example is typical of all the examples and exercises to be discussed in this section, and Figure 8.7(b) exemplifies the kind of solution we seek to these exercises: a simple graph of how the system evolves over time (warms or cools, or stays the same). We now outline a method that will enable us to draw such time-evolution graphs easily and quickly for a variety of differential equations.

We must first return to the matter of constant solutions. Consider the general autonomous differential equation

$$y' = g(y)$$

If M is a *number* such that $g(M) = 0$, then $y(t) = M$ is a *constant solution* to $y' = g(y)$, because (1) $y'(t) = 0$ since y is constant and (2) $g(y(t)) = g(M) = 0$. Therefore

---

Every root M of the algebraic equation $g(y) = 0$ yields a constant solution $y(t) = M$ of the differential equation $y' = g(y)$, and conversely.

---

456 Chapter 8 Differential Equations

[We saw this earlier in the particular case $y' = k(68 - y)$, where $g(y) = k(68 - y)$. Here $M = 68$, since $g(68) = 0$.]

EXAMPLE 1

Graph all constant solutions of the equation

$$y' = 3y(2 - y)$$

Solution Here $g(y) = 3y(2 - y)$. The *algebraic* equation $3y(2 - y) = 0$ has two (number) solutions, $M = 0$ and $M = 2$. There are then *two* constant solutions to the corresponding differential equation:

$$y(t) = 0 \quad \text{and} \quad y(t) = 2$$

We graph these in Figure 8.8. ∎

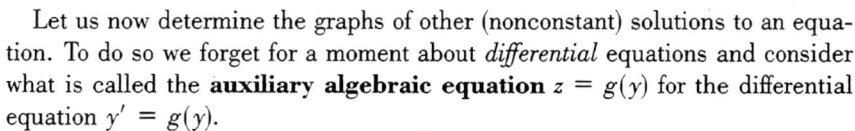

Figure 8.8

Let us now determine the graphs of other (nonconstant) solutions to an equation. To do so we forget for a moment about *differential* equations and consider what is called the **auxiliary algebraic equation** $z = g(y)$ for the differential equation $y' = g(y)$.

For example, the auxiliary algebraic equation for the differential equation $y' = 3y(2 - y)$ is the equation $z = 3y(2 - y)$. The auxiliary equation for $y' = k(68 - y)$ is $z = k(68 - y)$. The following example shows how we use an auxiliary equation to graph a solution to the corresponding differential equation with a given initial value.

EXAMPLE 2

Sketch the solution to the IVP $y' = 0.02(68 - y)$ and $y(0) = 45$.

Solution The auxiliary equation is $z = 0.02(68 - y) = 1.36 - 0.02y$. We first graph this linear algebraic equation, with slope -0.02 and vertical intercept 1.36, in a yz-coordinate system, as indicated in Figure 8.9.

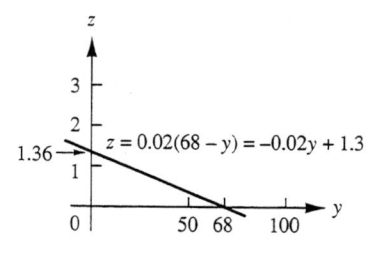

Figure 8.9

Here is the key point: Since $y' = 0.02(68 - y)$ and $z = 0.02(68 - y)$, then $z = y'(t)$ and therefore the values indicated on the z-axis in Figure 8.9 *represent the slope of the tangent lines to the solution $y(t)$*.

In Figure 8.10(a–h), we indicate how to use this idea. We first draw the auxiliary graph on a yz-axis system and then draw a second ty-axis system on which to represent solutions to the given differential equation.

8.3 Evolution of Time-Autonomous Systems

We initially sketch the one constant solution $y(t) = 68$ on this axis system in Figure 8.10(b). Notice how this one constant solution corresponds to the zero value on the auxiliary graph.

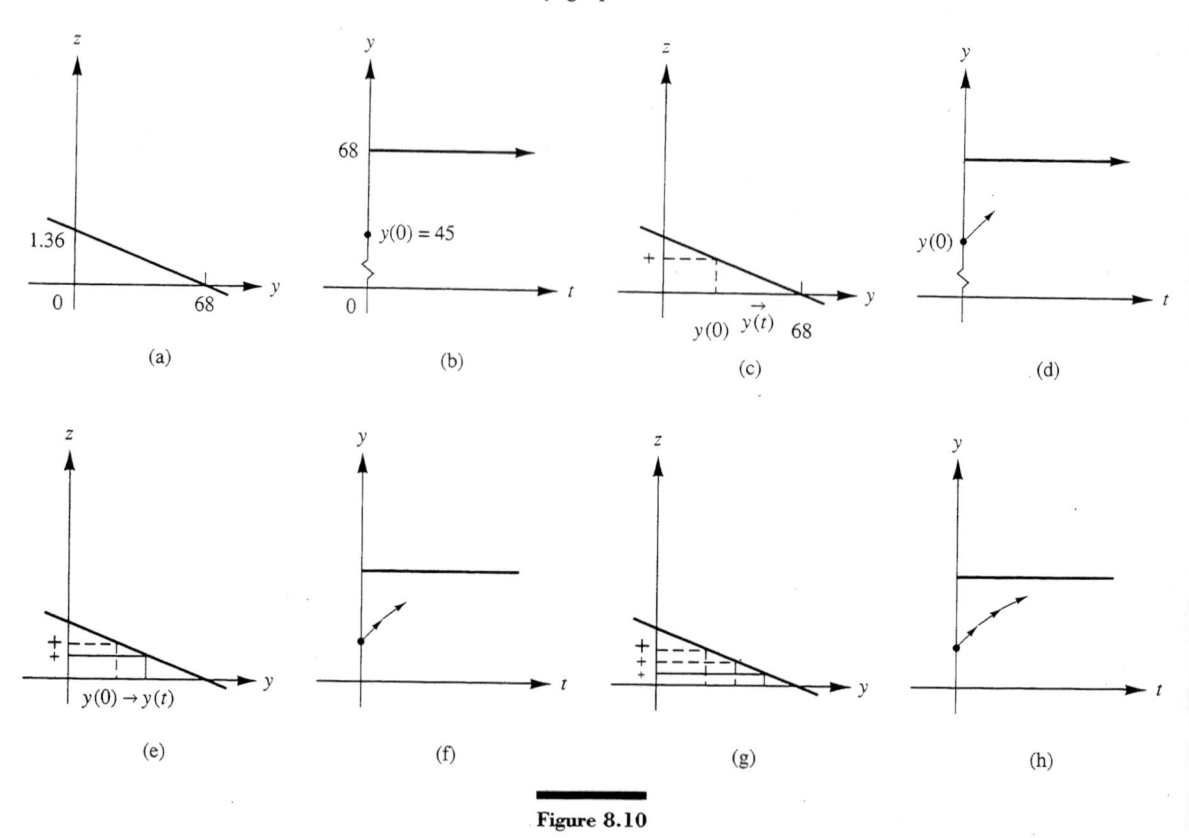

Figure 8.10

Now imagine a solution starting at $y(0) = 45 < 68$, as indicated in Figure 8.10(b). In the yz-coordinate system the corresponding z-value is positive $(+)$, as we see in Figure 8.10(c). Therefore the slope of the solution near $t = 0$ is *positive*. We indicate this in Figure 8.10(d), showing a positive tangent.

What happens as time moves on? "Later" values $y(t)$ must be larger [see Figure 8.10(c, d)] and therefore must be indicated on the y-axis in Figure 8.10(e) to the *right* of the initial value $y(0)$. The corresponding z-value, representing the slope of the solution at such later times, is therefore positive but *less* positive. Thus, at such a later time t, the slope of the tangent to the solution should be indicated as in Figure 8.10(f), showing a smaller, but still positive, slope.

Imagine time moving further on. Then Figure 8.10(f) tells us that a yet later value $y(t)$ moves even further to the right [Figure 8.10(g)]; the (tangent) slope remains positive but even less so (on the z-axis), yielding the graph shown in Figure 8.10(h).

458   Chapter 8   Differential Equations

Finally, it can be shown that the solution curve *cannot cross* the graph of the constant solution. (For example, in Newton's law of heating or cooling, a cool object cannot warm *above* the temperature of the room; it can only approach the room temperature.) Thus we draw a final graph of the solution as shown in Figure 8.11. ∎

Let us give a second example and then summarize the method.

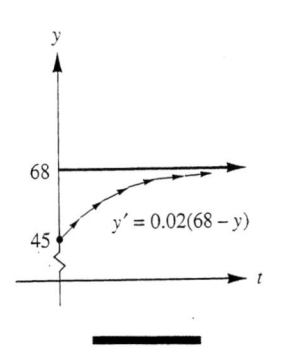

Figure 8.11

EXAMPLE 3

Graph *all* solutions to $y' = 3y(1 - y)$.

Solution The auxiliary equation is $g(y) = 3y(1 - y) = 3y - 3y^2$, with two roots $y = 0$ and $y = 1$. This quadratic function has the graph indicated in Figure 8.12(a), with a maximum value occurring at $y = \frac{1}{2}$. We first show *two* constant solutions corresponding to the two roots on the ty-graph in Figure 8.12(b). We are asked to graph *all* solutions.

(a)

(b)

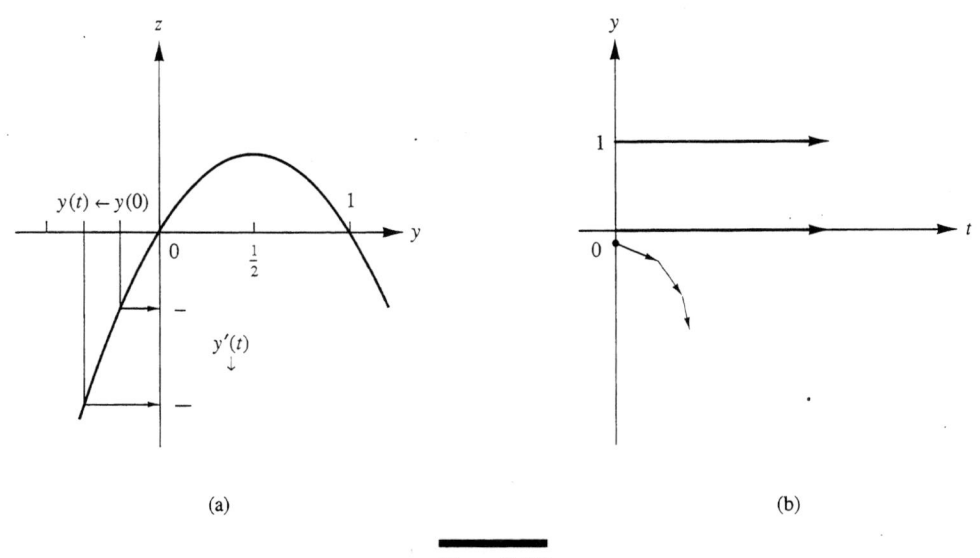

Figure 8.12

Case 1. Suppose that $y(0) < 0$. From Figure 8.12(a), the solution must begin with a negative $(-)$ slope, indicated in Figure 8.12(b). Later values of y therefore decrease and so *move to the left* in Figure 8.12(a), yielding a *more negative* slope, correspondingly indicated in Figure 8.12(b). This pattern can only continue as time goes on.

Case 2. Suppose that $y(0) > 1$. We illustrate the sign of the slope of tangents to the solutions and therefore the time evolution of the solution, as indicated in Figure 8.13(a, b).

8.3 Evolution of Time-Autonomous Systems   **459**

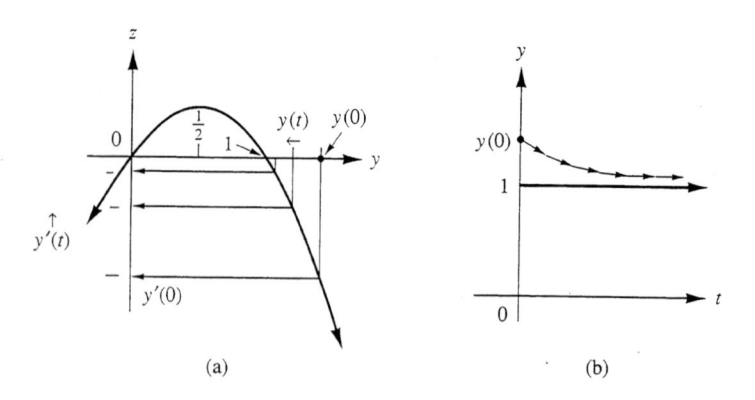

Figure 8.13

(a)   (b)

Case 3. Suppose that $y(0) < 1$ but $y(0) \geq \frac{1}{2}$. We then see in Figure 8.14(a) that the slope of the solution is positive (+) and that later values of $y(t)$ *move to the right* in the zy-axis system, yielding *less positive* slopes as time goes on. We indicate the flow of the solution in Figure 8.14(b).

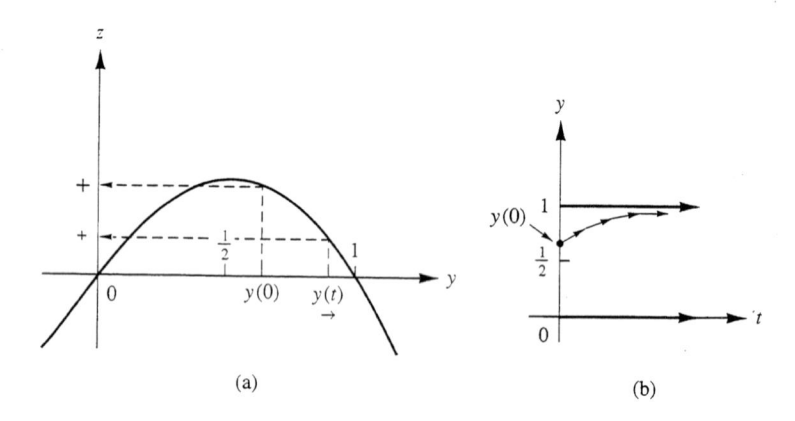

Figure 8.14

(a)   (b)

Case 4. This is an interesting case! Suppose that $0 < y(0) < \frac{1}{2}$. We first indicate such an initial value as in Figure 8.15(a, b).

From Figure 8.15(a), we observe a positive slope and therefore movement to the right, resulting in a *more* positive slope [Figure 8.15(b)]. But now the interesting point. As time goes on and the solution $y(t)$ passes the $\frac{1}{2}$ mark [Figure 8.15(c)], the slope of the solution *remains positive* but (here's the new "wrinkle") *less* so. We must indicate this as in Figure 8.15(d). In short, the peak on the auxiliary graph indicates a *peak rate of change* in the time-evolution (ty) graph [Figure 8.15(d)], resulting in an inflection point on the solution curve of the differential equation.

460   Chapter 8   Differential Equations

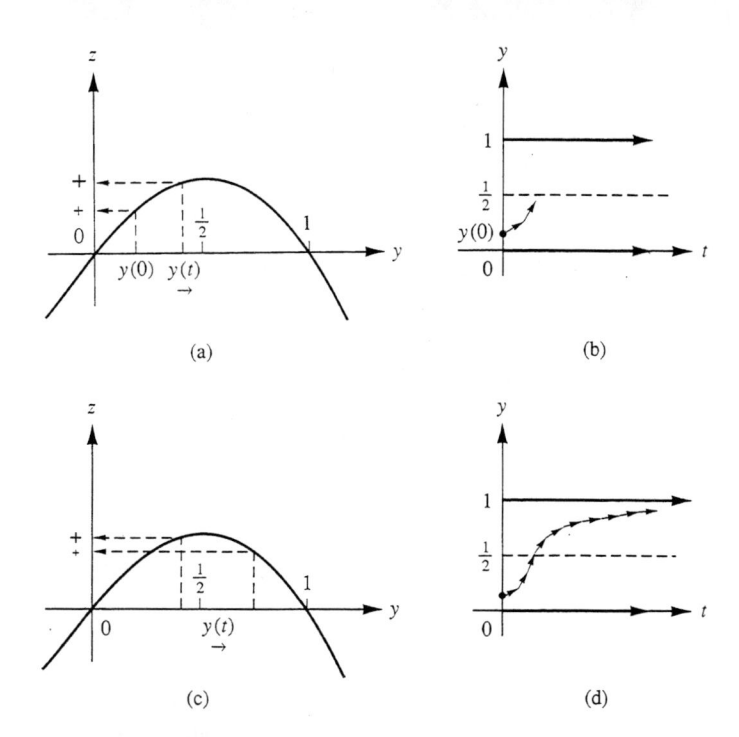

Figure 8.15

(a) (b) (c) (d)

These four cases indicate the *generic solutions* to the differential equation $y' = 3y(1 - y)$. No other cases, or geometric "shapes," are possible. We indicate then the solution to this example as an auxiliary graph and the associated four generic solutions, as shown in Figure 8.16. At the same time we realize that we are showing only a sketch of the general shape of all possible solutions and not a geometrically precise graph of any but the constant solutions.

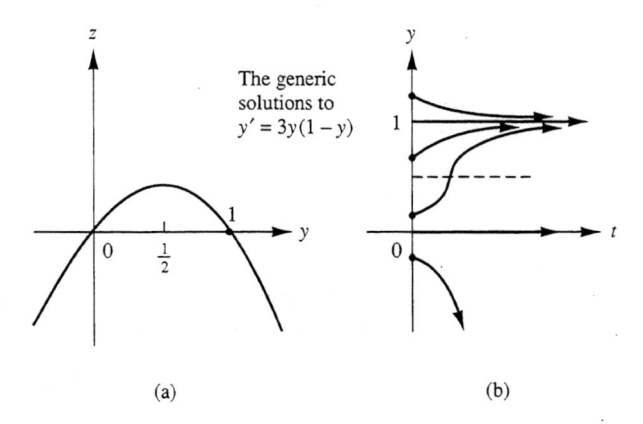

Figure 8.16

The generic solutions to $y' = 3y(1-y)$

(a) (b) ∎

Examples 1, 2, and 3 illustrate the method of this section entirely. Only a more involved algebraic formula for $g(y)$ in $y' = g(y)$ can create any difficulties for you in applying this method.

8.3 Evolution of Time-Autonomous Systems

The result of this method is known as a **qualitative solution** of a differential equation, since it yields only the shape of the solution curve, as distinct from the *quantitative* methods of Sections 8.1 and 8.2 (and later sections), which yield exact numerical values of the solution. If all we need is a picture of the evolution of a system over time, the qualitative method, summarized in the accompanying box, can be quite useful.

Time-Evolution Graphing (Qualitative Solution) for the Autonomous Differential Equation $y' = g(y)$

To graph the generic solutions to the autonomous differential equation $y' = g(y)$, take the following steps:

Step 1. Let $z = g(y)$ and graph this auxiliary algebraic equation on a yz-coordinate system.

Step 2. Locate all zero values of $g(y)$. Graph these as constant solutions to $y' = g(y)$ on a ty-coordinate system.

Step 3. Plot an initial value $y(0)$ on both coordinate systems.

Step 4. Determine the *sign* of $z = g(y)$ at this value. Use this sign to indicate the *direction* of the solution values in both coordinate systems.

Step 5. Use this direction to indicate the position of a later value $y(t)$ in the zy-system. Use this later value to determine if the slope of the solution $y(t)$ becomes greater or less, indicating the concavity of the solution graph in ty.

Step 6. Remember that peaks (maxima and minima) on the zy-graph suggest inflection points on the ty-graph. Also remember that a solution cannot cross a constant solution and can otherwise grow or decline without bound [see Figure 8.12(a)] over time.

We conclude with an example close to home.

EXAMPLE 4

Suppose that you have one unit of material to learn; we will denote this unit by 1. Let $y(t)$ be the proportion of the material that you know after t hours of study, and suppose that initially you know 25% of the material; thus $y(0) = \frac{1}{4}$.

How fast you learn, represented by $y'(t)$, ought to depend on two things: (1) how much you already know, $y(t)$, and (2) how much is yet to be learned, $1 - y(t)$. [For example, if you didn't know how to read, it would take you a long time to learn this subject. On the other hand, you know from experience that it is very hard to pin down the last few facts about a subject, represented by $1 - y(t)$.]

Suppose that you learn at the rate of 10% of the amount you already know, $y(t)$, and the amount you do not yet know, $1 - y(t)$. That is, suppose that the law governing the change in your state of knowledge is

$$y' = 0.10y(1 - y)$$

Graph your learning curve.

Solution This is essentially Example 3 with one number changed. The auxiliary graph is indicated in Figure 8.17, where again the max value occurs at $\frac{1}{2}$, representing a knowledge of half of what is to be learned.

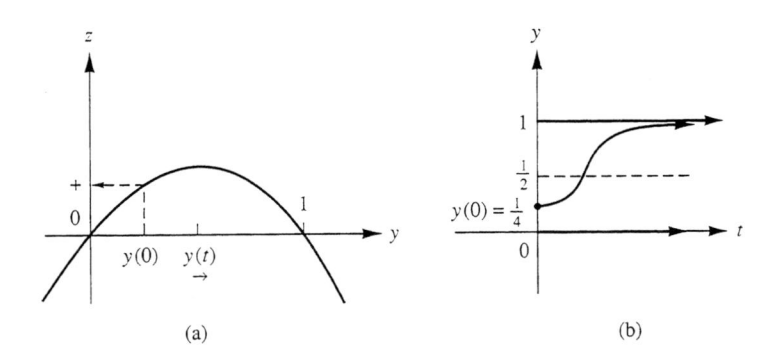

Figure 8.17

(a) (b)

Since $y(0) = \frac{1}{4}$, the solution must begin below $\frac{1}{2}$ [Figure 8.17(b)] and your knowledge increases [moves to the right in Figure 8.17(a)] at an increasing rate (larger z-values) until you have learned half the material, whereupon $y = \frac{1}{2}$ is crossed and your learning *rate* begins to slow down (less positive values of z), resulting in the graph shown in Figure 8.17(b). ∎

The dismal conclusion of this very general model, if it applies to you, is that the more you know, the slower you will learn what remains—but only *after* you already know a great deal!

Exercises 8.3

Specify the constant solutions (if any) to each of the differential equations in Exercises 1–6.

1. $y' = 30 - 6y$
2. $y' = 2y(y - 3)$
3. $y' = (y - 1)(y + 2)$
4. $y' = y^2 + 2y + 1$
5. $y' = \dfrac{1}{y}$
6. $y' = y^2 + 1$

7. Graph the solution to the IVP $y' = 0.02(68 - y)$, $y(0) = 98°$.

8. Graph the solution to the IVP $y' = 3y(1 - y)$, $y(0) = \frac{1}{2}$.

Graph the auxiliary graph and then graph the solution to each of the IVP's in Exercises 9–16. Show all constant solutions in each case.

9. $y' = 3y - 2$; $y(0) = 0$; $y(0) = 1$; $y(0) = \frac{2}{3}$

10. $y' = 2y + 1$; $y(0) = 0$; $y(0) = -1$; $y(0) = 1$
11. $y' = 9 - y^2$; $y(0) = 1$; $y(0) = -2$; $y(0) = 4$
12. $y' = y^2 - 4$; $y(0) = 1$; $y(0) = -2$; $y(0) = -3$
13. $y' = y^2 + 1$; $y(0) = -2$; $y(0) = 0$; $y(0) = 1$
14. $y' = (y - 1)(y + 2)$; $y(0) = 2$; $y(0) = -1$; $y(0) = 1$
15. $y' = (2y - 1)(y - 3)$; $y(0) = 0$; $y(0) = 1$; $y(0) = 2$
16. $y' = y^2 - y - 2$; $y(0) = 4$; $y(0) = -\frac{1}{2}$; $y(0) = 1$

In Exercises 17–20, you are given a graph of $z = g(y)$ and an initial value $A = y(0)$ for a solution to $y' = g(y)$. Graph the solution; show constant solutions if they exist.

17.

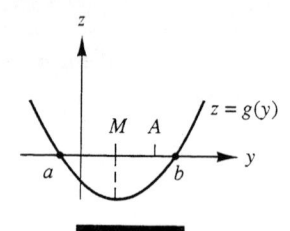

Figure 8.18

18.

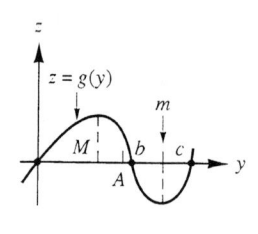

Figure 8.19

19.

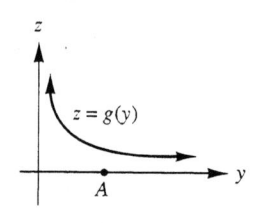

Figure 8.20

20.

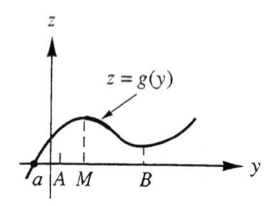

Figure 8.21

Sketch the graphs of all constant solutions and the generic solutions to each of the differential equations in Exercises 21–44.

21. $y' = 1 - y$
22. $y' = 1 + y$
23. $y' = 6 - 2y$
24. $2y' = 4 - 2y$
25. $y' = 3 - \frac{1}{2}y$
26. $y' = 1 - y^2$
27. $y' = y^2$
28. $y' = e^y$
29. $y' = \ln y$
30. $y' = y(3 + y)$
31. $y' = y(5 - y)$
32. $y' = e^{-y}$
33. $y' = y^2 - y - 6$
34. $y' = \frac{1}{4}$
35. $y' = 2 + y^2$
36. $y' = \frac{1}{2}y^2 - 8$
37. $y' = ye^{-y}$
38. $y' = y^2 - 2y$
39. $y' = y^3$
40. $y' = y^2 - 2y + 2$
41. $y' = y^3 - y^2 - 6y$
42. $y' = y(y^2 - 1)$
43. $y' = (y^2 - 1)(y^2 - 4)$
44. $y' = e^{-y^2}$

45. The differential equation of Newton's law of heating or cooling has the general form $y' = k(M - y)$. This form occurs repeatedly in widely differing applications. Graph the generic solutions to this equation if $M > 0$ and $k > 0$.

46. Repeat Exercise 45 with $k < 0$.

47. Repeat Exercise 45 for the equation $y' = ky - A$, writing it as $y' = k[y - (A/k)]$, where $k, A > 0$.

48. Repeat Exercise 47 for $y' = ky + A$, where $k, A > 0$.

49. The logistic equation $y' = k(M - y)$, with $M > 0$, is very useful.

   a. Show that $g(y) = ky(M - y)$ has a maximum value at $M/2$ if $k > 0$. [*Hint:* Use $g'(y)$ and $g''(y)$.] Note then that this max is independent of k.

   b. Show that g has a min value at $M/2$ if $k < 0$.

   c. Graph the generic solutions to this equation when $k > 0$.

   d. Repeat (c) for $k < 0$.

50. Graph the generic solutions to $y' = 0$.

51. Graph the generic solutions to $y' = 2$.

52. Graph the generic solutions to $y' = ky$ and $y' = ky^2$ for $k > 0$. Repeat this exercise for $k < 0$.

53. Let $y(t)$ denote the amount of money in a bank account paying 8% interest. Suppose that $y(0) = \$5,000$ and that money is being withdrawn from the account at the rate of $500 per year. The change-of-state of this system can be shown to be ruled by the differential equation

$$y' = 0.08y - 500$$

Indicate by a graph the evolution over time of the amount of money in this account.

54. If in Exercise 53 $200 were deposited in the account each year, the differential equation would become $y' = 0.08y + 200$. Indicate the time evolution of this account.

55. If in Exercise 53 withdrawals occurred at the rate of $300 per year and the ruling differential equation were $y' = 0.08y - 300$, indicate the time evolution of this account.

56. What if in Exercises 53 and 55 the withdrawal rate were $400 per year?

57. Many species grow quickly when young but more slowly when older. Let L_∞ be the mature length of a salmon, and let $L(t)$ be its actual length in year t. If we assume that the percentage rate of growth of a salmon is in direct proportion at any time to the percentage yet to be grown, we have $L'(t) = kL(t)(L_\infty - L(t))$ as the law of growth. At what size will the salmon be growing fastest?

58. Let $S(t)$ denote the percentage of the market share of a company producing a certain item. Thus $0 \leq S(t) \leq 100$.

    Since the successful sale of an item draws attention to it, we can reasonably expect that the *relative* rate of growth of this company's share of the market is proportional to the remaining share of the market that it does not yet have. Thus

    $$S'(t) = kS(t)[100 - S(t)]$$

    When $S(t) = 50\%$, the management of this company asserts, "If things keep going like this, we will soon control 90% of the market." Should you be skeptical?

Challenge Problems

59. A population of N interbreeding individuals will contain $2N$ genes of a single type that occur in pairs AA, aA, or aa in each individual. Let $p(t)$ be the proportion of (dominant) genotype A in the whole population; $p(t)$ is called the *gene frequency* of A in the population at time t. Environmental and random influences along with mating will cause p to vary over time according to a differential equation of the form

    $$p'(t) = \alpha(1 - p(t)) - \beta p(t)$$

    where $1 - p(t)$ is the proportion of (subdominant) genotype a in the population, α is the proportion of a mutating to A, and β is the proportion of A mutating to a. Graph the generic solutions to this differential equation: $p' = \alpha - (\alpha + \beta)p$.

60. Use the graph of Exercise 59 to find $\lim_{t \to \infty} p(t)$, the long-term distribution of genotype A in the population. What does this long-term proportion depend on?

8.4 Solution by Separation of Variables

In this section we learn a step-by-step procedure for solving a differential equation of the form

$$y' = p(t)q(y)$$

To go beyond a graphical picture of the time evolution of a system ruled by a differential equation and to know particular states that the system will be in at particular times, we must explicitly solve the equation for $y(t)$. A basic method for doing so is called **separation of variables**. This method can be used when the differential equation occurs in the *form*

$$y' = p(t)q(y)$$

of a function of t alone *multiplied* by a function of y alone. For example, the equation $y' = t^2 y^3$ is in this form. However, the equation $y' = t + y$ is not in this form, and the method of this section will not yield its solution. The equation $y' = 6 - 3y$ can be put into this form by writing it, for example, as $y' = 3(2 - y)$ and regarding $p(t) = 3$ and $q(y) = 2 - y$.

Before we show how to solve differential equations in this general form, we need a particular integral equality. Consider the integral

$$\int h(y(t))y'(t)\, dt$$

where h is some function. Make the change of variable

$$u = y(t) \quad \text{and} \quad du = y'(t)\, dt$$

to obtain

$$\int h(y(t))y'(t)\, dt = \int h(u)\, du$$

Recall that in such an indefinite integral "u" is a "dummy variable" and can be replaced by any letter. It is most efficient for our purposes to replace "u" by the letter "y" itself. We then have

$$\int h(y(t))y'(t)\, dt = \int h(y)\, dy \tag{1}$$

Let us use Equation 1 to solve a differential equation by the method of separation of variables.

EXAMPLE 1

Find the solutions of $y' = t^3/y^2$ by separation of variables.

Solution This equation is in the form $y' = p(t)q(y)$, where $p(t) = t^3$ and $q(y) = 1/y^2$. We first rewrite the equation, by multiplying both sides by y^2, as

$$y^2 y' = t^3$$

The variables y and t are now *separated* (by the equality). Of course y is an unknown function of t. Thus the equation is properly written as $y(t)^2 y'(t) = t^3$, and we can then say that

$$\int y(t)^2 y'(t)\, dt = \int t^3\, dt$$

Applying Equation 1 to the left side of this equation with $h(y) = y^2$ yields

$$\int y^2\, dy = \int t^3\, dt$$

Hence

$$\frac{y^3}{3} = \frac{t^4}{4} + C$$

Consequently,

$$y^3 = \tfrac{3}{4} t^4 + 3C$$

or

$$y = y(t) = \sqrt[3]{\tfrac{3}{4} t^4 + 3C}\ \blacksquare$$

You can establish by substitution that this function, with the unknown constant C included, represents (all) the solutions to the differential equation $y' = t^3/y^2$.

Remark. Since C is an unknown constant, so is $3C$, and we can simply write $y^3 = \tfrac{3}{4} t^4 + C$ with no loss. That is, once a constant is introduced into the solution, it can be used to "absorb" all constant multiples applied to it. Notice also that only one constant need be introduced in the course of the solution.

Notice also that the application of Equation 1 effectively replaces y' in the differential equation by "dy" in the integral.

Example 1 is a prototype for the solution of the exercises of this section but for one additional concern illustrated in the next example.

EXAMPLE 2

Solve $y' = ty^2$.

Solution This equation is in the form $y' = p(t)q(y)$ with $p(t) = t$ and $q(y) = y^2$. We separate variables by writing $y' = ty^2$ as

$$\frac{y'}{y^2} = t \quad \text{if } y \neq 0$$

We then have

$$\int \frac{y'}{y^2} dt = \int t \, dt$$

or

$$\int \frac{dy}{y^2} = \int t \, dt \quad \text{By Equation 1}$$

Hence

$$\frac{-1}{y} = \frac{t^2}{2} + C$$

so

$$y = \frac{-1}{(t^2/2) + C}$$

or

$$y(t) = \frac{-2}{t^2 + C}$$

This formula represents all but one of the solutions to the given differential equation. Notice that $y(t) = 0$ is a constant solution for the given equation but is omitted by our solution procedure in the very first step when we divide both sides of the equation by y^2. Notice further that no choice of C in the resulting formula can produce the value 0. Therefore the solutions to this equation must be given as

$$y(t) = \frac{-2}{t^2 + C} \quad \text{and} \quad y(t) = 0 \quad \blacksquare$$

With these examples in mind we can outline a general method.

The Solution of $y' = p(t)g(y)$ by the Method of Separation of Variables

To solve a differential equation in the *form*

$$y = p(t)q(y)$$

make the following steps:

Step 1. Rewrite the equation as

$$\frac{y'}{q(y)} = p(t) \quad \text{for } q(y) \neq 0$$

8.4 Solution by Separation of Variables

Step 2. Set integrals with respect to t equal to obtain

$$\int \frac{y'}{q(y)} \, dt = \int p(t) \, dt$$

but use Equation 1 to write the integral on the left instead as

$$\int \frac{dy}{q(y)} = \int p(t) \, dt$$

Step 3. Complete the indefinite integration on both sides, introducing an arbitrary constant C on the right.

Step 4. Solve the resulting *algebraic* equation for y as a function of t to obtain a general formula, with an unknown constant C, for the solution.

Step 5. In addition, list as additional solutions all constant functions $y(t) = A$, where $q(A) = 0$, since these constant solutions are necessarily omitted in performing step 1.

In many cases step 5 is unnecessary, but this can be established only on a case-by-case basis.

We now solve an IVP using separation of variables.

EXAMPLE 3

Solve the IVP $y' = te^y$, $y(0) = 1$.

Solution We first find the general solution with an unknown constant and then determine this constant by imposing the initial value on the solution.

Step 1. We write $y' = te^y$ as $e^{-y}y' = t$.

Step 2.
$$\int e^{-y} \, dy = \int t \, dt$$

or
$$-e^{-y} = \frac{t^2}{2} + C$$

Step 3. Applying the logarithm to both sides yields

$$-y = \ln e^{-y} = \ln(-t^2/2 + C)$$

or
$$y(t) = -\ln\left(\frac{C - t^2}{2}\right) = \ln\left(\frac{2}{C - t^2}\right)$$

Step 4. Now since

$$1 = y(0) = -\ln\left(\frac{2}{C - 0^2}\right) = \ln\left(\frac{2}{C}\right)$$

we have $2/C = e$, since $\ln e = 1$. Therefore $C = 2/e$ and so

$$y(t) = \ln\left(\frac{2}{(2/e) - t^2}\right) = \ln\left(\frac{2e}{2 - et^2}\right)$$

is the solution to this IVP. ∎

Note that we may omit step 5 of the method because in step 1 e^y is never 0.

EXAMPLE 4

Solve $y' = \pi/y$, $y(0) = -3$.

Solution

Step 1. We write $y' = \pi/y$ as $yy' = \pi$ (note that $y \neq 0$).

Steps 2 and 3.
$$\int y \, dy = \int \pi \, dt$$
$$\frac{y^2}{2} = \pi t + C$$

Step 4.
$$y = \pm\sqrt{2\pi t + C}$$

Now $-3 = y(0) = \pm\sqrt{2\pi t \cdot 0 + C}$ or $-3 = \pm\sqrt{C}$. Hence $C = 9$, and we have the solution $y(t) = -\sqrt{2\pi t + 9}$, choosing the minus sign on the radical to meet the given initial value. ∎

Solution by separation of variables can be made more complicated only by difficulties with integration. We conclude with two examples.

EXAMPLE 5

Find all solutions to $y' = tye^t$.

Solution

Step 1.
$$\frac{y'}{y} = te^t \qquad y \neq 0$$

Steps 2 and 3.
$$\int \frac{dy}{y} = \int te^t \, dt$$

Using integration by parts (see Section 7.1) on the right, we have
$$\ln y = \int t \, d(e^t) = te^t - \int e^t \, dt = te^t - e^t + C$$

Step 4. Hence
$$e^{\ln y} = e^{(te^t - e^t + C)}$$
or
$$y(t) = e^{(te^t - e^t + C)}$$

Step 5. We must include the constant solution $y(t) = 0$ as well, since no choice of C makes the function y of step 4 identically zero. ∎

Our last example illustrates how to solve the general logistic differential equation.

EXAMPLE 6

Find all solutions to $y' = ky(M - y)$.

Solution

Step 1.
$$\frac{y'}{y(M-y)} = k \qquad y \neq 0, y \neq M$$

Steps 2 and 3.
$$\int \frac{dy}{y(M-y)} = \int k\,dt = kt + C$$

Notice from algebra that

$$\frac{1}{M}\left(\frac{1}{y} + \frac{1}{M-y}\right) = \frac{1}{y(M-y)}$$

We therefore can write the integral on the left as

$$\frac{1}{M}\int \left(\frac{1}{y} + \frac{1}{M-y}\right) dy = kt + C$$

Hence
$$\frac{1}{M}(\ln|y| - \ln|M-y|) = kt + C$$

or
$$\ln\left|\frac{y}{M-y}\right| = Mkt + C$$

Therefore, taking exponentials and using $e^{\ln x} = x$, we have

$$\frac{y}{M-y} = \pm e^{Mkt+C} = \pm e^C e^{Mkt} = Ae^{Mkt}$$

Step 4. We must now solve for y in the equation $y/(M-y) = B$, where $B = Ae^{Mkt}$. We have $y = B(M-y)$, or $y + By = BM$, or $y(1+B) = BM$, or $y = BM/(1+B)$. That is,

$$y(t) = \frac{MAe^{Mkt}}{1 + Ae^{Mkt}} = \frac{MA}{e^{-Mkt} + A}$$

The solution is then usually written as

$$y(t) = \frac{M}{Ce^{-Mkt} + 1} = \frac{M}{1 + Ce^{-Mkt}} \qquad (2)$$

upon division by A and replacing $1/A$ by C.

Step 5. Notice that the constant solution $y(t) = M$ omitted in step 1 results by taking $C = 0$ and so is included in the solution. The constant solution $y(t) = 0$ is *not* included in the formula and should be listed as an additional solution. ∎

The next example reminds us again that, having a formula for the solution of a differential equation, we are in a position to determine future states of a system.

EXAMPLE 7

A manufacturer expects to sell a product to 150 thousand potential customers. Let $y(t)$ be the number of customers buying the product by month t after its introduction. The manufacturer knows that sales will be determined (not only by advertising but) by how many customers are already using the product and how many are not. Marketing surveys suggest that the rate of customers attracted to the product is ruled by the differential equation

$$y' = 0.004y(150 - y)$$

representing both the number of customers buying the product, y, and the number of potential customers yet to buy it, $150 - y$. The manufacturer initially gives away 250 copies of the product, so that $y(0) = 0.25$ (thousand).

In how many months will the company have sold products to 90% of the anticipated market?

Solution In Section 8.2, we learned that the growth of sales ruled by the logistic differential equation will have the pattern shown in Figure 8.22.

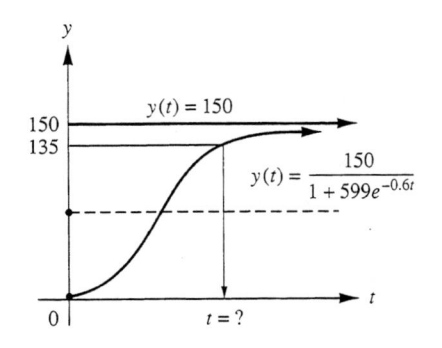

Figure 8.22

Since 90% of 150 is 135, we wish to determine that month t for which $y(t) = 135$. From Equation 2 of Example 6, with $M = 150$ and $k = 0.004$,

$$y(t) = \frac{150}{1 + Ce^{-0.6t}} \qquad \text{since } (0.004)(150) = 0.6$$

Now

$$0.25 = y(0) = \frac{150}{1 + C} \qquad \text{so } C = 599$$

Therefore $y(t) = 135$ when

$$135 = \frac{150}{1 + 599e^{-0.6t}}$$

or when

$$1 + 599e^{-0.6t} = 1.11$$

$$e^{-0.6t} = 0.000184$$

$$-0.6t = \ln(0.000184)$$

or

$$t = 14.33 \text{ months} \blacksquare$$

Exercises 8.4

Find all solutions to the differential equations in Exercises 1–20.

1. $y' = ty$
2. $y' = \dfrac{t}{y}$
3. $y' = 1 - y$
4. $y' = e^y(t + 1)$
5. $y' = e^t(1 - y)$
6. $y' = 2y + 1$
7. $y' = y + 3$
8. $y' = \dfrac{ty}{t^2 + 1}$
9. $y' = \dfrac{t^2}{y}$
10. $y' = \dfrac{y}{t}$
11. $u' = (t/u)^2$
12. $y' = \dfrac{y \ln t}{t}$
13. $ty' = \dfrac{1}{y}$
14. $y'y + ty^2 = 0$
15. $yy' = t^2 y^2$
16. $z' = e^{t-z}$
17. $u' = ute^{-t^2}$
18. $y' = y(1 - y)$
19. $A' = \dfrac{A}{x} + bA$
20. $f' = \sqrt{xf}$

Solve the initial value problems in Exercises 21–36.

21. $y' = \tfrac{1}{2}y - 4$; $y(0) = 1$
22. $ty' = 1 - y$; $y(1) = 2$
23. $y' = 21 - 7y$; $y(0) = 3$
24. $u' = \dfrac{t}{u}$; $u(0) = 5$
25. $y' = 2t - 3ty$; $y(0) = 1$
26. $y' = t^2 y$; $y(2) = 4$
27. $y' = \dfrac{t + 1}{y + 1}$; $y(0) = -4$
28. $D_t y = -3ty$; $y(0) = 4$
29. $y' = \dfrac{t^{1/3}}{\sqrt{y}}$; $y(0) = 1$
30. $zz' = 1$; $z(2) = -4$
31. $y' = \dfrac{t^2(y + 1)}{t^2 + 2}$; $y(0) = 3$
32. $tu' = 3$; $u(1) = 1$
33. $y' = \dfrac{t^2 e^{-t}}{y}$; $y(0) = 8$
34. $y' = t^3 y^2 - y^2$; $y(0) = 3$
35. $y' = y \ln t$; $y(1) = 1$
36. $y' = 2te^{-y} - e^{-y}$; $y(0) = 2$

37. In chemistry it often happens that a chemical compound in the amount of $y(t)$ molecules at time t decomposes into two other compounds at a rate proportional to $y(t)^2$ (the total possible number of molecular collisions). This rule yields a differential equation of the *form* $y' = -ky^2$. If $y(0) = 10^6$ and $k = 0.05$, solve the associated IVP.

38. In our first encounter with the logarithmic function (see Section 5.3), we saw that, in a natural way, a person's perception P of a stimulus is a logarithmic function of the strength S of the stimulus. This law of mathematical psychology is more generally known as the Weber–Fechner law and is expressed via the differential equation

$$\dfrac{dP}{dS} = k/S$$

That is, the Weber–Fechner law says that the rate of change of a person's perception P of a stimulus is inversely proportional to the strength of the stimulus. If the "threshold" stimulus level is S_0, then $P(S_0) = 0$; that is, a person does not notice the stimulus at this level. Solve the IVP $P' = k/S$ when $P(S_0) = 0$.

39. Imagine dropping a lead weight off the top of a 100-ft-high building, neglecting air resistance. Let $y(t)$ denote the height (in feet) of the weight at time t. Thus by time t the weight has fallen $100 - y(t)$ ft. Show that the speed at which the weight is falling *cannot* be proportional to how far it has fallen. [*Hint:* Assume it is. Write this as a differential equation, and with $y(0) = 100$, solve this IVP.]

40. Sociologists have found that within a group of size P the spread of a rumor is in proportion to the number $y(t)$ of the group who have heard the rumor by time t and the number $P - y(t)$ who have not. If $y' = 0.2y(P - y)$ and $y(0) = 1$, determine when

   a. 90% of the group has heard the rumor.

   b. Everyone except the person who is the subject of the rumor has heard it.

41. In DNA hybridization, double-strand DNA is broken down into single-strand DNA and then reannealed to double-strand DNA by a cooling process. This gives

472   Chapter 8   Differential Equations

rise to a differential equation of the form

$$y' = \frac{0.01}{y} - 0.5y$$

where $y(t)$ is the percent of single-strand DNA at time t in hours. With $y(0) = 99$ (percent), how much single-strand DNA remains after 1 hour?

42. Fick's principle for passive diffusion says that if a compartment is divided by a permeable membrane and on one side of the membrane a solute (such as salt in water) is kept at a constant *concentration* C_0, then the solute will diffuse across the membrane at a rate proportional to the difference between the constant concentration C_0 and the *concentration* $C(t)$ on the other side of the membrane at time t (see Figure 8.23). Express Fick's principle as a differential equation in $C(t)$, and solve this differential equation.

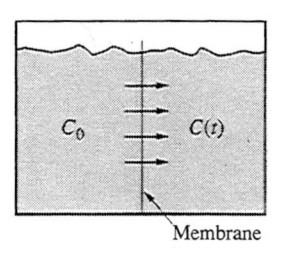

Figure 8.23

43. Using Example 6 as a (more involved) model, solve the differential equation of Newton's law of heating or cooling: $y' = k(M - y)$.

8.5 Applications to Compartment-Mixing Processes

This section uses rate-in less rate-out analysis of a compartment to model a wide variety of applications of the basic linear differential equation $y' = ay + b$.

The applications of this section span a variety of fields. We will divide these applications into two distinct types, first to business and economics and then to the environmental and life sciences. At the same time, such is the power of abstract mathematics to synthesize diversity into a single pattern that all applications in the section fall into the single category of *one-compartment mixing processes*. The key to modeling such systems is the concept of **rate-in less rate-out analysis.** This method uses a differential equation to state the law of change inside a compartment via analysis of naturally observable external rates.

Imagine an inaccessible compartment in which we are asked to describe an unknown amount $y(t)$ present in the compartment at time t (see Figure 8.24). For example, the compartment may be a city reservoir, and $y(t)$ could be the amount of a certain pesticide in the water supply. Or the compartment could be an investment account subject to earnings and losses, with $y(t)$ being the amount of money in the account at time t. In both examples, we want to initially forget about what is in the compartment and instead focus on what is going into and what is leaving the compartment. Although we cannot "see" inside the compartment, we can often observe what is entering and leaving, especially the *rate* at which this occurs. As we will soon see, this is a very natural approach.

Figure 8.24

What we want to imagine, then, is a system consisting of an inaccessible compartment which some quantity is flowing into, and out of, at known rates (see Figure 8.25). We wish to determine the state of this system $y(t)$—what *is in* the compartment at time t—through an analysis of its change-of-state $y'(t)$.

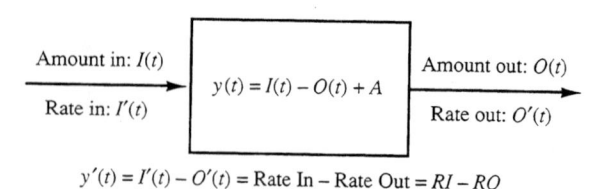

Figure 8.25

Amount in: $I(t)$
Rate in: $I'(t)$

$y(t) = I(t) - O(t) + A$

Amount out: $O(t)$
Rate out: $O'(t)$

$y'(t) = I'(t) - O'(t) =$ Rate In $-$ Rate Out $= RI - RO$

If $I(t)$ is the amount that has gone into the compartment by time t and $O(t)$ is the amount that has left the compartment by time t, the amount in the compartment at time t, $y(t)$, is given by the equation

$$y(t) = I(t) - O(t) + A$$

where A is an initial, fixed amount in the compartment. Therefore

$$y'(t) = I'(t) - O'(t)$$

That is,
$$y'(t) = \text{Rate in} - \text{Rate out}$$

This last equation is the single principle that we will repeatedly apply in this section to determine the state $y(t)$ of such systems at time t. We will often write the equation as

$$y' = RI - RO$$

Application of the concept represented by this equation in this section will invariably result in a differential equation of the *form*

$$y' = ay + b$$

where a and b are both constant and $a \neq 0$. Let us solve this equation by separation of variables. We will use this solution throughout this section. We have, by factoring a, that

$$y' = a\left(y + \frac{b}{a}\right)$$

and, by separating variables, that

$$\frac{y'}{y + \frac{b}{a}} = a \qquad y \neq -\frac{b}{a}$$

Hence
$$\int \frac{dy}{y + \frac{b}{a}} = \int a\, dt$$

or
$$\ln\left|y + \frac{b}{a}\right| = at + C_1$$

and therefore
$$y + \frac{b}{a} = \pm C_1 e^{at}$$

In conclusion

Theorem 8.1   The general solution of $y' = ay + b$ is
$$y(t) = \frac{-b}{a} + Ce^{at}$$

---

where C may be positive or negative and the constant solution $y = -b/a$ is obtained when $C = 0$.

Notice that the familiar equation of Newton's law of heating or cooling, $y' = k(M - y)$, is but a special case with $a = -k$ and $b = kM$, since $y' = -ky + kM$.

The remainder of this section is subdivided into two areas of application. All solutions rely on rate-in less rate-out analysis of a compartment to write a differential equation whose solution is given by Theorem 8.1.

Subsection A: Applications to Business and Economics

Our first and simplest applications are in the field of business and economics.

EXAMPLE 1

An account initially contains \$10,000 and draws 8% interest compounded continuously. Frequent withdrawals are made from the account at the overall rate of \$1,000 per year. When will there be only \$1,500 remaining in the account? How much will have been withdrawn from the account by that time?

Solution   Let us think of this problem as a compartment-mixing problem in which the compartment is the account itself and $y(t)$ is the amount of money in the account at time t, with $y(0) = \$10,000$. We will model the problem as an IVP and obtain its solution (see Figure 8.26).

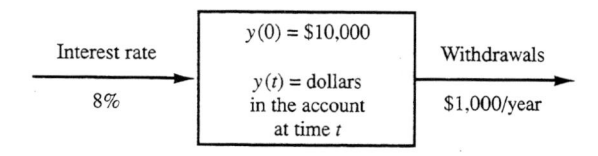

Figure 8.26

Interest rate 8% → [$y(0) = \$10{,}000$; $y(t) =$ dollars in the account at time t] → Withdrawals \$1,000/year

From the preceding discussion,
$$y'(t) = \text{Rate in} - \text{Rate out}$$

We are given that the rate of money leaving the account is \$1,000 per year. Thus
$$y'(t) = \text{Rate in} - 1{,}000$$

Now, at what rate is money entering the account? We can answer this by recalling the meaning of an 8% interest rate. To say that the account earns 8% interest is to say that money enters the account at the rate of 8% *of the amount in the account*. That is, the rate of money flowing into the account is $0.08y(t)$, since $y(t)$ is the amount in the account at time t. Therefore the rate-in less rate-out equation $y'(t) = RI - RO$ becomes the differential equation

$$y' = 0.08y - 1{,}000$$

From Theorem 8.1, we have the solution

$$y(t) = \frac{-(-1{,}000)}{0.08} + Ce^{0.08t}$$

or

$$y(t) = 12{,}500 + Ce^{0.08t}$$

Since $y(0) = 10{,}000$, we find that $C = -2{,}500$. Hence

$$y(t) = 12{,}500 - 2{,}500e^{0.08t}$$

There will be $1,500 remaining in the account when

$$1{,}500 = 12{,}500 - 2{,}500e^{0.08t}$$

$$\frac{-11{,}000}{-2{,}500} = e^{0.08t}$$

$$\ln \frac{110}{25} = 0.08t$$

or at

$$t = 18.52 \text{ years}$$

By this point in time, $1,000 × 18.52 = $18,520 will have been withdrawn from the account. ∎

Remark. Our model of this account treats interest and withdrawals in a continuous fashion, although in practice they occur at discrete intervals. Despite this discrepancy between the discrete and continuous in such examples from economics, this simple differential equation model is remarkably accurate. You will find models like it useful in your personal dealings with your bank, loan company, or insurance agent. The exercises at the close of this section will help teach you how.

This differential model also allows a quick analysis of a naturally related question.

EXAMPLE 2

What initial amount $y(0)$ deposited in an account subject to 8% interest and yearly withdrawals of $1,000 would enable the account to fund such withdrawals forever?

Solution The differential equation model $y' = 0.08y - 1{,}000$ found in Example 1 is independent of the initial condition $y(0)$ since it governs the *change* in the account. Notice that, by factoring 0.08, $y'(t) = 0.08(y - 12{,}500)$.

Therefore $y(t) = 12{,}500$ is the constant solution to this equation. Hence if the account initially contains $y(0) = \$12{,}500$, it can fund such withdrawals forever. ∎

The solutions to Examples 1 and 2 are illustrated in Figure 8.27.

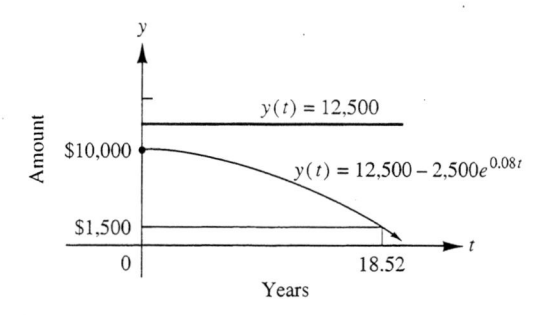

Figure 8.27

In our next example the "rate out" itself is an unknown and must be determined so that the compartment is empty by a certain time.

EXAMPLE 3

Suppose that you have contracted with a loan company to borrow \$8,000 on the purchase of a new car at an interest rate of 1% per month (or 12% per year). What must your monthly payment be so as to pay off the loan in 48 months? What will your interest charges be for the 4-year period?

Solution We regard this as a compartment problem in which the amount $y(t)$ in the compartment is the amount (principal) yet unpaid by time t (see Figure 8.28). Thus $y(0) = 8{,}000$.

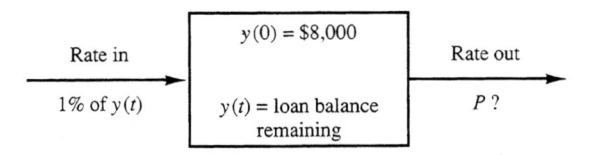

Figure 8.28

Further debt is accumulating in the compartment at the rate of 1% per month of the amount yet due on the loan. Thus the rate into the compartment is $0.01y(t)$. The "rate out" of the compartment will be the monthly payment on the loan, which we are to determine. Let us denote it by P. From $y' = RI - RO$, we have

$$y' = 0.01y - P$$

From Theorem 8.1, using $a = 0.01$ and $b = -P$, we have the solution

$$y(t) = 100P + Ce^{0.01t}$$

Since $y(0) = 8{,}000$, we then have

$$8{,}000 = y(0) = 100P + Ce^{(0.01)(0)} = 100P + C$$

so that $C = 8{,}000 - 100P$. Therefore

$$y(t) = 100P + (8{,}000 - 100P)e^{0.01t}$$
$$= 100P(1 - e^{0.01t}) + 8{,}000e^{0.01t}$$

To pay off the loan in 48 months, we must determine P so that

$$0 = y(48) = 100P(1 - e^{0.01(48)}) + 8{,}000e^{0.01(48)}$$

This yields the equation

$$0 = -61.61P + 12{,}928.60$$

or

$$P = \frac{12{,}928.60}{61.61} = \$209.85 \text{ per month}$$

In conclusion, the total amount paid on the loan will be ($209.85)(48) = $10,073 (against an original loan of $8,000), leading to a 4-year interest charge of $2,073. ∎

The next example shows how compartment analysis can be coupled with the qualitative methods of Section 8.3 to predict a trend within a system.

EXAMPLE 4

"Swords versus Plowshares" At time t a nation has a certain amount of resources $R(t)$. In general these resources can be spent in two ways: on production that leads to further production, such as the use of steel to produce a plow, which in turn produces food; or in production that ends with itself, such as the use of steel to produce a sword, which produces nothing further and is a loss against available economic resources.

Let $P(t)$ and $W(t)$ be the amount of resources spent on productive and nonproductive activity, respectively. Note that $R(t) = P(t) + W(t)$. Suppose that the nation's resources grow at the rate of 10% of the amount spent on productive activity per year. At the same time, some portion of even a nonproductive expenditure is returned to the economy, for example, through wages paid in its production. Suppose then that resources decline at a rate of only 25% of nonproductive expenditures.

Suppose that $R(0) = \$1{,}000$ billion and that spending on productive activity is held constant at $700 billion. Analyze the trend of the nation's total resources and its ability to support nonproductive spending in the long run.

Solution Here the compartment is the nation's total spendable resources (see Figure 8.29).

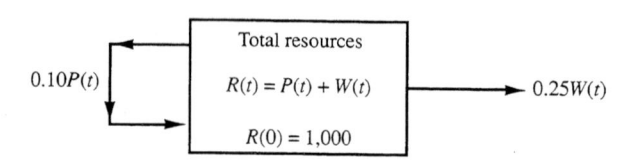

Figure 8.29

Further resources are flowing into the compartment at the rate of 10% of resources spent on productive activity. Losses occur at the rate of 25% spent on nonproductive spending. Since

$$R'(t) = \text{Rate in} - \text{Rate out}$$

we have
$$R'(t) = 0.10P(t) - 0.25W(t)$$

We must now write this differential equation in terms of R alone. Since $R = W + P$, then

$$R' = 0.10P - 0.25(R - P)$$

or
$$R'(t) = 0.35P(t) - 0.25R(t)$$

We are given further that $P(t) = 700$ in each year t—that is, productive spending is held constant. With this substitution we have

$$R' = 0.35(700) - 0.25R = 245 - 0.25R = 0.25(980 - R)$$

This differential equation has the constant solution $R(t) = 980$. Using the methods of Section 8.3, we see that, with $R(0) = 1{,}000$, the trend of the nation's resources is illustrated by the graph in Figure 8.30.

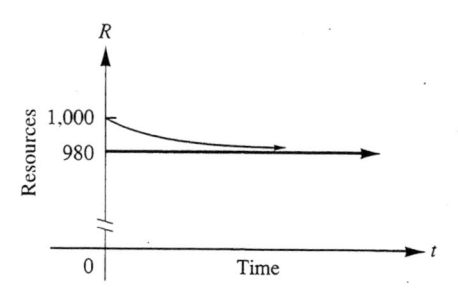

Figure 8.30

The graph indicates that, in the long run, the nation's resources will slowly decline and its spending on nonproductive activity $W(t)$ must decline from \$300 billion to \$280 billion, unless spending on productive activity is allowed to decline or the nation borrows to meet its deficit. ■

Subsection B: Applications to Natural Systems

Compartmental rate analysis is particularly useful in systems found throughout the natural world, from the interior of your body to the ecology of a planet. Space does not permit illustration of all the possibilities, but you will soon realize that the method applies to many subjects.

Our first example analyzes the lifetime of a dam constructed across a river to form a lake above the dam. As water flows downstream into the lake, particles of soil, organic matter, and man-made products flow in and settle behind the dam as sediment. As water is allowed to move through the dam, some part of this sediment leaves. Not all of it does, however, and over many years the sediment can build up to such an amount that some dredging must be done to prevent the dam from becoming useless (see Figure 8.31).

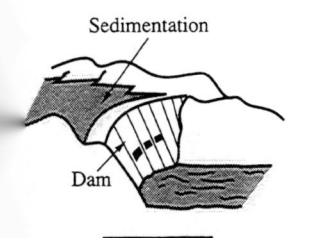

Figure 8.31

EXAMPLE 5

Sediment is accumulating behind a newly constructed dam at the rate of 25,000 tons per year. Suppose that in the course of a year 4% of the sediment behind the dam leaves with the water, which flows through the dam's electricity-generating turbines. It is estimated that when 600,000 tons of sediment have accumulated behind the dam, the turbines will no longer be usable. When will 90% of this accumulation occur so that dredging must be considered?

Solution Let us think of the lake behind the dam as a compartment in which $y(t)$ denotes the amount of sediment (in tons) behind the dam in year t (see Figure 8.32). As before, $y'(t) = RI - RO$.

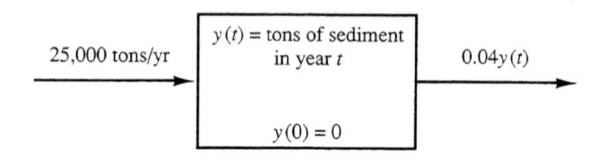

```
                    ┌─────────────────────┐
   25,000 tons/yr   │ y(t) = tons of      │   0.04y(t)
   ───────────────▶ │    sediment         │ ───────────▶
                    │    in year t        │
                    │                     │
                    │    y(0) = 0         │
                    └─────────────────────┘
```

Figure 8.32

Since new sediment accumulates at the rate of 25,000 tons per year, we have $RI = 25{,}000$. Additionally, in year t, 4% of the amount of sediment in the lake leaves as electricity is generated. Thus $RO = -0.04y(t)$. Consequently, the sedimentation process is modeled by the differential equation

$$y'(t) = 25{,}000 - 0.04y(t)$$

According to Theorem 8.1, we have

$$y(t) = 625{,}000 + Ce^{-0.04t}$$

Since $y(0) = 0$, we have $0 = 625{,}000 + C$ or

$$y(t) = 625{,}000(1 - e^{-0.04t})$$

Since $0.90(600{,}000) = 540{,}000$, dredging of sediment behind the dam will begin when

$$540{,}000 = 625{,}000(1 - e^{-0.04t})$$

or when

$$e^{-0.04t} = 1 - 0.864 = 0.136$$

This will occur when

$$-0.04t = \ln 0.136$$

or $t = 49.88$ years after the dam is built (see Figure 8.33).

Remark. Notice that $y' = 0.04(625{,}000 - y)$ so that this equation has the constant solution of $y(t) = 625{,}000$ tons. Thus if the dam could be constructed so as to maintain the 4% clearing rate and still function properly at a sedimentation level of 625,000 tons, dredging would never be necessary (provided that the input rate remains constant at 25,000 tons/year).

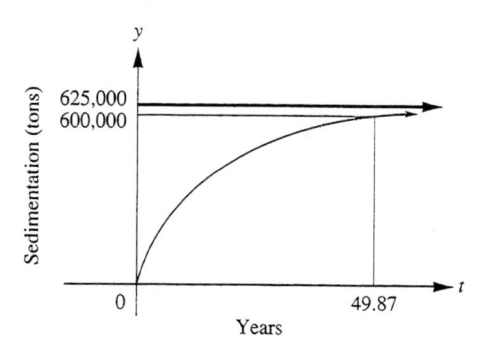

Figure 8.33

Our next example takes place at the microscopic level. But before we discuss it we must understand a preliminary calculation that naturally arises in compartment-mixing analysis. This calculation involves the concept of *concentration*, or proportional amounts.

Consider a compartment containing a liquid that is flowing out of the compartment at a (constant) rate of F liters per hour. We ignore any input rates in this initial discussion. Suppose additionally that the compartment has a volume of V liters. Finally, let $y(t)$ be the amount in grams of some substance S in the compartment that we wish to model. We need to determine RO, the rate at which this substance is leaving the compartment (see Figure 8.34).

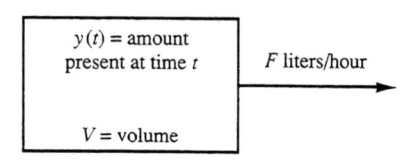

Figure 8.34

This is not hard to do. First, realize that each of the F liters per hour leaving the compartment contains the liquid in the compartment *and* some amount of the substance S. If thorough mixing occurs in the compartment, then the amount of S *in each liter* is in the proportion

$$\frac{y(t)}{V} \quad \frac{\text{grams}}{\text{liter}}$$

Therefore the amount of S leaving the compartment in the combined flow of F liters per hour is

$$RO = \frac{y(t)}{V}\frac{\text{g}}{\text{L}} \times F\frac{\text{L}}{\text{hr}} = \left(\frac{F}{V}\right) y(t) \frac{\text{g}}{\text{hr}}$$

Note that the units (grams, liters, hours) used in this discussion are not essential; any like system of units applies just as well.

EXAMPLE 6

Suppose that the room in which you are seated has a volume of 3,600 ft³. Suppose additionally that the room is vented so that 250 ft³ of air leave the room per minute.

a. If there are 75 ft³ of CO₂ (carbon dioxide) in the room, at what rate is this CO₂ leaving the room per minute?

b. If the amount of CO₂ in the room is a function $y(t)$ of time, at what rate is CO₂ leaving the room at time t?

Solution

a. Here $V = 3{,}600$ ft³, and $F = 250$ ft³/min is the flow rate out of the room. Since there are 75 ft³ of CO₂ in the room, the amount of CO₂ per cubic foot of volume (air) is

$$\frac{75}{3{,}600} = 0.02083$$

Each of the 250 ft³ of air leaving the room carries then 0.02083 ft³ of CO₂. Therefore, CO₂ is leaving the room at the rate of

$$RO = \frac{75}{V} \times F = (0.02083)(250) \approx 5.208 \frac{\text{ft}^3\text{-CO}_2}{\text{min}}$$

b. The analysis is exactly the same as in (a), except that the amount of CO₂ in the room is replaced by $y(t)$, yielding a CO₂ concentration of

$$\frac{y(t)}{3{,}600} \text{CO}_2/\text{ft}^3$$

Consequently, CO₂ is leaving the room at the rate of

$$RO = \frac{y(t)}{V} \times F = \frac{y(t)}{3{,}600} \times 250 = 0.0694 y(t) \frac{\text{ft}^3\text{-CO}_2}{\text{min}} \quad \blacksquare$$

This kind of reasoning is naturally needed in many compartment-mixing processes; the following is an example.

EXAMPLE 7

A food source can be partitioned into its various proteins by *ion-exchange chromatography* in which the substance is bathed in an acidic solution containing salt. A specific protein is "fixed" in accord with a pH and salt strength characteristic of that protein. We need to know the amount of salt in solution over the time of the "fixing" process.

Suppose that a 200-mL container is initially free of salt. At time $t = 0$, an acidic solution containing 0.1 g/mL of salt is pumped through the container at the rate of 50 mL per hour.

a. Model the amount of salt in the container at time t, and illustrate this amount by a graph.

b. What will be the salt concentration in the container at the end of 2 hours?

c. At what time will the amount of salt reach half its maximum level?

Solution We regard the container as a compartment in which the amount of salt at time t is $y(t)$ (see Figure 8.35). Note that $y(0) = 0$.

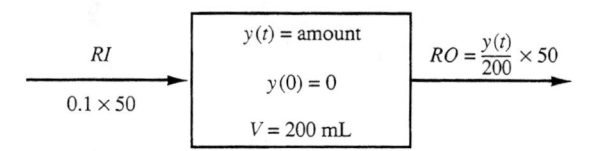

Figure 8.35

$$RI \quad \boxed{\begin{array}{c} y(t) = \text{amount} \\ y(0) = 0 \\ V = 200 \text{ mL} \end{array}} \quad RO = \frac{y(t)}{200} \times 50$$

$$0.1 \times 50$$

We must be careful in this analysis to distinguish between the *amount* of salt, measured in grams, and the *concentration* of salt, measured in grams per liter.

a. Note first that since 50 mL of solution enter the container per hour, at a concentration of 0.1 g/mL of salt, then (0.1) g/mL × 50 mL = 5 g of salt enter the container per hour. This is the rate-in (*RI*) for this system.

Note next that 50 mL/hr are *flowing through* the compartment. Reasoning as in Example 2(b), with $F = 50$, $V = 200$, and $y(t)$ being the amount of salt in the container at time t, we see that

$$RO = \frac{y(t)}{200} \frac{\text{g}}{\text{mL}} \times 50 \frac{\text{mL}}{\text{hr}} = 0.25 y(t) \text{ g/hr}$$

at time t. As before, $y' = RI - RO$. Consequently, the differential equation

$$y' = 5 - 0.25y$$

models this system. Writing the equation as

$$y' = 0.25(20 - y)$$

and using the qualitative methods of Section 8.3, we see that this system evolves over time toward the constant solution 20 g, as shown in Figure 8.36.

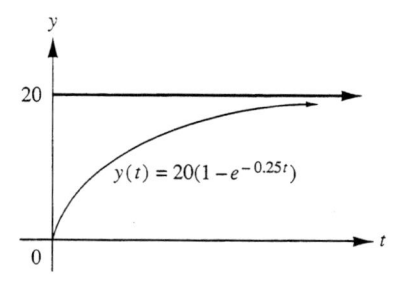

Figure 8.36

$y(t) = 20(1 - e^{-0.25t})$

b. With $y(0) = 0$, the equation has the solution

$$y(t) = 20(1 - e^{-0.25t})$$

At the end of 2 hours, there will be

$$y(2) = 20(1 - e^{-0.25(2)}) = 7.869 \text{ g}$$

of salt in the container. The *concentration* of salt at that time will be

$$\frac{7.869}{200} = 0.0393 \text{ g/mL}$$

c. As noted in (a) (and see Figure 8.36), the maximum level of salt ever to be obtained is 20 g. Half this amount is attained when

$$10 = 20(1 - e^{-0.25t}) = y(t)$$

which yields the solution $t = 2.77$ hours. ∎

In the next example, we are asked to determine an input rate necessary to accomplish a given goal.

EXAMPLE 8

A cell culture is to be grown in an incubator of volume 10,000 cc. The air inside the incubator must contain 5% CO_2 gas. Initially the air in the incubator contains 1.2% CO_2. Air containing 80% CO_2 is to be pumped into the incubator, and an equal volume of air inside the incubator is to be released. At what rate should the air be pumped into the incubator so as to reach the 5% level within 3 minutes?

Solution We regard the incubator as a compartment of volume $V = 10^4$ cc and let $y(t)$ be the amount of CO_2 (in cc) in the compartment at time t (see Figure 8.37). Since the air in the incubator initially contains 1.2% CO_2, then $y(0) = 0.012(10,000) = 120$ cc.

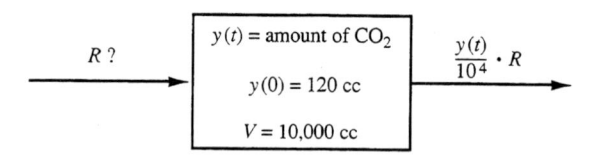

Figure 8.37

We are to determine an as yet unknown rate R cc/min (containing 80% CO_2) at which air is to be pumped through the incubator so that at $t = 3$, $y(t)/10,000 = 0.05$, that is, so that the percentage of CO_2 in the compartment is 5%.

Since $y'(t) = RI - RO$, we first write $y' = 0.80R - RO$ since 80% of the input rate R consists of CO_2. Now air leaves the compartment at the same rate of R cc/min. However, the amount of CO_2 per cc of exiting air is in the proportion $y(t)/10^4$. Therefore the $RO = (y(t)/10^4)R$. Consequently, the needed differential equation is

$$y' = 0.80R - \frac{R}{10^4}y(t)$$

By Theorem 8.1, the solution of this equation is

$$y(t) = \frac{-0.80R}{(-R/10^4)} + Ce^{(-R/10^4)t}$$

or

$$y(t) = 8{,}000 + Ce^{(-R/10^4)t}$$

Since $y(0) = 120$, we obtain $C = -7{,}880$. Consequently,

$$y(t) = 8{,}000 - 7{,}880e^{-Rt/10^4}$$

We must determine R so that

$$\frac{y(3)}{10^4} = 0.05$$

This yields the equation

$$8{,}000 - 7{,}880e^{-3R/10^4} = 500$$

or

$$e^{-3R/10^4} = 0.95178$$

Hence

$$\frac{-3R}{10^4} = \ln 0.95178 = -0.04942$$

The needed rate R is then

$$R = 164.73 \text{ cc/min} \quad \blacksquare$$

In closing we point out how we would treat *multiple* input/output rates in compartment analysis. If Figure 8.38 represents a system with multiple rates-in R_1, R_2 and rates-out P_1, P_2, we have only to add these to model this system. Thus $y' = RI - RO = (R_1 + R_2) - (P_1 + P_2)$.

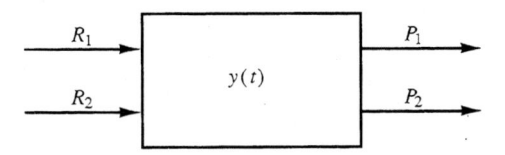

Figure 8.38

The principles of compartment-mixing analysis may also be used in applications where the rates-in and rates-out are not constant (see Exercise 35). We study these in Section 8.7.

Exercises 8.5

Subsection A: Applications to Business and Economics

In Exercises 1–4:
a. Write the solution to the differential equation using Theorem 8.1.   b. Determine the constant C.
c. Determine $y(5)$.   d. Determine the constant solution.
e. Determine when $y(t)$ is half the constant value found in (d), if this happens at all.

1. $y' = 1 - 2y$; $y(0) = 0.25$
2. $y' = 25 - 0.5y$; $y(0) = 30$
3. $y' = 8 - 0.04y$; $y(0) = 200$
4. $y' = 5 + 0.02y$; $y(0) = 4$

5. An account contains \$5,000 and draws 10% interest compounded continuously. Withdrawals are made at the rate of \$450 per year. When will the account contain \$6,000?

6. What will happen to the account in Exercise 5 if withdrawals are made at the rate of \$500 per year?

7. Suppose that you have a savings account, paying 9% yearly interest, to which you add \$500 per year. If the account initially contains \$1,000, how much will it hold at the end of 5 years?

8. At what rate should a company make deposits into an account paying 10% interest so as to have \$80,000 in the account at the end of 5 years, assuming that the initial balance in the account is zero?

9. An annuity is a repeated payment at definite intervals by one party of the annuity contract to another party. Suppose the annuity is to be $25,000 per year, funded by an investment account in Treasury Bills paying 9% per year.

   a. What is the smallest initial amount that will fund this annuity forever?

   b. What initial investment will fund the annuity for *exactly* 20 years?

10. Please refer to Example 3. Part of the experience of buying a new car is the last-minute decision to buy a few "options." How is the monthly payment rate R affected by an additional $800 expenditure for one more option? What will be the extra cost of this $800 over the life of the loan?

11. Compare the extra cost of the 10% increase of $800 made to a total loan of $8,800 found in Exercise 10, to the extra cost if instead you had to pay an additional 10% in interest costs to 1.1% per month on the original $8,000 loan.

12. Referring to Example 3 (and Section 7.4), what is the *present value* of this stream of loan payments at an interest rate of 8%? How may similar thinking be used to understand the effect of an 8% inflation rate over the life of such a loan?

13. This exercise refers to Example 4.

    a. What is the trend of the nation's resources if instead $P(t)$ is held constant at $500 billion? $750 billion? $850 billion? Which of these allows a larger nonproductive expenditure *in the long run*?

    b. Example 4 indicates that total resources would decline from $1,000 billion to $980 billion if $P(t)$ is held constant at $700 billion. What would happen if, when $R(t)$ drops to the level of $980 billion, the nation's leaders decide to reduce $P(t)$ to $680 billion? Raise $P(t)$ to $720 billion?

    c. What would happen in Example 4 if the nation instead opted for a constant level of $W(t) = 300$ in nonproductive spending?

14. a. Suppose that in Exercise 13 and Example 4 $P(t)$ were allowed to be a constant proportion $kR(t)$ of total resources. What proportion k would ensure that the nation's total resources never decline? That they never expand?

    b. Suppose that the rates of gain and loss in the nation's economy are unknown numbers α and β so that $R' = \alpha P - \beta W$. Show that if $P(t) = e^{-kt}R(t)(k > 0)$—that is, if spending on productive activity shares an exponentially declining part of the nation's resources—then the nation is headed for long-term ruin, by showing that $\lim_{t \to \infty} R(t) = 0$.

15. Which is the better offer for the purchase of the same item: (a) store A offers the item at a cost of $5,000 financed at 1% interest per month, or (b) store B offers the same item at a cost of $4,500 at 1.1% per month, both over a 36-month period? [*Hint:* Determine the monthly payment in each case.]

16. A certain country has $5 billion of paper currency in circulation. Each day $3 million flows through the nation's banks. It is decided to replace the paper dollars with new paper dollars as each appears in one of the banks. How long will it take for 50% of the old paper dollars to be replaced by new paper dollars? [**Remark.** If $y(t)$ is the amount of new currency in circulation on day t, then $(5 - y(t))/5$ is the *proportion* of old currency to total currency on day t.]

Subsection B: Applications to Natural Systems

In Exercises 17–18, imagine a compartment of volume V through which material is flowing at a constant rate F. Additionally the compartment contains a second, different substance A, in the amount $A(t)$ at time $t \geq 0$, which is thoroughly mixed with the substance flowing through the compartment. Answer (a)–(e).

17. $V = 30$ ft^3; $A(0) = 2$ g; $F = 5$ ft^3/hr

    a. What is the concentration of the substance A in the compartment at time t?

    b. At what rate per hour is the substance A leaving the compartment?

    c. How much of the substance A leaves the compartment in a 2-hour period?

    d. In a half hour?

    e. In 10 minutes?

18. $V = 50$ L; $F = 3$ L/min; $A(t) = te^{-0.02t}$ cc

    a. What is the concentration of A at time t?

    b. At what rate per minute is A leaving the compartment at time t?

    c. *Approximately* how much of A leaves the compartment during a 1-minute period at the end of $t = 2$ hours?

    d. During a 1-minute period after $t = 105$ minutes?

    e. Does more, or less, of A leave the compartment in the first quarter hour or the fourth quarter hour?

19. Research has shown that memory loss occurs as more is memorized. If a person is memorizing 30 words per hour in a foreign language and forgets 20% of the words learned each hour,

   a. How many words will be known at the end of 3 hours?

   b. What is the maximum number of words that will be retained "in the long run"?

20. A cancerous tumor initially weighs 3.5 g and is growing at the rate of 20% of its size (as mass) per week. Chemical treatment kills tumor cells at the rate of 0.8 g per week. Will the tumor be destroyed? [*Hint:* A qualitative solution (see Section 8.3) will suffice to answer this question.]

21. The total amount of CO_2 in the atmosphere today is 3,000 billion tons (a 13% increase over 100 years ago). It is estimated that the "greenhouse effect" on the earth's atmosphere will cause the polar ice caps to melt (flooding Manhattan) if the amount of CO_2 reaches 4,000 billion tons. The burning of fossil fuels plus natural sources is adding CO_2 to the atmosphere at the rate of 45 billion tons per year. Only 1% of the CO_2 is removed by the earth's oceans. If this continues, when will the polar ice caps melt?

22. Suppose that 100 years from now the CO_2 level in the atmosphere (see Exercise 21) has reached 3,800 billion tons and humanity has become organized enough to enforce an immediate reduction to 35 billion tons per year in the amount of CO_2 added to the atmosphere. Additionally, worldwide emergency measures are undertaken to remove R billion tons of CO_2 per year from the atmosphere. If the oceans continue to eliminate 1% of atmospheric CO_2 as well, how large must R be so as to reduce atmospheric CO_2 to 3,000 billion tons within two generations (about 50 years)?

23. Agribusiness has led to the planting of huge acreages of crops of a single genetic variety with a high yield. When a fungus or other blight specific to this variety strikes, the result can be devastating. A cautious small farmer who wishes to guard against such a disaster decides to plant corn in two genetic types A and B, where B has a smaller yield but is less susceptible to the same diseases. The farmer's machinery allows for mixing and stirring 100 lb of corn. The farmer fills the machine with type A and then begins to introduce type B into the machine at the rate of 5 lb/min, removing 5 lb of mixed type at the same time. When will there be a 50-50 mixture of types A and B in the machine?

24. A hundred acres of fields and forest have been destroyed in order to build a new subdivision complete with plastic stores and acres of pavement. People begin moving into the new subdivision at the rate of 20 individuals per year. The birthrate for the arriving and then existing population is 2% per year, and 0.5% of the inhabitants leave the area due to sensory deprivation each year. When will the population (which begins at zero) reach 200 individuals?

25. An island in the Pacific initially has a population of 1,800 giant tortoises. Their birthrate is 12% per year, and their natural death rate is 6% per year. An additional 10% die due to predation. Concerned scientists begin bringing new tortoises to the island at the rate of 8 tortoises per year.

   a. How many tortoises will be on the island in 7 years?

   b. At what rate should new tortoises be brought to the island to maintain the population near 1,800?

26. Suppose that in Exercise 25 the tortoises occur in two varieties—one variety having green eyes, the other blue. Initially there are 600 green-eyed tortoises, and the birth, death, and predation rates apply equally to green- or blue-eyed tortoises. New tortoises available to the scientists are found only in the green-eyed variety, and these are brought to the island at the rate of 30 per year. When will 50% of the tortoises on the island have green eyes? [*Hint:* Determine (separately) the total number of tortoises and the total number of green-eyed tortoises in year t.]

27. Protein deposits naturally build up behind soft contact lenses, causing irritation after a time. An enzymatic cleaner is used to remove these deposits, but it is not 100% effective. If a user deposits 0.0007 mg/day of protein on the lens, and the cleaner is used at weekly intervals and removes 80% of the protein on the lens in each application, and eye irritation occurs when 0.006 mg of protein is present, when will problems begin to occur? How can the user deal with this difficulty?

28. Model the population of muskrats in a pond in which there are initially 25 muskrats with a birthrate of 25% and a death rate of 20% and to which 8 new muskrats migrate in each year and 6 leave. Will the population rise, decline, or reach a steady state?

29. Suppose that a drug is being administered intravenously and continuously at the rate of 100 mg/hr and

that the patient's body is able to eliminate the drug at the rate of 20% of the amount present.

   a. Set up a mathematical model (differential equation) for the amount of the drug present in the patient's body at any time t.

   b. If the initial amount of the drug is zero, solve the equation in (a).

   c. When will the drug concentration (amount) reach 400 mg?

   d. Suppose that an amount of 400 mg is dangerous. Once that level is reached, when can the doctor begin readministering the drug safely (that is, when is the level at 300 mg again)?

30. Suppose that in Exercise 29 the elimination rate remains at 20%, that 200 mg is dangerous but that 150 mg is necessary (after the drug has built up to this level) to do any good, that the initial amount present is 0, and that the 150-mg level must be reached within 4 hours.

   a. At what rate would you prescribe this drug?

   b. When would you cut off and then resume administration over the next 2 days?

   c. If you can change the prescribed dosage per hour at some later time, when and how would you do it so that the patient can be maintained at an equilibrium level of 150 mg?

31. A new method of insulin treatment allows infusion at a constant rate rather than through periodic injection. Suppose a diabetic consumes 30% per hour of insulin present in his bloodstream and his pancreas secretes insulin at the rate of 25 cc/hr.

   a. If the insulin level should be 100 cc, at what rate should insulin be administered if the diabetic's insulin level has fallen to 80 cc and the level of 100 cc must be reached within 1 hour?

   b. Once the 100 cc level is restored, at what rate should insulin be administered so as to maintain this level?

   c. If insulin shock occurs at a level of 300 cc and the doctor erroneously infuses insulin at the rate of 50 cc/hr, is the patient in danger?

32. A city reservoir holds 100,000 gallons of water. Residents drink 1,000 gallons per day. On the average, 300 gallons a day are replaced by pure rainfall, and 700 gallons a day flow in from a stream. The stream water has a concentration of 0.002% PCB's, and the reservoir is initially free of PCB's.

   a. Write a differential equation for the amount of PCB's in the reservoir on day t.

   b. When will each resident consume 0.000005 gallon of PCB's with each gallon he or she drinks?

   c. If a railway accident near the stream results in 160 gallons of PCB's running into the reservoir initially (when it is free of PCB's), how will this affect the long-run problem?

33. A town of 100 individuals gets its water from a reservoir filled by runoff from the surrounding hills and streams. With the introduction of nitrate fertilizers into the soil of the surrounding area in the 1950s, the runoff water now contains nitrates at a level of 0.2% of volume. The water in the reservoir was free of nitrates in 1960 (year 0). Suppose the reservoir holds 10,000,000 gallons of water, that it is replenished at the rate of 500,000 gallons per month, that each individual in the town uses 3,000 gallons per month, and that 200,000 gallons per month are lost by evaporation. (The nitrates remain in the reservoir and are not lost by evaporation.)

   a. Write a differential equation for the amount of *water* (including nitrates) in the reservoir at time t.

   b. If the initial amount of water is 10,000,000 gallons, solve the equation in (a).

   c. Now write a differential equation for the amount of *nitrates* in the water at time t (in years, $t = 0$ is 1960).

   d. Solve the equation in (c).

   e. Sketch the solution (to help guide you through the remaining questions).

   f. What is the maximum number of gallons of nitrates ever to be reached in the reservoir?

   g. How many years does it take for the amount of nitrates in the lake to reach 90% of the level in (f)?

   h. What is the percent of nitrates in a gallon of water at the time t found in (g)?

   i. Assume that an individual drinks a gallon of water a day. How much nitrate does she drink each month?

34. It is known that 150 mg of Valium in a person's bloodstream is fatal and that amounts above 10 mg induce drowsiness, slow reaction times, and so on. The half-life of Valium is 12 hours (that is, within 12 hours the body metabolizes half of whatever Valium is present at a given time). From Section 5.5, this yields a decay rate of $(\ln 2)/12$, or 5.8% per hour. Suppose that a patient is given the following prescription: Take one Valium tablet every 8 hours; each contains 10 mg of Valium.

a. Under this prescription, when will the amount of Valium in the person's body reach 20 mg?

b. Draw a graph of the amount of Valium in the person's body during the first 24 hours of this prescription.

c. Suppose that instead the patient is continuously administered Valium at the rate of $\frac{10}{8} = 1.25$ mg/hr. When will the amount in the patient's body reach 20 mg?

Challenge Problems

35. When a drug is administered to a patient, the amount in the bloodstream can be modeled by exponential decay Ce^{-kt}, before entering an intended organ. This process gives rise to the linear differential equation with *nonconstant coefficients* treated in this exercise, where the organ is considered to be the compartment and the drug leaves the bloodstream and *enters* the organ at the *rate of* $I'(t) = ake^{-kt}$ g/min and where $y(t)$ is the amount of drug present in the organ at time t (see Figure 8.39). Part (c) is particularly important because it tells the physician when maximum concentration in the organ occurs. Show that if $I'(t) = ake^{-kt}$ is the rate at which a substance S is entering a compartment and if $b = F/V$ is the ratio of (constant) flow rate to volume through the compartment, then:

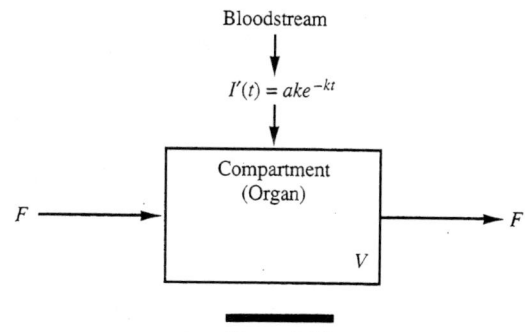

Figure 8.39

a. $y' = ake^{-kt} - by$ models this system.

b. $y(t) = [ak/(b - k)][e^{-bt} - e^{-kt}]$ is the solution of the differential equation in (a) when $y(0) = 0$.

c. The maximum value (maximum amount in the compartment) occurs at $T = (\ln b - \ln k)/(b - k)$.

36. Graph the solution to Exercise 35 when $a = 6$, $b = 0.03$, and $k = 0.02$, and determine the value of T in part (c).

Special Assignment

At this point in your study of calculus you have acquired the tools to make realistic applications to topics that interest you outside mathematics. Applications of the methods of this section are found throughout the literature in your field of study. As a special assignment, go to this literature and find a process that you can model using these methods and use this model to reach a conclusion about the process you have researched.

8.6 Euler's Method of Approximate Solution

This section exhibits a method of approximating the numerical value $y(t)$ of a solution to a differential equation at a particular point T when an explicit solution $y(t)$ in the variable t is unknown.

An attempt to solve the simple differential equation

$$y' = 2ty + 1$$

by the methods of this chapter will quickly lead to the need to find the integral $\int e^{t^2}\, dt$, which cannot be found in terms of known functions. The Swiss mathematician Leonhard Euler (1707–1783) (for whom the number e is named) invented a method of *approximating* the numerical value of a solution at a future time T. An understanding of **Euler's method** begins with an idea first seen in Section 8.3. If we are given a differential equation

$$y' = g(t, y)$$

then the numerical value of $g(t, y)$ is the slope, and hence the direction, of the solution y at time t.

Consider then a system modeled by a differential equation $y' = g(t, y)$, which begins in some initial state y_0 at time t_0 (see Figure 8.40). That is, we know that $y(t_0) = y_0$. Let T be some later time. If we are unable to actually solve the equation for $y(t)$, how can we *approximate* the number $y(T)$?

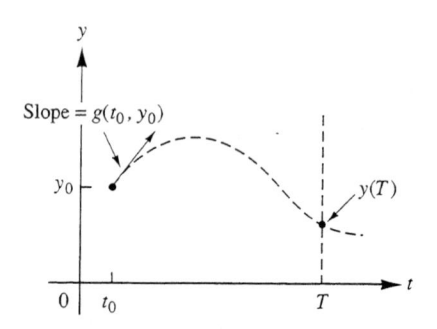

Figure 8.40

In attempting to approximate $y(T)$, we begin with the only thing that we have—that $y'(t_0) = g(t_0, y_0)$ is the slope of the solution $y(t)$ at t_0.

EXAMPLE 1

If $t_0 = 1$ and $y(t_0) = 2$, what is the slope of the solution to $y' = ty + 1$ at $t = t_0$?

Solution Here $g(t, y) = ty + 1$. Since $t_0 = 1$ and $y_0 = y(t_0) = 2$, we have that the slope of the solution is $y'(t_0) = g(1, 2) = 1(2) + 1 = 3$. We illustrate this answer in Figure 8.41.

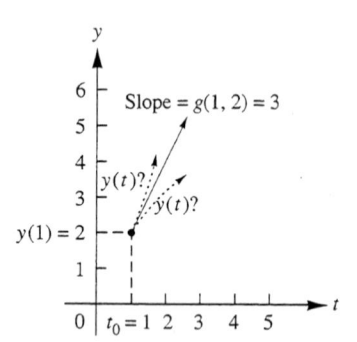

Figure 8.41

Either of the dotted-line curves could be the path of the solution to the given equation. Being unable to solve the equation, we cannot know this. However, we can assert that at, say, $t_1 = 1 + \frac{1}{10}$ the solution cannot be very far from the tangent line (of slope 3). This tangent has equation $y - y_0 = y'(t_0)(t - t_0)$ or $y - 2 = 3(t - 1)$. Since $y\left(1 + \frac{1}{10}\right)$ must be near this straight line, we can say that at the later time $t_1 = 1 + \frac{1}{10}$

$$y\left(1 + \tfrac{1}{10}\right) = y(t_1) \approx 3(t_1 - t_0) + 2 = 3\left(\left(1 + \tfrac{1}{10}\right) - 1\right) + 2 = \tfrac{23}{10}$$

490   Chapter 8   Differential Equations

Euler's method for finding an approximation of $y(T)$ repeats this process over and over, moving from $t = t_0$ [where an exact value $y(t_0)$ is known] in small steps to $t = T$, where an approximate value is found.

Explicitly, let us imagine splitting the interval from $t = t_0$ to $t = T$ into n short pieces, each $h = (T - t_0)/n$ units long, as illustrated in Figure 8.42. We then place a sequence of n points $t_1, t_2, \ldots, t_{n-1}, t_n$ between t_0 and T, each uniformly separated by the distance h. That is, $t_1 = t_0 + h$, $t_2 = t_1 + h, \ldots, t_{k+1} = t_k + h$, and so on, for $k = 1, 2, \ldots, n - 1$, where $T = t_n = t_{n-1} + h$.

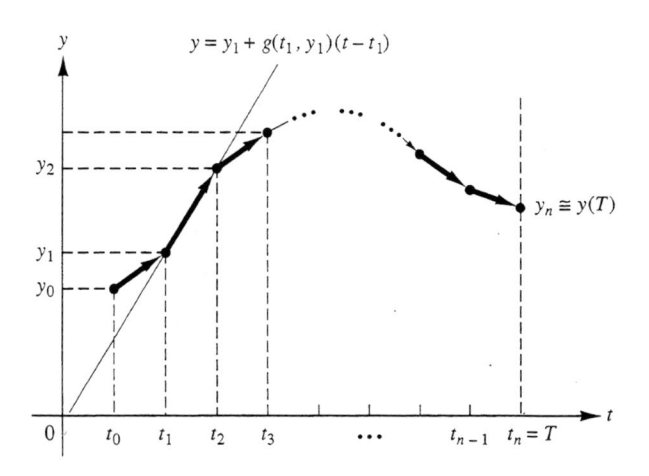

Figure 8.42

The solution to the given differential equation $y' = g(t, y)$ begins at the point (t_0, y_0). We will approximate this solution by a polygonal path like that indicated in Figure 8.42, stopping at each point t_k to redirect the line segment in the direction indicated by $y'(t_k)$. That is, the slope of each straight line is to be the slope of the solution $y'(t_k) = g(t_k, y_k)$ at the point t_k. In doing so we must also use a linear equation (as earlier) to determine an approximate value y_k to the solution at the point t_k to be used in the evaluation of $g(t_k, y_k)$. We do this as follows:

Since $g(t_0, y_0)$ is the slope given by the initial condition, as in Example 1, we have that

$$y - y_0 = g(t_0, y_0)(t - t_0)$$

is the equation of the first straight-line segment indicated in Figure 8.42. The y-value of this straight line at t_1 is

$$\begin{aligned} y_1 &= y_0 + g(t_0, y_0)(t_1 - t_0) \\ &= y_0 + g(t_0, y_0)h \\ &= y_0 + y'(t_0)h \end{aligned} \qquad (1)$$

since t_0 and t_1 are h units apart.

The slope of the second straight line, now indicated on an expanded scale in Figure 8.43, is $g(t_1, y_1)$, obtained from t_1 and the value y_1 just obtained. The equation of this second line is

$$y - y_1 = g(t_1, y_1)(t - t_1)$$

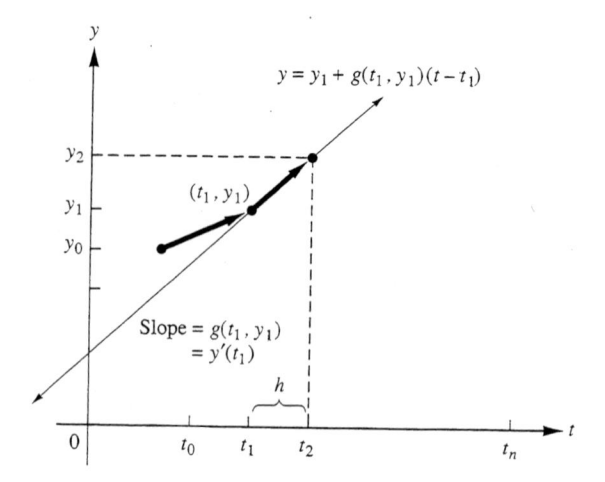

Figure 8.43

The y-value of this straight line at $t = t_2$ is then

$$y_2 = y_1 + g(t_1, y_1)(t_2 - t_1) = y_1 + g(t_1, y_1)h \qquad (2)$$

We can continue in this fashion, moving from each point to the next.

If we compare Equations 1 and 2, we can see a pattern. Given the kth time point t_k and the corresponding y-value y_k, we see that

$$y_{k+1} = y_k + g(t_k, y_k)h \qquad (3)$$

Thus the "next" y-value, y_{k+1}, is obtained by using previously determined values t_k and y_k. In this way we work our way across the interval from t_0 to T in steps of length h until, after n steps, we produce the desired approximation

$$y(T) \simeq y_n = y_{n-1} + g(t_{n-1}, y_{n-1})h$$

Finally, we should remember that at each step we have only approximate equality, $g(t_k, y_k) \simeq y'(t_k)$, since y_k is only an approximation to the actual solution curve $y(t)$ at $t = t_k$. This is all a lot easier to do than the notation suggests!

EXAMPLE 2

Approximate the solution to

$$y' = ty + 1 \qquad y(1) = 2$$

at $T = 3$, using $n = 4$ steps in this approximation.

492   Chapter 8   Differential Equations

Solution We begin by setting
$$h = \frac{T - t_0}{n} = \frac{3 - 1}{4} = \frac{1}{2}$$

This yields four points following $t_0 = 1$, up to (and including) $T = 3$—namely, $t_1 = \frac{3}{2}$, $t_2 = 2$, $t_3 = \frac{5}{2}$, and $t_4 = 3 = T$—as indicated in Figure 8.44.

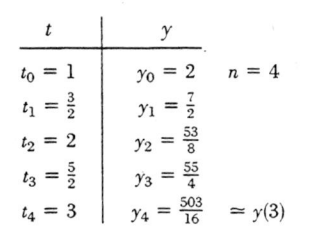

t	y
$t_0 = 1$	$y_0 = 2$
$t_1 = \frac{3}{2}$	$y_1 = \frac{7}{2}$
$t_2 = 2$	$y_2 = \frac{53}{8}$
$t_3 = \frac{5}{2}$	$y_3 = \frac{55}{4}$
$t_4 = 3$	$y_4 = \frac{503}{16} \approx y(3)$

$n = 4$

Figure 8.44

We next make a table of the values of t_k in which we will later place corresponding values y_k. This table is shown alongside Figure 8.44. According to Equation 3, each successive y-value is given by

$$y_{k+1} = y_k + g(t_k, y_k)h$$

Since $g(t, y) = ty + 1$ and $h = \frac{1}{2}$, we have that

$$y_{k+1} = y_k + (t_k y_k + 1)\frac{1}{2}$$

$$= y_k + \frac{t_k}{2}y_k + \frac{1}{2} = \left(1 + \frac{t_k}{2}\right)y_k + \frac{1}{2}$$

Consequently, using this last expression,

$$y_1 = \left(1 + \frac{t_0}{2}\right)y_0 + \frac{1}{2} = \left(1 + \frac{1}{2}\right)(2) + \frac{1}{2} = \frac{7}{2}$$

Since $t_1 = \frac{3}{2}$, we then have, using $y_1 = \frac{7}{2}$, that

$$y_2 = \left(1 + \frac{t_1}{2}\right)y_1 + \frac{1}{2} = \left(1 + \frac{\frac{3}{2}}{2}\right)\left(\frac{7}{2}\right) + \frac{1}{2} = \frac{53}{8}$$

With $t_2 = 2$, $y_2 = \frac{53}{8}$, we then obtain

$$y_3 = \left(1 + \frac{2}{2}\right)\left(\frac{53}{8}\right) + \frac{1}{2} = \frac{55}{4}$$

Hence

$$y_4 = \left(1 + \frac{\frac{5}{2}}{2}\right)\left(\frac{55}{4}\right) + \frac{1}{2} = \frac{503}{16}$$

Therefore

$$y(T) = y(3) \approx y_4 = 31.4375 \blacksquare$$

There is no reason to expect that Euler's method, using only a four-step approximation process, will yield a truly accurate approximation. If we used, say, $n = 10$ steps, the approximation may be a more accurate one. We illustrate this in Figure 8.45, where we show three polygonal path approximations to the solution curve of the equation

$$y' = \frac{-y}{2} + \frac{y}{t}, \quad y(0.5) = 1, \quad \text{and} \quad T = 4.5$$

using $n = 4$, $n = 8$, and $n = 16$. We also indicate in Figure 8.45 the path of the exact solution, the smooth curve $y(t) = te^{-t/2}$.

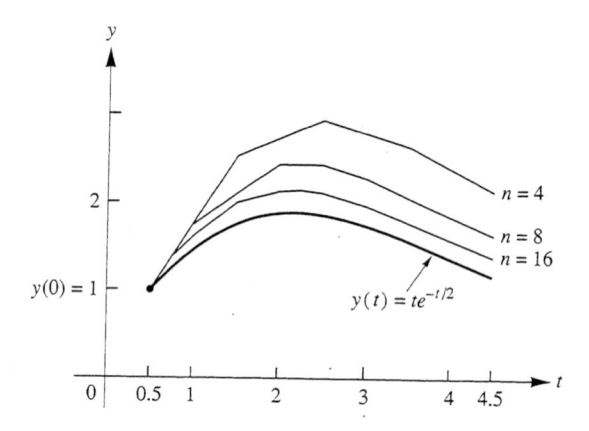

Figure 8.45

Here is a summary of Euler's method.

Euler's Method of Approximation

Given $y(t_0) = y_0$ and wishing to approximate the value $y(T)$ of the solution to $y' = g(t, y)$ at $t = T$ in n steps,

1. Find $h = \dfrac{T - t_0}{n}$.
2. Write $t_1 = t_0 + h$, $t_2 = t_1 + h$, ..., $t_n = t_{n-1} + h = T$.
3. Write $y_{k+1} = y_k + g(t_k, y_k)h$.
4. Simplify the expression on the right in step 3 as much as possible.
5. Use the equation in steps 3 and 4 along with t_0, y_0 to compute y_1. Then use t_1, y_1 and the equation to find y_2.
6. Continue this process until y_n is obtained. Conclude that $y(T) \simeq y_n$.

This procedure is well adapted to a simple loop program on a computer, which can be programmed to do step 5 repeatedly. (See Exercises 13–16.) Be aware, however, that if n is large, then roundoff error in the computer can be considerable. More sophisticated approximation methods that improve the accuracy of the calculations are available.

Exercises 8.6

1. If the point $(1, 2)$ is on the graph of the solution to $y' = t + y$, what is the slope of the solution through this point?

2. What is the slope of the solution to $y' = ty - t^2$ passing through the point $(0, 1)$?

3. Is the solution to $y' = ty + 1$ increasing or decreasing at the point $(0, -1)$?

4. Consider the IVP $y' = ty - e^t$, $y(0) = 2$. Is this solution increasing or decreasing at $t = 0$?

5. Solve the IVP $y' = ty$, $y(0) = 1$, by the method of separation of variables. Compute $y(2)$. Now use Euler's method to approximate $y(2)$ in $n = 4$ steps. Compare the two answers.

6. Approximate the solution to $y' = t + y$, $y(0) = 1$, at $T = 2$, using $n = 8$ points.

7. Approximate the solution to $y' = t - y$, $y(0) = 2$, at $T = 2$, using $n = 4$ points.

8. Approximate the solution to $y' = e^t y$, $y(0) = e^2$, at $T = 1$, using $n = 2$ steps. Then solve the equation and evaluate $y(1)$. Is the approximation accurate?

9. Use Euler's method with $n = 3$ to approximate the solution to the IVP $y' = e^{-t^2} y$, $y(0) = 1$, at $T = 1$.

10. Approximate the solution to $y' = y - ty^2$, $y(0) = 1$, at $T = 2$, in $n = 4$ steps.

11. In Example 1, Section 8.5B, sediment is accumulating in the lake above the dam at the rate of 25,000 tons per year. Suppose that, initially, water exiting the lake through turbines built in the dam removes 4% of the sediment per year. As sediment levels in the lake rise, however, a greater percentage of sediment is removed in this process, so that in year t, the proportion $(0.04 + 0.001t)$ of the sediment is removed. If $y(t)$ is the amount of sediment above the dam, then by RI/RO compartment analysis

$$y' = 25{,}000 - (0.04 + 0.001t)y$$

If $y(0) = 0$, approximate the amount of sediment above the dam in 4 years, using year-long steps in the approximation.

12. Suppose that in a lake of volume V, precipitation equals evaporation over the course of a year. Assume that a particular kind of pollutant has entered the lake via rainfall at the rate of 50 tons per year for 20 years. Assume that the pollutant thoroughly mixes in the water and then is lost with water through evaporation. Regard rainfall and evaporation as a constant flow rate F through the lake, and suppose that $F/V = 0.03$. Let $A(t)$ be the amount of pollutant in the lake in year t, with $A(0) = 0$. Answer (a)–(e) under these assumptions.

   a. Explain why $A'(t) = 50 - (F/V)A(t)$.

   b. How many tons of pollutant are in the lake at the end of 20 years?

   c. Suppose that a pollution control program is begun at that time and that pollutants enter the lake at the rate of $50 - 0.05t$ tons per year for the next 10 years. Model the amount of pollutants in the lake at time t.

   d. Approximate the number of tons of pollutant in the lake at the end of 10 years using year-long steps in the approximation.

   e. Suppose at that point no further pollutants enter the lake. How many years will it take for the amount of pollutants in the lake to reach 50% of the amount found in (b)?

Computer Application Problems

Use the BASIC program EULER to estimate the value of the solution to $y' = ty + 1$ for the intervals and initial values given in Exercises 13–16. In each case execute the program for $N = 8$, 16, and then 32 subintervals.

13. $y(0) = 1$; estimate $y(1)$.

14. $y(0) = 0$; estimate $y(2)$.

15. $y(1) = -2$; estimate $y(2)$.

16. $y(1) = -1$; estimate $y(4)$.

17. In Example 4, "Swords versus Plowshares," of Section 8.5A, suppose that the discovery of natural resources (oil, silver, and so on) allows the nation to have an input $\beta(t) = 100te^{-0.3t}$ of resources (billions of dollars per year) that rise and then decline over time. The differential equation then becomes (see Example 4, Section 8.5A) $R'(t) = 0.35P(t) - 0.25R(t) + \beta(t)$.

   a. Approximate the nation's resources 30 years later and 50 years later if $P(t) = 0.7R(t)$ using 5-year steps in the approximation. [Hint: You may need to reconsider the original equation $R' = 0.10P - 0.25W$.]

b. Modify the BASIC program EULER (line 30) and use it to find approximations at $t = 30$, 50, and 100 years using year-long steps.

8.7 Compartment-Mixing Processes with Variable Rates

In this section we use the first-order nonhomogenous linear differential equation with nonconstant coefficients to model a compartment subject to variable input and output rates.

In many systems it is both reasonable and natural that the rates in and out of a compartment not be constant, and one is led by compartmental analysis of an evolving system to a differential equation of the form

$$y'(t) = \alpha(t)y(t) + \beta(t)$$

An example of such an equation is, say, $y' = t^2 y + t$, with $\alpha(t) = t^2$ and $\beta(t) = t$. This differential equation is like the equation $y' = ay + b$ of Section 8.5 except that the coefficients α and β (determined by the input/output rates) are allowed to vary with time. As such, the equation is not of the autonomous type of Section 8.3, nor can the variables be separated as in Section 8.4. We first give a formula for the general solution of such an equation and then see how this formula can be obtained.

Theorem 8.2

The general solution of the first-order linear differential equation

$$y' = \alpha(t)y + \beta(t)$$

is

$$y(t) = (B(t) + C)e^{A(t)}$$

where

$$A(t) = \int \alpha(t)\, dt$$

and

$$B(t) = \int \beta(t) e^{-A(t)}\, dt$$

where C is an unknown constant.

Note again the appearance of the ubiquitous number e!

We can derive this solution as follows. We begin with the equation as given:

$$y' = \alpha(t)y + \beta(t)$$

Then we rewrite it as

$$y'(t) - \alpha(t)y(t) = \beta(t)$$

Now let $A(t) = \int \alpha(t)\, dt$ and multiply both sides of the last equation by $e^{-A(t)}$ to obtain

$$y'(t)e^{-A(t)} - y(t)\alpha(t)e^{-A(t)} = \beta(t)e^{-A(t)}$$

496 Chapter 8 Differential Equations

Note that $D_t e^{-A(t)} = e^{-A(t)} D_t(-A(t)) = -\alpha(t)e^{-A(t)}$. Therefore the equation can be written as

$$e^{-A(t)} D_t y(t) + y(t) D_t e^{-A(t)} = \beta(t)e^{-A(t)}$$

or, using the product rule for differentiation, as

$$D_t(y(t)e^{-A(t)}) = \beta(t)e^{-A(t)}$$

Consequently,

$$\int D_t(y(t)e^{-A(t)}) \, dt = \int \beta(t)e^{-A(t)} \, dt \qquad (1)$$

Now, from the Fundamental Theorem of Calculus,

$$\int D_t(y(t)e^{-A(t)}) \, dt = y(t)e^{-A(t)}$$

Let us write

$$\int \beta(t)e^{-A(t)} \, dt = B(t) + C$$

We then have, by substitution in Equation 1,

$$y(t)e^{-A(t)} = B(t) + C$$

or

$$y(t) = (B(t) + C)e^{A(t)}$$

as the solution.

EXAMPLE 1

Solve the IVP $y' = t^3 - ty$, $y(0) = 5$, using Theorem 8.2.

Solution Comparing this particular equation with the general form found in Theorem 8.2, we see that $\alpha(t) = -t$ and $\beta(t) = t^3$.

Consequently,

$$A(t) = \int -t \, dt = \frac{-t^2}{2} \quad \text{and} \quad B(t) = \int t^3 e^{+t^2/2} \, dt$$

By the method of integration by parts (Section 7.1), since $D_t e^{t^2/2} = te^{t^2/2}$, we then have

$$B(t) = \int t^3 e^{t^2/2} \, dt = \int t^2(te^{t^2/2}) \, dt = \int t^2 d(e^{t^2/2})$$
$$= t^2 e^{t^2/2} - \int e^{t^2/2} d(t^2)$$
$$= t^2 e^{t^2/2} - 2 \int te^{t^2/2} \, dt$$
$$= t^2 e^{t^2/2} - 2e^{t^2/2}$$

Hence
$$y(t) = (B(t) + C)e^{A(t)} = ((t^2 - 2)e^{t^2/2} + C)e^{-t^2/2}$$
$$= t^2 - 2 + Ce^{-t^2/2}$$

Therefore $5 = y(0) = -2 + C$, or $C = 7$. The solution to the given IVP is $y(t) = t^2 + 7e^{-t^2/2} - 2$. ∎

While the method of solution in Example 1 is technically more complex than that of Section 8.5, the method of application is not.

EXAMPLE 2

The runoff of nitrate fertilizer into a reservoir is $1{,}500e^{-0.25t}$ lb/week, t weeks after application on surrounding farmland in the spring of the year. Drinking water is removed from the reservoir at the rate of 100,000 gallons per week, with equal replenishment by rainfall. If the reservoir contains 10^6 gallons, how much nitrate fertilizer will be in the reservoir 8 weeks after application, if the reservoir is initially free of nitrate?

Solution As in Section 8.5, we imagine the reservoir as a compartment, but now with $RI = 1{,}500e^{-0.25t}$, a nonconstant input rate (see Figure 8.46).

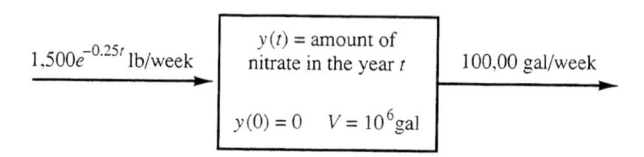

Figure 8.46

[Diagram: $1{,}500e^{-0.25t}$ lb/week → $y(t)$ = amount of nitrate in the year t; $y(0) = 0$, $V = 10^6$ gal → 100,00 gal/week]

With $y(t)$ the amount (in lb) of nitrate in the reservoir in week t, we immediately have

$$y'(t) = RI - RO = 1{,}500e^{-0.25t} - RO$$

Now, since the volume of the reservoir is $V = 10^6$, the concentration of nitrate per gallon is $y(t)/10^6$. Since the flow rate from the reservoir is $F = 100{,}000$ gal/week, we conclude that

$$y'(t) = 1{,}500e^{-0.25t} - \frac{y(t)}{10^6}(100{,}000)$$

or

$$y'(t) = 1{,}500e^{-0.25t} - 0.10y(t)$$

In order to apply Theorem 8.2, we set $\alpha(t) = -0.10$ and $\beta(t) = 1{,}500e^{-0.25t}$. Then $A(t) = \int \alpha(t)\,dt = \int -0.10\,dt = -0.10t$. Hence

$$B(t) = \int \beta(t)e^{-A(t)}\,dt = \int 1{,}500e^{-0.25t}e^{+0.10t}\,dt$$

$$= 1{,}500 \int e^{-0.15t}\,dt = -10{,}000e^{-0.15t}$$

Therefore, from Theorem 8.2,

$$y(t) = (C - 10{,}000e^{-0.15t})e^{-0.10t}$$

Since $y(0) = 0$, we have $C = 10{,}000$ and therefore

$$y(t) = 10{,}000(e^{-0.10t} - e^{-0.25t})$$

Hence, 8 weeks after application there will be

$$y(8) = 10{,}000(0.4493 - 0.1353)$$
$$= 3{,}140 \text{ lb}$$

of nitrate fertilizer in the reservoir.

The graph of this solution is indicated in Figure 8.47.

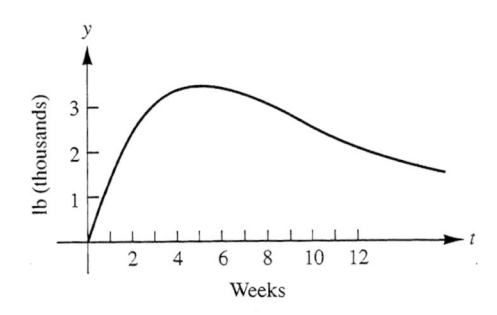

Figure 8.47

That $\beta(t)$ in Example 2 is an exponential decay function is rather natural. Recall from Section 4.5 that many substances dissipate exponentially; if A_0 is an initial amount present, then at a later time the amount present is $A_0 e^{-kt}$. Let us also note that in such a case the *rate* of loss is the derivative $kA_0 e^{-kt}$.

EXAMPLE 3

A fixed dose of 12 mg of a certain drug is injected into a patient's bloodstream at time $t = 0$. The drug enters an organ by leaving the bloodstream at the rate of 15% of the amount present in the blood. The organ in turn metabolizes 25% of the amount of drug it contains at any time. Model the amount of drug *in the organ* at time t.

Solution What we have here is one compartment leading into another, for the drug in the bloodstream is removed via the organ, as shown in Figure 8.48.

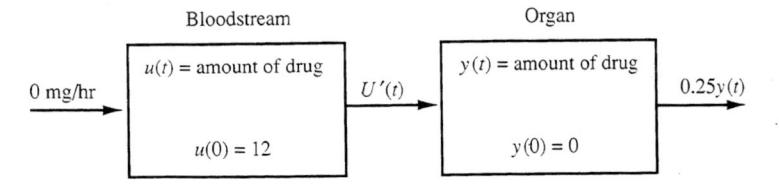

Figure 8.48

With $y(t)$ the amount of drug in the organ we have $y'(t) = RI - RO = RI - 0.25y(t)$ [in the second (organ) compartment].

The discussion preceding this example allows us to model the first (bloodstream) compartment and so determine the RI for the second compartment. For, if $u(t)$ is the amount of drug in the bloodstream, we have the IVP $u' = -0.15u$ and $u(0) = 12$, with the solution $u(t) = 12e^{-0.15t}$. Therefore the *rate* at which the drug is *leaving* the blood and (entering the organ) is $u'(t) = -12(0.15)e^{-0.15t} = -1.8e^{-0.15t}$.

Consequently, $y' = RI - 0.25y = +1.8e^{-0.15t} - 0.25y$. With $\alpha(t) = -0.25$ and $\beta(t) = 1.8e^{-0.15t}$ we then have $A(t) = \int \alpha(t)\, dt = -0.25t$. Hence

$$B(t) = \int \beta(t) e^{-A(t)}\, dt = \int 1.8 e^{-0.15t}(e^{+0.25t})\, dt$$

$$= 1.8 \int e^{0.10t}\, dt = 18 e^{0.10t}$$

According to Theorem 8.2, the amount of drug in the organ is

$$y(t) = (18 e^{0.10t} + C) e^{-0.25t}$$

Since $y(0) = 0$, $C = -18$ and

$$y(t) = 18(e^{-0.15t} - e^{-0.25t})$$

The graph of this solution is shown in Figure 8.49.

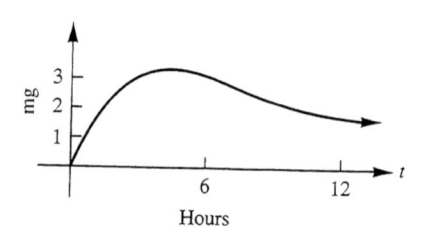

Figure 8.49

The general one-compartment mixing process always gives rise to the general first-order linear differential equation found in this section. Let V be the volume of the compartment, $F(t)$ the rate at which material is flowing through the compartment at time t, and $I(t)$ the rate at which a substance S is entering the compartment at time t (see Figure 8.50).

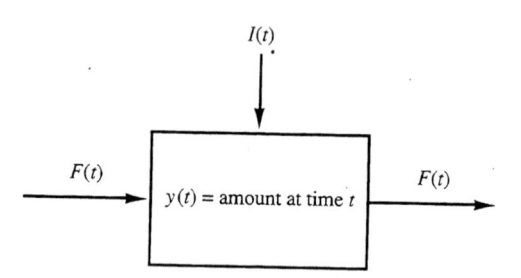

Figure 8.50

If $y(t)$ is the amount of substance S at time t, then $y(t)/V$ is the concentration of S per unit of volume. Therefore $[y(t)/V] \times F(t)$ is the rate at which S is leaving the compartment. Consequently,

$$y' = RI - RO = I(t) - \frac{y(t)}{V} F(t)$$

or

$$y' = \frac{-F(t)}{V} y + I(t)$$

500 Chapter 8 Differential Equations

If $F(t) = F$ happens to be constant, the equation is easy to solve, for $\alpha(t) = -F/V$, $\beta(t) = I(t)$, and therefore $B(t) = \int I(t)e^{(F/V)t}\,dt$ can often be found. If F or V is not constant, then this last integral may contain an exponential function of nonlinear degree in t and can be difficult to calculate. In such a case one can use the approximation methods of Section 8.6 to estimate the value of a solution. ∎

Exercises 8.7

Solve each differential equation in Exercises 1–8.
1. $y' = 2y + e^t$
2. $y' = y + e^t$
3. $y' = y + t$
4. $y' = y + (2 - t)$
5. $y' = ty$
6. $y' = y - t^2$
7. $y' = \frac{1}{t}y + t$
8. $y' = \frac{-1}{t^2}y + \frac{1}{t^3}$

Solve each IVP in Exercises 9–12.
9. $y' = y - t$, $y(0) = 3$
10. $y' = 2y + e^{3t}$, $y(0) = 1$
11. $ty' = y + t^3$, $y(1) = 5$
12. $y' = y - e^{-2t}$, $y(0) = 4$

13. Inflation may be considered as decay in the purchasing power of money. At an inflation rate of 6%, N dollars today has the purchasing power of $Ne^{-0.06t}$, t years from now. Suppose that Johnny B. Good deposits $800 into a savings account paying 9% interest and that the inflation rate is 6% per year.

   a. Model this savings plan by a differential equation, treating the inflation rate as a rate out of the compartment. Solve the equation. What is the real dollar value of the account 10 years later?

   b. Suppose that Johnny additionally deposits $500 a year into the account. Model and solve for this savings plan, determining the real dollar value 10 years later.

   c. Plan (b) is illusory in that $500 deposited t years from now has a purchasing power of $500e^{-0.06t}$. Regard this last figure as the real dollar rate at which money is deposited into the account, and model and solve this savings plan, determining the real dollar value 10 years later. How many actual dollars did Johnny put into the account over this 10-year period?

14. During the 120-day summer season, visitors canoeing down the Current River leave cans, bottles, Styrofoam™ containers, and other trash alongside the river at the rate of $360 - 0.10(t - 60)^2$ pieces per day. Five percent of such trash is removed by other more caring visitors each day. How many pieces of trash are left on the river at the end of the season?

15. A pregnant woman decides to have one glass of wine containing 18 mL of alcohol. Her body metabolizes the alcohol at the rate of 30% per hour. The baby in her womb absorbs alcohol from her body at the rate of 5% of the amount present and metabolizes the alcohol in its system at the rate of 20% per hour. How much alcohol has the baby consumed at the end of 2 hours? How much remains in the baby's system at the end of 2 hours? [Hint: If $m(t)$ is the amount of alcohol present in the mother at time t, then alcohol is consumed by the baby at the rate of $0.05m(t)$ at time t (see Example 3).]

16. This exercise asks you to recall Example 4, "Swords versus Plowshares," of Section 8.5A. Suppose that in that example the discovery of natural resources (oil, silver, and so on) allows the nation to have an input $\beta(t) = 100te^{-0.3t}$ of resources (billions of dollars per year) that rise and then decline over time. The differential equation then becomes (see Example 4) $R'(t) = 0.35P(t) - 0.25R(t) + \beta(t)$. Determine the nation's resources t years later if (a) $P(t) = 0.7R(t)$ and, alternatively, if (b) $W(t) = 0.3R(t)$. [Hint: You may need to reconsider the original equation $R' = 0.10P - 0.25W$.]

Challenge Problem

17. *A general logistic equation.* We have seen the importance of the logistic equation $y' =$

$ky(M - y) = kMy - ky^2$ many times. This equation can take the more general form of $y' = \alpha(t)y + \beta(t)y^2$. Bernoulli's method transforms this equation into a new equation that can be solved by the methods of this section. Here is an outline of Bernoulli's method.

a. Write $y(t) = 1/u(t)$. Show that, upon the substitution $y = 1/u$, the equation $y' = \alpha(t)y + \beta(t)y^2$ takes the form $u' = -\alpha(t)u - \beta(t)$.

b. Use Theorem 8.2 to write the solution of $u' = -\alpha(t)u - \beta(t)$.

c. Write $u = 1/y$ and use (b) to obtain the solution of the original (general logistic) equation.

d. Apply (a), (b), and (c) to find the solution of $y' = 0.05y - e^{0.6t}y^2$.

8.8 Two-Compartment Mixing and Linear Systems

In this section we consider mixing processes in which exchange occurs between two compartments at known rates. This leads to a linear system of differential equations.

In this section we consider the basic two-compartment mixing process in which each compartment contains distinct amounts $x(t)$ and $y(t)$ that feed into each other (see Figure 8.51). Examples abound, from the movement of currency back and forth across international boundaries to the process of soil nitrification in which the basic plant food element nitrogen moves back and forth between plant and soil.

A bit of reflection should convince you that it could easily happen that amounts in one compartment can rise while in the other they fall, and then vice versa. A predator–prey relationship, such as between foxes (one compartment) and rabbits (the second compartment), has this character. Such periodic rising and falling *cannot* be represented by the functions now at our disposal and demands the use of the trigonometric functions of Chapter 11 and the use of complex numbers $\alpha \pm \beta i$, where $i^2 = -1$. We will only consider the class of two-compartment processes in which periodic exchange does *not* occur.

Compartment 1 — $x(t) =$ amount present

Compartment 2 — $y(t) =$ amount present

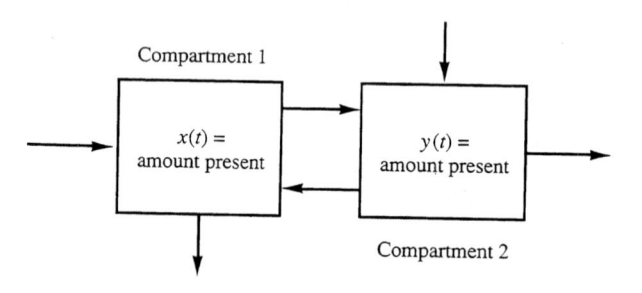

Figure 8.51

The Mathematical Form and Solution of a Two-Compartment Process

Let $x(t)$ and $y(t)$ be two quantities that are related in such a way that

$$x' = ax + by \qquad b \neq 0$$

and

$$y' = cx + dy$$

Two such equations taken together are called a **system of differential equations.** A solution to this system of equations consists of *two* functions $x(t)$ and $y(t)$ which, upon substitution in both expressions, yield equality.

EXAMPLE 1

Show that
$$x(t) = e^{-t} + e^{-2t} \quad \text{and} \quad y(t) = \frac{-e^{-t}}{2} - \frac{2e^{-2t}}{3}$$

are solutions to the linear system
$$x' = 2x + 6y$$
$$y' = -2x - 5y$$

Solution Consider the equation $x' = 2x + 6y$. On the left we have
$$x'(t) = D_t(e^{-t} + e^{-2t}) = -e^{-t} - 2e^{-2t}$$

On the right we have
$$2x + 6y = 2(e^{-t} + e^{-2t}) + 6\left(\frac{-e^{-t}}{2} - \frac{2e^{-2t}}{3}\right)$$
$$= 2e^{-t} - 3e^{-t} + 2e^{-2t} - 4e^{-2t}$$
$$= -e^{-t} - 2e^{-2t}$$

Therefore $x' = 2x + 6y$. The second equation can be verified similarly. ∎

We will not consider the lengthy mathematical reasoning needed to justify the following procedure for finding the solutions to a linear system, but will only state the procedure.

The Solution of a Linear System of Differential Equations with Constant Coefficients

In order to obtain the solutions to
$$x' = ax + by \quad\quad b \neq 0$$
$$y' = cx + dy$$
we must perform the following steps.

Step 1. Form the quadratic polynomial
$$z^2 - (a + d)z + (ad - bc) = 0$$

Step 2. Find the two roots r_1 and r_2 of this equation by factoring or by the quadratic formula.

8.8 Two-Compartment Mixing and Linear Systems   503

Step 3. If $r_1 \neq r_2$ and these roots are real numbers, the general solution to the linear system has the form

$$x(t) = C_1 e^{r_1 t} + C_2 e^{r_2 t}$$

and

$$y(t) = \frac{x'(t) - ax(t)}{b}$$

Step 4. If both roots are real and equal then with $r = r_1 = r_2$,

$$x(t) = (C_1 + C_2 t)e^{rt}$$

and

$$y(t) = \frac{x'(t) - ax(t)}{b}$$

Remark. If r_1 and r_2 are complex numbers $\alpha \pm \beta i$, the solution must be given in terms of trigonometric functions and can be found in more advanced texts.

Notice that the *two* given equations result in *two* unknown constants C_1 and C_2. This allows us to properly deal with two initial conditions $x(0)$ and $y(0)$.

EXAMPLE 2

Solve the linear system with initial values

$$x' = 2x - y \qquad x(0) = 1$$
$$y' = x - y \qquad y(0) = 2$$

Solution Here $a = 2$, $b = -1$, $c = 1$, and $d = -1$.

Step 1. We have $a + d = 1$ and $ad - bc = -1$. The needed quadratic is $z^2 - z - 1 = 0$.

Step 2. By the quadratic formula the two roots of this equation are

$$r_1 = \frac{1 + \sqrt{5}}{2} = 1.62 \quad \text{and} \quad r_2 = \frac{1 - \sqrt{5}}{2} = -0.62$$

Step 3. The solution for the unknown function $x(t)$ is

$$x(t) = C_1 e^{1.62t} + C_2 e^{-0.62t}$$

From the equation $x' = 2x - y$, we have $y = 2x - x'$ and

$$2x(t) - x'(t) = 2(C_1 e^{1.62t} + C_2 e^{-0.62t}) - 1.62 C_1 e^{1.62t} + 0.62 C_2 e^{-0.62t}$$

Therefore

$$y(t) = 0.38 C_1 e^{1.62t} + 2.62 C_2 e^{-0.62t}$$

We must now use the initial conditions to determine the constants C_1 and C_2. We have

$$1 = x(0) = C_1 + C_2$$

and

$$2 = y(0) = 0.38 C_1 + 2.62 C_2$$

From the first equation $C_1 = 1 - C_2$. In the second

$$2 = 0.38(1 - C_2) + 2.62C_2 = 0.38 + 2.24C_2$$

or
$$C_2 = \frac{1.62}{2.24} \simeq 0.72$$

Hence $C_1 \simeq 1 - 0.72 = 0.28$.

In conclusion, the solution to the given IVP is

$$x(t) = 0.28e^{1.62t} + 0.72e^{-0.62t}$$

and
$$y(t) = 0.11e^{1.62t} + 1.88e^{-0.62t} \quad \blacksquare$$

Remark. Note that determining C_1 and C_2 is equivalent to solving two algebraic equations in two unknowns. This may be done by any of several methods learned in algebra.

Both the small farmer and the agribusiness corporation must be concerned with the cost of growing the next crop. A part of this cost is the application of nitrogen fertilizer. The cost can be trimmed by knowledge of the process of soil nitrification whereby decomposition in the soil of the residue of the previous year's crop supplies nitrogen for the next year's crop. The next example is a simplified version of this process.

EXAMPLE 3

Consider the growing cycle on an acre of farmland. Suppose that growing plants take up nitrogen at the rate of 80% of the nitrogen in the soil during the growing season. The salable crop is harvested and the residue plowed back into the soil, returning nitrogen to the soil at the rate of 90% of the nitrogen left in the decaying plants (we assume the salable crop contains no nitrogen). During each month rainfall leaches away the soil's nitrogen at the rate of 5% of its level in the soil. If the soil initially contains 300 lb of nitrogen, and the plants none, model this system as a two-compartment mixing process and obtain the solution. How much nitrogen will be in the soil 1 year later? In the plants?

Solution We regard the soil as one compartment containing $x(t)$ lb of nitrogen and the growing crop as a second containing $y(t)$ lb of nitrogen. Initially, $x(0) = 300$ and $y(0) = 0$.

Figure 8.52 shows the **transfer coefficients** of nitrogen intake into both compartments and the rate out of the soil due to leaching. (**Remark.** These transfer coefficients are not readily found; see Exercise 18.) Consider first the rate $x'(t)$ of soil nitrification. We have

$$x'(t) = \text{Rate in} - \text{Rate out}$$
$$= 0.90y(t) - (0.80x(t) + 0.05x(t))$$

or
$$x' = -0.85x + 0.9y$$

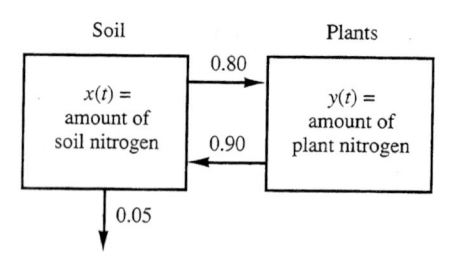

```
        Soil                      Plants
         0.80
   ┌──────────┐    →    ┌──────────┐
   │   x(t) = │         │   y(t) = │
   │ amount of│         │ amount of│
   │soil nitrogen│ 0.90 │plant nitrogen│
   └──────────┘    ←    └──────────┘
         │ 0.05
         ↓
```

Figure 8.52

Similarly, the rate of change of nitrogen in the plants is

$$y'(t) = RI - RO$$
$$= 0.80x(t) - 0.90y(t)$$

or
$$y' = 0.8x - 0.9y$$

We must now solve this linear system with $x(0) = 300$, $y(0) = 0$, $a = -0.85$, $b = 0.9$, $c = 0.8$, and $d = -0.9$.

Step 1. Since $a + d = -1.75$ and $ad - bc = 0.045$, the associated quadratic is

$$x^2 + 1.75z + 0.045 = 0$$

with solutions

$$z = \frac{-1.75 \pm \sqrt{3.063 - 0.18}}{2} = -0.03, -1.72$$

Step 2. $x(t) = C_1 e^{-0.03t} + C_2 e^{-1.72t}$

Step 3. $y(t) = \dfrac{x'(t) + 0.85x(t)}{0.90}$

$$= \frac{-0.03C_1 e^{-0.03t} - 1.72C_2 e^{-1.72t} + 0.85C_1 e^{-0.03t} + 0.85C_2 e^{-1.72t}}{0.90}$$

$$= 0.91 C_1 e^{-0.03t} - 0.97 C_2 e^{-1.72t}$$

Step 4. Therefore

$$x(t) = C_1 e^{-0.03t} + C_2 e^{-1.72t}$$

and
$$y(t) = 0.91 C_1 e^{-0.03t} - 0.97 C_2 e^{-1.72t}$$

Since $x(0) = 300$ and $y(0) = 0$, we obtain

$$300 = C_1 + C_2$$
$$0 = 0.91 C_1 - 0.97 C_2$$

Thus $C_1 = 1.01 C_2$ and $300 = 2.01 C_2$, and therefore

$$C_2 \simeq 149 \quad \text{and} \quad C_1 \simeq 151$$

by rounding off to unit coefficients.

Step 5. The solution is then
$$x(t) = 151e^{-0.03t} + 149e^{-1.72t}$$
and
$$y(t) = 137e^{-0.03t} - 145e^{-1.72t}$$

One year (12 months) later there will be $x(12) \simeq 105$ lb of nitrogen in the soil available for the next crop and $y(12) \simeq 86$ lb in decaying crop residue and fertilizer may be applied accordingly. The graphs of these solutions are shown in Figure 8.53.

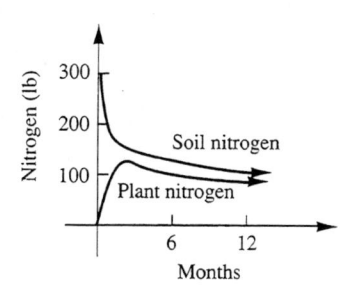

Figure 8.53

Fick's Law for Passive Diffusion

Some hint of how the solutions to linear compartment systems naturally take an exponential form can be obtained by a study of Fick's law for passive diffusion.

Figure 8.54 illustrates two compartments of volumes V_1 and V_2 containing differing amounts of the same substance at time t. The compartments are separated by a permeable membrane. In **passive diffusion,** substances flow through the membrane in both directions in direct proportion k to the *difference* in *concentration* of the substance on either side of the membrane (a bit like Newton's law of heating or cooling). Consequently, we obtain the equations

$$x' = \frac{k}{V_1}(x - y)$$

$$y' = \frac{k}{V_2}(y - x)$$

where x and y are the *concentrations* (amount divided by volume) in each compartment. [*Caution:* This means, for example, that $V_1 x(t)$ is the *amount* in the compartment at time t and $V_1 x'$ is the rate of change of this amount.]

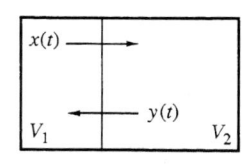

Figure 8.54

EXAMPLE 4

Solve the linear system

$$x' = \alpha(x - y)$$
$$y' = \beta(y - x)$$

directly.

Solution Let $u(t) = x(t) - y(t)$. Then $u' = x' - y'$. Subtracting the two equations we have

$$u' = x' - y' = \alpha x - \alpha y - \beta y + \beta x = (\alpha + \beta)(x - y) = (\alpha + \beta)u$$

That is, the function u satisfies the differential equation of exponential growth $u' = (\alpha + \beta)u$. Consequently, $u(t) = Ce^{(\alpha+\beta)t}$ Therefore

$$x(t) - y(t) = Ce^{(\alpha+\beta)t} \tag{1}$$

Substituting this in $x' = \alpha(x - y)$ we obtain

$$x' = \alpha Ce^{(\alpha+\beta)t}$$

Consequently,

$$x(t) = \int \alpha Ce^{(\alpha+\beta)t} \, dt = \frac{\alpha}{\alpha + \beta} Ce^{(\alpha+\beta)t} + C_1$$

Therefore, from Equation 1,

$$y(t) = \frac{\alpha}{\alpha + \beta} Ce^{(\alpha+\beta)t} + C_1 - Ce^{(\alpha+\beta)t}$$

or

$$y(t) = C_1 - \left(\frac{\beta}{\alpha + \beta}\right) Ce^{(\alpha+\beta)t} \quad \blacksquare$$

If we now apply this calculation to Fick's law for passive diffusion, with $\alpha = k/V_1$ and $\beta = k/V_2$, we obtain

$$x(t) = \left(\frac{V_2}{V_2 + V_1}\right) Ce^{k[(1/V_1)+(1/V_2)]t} + C_1$$

and

$$y(t) = C_1 - \left(\frac{V_1}{V_2 + V_1}\right) Ce^{k[(1/V_1)+(1/V_2)]t}$$

as the state equations for this system.

In this brief study we have not begun to touch on naturally related applications involving variable rates of exchange and external inputs, or external drains, on the two-compartment system. Such processes give rise to linear systems of the form

$$x'(t) = \alpha(t)x(t) + \beta(t)y(t) + \gamma(t)$$
$$y'(t) = \mu(t)x(t) + \mu(t)y(t) + \sigma(t)$$

Methods of solution can be found in advanced texts dedicated to the subject of ordinary differential equations.

Exercises 8.8

1. Show that the equation $y' = -2x - 5y$ of Example 1 is satisfied by the given functions.

2. Show that the two functions given in step 5 of Example 3 satisfy the given equations.

Solve the systems of linear differential equations in Exercises 3–8.

3. $x' = 2x - y$
   $y' = -x + 2y$

4. $x' = -x + y$
   $y' = x - y$

5. $x' = -2x + 4y$
   $y' = x + y$

6. $u' = \dfrac{u + v}{2}$
   $v' = \dfrac{u - v}{2}$

7. $x' = 2x + y$
   $y' = x + 2y$

8. $x' = x - y$
   $y' = 3(y - x)$

Solve the systems with given initial values in Exercises 9–12.

9. $x' = -y$
   $y' = x + 2y$
   $x(0) = 0$ and $y(0) = 1$

10. $2x' = 4x - 6y$
    $y' = -x$
    $x(0) = -1$ and $y(0) = 0$

11. $x' = \sqrt{3}\, x + \sqrt{3}\, y$
    $y' = \dfrac{x}{2} - y$
    $x(0) = 1$ and $y(0) = 1$

12. $y' = x - 3y$
    $x' = -x + y$
    $x(0) = 0$ and $y(0) = 0$

13. When Americans buy products manufactured abroad, U.S. dollars leave the domestic economy and enter the international economy. These dollars are in turn attracted back into the domestic economy by the sale of American products abroad as well as by higher interest rates paid on deposits in U.S. banks. In this way, high interest rates can be used to finance a balance of payments deficit to the international economy without immediate harm to the domestic economy. Use the two-compartment model in Figure 8.55 with indicated transfer rates to trace the distribution of a newly minted $1 billion of U.S. currency. What is $\lim_{t \to \infty} x(t)$?

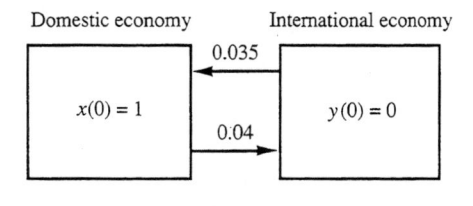

Figure 8.55

14. When a certain drug is consumed, its ingredients enter the bloodstream first. A large portion is eliminated rather quickly through the urine, but another portion is stored as body fat, which in turn feeds back into the blood system. Assume the transfer and elimination rates shown in Figure 8.56 and determine the amount of the substance still in the body 12 hours later of an initial amount of 75 mg.

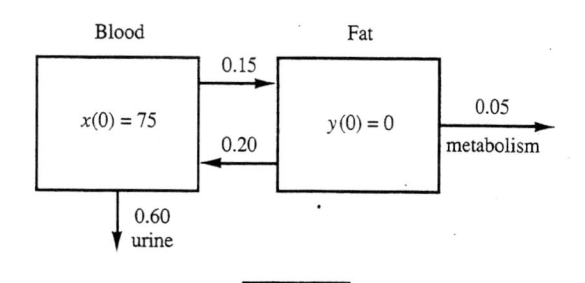

Figure 8.56

15. Two key components of the kidneys are collecting tubules and paratubular capillaries. Blood is filtered between these structures in order to purify it. This filtering is controlled by antidiuretic hormone (ADH), which is produced in precise consistent amounts each time the body has excreted the ADH previously produced.

    Let $x(t)$ and $y(t)$ be the amount (in mg) of ADH in the tubules and capillaries, respectively, with $x(0) = 10$ and $y(0) = 0$. Given the rate of liquid flow through the system as indicated in Figure 8.57, find a formula for $x(t)$ and $y(t)$, t days later. When is $x(t) = 0$?

8.8 Two-Compartment Mixing and Linear Systems

Collecting tubules   Paratubular capillaries

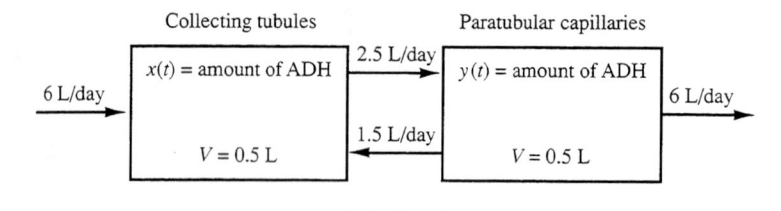

$x(t)$ = amount of ADH   2.5 L/day $y(t)$ = amount of ADH

6 L/day   6 L/day

1.5 L/day

$V = 0.5$ L $V = 0.5$ L

Figure 8.57

16. Two adjacent lakes in the Snowy Range of southeastern Wyoming are joined by a small stream. One lake has three times the volume of the other. The Wildlife and Fisheries Department stocks the larger lake with 20,000 brown trout. What does Fick's law, applied to the migration of trout between the two lakes, predict about the ultimate distribution of these 20,000 trout between the two lakes? [*Hint:* Let V be the volume of the smaller lake so that $3V$ is the volume of the larger lake.]

17. Consider a system consisting of two tanks connected as in Figure 8.58 with the indicated rates of flow and indicated volumes. At time $t = 0$, when liquid begins flowing through this system, 50 g of a red dye is deposited in the 40-liter tank and none is present in the 30-liter tank. Find the amount of dye in each tank t minutes later.

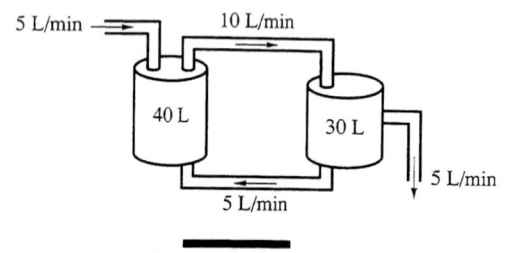

5 L/min   10 L/min

40 L   30 L

5 L/min

5 L/min

Figure 8.58

Challenge Problems

In Exercise 17, the rate of transfer (*transfer coefficients*) between two compartments is a known quantity. In some applications the transfer coefficients themselves are the main interest, but difficult to measure, while the amount present in each compartment can be measured intermittently. In such a case these measurements yield data from which one can obtain a formula for $x(t)$ and $y(t)$, the amounts in each compartment at time t. Such information can be used to determine the transfer coefficients. The next two exercises indicate how this can be done.

18. Seven mg of a harmless radioactive ion has been injected into the bloodstream of a patient. This ion moves back and forth between two compartments (denoted by I and II) of the blood at unknown rates r_1 and r_2 (see Figure 8.59).

I   II

$x(t)$ = amount of ion r_1

$x(t) = a + be^{kt}$ $y(t)$ = amount of ion

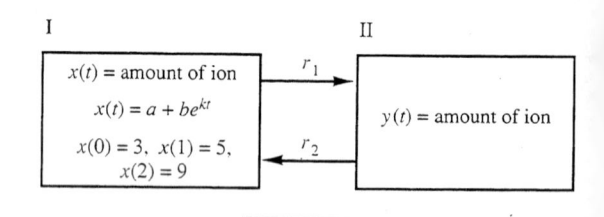

$x(0) = 3$, $x(1) = 5$, r_2

$x(2) = 9$

Figure 8.59

If $x(t)$ is the amount of ion in compartment I, it is known that $x(t)$ has the form $x(t) = a + be^{kt}$. At times $t = 0$, 1, and 2 minutes, a blood sample is taken and the amount of radioactive ion in compartment I is determined. The data $x(0) = 3$, $x(1) = 5$, and $x(2) = 9$ are obtained. Determine the values of a, b, and k. [*Hint:* Use the data to write three equations in the three unknowns a, b, and k. Determine b in terms of a first. Recall that $e^{2k} = (e^k)^2$.]

19. Use the values of a, b, and k found in Exercise 18 and Example 4 to determine the transfer coefficients r_1 and r_2, assuming that $x(0) = 5$ in compartment I. [*Hint:* $x' = r_2 y - r_1 x$; $y' = r_1 x - r_2 y$.]

Chapter 8 Summary

1. The three basic *differential equations*—$y'/y = g(t)$, $y' = k(M - y)$, and $y' = ky(M - y)$—and their solutions illustrate the major points of this chapter. Their solutions are, respectively:

$$y(t) = e^{\int g(t)} \qquad y(t) = M - Ce^{-kt} \qquad y(t) = \frac{M}{1 + Ce^{-Mkt}}$$

   The first equation models the percentage rate of growth of y; the second is a generalization of *Newton's law of heating or cooling* and models growth in proportion to remaining capacity; and the third is the *logistic equation*, modeling percentage rate of growth in proportion to the percentage of remaining capacity.

2. The solutions of a *time-autonomous differential equation* $y' = g(y)$ can often be graphed without explicitly solving the equation via the use of an auxiliary graph $z = g(y)$, which illustrates the slope (direction) of a solution at any point.

3. An equation in the form $y' = p(t)q(t)$ can often be solved by separating the variables as

$$\frac{y'}{q(y)} = p(t) \qquad q(y) \neq 0$$

   and integrating. Be sure to include constant solutions $y(t) = A$, where $g(A) = 0$, as well.

4. Compartment-mixing analysis of systems subject to constant rates of growth and/or decay, and constant inputs and/or depletions, gives rise to differential equations in the form

$$y' = ay + b$$

   with solutions $y(t) = (-b/a) + Ce^{at}$, whose graphs become either asymptotic to the constant solution $y(t) = -b/a$ or diverge in either direction from it.

5. If $y' = g(t, y)$, $y(t_0) = y_0$ cannot be solved exactly, *Euler's method*, based on the repetitive use of the equation $y_{k+1} = y_k + g(t_k, y_k)h$, can be used to approximate a value $y(T)$ of the solution [where $h = (T - t_0)/n$ and n is the number of repetitions].

6. In many applications, the differential equation $y' = \alpha(t)y + \beta(t)$ may be solved and yields an explicit model of a compartment-mixing system with *variable* rates in and out.

7. The most basic two-compartment mixing system is modeled by a system $x' = ax + by$ and $y' = cx + dy$ of differential equations whose solution is found via the roots of the algebraic equation $z^2 - (a + d)z + (ad - bc) = 0$.

Chapter 8 Summary Exercises

1. Show that $y(t) = 2e^t - t - 3$ is the solution to the IVP
$$y' = y + t + 2 \qquad y(1) = 2(e-1)$$

2. Show that $y(t) = e^{e^{2t}-1}$ is the solution to the IVP
$$y' = 2e^{2t}y \qquad y(0) = 1$$

3. Determine C so that $y(t) = 1 - Ce^{2t}$ is the solution of
$$y' = 2(y - 1) \qquad y(0) = 7$$

4. Show that $y(t) = 1 + Ce^{kt}$ satisfies $y' = k(y - 1)$.

5. Solve the IVP $y'/y = -t$, $y(0) = 2$.

6. State Newton's law of heating or cooling as a differential equation.

7. In a population of 500 individuals, a contagious infection is spreading in such a way that the percentage rate of growth of new infections is directly proportional to the percentage of individuals not yet affected. Express this statement as a differential equation. What is the name applied to such an equation?

Graph all constant solutions and the generic solutions to the differential equations in Exercises 8–11.

8. $y' = \frac{1}{3}y$
9. $u' = 2(1 - u)$
10. $y' = y(2 + y)$
11. $y' = (y + 1)(y + 2)$

12. What role do constant solutions to a differential equation play? How are they related, if at all, to the unknown constants that arise in the solution of a differential equation?

Solve the IVP's in Exercises 13–18.

13. $y' = \frac{y}{t^2}$; $y(1) = 1$
14. $y' = te^{-y}$; $y(0) = 0$
15. $y' = 7 - y$; $y(0) = 6$
16. $y' = t(y - 2)$; $y(0) = 2$
17. $ty' = t + 1$; $y(1) = 2$
18. $uu' = 3$; $u(-1) = -2$

19. Discuss the logistic differential equation and its solutions and applications.

20. Find all solutions to $y' = (1 - y)(1 - t)$.
21. Find all solutions to $y' = (t + 1)/y$.

22. If $z = g(y)$ has the graph in Figure 8.60, graph all solutions to $y' = g(y)$.

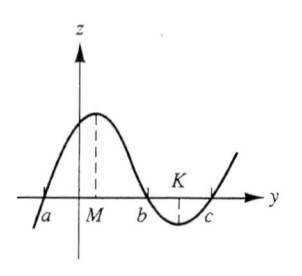

Figure 8.60

23. What must the monthly payment be so as to pay back a loan of $1,500 in 36 months at an interest rate of $1\frac{1}{2}\%$ per month? Graph the function representing the amount still owed on this loan at time t (months).

24. You are able to borrow money at 1% interest per month and can afford a monthly payment of $50 per month for the next 3 years. How much can you afford to borrow?

25. If annual withdrawals of $12,000 are made from an account initially containing $60,000 and earning 10% interest, when will there be only $24,000 left in the account?

26. Water is flowing through a 30-gal tank at the rate of 5 gal per half-hour. If 4 gal of salt are put into the tank at time $t = 0$, when will only 1 gal of salt remain?

27. Water is flowing through a 50-L mixing tank at the rate of 10 L per minute. If salt is added to the water entering the tank at a concentration of 30 g/L, when will the concentration of salt in the tank reach a concentration of 25 g/L, assuming that the tank is initially free of salt?

28. A patient metabolizes a certain drug at the ratio of 25% of the amount present per hour. At what rate should the drug be prescribed so as to reach a level of 40 mg within 5 hours?

29. Use Euler's method with $n = 4$ steps to approximate the solution of the IVP $y' = y - t$, $y(1) = 1$, at $T = 3$.

30. Use Euler's method with $n = 3$ steps to approximate the solution of the IVP $y' = e^{-t^2}$, $y(0) = 2$, at $T = 1$.

Solve the IVP's in Exercises 31 and 32.

31. $y' = \frac{t}{2}y + t^3$, $y(0) = 1$

32. $y' = 3t^2 y + t^2$, $y(1) = 0$

33. Suppose that in a lake of volume V, precipitation equals evaporation over the course of a year. Assume that a particular kind of pollutant has entered the lake via rainfall at the rate of 50 tons per year for 20 years. Assume that the pollutant thoroughly mixes in the water and then is lost with water through evaporation. Regard rainfall and evaporation as a constant flow rate F through the lake and suppose that $F/V = 0.03$. Let $A(t)$ be the amount of pollutant in the lake in year t, with $A(0) = 0$. Answer (a)–(e) under these assumptions.

  a. Explain why $A'(t) = 50 - (F/V)A(t)$.

  b. How many tons of pollutant are in the lake at the end of 20 years?

  c. Suppose that a pollution control program is begun at that time and pollutants enter the lake at the rate of $50 - 0.05t$ tons per year for the next 10 years. Model the amount of pollutants in the lake at time t.

  d. How many tons of pollutant are in the lake at the end of 10 or more years?

  e. Suppose that, at that point, no further pollutants enter the lake. How many years will it take for the amount of pollutants in the lake to reach the level of 50% of the amount found in (b)?

Solve the systems of differential equations in Exercises 34 and 35.

34. $x' = \frac{x}{2} + y$

$y' = x + \frac{y}{2}$

35. $u' = \frac{u+v}{2}$

$v' = \frac{u-v}{2}$

Solve the systems with given initial values in Exercises 36 and 37.

36. $x' = x - y$
$y' = y$
$x(0) = 0$ and $y(0) = 1$

37. $x' = 25x + 3y$
$y' = x + 23y$
$x(0) = -1$ and $y(0) = -1$

38. Glucose is the basic molecule of metabolism and is found in the blood and lymph systems of the body in the normal amounts of 7,000 mg and 3,000 mg, respectively. Additionally, glucose in either system is used throughout the day for energy, and what is unused is stored as adipose tissue in the body, at the rate of 1 lb per 0.85 kg of glucose. In a well-balanced diet little excess glucose remains for such storage. The cells of the body are supplied with glucose from both the blood and lymph systems. The diagram in Figure 8.61 shows the transfer rates of glucose between the blood and lymph systems and the rates per day at which glucose leaves each system for use by the cells.

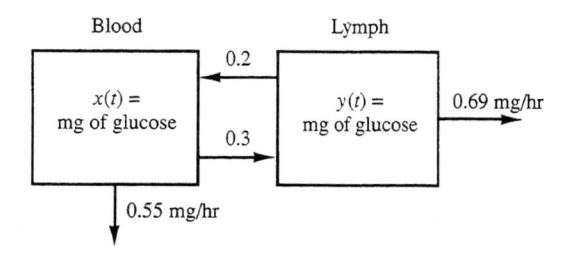

Figure 8.61

Suppose that a meal has been eaten that supplies the blood system with 1,000,000 mg of glucose. Thus $x(0) = 1,007,000$ and $y(0) = 3,000$. Determine the amount of glucose in the blood and lymph systems t days later. How much of this meal will be stored as adipose tissue 2 days later? [*Hint:* Find $[x(2) - 7,000] + [y(2) - 3,000]$.]

Chapter 9

Probability

Continuously Distributed Random Variables

The exact moment that you arrive for a 10 A.M. class is a number that cannot be predicted in advance and is subject to random influences, such as crowding in the hallways as you approach the classroom or friends you encounter on your way. Moreover, your arrival time can be any one of infinitely many values in the continuum of time from, say, 9:55 A.M. to 10:00 A.M. Calculus offers us the simplest model for computing the probability that such a *continuously distributed random variable* has any particular value. The Fundamental Theorem of Calculus tells us precisely how such probabilities are related to the frequency with which an event occurs. Consequently, a manufacturer can use the calculus of probability to determine the expected lifetime of a product, as well as the probability that the product will differ from its expected lifetime by a chosen amount, thus allowing the manufacturer to offer a competitive warranty for the product that will not lead to economic ruin in the long run. And the manufacturer can be as sure of this outcome as you are of the most likely time you will arrive for class, despite unpredictable encounters on your journey.

9.1 Improper Integrals

The purpose of this section is to define and calculate an integral over an unbounded interval.

Before we study the calculus of probability, we must learn to compute an integral over an interval of infinite length. Such an integral, called an **improper integral,** is needed in the study of probability for the following reason: If we are uncertain when an event may occur, we must use a model that allows the event to occur in an unbounded, or infinitely long, interval.

Each cell in your body, whose area is about 0.0003 cm^2, contains an identical strand of DNA that is itself about 200 cm long! Thus a very small two-dimensional area can contain a very long one-dimensional line. In fact, a region of finite area can contain an infinitely long line.

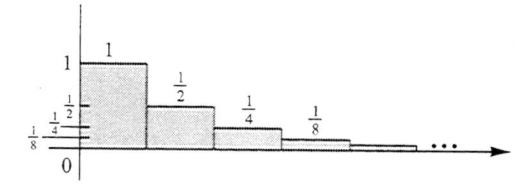

Figure 9.1

Consider Figure 9.1, where we draw an unending series of rectangles, all one unit long, but each successive rectangle only half as high as the previous rectangle. It is a surprising fact, at first, that the total area of the indicated shaded region is finite, even though the region itself is unending. Such a surprising outcome has real application, particularly to statistics and probability. But let us first consider a simple argument that "shows" that this shaded region indeed has finite area.

Suppose you walk across a 10-ft-wide room at 5 ft/sec. It will take you 2 sec to cross the room. On the other hand it is equally true that it will take you 1 sec to cross the first half of the room, plus $\frac{1}{2}$ sec to cross half of the remaining half of the room, plus an additional $\frac{1}{4}$ sec to cross half of the (yet) remaining half, and so on. Figure 9.2 shows a graph representing the amount of time it takes you to cross the room in "half-stages."

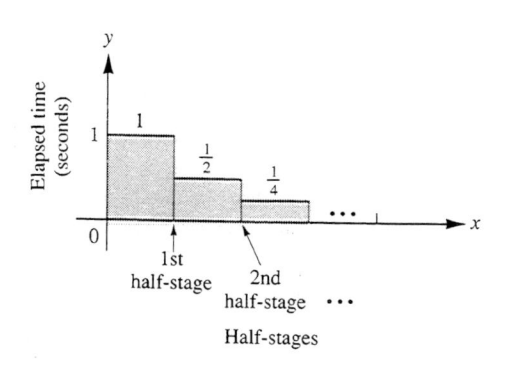

Figure 9.2

The area of the shaded region coincides exactly with the amount of time it takes you to cross the room in half-stages: $1 + \frac{1}{2} + \frac{1}{4} + \frac{1}{8} + \ldots$. Since the total time to cross the room is 2 sec, the area of the shaded region in Figure 9.2 (and therefore also in Figure 9.1) is 2 units. This unending region has finite area!

Consider now the graph of $y = 2^{-x}$, shown in Figure 9.3. Since $2^{-x} = 1/2^x \leq 1/2^n$ for $x \geq n$, this graph lies *within* the shaded region carried over from Figure 9.2. It follows that the area beneath the graph of $y = 2^{-x}$, from $x = 0$ and continuing without end, is also finite, since it is smaller than the shaded region having an area of 2 units.

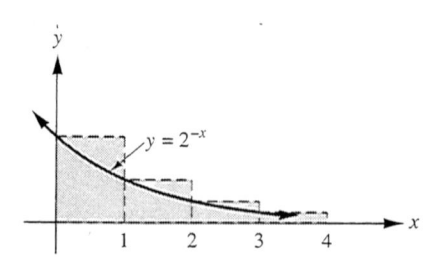

Figure 9.3

We use an *improper integral* to compute the area of such an unbounded region. The area beneath the graph of $y = 2^{-x}$ over the unending interval to the right beginning at 0 is denoted by

$$\int_0^\infty 2^{-x}\, dx$$

The general definition of an improper integral also tells us how to find the numerical value of an improper integral.

Definition

Let f be a continuous *nonnegative* function defined for all numbers $x \geq a$, where a is some fixed number. We define the improper integral of f over the interval $[a, \infty)$, denoted by $\int_a^\infty f(x)\, dx$, by the equation

$$\int_a^\infty f(x)\, dx = \lim_{b \to \infty} \int_a^b f(x)\, dx$$

when this limit exists as a real number. If this limit does not exist, the integral is said to *diverge*.

That is, the improper integral is the limit of a sequence of *definite* integrals as the upper endpoint is allowed to increase without bound.

EXAMPLE 1

Calculate $\int_0^\infty e^{-x}\, dx$, which measures the area beneath the entire graph of $y = e^{-x}$ to the right of $x = 0$.

Solution According to the definition,

$$\int_0^\infty e^{-x}\,dx = \lim_{b\to\infty} \int_0^b e^{-x}\,dx$$

To make the calculation on the right, we have to do two things in a specific order: (1) Calculate the definite integral $\int_0^b e^{-x}\,dx$, and *then* (2) evaluate the limit of this calculation as b increases without bound. We have:

Step 1. $\qquad \int_0^b e^{-x}\,dx = -e^{-x}\Big]_0^b = -e^{-b} - (-e^{-0}) = 1 - e^{-b}$

Therefore

Step 2. $\qquad \int_0^\infty e^{-x}\,dx = \lim_{b\to\infty}\int_0^b e^{-x}\,dx$

$\qquad\qquad\qquad\qquad = \lim_{b\to\infty}(1 - e^{-b})$

$\qquad\qquad\qquad\qquad = 1 - \lim_{b\to\infty}\frac{1}{e^b} = 1 - 0$

$\qquad\qquad\qquad\qquad = 1 \blacksquare$

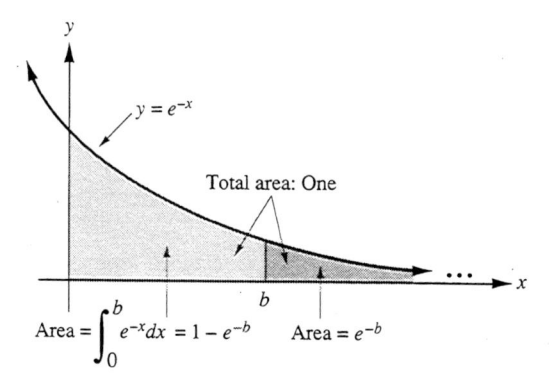

Figure 9.4

Figure 9.4 helps us visualize the process of calculating an improper integral. We approximate the area of the unending region below the graph by the area below this graph to point b. We then let b "approach infinity" to make the calculation. In step 1, we obtain an answer of $1 - e^{-b}$ for the area of the region up to b. In step 2, we find that the area of the total unending region is 1. (In essence, then, the term e^{-b}, which appears in the calculation, measures the area beneath the graph to the right of the number b.)

The next example shows that not just *any* unending region has a finite area.

EXAMPLE 2

Show that the improper integral $\int_1^\infty (1/x)\,dx$ does not exist, or *diverges*. That is, show that the area beneath the graph of $y = 1/x$ to the right of $x = 1$ cannot be finite.

9.1 Improper Integrals   517

Solution We indicate the region whose area we are to calculate in Figure 9.5(a).

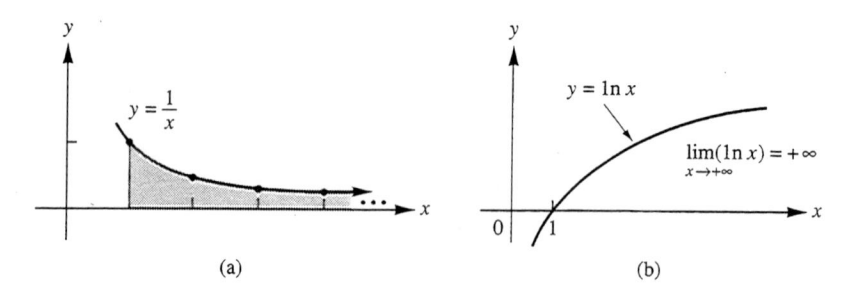

Figure 9.5

(a)   (b)

According to the definition,

$$\int_1^\infty \frac{1}{x}\,dx = \lim_{b\to\infty} \int_1^b \frac{1}{x}\,dx$$

$$= \lim_{b\to\infty} (\ln x]_1^b$$

$$= \lim_{b\to\infty} (\ln b - \ln 1)$$

$$= \lim_{b\to\infty} \ln b \quad \text{Since } \ln 1 = 0$$

Therefore $\quad \displaystyle\int_1^\infty \frac{1}{x}\,dx = \lim_{b\to\infty} \ln b = \infty \quad$ See Figure 9.5(b)

since $y = \ln x$ becomes larger and larger as x increases without bound. ∎

The next example considers an integral whose *form* is frequently encountered. The closing steps in its evaluation demand special attention.

EXAMPLE 3

Evaluate $\int_0^\infty xe^{-x}\,dx$.

Solution By definition, $\displaystyle\int_0^\infty xe^{-x}\,dx = \lim_{b\to\infty} \int_0^b xe^{-x}\,dx.$

As before we have

Step 1.

$$\int xe^{-x}\,dx = \int x\,d(-e^{-x}) = -xe^{-x} - \int -e^{-x}\,d(x)$$

$$= -xe^{-x} + \int e^{-x}\,dx = -xe^{-x} - e^{-x} + C \quad \text{By the method of integration by parts}$$

Therefore

$$\int_0^b xe^{-x}\,dx = -xe^{-x} - e^{-x}\Big]_0^b = (-be^{-b} - e^{-b}) - (-0 \cdot e^{-0} - e^{-0})$$

$$= 1 - e^{-b} - be^{-b}$$

We now move to
Step 2.

$$\int_0^\infty xe^{-x}\,dx = \lim_{b\to\infty}(1 - e^{-b} - be^{-b})$$

$$= 1 - \lim_{b\to\infty}\frac{1}{e^b} - \lim_{b\to\infty}\frac{b}{e^b}$$

Since e^b increases without bound as $b\to\infty$, $\lim_{b\to\infty}(1/e^b) = 0$. Turning to $\lim_{b\to\infty} b/e^b$, we see that both numerator and denominator approach infinity. Nonetheless the graph in Figure 9.6 (which you may recall from Example 9, Section 5.2) tells us that $\lim_{b\to\infty} b/e^b = 0$. Combining these conclusions, we have $\int_0^\infty xe^{-x}\,dx = 1 - 0 - 0 = 1$.

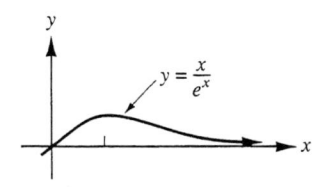

Figure 9.6

Remark. L'Hôpital's rule (Section A.2) can be used to show that

$$\lim_{x\to\infty}\frac{x^n}{e^x} = 0 \qquad \text{for any number } n > 0$$

since, according to L'Hôpital's rule,

$$\lim_{x\to\infty}\frac{x^n}{e^x} = \lim_{x\to\infty}\frac{nx^{n-1}}{e^x} = \ldots = \lim_{x\to\infty}\frac{n!}{e^x} = 0$$

Our next example shows how an improper integral is used to measure a long-term trend.

EXAMPLE 4

A business that is planning to introduce a new product into the market is interested in predicting the number of long-term "faithful" customers for the new product. The product will be introduced with a heavy advertising campaign that will continue until the rate of new customers for the product begins to decline. After that the reputation of the product will be relied on to attract new customers. Marketing surveys predict the following:

1. Advertising and the reputation of the product will attract new customers to the product at the rate of $15{,}000te^{-t}$ new customers per week during week t.

2. Of those new customers who try the product, only a certain proportion can be expected to remain customers. It is found that the later a customer tries the product, the less likely it is that he or she will be a long-term customer. The marketing survey predicts that, of those new customers who try the product in week t, only the proportion $e^{-0.30t}$ will remain faithful customers of the product in the long run.

How many customers can be expected to remain faithful to the product in the long run?

Solution Let T be some fixed time in the future. Imagine a point in time t between 0 and T and a short time span Δt about t. The number of new customers over this time span will be $15{,}000te^{-t}\Delta t$. Of these, only $e^{-0.30t}[(15{,}000te^{-t})\Delta t]$ will remain faithful customers, and the *sum* of all such terms, for all such intervals of length Δt between 0 and T, is a Riemann sum for $f(t) = 15{,}000te^{-1.3t}$, approximating the *total* number of faithful customers. Applying the approximation principle (AP) for the definite integral (Section 4.5), we have that the total number of faithful customers attracted to the product over the time span $[0, T]$ is then

$$\int_0^T 15{,}000te^{-1.3t}\, dt$$

The long-term trend, the number of faithful customers obtained over the long run, is measured by

$$\lim_{T\to\infty} \int_0^T 15{,}000te^{-1.3t}\, dt = \int_0^\infty 15{,}000te^{-1.3t}\, dt$$

Using integration by parts, we have

$$\int_0^T 15{,}000te^{-1.3t}\, dt = 15{,}000\left(\frac{-te^{-1.3t}}{1.3} - \frac{e^{-1.3t}}{(1.3)^2}\right)\bigg|_0^T$$

$$= \frac{15{,}000}{1.3}\left[-Te^{-1.3T} - \frac{e^{-1.3T}}{1.3} + \frac{1}{1.3}\right] \qquad (1)$$

Consequently, the total number of long-term customers is

$$\int_0^\infty 15{,}000te^{-1.3t}\, dt = \frac{15{,}000}{(1.3)^2} \approx 8{,}875.74 \quad \blacksquare$$

A graph of the arrival rate function, shown in Figure 9.7, indicates that the peak arrival rate is $15{,}000/e$, with a retention of $e^{-0.3}(15{,}000/e) \approx 4{,}088$ faithful customers from that point. Notice from Equation 1 that the numbers 15,000 and 0.30 essentially determine the final calculation, because all other terms vanish when the limit is taken. That is, the long-term trend is closely governed by the peak rate and the retention rate constant (0.30)—and nothing more. In such applications, the improper integral is said to calculate the **steady state** of the system—the state that the system will settle down to in the "long run."

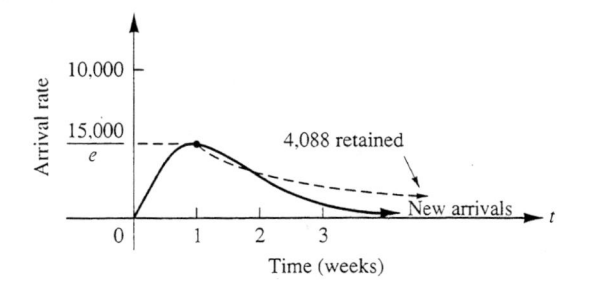

Figure 9.7

Remarks. Although we will not emphasize it, you may also wish to use the improper integral of a nonnegative function *up to* a point a (see Figure 9.8), defined as follows:

$$\int_{-\infty}^{a} f(x)\, dx = \lim_{b \to -\infty} \int_{b}^{a} f(x)\, dx$$

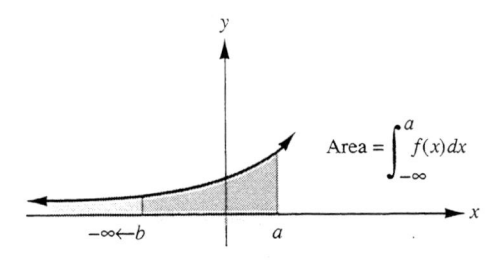

Figure 9.8

We can now define the improper integral of a nonnegative function over the entire real line by

$$\int_{-\infty}^{\infty} f(x)\, dx = \int_{-\infty}^{0} f(x)\, dx + \int_{0}^{\infty} f(x)\, dx$$

Exercises 9.1

Evaluate the limits in Exercises 1–10.

1. $\lim_{b \to \infty} \dfrac{1}{b^2}$

2. $\lim_{b \to \infty} \dfrac{1}{\sqrt{b}}$

3. $\lim_{b \to \infty} \dfrac{1}{\ln b}$

4. $\lim_{b \to \infty} e^{1/b}$

5. $\lim_{b \to \infty} \ln\left(\dfrac{1}{b}\right)$

6. $\lim_{b \to \infty} \dfrac{1}{b} - e^{-b^2}$

7. $\lim_{b \to \infty} b + e^{-b}$

8. $\lim_{b \to \infty} \dfrac{b^3}{e^b} + 2e^{1/b}$

9. $\lim_{b \to \infty} \ln(\ln b)$

10. $\lim_{b \to \infty} 3\left(1 - \dfrac{1}{\sqrt{b+1}}\right)$

11. Find the area beneath the graph of $y = 1/x^2$ for $x \geq 1$.

12. Find the area beneath the graph of $y = 1/x^3$ for $x \geq 1$.

13. Compare the values of the following three improper integrals.

   a. $\displaystyle\int_{1}^{\infty} \dfrac{1}{x^{0.999}}\, dx$ b. $\displaystyle\int_{1}^{\infty} \dfrac{1}{x}\, dx$ c. $\displaystyle\int_{1}^{\infty} \dfrac{1}{x^{1.001}}\, dx$

14. Find the area beneath the graph of $y = 1/(2x + 1)^{3/2}$ for $x \geq 0$.

15. Find the area beneath the graph of $y = 2xe^{-x/2}$ for $x \geq 0$.

16. a. Find the area *between* the graphs of $y = e^{-x}$ and $y = xe^{-x}$ for $x \geq 0$.

    b. What is the smallest value of x for which $xe^{-x} \geq e^{-x}$?

Evaluate the improper integrals in Exercises 17–34.

17. $\int_3^\infty \frac{1}{x^2}\, dx$

18. $\int_0^\infty \frac{x}{x^2 + 1}\, dx$

19. $\int_1^\infty e^{-x/2}\, dx$

20. $\int_1^\infty e^{-0.0001x}\, dx$

21. $\int_0^\infty e^{4-x}\, dx$

22. $\int_1^\infty xe^{-x^2}\, dx$

23. $\int_{-5}^\infty 3e^{-2x}\, dx$

24. $\int_1^\infty \frac{1}{(1-2x)^2}\, dx$

25. $\int_1^\infty x\, dx$

26. $\int_0^\infty 2\, dx$

27. $\int_0^\infty xe^{-2x}\, dx$

28. $\int_0^\infty x^3 e^{-x/3}\, dx$

29. $\int_1^\infty \frac{\ln x}{x}\, dx$

30. $\int_1^\infty \frac{n}{x^{n+1}}\, dx,\ n > 0$

31. $\int_0^\infty \alpha e^{-\alpha x}\, dx,\ \alpha > 0$

32. $\int_0^\infty x^2 e^{-x^3}\, dx$

33. $\int_{-\infty}^1 e^x\, dx$

34. $\int_{-\infty}^2 \frac{1}{(x-3)^2}\, dx$

35. Notice that the area beneath the graph of $y = 1/x$, $x \geq 1$, is not finite. How could you attempt to calculate the volume of the figure obtained by revolving the graph of $y = 1/x$ about the x-axis for $x \geq 1$? What answer do you obtain? [*Hint:* See Section 4.5.]

36. Consider the rectangle in Figure 9.9 of finite area (2 units).

    a. Draw a sequence of connected straight lines that lies inside this rectangle and that is 3 units long; 6 units long; 7 units long; 100 units long. [*Hint:* The sequence of straight lines indicated in Figure 9.9(b) is 5 units long.]

    b. Is there any apparent limit on how long a sequence of straight lines can be drawn inside this rectangle? What is the *area* of a line? What is the area of the sequence of lines you drew in each of the parts of (a)?

    c. Indicate how this rectangle could be "cut up" into infinitely many pieces, each 1 unit long, and these pieces then placed end-to-end so as to make up an infinitely long region of the same finite area.

37. Assuming that the graph of the function in Figure 9.10 continues the indicated pattern "out to infinity," what can you conclude about $\lim_{b \to \infty} \int_0^b f(x)\, dx$? [*Hint:* Consider when b is an even whole number and then when b is an odd whole number.]

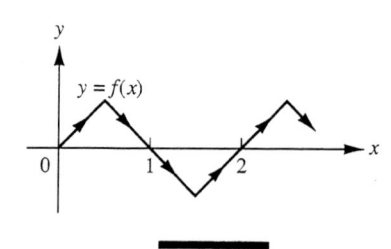

Figure 9.10

38. A new product that has no customers initially is attracting 12,000 new customers a month. However, t months later only $e^{-0.04t}$ of these remain customers for the product.

    a. Of those new customers arriving over an interval of time Δt about time t, how many are still customers T years later?

    b. Of new customers arriving in the interval of time $[0, T]$, how many are still customers at time T? Compute this value when $T = 60$ months.

    c. What is the long-run expectation of faithful customers (the steady state of sales for this product)?

39. The Boom-and-Bust Gold Mining Company is a slipshod operation run by a former oil wildcatter who locates new lodes of gold at the rate of $50t$ lb of gold per month. However, only $e^{-0.50t}$ of the newly discovered gold ever makes it to sale. The rest is lost, stolen, or has its location misplaced since the

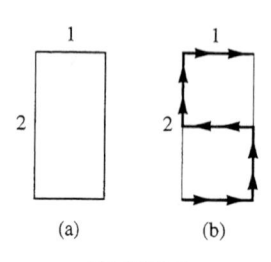

Figure 9.9

manager keeps all records inside his hat. What is the long-term production of the Boom-and-Bust Company "as time goes on"?

40. Use the graph in Figure 9.11 to conclude that $\int_1^\infty (1/x)\,dx = \infty$ by finding the sum of the areas of the indicated rectangles as $x \to \infty$.

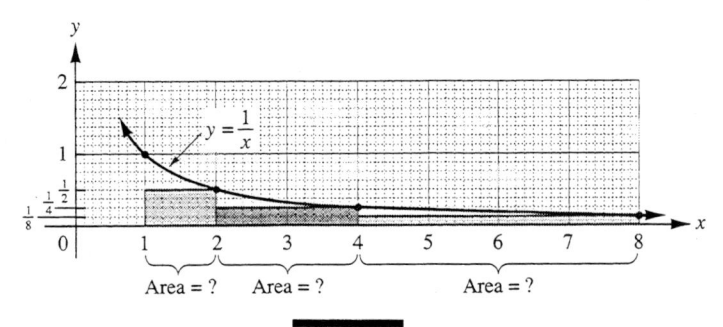

Figure 9.11

9.2 The Probability Distribution of a Random Variable

In this section we see how the probability that a random variable has particular values may be studied as a function of a real number.

A mathematical model is called **deterministic** if it predicts the future given exact information about the present and a rule telling how change is occurring. The differential equation of exponential change, $y' = ky$, with initial condition $y(0)$, of Section 5.5 is a deterministic model with many applications. Another mathematical model of comparable strength in applications has an opposite nature. In a **nondeterministic**, or **probabilistic, model** we view the world as made up of *events of an unpredictable nature* and find considerable accuracy not in precise measure of detail but via large-scale, average effects. Some of the most profound applications of mathematics bind these apparently opposed views of a deterministic versus a random world into a unified view. For example, the mathematical theory of Brownian motion explains how the random collisions of molecules in a heated body yield the (deterministic) solution of the differential equation that models heat flow through a solid. While well beyond this text, this example indicates that random models are not as far from the topics of calculus and the familiar deterministic world as might at first appear.

The Probability of a Random Variable

A measurement X whose value cannot be known exactly, or in advance, and whose determination is subject to "random" influences is called a **random variable.** For example, if X is the number that appears on the top face of an ordinary six-sided die after it is cast, then X is a random variable whose value can be any one of the numbers 1, 2, 3, 4, 5, or 6. The exact time X that you arrive at an appointment is a random variable subject to traffic, encounters with friends along the way, and so on. We will discuss the *probability* that a random variable will have particular values.

In doing so we will avoid a detailed discussion of probability itself and rely instead on your intuitive sense. For example, you would say that the probability of tossing "heads" with a "fair" coin is $\frac{1}{2}$. Similarly, when a fair die is cast, any one of six numbers could equally well appear. If you were asked, "What is the probability that the number '3' results?", you would reply, "$\frac{1}{6}$." In both of these examples you are thinking of probability as a *proportion*, $\frac{1}{2}$ or $\frac{1}{6}$, *of the whole* of equally likely possibilities, 2 and 6, respectively. Going one step further, if you were asked, "What is the probability of the event E that a '3' or a '5' results when the die is cast?", you would reply, "$\frac{2}{6}$" or "$\frac{1}{3}$," because either of *two* outcomes among the six equally likely possibilities favors the event E, and its probability is then the ratio $\frac{2}{6} = \frac{1}{3}$. The following definition is suitable for our purposes.

Definition

The **probability $P(E)$ of an event E** is the ratio

$$P(E) = \frac{\text{measure of favorable outcomes}}{\text{measure of total possible outcomes}}$$

Consequently, $P(E)$ is a fraction and $0 \leq P(E) \leq 1$ for any event E.

In the toss of a coin or the cast of a die, the equally possible outcomes are *discrete* and *finite* in number, and "the measure" of favorable outcomes and of total possibilities is a numerical count of each of these. On the other hand, if you were asked the probability that you will be at most 2 minutes late for an appointment, you would be unsure how to answer based on this definition because there are too many (indeed infinitely many!) possible arrival times and the favorable outcomes and total possible outcomes cannot be counted. As we shall see shortly, calculus allows us to assign a useful "measure" to such events and to model other such "continuously distributed" events.

Let X be a random variable. We are interested in questions like What is the probability that $X = 2$? or What is the probability that X is greater than 2 and less than or equal to 4? We will regard each of these requirements placed on the variable X as an event E and write

$P(X = a)$ for "the probability that X equals the number a"

$P(a < X \leq b)$ for "the probability that X is greater than a and at most b"

$P(X \leq t)$ for "the probability that X has a value less than or equal to t"

For example, we regard the inequality $a < X \leq b$ as defining the event E that, when X is found, its value will be greater than a and less than or equal to b; thus $P(a < X \leq b) = P(E)$.

Before turning to continuously distributed events like an arrival time, let us return to the discrete outcomes resulting from the cast of a die and gain familiarity with this new notation and the calculation of probabilities.

EXAMPLE 1

Let X be the number that appears on the top face of a six-sided die after it is cast.
a. Find $P(X = 2)$.
b. Find $P(3 < X \leq 5)$.
c. Find $P(X \leq 3)$ and $P(X < 5)$.

Solution Remember that six equally likely events are possible when the die is cast.
a. $X = 2$ is the event that the number "2" appears on the die. This is one event among six equally likely events. Thus $P(X = 2) = \frac{1}{6}$.
b. $3 < X \leq 5$ is the event that a number greater than "3" and at most "5" appears. This event is represented by two outcomes: that a "4" or a "5" appears. Thus $P(3 < X \leq 5) = \frac{2}{6} = \frac{1}{3}$.
c. $X \leq 3$ is the event that a number no larger than "3" appears on the die. There are three such possibilities: "1," "2," or "3" among the whole of six. Thus $P(X \leq 3) = \frac{3}{6} = \frac{1}{2}$. Similarly, $P(X < 5) = \frac{4}{6} = \frac{2}{3}$, since $X < 5$ allows for four possible outcomes. ∎

The concept of a random variable allows for wider application than might at first be imagined. Let us begin with an example where X can have infinitely many outcomes.

EXAMPLE 2

Imagine the following game. A pointer is placed at the center of a circle of radius 1. You will spin the pointer. If it stops in the first third of the circle, I will pay you \$1. If it stops in the last third, you will pay me \$1. If it stops in the middle third, no one will pay (see Figure 9.12).

Let X be your winnings. Then X is a random variable having any one of the values 0, 1, or -1.
a. Find $P(X \leq 0)$.
b. Find $P(X = 1)$.
c. Find $P(-1 < X \leq 1)$.
d. Find $P(X \leq 1)$.

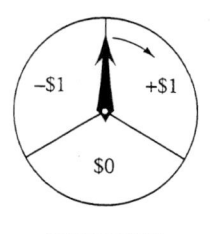

Figure 9.12

Solution There are three equally likely possibilities for the pointer.
a. The event $X \leq 0$ means that you lose \$1 or lose nothing. This is represented by two outcomes—the pointer stops in either the last or the middle third of the circle. Thus $P(X \leq 0) = \frac{2}{3}$.
b. Here you win \$1 and the pointer must stop in the first third of the circle. Thus $P(X = 1) = \frac{1}{3}$.
c. This event is that you win \$1 or pay nothing. These two possibilities mean that $P(-1 < X \leq 1) = \frac{2}{3}$.
d. The event $X \leq 1$ is certain, since X can be -1, 0, or 1 and is represented by all three possibilities. Thus $P(X \leq 1) = \frac{3}{3} = 1$. ∎

What should be done if the pointer stops exactly on the one-third (or two-thirds) mark of the circle? This question brings us to an example of a "continuously distributed" random variable.

EXAMPLE 3

In a circle of radius 1, mark the 12 o'clock position as 0, and attach a pointer to the center of the circle. Spin the pointer and let X be the distance of the end of the pointer from 0, when it comes to rest, measured *along the arc* of the circle in a clockwise direction [see Figure 9.13(a)].

a. What values can this random variable have?

b. Explain how we can conclude that $P(X = a) = 0$, where a is any one value of X specified *in advance* before the pointer is spun.

Solution

a. Since the circle has radius 1, its circumference is $2\pi \cdot 1 = 2\pi$ units long. Thus the end of the pointer can stop at any distance from 0 to 2π units from the 12 o'clock position. That is, $0 \leq X \leq 2\pi$.

b. Let a be a number between 0 and 2π. The event $X = a$ means that the pointer stops *exactly* a units from the 12 o'clock position where a is specified *in advance, before the pointer is spun*. Imagine a short interval I that is h units long, centered at the point a [see Figure 9.13(b)]. Let E be the event that the pointer stops in the interval I. We then measure the outcome favorable to E as the length h of the interval I, and we measure the total of possible outcomes as the length, 2π, of the whole circle. Consequently, $P(E) = h/2\pi$.

Now, the event $X = a$ is that the pointer stops exactly at a; thus the pointer would stop in *any* such interval I containing a. This means that the event $X = a$ is *less likely* than the event E no matter how short the interval I is. Thus $P(X = a) \leq P(E) = h/2\pi$ no matter how small h is. Hence

$$0 \leq P(X = a) = \lim_{h \to 0} P(X = a) \leq \lim_{h \to 0} \frac{h}{2\pi} = 0$$

since the number $P(X = a)$ is independent of h. Thus we conclude that $P(X = a) = 0$. ∎

You may be thinking that the pointer must come to rest at some point a and hence $P(X = a) = 1$. But that is not correct; if the point a is selected *in advance*, then the probability that the pointer stops at this *preselected point* is 0, as the calculation shows. We see then that in the game played in Example 2 the event that the pointer stops on exactly the one-third mark of the circle (that is, that $a = 2\pi/3$) has probability $P(X = 2\pi/3) = 0$. Thus the probability that the game played in Example 2 will end inconclusively is 0. At the same time it remains true that this event is possible; it only has mathematical probability 0.

We may use Example 3 to exhibit another essential property of probability. Consider the event E that the pointer stops in either the first or the third quadrant (see Figure 9.14). Since these two quadrants make up half the circle, we conclude that $P(E) = \frac{1}{2}$. Now let E_1 be the event that the pointer stops in the first quadrant, and let E_3 be the event that it stops in the third. Each of these quadrants makes up a fourth of the circle. Thus $P(E_1) = P(E_3) = \frac{1}{4}$.

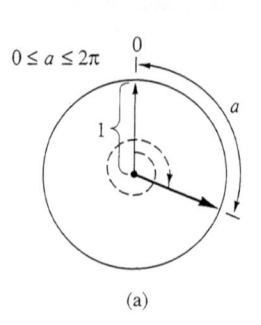

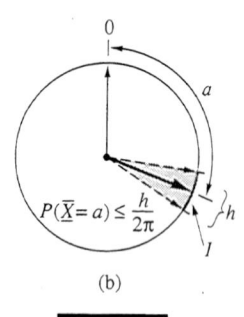

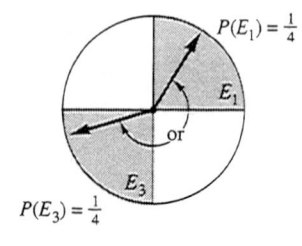

Figure 9.13

Figure 9.14

Notice now that E is the event that E_1 or E_3 occurs. Since $\frac{1}{2} = \frac{1}{4} + \frac{1}{4}$, we can write this as

$$P(E) = P(E_1) + P(E_3)$$

or, as we wish to observe,

$$P(E_1 \text{ or } E_3) = P(E_1) + P(E_3)$$

Such an equation is generally true about probability. If A and B are two events *that have nothing in common*, then

$$P(A \text{ or } B) = P(A) + P(B) \tag{1}$$

In particular, for a random variable X with $P(X = a) = 0$, we have that

$$P(a \leq X \leq b) = P(\ (X = a) \quad \text{or} \quad (a < X \leq b)\ )$$
$$= P(X = a) + P(a < X \leq b)$$

or
$$P(a \leq X \leq b) = P(a < X \leq b) \tag{2}$$

because the events $X = a$ and $a < X \leq b$ clearly have nothing in common and $P(X = a) = 0$.

Second, since the events $X \leq a$ and $a < X \leq b$ have nothing in common, we can also conclude that

$$P(X \leq b) = P(\ (X \leq a) \text{ or } (a < X \leq b)\ )$$
$$= P(X \leq a) + P(a \leq X \leq b) \quad \text{From (1) and (2)}$$
$$= P(X \leq a) + P(a \leq X \leq b)$$

Thus, subtracting,

$$P(a \leq X \leq b) = P(X \leq b) - P(X \leq a) \tag{3}$$

when $P(X = a) = 0$. In the next section we will see, via Equation 3, how calculus enters probability.

The Distribution of a Random Variable

A rigorous theory of probability defines a random variable as a *function* but not quite as you have grown to use this term. A random variable is a "function" of the result of an experiment. Let X be the number of heads that result when a coin is tossed three times in succession. Then X is a random variable with the values 0, 1, 2, or 3. The value of X can be determined only by actually tossing the coin three times and counting the number of heads that appear. Here is a key idea: If we toss the coin again, we will likely get a different number of heads—a different value for X. The domain of a random variable X is a list of *all* possible outcomes of an experiment and is called the **sample space** (see Exercise 16) for X. A value of X is obtained by actually running the experiment and observing the outcome.

This description is clearly quite different from the familiar idea of a function as a relation whose value is determined by putting a number into a formula and working out the result. Thus mathematicians do not study a random variable directly as a function represented by a formula with an explicit domain but rather study the (*cumulative*) *distribution* of a random variable. This allows a treatment of probability via a function of a real number in the familiar sense.

Definition

The **distribution** of a random variable X is the function F whose value at t is

$$F(t) = P(X \leq t)$$

the probability that X has at most the value t. We will henceforth write r.v. for the words *random variable* and F for its distribution.

The distribution of an r.v. is a key idea in probabilistic reasoning but, at least initially, is not intuitively meaningful.

EXAMPLE 4

Find the distribution of the r.v. X of Example 3.

Solution If $0 < t \leq 2\pi$, the event $X \leq t$ is the event that the pointer stops within a distance t of the 12 o'clock position, as measured *along the arc* of the circle and shown in Figure 9.15(a). The proportion of the circle represented by such an arc is $t/2\pi$. Therefore $F(t) = P(X \leq t) = t/2\pi$ for $0 < t \leq 2\pi$. ∎

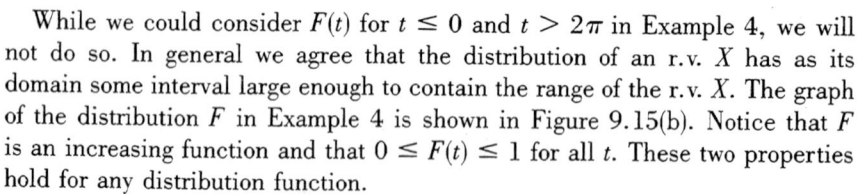

While we could consider $F(t)$ for $t \leq 0$ and $t > 2\pi$ in Example 4, we will not do so. In general we agree that the distribution of an r.v. X has as its domain some interval large enough to contain the range of the r.v. X. The graph of the distribution F in Example 4 is shown in Figure 9.15(b). Notice that F is an increasing function and that $0 \leq F(t) \leq 1$ for all t. These two properties hold for any distribution function.

Discrete random variables, like the roll of a die, may seem easier to deal with than the stopping point of a spinning pointer. In certain ways this is not so, as the next example indicates.

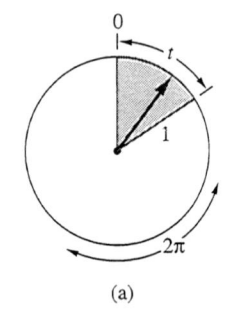

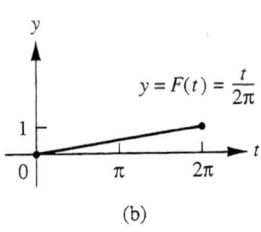

Figure 9.15

EXAMPLE 5

Cast a die and let X be the number appearing on its top face. Describe the (cumulative) distribution function for this random variable and sketch its graph.

Solution The random variable X has six values: 1, 2, 3, 4, 5, 6. Therefore, if $t < 1$, then

$$F(t) = P(X \leq t) = 0$$

since a value for X less than 1 is impossible. Now, since $F(1) = P(X \leq 1)$ is the probability that X is 1 or less, we see that $F(1) = \frac{1}{6}$. It follows in turn that if $t < 2$, then $F(t) = P(X \leq t)$ is again $\frac{1}{6}$.

Now, however (for $t = 2$), $F(2) = P(X \leq 2) = \frac{1}{3}$, because there are two of six ways in which the cast can result in a number less than *or equal to* 2—namely, a "1" or a "2" results.

Continuing this reasoning, we see that

$$F(t) = P(X \leq t) = \begin{cases} 0 & \text{if } t < 1 \\ \frac{1}{6} & \text{if } 1 \leq t < 2 \\ \frac{1}{3} & \text{if } 2 \leq t < 3 \\ \frac{1}{2} & \text{if } 3 \leq t < 4 \\ \frac{2}{3} & \text{if } 4 \leq t < 5 \\ \frac{5}{6} & \text{if } 5 \leq t < 6 \\ 1 & \text{if } 6 \leq t \end{cases}$$

which has the graph shown in Figure 9.16, where the circle on the end of each line indicates that F has its value at the next highest line—that is, that F has a discontinuity at each of the possible outcomes $X = 1, 2, \ldots, 6$.

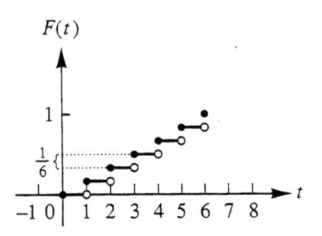

Figure 9.16

Discrete and Continuous Random Variables

We have seen two principal types of random variables: the discrete and the continuous. Discrete random variables, such as the result of the cast of a die, have distinct, clearly separated outcomes. Continuous random variables, like the stopping point of a spinning pointer, while having infinitely many possibilities, yield *continuous* distribution functions [see Figure 9.17(a, b) page 530]. In this case, calculus may be readily applied to probability, and as a result continuous random variables are, in a very real sense, easier to deal with than discrete random variables.

The simplest continuous probability distribution is the **uniform distribution**. A random variable X having all its values between two numbers $a < b$ is said to be uniformly distributed if

$$P(X \leq t) = \frac{t - a}{b - a}$$

9.2 The Probability Distribution of a Random Variable

(a) Discrete distribution (3 outcomes)

(b) General continuous distribution

(c) Uniform distribution

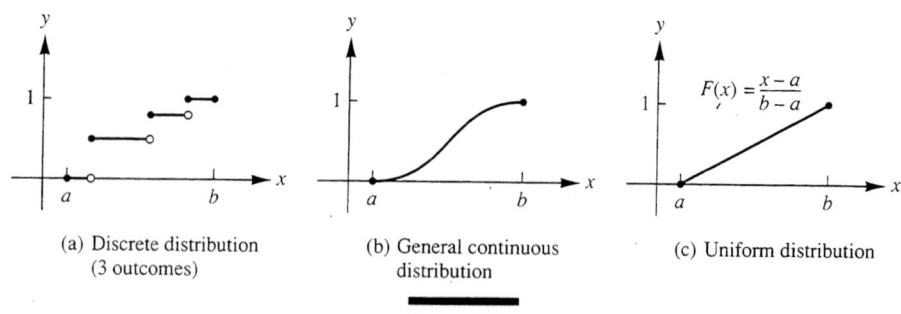

Figure 9.17

Thus if $c < d$, then using Equation 3,

$$P(c \leq X \leq d) = P(X \leq d) - P(X \leq c) = \frac{d-a}{b-a} - \frac{c-a}{b-a} = \frac{d-c}{b-a}$$

This equation tells us that the probability that X occurs in the interval $[c, d]$ is the ratio of the length, $d - c$, of that subinterval to the length, $b - a$, of the whole interval. This means that if $[c', d']$ is another interval *of the same length*, then X has the same probability of being in $[c', d']$ as of being in $[c, d]$. That is, its probabilities are "uniformly distributed" over the interval $[a, b]$, having just as good a chance to be in any one place as in any other [see Figure 9.17(c)].

The stopping point X of the spinning pointer of Example 5 has exactly this character—the pointer is likely to stop anywhere on the circle, with no particular preference. Moreover, as we have seen,

$$P(X \leq t) = t/2\pi = (t - 0)/(2\pi - 0)$$

the ratio of the length of arc up to the point t, to that of the length of the whole circle, 2π, regarded as the interval of numbers $[a, b] = [0, 2\pi]$.

The concept of probability as a measure of the random nature of events that cannot be predicted in advance began in the gambling halls of Europe in the seventeenth century. Although the first written work on probability was due to a mathematician, Pierre LaPlace, probability did not take its place alongside geometry, algebra, and calculus as a part of mathematics until this century, when physicists were shocked to realize that apparently "God does . . . play dice with the universe," despite Einstein's initially contrary opinion. The uncertainty principle, proposed by Werner Heisenberg, showed that one cannot know where an electron is *and* how fast it is moving. It was realized that in order to understand the most fundamental building blocks of the universe, it is necessary to model the atom in terms of the probability that its electrons are here *or* there. The immense speed and nearly infinitesimal masses involved called, however, for a precise mathematical description of such phenomena that would allow one to "see" what could not be observed. The Russian mathematician A. Kolmogorov, building on Henri Lebesgue's 1906 theory of measure, responded with a suitable mathematical theory of probability in the early 1930s.

Exercises 9.2

1. The notation $P(a < X \leq b)$ is a brief way of writing the English phrase "the probability that (the random variable) X is greater than a and less than or equal to b." Express each of the following in similar English.
   a. $P(0 < X < 9)$
   b. $P(X \leq 0)$
   c. $P(X \geq 8)$
   d. $P(3.5 \leq X \leq 5)$
   e. $P(X \leq 5) - P(X \leq 3)$
   f. $P(X \leq 2) + P(X \geq 4)$

2. Write each of the following phrases using notation like $P(X \leq a)$, $P(a \leq X \leq b)$, and so on.
   a. The probability that X is less than 2
   b. The probability that X exceeds 3
   c. The probability that X is no larger than 1
   d. The probability that X is larger than 5 and smaller than 7
   e. The probability that X does not exceed 3
   f. The probability that X is greater than 0 or less than -1
   g. The probability that X is greater than 2 plus the probability that X is less than 1
   h. The probability that X is 7

3. Imagine four cards numbered 1, 2, 3, and 4 on one side. Place these face down in front of you so that the numbers cannot be seen. Choose one card, and let X be the number appearing on this card.
   a. What is $P(X = 2)$?
   b. What is $P(X \leq 1)$?
   c. What is $P(X \leq 3)$?
   d. What is $P(1 \leq X \leq 3)$?
   e. Let $F(t) = P(X \leq t)$. Describe F as in Example 5, and draw the graph of this distribution.
   f. Write $P(1 < X \leq 3)$ in terms of $F(1)$ and $F(3)$.

4. Suppose that in Exercise 3 a friend agrees to pay you (in dollars) twice the number that appears on the face of the card you draw. Let Y be the amount that you win on one draw.
   a. What values can Y have?
   b. What is $P(Y \geq 6)$?
   c. What is $P(Y \leq 2)$?
   d. What is $P(Y \leq 1)$?
   e. Draw the graph of $F(t) = P(Y \leq t)$.

5. Flip a quarter and write 1 if a head appears, 0 if a tail appears. Let X be the number that you write down. What is $P(X \leq \frac{1}{4})$? $P(X \leq 1)$? Graph $F(t) = P(X \leq t)$.

6. Flip both a quarter and a nickel, and count the number of heads that appear. Let X be this number.
   a. What values can X have?
   b. How many outcomes are possible when you observe both coins? [Distinguish between a head on the quarter and a head on the nickel. List the possibilities as (H, H), (H, T), and so on.]
   c. What is $P(X = 0)$, $P(X = 1)$, $P(X = 2)$?
   d. Graph $F(t) = P(X \leq t)$. (Remember that X is a discrete random variable like that in Example 5.)

7. Imagine the following game. A pointer is spun around a circle. You will win the amounts shown in Figure 9.18 if the pointer stops in the region corresponding to each of the indicated payoffs of -1, 0, 1, or 2 dollars. Let X be the amount that you win. Thus X is a random variable with any one of these four values.

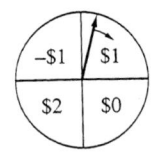

Figure 9.18

   a. Find $P(X \leq 1)$, $P(X = 2)$, $P(X \leq 0)$, $P(X > 1)$, and $P(X \geq \frac{3}{2})$.
   b. Find $P(X \leq 0$ or $X > 1)$ and $P(0 \leq X \leq 2)$.
   c. Find the distribution function $F(t) = P(X \leq t)$ and draw its graph.

8. Please refer to the spinning pointer of Example 3 and think of half the circle as having length π, a quarter as having length $\pi/2$, a sixth as having length $2\pi/6 = \pi/3$, and, conversely, think of a number

like, say, $\pi/4$ as defining the first eighth of the circle. Let X be the stopping point of the spinning pointer.

a. Find the probability that the pointer stops in the first quadrant of the circle.

b. Find $P\left(0 \leq X \leq \dfrac{\pi}{2}\right)$.

c. Find $P\left(\dfrac{\pi}{2} \leq X \leq \pi\right)$.

d. Find the probability that the pointer stops in the first one-sixth of the circle.

e. Find $P\left(\dfrac{\pi}{3} \leq X \leq 2\pi\right)$.

f. Find $P\left(\dfrac{\pi}{4} \leq X \leq \dfrac{\pi}{3}\right)$.

Find the distribution F for a random variable that is uniformly distributed over the indicated interval $[a, b]$ in Exercises 9–12, with this interval in each case being the domain of F. [*Hint:* In Example 4, we can regard $[a, b] = [0, 2\pi]$ and X as a stopping point in this interval that is uniformly distributed.]

9. $[a, b] = [0, 1]$ **10.** $[a, b] = [0, 2\pi]$

11. $[a, b] = [1, 4]$ **12.** $[a, b] = [-1, 1]$

13. Find $P(1 \leq X \leq 2)$, $P\left(\dfrac{1}{2} \leq X \leq \dfrac{3}{2}\right)$, and $P(2 \leq X \leq 3.5)$ in Exercise 11.

14. The piston in a gasoline-powered chain saw moves a total of 1 in. *up and down* during its stroke. The owner randomly shuts the saw off, and the piston stops somewhere along its 1-in. stroke length. If it stops within $\frac{1}{8}$ in. of the top of its stroke *while moving upward*, the saw will be hard to start again because the owner will have to tug the starting cord against the full compression of the engine. What is the probability of this happening?

15. Even the smallest computers have a *random number generator* that picks a number X "at random" from the unit interval $[0, 1]$. More precisely, this means that X is a uniformly distributed r.v., with outcomes in subintervals of equal length being equally likely. That is, $P(X \leq t) = t$, since the proportion of $[0, t]$ to $[0, 1]$ is t.

a. Find $P(0.20 \leq X \leq 0.40)$.

b. Find the probability that the first digit of X is a "3."

c. Find $P(0.45 \leq X \leq 0.50)$.

d. Find $P(X \leq 0.75) - P(X \leq 0.50)$.

e. Find $P(0.950 \leq X \leq 0.957)$.

f. Find the probability that the second digit of X is a "2."

g. Find the probability that the first digit of X is a "5" and the second is a "3."

[*Hint:* Try first to draw a subinterval(s) of $[0, 1]$, which represents each of these outcomes.]

16. We understand probability as a proportion of a whole. If we can describe the whole in some fashion, we have seen that it is possible to find probabilities directly. The "whole" we refer to is called a *sample space*. Consider the following problem. You agree to meet a friend at the bench beside the library sometime between 1:00 and 2:00, but you don't know when you'll arrive. Your friend has a similar problem of uncertainty. You both agree to wait 20 minutes and, if the other party does not arrive, to go on your way.

Let X represent your arrival time (in minutes after 1:00), and let Y represent your friend's arrival time. Use a unit interval $[0, 1]$ drawn horizontally to represent your arrival time and another unit interval $[0, 1]$ drawn vertically to represent your friend's arrival time (see Figure 9.19).

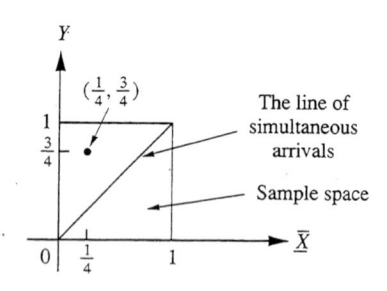

Figure 9.19

The resulting *square* represents the whole of possibilities—an appropriate sample space for this problem in probability. For example, the point $(X, Y) = \left(\dfrac{1}{4}, \dfrac{3}{4}\right)$ means that you arrive at 1:15 (a quarter hour after 1:00) and your friend at 1:45, thus missing each other.

a. You will meet your friend if and only if $|X - Y| \leq \dfrac{1}{3}$, since 20 minutes is $\dfrac{1}{3}$ of an hour. Draw a picture of the set of points in the square satisfying this condition.

b. What is the area of the region you drew in (a)? What is the area of the whole square? What is the probability that you and your friend meet?

532   Chapter 9   Probability

c. Repeat (a) and (b) for an arbitrary waiting time w. What minimum length of time w ensures that there is a 50-50 chance that you two will meet?

17. This exercise indicates why a random variable is really a function of an experiment. Imagine flipping a coin four times in succession and recording the outcome as a 1 if a head appears and a 0 if a tail appears. One possible outcome is $(0, 1, 1, 0)$, indicating a tail, then a head, a head again, then a tail. Thus we can indicate any possible outcome of this experiment as a quadruple: (w_1, w_2, w_3, w_4), where w_i is a 0 or a 1. For example, if four heads appear in succession, then $w_1 = w_2 = w_3 = w_4 = 1$.

   a. Make a list of the 16 possible outcomes of this experiment. That is, list 16 different quadruples of 0's and 1's.

   b. Let X be the number of heads that appear in four tosses. Explain why
   $$X(w_1, w_2, w_3, w_4) = w_1 + w_2 + w_3 + w_4$$
   is a function, and a formula, representing this random variable as a function of this experiment.

   c. How many outcomes consist of exactly two heads? What is $P(X = 2)$?

   d. Suppose a friend agrees to pay you $1 if a tail appears and that you will pay him $1 if a head appears, for each of the four tosses of the coin. Let Y represent your winnings. Explain why
   $$Y(w_1, w_2, w_3, w_4) = (-1)^{w_1} + (-1)^{w_2} + (-1)^{w_3} + (-1)^{w_4}$$
   represents your winnings for any possible outcome and thus is a formula for the r.v. Y.

   e. What is $P(Y = 0)$? $P(Y = 1)$? $P(Y \geq 3)$?

   f. Graph the distribution functions $F(t) = P(X \leq t)$ and $G(t) = P(Y \leq t)$ after you have first determined the possible values of each of these r.v.'s.

Challenge Problem

18. In the examples and exercises of this section, each of the various outcomes taken on by the random variable was *equally likely*; each single outcome had the same probability. In this exercise the outcomes are not equally likely, but the reasoning is as before.

    Divide a circle as in Figure 9.20. Let X be the random variable that pays $-\$1$ if the pointer stops in the region A, pays $\$0$ if the pointer stops in the region B, and pays $\$1$ if the pointer stops in the region C.

    a. Find $P(X = 1)$; $P(X = 0)$; $P(X = -1)$.
    b. Find $P(-1 \leq X \leq 0)$; $P(0 \leq X \leq 1)$.
    c. Find $P(-1 < X < 1)$; $P(0 < X < 1)$.
    d. Find $P(|X| = 1)$.

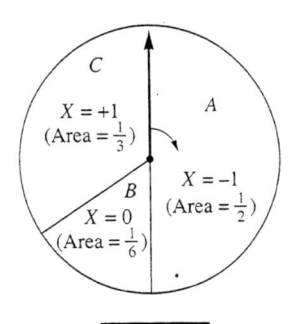

Figure 9.20

9.3 The Probability Density Function of a Continuous Random Variable

In this section we see how calculus can be used to calculate probabilities via the representation of the probability distribution of a random variable as the integral of its density function.

The distribution of a random variable does not in general offer the easiest or most efficient way to calculate probability. We need a step-by-step procedure, offered by the concept of a **probability density function**.

A random variable X is *continuous* if its distribution $F(t) = P(X \leq t)$ is a continuous function on its domain. For our purposes a continuous r.v. has the property $P(X = a) = 0$ encountered in Example 3, Section 9.2. This property, and the Fundamental Theorem of Calculus, allows calculus to enter probability via the following concept.

Definition

Let X be a random variable, and let $F(t) = P(X \le t)$ be its distribution. The function

$$\rho(t) = F'(t)$$

(if this derivative exists) is called the *probability density function* (p.d.f.) for the random variable X.

Because F itself is an increasing function, the density function ρ of an r.v. is always nonnegative. Since a differentiable function is necessarily continuous (Section A.1), an r.v. X is a continuous r.v. whenever it has a density function defined at every value of t in the domain of F.

EXAMPLE 1

Find the density function for the stopping point of a spinning pointer of Example 4, Section 9.2.

Solution In Example 4, we found that $F(t) = t/2\pi$. Therefore $\rho(t) = F'(t) = 1/2\pi$ and is a constant function. We show F and ρ in Figure 9.21(a, b).

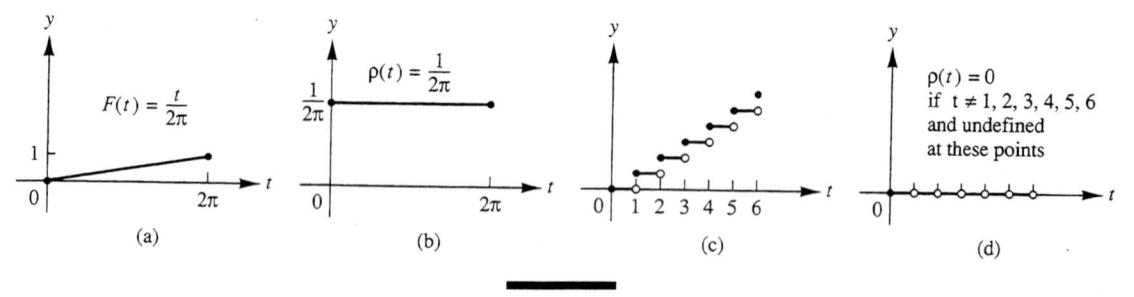

Figure 9.21

A density function is essential for studying continuous r.v.'s but is largely useless in dealing with discrete r.v.'s. For example, if X is the number resulting from the roll of a die, Figure 9.21(c) shows that $F(t) = P(X \le t)$ is constant, and hence has $\rho(t) = F'(t) = 0$ at every point t, except those t where F is interesting! Where F has a discontinuity representing an actual outcome of casting the die, $F'(t)$ cannot exist [Figure 9.21(d)].

As we will see, the importance of density functions resides in two things: (1) the Fundamental Theorem of Calculus and (2) the fact that direct observation and statistical measurement of random phenomena often suggest the form of an appropriate density function even when F and the domain of the r.v. X are uncertain (see Exercise 19). The principal fact of this section and of the *calculus of probability* is the following theorem, which tells us that the probabilities of a continuous r.v. may be found directly by integration of its density function. (See Figure 9.22.)

Theorem 9.1   If X is a continuously distributed random variable having a probability density function ρ, then

$$P(a \leq X \leq b) = \int_a^b \rho(x)\, dx \qquad (1)$$

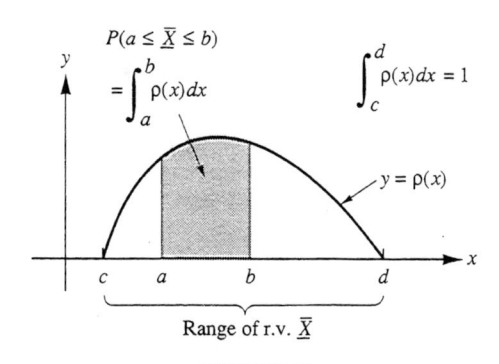

Figure 9.22

In particular, if the entire range of the r.v. X lies between two numbers c and d, then $P(c \leq X \leq d)$ has probability 1 and so

$$P(c \leq X \leq d) = \int_c^d \rho(x)\, dx = 1 \qquad (2)$$

It is easy to see why Theorem 9.1 is true. Since $F'(t) = \rho(t)$, F is an antiderivative of ρ and the Fundamental Theorem of Calculus tells us that

$$F(b) - F(a) = \int_a^b \rho(x)\, dx$$

But $\qquad F(b) - F(a) = P(X \leq b) - P(X \leq a) = P(a \leq X \leq b)$

Thus $\qquad P(a \leq X \leq b) = \int_a^b \rho(t)\, dt$

Let us observe the validity of Theorem 9.1 in the case of a spinning pointer.

EXAMPLE 2

It is visually clear that the probability that a spinning pointer stops somewhere in the second quarter of a circle is $\frac{1}{4}$. Show that this answer coincides with

$$\int_{\pi/2}^{\pi} \rho(t)\, dt$$

where, from Example 1, $\rho(t) = 1/2\pi$ is the p.d.f. for this r.v. X (see Figure 9.23).

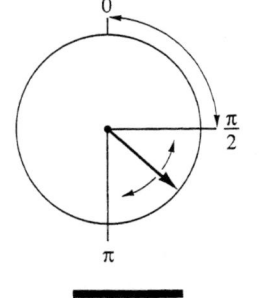

Figure 9.23

9.3 The Probability Density Function of a Continuous Random Variable   535

Solution Since half the circle is π units long, the second quarter of the circle extends from $a = \pi/2$ to $b = \pi$. From Example 1, we have that $\rho(t) = 1/2\pi$. Therefore

$$P\left(\frac{\pi}{2} \leq X \leq \pi\right) = \int_{\pi/2}^{\pi} \rho(t)\, dt = \int_{\pi/2}^{\pi} \frac{1}{2\pi}\, dt = \frac{1}{2\pi}(t)\Big]_{\pi/2}^{\pi} = \frac{1}{2\pi}\left(\pi - \frac{\pi}{2}\right) = \frac{1}{2\pi}\left(\frac{\pi}{2}\right) = \frac{1}{4}$$

Thus the computation of probability via the integral, according to the basic Theorem 9.1, coincides with our visual calculation. ∎

In practice statisticians determine appropriate probability distributions and/or probability density functions for an r.v. of interest. This often involves the determination of appropriate constants called **parameters.** The most basic property that a p.d.f. must have is that its integral over the whole range of the r.v. *equal 1* since there is probability 1 of at least some outcome in this *whole* range of X. We use this fact to determine the value of such a parameter, as in the next example.

EXAMPLE 3

A random variable is known to have range in the interval [0, 2]. A statistician wishes to model its probabilities by a p.d.f. of the form

$$\rho(x) = Cx(2 - x)$$

Determine the parameter C so that $\int_0^2 \rho(x)\, dx = 1$.

Solution We want C to satisfy the equation

$$1 = \int_0^2 \rho(x)\, dx = \int_0^2 Cx(2 - x)\, dx$$

$$= C\int_0^2 (2x - x^2)\, dx = C\left[x^2 - \frac{x^3}{3}\right]_0^2 = C\left(\frac{4}{3}\right)$$

Therefore we conclude that $C = \frac{3}{4}$. The statistician will thus use $\rho(x) = \frac{3}{4}x(2 - x)$ as the appropriate p.d.f. for this r.v. ∎

Let us now take Formula 2 one step further and find a formula for the distribution function $F(t) = P(X \leq t)$. If X has *all* of its values in the interval $[a, b]$, then $F(a) = P(X \leq a) = 0$ and therefore $F(t) = F(t) - F(a) = \int_a^t \rho(x)\, dx$. That is, for any number t

$$F(t) = P(X \leq t) = \int_a^t \rho(x)\, dx \tag{3}$$

EXAMPLE 4

Suppose that you make a running, blindfolded jump from point A and land in the interval between points B and C, which are 2 ft apart, as indicated in Figure 9.24. Let X be the point where you land in this interval, and suppose that $\rho(t) = -\frac{3}{4}(t - 6)^2 + \frac{3}{4}$ is the density function for your landing point.
  a. Find $P(5.5 \leq X \leq 6.5)$, the probability that you will land within 6 in. of the center of the interval $[B, C]$.

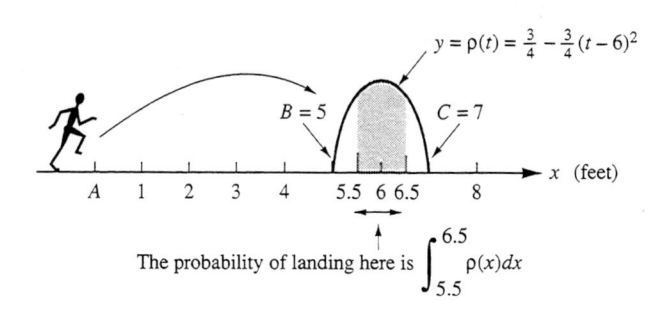

Figure 9.24

[Figure: A runner jumping with landing positions marked on x-axis (feet). Points labeled A, 1, 2, 3, 4, 5.5, 6, 6.5, 8. B = 5, C = 7. Curve $y = \rho(t) = \frac{3}{4} - \frac{3}{4}(t-6)^2$. The probability of landing here is $\int_{5.5}^{6.5} \rho(x)\,dx$.]

b. Find the probability distribution $F(t) = P(X \leq t)$, for $5 \leq t \leq 7$, using Equation 3.

c. Use (b) to find $P(X \leq 5.5)$, the probability that you land within 6 in. of point B.

Solution

a. According to Theorem 9.1

$$P(5.5 \leq X \leq 6.5) = \int_{5.5}^{6.5} \left[-\frac{3}{4}(t-6)^2 + \frac{3}{4} \right] dt$$

$$= -\frac{3}{4} \frac{(t-6)^3}{3} + \frac{3}{4} t \Big]_{5.5}^{6.5}$$

$$= -\frac{1}{4}[(0.5)^3 - (-0.5)^3] + \frac{3}{4}(6.5 - 5.5)$$

$$= -\frac{2(0.125)}{4} + \frac{3}{4} = 0.6875$$

Thus there is a better than $\frac{2}{3}$ probability that you will land within 6 in. of the center of this interval. Notice how in Figure 9.24 the graph of $y = \rho(t)$ concentrates most of its area (which determines probability, according to Theorem 9.1) toward the center of the interval and less near the edge.

b. Since $F(5) = P(X \leq 5) = 0$, using Equation 3, we have

$$F(t) = P(X \leq t) = \int_5^t \rho(x)\,dx$$

$$= \int_5^t \left[-\frac{3}{4}(x-6)^2 + \frac{3}{4} \right] dx$$

$$= \frac{3}{4}\left[x - \frac{(x-6)^3}{3} \right]_5^t$$

$$= \frac{3}{4}\left[\left(t - \frac{(t-6)^3}{3} \right) - \left(5 - \frac{(5-6)^3}{3} \right) \right]$$

or $$F(t) = \frac{3}{4}\left[t - \frac{(t-6)^3}{3} - \frac{16}{3} \right] \tag{4}$$

9.3 The Probability Density Function of a Continuous Random Variable

c. The probability, $P(X \leq 5.5)$, that you land within 6 in. of the left end of the interval is, upon substitution in Equation 4,

$$P(X \leq 5.5) = F(5.5) = \frac{3}{4}\left[5.5 - \frac{(5.5-6)^3}{3} - \frac{16}{3}\right]$$

$$\simeq \frac{3}{4}\left[5.5 + \frac{0.125}{3} - 5.333\right] \simeq 0.157 \quad \blacksquare$$

Your landing point in Example 4 is a continuously distributed random variable with infinitely many possible outcomes—you could land anywhere in the interval—and these are far too numerous to count. Given our original definition of probability, the concept of a probability density function allows us to "measure" the outcomes favoring an event E, even when there are infinitely many such outcomes, by integration of ρ over the interval representing E.

What does a *single function value* $\rho(x)$ represent, or measure? This is not easy to answer. Returning to Figure 9.24 notice that, intuitively, possible outcomes are not "equally likely"; landing near the center of the interval is more likely than landing near the edge, and the calculations in Example 4 support this. The values $\rho(x)$ allow us to "weight," or give more "measure" to, some events over others. When $\rho(c)$ is large, events *near* $X = c$ are more likely than events *near* $X = d$, where $\rho(d)$ is small. Thus, in Example 4, the event $6 \leq X \leq 6.5$ is much more likely than the event $5 \leq X \leq 5.5$ and, correspondingly, $\rho(6) = \frac{3}{4}$ whereas $\rho(5) = 0$.

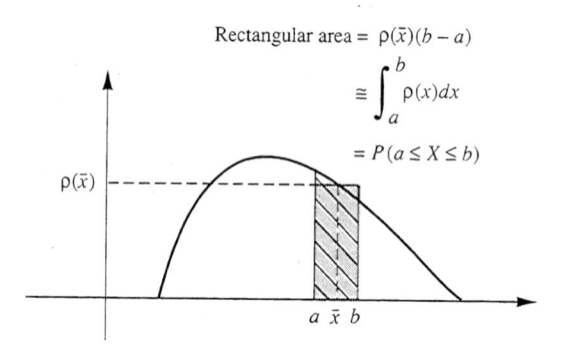

Figure 9.25

Figure 9.25 can help us interpret $\rho(x)$ more generally and more precisely. We see there that

$$\rho(\bar{x})(b - a) \simeq \int_a^b \rho(x)\,dx = P(a \leq X \leq b)$$

That is, $P(a \leq X \leq b)$ is *(approximately)* directly *proportional to the length* $b - a$ *of the interval* $I = [a, b]$ measuring the event, *by the proportionality factor* $\rho(\bar{x})$, where $\bar{x}$ lies between a and b.

Figure 9.25 reminds us too of our original definition of probability as a proportion of a whole. We see that $P(a \leq X \leq b)$ is the proportion of the area beneath the graph of the p.d.f. ρ between a and b to the area beneath the *entire* graph of ρ, this entire area being exactly 1.

In each of the examples of this section, the r.v. X has a limited range. If the range of X is unknown or if we allow $-\infty < X < \infty$, we then regard $F(t) = P(X \le t)$ as an increasing function defined on the entire real line with $\lim_{t \to \infty} F(t) = 1$ and $\lim_{t \to -\infty} F(t) = 0$ [Figure 9.26(a)]. We then consider the associated density function ρ (when it exists) on the same domain and have that

$$F(a) = P(X \le a) = \int_{-\infty}^{a} \rho(x)\, dx$$

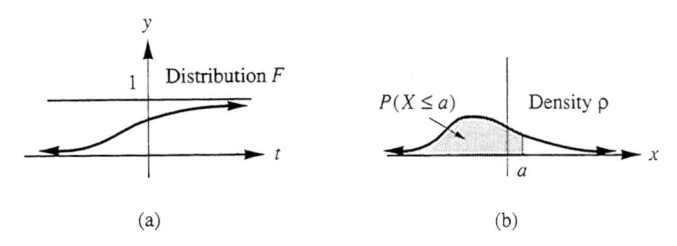

Figure 9.26

(a)   (b)

Moreover, if X has range between 0 and $+\infty$, then a complementary formula holds [see Figure 9.26(a)]:

$$P(X \ge a) = \int_{a}^{\infty} \rho(x)\, dx$$

EXAMPLE 5

A random variable X can have any value between 0 and $+\infty$ and is known to have p.d.f. $\rho(x) = 2e^{-2x}$. Find $P(X \ge 1)$ (see Figure 9.27).

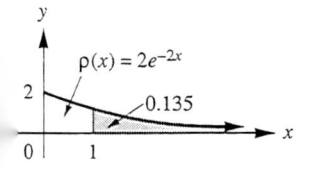

Figure 9.27

Solution We have

$$P(X \ge 1) = \int_{1}^{\infty} 2e^{-2x}\, dx = \lim_{b \to \infty}(-e^{-2x})\Big]_{1}^{b} = \lim_{b \to \infty}(e^{-2} - e^{-2b}) = \frac{1}{e^2} \approx 0.135 \quad \blacksquare$$

Exercises 9.3

Find the number C so that each of the functions in Exercises 1–4 is a p.d.f. Graph each function.

1. $\rho(x) = C$ on $[a, b] = [1, 3]$
2. $\rho(x) = Cx + 1$ on $[a, b] = [0, 2]$
3. $\rho(x) = Cx(1 - x)$ on $[a, b] = [0, 1]$
4. $\rho(x) = Ce^{-2x}$ on $[a, b] = [0, \infty)$ (Use an improper integral.)

Find the following probabilities by integration of the density function obtained in Exercises 1–4.

5. $F(\frac{3}{2}) = P(X \le \frac{3}{2})$ in Exercise 1
6. $F(0) = P(X \le 0)$ in Exercise 2
7. $F(\frac{1}{3}) = P(X \le \frac{1}{3})$ in Exercise 3
8. $F(1) = P(X \le 1)$ in Exercise 4

In Exercises 9–12, find a formula for $F(t) = P(X \leq t) = \int_a^t \rho(x)\,dx$ for each function ρ of Exercises 1–4. Graph F.

9. In Exercise 1, $a = 1$ and $1 \leq t \leq 3$.
10. In Exercise 2, $a = 0$ and $0 \leq t \leq 2$.
11. In Exercise 3, $a = 0$ and $0 \leq t \leq 1$.
12. In Exercise 4, $a = 0$ and $0 \leq t < \infty$.
13. Let X be an r.v. with values in the interval $[0, 5]$ whose p.d.f. is $\rho(x) = 0.048x(5 - x)$. Refer to Example 4.

    a. Graph ρ (using an exaggerated scale on the y-axis for clarity).

    b. Which event is more likely: $X \leq 1$ or $2 \leq X \leq 3$? [Answer by using (a) and *without* making an exact calculation.]

    c. Find $P(1 \leq X \leq 2)$; $P(1 \leq X \leq 3)$; $P(2 \leq X \leq 3)$; $P(4 \leq X \leq 5)$.

    d. Find $P(X \leq 2.5)$ and $P(X \leq 3)$.

    e. Indicate each answer in (c) and (d) as an area on the graph found in (a).

    f. Compare $\rho(2)(\frac{1}{2})$ with $P(2 - \frac{1}{4} \leq X \leq 2 + \frac{1}{4})$. (Note that the length of this interval is $\frac{1}{2}$.)

14. Let X be the number of minutes that you are late for an appointment, and suppose that the function ρ of Exercise 13 is the p.d.f. for this random variable X.

    a. What is the probability that you will be 2 to 4 minutes late for your appointment?

    b. What is the probability that you will be at least 3 minutes late?

    c. What is the probability that you will be at most 1 minute late?

    d. What is the probability that you will be at most 2 minutes late or at least 4 minutes late?

    e. Find a formula for $F(t) = P(X \leq t)$, and graph F.

Find $P(X \geq 1)$ for the p.d.f. ρ in Exercises 15–18.

15. $\rho(x) = 3e^{-3x}$, $x \geq 0$
16. $\rho(x) = \dfrac{1}{(x+1)^2}$, $x \geq 0$
17. $\rho(x) = 4xe^{-2x}$, $x \geq 0$
18. $\rho(x) = \dfrac{2}{(x+1)^3}$, $x \geq 0$

The next exercise indicates how a p.d.f. could be chosen based on *frequency data* and how it may be tested against this data.

19. Mary claims that John is "always" late for an appointment. She has gone so far as to keep some records of John's tardiness. These are indicated in Figure 9.28 and show that of 34 appointments John was on time once, 1 minute late 6 times, 2 minutes late 7 times, and so on. Since the graph of $y = 10xe^{-0.5x}$ closely matches this data, Mary decides to use such a function as a p.d.f. to predict John's tardiness.

    a. Find a suitable constant k so that $\int_0^8 k(10xe^{-0.5x})\,dx = 1$. The resulting function $\rho(x) = 10kxe^{-0.5x}$ on $[0, 8]$ is then a usable p.d.f. Use this value of k in (b)–(e).

    b. The graph indicates that John has a probability of $\frac{7}{34} = 0.205$ of being 2 minutes late. Compare this value with that of the integral $\int_{1.5}^{2.5} \rho(x)\,dx$.

    c. Find $\int_{3.5}^{7.5} \rho(x)\,dx$, and compare this number with the observed probability of $\frac{14}{34} = 0.411$ that John will be 4 to 7 minutes late.

    d. What is the probability predicted by ρ that John will be 1 to 3 minutes late?

    e. Is ρ a good predictor of John's arrival times?

20. Let X be uniformly distributed on the interval $[1, 5]$.

    a. Find the formula for $F(t) = P(X \leq t)$, $1 \leq t \leq 5$.

    b. Find $\rho(t) = F'(t)$.

    c. Compute $P(2 \leq X \leq 3.25)$ using ρ.

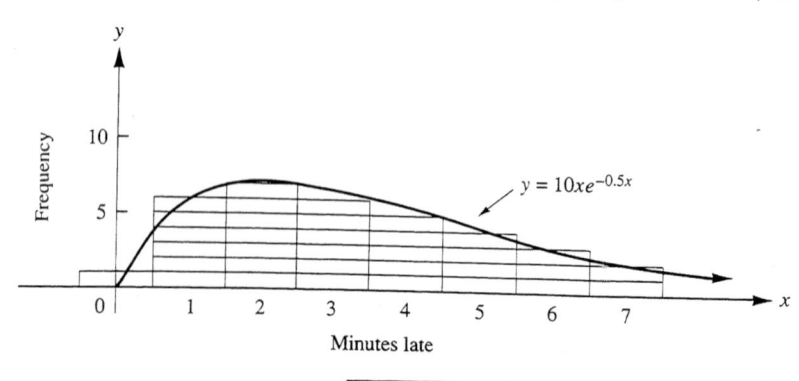

Figure 9.28

9.4 The Expectation and Variance of a Random Variable

In this section we learn how to find the expected value of a random variable and study two exponential distributions.

To be competitive, a manufacturer of roof shingles must guarantee that the shingles will last a certain number of years. At the same time, if the guarantee is too good, huge replacement costs may be incurred in the future. Given that it is impossible to know in advance when a roof will begin leaking, the manufacturer must regard the roof's lifetime as a continuous random variable X and needs to know what value X can be "expected" to have and guarantee the shingles to last no longer than that value. The ability to compute such an *expected value* has a variety of applications and is the subject of this section.

The Expectation of a Random Variable

Let us introduce the concept of expected value by considering the following game. You will cast a six-sided die and I will give you half the number of dollars that appear on its face; if you roll a "3," I will give you $1.50. How much can you "expect" to win in this game?

Let X be the number of dollars that you win on the roll of the die. Then X is an r.v. that can have any one of the values $\frac{1}{2}$, 1, $\frac{3}{2}$, 2, $\frac{5}{2}$, and 3 (dollars). Each of these outcomes has probability $\frac{1}{6}$. It would seem reasonable that you would win $\frac{1}{2}$ dollars with probability $\frac{1}{6}$, or 1 dollar with probability $\frac{1}{6}$, or $\frac{3}{2}$ dollars with probability $\frac{1}{6}$, and so on. That is, it seems reasonable that you should "expect" to win

$$\left(\tfrac{1}{2} \cdot \tfrac{1}{6}\right) + \left(1 \cdot \tfrac{1}{6}\right) + \left(\tfrac{3}{2} \cdot \tfrac{1}{6}\right) + \left(2 \cdot \tfrac{1}{6}\right) + \left(\tfrac{5}{2} \cdot \tfrac{1}{6}\right) + \left(3 \cdot \tfrac{1}{6}\right) = \tfrac{21}{12} \quad (1)$$

or $1.75, in this game. Equation 1 is called a **weighted average**—each possible value for X is "weighted" (multiplied) by its probability, and the sum of these is a reasonable measure of the **expected value** of X. (It is interesting to note that this expected value cannot actually occur, since X is never equal to $1.75!)

How can we make a general definition for the *expected value* of a random variable? To begin, notice that we can regard Equation 1 as follows: Each outcome $d = \frac{1}{2}, 1, \ldots, 3$ is multiplied by its probability. That is, each term has the *form* $d \cdot P(X = d)$. Thus Equation 1 can be written as

$$\tfrac{1}{2}P\!\left(X = \tfrac{1}{2}\right) + 1 \cdot P(X = 1) + \tfrac{3}{2}P\!\left(X = \tfrac{3}{2}\right)$$
$$+ 2 \cdot P(X = 2) + \tfrac{5}{2}P\!\left(X = \tfrac{5}{2}\right) + 3 \cdot P(X = 3) = \tfrac{21}{12} \quad (2)$$

This expression is a *sum of outcomes by the probability of each outcome.*

Let us form a sum of the same *form* for a continuous r.v. X with p.d.f. ρ, as in Figure 9.29 (page 542), having all its values between a and b, as follows. Partition the interval $[a, b]$ into many short intervals of length Δt, and let t_k be, say, the midpoint of the kth interval $I_k = [a_k, b_k]$. [Since $P(X = t_k) = 0$, we can only make use of $P(a_k \le X \le b_k)$, the probability that X is near t_k, that is, the probability that X occurs in the interval I_k.]

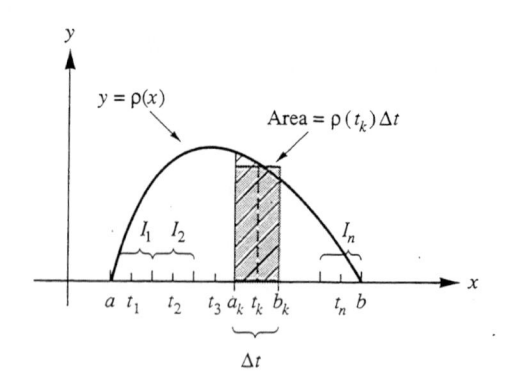

Figure 9.29

The sum

$$t_1 P(a_1 \leq X \leq b_1) + t_2 P(a_2 \leq X \leq b_2) + \cdots + t_n P(a_n \leq X \leq b_n) \quad (3)$$

is, like Equation 2, a sum of outcomes t_k times the probability of an outcome near t_k. Because of its similarity to Equation 2, the sum (Expression 3) can be taken as an *approximation* to an expected value for X.

At the same time, our discussion of the meaning of $\rho(t_k)$ in Section 9.3 tells us that $P(a_k \leq X \leq b_k) \simeq \rho(t_k)\Delta t$. Thus, upon substitution in Equation 3, an expected value for X is also approximated by a Riemann sum of the form

$$t_1 \rho(t_1) \Delta t + t_2 \rho(t_2) \Delta t + \cdots + t_n \rho(t_n) \Delta t \quad (4)$$

This is a Riemann sum for the integral $\int_a^b t\rho(t)\,dt$. This leads us to make the following definition, based on the approximation principle (AP) of Section 4.5.

Definition

The *expected value* $E(X)$ of a random variable X with density function ρ is the number

$$E(X) = \int_a^b x\rho(x)\,dx \quad (5)$$

where X has all its (possible) values between a and b. If X has unlimited range, we allow $a = -\infty$ and $b = +\infty$.

EXAMPLE 1

Find $E(X)$ for the r.v. X (the landing point in a blindfolded jump of Example 4, Section 9.3) whose p.d.f. is

$$\rho(x) = -\frac{3}{4}(x-6)^2 + \frac{3}{4} \quad \text{for } 5 \leq x \leq 7$$

indicated in Figure 9.30.

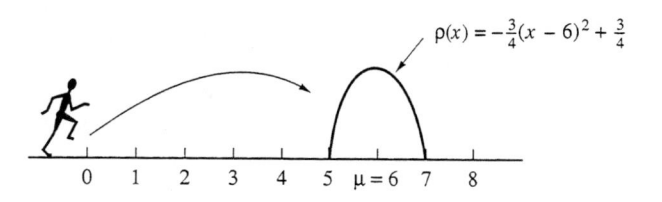

Figure 9.30

$$p(x) = -\tfrac{3}{4}(x-6)^2 + \tfrac{3}{4}$$

0 1 2 3 4 5 μ = 6 7 8

Solution According to the definition,

$$E(X) = \int_5^7 x\rho(x)\,dx = \int_5^7 x\left[-\frac{3}{4}(x-6)^2 + \frac{3}{4}\right]dx$$

$$= \frac{3}{4}\int_5^7 x(1 - (x-6)^2)\,dx = \frac{3}{4}\left(\frac{x^2}{2}\right)\Big|_5^7 - \frac{3}{4}\int_5^7 x(x-6)^2\,dx$$

$$= 9 - \left(\frac{3}{4}\right)(4) = 6 \quad \blacksquare$$

A glance at the graph of the density function ρ suggests that the most likely landing spot X in this blindfolded jump is in fact (centered) at 6 ft. Thus $E(X) = 6$ is consistent with our intuition.

The expected value, or expectation, $E(X)$, of an r.v. X is also referred to as its **mean** or **average value**. In this case it is common to use the Greek letter μ ("mu") to denote $E(X)$. Thus, in Example 1, we would say that X is an r.v. with p.d.f. $\rho(x) = -\tfrac{3}{4}(x-6)^2 + \tfrac{3}{4}$ and mean (or expectation) $\mu = 6$.

The Variance and Standard Deviation of a Random Variable

While in Example 1 we can reasonably "expect" an outcome of $X = 6$, in fact $P(X = 6) = 0$. It is the essence of this subject that X will not have any single value with much probability, even though $X = 6$ is "expected." These two superficially contradictory comments are resolved by defining a measure of dispersion—the tendency of a random outcome to be "near" its expected value.

Definition

The **variance** $V(X)$ of a random variable X (with p.d.f. ρ) about its mean $\mu = E(X)$ is the number

$$V(X) = \int_a^b (x - \mu)^2 \rho(x)\,dx \tag{6}$$

where X can have values between a and b. The **standard deviation** σ of X is the number

$$\sigma = \sqrt{V(X)}$$

If X has unlimited range, then $a = -\infty$ and $b = +\infty$.

Essentially, the concept of standard deviation measures how far an outcome $X = x$ is from the mean μ—thus the factor $(x - \mu)^2$— and weights this number by the probability density $\rho(x)$ at x. The next example gives some sense of how σ is a measure of the tendency of a random outcome to be near its expected outcome μ.

EXAMPLE 2

Let X be the distance a spinning pointer stops when measured along the arc of a circle of radius 1 of Example 4, Section 9.2. Example 1, Section 9.3, gives $\rho(x) = 1/2\pi$ as the p.d.f. for X.
a. Find $\mu = E(X)$.   **b.** Find $V(X)$ and σ.

Solution

a. By definition, since X ranges between 0 and 2π,

$$\mu = E(X) = \int_0^{2\pi} x\rho(x)\, dx = \int_0^{2\pi} x\left(\frac{1}{2\pi}\right) dx = \frac{1}{2\pi} \frac{x^2}{2}\Big]_0^{2\pi}$$

$$= \frac{(2\pi)^2}{2\pi(2)} = \pi$$

We should "expect" the pointer to stop at the 6:00 position, the midposition of the circle, as indicated in Figure 9.31.

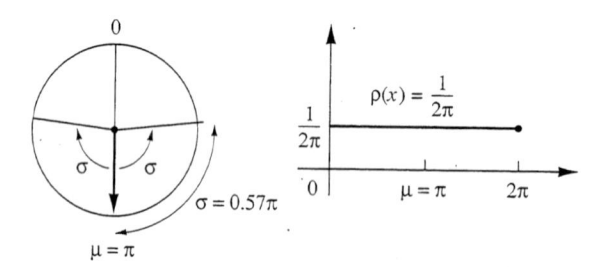

Figure 9.31

b. By definition,

$$V(X) = \int_0^{2\pi} (x - \mu)^2 \rho(x)\, dx$$

$$= \int_0^{2\pi} (x - \pi)^2 \left(\frac{1}{2\pi}\right) dx = \frac{1}{2\pi} \int_0^{2\pi} (x^2 - 2\pi x + \pi^2)\, dx$$

$$= \frac{1}{2\pi}\left(\frac{x^3}{3} - \pi x^2 + \pi^2 x\right)\Big]_0^{2\pi} = \frac{1}{2\pi}\left(\frac{(2\pi)^3}{3} - 4\pi^3 + 2\pi^3\right)$$

$$= \frac{1}{2\pi}\left(\frac{8\pi^3 - 6\pi^3}{3}\right) = \frac{\pi^2}{3} \simeq (1.0472)\pi$$

Therefore $\sigma = \sqrt{V(X)} \simeq 1.023\sqrt{\pi} \simeq 1.81 \simeq 0.58\pi$. ∎

Thus while we "expect" the pointer to stop at $\mu = \pi$, it has a tendency to deviate about this stopping point by more than a quarter circle (which has length 0.5π) on either side, as indicated in Figure 9.31.

EXAMPLE 3

Find $V(X)$ and σ for the r.v. X of Example 1, with p.d.f. $\rho(x) = \frac{3}{4}[-(x-6)^2 + 1]$ and $\mu = 6$.

Solution By definition,

$$V(X) = \int_5^7 (x-6)^2 \left(\frac{3}{4}(1 - (x-6)^2)\right) dx$$

$$= \frac{3}{4}\left[\int_5^7 (x-6)^2 \, dx - \int_5^7 (x-6)^4 \, dx\right]$$

$$= \frac{3}{4}\left[\frac{(x-6)^3}{3}\bigg]_5^7 - \frac{(x-6)^5}{5}\bigg]_5^7\right]$$

$$= \frac{3}{4}\left(\frac{2}{3} - \frac{2}{5}\right) = \frac{1}{5} = 0.2$$

Therefore $\sigma = \sqrt{1/5} \approx 0.447$. ∎

Recall that in this model events near the mean $\mu = 6$ are much more likely since the p.d.f. ρ bounds most of its area near $\mu = 6$ (see Figure 9.30). In this blindfolded jump we should expect to land 6 ft away with a deviation of about 5 in. (approximately 0.447×12 in.).

It may be shown (Exercise 34) that $V(X)$ is also given by

$$V(X) = \int_a^b x^2 \rho(x) \, dx - \mu^2 \qquad (7)$$

This formula is sometimes easier to use when finding $V(X)$ and σ.

Exponentially Distributed Random Variables

A manufacturer is always concerned with reliability whether the product is a light bulb or an automobile. The useful life X of a product can sometimes be modeled as a random variable that is *exponentially distributed*.

Definition

A random variable X is said to be **exponentially distributed** if $X \geq 0$ and the p.d.f for X is $\rho(x) = ke^{-kx}$, for $x \geq 0$ and for some $k > 0$. That is,

$$P(X \leq t) = \int_0^t ke^{-kx} \, dx$$

Figure 9.32 (page 546) indicates the graph of an exponential density function. Since

$$\int_0^\infty ke^{-kx} \, dx = \lim_{b \to \infty}\left(k \int_0^b e^{-kx} \, dx\right) = \lim_{b \to \infty} k\left[\frac{e^{-kx}}{-k}\right]_0^b = \lim_{b \to \infty}(1 - e^{-kb}) = 1$$

no matter what number k is, we are always assured that $0 \leq F(t) = P(X \leq t) \leq 1$. Moreover, since $P(X > t) = 1 - P(X \leq t)$ (because the events

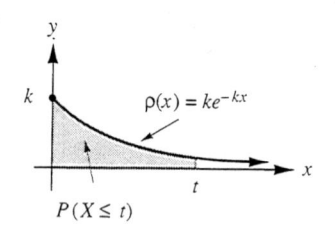

$P(X \le t)$

Figure 9.32 The exponential density function

$X > t$ and $X \le t$ can have nothing in common), we see that the probability that X exceeds large values of t must be small, because most of the area beneath the graph in Figure 9.32 is concentrated to the left side of the region bounded by the graph.

It may be shown (see Exercise 16) that if X is exponentially distributed, then

$$\mu = E(X) = \int_0^\infty x(ke^{-kx})\, dx = \frac{1}{k} \tag{8}$$

The number k is again a *parameter* and must be adapted to the particular case at hand, as in the next example.

EXAMPLE 4

A 75-watt light bulb has an expected lifetime of 1,200 hours before it burns out. Let X be the lifetime (in days) of such a bulb that is left burning 24 hours a day, and suppose that X is exponentially distributed.
  a. Find $P(X \le 40)$, the probability that the bulb will fail within 40 days (960 hours).
  b. Find $P(50 \le X)$, the probability that the bulb will last at least 50 days.
  c. Find the standard deviation σ of the bulb's lifetime about its mean lifetime.

Solution First we realize that $\mu = E(X) = \frac{1,200}{24} = 50$ days. From Equation 8, this means that $1/k = 50$ (or $k = 0.02$) in the formula $\rho(x) = ke^{-kx}$ for this particular r.v.
  a. Since X is exponentially distributed,

$$P(X \le 40) = \int_0^{40} ke^{-kx}\, dx \quad \text{where } k = \tfrac{1}{50}$$

$$= -e^{-kx} \Big]_0^{40}$$

$$= -e^{-40/50} + e^0 = 1 - e^{-0.8} \approx 0.5507$$

There is a better than even (55.07%) chance that the bulb will fail within 40 days of use.

  b. $$P(X \ge 50) = \int_{50}^{\infty} ke^{-kx}\, dx = \lim_{b \to \infty} (-e^{-kx}) \Big]_{50}^{b}$$

$$= \lim_{b \to \infty} -e^{-kb} + e^{-50/50} \approx 0 + 0.3679 = 0.3679$$

546 Chapter 9 Probability

There is a better than one-third (36.78%) chance that the bulb will last longer than its expected lifetime of 50 days.

c. From Equation 7,

$$V(X) = \int_0^\infty x^2 f(x)\, dx - \mu^2 = \lim_{b \to \infty} \int_0^b x^2 (ke^{-kx})\, dx - \mu^2$$

$$= \lim_{b \to \infty} k \left[\left(\frac{x^2 e^{-kx}}{-k} \right)_0^b + \frac{2}{k} \int_0^b x e^{-kx}\, dx \right] - \mu^2 \quad \text{Using integration by parts}$$

$$= \lim_{b \to \infty} \frac{2}{k} \int_0^b x e^{-kx}\, dx - \mu^2 \quad \text{Since } \lim_{b \to \infty} b^2 e^{-kb} = 0$$

$$= \frac{2}{k^2} - \mu^2 \quad \text{Using Equation 8 to evaluate the integral}$$

$$= \frac{2}{k^2} - \frac{1}{k^2} = \frac{1}{k^2} \quad \text{Since } \mu = \frac{1}{k}$$

Thus $V(X) = 1/k^2 = (1/k)^2 = \mu^2 = (50)^2 = 2{,}500$ and therefore $\sigma = \sqrt{2{,}500} = 50$. ∎

This large deviation indicates that while the light bulb is expected to last 50 days, it could last a great deal longer, or a great deal less, and that failure the moment the bulb is turned on is within one standard deviation of its mean! We graph this density function p in Figure 9.33.

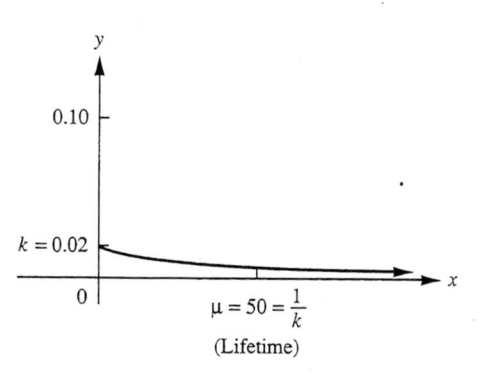

Figure 9.33

(Lifetime) $\mu = 50 = \frac{1}{k}$, $k = 0.02$

In general it can be shown that when an r.v. X has a certain "memoryless property," then the exponential distribution is often a good model for its behavior. This memoryless property is that if X has not occurred by time a, the probability that X will not occur by time $a + b$ is the same as the probability that X would not occur by time b, if put in service at time 0. A light bulb has this property: It has no moving parts and only conducts electricity through a filament that remains essentially stable while in use and is in much the same state each time it is turned on; that is, once turned off the light bulb "forgets" that it has been used and burns just as brightly as before when turned on again.

The exponential distribution is a special case of a general class of probability density functions known as **"gamma densities."** The next simplest member of this class is a random variable whose p.d.f. is the function $\rho(x) = b^2 x e^{-bx}$, shown in Figure 9.34. This graph indicates that there is a low probability of an event near 0, a high probability for events near $1/b$, and a slowly dissipating probability thereafter.

Let us first verify that $\rho(x) = b^2 x e^{-bx}$, $x \geq 0$, has unit area below its graph, as it must if it is to represent a probability distribution. We have, using integration by parts,

$$\int_0^\infty b^2 x e^{-bx}\, dx = b^2 \left(\lim_{r \to \infty} \int_0^r x e^{-bx}\, dx \right)$$

$$= b^2 \cdot \lim_{r \to \infty} \int_0^r x\, d\!\left(\frac{e^{-bx}}{-b}\right)$$

$$= b^2 \lim_{r \to \infty} \left[\left(\frac{x e^{-bx}}{-b}\right)_0^r - \left(\frac{1}{-b}\right) \int_0^r e^{-bx}\, dx \right]$$

$$= b^2 \lim_{r \to \infty} \frac{1}{b} \int_0^r e^{-bx}\, dx \qquad \text{Since } \lim_{r \to \infty} r e^{-br} = 0 \text{ and } 0 \cdot e^{-b \cdot 0} = 0$$

$$= b^2 \lim_{r \to \infty} \frac{1}{b} \left(\frac{e^{-bx}}{-b}\right)_0^r$$

$$= \lim_{r \to \infty} (e^0 - e^{-br}) = 1$$

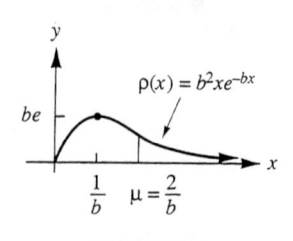

Figure 9.34

$\rho(x) = b^2 x e^{-bx}$

$\dfrac{1}{b}\quad \mu = \dfrac{2}{b}$

It may similarly be shown (Exercise 18) that if X has $\rho(x) = b^2 x e^{-bx}$ as its p.d.f., then

$$\mu = E(X) = \frac{2}{b} \quad \text{and} \quad \sigma = \frac{\sqrt{2}}{b} \qquad (9)$$

EXAMPLE 5

A filter for the fuel injection system of an automobile is known to have a useful life X (in thousands of miles) that is distributed with p.d.f. $\rho(x) = b^2 x e^{-bx}$. Moreover, the filter has an expected useful life of 6,000 miles.

a. Find $P(X \geq 8)$, the probability that the filter will last at least 8,000 miles.
b. Find $P(X \leq 2)$ and explain what this number means to a car dealer who has sold 150 new cars equipped with this filter.

Solution We are given that $\mu = E(X) = 6$. Therefore from Equation 9 $2/b = 6$ or $b = 1/3$. Hence $\rho(x) = x e^{-x/3}/9$ for this particular product.

a. Let us first find $P(X \leq 8)$, the probability that the filter fails *within* 8,000 miles. We have, using integration by parts,

$$P(X \leq 8) = \int_0^8 \frac{x e^{-x/3}}{9}\, dx = \frac{1}{9}[-3(x e^{-x/3} + 3 e^{-x/3})]_0^8$$

$$= -\frac{1}{3}(8 e^{-8/3} + 3 e^{-8/3} - 3) \approx 0.745$$

Now $P(X \leq 8) + P(X > 8) = 1$ so $P(X > 8) = 1 - P(X \leq 8) \approx 1 - 0.745 = 0.255$. Remembering that $P(X = 8) = 0$, we have $P(X \geq 8) \approx 0.255$. There is only about a 25% chance of the filter lasting 2,000 miles more than its expected 6,000-mile life.

b. We have

$$P(X \leq 2) = \frac{1}{9} \int_0^2 xe^{-x/3} \, dx \approx 0.144$$

integrating as before. Thus there is about a 14% chance of failure within 2,000 miles. The dealer can expect to see $0.144 \, (150) = 21.6$ of the 150 original customers back shortly, wanting to know why their cars are not running properly. ∎

How do we know what kind of p.d.f. to use in modeling a given r.v.? The physical characteristics and operating environment of a light bulb versus an automotive filter suggest that they could not be modeled by the same kind of distribution (the filter must remove and store dirt from the fuel before it is burned in the engine and is subject to an increasing buildup of dirt over time). A statistician begins with data on such devices. The data, which will necessarily be discrete, and not continuous, are often represented by a graph called a **frequency histogram,** showing the frequency of success (or failure) for a device. The shape of this histogram is then compared to the graphs of various possible p.d.f.'s. The statistician may then choose one of these because its shape is most similar to that shown by the data (see Exercise 19, Section 9.3). Probabilistic predictions may then be made using techniques like those in this section. If these predictions yield useful information consistent with future events, the chosen p.d.f. can be regarded as a useful working model. If they do not, the statistician must analyze the system further, find better data, try a different p.d.f., and test its predictions against reality.

Exercises 9.4

Find the expected value μ and the standard deviation σ of the r.v. whose p.d.f. is given on the indicated interval in Exercises 1–14. Graph each function and indicate the point μ on the given interval. Then indicate σ as a distance measurement about each side of μ.

1. $\rho(x) = \frac{1}{3}$ on $[1, 4]$ **2.** $\rho(x) = \frac{x+1}{2}$ on $[-1, 1]$

3. $\rho(x) = \frac{2}{9}x(3 - x)$ on $[0, 3]$

4. $\rho(x) = \frac{1 - x^2}{2}$ on $[-1, 1]$

5. $\rho(x) = \frac{1}{x \ln 3}$ on $[1, 3]$

6. $\rho(x) = \frac{3}{4}x^2(2 - x)$ on $[0, 2]$

7. $\rho(x) = 1.02e^{-x}$ on $[0, 4]$

8. $\rho(x) = \left(\frac{e}{e - 2}\right)xe^{-x}$ on $[0, 1]$

9. $\rho(x) = 2e^{2(1-x)}$ on $[1, \infty]$

10. $\rho(x) = \left(\frac{e}{2}\right)xe^{-x}$ on $[1, \infty]$

11. $\rho(x) = \dfrac{9}{4x^3}$ on $[1, 3]$ 12. $\rho(x) = \dfrac{2}{x^3}$ on $[1, \infty]$

13. $\rho(x) = |x|$ on $[-1, 1]$

14. $\rho(x) = \tfrac{1}{4}|x - 2|$ on $[0, 4]$

15. Find the mean μ and standard deviation σ of the r.v. with p.d.f. $\rho(x) = \tfrac{1}{2}e^{-0.5x}$ on $[0, \infty)$.

16. Show that $\int_0^\infty x(ke^{-kx})\, dx = 1/k$, establishing that $E(X) = 1/k$ for an exponentially distributed r.v. X.

17. Find the mean μ, variance, and standard deviation σ of a random variable with p.d.f. $\rho(x) = 4xe^{-2x}$ on $[0, \infty)$.

18. Show that if X is distributed according to the p.d.f. $\rho(x) = b^2xe^{-bx}$ then $E(X) = 2/b$ and $\sigma = \sqrt{2}/b$, establishing Equation 9.

The *median* of a probability distribution is the number M such that $P(X \le M) = \tfrac{1}{2}$; that is, there is a 50-50 chance of an outcome on either side of M, and a vertical line drawn at M divides the area beneath the graph of ρ in half. Find the median M for each p.d.f. in Exercises 19–22 by solving for M in the equation $\int_0^M \rho(x)\, dx = \tfrac{1}{2}$ after performing the indicated integration. Sketch the graph of ρ, and indicate the position of the number M on the x-axis.

19. $\rho(x) = \tfrac{1}{5}$ on $[0, 5]$

20. $\rho(x) = 6x(1 - x)$ on $[0, 1]$

21. $\rho(x) = \tfrac{1}{2}e^{-0.5x}$ on $[0, \infty)$

22. $\rho(x) = \dfrac{1}{(\ln 3)(x + 1)}$ on $[0, 2]$

23. Find a formula for the cumulative distribution F of an r.v. X that is exponentially distributed. Find $P(X \le 2)$, $P(3 \le X \le 5)$, $P(X \ge 6)$, and $P(X \ge 1/k)$ using this formula.

24. Find a formula for the cumulative distribution F of an r.v. X that has p.d.f. $\rho(x) = b^2xe^{-bx}$ on $[0, \infty)$. Use this formula to find $P(X \le 1)$, $P(1 \le X \le 3)$, $P(X \ge 3)$, $P(X \le 2/b)$, and $P[X \le (2/b) - (2/b^2)]$.

25. Assume that the expected lifetime X of a U.S. citizen born in 1941 is 65 years and that X is exponentially distributed. Find $P(X \ge 70)$ and $P(X \ge 80)$. Is this a good model? That is, are these answers reasonable?

26. Let X be the percentage of intended impurities in a transistor. If X is less than 30% or greater than 50%, the transistor will not function properly. Suppose that X has p.d.f. $\rho(x) = 12(x - 1)^2 x$, $0 \le x \le 1$, by writing the percentage X as a decimal. What is the probability that a transistor will be defective?

27. Assume that the lifetime of the blue crab is exponentially distributed with mean $\mu = 5$ years and that crab traps are designed so that crabs younger than 2 years can escape. What proportion of the crab population is subject to being caught?

28. Suppose that the number of tons X of raw material used in a manufacturing process each day is exponentially distributed with $\mu = 20$ tons per day. What is the probability that 23 tons will be used in 1 day? That between 18 and 22 tons will be used? What is the variance in use?

29. Please refer to Exercise 28. How much of the raw material should be stocked so that the probability of running out of the material on a given day is 5%? [*Hint:* Find x so that $P(X \ge x) = 0.05$.]

30. The time X taken to manufacture a certain product is a random variable with mean time $\mu = 5$ hours and p.d.f. $\rho(x) = b^2xe^{-bx}$. What is the probability that the product can be manufactured in 4 hours? That it will take at least 5 hours to manufacture it? Between 4.5 and 5.5 hours to manufacture it?

Challenge Problems

31. Let X be the *time between arrivals* of buses at a bus stop, and assume that X is a random variable with p.d.f. $\rho(x) = (x/4,500)(30 - x)$, $0 \le x \le 30$ minutes. Find the average (mean) time between arrivals. What is the probability that successive buses will arrive within 2 minutes of this expected "interarrival" time?

32. Please refer to Exercise 31. Suppose that just as you walk up to the bus stop you see the bus leaving. What is the probability that you will catch the next bus if you wait 12 minutes? How long should you wait so as to have a 95% chance of catching the next bus?

33. Aircraft engines are routinely rebuilt after a certain specified number of operating hours so as to ensure that the engine will not fail in use. Assume that the average number of trouble-free hours of operation of a certain aircraft engine is 500 hours and that the number of hours X of trouble-free operation has p.d.f. $\rho(x) = b^2xe^{-bx}$. Assume further that the carrier rebuilds the engine after 75 hours of operation. What is the probability of a failure during flight under this policy?

34. Remembering that $\int_a^b \rho(x)\, dx = 1$ for p.d.f. ρ and that μ is a constant, show that $V(X) = \int_a^b x^2 \rho(x)\, dx - \mu^2$ by squaring $(x - \mu)^2$ and then integrating.

9.5 The Normal Distribution

In this section we study the normal distribution found throughout the statistical analysis of real systems.

Many phenomena have the character of resulting from repeated "either-or" outcomes. The inheritance of physical characteristics such as height and hair color is among these. It can be shown that a random variable representing the outcome of a long series of repeated either-or outcomes has a density function that is well approximated by the "normal, bell-shaped curve" of Figure 9.35. Indeed, Sir Francis Galton, who initiated statistical analysis, first developed the normal curve in order to explain the inheritance of physical characteristics.

The Standard Normal Curve

If a continuously distributed random variable X has a probability density function N whose graph is like that shown in Figure 9.35, then an appropriate formula for N is the **standard normal density function**

$$N(x) = \frac{1}{\sqrt{2\pi}} e^{-x^2/2}$$

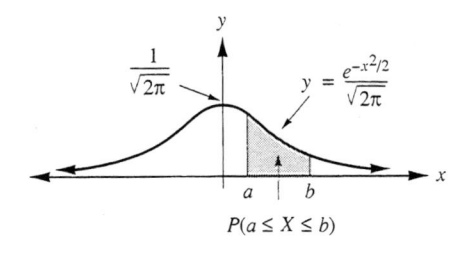

Figure 9.35

The standard normal curve

Therefore,
$$P(a \le X \le b) = \frac{1}{\sqrt{2\pi}} \int_a^b e^{-x^2/2} \, dx$$

Such a random variable X is said to be a **standard normal random variable**, and it can be shown that X has mean $\mu = 0$ and standard deviation $\sigma = 1$. The total area beneath the normal curve of Figure 9.35, an improper integral, must of course be 1. The appropriate value of a parameter to ensure this is the number $1/\sqrt{2\pi}$, which appears in the formula for the normal density function N.

Thus, to determine normal probabilities, we have to compute the integral

$$\int e^{-z^2/2} \, dz$$

However, as we have noted before, it is not possible to find an antiderivative for e^{-z^2}. Instead, techniques of approximate integration must be used to approximate the values of this integral.

A table of (approximate) values for the integral

$$\frac{1}{\sqrt{2\pi}} \int_0^a e^{-z^2/2} \, dz$$

for $a \geq 0$ is given at the close of this section (see Table 9.1, p. 558). Let us now see how standard normal probability values can be determined from Table 9.1.

EXAMPLE 1

Find the value of

$$\frac{1}{\sqrt{2\pi}} \int_0^{1.57} e^{-z^2/2} \, dz$$

using Table 9.1.

Solution We determine this value by reading *down the left column* in Table 9.1 to the number 1.5 and then *across that row* to column .07 to obtain the value

$$\frac{1}{\sqrt{2\pi}} \int_0^{1.57} e^{-z^2/2} \, dz = 0.4418 \quad \blacksquare$$

This value is the area beneath the curve of the shaded region shown in Figure 9.36.

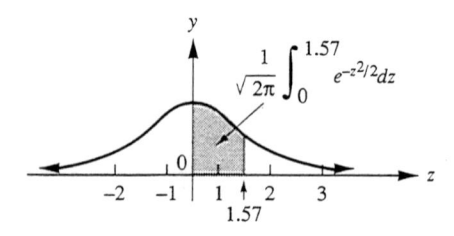

Figure 9.36

Let us next see how all other areas beneath the standard normal curve can be found using Table 9.1. This method involves no more than the subtraction (and sometimes addition) of the areas of appropriate regions whose areas can be found in Table 9.1 (see Figure 9.37).

Let $A(x) = (1/\sqrt{2\pi}) \int_0^x e^{-z^2/2} \, dz$. The number $A(x)$ can be found in Table 9.1 and represents the area of the region shown in Figure 9.38.

Now, returning to Figure 9.37, we see that the indicated areas are

Figure 9.37

$A = A(b) - A(a)$ (a)
$B = \frac{1}{2} - A(b)$ (b)
$C = A(b)$ (c)
$D = A(a) + A(b)$ (d)

(a)

(b)

(c)

(d)

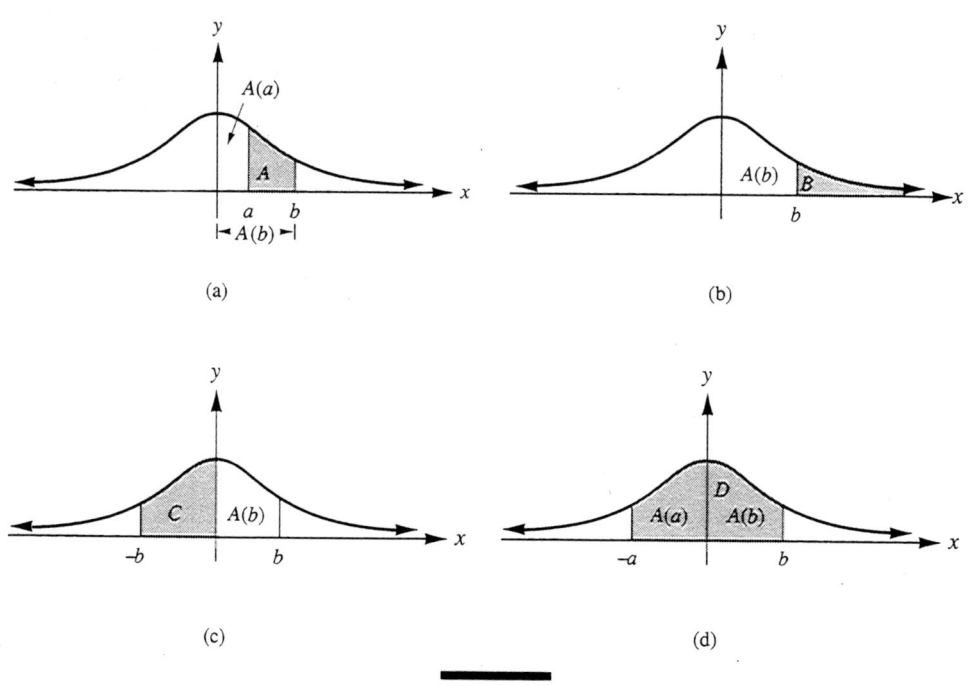

Figure 9.37

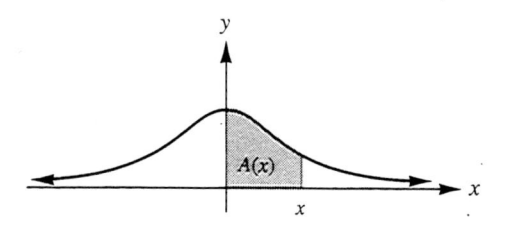

Figure 9.38

Notice that the formula $B = \frac{1}{2} - A(b)$ follows from the fact that since the area below the entire normal curve is one unit, the area beneath the curve to the right of the y-axis is one-half unit.

EXAMPLE 2
Use Table 9.1 to find

a. $\dfrac{1}{\sqrt{2\pi}} \displaystyle\int_{-2}^{1/2} e^{-z^2/2}\, dz$ **b.** $\dfrac{1}{\sqrt{2\pi}} \displaystyle\int_{0.42}^{1.5} e^{-z^2/2}\, dz$

Solution
a. From Figure 9.37(d), this integral is
$$A(2) + A(\tfrac{1}{2}) = 0.4772 + 0.1915 = 0.6687$$
b. From Figure 9.37(a), this integral is
$$A(1.5) - A(0.42) = 0.4332 - 0.1628 = 0.2704 \quad\blacksquare$$

9.5 The Normal Distribution

The General Normal Curve with Mean μ and Standard Deviation σ

It can be shown that if a continuously distributed random variable X has a density function whose graph is like that in Figure 9.39, then an appropriate probability density function for X is the function

$$N_{\sigma,\mu}(x) = \frac{1}{\sigma\sqrt{2\pi}} e^{(-1/2)[(x-\mu)/\sigma]^2}$$

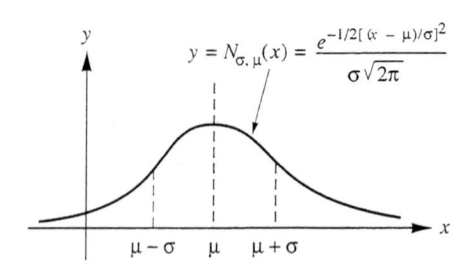

Figure 9.39

Such a random variable X is then said to be a **normally distributed random variable,** and it may be shown (see Exercises 24 and 25) that $E(X) = \mu$ and $V(X) = \sigma^2$. This means that the numbers μ and σ, which arise in the formula for N in order to give the desired graph for algebraic and geometric reasons, coincide in the end with the probabilistic mean and standard deviation.

To compute probabilities of the general normal r.v. X, we must perform a change of variable that reduces the integral

$$P(a \leq X \leq b) = \frac{1}{\sigma\sqrt{2\pi}} \int_a^b e^{(-1/2)[(x-\mu)/\sigma]^2} \, dx \tag{1}$$

to an integral of the *standard* normal r.v. of mean $\mu = 0$ and standard deviation $\sigma = 1$ already studied. This is not difficult.

Consider what happens to the indefinite integral

$$\int \frac{1}{\sigma\sqrt{2\pi}} e^{(-1/2)[(x-\mu)/\sigma]^2} \, dx$$

under the change of variable

$$z = \frac{x-\mu}{\sigma} = \frac{x}{\sigma} - \frac{\mu}{\sigma}$$

We have $dz = (1/\sigma) \, dx$ and therefore

$$\int \frac{1}{\sigma\sqrt{2\pi}} e^{(-1/2)[(x-\mu)/\sigma]^2} \, dx = \int \frac{e^{-z^2/2}}{\sqrt{2\pi}} \, dz$$

This second integral is the integral for the standard normal curve. Moreover, when $x = a$, then $z = (a - \mu)/\sigma$; and when $x = b$, then $z = (b - \mu)/\sigma$. It can then be shown that for the *definite* integral, we have

554 Chapter 9 Probability

$$\frac{1}{\sigma\sqrt{2\pi}} \int_a^b e^{(-1/2)[(x-\mu)/\sigma]^2} \, dx = \frac{1}{\sqrt{2\pi}} \int_{(a-\mu)/\sigma}^{(b-\mu)/\sigma} e^{-z^2/2} \, dz \qquad (2)$$

EXAMPLE 3

Let X be normally distributed with mean $\mu = 5$ and standard deviation $\sigma = 2$. Find $P(3 \leq X \leq 6)$, the probability that X occurs between $a = 3$ and $b = 6$.

Solution From Equation 1, with $\mu = 5$ and $\sigma = 2$,

$$P(3 \leq X \leq 6) = \frac{1}{2\sqrt{2\pi}} \int_3^6 e^{-1/2[(x-5)/2]^2} \, dx$$

Let $z = (x - 5)/2$. When $x = 3$, we have $z = (3 - 5)/2 = -1$; when $x = 6$, we have $z = (6 - 5)/2 = \frac{1}{2} = 0.5$. Therefore, from Equation 2,

$$P(3 \leq X \leq 6) = \frac{1}{\sqrt{2\pi}} \int_{-1}^{0.5} e^{-z^2/2} \, dz$$

Using Table 9.1, as in Examples 1 and 2, gives

$$P(3 \leq X \leq 6) = 0.3413 + 0.1915 = 0.5328 \quad\blacksquare$$

It may be shown that the inflection points on the normal curve defined by $y = N_{\sigma,\mu}(x)$ occur at precisely the points $x = \mu \pm \sigma$ (see Exercises 21 and 22). That is, the inflection points occur at exactly one standard deviation from that mean. The interval $[\mu - \sigma, \mu + \sigma]$ has an interesting probabilistic property as well.

EXAMPLE 4

Show that, no matter what the mean μ and standard deviation σ of a normal r.v. are, we always have

$$P(\mu - \sigma \leq X \leq \mu + \sigma) = 0.6826$$

That is, there is always a better than $\frac{2}{3}$ chance that X will occur within one standard deviation σ of its mean μ.

Solution Let $z = (x - \mu)/\sigma$, $a = \mu - \sigma$, and $b = \mu + \sigma$. Using Equations 1 and 2, we have

$$P(\mu - \sigma \leq X \leq \mu + \sigma) = \frac{1}{\sqrt{2\pi}} \int_{(b-\mu)/\sigma}^{(a-\mu)/\sigma} e^{-(z^2/2)} \, dz$$

But

$$\frac{a - \mu}{\sigma} = \frac{(\mu + \sigma) - \mu}{\sigma} = -\frac{\sigma}{\sigma} = -1$$

$$\frac{b - \mu}{\sigma} = \frac{(\mu + \sigma) - \mu}{\sigma} = \frac{\sigma}{\sigma} = +1$$

and therefore $P(\mu - \sigma \le X \le \mu + \sigma) = \dfrac{1}{\sqrt{2\pi}} \displaystyle\int_{-1}^{1} e^{-z^2/2}\, dz$

$$= 0.3413 + 0.3413$$
$$= 0.6826 \quad \blacksquare$$

The next example may have "everyday" meaning.

EXAMPLE 5

Your faithful friend promises to call you every evening at 7 P.M. After some time has passed, you begin to realize that the time the call actually arrives is a normal random variable X with mean $\mu = 7$ and standard deviation 5 minutes. On any given evening, what is the probability that your friend's call will arrive 6 to 9 minutes late?

Solution Since 5 minutes is $\frac{5}{60} \approx 0.08$ of an hour, the numerical standard deviation σ for this random variable is $\sigma = 0.08$. Since $\frac{6}{60} = 0.10$ and $\frac{9}{60} = 0.15$, we are asked for $P(7.10 \le X \le 7.15)$, the probability that the call arrives between 7:06 and 7:09.

Let $z = (x - 7)/0.08$, $a = 7.10$, and $b = 7.15$. Using Equations 1 and 2, we have

$$P(7.10 \le X \le 7.15) = \dfrac{1}{0.08\sqrt{2\pi}} \int_{7.10}^{7.15} e^{-1/2[(x-7)/0.08]^2}\, dx$$

$$= \dfrac{1}{\sqrt{2\pi}} \int_{1.25}^{1.875} e^{-(z^2/2)}\, dz$$

since $(7.10 - 7)/0.08 = 1.25$ and $(7.15 - 7)/0.08 = 1.875$. Consequently, using Table 9.1,

$$P(7.10 \le X \le 7.15) \approx 0.4699 - 0.3944 = 0.075$$

where we read the normal probability for $z = 1.88 \approx 1.875$. Thus there is about a 7.5% chance that your friend will call 6 to 9 minutes late. $\blacksquare$

The graph in Figure 9.40 suggests a needed fact about the normal curve that can be proven, although we will not do so.

$$\dfrac{1}{\sigma\sqrt{2\pi}} \int_{a}^{\infty} e^{(-1/2)[(x-\mu)/\sigma]^2}\, dx = \dfrac{1}{\sqrt{2\pi}} \int_{(a-\mu)/\sigma}^{\infty} e^{-(z^2/2)}\, dz \qquad (3)$$

That is, the entire area to the right of a beneath the general normal curve equals the entire area to the right of $(a - \mu)/\sigma$ beneath the standard normal curve.

EXAMPLE 6

A roofing contractor installs shingles with an expected lifetime X that is normally distributed with mean $\mu = 15$ years and standard deviation $\sigma = 2$ years.

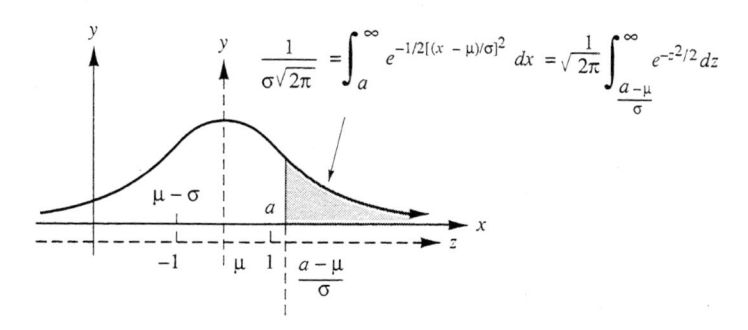

Figure 9.40

$$\frac{1}{\sigma\sqrt{2\pi}} = \int_a^\infty e^{-1/2[(x-\mu)/\sigma]^2}\,dx = \frac{1}{\sqrt{2\pi}}\int_{\frac{a-\mu}{\sigma}}^\infty e^{-z^2/2}\,dz$$

a. What is the probability that the shingles installed on a given roof will last more than 16 years?
b. The contractor shingles 120 roofs in 1 year. How many of these jobs can be expected to last at least 16 years?

Solution

a. Here X is normally distributed with p.d.f.

$$N_{2,15}(x) = \frac{1}{2\sqrt{2\pi}} e^{(-1/2)[(x-15)/2]^2}$$

We are asking for $P(16 \leq X)$. Let $z = (x-15)/2$. When $x = 16$, $z = (16-15)/2 = \frac{1}{2}$. According to Equation 3, we have

$$P(16 \leq X) = \frac{1}{2\sqrt{2\pi}} \int_{16}^\infty e^{(-1/2)[(x-15)/2]^2}\,dx$$

$$= \frac{1}{\sqrt{2\pi}} \int_{1/2}^\infty e^{-z^2/2}\,dz$$

$$= \frac{1}{2} - 0.1915 = 0.3085$$

using Table 9.1.

b. From (a), we can expect that the proportion $(0.3085)(120) = 37$ of these roofing jobs will last at least 16 years. ∎

How do we know when to model a random variable X by a normal distribution? In practice we record the outcomes of X over many trials in a frequency histogram. If X "seems" to be normally distributed, we use the data to estimate μ and σ. Predictions based on a then assumed normal distribution (with these values for μ and σ) are then tested against reality.

In Section A.4, you will find an exposition of how M. King Hubbert used the shape of the normal curve to predict the peak output and then the decline of U.S. oil reserves 15 years before it occurred and contrary to all prevailing opinion at the time.

$$A(a) = \frac{1}{\sqrt{2\pi}} \int_0^a e^{-z^2/2} dz$$

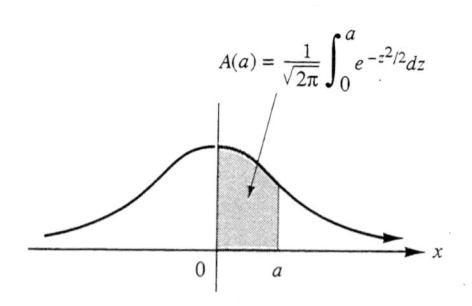

Table 9.1

a	.00	.01	.02	.03	.04	.05	.06	.07	.08	.09
0.0	.0000	.0040	.0080	.0120	.0160	.0199	.0239	.0279	.0319	.0359
0.1	.0398	.0438	.0478	.0517	.0557	.0596	.0636	.0675	.0714	.0753
0.2	.0793	.0832	.0871	.0910	.0948	.0987	.1026	.1064	.1103	.1141
0.3	.1179	.1217	.1255	.1293	.1331	.1368	.1406	.1443	.1480	.1517
0.4	.1554	.1591	.1628	.1664	.1700	.1736	.1772	.1808	.1844	.1879
0.5	.1915	.1950	.1985	.2019	.2054	.2088	.2123	.2157	.2190	.2224
0.6	.2257	.2291	.2324	.2357	.2389	.2422	.2454	.2486	.2517	.2549
0.7	.2580	.2611	.2642	.2673	.2704	.2734	.2764	.2794	.2823	.2852
0.8	.2881	.2910	.2939	.2967	.2995	.3023	.3051	.3078	.3106	.3133
0.9	.3159	.3186	.3212	.3238	.3264	.3289	.3315	.3340	.3365	.3389
1.0	.3413	.3438	.3461	.3485	.3508	.3531	.3554	.3577	.3599	.3621
1.1	.3643	.3665	.3686	.3708	.3729	.3749	.3770	.3790	.3810	.3830
1.2	.3849	.3869	.3888	.3907	.3925	.3944	.3962	.3980	.3997	.4015
1.3	.4032	.4049	.4066	.4082	.4099	.4115	.4131	.4147	.4162	.4177
1.4	.4192	.4207	.4222	.4236	.4251	.4265	.4279	.4292	.4306	.4319
1.5	.4332	.4345	.4357	.4370	.4382	.4394	.4406	.4418	.4429	.4441
1.6	.4452	.4463	.4474	.4484	.4495	.4505	.4515	.4525	.4535	.4545
1.7	.4554	.4564	.4573	.4582	.4591	.4599	.4608	.4616	.4625	.4633
1.8	.4641	.4649	.4656	.4664	.4671	.4678	.4686	.4693	.4699	.4706
1.9	.4713	.4719	.4726	.4732	.4738	.4744	.4750	.4756	.4761	.4767
2.0	.4772	.4778	.4783	.4788	.4793	.4798	.4803	.4808	.4812	.4817
2.1	.4821	.4826	.4830	.4834	.4838	.4842	.4846	.4850	.4854	.4857
2.2	.4861	.4864	.4868	.4871	.4875	.4878	.4881	.4884	.4887	.4890
2.3	.4893	.4896	.4898	.4901	.4904	.4906	.4909	.4911	.4913	.4916
2.4	.4918	.4920	.4922	.4925	.4927	.4929	.4931	.4932	.4934	.4936
2.5	.4938	.4940	.4941	.4943	.4945	.4946	.4948	.4949	.4951	.4952
2.6	.4953	.4955	.4956	.4957	.4959	.4960	.4961	.4962	.4963	.4964
2.7	.4965	.4966	.4967	.4968	.4969	.4970	.4971	.4972	.4973	.4974
2.8	.4974	.4975	.4976	.4977	.4977	.4978	.4979	.4979	.4980	.4981
2.9	.4981	.4982	.4982	.4983	.4984	.4984	.4985	.4985	.4986	.4986
3.0	.4987	.4987	.4987	.4988	.4988	.4989	.4989	.4989	.4990	.4990
3.1	.4990	.4991	.4991	.4991	.4992	.4992	.4992	.4992	.4993	.4993
3.2	.4993	.4993	.4994	.4994	.4994	.4994	.4994	.4995	.4995	.4995
3.3	.4995	.4995	.4995	.4996	.4996	.4996	.4996	.4996	.4996	.4997
3.4	.4997	.4997	.4997	.4997	.4997	.4997	.4997	.4997	.4997	.4998
3.5	.4998	.4998	.4998	.4998	.4998	.4998	.4998	.4998	.4998	$A(a)$

Exercises 9.5

Let X be a standard normal r.v. Write each of the probabilities in Exercises 1–12 as an integral, and determine its value using Table 9.1.

1. $P(0 \leq X \leq 1.5)$
2. $P(-1 \leq X \leq 0.5)$
3. $P(0.5 \leq X \leq 1.5)$
4. $P(-1 \leq X \leq 0)$
5. $P(-1 \leq X \leq -0.5)$
6. $P(X \geq 1.3)$
7. $P(X \leq -0.55)$
8. $P(X \leq 1.5)$
9. $P(X \geq 0.33)$
10. $P(X \geq -0.85)$
11. $P(X \leq -1) + P(X \geq 1)$
12. $P(|X| \leq 0.5)$

13. A *binomial random variable* is one with an "either-or" outcome, such as observing whether "heads" results on the toss of a coin. Imagine tossing a coin four times. The event that "one head" occurs in four tosses can occur in four different ways, showing on the first, second, third, or fourth toss. Explain how the graph in Figure 9.41 is an appropriate probabilistic model of the total number of heads that can occur in four tosses, first convincing yourself that 16 different outcomes are possible in four tosses. Note how the graph distributes these outcomes in analogy with the normal curve. A similar graph, representing all the possibilities in, say, 100 repeated tosses, would look very much like the normal curve. In a precise sense the normal density is the limiting case of such *repeated* binomial trials, accounting for its persistence in applications.

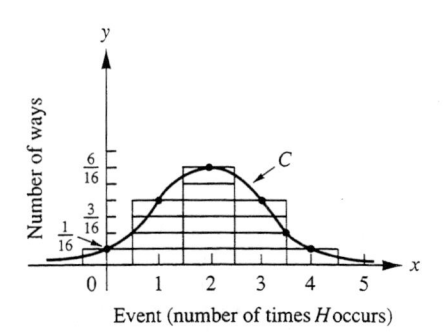

Figure 9.41

14. Repeat Exercise 13 for three tosses, having eight possible outcomes.

Use a change of variable and Table 9.1 to find the value of each integral in Exercises 15–20. Round off the limits of integration before using Table 9.1.

15. $\dfrac{1}{3\sqrt{2\pi}} \displaystyle\int_{19}^{22} e^{(-1/2)[(x-21)/3]^2}\,dx$

16. $\dfrac{1}{\sqrt{2\pi}} \displaystyle\int_{4}^{5.5} e^{(-1/2)(x-5)^2}\,dx$

17. $\dfrac{1}{3\sqrt{2\pi}} \displaystyle\int_{0}^{2} e^{(-1/2)(z/3)^2}\,dz$

18. $\dfrac{2}{\sqrt{2\pi}} \displaystyle\int_{0}^{3} e^{(-1/2)[2(x-1)]^2}\,dx$

19. $\dfrac{1}{3\sqrt{2\pi}} \displaystyle\int_{-\infty}^{7} e^{(-1/2)[(x-8)/3]^2}\,dx$

20. $\dfrac{1}{\sqrt{2\pi}} \displaystyle\int_{1}^{\infty} e^{(-1/2)(x-0.5)^2}\,dx$

21. Show that the only two inflection points of
$$f(x) = \frac{e^{-1/2(x/\sigma)^2}}{\sigma\sqrt{2\pi}}$$
are at $x = \pm\sigma$. [*Hint:* Find where $f''(x) = 0$.]

22. Show that the points $x = \mu \pm \sigma$ are the inflection points of
$$f(x) = \frac{1}{\sigma\sqrt{2\pi}} e^{(-1/2)[(x-\mu)/\sigma]^2}$$

23. This exercise is intended to show how the size of the standard deviation σ of a normal r.v. reflects the dispersion of possible events. Suppose that X is a normal r.v. with standard deviation σ. Compute $P(0 \leq X \leq 1)$ when **a.** $\sigma = 1$, **b.** $\sigma = 2$, **c.** $\sigma = \frac{1}{2}$. In each case sketch the graph of
$$f(x) = \frac{1}{\sigma\sqrt{2\pi}} e^{-1/2(x/\sigma)^2}$$
and the region corresponding to this probability.

24. Show that if X is normal with mean μ and variance σ^2, then indeed $E(X) = \mu$ is the true mean, or expected value, of X. That is, evaluate $\int_{-\infty}^{\infty} x N_{\sigma,\mu}(x)\,dx$ by integration by substitution.

25. Show that if X is normal with mean $\mu = 0$ and standard deviation σ, then $V(X) = \sigma^2$. That is, evaluate
$$\int_{-\infty}^{\infty} x^2 N_{\sigma,0}(x)\,dx$$

26. Show that

$$\frac{1}{\sigma\sqrt{2\pi}} \int_{\mu-2\sigma}^{\mu+2\sigma} e^{(-1/2)[(x-\mu)/\sigma]^2} \, dx = 0.9544$$

27. Suppose that in Example 5 the standard deviation σ of X is $\sigma = 8$ minutes. What is $P(X \geq 7{:}05)$? If $\sigma = 12$ minutes, what is $P(X \geq 7{:}05)$?

28. What is the probability in Example 5 that your phone call will arrive 3 to 5 minutes early? 2 to 6 minutes late?

29. Let X be the time that you arrive for this class. Suppose that X is a normal r.v. with standard deviation $\sigma = 2$ minutes and mean μ equal to 1 minute *before* the scheduled time for class. What is the probability that you arrive for class

   a. 1 to 2 minutes early?
   b. Before class begins?
   c. At least 1 minute late?
   d. Exactly on time?
   e. 3 to no minutes early?
   f. 1 minute early to right on time?
   g. 6 to no *seconds* early?

30. Virtually any measurement is subject to some error. Minute quantities of a drug in the bloodstream can have pronounced effects on a patient. Suppose that a physician expects to find 30 mg of a certain drug in a patient's bloodstream and knows that 32 mg can be dangerous. A measuring instrument can measure the amount in the patient's bloodstream with a certain level of accuracy. Assume that this measurement X is a normal r.v. with mean 30 and standard deviation 0.5 mg. What is $P(X \geq 32)$? How should the physician interpret the answer?

31. The shingles used on the roof of a house have an average lifetime of 15 years with a standard deviation σ of 2 years. If the roofer guarantees the shingles to last at least 15 years, what is the probability that they will? What is the probability that they will last between 16 and 17 years? That the roofer will have to deal with a customer's complaint 12 to 13 years later? That the shingles will fail to protect the roof within 10 years?

32. Show that

$$\frac{1}{\sigma\sqrt{2\pi}} \int_{-2\sigma}^{2\sigma} e^{-1/2(x/\sigma)^2} \, d\sigma = 0.9544$$

Explain in English what this answer means.

33. Suppose that the average math SAT score of the freshman class at a certain university was 500 with a standard deviation of $\sigma = 100$. What percentage of the scores were between 500 and 600? Between 550 and 600? What was the minimum score needed to be in the top 25% of the class?

34. A gardening supplier sells grass seed with an expected germination rate of 80% and a standard deviation of 10%. Assume that the germination rate is a normal r.v. What is the probability that less than 70% of the seed will germinate? More than 95%? At least 15% will not germinate? (Notice that this is a *truncated* normal distribution because no more than 100% of the seed can germinate.)

Chapter 9 Summary

1. For a continuous function $f \geq 0$ on the interval $[a, \infty)$, the improper integral of f is defined by

$$\int_a^\infty f(x) \, dx = \lim_{b \to \infty} \int_a^b f(x) \, dx$$

2. Probability is a proportion of the whole of possibilities. If X is a random variable, the function $F(t) = P(X \leq t)$, denoting the probability that X does not exceed t, is the (cumulative) *distribution* of X.

3. The function $\rho(t) = F'(t)$ is called the *probability density function* for X and

$$P(a \leq X \leq b) = \int_a^b \rho(x) \, dx$$

provided that X is continuously distributed—that is, that $P(X = a) = 0$ for any number a. Moreover, $F(t) = \int_a^t \rho(x)\,dx$ provided that $P(X \le a) = 0$.

4. The *mean* μ, or *expected value* $E(X)$, of a random variable X with range between a and b is defined by

$$\mu = E(X) = \int_a^b x\rho(x)\,dx$$

5. The *variance* $V(X)$ and *standard deviation* σ of an r.v. X with range in $[a, b]$ and mean μ are defined by

$$V(X) = \int_a^b (x - \mu)^2 \rho(x)\,dx \quad \text{and} \quad \sigma = \sqrt{V(X)}$$

The number σ is a measure of the "tendency" of X to occur near its mean.

6. A random variable X is exponentially distributed if its p.d.f. has the form $\rho(x) = ke^{-kx}$; a second density of exponential type is $\rho(x) = b^2 x e^{-bx}$.

7. A random variable X is normally distributed with mean μ and variance σ^2 if

$$P(a \le X \le b) = \frac{1}{\sigma\sqrt{2\pi}} \int_a^b e^{(-1/2)[(x-\mu)/\sigma]^2}\,dx = \frac{1}{\sqrt{2\pi}} \int_{(a-\mu)/\sigma}^{(b-\mu)/\sigma} e^{-z^2/2}\,dz$$

Chapter 9 Summary Exercises

Find the integrals in Exercises 1–6.

1. $\displaystyle\int x \ln x\,dx$

2. $\displaystyle\int xe^{kx}\,dx$

3. $\displaystyle\int_0^1 x^2 e^x\,dx$

4. $\displaystyle\int x(1 + x)^3\,dx$

5. $\displaystyle\int xe^{x^2}\,dx$

6. $\displaystyle\int_1^e \ln 2x\,dx$

Find the improper integrals in Exercises 7–10.

7. $\displaystyle\int_1^\infty \frac{1}{x^{3/2}}\,dx$

8. $\displaystyle\int_0^\infty xe^{-x}\,dx$

9. $\displaystyle\int_{-\infty}^{-1} e^x\,dx$

10. $\displaystyle\int_0^{32} \frac{1}{x^{4/5}}\,dx$

11. **a.** If an assembly-line worker blindly reaches into a box containing 50 bolts, 30 nuts, and 20 washers and removes one object, what is the probability that it will be a washer?

    b. Each minute the worker removes one bolt, one washer, and one nut from the box. Ten minutes later, what is the probability when blindly reaching into the box of removing a bolt?

12. Suppose that a random variable X having all its values between 0 and 3 has the p.d.f. ρ whose graph is shown in Figure 9.42. Think of $P(a \le X \le b)$ directly as the area beneath the graph of ρ. Use the triangular area calculations in a–e.

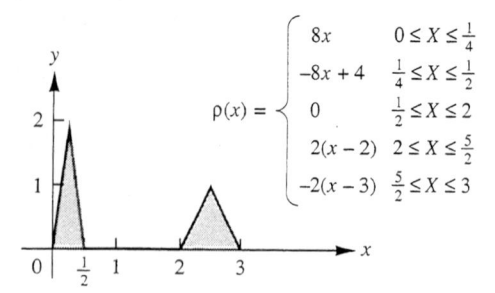

$$p(x) = \begin{cases} 8x & 0 \leq x \leq \frac{1}{4} \\ -8x + 4 & \frac{1}{4} \leq x \leq \frac{1}{2} \\ 0 & \frac{1}{2} \leq x \leq 2 \\ 2(x-2) & 2 \leq x \leq \frac{5}{2} \\ -2(x-3) & \frac{5}{2} \leq x \leq 3 \end{cases}$$

Figure 9.42

a. Find the total area beneath the graph of ρ.
b. Find $P(X \leq 1)$; $P(1 \leq X \leq 3)$; $P\left(X \leq \frac{1}{2}\right)$; $P\left(2 \leq X \leq \frac{5}{2}\right)$.
c. Find $P\left(X \leq \frac{1}{4} \text{ or } X \geq \frac{3}{2}\right)$.
d. What is $P\left(X = \frac{5}{2}\right)$?
e. Find $P\left(\frac{1}{4} \leq X \leq \frac{3}{2}\right)$.
f. Find a formula for $F(t) = P(X \leq t)$.
g. What is the expected value $\mu = E(X)$ of X?

13. A random variable X with range $0 \leq X \leq 2$ is hypothesized to have a p.d.f. of the form $\rho(x) = Cx(2-x)$. Since we must have $\int_0^2 \rho(x)\,dx = 1$, determine the value that C must have.

14. Find the mean μ and standard deviation σ of the r.v. in Exercise 13.

15. a. Find the mean $\mu = E(X)$ and standard deviation σ of an r.v. X that is exponentially distributed on $[0, \infty)$ with p.d.f. $\rho(x) = 3e^{-3x}$.
b. Find $P(0 \leq X \leq 2)$; $P(0 \leq X \leq \mu)$; $1 - P(\mu \leq X)$.

16. What is $P(X = a)$ equal to for a continuously distributed random variable?

17. Find the distribution of an r.v. X with p.d.f. $\rho(x) = 2e^{-2x}$ for $x \geq 0$. Then find $P(X \leq 4)$ and $P(X \geq 5)$.

18. Let X be a standard normal random variable.
a. Find $P\left(0 \leq X \leq \frac{1}{2}\right)$.
b. Find $P(-1 \leq X \leq 2)$.
c. Find $P(-\infty < X \leq -1)$.
d. Find $P(2 \leq X < \infty)$.

19. Let X be normally distributed with mean $\mu = 0$ and standard deviation σ. Find $P(-\sigma \leq X \leq 2\sigma)$.

20. Sketch the graphs of two normally distributed random variables X and Y, both with mean $\mu = 3$, but where X has standard deviation $\sigma = 1$ and Y has standard deviation $\sigma = 2$.

21. An r.v. X is known to be exponentially distributed with mean $\mu = 5$. Find its p.d.f. ρ.

22. The number of days from planting to harvest of a certain variety of corn is a normal random variable with mean (growing time) $\mu = 75$ days and standard deviation $\sigma = 5$ days.. Seventy-eight days after a certain 10 acres of corn was planted, harvesting is begun and completed (that day), yielding 2,000 bushels of corn. What proportion of these bushels is likely to be not fully ripened?

23. A pointer is attached to the center of a circle and spun until it stops. Let X be the number of revolutions it makes before stopping, times 2π, plus the portion of the length of the circle that it travels in its last incomplete revolution. Then $0 < X < \infty$. Suppose that the p.d.f. ρ for X is $\rho(x) = (x/100\pi^2)e^{-x/10\pi}$.

a. Graph ρ.
b. Find the probability $P(16\pi \leq X \leq 22\pi)$ that the pointer rotates 8 to 11 times before stopping.
c. Find $P(20\pi \leq X \leq 23\pi)$.
d. Find the mean μ and standard deviation σ of X.
e. How many times should you expect the pointer to rotate before stopping?

24. Let X be uniformly distributed on the interval $[-1, 2]$.
a. Find $P(0 \leq X \leq 1)$.
b. Find a formula for F and ρ, the probability distribution and density for X.

25. Let X be normally distributed with mean $\mu = 5$ and standard deviation $\sigma = \frac{3}{2}$. Find $P(4 \leq X \leq 7)$; $P(X \geq 5.5)$; $P(X \leq 3.5)$.

26. Discuss the relationships between a probability distribution function, a probability density function, and the probabilities that the associated random variable has. Use complete sentences and the minimum amount of notation needed for precision in your discussion.

Chapter 10

Sequences and Series

Discrete Change

The concepts of differentiation and integration yield powerful tools for analyzing change that takes place continually over time, like the changing position of an arrow in flight. But sometimes change occurs at distinct points, such as the successive ticks of a clock, marked by the sudden jump of the second hand. Such change is referred to as *discrete*. That discrete analysis can be as precise and as useful as continuous analysis is made apparent by the success of digital recording, wherein the frequency of sound produced by a singer or instrument is recorded as a number. This is done repeatedly at exceedingly close, but still discrete, time intervals with the result that the smoothest change of all, the sound of music itself, is reproduced with a quality unmatched by "continuous" recording. The recorded set of numbers, listed in the order in which they occur as the sound is produced, is called a *sequence*. The sequential modeling of data is very natural and occurs in a wide variety of fields, from finance to medical care. Curiously, while we seem to more readily greet sequential, rather than continuous, data, the processing of such data is often guided by experience in the continuous case, where the analysis, if not always the understanding, is typically easier.

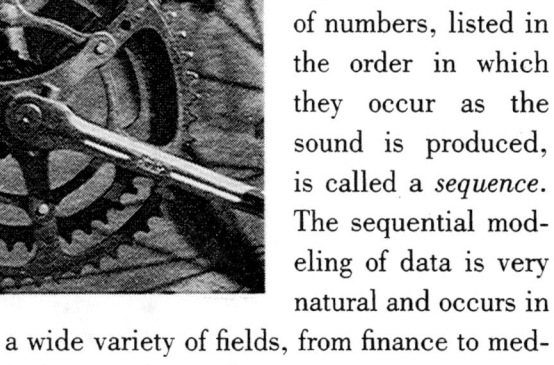

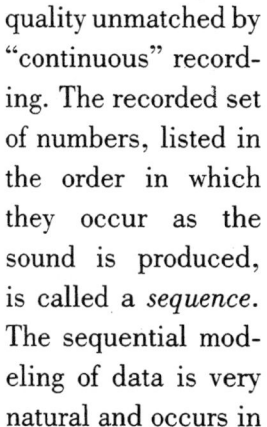

10.1 Sequences

In this section we study the limit of a sequence of numbers.

A sequence is an unending list of numbers *in the order* in which it is given. Thus

$$1, 0, 1, 3, \sqrt{2}, -3, 7, \pi, \tfrac{1}{2}, 3, \ldots$$

is a sequence. But such a random list of numbers is not really typical of the sequences we will encounter. Most often a sequence is a list of numbers that follows some pattern, such as

$$1, \tfrac{1}{2}, \tfrac{1}{3}, \tfrac{1}{4}, \tfrac{1}{5}, \ldots$$

or

$$+1, -1, +1, -1, +1, -1, \ldots$$

The simplest sequence is the list of whole numbers

$$1, 2, 3, \ldots$$

For this reason a **sequence** is defined as a list of numbers *in correspondence with the sequence of whole numbers* (that is, as a function whose domain is the set of whole numbers). When the list has some evident pattern, such as $1, \tfrac{1}{2}, \tfrac{1}{3}, \ldots$, we say that the **nth term** of the sequence is $1/n$, because this expression formulates the pattern of the sequence and corresponds to the whole number n.

This view leads to a general notation for sequences. We use the natural numbers as *subscripts* for the terms of the sequence in order and write

$$a_1, a_2, a_3, \ldots$$

as the general notation for a sequence whose nth term is a_n. With this notation we then describe the sequence $1, \tfrac{1}{2}, \tfrac{1}{3}, \ldots$ by the single equation $a_n = 1/n$. That is, a sequence may be defined by simply giving a prescription for its nth term. For example, the statement Let $\{a_n\}$ be a sequence whose nth term is

$$a_n = \frac{1}{2^n} \quad \text{for } n \geq 1$$

describes the (infinite) sequence

$$\frac{1}{2^1}, \frac{1}{2^2}, \frac{1}{2^3}, \ldots$$

That is, the sequence $\tfrac{1}{2}, \tfrac{1}{4}, \tfrac{1}{8}, \tfrac{1}{16}, \tfrac{1}{32}, \tfrac{1}{64}, \ldots$.

EXAMPLE 1

Find the first five terms of the sequences $\{a_n\}$ and $\{b_n\}$ given by

a. $a_n = \dfrac{n}{n+1}$ **b.** $b_n = (-1)^n \left(1 - \dfrac{1}{n}\right)$

Solution

a. The first term is $a_1 = 1/(1+1) = 1/2$. The second is $a_2 = 2/(2+1) = 2/3$; the third is $a_3 = 3/(3+1) = 3/4$. Clearly now, $a_4 = 4/5$ and $a_5 = 5/6$.

b. The first term is $b_1 = (-1)^1\left(1 - \frac{1}{1}\right) = 0$. The second is $b_2 = (-1)^2\left(1 - \frac{1}{2}\right) = \frac{1}{2}$; the third is $b_3 = (-1)^3\left(1 - \frac{1}{3}\right) = -\frac{2}{3}$. Similarly, $b_4 = (-1)^4\left(1 - \frac{1}{4}\right) = \frac{3}{4}$ and $b_5 = (-1)^5\left(1 - \frac{1}{5}\right) = -\frac{4}{5}$. ∎

EXAMPLE 2

Write a formula for the general term of the sequences

a. 3, 5, 7, 9, . . .   **b.** $1, -\frac{1}{2}, \frac{1}{3}, -\frac{1}{4}, \frac{1}{5}, -\frac{1}{6}, \ldots$

Solution

a. This sequence lists every odd number, except the number 1. Since every *even* number has the form $2n$, the odd numbers have the form $2n + 1$. If we call this sequence $\{a_n\}$, we can then write $a_n = 2n + 1$.

b. Let $\{b_n\}$ denote this sequence. A little inspection convinces us that b_n is $1/n$ except for an alternating sign. Thus $b_n = (-1)^{n+1}/n$; the exponent $n + 1$ ensures that the first term and then every other term is positive. ∎

We have seen that the graph of a function is a useful tool. This is less true for a sequence, but we can illustrate a sequence by a graph, as we indicate in Figure 10.1.

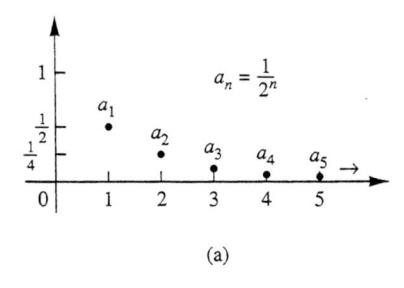

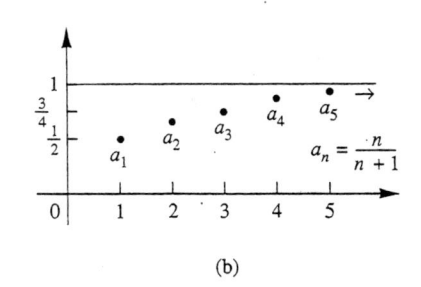

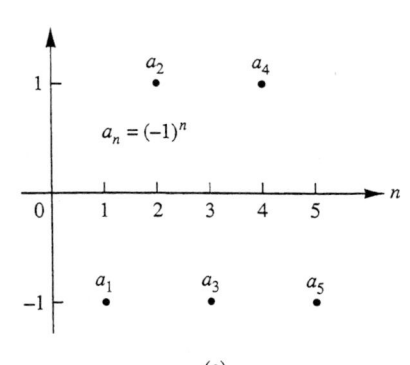

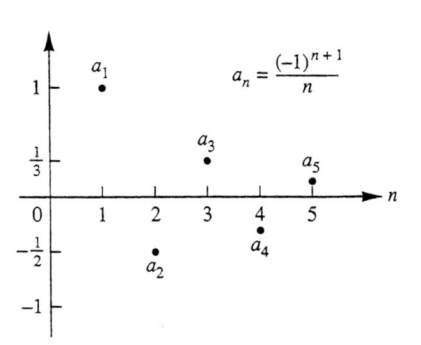

Figure 10.1

(a) $a_n = \dfrac{1}{2^n}$

(b) $a_n = \dfrac{n}{n+1}$

(c) $a_n = (-1)^n$

(d) $a_n = \dfrac{(-1)^{n+1}}{n}$

10.1 Sequences   **565**

In Figure 10.1(a, b, d), we see a "long-run trend." The terms of the sequence "approach" a particular value. We identify this value as the *limit* of the sequence.

Definition

We say that the **limit of a sequence** $\{a_n\}$ is the number L, written

$$\lim_{n \to \infty} a_n = L$$

if terms of the sequence are as close to L as we demand if n is a large enough number. We say then that the nth term a_n **approaches** L and that the sequence **converges** to L.

This definition is entirely analogous to that of Section 1.3 for the limit of a function at infinity.

EXAMPLE 3

Find

a. $\lim_{n \to \infty} \dfrac{1}{n}$ **b.** $\lim_{n \to \infty} \dfrac{(-1)^n}{n}$ **c.** $\lim_{n \to \infty} (-1)^n$

Solution

a. If n is taken to be very large, then $a_n = 1/n$ is approximately 0. (For example, $a_{9,999} = 0.000100010 \ldots$.) Thus $\lim_{n \to \infty}(1/n) = 0$.

b. Here $\lim_{n \to \infty}(-1)^n/n = 0$ again, since whether n is odd or even, $(-1)^n/n$ is approximately 0 if n is very large. For example, $(-1)^{9,999}/9,999 = -0.000100010 \ldots$.

c. This sequence *has no limit* L, since the terms of the sequence alternate between $+1$ and -1 and are not *all* close to any one number L no matter how large n is. ∎

The next example reminds us of methods found in Example 7, Section 1.3.

EXAMPLE 4

Evaluate the limits

a. $\lim_{n \to \infty} \dfrac{n}{n+1}$ **b.** $\lim_{n \to \infty} \dfrac{n^2}{n^2 - n + 1}$

Solution

a. Since

$$\frac{n}{n+1} = \frac{\frac{n}{n}}{\frac{n+1}{n}} = \frac{1}{1 + \frac{1}{n}}$$

we see that

$$\lim_{n \to \infty} \frac{n}{n+1} = \lim_{n \to \infty} \frac{1}{1 + \frac{1}{n}} = \frac{1}{1+0} = 1$$

because $1/n$ approaches 0 as n becomes larger.

b. Dividing both numerator and denominator by the largest power of n that appears, we see that

$$\lim_{n \to \infty} \frac{n^2}{n^2 - n + 1} = \lim_{n \to \infty} \frac{\frac{n^2}{n^2}}{\frac{n^2 - n + 1}{n^2}} = \lim_{n \to \infty} \frac{1}{1 - \frac{1}{n} + \frac{1}{n^2}}$$

Since $1/n$ and $1/n^2$ approach 0 as n grows larger, the limit is

$$1/(1 - 0 + 0) = 1 \quad \blacksquare$$

We frequently encounter a sequential limit where the index (or counter) letter n appears in the exponent.

EXAMPLE 5

Evaluate

a. $\lim_{n \to \infty} \left(\frac{1}{2}\right)^n$ **b.** $\lim_{n \to \infty} 2^n$ **c.** $\lim_{n \to \infty} 1^n$

Solution

a. Since $(1/2)^n = 1/2^n$ and 2^n is very large when n is large, we conclude that $(1/2)^n$ is approximately 0. For example, $1/2^{16} = 1/65,536 \simeq 0.000015$. Thus $\lim_{n \to \infty}(1/2)^n = 0$.

b. Here the opposite happens. The nth term 2^n becomes ever larger as n increases and so no limit exists. However, since 2^n "approaches infinity," we do write

$$\lim_{n \to \infty} 2^n = +\infty$$

c. Here $1^n = 1$ no matter how large n is. Thus $\lim_{n \to \infty} 1^n = 1$. $\quad \blacksquare$

The following rules allow us to find sequential limits just as we find limits for a function at a point (Section 1.3).

The Limit Properties for Sequences

If $\lim_{n\to\infty} a_n = L$ and $\lim_{n\to\infty} b_n = K$ (and neither L nor K is infinite), then

1. $\lim\limits_{n\to\infty} (a_n \pm b_n) = L \pm K.$

2. $\lim\limits_{n\to\infty} a_n b_n = LK.$

3. If $K \neq 0$, $\lim\limits_{n\to\infty} \dfrac{a_n}{b_n} = \dfrac{L}{K}.$

4. $\lim\limits_{n\to\infty} f(a_n) = f(L)$ for any function f that is continuous at L.

EXAMPLE 6

Find

a. $\lim\limits_{n\to\infty} \sqrt{\dfrac{n}{4n+1}}$ **b.** $\lim\limits_{n\to\infty} e^{n/(1-n)}$

Solution

a. Write $n/(4n+1) = 1/[4 + (1/n)]$ by division of numerator and denominator by n. Property 3 yields

$$\lim_{n\to\infty} \frac{n}{4n+1} = \lim_{n\to\infty} \frac{1}{4 + \dfrac{1}{n}} = \frac{1}{4}$$

Now since $y = \sqrt{x}$ is continuous on its domain, property 4 yields

$$\lim_{n\to\infty} \sqrt{\frac{n}{4n+1}} = \sqrt{\lim_{n\to\infty}\left(\frac{n}{4n+1}\right)} = \sqrt{\frac{1}{4}} = \frac{1}{2}$$

b. Since

$$\lim_{n\to\infty}\left(\frac{n}{1-n}\right) = \lim_{n\to\infty}\left(\frac{1}{\dfrac{1}{n} - 1}\right) = -1$$

and since $y = e^x$ is continuous, property 4 again yields

$$\lim_{n\to\infty} e^{n/(1-n)} = e^{\lim_{n\to\infty}[n/(1-n)]} = e^{-1} \approx 0.367879 \quad \blacksquare$$

In Section 10.2, we will apply the concept of a sequence and its limit to discrete state compartment-mixing processes. In Section A.5, we use sequences to develop and apply the Newton-Raphson algorithm to finding the roots of an equation $f(x) = 0$.

Background Review 10.1

1. For a positive whole number
$$n! = n(n-1)(n-2) \cdot \ldots \cdot 3 \cdot 2 \cdot 1$$
For example, $5! = 5 \cdot 4 \cdot 3 \cdot 2 \cdot 1 = 120$. Additionally, as a special case, we define $0! = 1$.

2. From Section A.2, we state L'Hôpital's rule:
$$\lim_{x \to \infty} \frac{f(x)}{g(x)} = \lim_{x \to \infty} \frac{f'(x)}{g'(x)}$$
when f and g have limits at infinity that are *both* 0 or *both* infinite. If, for example, we wish to find $\lim_{n \to \infty}(n/e^{2n})$, we may equally well use L'Hôpital's rule to find $\lim_{x \to \infty}(x/e^{2x}) = \lim_{x \to \infty}(1/2e^{2x}) = 0$ and conclude that $\lim_{n \to \infty}(n/e^{2n}) = 0$ also.

Exercises 10.1

Determine the first three terms and the sixth and eleventh terms of the sequences given in Exercises 1–10.

1. $a_n = 2^n - 1$
2. $a_n = \frac{1}{n} - \frac{1}{n+1}$
3. $b_k = \frac{1}{3^k}$
4. $x_n = \frac{1 + (-1)^n}{2}$
5. $y_n = ne^{-n}$
6. $a_n = \frac{2^n}{n!}$, where $n! = n(n-1) \cdot \ldots \cdot 3 \cdot 2 \cdot 1$
7. $a_n = n \ln(\ln n)$
8. $a_n = r^n$, $0 \le r < 1$
9. $a_n = b_{n+3}$, where $b_k = \frac{1}{k}$
10. $a_n = b_{n+2}$, where $b_k = \sqrt{k+1} - \sqrt{k}$

Find a formula for the nth term of the sequence given in Exercises 11–18.

11. $2, 4, 6, 8, \ldots$
12. $1, \frac{1}{5}, \frac{1}{9}, \frac{1}{13}, \ldots$
13. $e^{-1}, e, e^{-1}, e, e^{-1}, e, \ldots$
14. $1, \frac{9}{10}, \frac{81}{100}, \frac{729}{1{,}000}, \ldots$
15. $1, -\frac{1}{2}, \frac{1}{3}, -\frac{1}{4}, \ldots$
16. $\frac{1}{2}, \frac{1}{6}, \frac{1}{12}, \frac{1}{20}, \ldots$
17. $0.5, 0.55, 0.555, \ldots$
18. $0, \frac{1}{2}, \frac{3}{4}, \frac{7}{8}, \frac{15}{16}, \ldots$
19. Let $f(x) = x(x+1)/(x^2+1)$. Find the first five terms of the sequence $a_n = f(n)$.
20. Let
$$s_n = 1 + \frac{1}{2} + \frac{1}{4} + \frac{1}{8} + \cdots + \frac{1}{2^{n-1}}$$
Find $s_1, s_2, s_3,$ and s_5.

Determine the limit of the sequences in Exercises 21–42.

21. $a_n = \frac{2n}{n+2}$
22. $b_n = \frac{1}{3^n}$
23. $c_n = \frac{1}{n^3}$
24. $a_n = 2 + \left(-\frac{1}{2}\right)^n$
25. $x_n = \frac{2^n}{3^n}$
26. $b_n = \frac{\sqrt{n+1}}{n}$
27. $a_n = \frac{2n^2 - n}{1 + 4n^2}$
28. $a_k = \frac{352 - k}{2k + 1}$
29. $a_n = (-2)^n$
30. $b_k = \frac{3k^2 - 2k + 1}{1 - 2k^2}$

31. $c_j = \dfrac{j}{(j+1)(j+2)}$ 32. $a_k = e^{1/k}$

33. $a_n = \sqrt{n+1} - \sqrt{n}$ [Hint: Multiply by $(\sqrt{n+1} + \sqrt{n})/(\sqrt{n+1} + \sqrt{n})$.]

34. $a_n = \sqrt{n^2 + 9} - n$

35. $a_n = \dfrac{e^n}{3^n}$

36. $b_n = ne^{-n}$ [Hint: Use L'Hôpital's rule.]

37. $b_n = 2 + \dfrac{(-1)^n}{2^n}$ 38. $a_n = \dfrac{1}{n^k}, k > 0$

39. $a_n = \dfrac{\ln n}{n}$ [Hint: Let $f(x) = \ln x$ and $g(x) = x$, and use L'Hôpital's rule.]

40. $x_n = ane^{-bn}, a, b > 0$

41. $c_n = \dfrac{n^2 + 9n}{n}$

42. $s_n = \dfrac{1 - r^{n+1}}{1 - r}, -1 < r < 1$

10.2 First-Order Difference Equations

In this section we explore the concept of a difference equation, which is the discrete analog of a differential equation.

In its appearances the world evolves continuously over time, changing imperceptibly from moment to moment. As we have seen in earlier chapters, calculus is superbly designed to model such evolution. But not all change is smooth; indeed we naturally tend to think of change in discrete packets. A merchant replaces inventory in timely batches. The second hand on my watch stops, waits, then *jumps* ahead—one more second gone—and waits, then jumps again. This is an example of **discrete change**, or change that is abrupt and occurs at specific intervals.

Measurements of discrete change give rise to a sequence of numbers. For example, suppose you are given this prescription by your doctor: Take one of these pills every 8 hours. The pill contains 6 mg and has a half-life of 8 hours. Thus, of the 6 mg ingested upon taking the first pill, only $\tfrac{1}{2}(6) = 3$ mg is present when you take the second pill, at which time the amount in your body suddenly jumps to $3 + 6 = 9$ mg. If you continue to follow the prescription, the amount of drug present in your body at 8-hour intervals is a sequence, as shown in the following table.

Hour	Amount Present
0	6 mg
8	$\tfrac{1}{2}(6) + 6 = 9$ mg
16	$\tfrac{1}{2}(9) + 6 = 10.5$ mg
24	$\tfrac{1}{2}(10.5) + 6 = 11.25$ mg
.	.
.	.
.	.

We will make a complete analysis of the cumulative effect of such a prescription later in this section.

Attempts to model discrete change in a system give rise to a **difference equation.** Let y_0 be the initial amount deposited in a savings account paying

6% interest that is applied once at the end of each year to the amount of money in the account. Let y_n be the state (amount) of this account at the end of year n. Assuming no other deposits are made, we have the difference equation

$$y_{n+1} - y_n = 0.06 y_n$$

as a model of the change in the account from one year (n) to the next year $(n + 1)$, since this change must be 6% of the amount y_n in the account in year n. This difference equation is the discrete analog of the familiar differential equation $y' = 0.06y$ of exponential growth (Section 5.5), which models 6% interest applied continuously.

Definition

A **first-order difference equation** is an equation involving an unknown sequence y_n, $n \geq 0$, in which the difference between successive terms of the sequence is a function of y_n alone—that is, an equation in which

$$y_{n+1} - y_n = g(y_n)$$

for some function g. Note that we allow the sequence to begin at $n = 0$; we call the number y_0 the *initial value* of the sequence.

In the preceding paragraph we would take $g(y_n) = 0.06 y_n$. In this section we are interested in systems that are naturally modeled by a discrete state analysis. Here the state, or condition, of a system after n changes is regarded as the nth term of a sequence. The appropriate law of change for the system is then written as a difference equation. A solution of the difference equation produces a model of the system from which measurements and predictions can be made.

Definition

A **solution** of a difference equation

$$y_{n+1} - y_n = g(y_n) \qquad n \geq 0$$

is an expression, or a formula, for y_n which, when substituted in the equation, yields equality for every value of n.

EXAMPLE 1

A single bacterium of *E. coli* reproduces by division every 20 minutes. Model this system as a difference equation and find its solution.

Solution Let $y_0 = 1$ be the initial number of *E. coli* bacteria, and let y_n be the number of bacteria present after n divisions. (Thus each integer n represents a 20-minute time gap between divisions.) It is apparent that

$$y_{n+1} = 2 y_n \tag{1}$$

We can write Equation 1 as a difference equation by subtracting y_n from each side to obtain

$$y_{n+1} - y_n = 2y_n - y_n = y_n \qquad (2)$$

This difference equation simply says that the *change* $y_{n+1} - y_n$ in the number of bacteria present is the previous number, as each bacterium divides into two.

Returning to Equation 1, we see that, since $y_0 = 1$, we have $y_1 = 2y_0 = 2 \cdot 1 = 2$ and

$$y_2 = 2y_1 = 2(2) = 4$$
$$y_3 = 2y_2 = 2(4) = 8$$

Continuing like this, we see that $y_n = 2^n$. That is, the formula $y_n = 2^n$ is the solution to this difference equation when $y_0 = 1$. To check this solution, we substitute it in the equation $y_{n+1} = 2y_n$ and observe that

$$y_{n+1} = 2^{n+1} \quad \text{and} \quad 2y_n = 2(2^n) = 2^{n+1}$$

so that, indeed, $y_{n+1} = 2y_n$. ∎

Geometric Sequences

The discrete analog of the differential equation $y' = ky$ of exponential change [with the solution $y(t) = Ce^{kt}$] is the **difference equation of geometric growth**

$$y_{n+1} - y_n = ky_n \qquad (3)$$

where k is constant.

The general solution to the difference equation of geometric change is the **geometric sequence**

$$y_n = C(1 + k)^n \qquad (4)$$

where C is an arbitrary constant. If y_0 is given as the initial term of the sequence (and if $k \neq -1$), then $C = y_0$.

We can verify the solution (Equation 4) by substitution in Equation 3. We have

$$y_{n+1} - y_n = C(1 + k)^{n+1} - C(1 + k)^n = C(1 + k)^n[(1 + k) - 1]$$
$$= C(1 + k)^n[k] = y_n \cdot k = ky_n$$

Moreover, if y_0 is specified (and if $k \neq -1$), then

$$y_0 = C(1 + k)^0 = C$$

EXAMPLE 2

If $300 is deposited (once) in a savings account paying 8% annual interest and y_n is the amount in the account n years later, find a formula for y_n. How much is in the account 9 years later?

Solution The difference $y_{n+1} - y_n$ of the amounts in the account between successive years is the interest earned on the amount in the account during the year. Since y_n is the amount in the account at year n, the earned interest by the end of the year is $0.08 y_n$. Therefore, with $k = 0.08$, the equation of geometric change

$$y_{n+1} - y_n = 0.08 y_n$$

models this system. Its solution is $y_n = C(1 + 0.08)^n = C(1.08)^n$. Since $y_0 = 300$, then $y_n = 300(1.08)^n$, and the account grows exponentially. Nine years later there will be $y_9 = 300(1.08)^9 \simeq \$599.70$ in the account. ∎

Arithmetic Sequences

The analog of the differential equation $y' = k$ [with linear solution $y(t) = kt + C$] is the **difference equation of arithmetic growth**

$$y_{n+1} - y_n = k \qquad (5)$$

The general solution of the difference equation of arithmetic growth is the **arithmetic sequence**

$$y_n = kn + C \qquad (6)$$

If y_0 is given, then $C = y_0$.

Let us derive this solution as follows. First, observe that

$$(y_n - y_{n-1}) + (y_{n-1} - y_{n-2}) + \cdots + (y_2 - y_1) + (y_1 - y_0)$$

$$= \underbrace{k + k + k + \cdots + k + k}_{n \text{ times}} = kn$$

Next, notice that all succeeding terms in the sum on the left cancel preceding terms—save for the first term, y_n, and the last, y_0. That is, the sum on the left equals $y_n - y_0$. Therefore $y_n - y_0 = kn$ or $y_n = kn + y_0$.

EXAMPLE 3

A manufacturer has 200 washing machines on hand and continues production at the rate of 7 machines per day but is unable to sell any machines. Model the company's miserable circumstance by a difference equation and write its solution. How many machines are clogging the firm's warehouse 8 days later?

Solution Let y_n be the number of machines on hand on day n. Then $y_0 = 200$ is the initial state of this system and

$$y_{n+1} - y_n = 7$$

models its changing state. This equation has the solution

$$y_n = 7n + 200$$

Eight days later there are $y_8 = 7(8) + 200 = 256$ washing machines in the warehouse. ∎

Compartment Mixing and the General Linear, First-Order Difference Equation

Each of the preceding difference equations may be written in the (more general) *standard form* of a first-order difference equation

$$y_{n+1} = h(y_n)$$

We are particularly interested in this difference equation when $h(y_n) = ay_n + b$, a linear function of y_n, since this difference equation arises in the discrete analysis of a compartment-mixing system subject to constant rates of change. It is also the analog of the familiar differential equation $y' = ay + b$ studied in Section 8.5. We have the following:

The general solution of the difference equation

$$y_{n+1} = ay_n + b$$

is

$$y_n = \begin{cases} y_0 + nb & \text{if } a = 1 \\ \dfrac{b}{1-a} + Ca^n & \text{if } a \neq 1 \end{cases} \quad (7)$$

With some effort we can derive these solutions. Notice first that if $a = 1$, the equation becomes that of arithmetic change: $y_{n+1} = ay_n + b = y_n + b$ or $y_{n+1} - y_n = b$ with solution $y_n = y_0 + nb$, from Equation 6.

If $a \neq 1$, consider the *constant* solution $y_n = b/(1-a)$ as a *possible* solution. Upon substitution in $y_{n+1} = ay_n + b$, we have

$$y_{n+1} = \frac{b}{1-a} \quad (8)$$

and

$$ay_n + b = a\left(\frac{b}{1-a}\right) + b = \frac{ab + (1-a)b}{1-a} = \frac{b}{1-a} \quad (9)$$

Thus, combining Equations 8 and 9, $y_{n+1} = ay_n + b$ and so $y_n = b/(1-a)$ is *a* solution.

Now imagine any other solution—say, x_n. We have that both the equations
$$y_{n+1} = ay_n + b$$
and
$$x_{n+1} = ax_n + b$$
hold. Therefore, upon subtraction of one equation from the other,
$$y_{n+1} - x_{n+1} = (ay_n + b) - (ax_n + b) = a(y_n - x_n) \qquad (10)$$
Let $z_n = y_n - x_n$. Then by substitution Equation 10 becomes the equation
$$z_{n+1} = az_n$$
Next, subtracting z_n from both sides,
$$z_{n+1} - z_n = (a - 1)z_n$$
This is the equation of geometric change with $k = a - 1$. If $a \neq 0$, then from Equation 4 we have the solution
$$z_n = C[1 + (a - 1)]^n = Ca^n$$
Thus $y_n - x_n = Ca^n$. Since $y_n = b/(1 - a)$ is the constant solution already found, then
$$x_n = y_n - Ca^n = \frac{b}{1 - a} + Ca^n$$
where we "absorb" the minus sign into a new constant C^*.

If $a \neq 1$ and an initial state y_0 is given, we may also solve for the constant C as follows:
$$y_0 = \frac{b}{1 - a} + Ca^0 = \frac{b}{1 - a} + C$$
or
$$C = \frac{(1 - a)y_0 - b}{1 - a} \qquad (11)$$

Let us apply (7) to the analysis of a system that is naturally discrete.

EXAMPLE 4

A Colorado ranch of 150,000 acres can support a herd of 1,250 head of cattle. A rancher who begins with a herd of 500 cattle is able to reproduce the herd at the rate of 30% in the spring of each year. The rancher plans to sell 80 cattle per year.
a. How large is the herd n years later?
b. How large is the herd after 4 years?
c. When will overgrazing occur?
d. At what rate can cattle be sold so as to reach a herd size of 1,250 in 10 years?

---

*If $a = 0$, the original equation is $y_{n+1} = b$ with the constant solution $y_n = b$.

Solution

a. Let us imagine the ranch as a compartment, where y_n is the size of the herd in year n.

$$0.30y_n \xrightarrow{\text{Amount in}} \boxed{\begin{array}{l} y_0 = 500 \\ y_n = \text{State in year } n \\ y_{n+1} = \text{Next state} \end{array}} \xrightarrow{\text{Amount out}} 80$$

The amount y_{n+1} is the size one year later, and the difference $y_{n+1} - y_n$ must be the number of cattle added to the herd, less the number of cattle removed, during the year. That is, we have an "amount-in/amount-out" difference equation

$$y_{n+1} - y_n = \text{Amount in} - \text{Amount out}$$

modeling the changing size of the herd. Since the herd increases by 30% of its size, y_n, over the year, the *amount in* is therefore $0.30y_n$. In the same year, 80 cattle are sold and the *amount out* is 80. Therefore

$$y_{n+1} - y_n = 0.30y_n - 80$$

Hence
$$y_{n+1} = 1.3y_n - 80$$

with solution
$$y_n = \frac{-80}{1 - 1.3} + C(1.3)^n \qquad \text{From (7)}$$
$$= C(1.3)^n + 266.67$$

Additionally, since $y_0 = 500$, then

$$C = \frac{(1 - 1.3)(500) - (-80)}{1 - 1.3} = 233.33$$

from Equation 11. Thus $y_n = 233.33(1.3)^n + 266.67$ is the (approximate) size of the herd n years later.

b. After $n = 4$ years, the herd contains (approximately) $y_4 = 233.33(1.3)^4 + 266.67 \simeq 933$ cattle.

c. Overgrazing will occur when $y_n \geq 1{,}250$. Solving for n in $233.33(1.3)^n + 266.67 = 1{,}250$, we have

$$(1.3)^n \simeq 4.214$$
$$\ln(1.3)^n \simeq \ln 4.214$$

or
$$n \simeq \frac{\ln 4.214}{\ln 1.3} \simeq 5.48$$

We conclude that by the end of $n = 6$ years the rancher would be attempting to raise more cattle than the land can support.

d. Here we must assume an *unknown amount out* of the compartment. Let us denote it by b. The difference equation then becomes

$$y_{n+1} = 1.3y_n - b$$

with solution

$$y_n = C(1.3)^n + \frac{-b}{1-(1.3)} = C(1.3)^n + 3.33b$$

where (from Equation 11)

$$C = \frac{(1-1.3)(500)-(-b)}{1-1.3} = 500 - 3.33b$$

Thus $y_n = (500 - 3.33b)(1.3)^n + 3.33b$. We want $y_{10} = 1{,}250$ or

$$1{,}250 = (500 - 3.33b)(1.3)^{10} + 3.33b$$

$$1{,}250 = 6{,}892.92 + 3.33b - 45.91b$$

or

$$b = \frac{-5{,}642.92}{-42.58} = 132.53$$

The rancher can sell about 133 cattle a year and support the herd for 10 years. ∎

A Drug Prescription Model

In the next example we study a common experience of discrete change that contains an element of continuous change (Section 5.5) as well. Here we are particularly interested not only in y_n but in the number $\lim_{n\to\infty} y_n$.

EXAMPLE 5

A doctor prescribes a 20-mg dosage of a certain drug to be taken in pill form every 8 hours. The patient's body metabolizes the drug at the rate of 3% of the amount present.

a. Using a difference equation, model the amount of drug in the patient's body each time a pill is taken.
b. Graph the solution to this difference equation, and show on the graph the amount of drug in the patient's body between doses.
c. What is the long-run behavior of this system? That is, to what level will the drug build up in the patient's body?

Solution

a. As in Example 4, let us imagine the patient's body as a compartment where y_n is the amount in the patient's body when the nth pill is taken.

20 mg —Amount in→ | $y_0 = 20$
 y_n = Amount (state) when nth pill is taken
 y_{n+1} = Next state | —Amount out→ $y_n - e^{-0.03(8)}y_n = 0.213y_n$

We may write the change in the patient's body from one pill to the next as $y_{n+1} - y_n =$ Amount in − Amount out. We know that the

$$\text{Amount in} = 20$$

Imagine now that the patient has just taken the nth pill and has y_n mg of the drug in the body; what happens next? The patient's body immediately begins to eliminate (metabolize) the drug according to the *differential equation* $y' = -0.03y$ of Section 5.5, where $y(t)$ is the continually decreasing amount of drug in the body. This differential equation has the solution $y(t) = y(0)e^{-0.03t}$. Since this continual elimination process began when there was $y(0) = y_n$ mg present, we have $y(t) = y_n e^{-0.03t}$. Therefore, 8 hours later, when $t = 8$ and (just before) the next pill is taken, only

$$y_n e^{-0.03(8)} \approx 0.787 y_n$$

remains in the patient's body. Consequently, the *amount out* during this 8-hour period is

$$\text{Amount out} = y_n - 0.787 y_n = 0.213 y_n$$

The change between successive pills is therefore

$$y_{n+1} - y_n = \text{Amount in} - \text{Amount out}$$
$$= 20 - 0.213 y_n$$

or
$$y_{n+1} = 0.787 y_n + 20$$

From (7), this equation has the solution

$$y_n = 93.897 + C(0.787)^n$$

where $\qquad C = -73.897 \qquad$ Using Equation 11, page 575

Thus $\qquad y_n = 93.897 - 73.897(0.787)^n$

is the amount in the body when the nth pill is taken.

b. Using a calculator, we have the following accumulation of the drug in the patient's body as each pill is taken.

$$y_0 = 20 \qquad\qquad y_3 \approx 57.88$$
$$y_1 \approx 35.74 \qquad\qquad y_4 \approx 65.55$$
$$y_2 \approx 48.13 \qquad\qquad y_5 \approx 71.59$$

Additionally, as we saw in (a), each of these amounts decays by the factor $e^{-0.03(8)} = 0.787$ between pills. A graph of this drug-dosage scheme appears in Figure 10.2.

c. In the long run, the drug in the patient's body builds up to a level of approximately

$$\lim_{n \to \infty} y_n = \lim_{n \to \infty} 93.897 - 73.896(0.787)^n = 93.897 \text{ mg} \quad\blacksquare$$

The Logistic Difference Equation

Because we instinctively perceive change in the world about us in a discrete fashion, difference equations are more natural as initial mathematical models whereas differential equations require a lengthy study of calculus before the

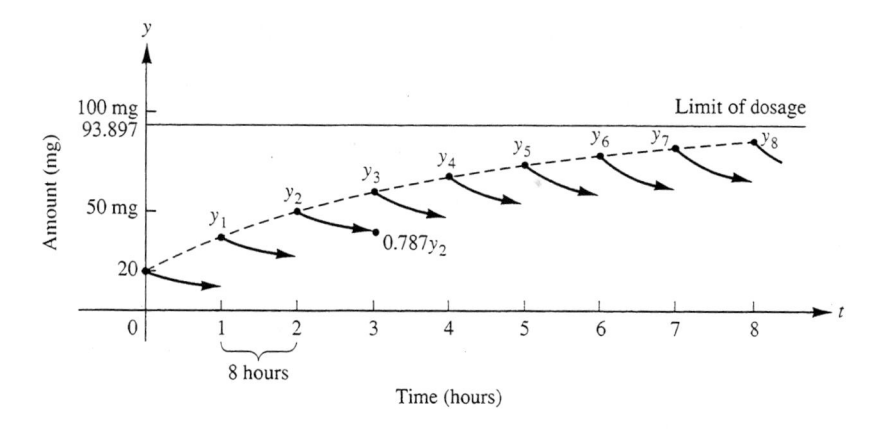

Figure 10.2

can even be discussed. On the other hand, even simple difference equations cannot always be solved by procedural techniques, such as integration, while basic differential equations often can be.

An important example is the difference equation of logistic growth. Logistic growth was studied in Section 8.2, where such growth is modeled by the differential equation $y' = ky(M - y)$. In this model, growth can occur up to a maximum capacity M (see, for example, Figure 8.17), but the solution $y(t)$ can never exceed M [if $y(0) \leq M$].

The logistic difference equation is the equation

$$y_{n+1} - y_n = ky_n(M - y_n)$$

Unlike the continuous logistic equation, no method of solution for the logistic difference equation is known, and a solution can oscillate above and below M (see Exercise 42). Example 5 suggests why the most important issue might be to know, not an entire solution y_n, but simply the number $\lim_{n \to \infty} y_n$. This limit, if it exists, describes the long-run trend of the system.

EXAMPLE 6

Suppose $\lim_{n \to \infty} y_n$ exists for a solution (if any) of the difference equation

$$y_{n+1} - y_n = \tfrac{1}{2} y_n(60 - y_n) \tag{12}$$

Find $\lim_{n \to \infty} y_n$.

Solution Let $\bar{y} = \lim_{n \to \infty} y_n$. Since the limit of a sequence is determined only by its "tail," $y_k, y_{k+1}, y_{k+2}, \ldots$, for any k, it follows that $\bar{y} = \lim_{n \to \infty} y_{n+1}$ as well. Applying the limit to both sides of Equation 12, we have

$$\lim_{n \to \infty} (y_{n+1} - y_n) = \lim_{n \to \infty} \left[\tfrac{1}{2} y_n(60 - y_n)\right]$$

$$\lim_{n \to \infty} y_{n+1} - \lim_{n \to \infty} y_n = \tfrac{1}{2}(\lim_{n \to \infty} y_n)(60 - \lim_{n \to \infty} y_n)$$

$$\bar{y} - \bar{y} = \tfrac{1}{2}\bar{y}(60 - \bar{y})$$

or

$$0 = \tfrac{1}{2}\bar{y}(60 - \bar{y})$$

This quadratic equation has the two solutions
$$\bar{y} = 0 \quad \text{or} \quad \bar{y} = 60$$
Thus in the long run the terms of a solution to the equation will approach either 0 or 60. ∎

Exercises 10.2

Find the solution and the first, second, third, fifth, and ninth values of the solution for each of the difference equations in Exercises 1–10, where $y_0 = 2$ in each case. You may need to use a calculator.

1. $y_{n+1} - y_n = \frac{1}{2}y_n$
2. $y_{n+1} - y_n = 0$
3. $y_{n+1} - y_n = -y_n$
4. $y_{n+1} - y_n = 1$
5. $y_{n+1} - y_n = 3$
6. $y_{n+1} - y_n = -\frac{1}{3}$
7. $y_{n+1} - y_n = -\frac{y_n}{2}$
8. $y_{n+1} - y_n = -3y_n$
9. $y_{n+1} - y_n = 0.25 y_n$
10. $y_{n+1} - y_n = -0.08 y_n$

Find the solution and the first, third, sixth, and tenth terms of the solution in Exercises 11–20.

11. $y_{n+1} = \frac{1}{2}y_n + 1; \quad y_0 = 1$
12. $y_{n+1} = y_n + 1; \quad y_0 = -1$
13. $y_{n+1} = 2y_n - 4; \quad y_0 = 0$
14. $y_{n+1} = 2 - y_n; \quad y_0 = 3$
15. $2y_{n+1} - 6y_n + 4 = 0; \quad y_0 = 1$
16. $y_{n+1} - y_n = 4; \quad y_0 = 1$
17. $y_n = -2y_{n+1} - 3; \quad y_0 = 1$
18. $y_{n+1} = 0.2 y_n + 100; \quad y_0 = 15$
19. $y_n - y_{n+1} = 2y_n + 1; \quad y_0 = -\frac{3}{2}$
20. $\dfrac{y_{n+1} - y_n}{y_n} = \dfrac{1}{2}; \quad y_0 = 1$

21. This exercise asks you to compare continuously compounded change with change compounded at discrete intervals. Two savings accounts are available. One pays 8.7% interest per year compounded continuously (Section 5.5), and the other pays 9% a year compounded yearly. Model the first by a differential equation, the second by a difference equation. Solve both equations. If $1,000 is initially deposited in each account, what amount is in each account 1, 2, 5, and 10 years later?

22. An account initially contains $5,000 and draws 9% interest. Each year, just after interest is applied to the account, the owner withdraws $400. Model the amount in the account n years later. How much is in the account after 5 years?

23. A stream initially contains 25,000 trout with a birthrate of 15%. Fishermen harvest 17% of the trout in the stream each year, and the stream is stocked with 3,000 trout each year. What is the trout population 5 years later?

24. A disease is occurring in a population at the rate of 50 new cases a year. Treatment results in a cure rate of 35% of infected individuals. Initially 75 individuals were affected. How many individuals will carry the disease 5 years later?

25. Suppose that you borrow $5,000 to buy a car at 1% interest per month with a monthly payment of $200. How much will you owe on the loan 2 years later? How much interest will you have paid by that time?

26. What should your monthly payment be so as to pay off a $50,000 loan in 25 years at an interest rate of 1% per month?

27. A deer population of a certain area reproduces at the rate of 12% a year, beginning with an initial population of 1,500. How many deer can be taken during the 2-week hunting season so as to allow the population to reach a level of 2,000 in 10 years? How large is this herd after 5 years? If the area can support only 1,850 healthy deer, when will overgrazing occur? At what rate should deer be taken each season so as to approach but never exceed 1,850?

28. A nerve cell acts like an electrical condenser. It will not "fire" (send a message on) until its electrical charge reaches a certain level—say, 6 mV.

   a. If the cell's charge is increasing at the rate of 1.5 mV per msec, when will the cell fire if it initially has a zero charge?

b. Suppose that the cell loses 20% of its existing charge each msec. When will it fire?

29. Rework Example 5 for a dosage of 10 mg taken every 6 hours at a metabolic rate of 20%.

30. A level of 150 mg of Valium is considered dangerous. How frequently can a 25-mg dose of Valium be taken if the body metabolizes Valium at the rate of 6% of the amount present? [*Hint:* Use Example 5. Let T be the unknown time. If y_n is the amount when the nth dose is taken, then $e^{-0.06T}y_n$ remains when the next dose is taken.]

31. Let $y_{n+1} = 0.56y_n + b$ and $y_0 = 50$. Find b so that $\lim_{n\to\infty} y_n = 100$.

Challenge Problems

32. A difference equation may also be given as $y_{n+1} = f(y_0, y_1, \ldots, y_n)$, where y_{n+1} depends on *all* preceding terms. The *second-order* difference equation (with $y_1 = 1$ and $y_2 = 1$)

$$y_{n+1} = y_n + y_{n-1} \qquad (13)$$

is a famous example yielding the **Fibonacci sequence.** A sequence given in this way is said to be a **recursive relation.** This means that previous terms are used to define the next term; this next term is then used with previous terms to define the "second-next" term. The concept of a recursive relationship is important in both mathematics and computer science, where a program loop is often a recursive relationship.

a. Find the first seven terms of the Fibonacci sequence.

b. Let $r_n = y_{n+1}/y_n$. Divide Equation 13 by y_n to obtain

$$\frac{y_{n+1}}{y_n} = \frac{y_n + y_{n-1}}{y_n} = 1 + \frac{y_{n-1}}{y_n} = 1 + \frac{1}{\frac{y_n}{y_{n-1}}}$$

or

$$r_n = 1 + \frac{1}{r_{n-1}}$$

Now find $r = \lim_{n\to\infty} r_n$ as a positive number. [*Hint:* See Example 6.] Conclude that if $\lim_{n\to\infty} r_n$ exists then $\lim_{n\to\infty} y_n$ cannot exist.

33. The Fibonacci sequence occurs throughout nature in such patterns as the sunflower and the spiral of branches forming a tree. Suppose that a certain insect species attains reproductive maturity 2 days after birth and that each mature pair of insects produces a new pair of insects each day. Assuming an initial population of 50 mature insect pairs, show that the insect population n days later is a (constant) multiple of the Fibonacci sequence. [*Hint:* List the population size in successive months in three columns, representing mature, young, and total pairs of insects in each day, respectively.]

34. Draw a rectangle of width w and height h of any size you choose, but draw the two sides so that the rectangle has what seems to you a pleasing proportion of width to height.

a. Measure the resulting sides and compare the ratio w/h to the limit r found in Exercise 32. The number r in Exercise 32 has been known for centuries as the "golden ratio" and matches the rectangular dimensions of the Athenian Parthenon; this ratio occurs in science, art, and nature far more often than should be expected by coincidence.

b. Draw a square, divide it in two, and rotate the line AB so as to lengthen the base and construct a new rectangle. Use a ruler to see that the larger rectangle has a w/h ratio of r in Exercise 32.

If you remove the square from the picture, the remaining rectangle is again a "golden rectangle" with $w_1/h_1 = r$ (see Figure 10.3).

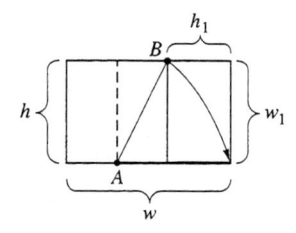

Figure 10.3

In Exercises 35–40, assume $\lim_{n\to\infty} y_n$ exists and find its value(s) without solving the difference equation (as in Example 6).

35. $y_{n+1} = 2 + 3y_n$

36. $y_{n+1} = -0.2\, y_n + \dfrac{n}{n+1}$

37. $y_{n+1} = -2 + y_n^2$

38. $y_{n+1} - y_n = y_n(3 - y_n)$

39. $y_{n+1} = \dfrac{y_n}{1 + y_n}$

40. $\dfrac{y_n}{y_{n+1} + 2} = \sqrt{\dfrac{n^2}{n^2 + n}}$

41. A local environment can support M individuals. Let $x_0 = M/4$ be the initial population. Assume that the yearly population change $x_{n+1} - x_n$ is proportional to $M - x_n$. Find $\lim_{n \to \infty} x_n$.

42. **a.** In Section 8.3, we saw that if y is a solution to the logistic differential equation $y' = ky(M - y)$ with $0 < y(0) < M$, then $y(t) < M$ for any value of t. Show that the same property is *not* true for the difference equation $y_{n+1} - y_n = 2y_n(100 - y_n)$, where $y_0 = 80$, by finding y_1. In general the solution to the logistic difference equation can move above and below M as n changes.

b. Show that if $\lim_{n \to \infty} y_n$ exists, it equals 0 or M.

Computer Application Problems

Computer-driven applications of difference equations may be easily found. Use the BASIC program LOANPAY, which uses no more than the solution to the difference equation $y_{n+1} = ay_n + b$, to determine the monthly payment for the life of the loan and the payment schedule for both principal and interest in the indicated year for the loans in Exercises 43 and 44. Explain and/or interpret the output of the program in relation to the elements of this difference equation and its solution.

43. Loan amount $40,000 for 20 years at 9% interest; payment schedule for year 1 of the loan.

44. Loan amount $4,000 for 36 months at 10% interest; payment schedule for year 2 of the loan.

Use the BASIC program DISDIFF in Exercises 45 and 46.

45. When will the trout population in Exercise 23 reach a level of 30,000?

46. Suppose that in Exercise 25 you borrow $5,500 instead. What will your monthly payment be if you wish to pay off the loan in 4 years? When will you still owe $2,500 on the loan? How much interest do you pay as part of your 12th loan payment?

10.3 Infinite Series

In this section we study when and how we may add infinitely many numbers.

An **infinite series** is a sum of numbers

$$a_1 + a_2 + a_3 + \cdots + a_n + \cdots$$

without end. Such unending sums can be given computational meaning, occur naturally, and provide convenient and useful mathematical models. Consider a person walking across a room 10 ft wide at a rate of 5 ft per second. This will take 2 sec. At the same time, 1 sec will be required to cross half the room, $\frac{1}{2}$ sec to cross half of what remains, $\frac{1}{4}$ sec to cross half of what yet remains, and so on, so that we should expect that

$$2 = 1 + \frac{1}{2} + \frac{1}{4} + \frac{1}{8} + \cdots + \frac{1}{2^n} + \cdots$$

We will verify that this is so after giving a definition of the sum of such an infinite series. We will also see that exactly the same kind of sum models the effect on an economy of savings rates, taxing policies, and the like.

We cannot actually add infinitely many numbers in the usual sense, but it is possible to add any finite list of numbers. This is the basis for the concept of the **sum of an infinite series.** Given an infinite series of numbers

$$a_1 + a_2 + a_3 + \cdots + a_k + \cdots$$

the number

$$S_n = a_1 + a_2 + a_3 + \cdots + a_n$$

obtained by adding only the first n terms is called the **nth partial sum** of the series.

EXAMPLE 1

a. Determine S_1, S_2, S_3, and S_5 for the infinite series

$$a_1 + a_2 + a_3 + \cdots = 1 + \frac{1}{2} + \frac{1}{4} + \frac{1}{8} + \frac{1}{16} + \frac{1}{32} + \cdots + \frac{1}{2^{n-1}} + \cdots$$

whose nth term is $a_n = 1/2^{n-1}$.

b. Determine S_n for any number (of summands) n.

Solution

a. By definition,

$$S_1 = a_1 = 1 \qquad S_2 = a_1 + a_2 = 1 + \tfrac{1}{2} = \tfrac{3}{2}$$

$$S_3 = a_1 + a_2 + a_3 = 1 + \tfrac{1}{2} + \tfrac{1}{4} = \tfrac{7}{4}$$

and $\quad S_5 = a_1 + a_2 + a_3 + a_4 + a_5 = 1 + \tfrac{1}{2} + \tfrac{1}{4} + \tfrac{1}{8} + \tfrac{1}{16} = \tfrac{31}{16}$

b. Now consider an arbitrary nth partial sum

$$S_n = a_1 + a_2 + \cdots + a_n = 1 + \frac{1}{2} + \frac{1}{4} + \frac{1}{8} + \cdots + \frac{1}{2^{n-1}}$$

It is hardly obvious what this sum is as a single fraction, but there is a clever way to find S_n. Notice first that S_n has the *form*

$$S_n = 1 + r + r^2 + r^3 + \cdots + r^{n-1}$$

where $r = \tfrac{1}{2}$. If we multiply both sides of this equation by $1 - r$, we have

$$(1 - r) S_n = (1 - r)(1 + r + r^2 + \cdots + r^{n-1})$$

$$= 1 + r + r^2 + \cdots + r^{n-1} - r - r^2 - r^3 - \cdots - r^n$$

or $\quad (1 - r) S_n = 1 - r^n$

by cancellation of positive and negative terms. Solving for S_n by division of both sides of this equation by $1 - r$ gives

$$S_n = \frac{1 - r^n}{1 - r}$$

Since $r = \tfrac{1}{2}$, we conclude that

$$S_n = \frac{1 - \left(\tfrac{1}{2}\right)^n}{1 - \tfrac{1}{2}} = 2 - \frac{1}{2^{n-1}} \quad\blacksquare$$

Notice from Example 1 that the partial sums $S_1, S_2, S_3, \ldots, S_n$ form an ordinary *sequence* of numbers

$$1, \frac{3}{2}, \frac{7}{4}, \frac{15}{8}, \frac{31}{16}, \ldots, 2 - \frac{1}{2^{n-1}}, \ldots$$

This suggests how we may define the sum of an infinite series (a definition analogous to that for an improper integral in Section 9.1).

Definition The **sum** S of an infinite series

$$a_1 + a_2 + a_3 + \cdots + a_n + \cdots$$

is the number

$$S = \lim_{n \to \infty} S_n$$

when this limit exists as a real number. We then write

$$S = a_1 + a_2 + a_3 + \cdots$$

and say that the series **converges** to S. If $\lim_{n \to \infty} S_n$ does not exist, or is infinite, we say that the series **diverges.**

EXAMPLE 2

Find the infinite sum

$$1 + \frac{1}{2} + \frac{1}{4} + \frac{1}{8} + \cdots + \frac{1}{2^{n-1}} + \cdots$$

Solution According to Example 1, $S_n = 2 - (1/2^{n-1})$. Thus $\lim_{n \to \infty} S_n = \lim_{n \to \infty} [2 - (1/2^{n-1})] = 2$. Therefore the given series converges with sum $S = 2$. That is,

$$2 = 1 + \frac{1}{2} + \frac{1}{4} + \frac{1}{8} + \cdots + \frac{1}{2^{n-1}} + \cdots \quad \blacksquare$$

The Geometric Series

Examples 1 and 2 are special cases of a singularly important infinite series called the **geometric series.** A geometric series is an infinite series of the form

$$1 + r + r^2 + r^3 + \cdots + r^n + \cdots$$

whose nth term $a_n = r^{n-1}$ is a power of the number r. In Examples 1 and 2, $r = \frac{1}{2}$. In the solution of Example 1, we saw that the nth partial sum G_n of the geometric series is

$$G_n = 1 + r + r^2 + \cdots + r^{n-1} = \frac{1 - r^n}{1 - r}$$

When then does the geometric series converge? The answer depends entirely on the value of r.

Case 1. If $-1 < r < 1$, then $\lim_{n \to \infty} r^{n+1} = 0$ and therefore

$$\lim_{n \to \infty} G_n = \lim_{n \to \infty} \frac{1 - r^{n+1}}{1 - r} = \frac{1}{1 - r}$$

is the sum of the converging geometric series.

Case 2. If $r = 1$, then $G_n = 1 + 1^2 + 1^3 + \cdots + 1^n = n$. Therefore $\lim_{n \to \infty} G_n$ does not exist and the series diverges. If $r = -1$, then $G_n = -1 + (-1)^2 + (-1)^3 + \cdots + (-1)^n$, which is alternatively -1 or 0. Again $\lim_{n \to \infty} G_n$ cannot exist and the series diverges.

Case 3. If $r > 1$ or $r < -1$, then $\lim_{n \to \infty} r^{n+1}$ is $\pm \infty$ and again the series diverges.

In conclusion:

The geometric series converges if and only if $|r| < 1$ and in this case

$$\frac{1}{1 - r} = 1 + r + r^2 + r^3 + \cdots + r^n + \cdots \tag{1}$$

is its sum.

Notice that this sum is valid even when r is negative (but $r > -1$); the geometric series can contain negative summands.

EXAMPLE 3

Suppose that the average individual spends 90% of each dollar earned. Conclude that each single dollar released into an economy generates $10 worth of economic activity in the long run.

Solution The single dollar is first earned by someone, who in turn spends 90 cents of this dollar. This 90 cents is received by someone else, who in turn spends $0.9(90 \text{ cents}) = 81$ cents. This process continues so that the total amount of economic activity in the long run is represented by

$$1 + 0.9 + (0.9)^2 + (0.9)^3 + \cdots + (0.9)^n + \cdots$$

This is a geometric series with $r = 0.9$. From Equation 1, its sum is

$$\frac{1}{1 - r} = \frac{1}{1 - 0.9} = 10 \quad \blacksquare$$

Economists refer to the principle revealed by Example 3 as the **multiplier effect**. Whether this one dollar is released into an economy by government spending or by a decrease in taxation the effect is the same.

The Divergent Harmonic Series

The series

$$1 + 1 + 1 + \cdots + 1 + \ldots$$

clearly *diverges* because $S_n = n$, and therefore $\lim_{n \to \infty} S_n = \infty$ is not a real number.

It is much less clear that the **harmonic series**

$$1 + \frac{1}{2} + \frac{1}{3} + \cdots + \frac{1}{n} + \cdots$$

also *diverges*, despite the decreasing size of its nth term, $a_n = 1/n$. Let us see why.

The harmonic series is a prototype of the subtlety and difficulty in determining the convergence of an infinite series. We cannot simply look at a series, or at its nth term, and determine convergence. But in this case a graph is appropriate.

It is apparent from Figure 10.4 that the nth partial sum $S_n = 1 + (1/2) + (1/3) + \cdots + (1/n)$ of the harmonic series is the sum of the areas of the indicated rectangles, each of width 1 unit and height $a_n = 1/n$ (between $x = n$ and $x = n + 1$). It is equally apparent from Figure 10.4 that this sum S_n of rectangular areas is *larger* than the area

$$\int_1^{n+1} \frac{1}{x}\, dx = \ln x \Big]_1^{n+1} = \ln(n+1)$$

beneath the graph of $y = 1/x$. This means that $S_n \geq \ln(n+1)$. Consequently, $\lim_{n \to \infty} S_n \geq \lim_{n \to \infty} \ln(n+1) = +\infty$, and the harmonic series *diverges*.

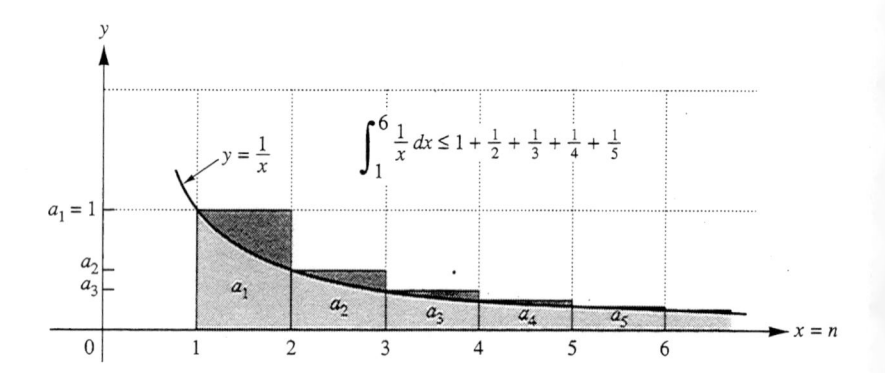

Figure 10.4

Notation and General Properties of Infinite Series

Before discussing a test procedure for the convergence of an infinite series, we must introduce some notation and a basic theorem. We will adopt the standard notation

$$\sum_{n=1}^{\infty} a_n = a_1 + a_2 + \cdots + a_n + \cdots$$

for an infinite series whose nth term is a_n, and correspondingly $\sum_{k=1}^{n} a_k = a_1 + a_2 + \cdots + a_n = S_n$ for its nth partial sum. When these are taken together, we typically write $S = \sum_{k=1}^{\infty} a_k$ when $S = \lim_{n\to\infty} S_n$. When no confusion is possible, we will write, more briefly and simply, Σa_n for the sum $\sum_{n=1}^{\infty} a_n$.

Because Σa_n is defined as a limit of a sequence, we have the following basic properties directly from the limit properties for sequences (Section 10.1).

Algebraic Properties of Infinite Series

If Σa_n and Σb_n both converge and if k is constant, then $\Sigma k a_n$ and $\Sigma (a_n + b_n)$ also converge and

$$\Sigma k a_n = k \Sigma a_n \tag{2}$$

$$\Sigma (a_n + b_n) = \left(\Sigma a_n\right) + \left(\Sigma b_n\right) \tag{3}$$

For example, using these properties and the sum of a geometric series, we say that

$$\sum_{n=0}^{\infty}\left(\frac{3}{2^n} + \frac{5}{9^n}\right) = \sum_{n=0}^{\infty}\frac{3}{2^n} + \sum_{n=0}^{\infty}\frac{5}{9^n} = 3\sum_{n=0}^{\infty}\frac{1}{2^n} + 5\sum_{n=0}^{\infty}\frac{1}{9^n}$$

$$= 3\left(\frac{1}{1-\frac{1}{2}}\right) + 5\left(\frac{1}{1-\frac{1}{9}}\right) = 6 + \frac{45}{8} = \frac{93}{8}$$

The Ratio Test for Convergence

The geometric series is only one of many infinite series. It may surprise you to learn that it is one of the few series for which we actually find a sum in "closed," concise form. Our emphasis will now change from finding the actual sum S of a series to determining only whether or not it has a sum—whether or not it converges. We take this view because in practice we cannot actually add infinitely many numbers $a_1 + a_2 + \cdots + a_n + \cdots$, but if we at least know that the series *has a sum* S, we may use the finite sum S_n as a very good approximation to S. For example, in the multiplier effect (Example 3), after (say) 50 such economic exchanges, we have

$$S - S_{50} = 10 - \frac{1 - 0.9^{50}}{1 - 0.9} \simeq 10 - 9.9485 = 0.0515$$

which, as a "dollar" figure, is only slightly more than a nickel.

The **ratio test** offers a direct test procedure for determining the convergence or divergence of an infinite series, all of whose terms are positive numbers.

Theorem 10.1   **The Ratio Test**

Suppose that $a_n > 0$ for all n and let

$$R = \lim_{n \to \infty} \frac{a_{n+1}}{a_n}$$

1. If $R < 1$, then $\Sigma\, a_n$ converges.
2. If $R > 1$, then $\Sigma\, a_n$ diverges.
3. If $R = 1$, this test is inconclusive.

EXAMPLE 4

Use the ratio test to show that

a. $\Sigma\, 1/n!$ converges   **b.** $\Sigma\, n^n/n!$ diverges

Solution

a. Here $a_n = 1/n!$ and

$$\frac{a_{n+1}}{a_n} = \frac{\frac{1}{(n+1)!}}{\frac{1}{n!}} = \frac{n!}{(n+1)!} = \frac{1}{n+1}$$

Thus $R = \lim_{n \to \infty} \frac{a_{n+1}}{a_n} = \lim_{n \to \infty} \frac{1}{n+1} = 0$

Since $R < 1$, the series converges.

b. Here $a_n = n^n/n!$ and

$$\frac{a_{n+1}}{a_n} = \frac{\frac{(n+1)^{n+1}}{(n+1)!}}{\frac{n^n}{n!}} = \frac{(n+1)^{n+1}}{n^n} \cdot \frac{n!}{(n+1)!} = \frac{(n+1)^{n+1}}{n^n} \cdot \frac{1}{n+1}$$

$$= \frac{(n+1)^n}{n^n} \cdot (n+1) \cdot \frac{1}{n+1} = \left(\frac{n+1}{n}\right)^n = \left(1 + \frac{1}{n}\right)^n$$

Consequently, $R = \lim_{n \to \infty} \frac{a_{n+1}}{a_n} = \lim_{n \to \infty} \left(1 + \frac{1}{n}\right)^n = e$

from Chapter 5. Since $R = e > 1$, $\Sigma\, a_n$ diverges. ∎

    Why is the ratio test valid? Suppose, *for simplicity*, that $a_{n+1}/a_n = R < 1$. Then, for all n, $a_{n+1} = R a_n$. Consequently,

$$a_2 = R a_1 \qquad a_3 = R a_2 = R(R a_1) = R^2 a_1 \qquad a_4 = R a_3 = R(R^2 a_1) = R^3 a_1$$

and so on. That is,

$$a_1 + a_2 + a_3 + \cdots = a_1 R + a_1 R^2 + a_1 R^3 + \cdots = a_1(R + R^2 + R^3 + \cdots)$$

from Equation 2. Since $R < 1$, this geometric series in R converges and therefore so does $\Sigma\ a_k$. A complete argument for the ratio test would have to deal with the fact that we only know that $a_{n+1}/a_n \simeq R$, using the limit hypothesis of the test.

Because the ratio test is inconclusive when $R = 1$, it does not allow us to determine the convergence of all positive term series. Indeed the ratio test is inconclusive for the harmonic series.

EXAMPLE 5

Show that $R = 1$ for the harmonic series $\Sigma\ 1/n$.

Solution Here $a_n = 1/n$ and

$$\frac{a_{n+1}}{a_n} = \frac{\frac{1}{n+1}}{\frac{1}{n}} = \frac{n}{n+1}$$

Therefore

$$R = \lim_{n \to \infty} \frac{a_{n+1}}{a_n} = \lim_{n \to \infty} \frac{n}{n+1} = \lim_{n \to \infty} \frac{1}{1 + \frac{1}{n}} = 1 \ \blacksquare$$

In Exercise 50, you are asked to show that $R = 1$ for the *convergent* series $\Sigma\ 1/n^2$. This exercise means that the *ratio test* truly is *inconclusive* when $R = 1$.

One might ask at this point, "If we are interested only in knowing whether a series converges so that, if it does, we can use a computer to approximate its sum by finding a partial sum S_n for n very large, why not just try adding the series directly on a computer and see if it 'adds up' to something?" Again the subtleties of convergence intrude, and only abstract mathematics is powerful enough to decide the issue of convergence (see Exercise 63).

Alternating Series

Except in the case of the geometric series, when $-1 < r < 0$, we have only considered series of positive numbers a_n. An **alternating series** is a series of the form

$$b_1 - b_2 + b_3 - b_4 + \cdots + (-1)^{n+1}\ b_n + \cdots$$

where b_n is itself a nonnegative number. The next theorem tells us that an alternating series behaves as we might wish all series behaved.

Theorem 10.2 If for each n, $b_n \geq b_{n+1}$, then the alternating series $\Sigma(-1)^{n+1}\ b_n$ converges if and only if $\lim_{n \to \infty} b_n = 0$.

EXAMPLE 6

Show that the alternating harmonic series

$$1 - \tfrac{1}{2} + \tfrac{1}{3} - \tfrac{1}{4} + \cdots + (-1)^{n+1}/n + \cdots$$

converges.

Solution Here $b_n = 1/n$. Since $\lim_{n\to\infty} b_n = \lim_{n\to\infty} (1/n) = 0$, the series $\Sigma(-1)^{n+1}/n$ converges. [Note, of course, that $1/n \geq 1/(n+1)$ for each n.] ∎

Absolute Convergence

The simplifying concept of **absolute convergence** helps to bring some order to the general study of infinite series. A series $\Sigma\, a_n$ is said to be *absolutely convergent* if the series of absolute values $\Sigma\, |a_n|$ converges. In general we most wish to know that a series is absolutely convergent for the following reason.

Theorem 10.3

If a series $\Sigma\, a_n$ is absolutely convergent, then it is convergent.

EXAMPLE 7

Use the ratio test to conclude that the series $\Sigma(-2)^n/n!$ converges absolutely and therefore converges.

Solution Here $a_n = (-2)^n/n!$, and therefore $|a_n| = 2^n/n!$. Let $b_n = |a_n|$. Then

$$\frac{b_{n+1}}{b_n} = \frac{\dfrac{2^{n+1}}{(n+1)!}}{\dfrac{2^n}{n!}}$$

$$= \frac{2^{n+1}}{(n+1)!} \cdot \frac{n!}{2^n} = \frac{2}{n+1}$$

Consequently,
$$R = \lim_{n\to\infty} \frac{b_{n+1}}{b_n} = \lim_{n\to\infty} \frac{2}{n+1} = 0$$

and so, by the ratio test, $\Sigma\, b_n = \Sigma\, |a_n|$ converges. Therefore $\Sigma\, a_n$ is absolutely convergent and so, by Theorem 10.3, it is convergent. ∎

Remarks.

1. The series in Example 7 is of course an alternating series $a_n = (-1)^n 2^n/n!$, but Theorem 10.2 is not easily applied to it.
2. The *converse* of Theorem 10.3 is *false*. Example 6 shows that the alternating harmonic series $\Sigma\,(-1)^n/n$ is convergent, but $\Sigma\,|(-1)^n/n| = \Sigma\, 1/n$ is the divergent harmonic series. Therefore the alternating harmonic series is convergent but *not* absolutely convergent.

This concludes a brief study of infinite series of numbers. This topic is challenging because it does not yield to "formula" solutions. The best first test for convergence is the ratio test. A *comparison test* (see Exercises 51–60) is available when the ratio test fails. After these, only ad hoc methods, and possibly a more advanced text, may help.

Exercises 10.3

Find the first, second, third, and sixth partial sums of the series whose nth term a_n is given in Exercises 1–8.

1. $a_n = \dfrac{1}{3^{n-1}}$
2. $a_n = (-1)^n$
3. $a_n = 2$
4. $a_n = \dfrac{1}{n}$
5. $a_n = 1/n - 1/(n+1)$
6. $a_n = 1/n + 1/2^{n-1}$
7. $a_n = \dfrac{3}{2^{n-1}}$
8. $a_n = \begin{cases} 0, & n \text{ odd} \\ \dfrac{1}{2^{n-1}}, & n \text{ even} \end{cases}$

9. Let $a_n = n$. Find $S_n = \sum_{k=1}^{n} a_k$ for $n = 2, 5, 15,$ and 100. [*Hint:* In the case $n = 100$, try adding the first and last terms (1 plus 100), then the second and second-to-last (2 plus 99), and continue this way until you "see the light." Can you now write a general formula for S_n in terms of n? Does the infinite series $\sum_{n=1}^{\infty} n$ converge?

10. Consider the harmonic series $1 + \tfrac{1}{2} + \tfrac{1}{3} + \tfrac{1}{4} + \tfrac{1}{5} + \cdots$, and group the terms of the series as

$$1 + \frac{1}{2} + \left(\frac{1}{3} + \frac{1}{4}\right) + \left(\frac{1}{5} + \frac{1}{6} + \frac{1}{7} + \frac{1}{8}\right) + \cdots$$
$$+ \left(\frac{1}{2^{n-1}+1} + \frac{1}{2^{n-1}+2} + \cdots + \frac{1}{2^n}\right) + \cdots$$

  a. Show that each sum in parentheses is equal to or larger than $\tfrac{1}{2}$. [*Hint:* Each of the *two* terms $\tfrac{1}{3}$ and $\tfrac{1}{4}$ is no smaller than the single term $\tfrac{1}{4}$.]

  b. Conclude from (a) that $S_1 = 1$, $S_2 = \tfrac{3}{2}$, $S_4 \geq 2$, $S_8 \geq \tfrac{5}{2}$, $S_{16} \geq 3$.

  c. Conclude from (a) that $S_{2^n} \geq 1 + (n/2)$ and again that the harmonic series diverges.

Write out the geometric series $1 + r + r^2 + \cdots + r^n + \cdots$, and find its sum when r is given as in Exercises 11–16.

11. $r = \tfrac{1}{3}$
12. $r = -\tfrac{1}{2}$
13. $r = 0$
14. $r = 0.25$
15. $r = -0.6$
16. $r = -0.01$

Use the series property $\sum k a_n = k \sum a_n$ to factor an appropriate constant in each of the series in Exercises 17–21 so as to express it as a multiple k of a geometric series and then find its sum.

17. $\dfrac{1}{2} + \dfrac{1}{4} + \dfrac{1}{8} + \cdots + \dfrac{1}{2^{n+1}} + \cdots$
    [*Hint:* Factor $k = \tfrac{1}{2}$.]

18. $\dfrac{3}{10} + \dfrac{3}{10^2} + \dfrac{3}{10^3} + \cdots$

19. $\dfrac{1}{4} - \dfrac{1}{8} + \dfrac{1}{16} - \dfrac{1}{32} + \cdots + \left(-\dfrac{1}{2}\right)^n + \cdots$

20. $9 - \dfrac{9}{10^2} + \dfrac{9}{10^4} + \cdots$

21. $\dfrac{5}{10^2} + \dfrac{5}{10^4} + \dfrac{5}{10^6} + \cdots$

22. Let $G = 1 + r + r^2 + \cdots = 1/(1-r)$, $|r| < 1$, and let $G_n = 1 + r + \cdots + r^{n-1}$.

  a. Show that $G - G_n = r^n/(1-r)$.

  b. Let $r = \tfrac{1}{3}$. How well does G_{10} approximate G? That is, how large is the difference $G - G_{10}$?

By definition an infinite decimal $0.a_1 a_2 a_3 \ldots$ equals the infinite series $(a_1/10) + (a_2/10^2) + (a_3/10^3) + \cdots$. Write each infinite decimal in Exercises 23–28 as an infinite series, and find its sum as a fraction using the geometric series.

23. $0.333\ldots$
24. $0.999\ldots$
25. $0.050505\ldots$
26. $2.666\ldots = 2 + 0.666\ldots$
27. $0.4999\ldots$
28. $0.252525\ldots$ [*Hint:* Use the property $\sum (a_n + b_n) = \sum a_n + \sum b_n$.]

10.3 Infinite Series

29. Calculate the multiplier effect (Example 3) when individuals spend 95 cents of every dollar they earn; when they spend 80 cents of every dollar they earn.

30. Perform long division on the fraction $\frac{123}{999}$ until you are able to "see" how to write this fraction as an infinite decimal $0.a_1 a_2 a_3 \ldots a_n \ldots$. Then show that this decimal representation is correct using the geometric series, as in Exercise 25 or 28.

Use the ratio test to determine convergence/divergence of the infinite series in Exercises 31–40.

31. $\sum \dfrac{1}{n!}$

32. $\sum \dfrac{2^n}{n!}$

33. $\sum \left(\dfrac{5}{4}\right)^n$

34. $\sum ne^{-n}$

35. $\sum \dfrac{n^{n/2}}{n}$

36. $\sum \dfrac{1}{\sqrt{2^n}}$

37. $\sum \dfrac{2^{2n}}{n!}$

38. $\sum \dfrac{n^2}{2^n}$

39. $\sum \dfrac{\ln n}{n!}$

40. $\sum \dfrac{n!}{n^n}$

In Exercises 41–46, determine (a) whether the series converges and (b) whether it converges absolutely.

41. $\sum (-1)^n \left(\dfrac{n}{n+1}\right)$

42. $\sum (-1)^n \left(\dfrac{n}{n^2+1}\right)$

43. $\sum (-1)^n \left(\dfrac{11}{10}\right)^n$

44. $\sum (-1)^n \left(\dfrac{9}{10}\right)^n$

45. $\sum \dfrac{(-5)^n}{n}$

46. $\sum \dfrac{(-1)^n}{\sqrt{n}}$ [Hint: $\sqrt{n} < n$.]

Use the property $\sum (a_n + b_n) = \sum a_n + \sum b_n$ to find the sum of the series in Exercises 47 and 48.

47. $\sum (-1)^n/2^n + (2/3)^n$

48. $\sum \dfrac{1 + (-1)^n}{2^n}$

49. Find the ninth and tenth partial sums of the series $\sum_{n=1}^{\infty} (-1)^n/10^n$ as a decimal. Find the sum of this series as an infinite decimal.

50. We have seen that $\sum_{k=1}^{n} 1/k \geq \ln(n+1)$ since $\sum_{k=1}^{n} 1/k \geq \int_1^{n+1} (1/x)\, dx$ (see Figure 10.4).

   a. Show that $R = 1$ for the harmonic series.

   b. Draw a similar graph and conclude that
   $$1 + \dfrac{1}{2^2} + \dfrac{1}{3^2} + \cdots + \dfrac{1}{n^2} \leq 1 + \int_1^n \dfrac{1}{x^2}\, dx$$

c. Let $S_n = \sum_{k=1}^{n} 1/k^2$. Use (b) to conclude that $S_n \leq 2 - (1/n)$.

d. Because $S_n \leq S_{n+1}$, it is known that $\lim_{n \to \infty} S_n$ exists or is infinite. From (c) conclude that $\lim_{n \to \infty} S_n \leq 2$ and hence that $\sum 1/n^2$ converges with sum at most 2.

e. Extend this argument to $\sum 1/n^p$, for any $p > 1$.

Challenge Problems

When the ratio test fails, a useful alternative is given by the following theorem, which is known as the *comparison test* for convergence and divergence.

1. If $0 \leq a_n \leq b_n$ for all n and $\sum b_n$ converges, then so does $\sum a_n$ converge.
2. If $0 \leq b_n \leq a_n$ and $\sum b_n$ diverges, then so does $\sum a_n$ diverge.

Do not let all the notation distract you; the idea is simple. In plain English, a *positive* term series that is (a) smaller than a convergent series is also convergent; (b) larger than a divergent series is also divergent.

For example, (a) $\sum 1/n2^n$ converges because $a_n = 1/n2^n \leq 1/2^n = b_n$ and $\sum 1/2^n$ converges; (b) $\sum 1/\sqrt{n}$ diverges because $a_n = 1/\sqrt{n} \geq 1/n = b_n$ and $\sum b_n$ diverges.

Use the *comparison test* along with the geometric series, the harmonic series, or the convergent series $\sum 1/n^2$ to determine the convergence or divergence of the series in Exercises 51–60.

51. $\sum \dfrac{1}{\ln 2^n}$

52. $\sum \dfrac{1}{n^{3/2}}$

53. $\sum \dfrac{1}{e^n}$

54. $\sum \dfrac{2^n}{3^n}$

55. $\sum \dfrac{1}{n!}$ [Hint: $n! \geq n^2$ if $n \geq 4$.]

56. $\sum \dfrac{n}{n+1}$ [Hint: $n^2 \geq n+1$ if $n \geq 2$.]

57. $\sum \dfrac{1}{n^{1/n}}$

58. $\sum \dfrac{n}{n^2+n}$

59. $\sum \dfrac{1}{n2^n}$

60. $\sum \dfrac{1}{\sqrt{n^2+n}}$

61. Show that $\sum 1/(2^n - n)$ converges using the fact that $2^n - n(n+1) > 0$ if $n \geq 5$.

62. Show by drawing a graph that $\sum_{k=2}^{n} 1/(k \ln k) \geq \int_2^{n+1} dx/(x \ln x)$ and conclude that $\sum_{k=2}^{\infty} 1/(k \ln k)$ diverges.

Computer Application Problems

For all its speed of computation the computer cannot be used to determine whether an infinite series converges; once the series is known to converge, the computer can be a useful tool for estimating its sum. Use the BASIC program SERIES in Exercises 63 and 64.

63. Find the partial sums of the harmonic series $\sum_{k=1}^{n} 1/k$ for $n = 100$, $n = 500$, and $n = 10,000$.

64. Repeat Exercise 63 for the convergent alternating harmonic series $\sum_{k=1}^{n} (-1)^k/k$.

10.4 Power Series

In this section we will see that many useful functions are "almost" polynomials by virtue of their Taylor power series representation.

If every function were a polynomial, mathematics would in some ways be a lot simpler! Calculus gives us a way to *approximate* most functions by polynomial functions.

Recall that a *polynomial function* P is given by a finite sum of powers of x:

$$P(x) = a_0 + a_1 x + a_2 x^2 + \cdots + a_n x^n$$

A **power series** in turn is such an expression but *without end*:

$$a_0 + a_1 x + a_2 x^2 + \cdots + a_n x^n + \cdots \quad (1)$$

The numbers $a_0, a_1, a_2, \ldots$ are called the **coefficients** of the power series. The theme of this section is that while functions like e^x, $1/(1 - x)$, $\ln(x + 1)$, and so on, are not exactly equal to a polynomial, they are equal to the sum of a power series over part, or all, of their domain and thus are approximately equal to a polynomial.

EXAMPLE 1

Write $f(x) = 1/(1 - x)$ as a power series for $-1 < x < 1$.

Solution We have already seen that the geometric series

$$1 + x + x^2 + \cdots + x^n + \cdots = 1/(1 - x) \quad \text{for } |x| < 1$$

Since the expression on the left is indeed a power series with all coefficients $a_0 = 1$, $a_1 = 1$, $a_2 = 1, \ldots, a_n = 1, \ldots$, we can write the function f as a power series:

$$f(x) = 1 + x + x^2 + \cdots + x^n + \cdots \quad \blacksquare$$

Convergence of a Power Series

Before we see in general how a function may be written as a power series, we must clearly understand the nature of the convergence of a power series. First, a power series is also an infinite series with nth term $b_n(x) = a_n x^n$ and nth partial sum (with nth degree polynomial)

$$S_n(x) = a_0 + a_1 x + \cdots + a_n x^n$$

What is new here is that the values of the nth term and of the nth partial sum depend on x. The power series will *converge for a particular value of x*, with sum $S(x)$, if (as for any infinite series) $\lim_{n\to\infty} S_n(x) = S(x)$ exists.

Example 1 exhibits all the relevant points for this section. The function $f(x) = 1/(1-x)$ is defined for all $x \neq 1$. On the other hand, the power series $1 + x + x^2 + \cdots$ converges only for values of x between -1 and 1, that is, for x in the *interval* $-1 < x < 1$. The sum

$$S(x) = \lim_{n\to\infty} S_n(x) = \lim_{n\to\infty} \frac{1-x^{n+1}}{1-x} = \frac{1}{1-x} = f(x) \qquad \text{for } |x| < 1$$

from Section 10.3. In short we cannot expect a power series to converge for just any x, and when, as is our goal, we write a function as a power series, equality of the series and the function may be restricted.

Since a power series is an infinite series, the ratio test offers a useful test for its convergence. For the series

$$a_0 + a_1 x + a_2 x^2 + \cdots + a_n x^n + \cdots$$

let $L = \lim_{n\to\infty} |a_{n+1}/a_n|$. Now let $r = 1/L$, when this limit exists and is not zero; if $L = 0$, let $r = +\infty$. The basic fact for this section is the following theorem:

Theorem 10.4

For any number x, with $|x| < r$, the power series (Equation 2) is absolutely convergent. If we write

$$S(x) = a_0 + a_1 x + a_2 x^2 + \cdots + a_n x^n + \cdots \qquad (2)$$

for the numerical sum of this convergent series at a number x in the interval $(-r, r)$, this formula defines a continuous and differentiable function S on $(-r, r)$ for which

1. $S'(x) = a_1 + 2a_2 x + 3a_3 x^2 + \cdots + n a_n x^{n-1} + \cdots$ and

2. $\displaystyle\int S(x)\, dx = a_0 x + a_1\left(\frac{x^2}{2}\right) + a_3\left(\frac{x^2}{3}\right)$
$$+ \cdots + a_n\left(\frac{x^{n+1}}{n+1}\right) + \cdots + C$$

are both valid for all x in $(-r, r)$.

There is a great deal of useful information in Theorem 10.4, and we will see that the apparent technicalities in it can be largely avoided. For example, parts 1 and 2 tell us that we may differentiate or integrate $y = S(x)$ by performing these operations on each power term $a_n x^n$ separately, and this is a simple operation. Although the proof of the theorem is well beyond this text, we can understand the role of the number r. By the ratio test, the series (2) converges absolutely if

$$1 > \lim_{n\to\infty} \left|\frac{a_{n+1} x^{n+1}}{a_n x^n}\right| = \lim_{n\to\infty} \left|\frac{a_{n+1}}{a_n}\right| |x| = |x| \cdot L$$

That is, the series is absolutely convergent if $|x| < 1/L = r$. For this reason the open interval $(-r, r)$ is called the **interval of convergence** of the power series, and the number r is called the **radius of convergence** of the power series. It may or may not happen that the series also converges at the endpoint $x = r$ (or $x = -r$). We will not consider this subtlety in this text.

EXAMPLE 2

Determine the radius and interval of convergence of the power series

$$x + \frac{x^2}{2} + \frac{x^3}{3} + \frac{x^4}{4} + \cdots + \frac{x^n}{n} + \cdots$$

Solution Here $a_0 = 0$ and for $n \geq 1$, $a_n x^n = x^n/n$ so $a_n = 1/n$. Consequently,

$$L = \lim_{n \to \infty} \left| \frac{a_{n+1}}{a_n} \right| = \lim_{n \to \infty} \frac{\frac{1}{n+1}}{\frac{1}{n}} = \lim_{n \to \infty} \frac{n}{n+1} = 1$$

The radius of convergence is $r = 1/L = 1$, and the interval of convergence is $(-1, 1)$, an open interval centered at $x = 0$ with radius 1. ∎

We will exploit Theorem 10.4 in the remainder of this section. The next example illustrates how Theorem 10.4 may be (1) interpreted as defining a function of x, (2) used to find this function, and (3) used to find the sum of a power series (in this case).

EXAMPLE 3

a. Find the derivative of the function defined by the sum of the power series in Example 2, on its interval of convergence $(-1, 1)$.
b. Write the series in Example 2 as a function of x for $|x| < 1$.
c. Find the (numerical) sum of this power series at $x = \frac{1}{2}$.

Solution

a. The series in Example 2 defines the function

$$S(x) = x + \frac{x^2}{2} + \frac{x^3}{3} + \cdots + \frac{x^n}{n} + \cdots \quad \text{on } (-1, 1) \quad (3)$$

Therefore from Theorem 10.4(1)

$$S'(x) = 1 + x + x^2 + \cdots + x^{n-1} + \cdots \quad (4)$$

since $D_x(x^n/n) = nx^{n-1}/n = x^{n-1}$.

b. Since the series (Equation 4) is again the geometric series on $(-1, 1)$, we see that $S'(x) = 1/(1 - x)$. Therefore $S(x) = \int S'(x)\, dx = \int 1/(1 - x)\, dx = -\ln(1 - x) + C$.

Let us determine C by comparing values at $x = 0$. Notice first that $S(0) = -\ln(1 - 0) + C = C$. Second, from Equation 3, $S(0) = 0 + (0^2/2) + (0^3/3) + \cdots = 0$. Therefore $C = 0$ and we have

$$S(x) = x + \frac{x^2}{2} + \frac{x^3}{3} + \cdots = -\ln(1 - x) = \ln\left(\frac{1}{1 - x}\right) \quad \text{for } |x| < 1$$

c. At $x = \frac{1}{2}$, the series

$$\frac{1}{2} + \frac{\left(\frac{1}{2}\right)^2}{2} + \frac{\left(\frac{1}{2}\right)^3}{3} + \cdots = \ln\left(\frac{1}{1 - \frac{1}{2}}\right) = \ln 2 \approx 0.69 \quad \blacksquare$$

The Taylor-MacLaurin Series Representation of a Function

We have seen in Examples 1 and 3 how a function and a power series represent each other. The important question now before us is this: Given a function $y = f(x)$, how can the function value $f(x)$ be written as a power series $a_0 + a_1 x + a_2 x^2 + \cdots + a_n x^n + \cdots$? In particular, how should the coefficients a_n be chosen? The answer is that

$$a_n = \frac{f^{(n)}(0)}{n!} \tag{5}$$

where $f^{(n)}(0)$ is the nth derivative of f evaluated at $x = 0$ (if this nth derivative exists).*

Why is this? Suppose that we could write $f(x)$ as a power series

$$f(x) = a_0 + a_1 x + a_2 x^2 + \cdots + a_n x^n + \cdots \tag{6}$$

in its interval of convergence. By substitution,

$$f(0) = a_0 + a_1 0 + a_2 0^2 + \cdots + a_n 0^n + \cdots = a_0$$

Thus Equation 5 is valid for $n = 0$ because the leading coefficient a_0 must be $f(0)$.

Now, by Theorem 10.4,

$$f'(x) = a_1 + 2a_2 x + 3a_3 x^2 + \cdots + na_n x^{n-1} + \cdots$$

Therefore, at $x = 0$,

$$f'(0) = a_1 + 2a_2 \cdot 0 + 3a_3 \cdot 0^2 + \cdots + na_n \cdot 0^{n-1} + \cdots = a_1$$

Thus $a_1 = f^{(1)}(0) = f^{(1)}(0)/1!$, and again Equation 5 must hold for $n = 1$.

---

*Where by $f^{(0)}(0)$ we mean $f(0)$; additionally, from algebra, $0! = 1$.

Again by Theorem 10.4,

$$f^{(2)}(x) = f''(x) = 2a_2 + 3 \cdot 2a_3 x + \cdots + n(n-1)a_n x^{n-2} + \cdots \quad (7)$$

so $f^{(2)}(0) = 2a_2 + 3 \cdot 2a_3 \cdot 0 + \cdots + n(n-1)a \cdot 0^{n-2} + \cdots = 2a_2$

Thus $a_2 = f^{(2)}(0)/2 = f^{(2)}(0)/2!$, and Equation 5 is valid for $n = 2$.

We can continue in this way. Notice how $n!a_n$ builds up with each successive differentiation in succeeding factors. In Equation 7, we see $3 \cdot 2 \cdot a_3 = 3!a_3$ as a factor of x, which upon the *next* differentiation *and* evaluation at $x = 0$ yields $f^{(3)}(0) = 3!a_3$ or $a_3 = f^{(3)}(0)/3!$. The $(n-2)$th coefficient in Equation 7 will become $n(n-1)(n-2)a_n$ in the next differentiation. After n successive differentiations, followed by evaluation at $x = 0$, the equation $f^{(n)}(0) = n!a_n$ results, showing that Equation 5 must hold in general.

The logic of our position is that to write a function as a power series we *must* use Equation 5 as a formula for the coefficients of this series. We do not yet know if such a series then converges or, in particular, that it converges to the given function.

EXAMPLE 4

Write $f(x) = e^x$ as a power series. What is the interval of convergence of this series?

Solution Since $f'(x) = e^x$, $f''(x) = e^x, \ldots, f^{(n)}(x) = e^x$, then $a_n = f^{(n)}(0)/n! = e^0/n! = 1/n!$. Thus we write, as the only possible power series that may converge to e^x,

$$a_0 + a_1 x + a_2 x^2 + \cdots + a_n x^n + \cdots$$

$$= 1 + x + \frac{1}{2!}x^2 + \cdots + \frac{1}{n!}x^n + \cdots$$

$$= 1 + x + \frac{x^2}{2!} + \cdots + \frac{x^n}{n!} + \cdots$$

To conclude we must first find where this series converges and then where it equals the function $y = e^x$.

We have

$$L = \lim \left| \frac{a_{n+1}}{a_n} \right| = \lim_{n \to \infty} \frac{\frac{1}{(n+1)!}}{\frac{1}{n!}}$$

$$= \lim_{n \to \infty} \frac{n!}{(n+1)!} = \lim_{n \to \infty} \frac{1}{n+1} = 0$$

Since $L = 0$, then $r = +\infty$ and the interval of convergence is the entire real line $(-\infty, +\infty)$.

Second, let $S(x) = 1 + x + (x^2/2!) + \cdots + (x^n/n!) + \cdots$ be this series. By Theorem 10.4,

$$S'(x) = 0 + 1 + \frac{2x}{2!} + \frac{3x^2}{3!} + \cdots + \frac{nx^{n-1}}{n!} + \cdots$$

$$= 1 + x + \frac{x^2}{2} + \cdots + \frac{x^{n-1}}{(n-1)!} + \cdots = S(x) \qquad (8)$$

That is, $S'(x) = S(x)$. From Chapter 5, $S(x) = Ce^x$. Moreover, $C = Ce^0 = S(0) = 1 + 0 + (0^2/2!) + \cdots + (0^n/n!) + \cdots = 1$. Since $C = 1$, we then have $S(x) = 1 \cdot e^x = e^x$ or

$$e^x = 1 + x + \frac{x^2}{2!} + \cdots + \frac{x^n}{n!} + \cdots$$

from Equation 8. ∎

The question of when the power series defined via Equations 5 and 6 by a function f converges at all, and then converges to the function f itself, is a complex one. Suppose that there is a number M such that $\left|f^{(n)}(x)\right| \leq M$ for all derivatives $f^{(n)}$ and all numbers x in some open interval I containing $x = 0$. We then have

Theorem 10.5

For each x in the interval I

$$f(x) = f(0) + f'(0)x + \frac{f''(0)}{2!}x^2 + \cdots + \frac{f^{(n)}(0)}{n!}x^n + \cdots$$

Additionally, the series on the right is unique: It is the only possible power series equal to f on its interval of convergence.

The series in Theorem 10.5 is called the **Taylor-MacLaurin series** for the function f. [A more general series expansion in terms of powers $(x - a)^n$ and $f^{(n)}(a)/n!$ is possible and is known as the Taylor series for f, centered at $x = a$.]

Let us see why the uniqueness property in Theorem 10.5 is useful.

EXAMPLE 5

Write $f(x) = e^{-x^2}$ as a power series. What is $f''(0); f^{(3)}(0); f^{(4)}(0)$?

Solution By Example 4, for any number t,

$$e^t = 1 + t + \frac{t^2}{2!} + \cdots + \frac{t^n}{n!} + \cdots$$

Therefore, for $t = -x^2$,

$$e^{-x^2} = 1 + (-x^2) + \frac{(-x^2)^2}{2!} + \cdots + \frac{(-x^2)^n}{n!} + \cdots$$

$$= 1 - x^2 + \frac{x^4}{2!} + \cdots + \frac{(-1)^n x^{2n}}{n!} + \cdots$$

Since this is a power series (whose odd indexed coefficients happen to be zero), then according to Theorem 10.5, this series must be *the* power series for $f(x) = e^{-x^2}$.

Consequently, since $a_n = f^{(n)}(0)/n!$, then $f^{(2)}(0)/2! = a_2 = -1$ and so $f^{(2)}(0) = -(2!) = -2$. Similarly, $f^{(3)}(0)/3! = a_3 = 0$, so $f^{(3)}(0) = 0$. Finally, $f^{(4)}(0)/4! = a_4 = 1/2!$, so $f^{(4)}(0) = 4!/2! = 12$. ∎

Applications of the Taylor-MacLaurin Series

The topic of Taylor-MacLaurin power series representation and approximation of a function is largely beyond the level of this text. We can give only the most basic applications. Recall that we have seen that it is impossible to find an antiderivative for e^{-x^2}.

EXAMPLE 6

Use the Taylor-MacLaurin series for $f(x) = e^{-x^2}$ to estimate $\int_0^1 e^{-x^2}\,dx$ using the fourth partial sum.

Solution From Example 4,

$$e^{-x^2} = 1 - x^2 + \frac{x^4}{2!} - \frac{x^6}{3!} + \cdots$$

Therefore, from Theorem 10.4(2),

$$\int_0^1 e^{-x^2}\,dx = \int_0^1 \left(1 - x^2 + \frac{x^4}{2!} - \frac{x^6}{3!} + \cdots \right) dx$$

$$= \left(x - \frac{x^3}{3} + \frac{x^5}{5 \cdot 2!} - \frac{x^7}{7 \cdot 3!} + \cdots \right)\Big|_0^1$$

$$\approx 1 - \frac{1}{3} + \frac{1}{5 \cdot 2!} - \frac{1}{7 \cdot 3!} = 0.74286$$

using only the first four terms of this series. ∎

Exercises 10.4

Determine the radius and interval of convergence of each power series in Exercises 1–8. First specify the nth coefficient a_n.

1. $x - x^3 + x^5 - x^7 + \cdots + (-1)^n x^{2n+1} + \cdots$

2. $1 + \dfrac{x}{2} + \dfrac{x^2}{4} + \dfrac{x^3}{8} + \dfrac{x^4}{16} + \cdots + \dfrac{x^n}{2^n} + \cdots$

3. $x - \dfrac{x^3}{3!} + \dfrac{x^5}{5!} - \dfrac{x^7}{7!} + \cdots + \dfrac{(-1)^n x^{2n+1}}{(2n+1)!} + \cdots$

4. $1 + x + 2x^2 + 3x^3 + \cdots + nx^n + \cdots$

5. $1 - x + \dfrac{x^2}{2!} - \dfrac{x^3}{3!} + \cdots + \dfrac{(-1)^n x^n}{n!} + \cdots$

6. $1 + ex + e^2 x^2 + e^3 x^3 + \cdots + e^n x^n + \cdots$

7. $1 + x + 4x^2 + 27x^3 + \cdots + n^n x^n + \cdots$
   [*Hint:* $(n+1)^{n+1} = (n+1)^n (n+1)$.]

8. $1 + 2x^2 + \dfrac{8x^3}{3!} + \dfrac{16x^4}{4!} + \cdots + \dfrac{2^n x^n}{n!} + \cdots$

Using $1/(1-x) = 1 + x + x^2 + \cdots$ for $|x| < 1$ and $e^x = 1 + x + (x^2/2!) + (x^3/3!) + \cdots$ for all x, represent each of the functions in Exercises 9–16 as a power series, as was done in Example 5. Determine the interval of convergence in each case.

9. $\dfrac{1}{1-x^2}$ 10. $\dfrac{1}{1-2x}$ 11. $\dfrac{x}{1-x}$

12. e^{-x} 13. $\dfrac{1}{1+x}$ 14. $\dfrac{x}{1+x^2}$

15. xe^{-x} 16. xe^{-x^2}

Express each of the power series in Exercises 17–20 as a familiar function by an appropriate substitution in the known series for e^x or $1/(1-x)$.

17. $1 - x^2 + x^4 - x^6 + \cdots + (-1)^n x^{2n} + \cdots$

18. $1 + 2x + \dfrac{4x^2}{2!} + \dfrac{8x^3}{3!} + \cdots + \dfrac{2^n x^n}{n!} + \cdots$

19. $1 + \dfrac{x}{2} + \dfrac{x^2}{4 \cdot 2!} + \dfrac{x^3}{8 \cdot 3!} + \cdots + \dfrac{x^n}{2^n n!} + \cdots$

20. $x - x^2 + x^3 - x^4 + \cdots + (-1)^{n+1} x^n + \cdots$
[Hint: First factor an x.]

In the manner of Example 3, use Theorem 10.4, part 1 or 2, along with the known series expansions for $1/(1-x)$ and e^x to write each power series in Exercises 21–26 as a function of x. Determine the sum $S(x)$ of the given series at $x = \tfrac{1}{2}$ in each case.

21. $S(x) = x - \dfrac{x^2}{2} + \dfrac{x^3}{3} - \dfrac{x^4}{4} + \cdots$

22. $S(x) = x - \dfrac{x^2}{2} + \dfrac{x^3}{3 \cdot 2!} - \dfrac{x^4}{4 \cdot 3!}$

23. $S(x) = 1 + 2x + 3x^2 + 4x^3 + \cdots$
[Hint: Factor an x at a suitable stage.]

24. $S(x) = x + \dfrac{x^3}{3} + \dfrac{x^5}{5} + \dfrac{x^7}{7} + \cdots$

25. $S(x) = \dfrac{x^2}{2} + \dfrac{x^4}{4} + \dfrac{x^6}{6} + \cdots$
[Hint: Factor an x at a suitable stage.]

26. $S(x) = 1 + 2x + \dfrac{3x^2}{2!} + \dfrac{4x^3}{3!} + \cdots$

Using Example 5 as a model, find $f^{(6)}(0)$ and $f^{(25)}(0)$ for each function in Exercises 27–30.

27. $f(x) = \dfrac{1}{1+x}$ 28. $f(x) = \dfrac{x}{1-x^2}$

29. $f(x) = xe^{-x}$ 30. $f(x) = -\ln|1-x|$

Write the Taylor-MacLaurin series expansion of each function in Exercises 31–38.

31. $f(x) = \ln(x+1)$ 32. $f(x) = x^7$

33. $f(x) = \dfrac{x+1}{1-x}$ 34. $g(x) = \dfrac{2x}{1-x}$

35. $g(x) = \dfrac{1}{(1-x)^2}$ 36. $f(x) = \dfrac{\ln(1+x)}{1-x}$

37. $h(x) = \sqrt{x+1}$ 38. $h(x) = \dfrac{1}{\sqrt{1+x}}$

39. Use the first five terms of the series expansion obtained in Exercise 37 to estimate $\sqrt{2}$.

40. Use the first four terms of the series expansion obtained in Exercise 35 to estimate $g(\tfrac{1}{2}) = 4$. How good an approximation is your answer?

41. Use Example 6 as a model and estimate the value of $\int_0^{1/2} dx/(1+x^2)$ by writing $1/(1+x^2) = 1/[1-(-x^2)] = 1 - x^2 + x^4 - x^6 + \cdots$ using the geometric series and the first four terms of the resulting series. Compare your answer with that in Example 1, Section 7.5.

42. Repeat Exercise 41 for

$$\int_{-1/4}^{1/2} \dfrac{1}{1+x^2}\, dx$$

Chapter 10 Summary

1. A sequence is a list of numbers a_n in a particular order in correspondence with the natural numbers $n = 1, 2, 3, \ldots$.

2. The limit properties and procedures for a sequence are much like those for finding the limit of a function at infinity.

3. A *difference equation* is the discrete analog of a differential equation and can be used to model discrete change compartment-mixing processes. The solution of the difference equation

$$y_{n+1} = ay_n + b$$

is the sequence

$$y_n = \begin{cases} ca^n + \dfrac{b}{1-a} & \text{if } a \neq 1 \\ y_0 + nb & \text{if } a = 1 \end{cases}$$

The *arithmetic/geometric sequences* arise as solutions to the two basic difference equations: $y_{n+1} = y_n + b$ and $y_{n+1} = ay_n$.

4. An infinite series is the sum of the terms of a sequence. The infinite series $\Sigma\, b_n$ is said to *converge* with sum S if S is the limit of the sequence of partial sums $S_n = \Sigma_{k=1}^{n} b_k$.

5. The *geometric series* $1 + x + x^2 + \cdots + x^n + \cdots = 1/(1-x)$ if $|x| < 1$, and the *harmonic series*

$$1 + \frac{1}{2} + \frac{1}{3} + \frac{1}{4} + \cdots + \frac{1}{n} + \cdots$$

diverges.

6. It is important to be able to determine whether an infinite series converges. When an infinite series $\Sigma\, a_n$ converges, then necessarily $\lim_{n \to \infty} a_n = 0$ but *not conversely*. The *ratio test* offers a basic method of testing the *convergence* of an infinite series. An absolutely convergent series (a series for which $\Sigma\, |a_n|$ converges) is itself convergent. An alternating series, $\Sigma\, (-1)^{n+1} a_n$, converges if $\lim_{n \to \infty} a_n = 0$ (and $a_n \geq a_{n+1}$) but, even in this event, may fail to be absolutely convergent.

7. A power series $\Sigma\, a_n x^n$ defines a continuous, integrable, differentiable function $S(x)$ as its sum, within its interval r of convergence, where $1/r = \lim_{n \to \infty} |a_{n+1}/a_n|$. Moreover, both the integral $\int S(x)\, dx$ and the derivative $D_x S(x)$ may be found by integrating, or differentiating, each term $a_n x^n$ of the power series.

8. A function f may be written as a Taylor-MacLaurin power series

$$f(x) = \sum_{n=1}^{\infty} \frac{f^{(n)}(0)}{n!} x^n$$

in the interval of convergence of this power series under suitable conditions (see Theorem 10.5).

Chapter 10 Summary Exercises

Determine the first two terms and the fifth and eighth terms of the sequences in Exercises 1 and 2.

1. $a_n = \ln\left(\dfrac{n}{10}\right)$ **2.** $b_n = \dfrac{2^{n-1}}{n(n+1)^2}$

Find a formula for each sequence given in Exercises 3 and 4.

3. $1, \frac{3}{2}, \frac{7}{4}, \frac{15}{8}, \ldots$ **4.** $1, -3, 5, -7, 9, \ldots$

Find the limit of each sequence in Exercises 5–10.

5. $a_n = \dfrac{1}{2n+3}$ **6.** $b_n = 100ne^{-n/1,000}$

7. $c_k = \dfrac{k^3 - 2k^2 + 1}{1 - 5k^3}$ **8.** $a_n = (-1)^n 2^{1/n}$

9. $a_p = e^{1/p}\left(1 - \dfrac{2}{p}\right)$

10. $b_k = \dfrac{1}{k} + \dfrac{(-1)^{k+1}}{2^k} - \dfrac{k}{1+2k}$

11. $a_n = \ln(n/n + 1)$

12. $b_n = \dfrac{1 - (\ln 2)^n}{1 - \ln 2}$

Find the solution and the first three terms of this solution to each difference equation in Exercises 13–20.

13. $y_{n+1} - y_n = \frac{3}{2}y_n;\ y_0 = 2$

14. $z_{n+1} - z_n = 7;\ z_0 = 1$

15. $y_{n+1} - y_n = -\dfrac{y_n}{4};\ y_0 = 0$

16. $x_{n+1} - x_n = \frac{1}{3}x_n;\ x_0 = 9$

17. $y_n - y_{n+1} = 1 - 2y_n;\ y_0 = 5$

18. $10y_n - 5y_{n+1} = 20y_n + 50;\ y_0 = 5$

19. $y_{n+1} = 0.3y_n - 2;\ y_0 = 0$

20. $\dfrac{y_{n+1} - 1}{y_n} = \dfrac{1}{2};\ y_0 = 1$

Assume $\lim_{n\to\infty} y_n$ exists and find its value in Exercises 21 and 22.

21. $2y_{n+1} = -1 - y_n^2$

22. $y_n + 1 = y_{n+1}^2 + \dfrac{n}{n+1}$

23. In January 1983, the United States had a cushion of $150 billion in investments abroad with which to finance imports. Suppose that these investments earn 1% per month and that Americans imported about $8 billion more per month than exported. Use a difference equation model to explain how, by early 1985, the United States quietly became a debtor nation, with its earnings abroad no longer paying for its excess imports.

24. Most sewage treatment facilities in the United States were not designed to deal with volatile organic compounds such as paint thinner, industrial solvents, and so on. At the same time such facilities became collecting points for the wastes of whole communities. Suppose that a community generates 30 tons of such waste per month, that 25% evaporates in the sewer lines, and that the sewage facility is able to safely dispose of 40% of the amount arriving at the facility. How many tons of these pollutants are released to the environment in the area of the sewage facility over a 1-year period?

Find the second, third, and sixth partial sums of each infinite series whose nth term is given in Exercises 25–28.

25. $a_n = \dfrac{1}{2n}$ **26.** $b_n = \dfrac{2}{3^n}$

27. $b_n = \dfrac{n+1}{n}$ **28.** $c_n = 1 + (-1)^n$

Find the sum of each infinite series in Exercises 29 and 30. Compare this value with that of the fifth partial sum.

29. $\displaystyle\sum_{n=0}^{\infty} \left(-\dfrac{1}{4}\right)^n$ **30.** $\displaystyle\sum_{n=1}^{\infty} 3\left(\dfrac{7}{10}\right)^{n-1}$

Write each infinite decimal in Exercises 31 and 32 as a fraction.

31. $0.4444\ldots$ **32.** $3.121212\ldots$

Determine whether each series in Exercises 33–42 converges and converges absolutely.

33. $\displaystyle\sum \left(\dfrac{5}{7}\right)^n$ **34.** $\displaystyle\sum \dfrac{1}{2n}$

35. $\displaystyle\sum \dfrac{1}{n!}$ **36.** $\displaystyle\sum \dfrac{n^2 + 2n}{n!}$

37. $\displaystyle\sum \dfrac{(-1)^n}{\sqrt{n}}$ **38.** $\displaystyle\sum e^{-n}$

39. $\displaystyle\sum \dfrac{n}{2^n}$ **40.** $\displaystyle\sum \dfrac{(-3)^n}{n!}$

41. $\displaystyle\sum \dfrac{(-1)^n}{5^n}$ **42.** $\displaystyle\sum \dfrac{2^n}{n^{1/n}}$

Determine the radius and interval of convergence of each power series in Exercises 43–46.

43. $x + x^3 + x^5 + x^7 + \cdots + x^{2n+1} + \cdots$

44. $1 + 2x + 4x^2 + 8x^3 + \cdots + 2^n x^n + \cdots$

45. $3 + x + \dfrac{x^2}{2!} + \dfrac{x^3}{3!} + \cdots + \dfrac{x^n}{n!} + \cdots$

46. $1 - \dfrac{x^2}{2!} + \dfrac{x^4}{4!} - \dfrac{x^8}{8!} + \cdots + \dfrac{(-1)^{n+1} x^{2^n}}{(2^n)!} + \cdots$

Write each function in Exercises 47–50 as a power series.

47. $\dfrac{1}{1 - 2x}$

48. e^{-3x}

49. $\dfrac{x}{1 + x} = x\left(\dfrac{1}{1 - (-x)}\right)$

50. $\dfrac{x}{e^x} = xe^{-x}$

51. Use the geometric series to write
$$1 - 4x^2 + 16x^4 + \cdots + (-1)^n 2^{2n} x^{2n} + \cdots$$
as a function.

52. Use the series for $y = e^x$ to write
$$1 + \dfrac{x}{2} + \dfrac{x^2}{4 \cdot 2!} + \dfrac{x^3}{8 \cdot 3!} + \cdots + \dfrac{x^n}{2^n n!} + \cdots$$
as a function.

53. Find a formula for the sum
$$S(x) = x + \dfrac{x^2}{2} + \dfrac{x^3}{3} + \dfrac{x^4}{4} + \cdots$$
by first differentiating the series (within its interval of convergence), recognizing the resulting series, and then integrating.

54. Find $f^{(7)}(0)$ for $f(x) = \dfrac{1}{1 + x^2}$.

Write the Taylor-MacLaurin series expansion of each function in Exercises 55 and 56.

55. $f(x) = x^3 + \ln(x + 1)$
56. $g(x) = \dfrac{1}{(x + 1)^2}$

57. $h(x) = xe^{-x}$
58. $f(x) = 1 - x^3$

59. Integrate the given power series within its interval of convergence so as to find its sum:
$$S(x) = 1 - 2x + 3x^2 - 4x^3 + \cdots$$

60. Use the fourth partial sum of the series expansion of $f(x) = \ln(x + 1)$ to estimate $\ln 2$.

Chapter 11

Trigonometric Functions

Repetitive Behavior: Periodic Systems

The ocean tide ebbs and flows, interest rates rise and fall, and your blood pressure fluctuates. A rabbit population grows with decline of the neighboring fox population, which in turn expands following the increasing rabbit population. On a smaller scale the fever you ran for a few days rises and falls with the strength and weakness of a virus invading your cells. Each of these is an example of *periodic*, or regularly repetitive, behavior. The mathematical study of periodic behavior has been vast and fruitful since the discoveries of the French physicist and mathematician Jean Fourier (1768–1830) who, while trying to describe how heat moves through a solid, showed that only two special functions suffice to describe most periodic phenomena. These two functions are modern versions of the ancient trigonometric functions of *sine* and *cosine*, invented more than 2,000 years earlier to describe quite different, and quite rigid, phenomena. In our day Fourier series are used to analyze the processing of computer-generated images of the surface of Mars, or, equally, of a medical patient's internal organs. While Fourier analysis is well beyond this text, it begins with the calculus of the trigonometric functions.

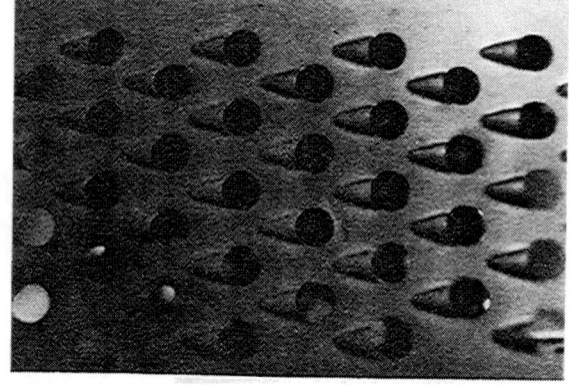

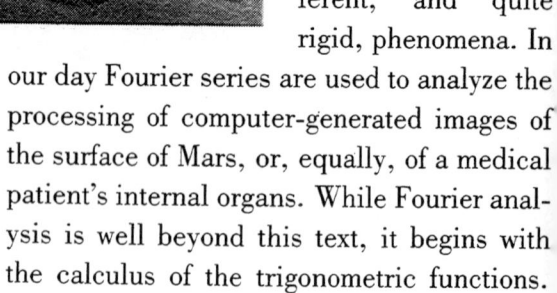

11.1 The Radian Measure of an Angle

In this section we define the radian measure of an angle and establish its relationship to the degree measure of an angle.

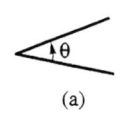

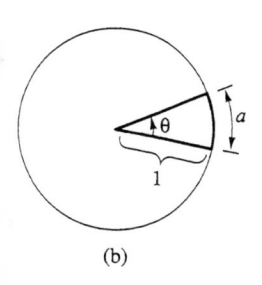

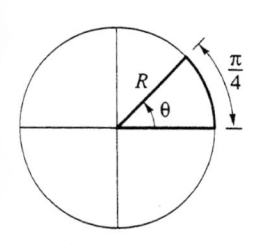

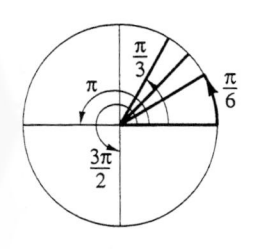

Mathematical constructions of a particular type are often built upon the simplest example of the same type. The simplest example of repetitive, or periodic, behavior is "going-around-in-a-circle," and this is where we begin the construction of functions capable of describing the most complex periodic phenomena.

Specifically we want to give a numerical measure—an ordinary number—to the size of an angle, rather than the more familiar degree measurement. This numerical measure is called the *radian measure* of an angle. Consider an angle θ, as in Figure 11.1(a), and inscribe it in a circle of radius 1, as in Figure 11.1(b).

The **radian measure** of the angle θ is defined to be the *length of arc* subtended by the angle θ when it is inscribed in a circle of radius 1.

It is not difficult to determine the radian measure of the most common angles if we think in terms of a *proportion of the circle*. The arc length of the *entire* circle—its circumference—is $2\pi \cdot 1 = 2\pi$, since the radius is 1. Therefore, for example, the angle θ in Figure 11.2, which takes up $\frac{1}{8}$ of the entire circle, must have a proportionate arc length—namely,

$$\frac{1}{8}(2\pi) = \frac{\pi}{4}$$

That is, the radian measure of θ is the number $\pi/4 \approx 3.1416/4 = 0.7854$.

If you understand, and can extend, this proportion principle, you will understand radian measure. In the ensuing discussion the line R in Figure 11.2 is called the **terminus** of the angle θ, or the **radial line** determined by θ.

EXAMPLE 1

The degree measurement of the angle indicated in Figure 11.2 is 45°. Thus the radian measure of an angle of 45° is $\pi/4$. The radian measure of other familiar angles is summarized in Table 11.1 and illustrated in Figure 11.3.

For example, an angle of 60° is $\frac{1}{6}$ of the 360° of the entire circle and so has a radian measure of

$$\frac{1}{6}(2\pi) = \frac{\pi}{3} \quad \blacksquare$$

The radian measure of certain other angles not given in Example 1 may be derived from Table 11.1. For example, an angle of 135° consists of three angles of 45° and so has radian measure $3\pi/4$, three times the arc length $\pi/4$ of an angle of 45°.

Table 11.1

Degrees	Proportion of the Circle	Radian Measure
0	0	0
30	$\frac{1}{12}$	$\frac{\pi}{6}$
45	$\frac{1}{8}$	$\frac{\pi}{4}$
60	$\frac{1}{6}$	$\frac{\pi}{3}$
90	$\frac{1}{4}$	$\frac{\pi}{2}$
180	$\frac{1}{2}$	π
270	$\frac{3}{4}$	$\frac{3\pi}{2}$

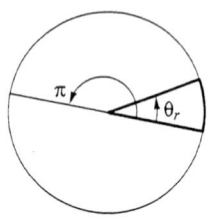

The proportion principle used in Example 1 can be extended to a general formula for the conversion of an angle in degree measure θ_d to (or from) its radian measure θ_r. This formula is

$$\frac{\theta_d}{180} = \frac{\theta_r}{\pi} \tag{1}$$

or, equivalently, $\quad \theta_r = \pi\left(\dfrac{\theta_d}{180}\right) \quad$ and $\quad \theta_d = 180\left(\dfrac{\theta_r}{\pi}\right) \tag{2}$

Formula 1 is valid because the two ratios

$$\frac{\theta_d}{360°} \quad \text{and} \quad \frac{\theta_r}{2\pi}$$

must be the same number because they measure the *same proportion* of an angle θ relative to the whole circle (see Figure 11.4). The first ratio is the proportion of θ relative to the 360 degrees of the circle, and the second ratio is the proportion of θ relative to the 2π units of arc length of the unit circle. Formula 2 is applied as follows.

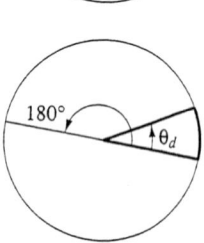

Figure 11.4

EXAMPLE 2
a. Find the radian measure of an angle of 120°.
b. Find the degree measure of an angle of $5\pi/6$ radians.

Solution
a. Using Formula 2, $\theta_r = \pi(120/180) = 2\pi/3$.
b. Using Formula 2 again,

$$\theta_d = 180\left(\frac{\frac{5\pi}{6}}{\pi}\right) = 180\left(\frac{5}{6}\right) = 150° \quad \blacksquare$$

Three important, but simple, questions remain to be answered.

1. What is 1 radian? That is, how large is an angle of 1 radian? Is it important to know this?

   Answer. If $\theta_r = 1$, then (from Formula 2) the degree measure of θ is

   $$\theta_d = 180\left(\frac{1}{\pi}\right) = \frac{180}{\pi} \simeq 57.296°$$

   This angle is indicated in Figure 11.5. An angle of 1 radian is unwieldy to deal with and is *not* really important. Think of an angle of π radians as the "basic unit" of radian measure, and notice how in Table 11.1 all radian measures are multiples of π.

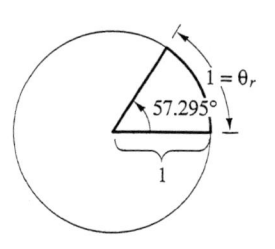

Figure 11.5

2. Is there such a thing as negative radian measure?

   Answer. Yes. The angle θ indicated in Figure 11.6 is said to have radian measure $-\pi/4$. The minus sign means no more than that the angle is to be indicated, or thought of, in the opposite, or clockwise, direction. For example, angles of $7\pi/4$ and $-\pi/4$ radians determine the same radial line R in the circle. They are not, however, considered to be the same *angle*, because we want to regard an angle as being directed in either the counterclockwise direction (with *positive* radian measure) or the clockwise direction (with *negative* radian measure).

   Negative radian measure is important for a number of reasons, as we will see throughout this chapter.

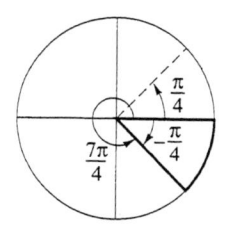

Figure 11.6

3. How can we compute the radian measure of an angle θ if this angle is found in a circle of radius *not* equal to 1?

   Answer. A simple formula for this computation rests on plane geometry. Just as the ratios a/b and A/B in the two triangles in Figure 11.7(a) are the same—that is, just as $a/b = A/B$—so also are the ratios a/r and A/R indicated in the circles in Figure 11.7(b) the same. That is, $a/r = A/R$.

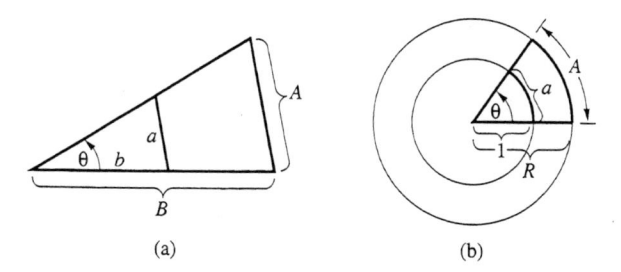

Figure 11.7

(a)   (b)

As a special case of this formula, if the smaller circle is drawn so as to have radius $r = 1$, then, since the radian measure θ_r of θ is defined as the arc length a, we have $\theta_r = a = a/1 = a/r = A/R$. Thus

$$\theta_r = \frac{A}{R}$$

That is, the radian measure of an angle is *the ratio of arc length to radius* in any size circle in which the angle is inscribed.

11.1 The Radian Measure of an Angle   607

EXAMPLE 3

Determine the radian measure of (a) the angle θ and (b) the angle γ in Figure 11.8.

Solution

a. Here the radius $R = \frac{3}{2}$ and the arc length $A = \frac{4}{3}$. Therefore

$$\theta_r = \frac{A}{R} = \frac{\frac{4}{3}}{\frac{3}{2}} = \frac{8}{9}$$

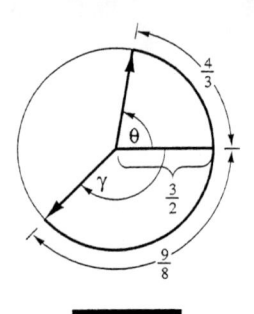

Figure 11.8

b. The radius R is again $\frac{3}{2}$, but the angle is observed to be in the counterclockwise direction. Therefore we take $A = -\frac{9}{8}$ and hence

$$\theta_r = \frac{-\frac{9}{8}}{\frac{3}{2}} = \frac{-3}{4} \quad \blacksquare$$

Exercises 11.1

Determine the radian measure of the indicated angles in Exercises 1–4.

1.

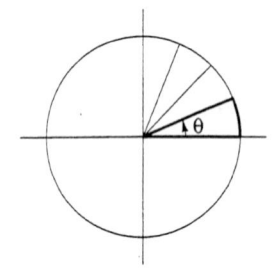

Figure 11.9

2.

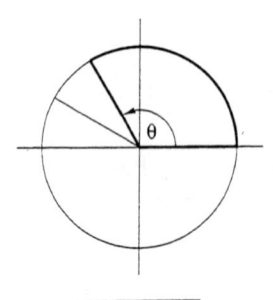

Figure 11.10

3.

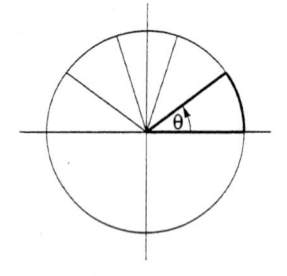

Figure 11.11

4.

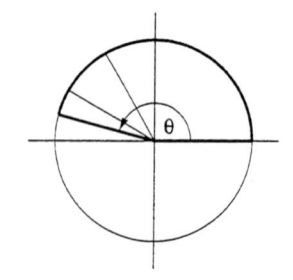

Figure 11.12

608 Chapter 11 Trigonometric Functions

Determine the radian measure of the angles in Exercises 5–12. First make this determination geometrically, and then use the formula $\theta_r = \pi(\theta_d/180)$.

5. 20°
6. 15°
7. 150°
8. 36°
9. 240°
10. 300°
11. 450° [*Hint:* Your answer should be greater than 2π.]
12. 390°

Convert the angles, given in radian measure, in Exercises 13–20 to degree measure. Draw a figure representing each angle.

13. $\dfrac{\pi}{9}$
14. $\dfrac{2\pi}{3}$
15. $\dfrac{7\pi}{6}$
16. $\dfrac{21\pi}{30}$
17. $\dfrac{5\pi}{4}$
18. 3π
19. $\dfrac{5\pi}{2}$
20. 0.6π

21. Find an angle θ of positive radian measure between θ and 2π having the same terminal side as an angle of

   a. $-\dfrac{\pi}{3}$ radians b. $-\dfrac{3\pi}{4}$ radians

   c. $-\dfrac{11\pi}{12}$ radians d. $-\dfrac{9\pi}{4}$ radians

22. Convert the following angles from radian/degree measure to degree/radian measure.

   a. $-30°$ b. $-\dfrac{2\pi}{3}$
   c. $-135°$ d. $-\dfrac{3\pi}{2}$
   e. $-2°$ f. $-\dfrac{7\pi}{12}$

23. Suppose you run around in a perfect circle of radius 30 yards through an angle of 25 radians. How far did you run?

Determine the radian measure of the angle θ with the dimensions indicated in Exercises 24–27. Then convert the radian measure to a multiple of π.

24.

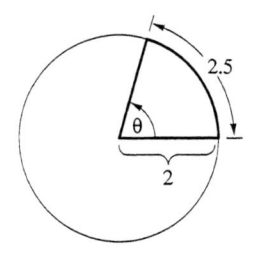

Figure 11.13

25.

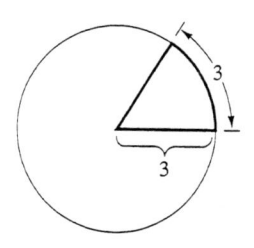

Figure 11.14

26.

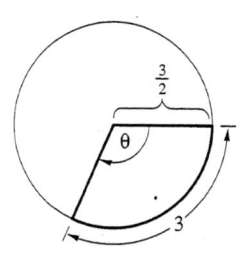

Figure 11.15

27.

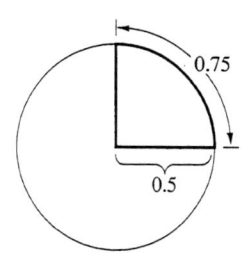

Figure 11.16

11.1 The Radian Measure of an Angle   609

11.2 The Basic Trigonometric Functions

In this section we define the sine, cosine, and tangent trigonometric functions by observing the y- and x-coordinates of a point moving around a circle.

A quantity Q that rises and falls over time t in a *periodic* manner could be represented by a graph similar to that in Figure 11.17. In this section we define the two basic periodic functions, sine and cosine, which, in combination, can describe many periodic phenomena. Their graphs will be seen to be periodic in a simple way. We will obtain these periodic functions again by beginning with the simple periodic phenomena of "going around in a circle." Later in this section we will relate this definition to the ancient definition of sine and cosine in terms of the angles within a right triangle.

Figure 11.17

Consider the angle θ indicated in Figure 11.18, where the indicated circle is of radius 1 drawn in an xy-coordinate system and the radial line R (determined by θ) meets the circle at point $P = (a, b)$. We define the **sine** and **cosine** of the angle θ by

$$\sin \theta = b$$

the y-coordinate of P, and

$$\cos \theta = a$$

the x-coordinate of P.

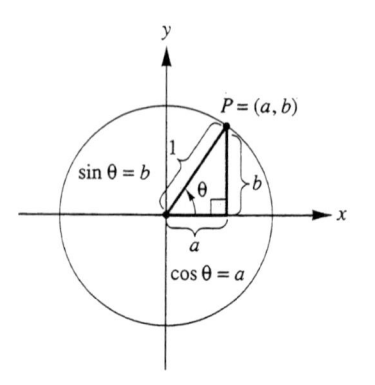

Figure 11.18

610   Chapter 11   Trigonometric Functions

Let us graph this new function, $y = \sin \theta$, so as to illustrate its pattern and periodicity. To obtain this graph we first imagine the circle as a straight line of length 2π; see Figure 11.19 (effectively, we cut the circle and unroll it into a line). We mark θ on this horizontal axis and $y = \sin \theta$ on a vertical axis. Next we imagine the point P moving around the circle as the angle θ changes, beginning at $\theta = 0$, and record the y-coordinate $\sin \theta$ of P. We then mark this number with respect to the y-axis in Figure 11.19, obtaining the values and points shown there.

θ	$\sin \theta$
0	0
$\frac{\pi}{2}$	1
π	0
$\frac{3\pi}{2}$	-1
2π	0
$\frac{5\pi}{2}$	1
3π	0
$\frac{7\pi}{2}$	-1
4π	0

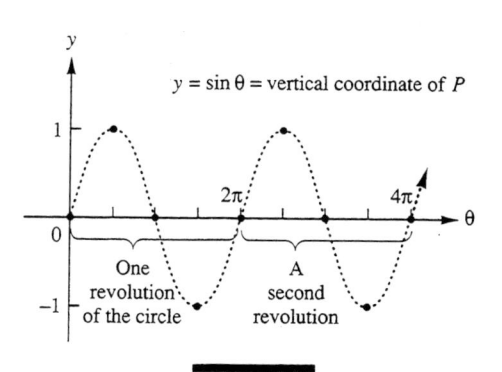

Figure 11.19

The dotted curve in Figure 11.19 indicates the likely position of the graph for angles θ other than multiples of $\pi/2$. We can establish these other points on the graph by using the definition of $y = \sin \theta$ to compute its values at other angles.

Let us compute $\sin(\pi/4)$ and $\sin(\pi/6)$. In Figure 11.20(a), we see that an angle of $\pi/4$ radians determines a radial line of slope 1, so that $a = b$. Since the radius is also 1, we have $1 = a^2 + b^2 = 2b^2$, or $b = 1/\sqrt{2} = \sqrt{2}/2$. Thus $\sin(\pi/4) = \sqrt{2}/2 \approx 0.707$. In Figure 11.20(b), where $\theta = \pi/6$, we know from plane geometry that the hypotenuse is twice the length of the side of height b. Thus $2b = 1$, or $b = \frac{1}{2}$, and therefore, $\sin(\pi/6) = 1/2 = 0.5$.

$\sin \frac{\pi}{4} = \frac{1}{\sqrt{2}} = \frac{\sqrt{2}}{2} \approx 0.707$

(a)

$\sin \frac{\pi}{6} = \frac{1}{2}$

(b)

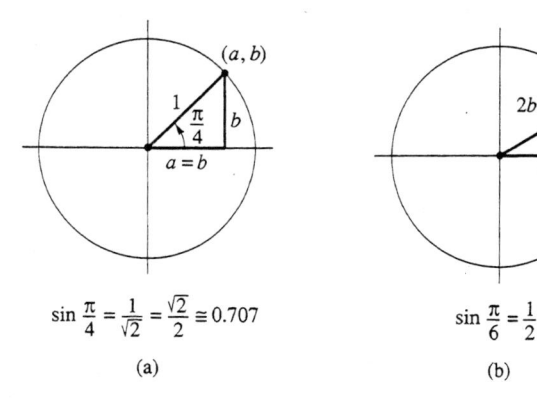

Figure 11.20

11.2 The Basic Trigonometric Functions

Using these two new values, we are led to illustrate the entire graph of $y = \sin \theta$ as in Figure 11.21. Notice that for angles greater than 2π (as we go around the circle a *second time* laying out a second line of length 2π, from $\theta = 2\pi$ to $\theta = 4\pi$), the values of $y = \sin \theta$ can only repeat themselves again. Thus the graph of $y = \sin \theta$ is *periodic*. We can also graph $y = \sin \theta$, for $\theta < 0$, by regarding θ as the radian measure of an angle drawn in the clockwise direction. We indicate this in Figure 11.22, where in a manner comparable to Figure 11.20, we find that $\sin(-\pi/4) = -\sqrt{2}/2$.

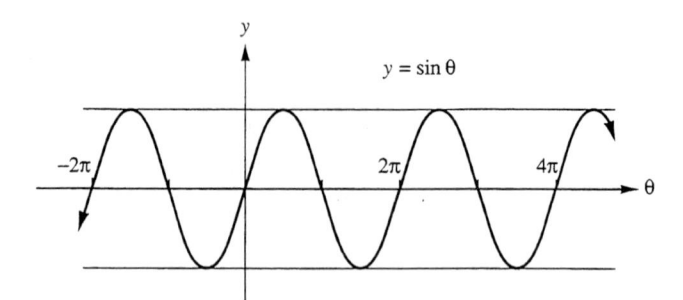

Figure 11.21

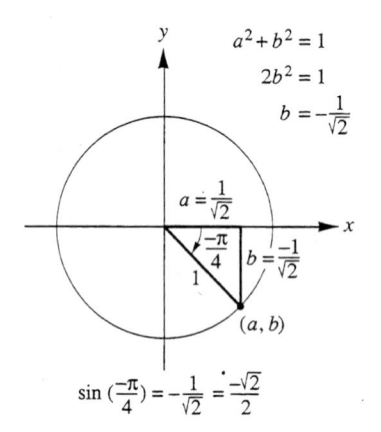

Figure 11.22

Similarly, we can graph $y = \cos \theta$ by determining the x-coordinate of the point P as P moves around the circle (see Figure 11.23). Notice that, in contrast to the sine function, $\cos(-\pi/4) = \sqrt{2}/2 = \cos(+\pi/4)$.

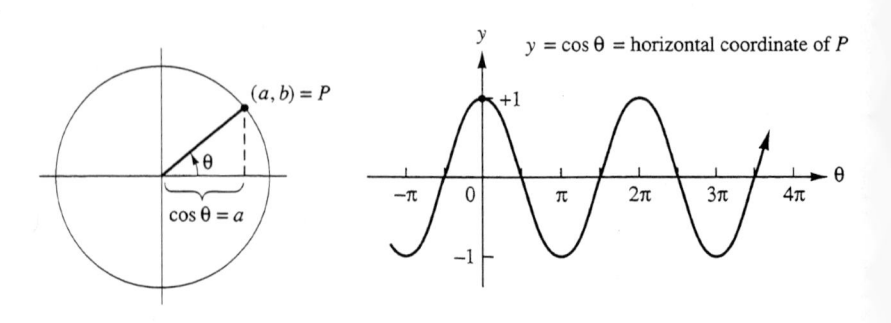

Figure 11.23

Some Trigonometric Identities

Identities are equations that are satisfied for all values of the variable. We have need for four particularly useful identities involving sine and cosine. The first is

$$\sin^2 \theta + \cos^2 \theta = 1 \tag{1}$$

Here you are to understand that $\sin^2 \theta$ means the square of $\sin \theta$—that is, $\sin^2 \theta = (\sin \theta)^2$. Thus, for example,

$$\sin^2 \frac{\pi}{4} = \left(\sin \frac{\pi}{4}\right)^2 = \left(\frac{\sqrt{2}}{2}\right)^2 = \frac{1}{2}$$

Equation 1 is an identity; it is true for *any* angle θ. The reason this is so is indicated in Figure 11.24 and depends on the Pythagorean theorem, which says that in a right triangle the square of the hypotenuse equals the sum of the squares of the remaining sides. That is, $1^2 = (\cos \theta)^2 + (\sin \theta)^2$, or $\sin^2 \theta + \cos^2 \theta = 1$.

Two additional identities are

$$\sin(-\theta) = -\sin \theta \quad \text{and} \quad \cos(-\theta) = \cos \theta \tag{2}$$

These are seen to be true by recalling that the sine and cosine are, respectively, the coordinates of the point P in Figure 11.25(a, b). Notice that in Figure 11.25(a) either θ or $-\theta$ determines the same x-coordinate, and this x-coordinate is the cosine of both θ and $-\theta$ at once. Thus $\cos \theta = \cos(-\theta)$. In Figure 11.25(b), the y-coordinates determined by θ and $-\theta$ are the *same length* but of different sign. Thus $\sin(-\theta) = -\sin \theta$.

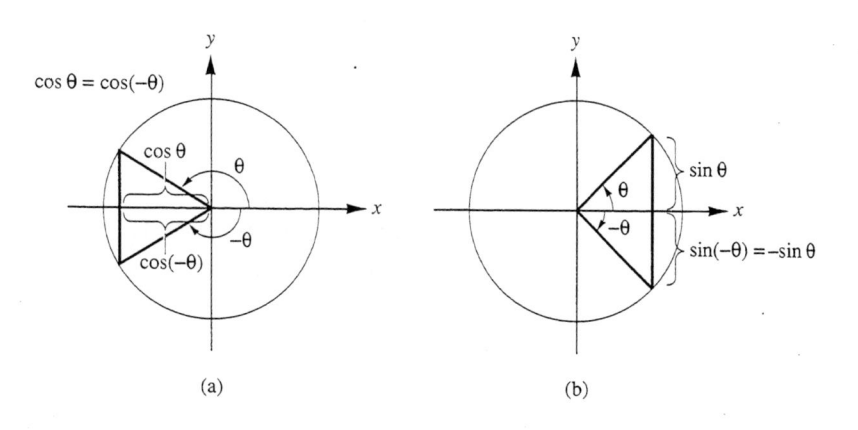

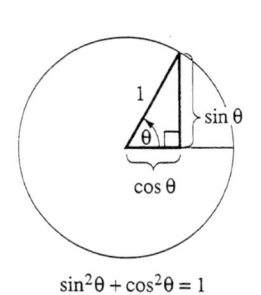

Figure 11.24

$\sin^2\theta + \cos^2\theta = 1$

Figure 11.25

(a) $\cos \theta = \cos(-\theta)$

(b) $\sin(-\theta) = -\sin \theta$

Another pair of identities results when we realize that the graph of the cosine is a **translation,** a shift to the right, of the graph of the sine function, as indicated in Figure 11.26. For example, at the angle θ indicated in the figure, we see that $\cos[\theta + (\pi/2)] = \sin \theta$.

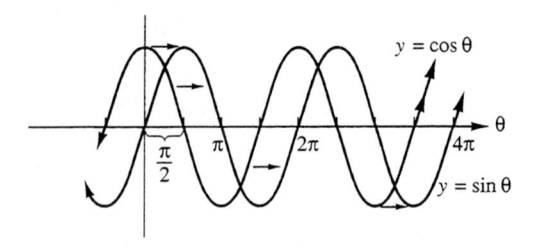

Figure 11.26

Similar reasoning about the angle $[\theta - (\pi/2)]$ yields the identities

$$\cos\left(\theta - \frac{\pi}{2}\right) = \sin\theta \qquad \sin\left(\theta + \frac{\pi}{2}\right) = \cos\theta \qquad (3)$$

A final formula is needed in the next section to compute the derivative of $y = \sin x$.

$$\sin(\theta + \gamma) = \sin\theta\cos\gamma + \cos\theta\sin\gamma \qquad (4)$$

Unlike the preceding identities, this formula is not easy to verify. Rather than do so we instead discuss its meaning. The numbers θ and γ represent the radian measure of the angles θ and γ. As numbers these may be added to obtain the number $\theta + \gamma$. We then regard $\theta + \gamma$ as the radian measure of a third angle (Figure 11.27). Formula 4 then allows us to compute the sine of this third angle $\theta + \gamma$ in terms of the sine and cosine of the original angles θ and γ.

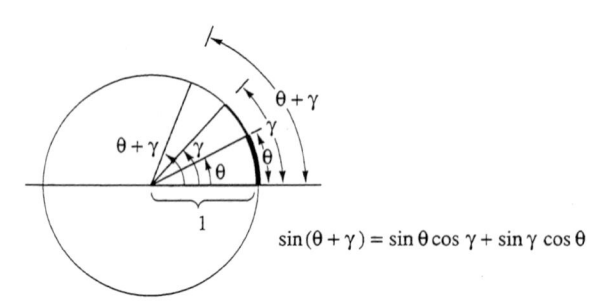

Figure 11.27

Frequency and Amplitude of Periodic Phenomena

The electrical current in an alternating 120-volt circuit typically cycles between a positive and negative potential of 120 volts exactly 60 times per second (see Figure 11.28).

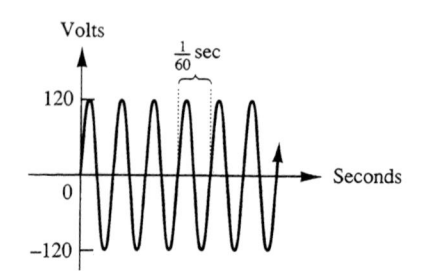

Figure 11.28
Voltage (potential) pattern repeats 60 times in each second.

614   Chapter 11   Trigonometric Functions

The blood pressure of an individual at rest rises and falls with the pumping of the heart from a high of 120 to a low of 80. The individual's pulse rate, which is typically 70 beats per minute, is a measure of how frequently this rise and fall occurs per minute (see Figure 11.29). Let us see how both these processes, alternating current and blood pressure, may be modeled using the sine (or cosine) function.

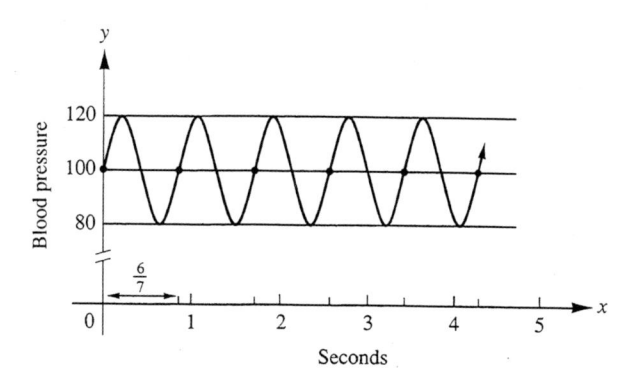

Figure 11.29
Pressure pattern repeats every $\frac{6}{7}$ seconds.

The ideas of frequency and amplitude are the key to such models. In the preceding examples, the **frequency** is the number of repetitions per unit (of time) of the periodic phenomena. This frequency is 60 cycles per second for alternating current and 70 beats per minute for blood pressure. The **amplitude** is one-half the magnitude, or absolute value, of the distance between the high and low values of the periodic phenomena. For alternating current this amplitude is 120 (volts), and for blood pressure the amplitude is $20 = \frac{1}{2}(120 - 80)$ (see Figure 11.29).

The sine function, on the other hand, has an amplitude of 1 unit and a frequency of 1 per 2π units of length (see Figure 11.30).

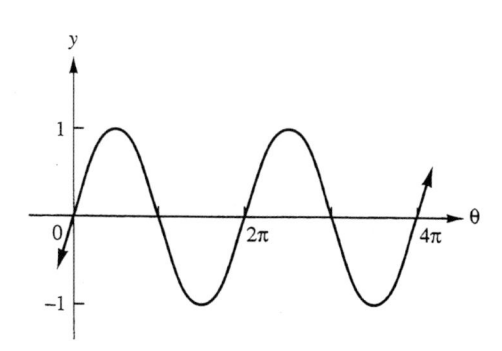

Figure 11.30

It is a simple matter to create a graph with greater (or smaller) amplitude. For example, the graph of $y = 2 \sin x$ will have amplitude 2, because each value of the sine function is doubled. The graph of $y = \frac{1}{4} \sin x$ will have amplitude $\frac{1}{4}$ unit, since each value of the sine is reduced by division by 4 (see Figure 11.31).

11.2 The Basic Trigonometric Functions

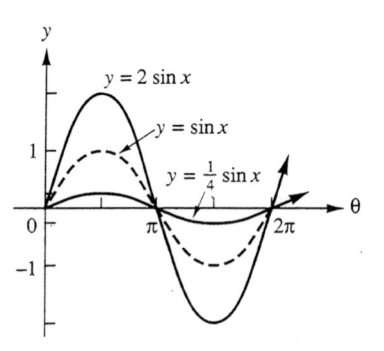

Figure 11.31

More involved is an adjustment for frequency. The basic principle, however, can be seen by a consideration of the graph of, say, $y = \sin 2x$. As the value of x changes from $x = 0$ to $x = 2\pi$, the values of $2x$ will range from 0 to 2π (at $x = \pi$) *and on up to* $2(2\pi) = 4\pi$. This means that the input $2x$ in the sine function will effectively represent angles varying through *two cycles* of the circle as x itself varies through only one cycle. The result is the graph shown in Figure 11.32. You may wish to evaluate $y = \sin 2x$ for, say, $x = \pi/4$, $\pi/2$, and $3\pi/4$ to convince yourself of the correctness of this graph.

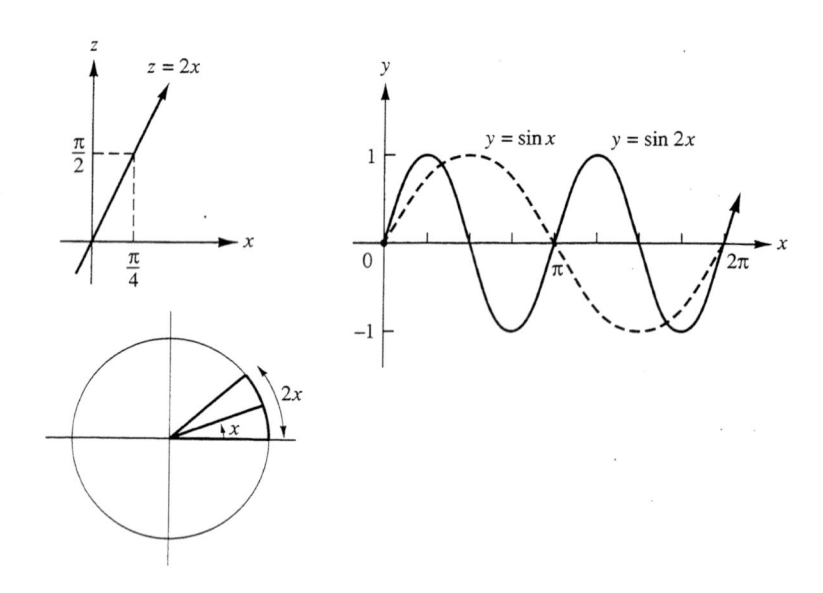

Figure 11.32

Combining the two ideas of adjusting the frequency and amplitude of $y = \sin x$, we see that the graph of $y = \frac{1}{2} \sin 3x$ will have frequency 3 (over 2π units) and amplitude $\frac{1}{2}$ and be drawn as in Figure 11.33.

616   Chapter 11   Trigonometric Functions

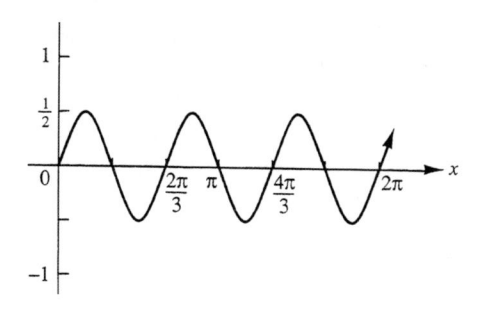

Figure 11.33

One more idea is needed to model alternating current or a person's blood pressure. We need to adjust frequency so that it occurs relative to *one unit* (of measure). The idea is this: Since the whole pattern of the curve of $y = \sin x$ is obtained when x varies from 0 to 2π, the entire curve can be obtained by the formula

$$y = \sin 2\pi t$$

when t varies from 0 to 1 (= *one unit*), because the input in the sine function, $2\pi t$, varies from 0 to 2π when t varies from 0 to 1 (see Figure 11.34).

t	$2\pi t$	$\sin 2\pi t$
0	0	0
$\frac{1}{8}$	$\frac{\pi}{4}$	$\frac{\sqrt{2}}{2}$
$\frac{1}{4}$	$\frac{\pi}{2}$	1
$\frac{1}{2}$	π	0
$\frac{5}{8}$	$\frac{5\pi}{4}$	$-\frac{\sqrt{2}}{2}$
$\frac{3}{4}$	$\frac{3\pi}{2}$	-1
1	2π	0

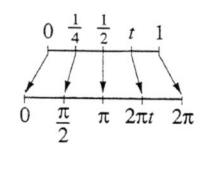

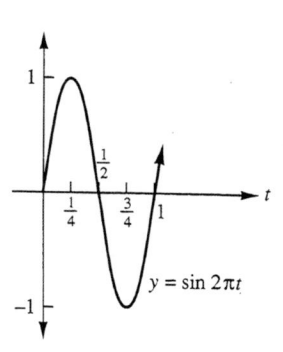

Figure 11.34

Putting all these ideas together, we see that alternating electrical current can be modeled by the graph of $y = 120 \sin[60(2\pi t)]$ (see Figure 11.35). The number 120 determines the amplitude, the number 60 determines the frequency of repetitions, and the number 2π causes this frequency to occur over one unit of the variable t.

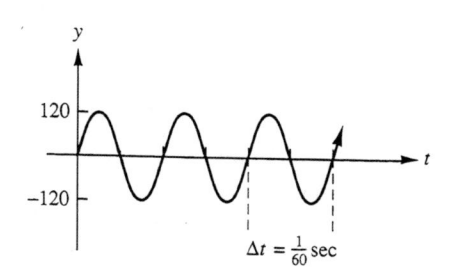

Figure 11.35

11.2 The Basic Trigonometric Functions

A model of blood pressure is complicated only by the need to have the amplitude "centered" about an average value of 100. A suitable model, then, is the function $y = 100 + 20 \sin 70(2\pi)t = 100 + 20 \sin 140\pi t$ (see Figure 11.36). The addition of "100" raises each value of the sine by 100 units; "20" ensures the needed amplitude; and "70," along with "2π," ensures a frequency of 70 beats per minute. Note, for example, that at values of t such that $2\pi t$ is an odd multiple of $\pi/2$ (say, $t = \frac{1}{240}$ or $\frac{3}{240}$) the value of y will be $y = 100 \pm 20 = 120$ or 80.

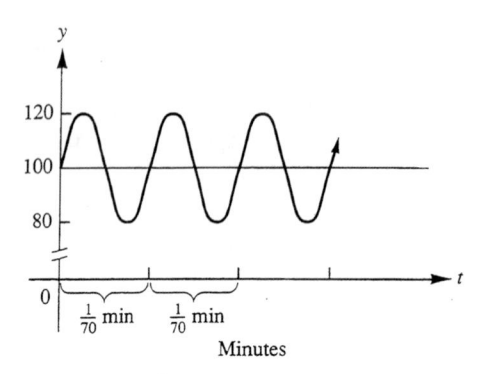

Figure 11.36

Fourier proved that much more complex phenomena than these can be modeled by combinations of the sine and cosine, but the extent of his work is well beyond this text. In certain cases we can readily see by *visually adding* the graphs of the basic sine and/or cosine curves that other periodic functions can be formed. For example, the graph of $y = \sin x + \frac{1}{2} \cos x$ is indicated in Figure 11.37.

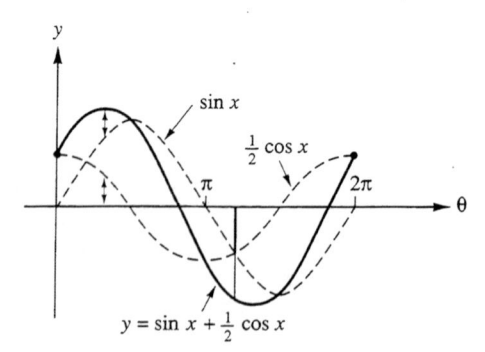

Figure 11.37

The Sine and Cosine via Triangular Measurements

The sine and cosine functions were originally defined over 2,000 years ago for reasons quite unrelated to periodic phenomena. Let us state these definitions and their relationship to those used at the start of this section. Consider Figure

11.38. By definition, $\sin \theta = b$ and $\cos \theta = a$ in Figure 11.38. But, from plane geometry,

$$b = \frac{b}{1} = \frac{B}{R} \quad \text{and} \quad a = \frac{a}{1} = \frac{A}{R}$$

since the indicated triangles are *similar triangles*. That is,

$$\sin \theta = \frac{B}{R} = \frac{\text{side opposite } \theta}{\text{hypotenuse}}$$

and

$$\cos \theta = \frac{A}{R} = \frac{\text{side adjacent to } \theta}{\text{hypotenuse}}$$

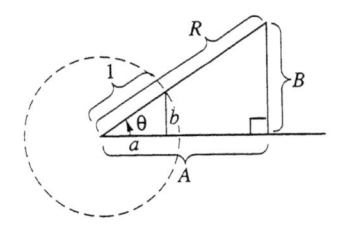

Figure 11.38

Thus, for example, if the angle θ is given in the triangle in Figure 11.39, then, by the Pythagorean theorem, $R^2 = 1^2 + 3^2 = 10$, so $R = \sqrt{10}$ and

$$\sin \theta = \frac{1}{\sqrt{10}} = \frac{\sqrt{10}}{10}$$

and

$$\cos \theta = \frac{3}{\sqrt{10}} = \frac{3}{10}\sqrt{10}$$

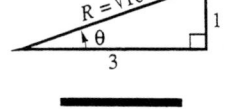

Figure 11.39

The Tangent Function

There is a third particularly useful trigonometric function. In the notation of Figure 11.40 we define the tangent of the angle θ

$$\tan \theta = \frac{b}{a} \quad \text{whenever } a \neq 0 \tag{5}$$

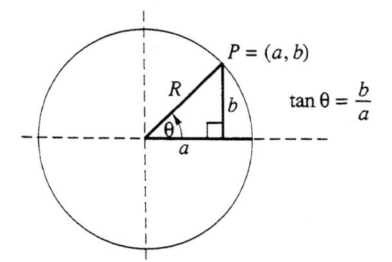

Figure 11.40

11.2 The Basic Trigonometric Functions

That is, $\tan \theta$ is the slope of the radial line R (however long R is, consistent with the preceding discussion). Note also that $a = 0$ whenever $\theta = \pi/2$ or $3\pi/2$. Therefore these angles are *not in the domain* of the tangent function.

It follows that since $\sin \theta = b/R$ and $\cos \theta = a/R$, we have the identity

$$\tan \theta = \frac{\sin \theta}{\cos \theta} \tag{6}$$

Many properties of the tangent function can accordingly be derived from those of the sine and cosine.

The function $f(x) = \tan x$, for $0 \leq x \leq 2\pi$ and $x \neq \pi/2$ or $3\pi/2$, also has a periodic graph, but it is different from that of the sine or cosine. If we think of the number $\tan \theta$ as the ratio of vertical to horizontal coordinates of P (Figure 11.40) as θ moves around the circle, we obtain the graph of the tangent function in Figure 11.41.

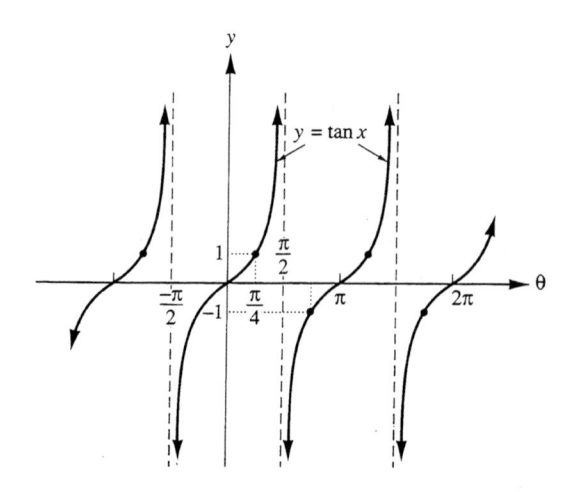

Figure 11.41

Notice, for example, that as θ approaches $\pi/2$, the y-coordinate b of P approaches 1, the x-coordinate a approaches 0, and the ratio b/a defining $\tan \theta$ becomes larger and larger without bound. Figure 11.41 illustrates that the tangent function has a vertical asymptote at every odd multiple of $\pi/2$. Note that beyond, but near, $\pi/2$ $\tan \theta$ is a large negative number.

Exercises 11.2

1. Make a table of values for $y = \cos \theta$ for

    $\theta = 0, \dfrac{\pi}{6}, \dfrac{\pi}{4}, \dfrac{\pi}{3}, \dfrac{\pi}{2}, \dfrac{2\pi}{3}, \dfrac{3\pi}{4}, \dfrac{5\pi}{6}, \dfrac{5\pi}{4}, \dfrac{3\pi}{2}, \dfrac{11\pi}{6}, 2\pi$

    and use these values to justify the graph of $y = \cos \theta$ shown in Figure 11.23.

2. For $x = \pi/3$ and $x = \pi/4$, compute the values of

    $\sin x$, $(\sin x)^2$, $\sin^2 x$, $\sin 2x$, $2 \sin x$,
    $2 \sin 3x$, $4 \sin x$, $\sin^2 x + \cos^2 x$

3. **a.** Use an accurate drawing of a circle of radius 1 and the best measurements your eye and a ruler can make to compute $\sin(x^2)$ for $x = \pi/3$. [*Hint:* Recall that radian measure must be measured along the arc of the circle. Be careful to distinguish between $(\sin x)^2$ and $\sin x^2$.]

   b. Now write $(\pi/3)^2 = \alpha\pi$ and solve for α; try again to evaluate $\sin(x^2)$ by drawing an angle of $\alpha\pi$ radians in the circle.

4. Using Exercises 2 and 3, what can you conclude about the equation $\sin^2 x = \sin(x^2)$?

Verify the validity of the formulas in Exercises 5–7.

5. $\sin(-\theta) = \sin\theta$ and $\cos(-\theta) = \cos\theta$, for $\theta = \pi/2$, π, $\pi/6$, and $5\pi/6$

6. $\sin[\theta + (\pi/2)] = \cos\theta$, for $\theta = 0$, $\pi/4$, $\pi/2$, $\pi/6$, π, and $7\pi/6$

7. $\sin(x+y) = \sin x \cos y + \cos x \sin y$, for $x, y = 0$, $\pi/2$; $\pi/4$, $\pi/4$; $\pi/6$, $\pi/3$

8. This exercise indicates that radian measure allows for a wider range of mathematical operations inside the sine function than could be done with degree measure.

   a. Use a drawing (as in Exercise 3) to make the best estimate you can of the value of $\sin xy$ for $x = \pi/3$ and $y = \pi/4$.

   b. Write $(\pi/3)(\pi/4) = \alpha\pi$ and solve for α, so as to draw $(\pi/3)(\pi/4)$ more accurately as an angle.

   c. How would you respond to this same question had it been asked in terms of degree measure, for $x = 60°$, $y = 45°$?

9. Compute values of $y = \tan\theta$ for $\theta = 0$, $\pi/3$, $\pi/4$, $\pi/6$, $3\pi/4$, $5\pi/4$, and $7\pi/6$.

10. Compute the values of $\sin x$, $\cos x$, $\sin 2x$, and $\cos 3x$ for $x = -\pi/4$, $-\pi/6$, $-\pi/3$, and $-\pi/2$.

11. Make a table of values for $y = \sin 2x$ and $y = \sin 4x$ for $x = 0$, $\pi/3$, $\pi/4$, $\pi/6$, $\pi/2$, $3\pi/4$, π, $7\pi/6$, $3\pi/2$, and 2π, and draw the graphs of both functions.

12. Repeat Exercise 11 for $\cos 2x$ and $\cos 3x$.

13. On the same axis system, draw the graphs of $\sin 2x$ and $2\sin x$, $0 \le x \le 2\pi$. What does this tell you about the equation $\sin 2x = 2\sin x$? What particular values of x make this a true equation?

14. Do Exercise 13 for $\cos\frac{1}{2}x$ and $\frac{1}{2}\cos x$, $0 \le x \le 2\pi$.

Graph the functions in Exercises 15–20 on $0 \le x \le 2\pi$.

15. $y = \sin 3x$
16. $y = 2\sin 2x$
17. $y = \frac{1}{4}\cos 4x$
18. $y = \sin^2 x = (\sin x)^2$
19. $y = -\cos x + 2$
20. $y = x + \sin x$

Graph the functions in Exercises 21–26 on $0 \le x \le 2\pi$.

21. $y = -\frac{1}{2}\sin 2x$
22. $y = \sqrt{\sin\frac{1}{2}x}$
23. $y = |\cos x|$
24. $y = \ln|\sin x + 2|$
25. $y = e^{\sin x}$
26. $y = (\sin x)(\cos x)$

On the same axis system draw the graph of the two functions given first in Exercises 27–32, and then visually add these graphs to obtain the graph of the third function. Do this on $0 \le x \le 2\pi$.

27. $\sin x$; x; $x + \sin x$
28. $\sin x$; e^{-x}; $e^{-x} + \sin x$
29. $\sin x$; $2\sin x$; $\sin x + 2\sin x$
30. $\frac{1}{2}\sin\frac{1}{2}x$; $\sin x$; $\frac{1}{2}\sin\frac{1}{2}x + \sin x$
31. $\frac{1}{4}\sin\frac{x}{4}$; $\sin x$; $\frac{1}{4}\sin\frac{x}{4} + \sin x$
32. $\frac{1}{4}\cos\frac{x}{4}$; $\sin x$; $\frac{1}{4}\cos\frac{x}{4} + \sin x$

Graph the functions in Exercises 33–36 for $0 \le t \le 1$.

33. $\sin 2\pi t$
34. $\cos 4\pi t$
35. $\cos\left(\frac{\pi}{2}\right)t + \frac{1}{2}\sin 4\pi t$
36. $\sin 6\pi t + \frac{1}{2}\sin 4\pi t$

37. For $0 < x < 1/2\pi$, graph the following:

   a. $\sin\frac{1}{x}$ [*Hint:* Compute values for $1/\left(n\frac{\pi}{2}\right)$, $n = 1, 2, 3, \ldots$]

   b. $x\sin\frac{1}{x}$

   c. $x^2\sin\frac{1}{x}$

In each of the following exercises be careful to specify the domain of each function.

38. A broker in precious metals is trying to determine when to advise her clients to buy gold. She notices that the price of gold fluctuates between $300 and $350 per ounce regularly at 6-day intervals. Model this periodic pattern by a trigonometric function for a 1-month (30-day) period.

39. Claire has decided to give up smoking cigarettes to save money. She measures the intensity of her craving to have a cigarette by the amount of money she is willing to spend at any moment for another pack, from a high of $1 to a low of $0. Sitting at her desk, trying to concentrate on her work, Claire is fighting off her periodic cravings for a cigarette, which rise to peak intensity every 40 seconds. Define a trigonometric function that models her struggle against her craving for a period of 3 minutes.

11.2 The Basic Trigonometric Functions   621

40. It is early December in Missouri and the temperature is varying between 33° and 29°F each day, with the low and high temperatures being reached at 3 A.M. and 3 P.M., respectively. Model the temperature for a 4-day period by a trigonometric function, beginning at 3 P.M. on the first day.

41. John has just purchased a stock for $10 and begins to watch its performance on the stock exchange. He notices that the price rises at the overall rate of $0.50 a day (noted at the close of each trading day). But John also notices that during the trading day the stock regularly falls from its previous closing price by $0.25 and rises to $0.25 above its final closing price each day. First, model the closing price of John's stock for a 5-day period by a linear function. Then add to this function a trigonometric function that models the movement of his stock's price between closing prices for this 5-day period.

42. In chemistry the pH of a solution is $-0.43(\ln H)$, where H is the hydrogen ion concentration of the solution. Suppose the hydrogen ion concentration in a medical patient's saliva is periodically varying between 9×10^{-8} and 10^{-7} moles per liter 4 times per hour. Write a function that *could* be used to model the pH of the patient's saliva for a 2-hour period.

Three additional functions provide a complete list of simple periodic functions. Each of these is defined as the reciprocal of the basic functions sine, cosine, and tangent:

$$\csc x = \frac{1}{\sin x} \qquad \sec x = \frac{1}{\cos x} \qquad \cot x = \frac{1}{\tan x}$$

and are known as the cosecant, secant, and cotangent, respectively.

43. Find $\sec \pi$, $\csc \pi/2$, $\cot \pi/4$, $\sec \pi/3$, and $\csc \pi/4$. Explain why $\sec \pi/2$ and $\csc 0$ are undefined.

44. Graph $y = \sec x$ and $y = \cot x$.

45. Show that $\cot x = \cos x / \sin x$.

46. Show that $\sec^2 x = 1 + \tan^2 x$.

11.3 Differentiation and Integration of Trigonometric Functions

In this section we apply the two operations of calculus to the basic trigonometric functions.

The trigonometric functions were invented hundreds of years before the development of calculus. Yet it turns out that

$$D_x(\sin x) = \cos x$$

To see why this formula is correct, we first need to evaluate a difficult limit. We wish to make a reasonably convincing argument that

$$\lim_{\theta \to 0} \frac{\sin \theta}{\theta} = 1$$

Notice that both numerator and denominator vanish as $\theta \to 0$. Since there is no apparent algebraic simplification that could help us compute the limit, we use a geometrical approach.

Notice that, relative to the circle of radius 1 in Figure 11.42, it appears that

$$\text{Length } AB \leq \text{ arc length } CB \leq \text{ length } CD \tag{1}$$

But recall that in the formula $\sin \theta / \theta$

$$\sin \theta = \text{length } AB \tag{2}$$

and radian measure

$$\theta = \text{arc length } CB \tag{3}$$

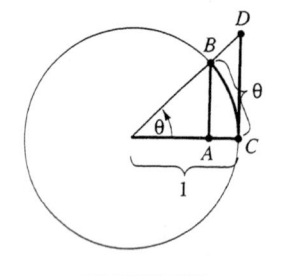

Figure 11.42

Furthermore,

$$\text{Slope of line } OD = \frac{\text{length } CD}{1} = \frac{\text{length } AB}{\text{length } OA} = \frac{\sin\theta}{\cos\theta} \quad (4)$$

Now, combining Equations 2, 3, and 4 in Equation 1, we have

$$\sin\theta \leq \theta \leq \frac{\sin\theta}{\cos\theta} \quad (5)$$

[Because $\cos(-\theta) = \cos\theta$ and $\sin(-\theta) = -\sin\theta$, Equation 6 remains true for $\theta < 0$ as well.]

For $\theta > 0$, this yields *two* inequalities:

$$\frac{\sin\theta}{\theta} \leq 1 \qquad \cos\theta \leq \frac{\sin\theta}{\theta}$$

That is, $\qquad \cos\theta \leq \dfrac{\sin\theta}{\theta} \leq 1 \qquad$ for $\theta > 0 \qquad$ Figure 11.43 $\qquad$ (6)

But now, recalling the graph of $y = \cos\theta$ in Figure 11.43, we have $\lim_{\theta\to 0}\cos\theta = 1$. Hence, using Equation 6,

$$1 = \lim_{\theta\to 0}\cos\theta \leq \lim_{\theta\to 0}\frac{\sin\theta}{\theta} \leq \lim_{\theta\to 0} 1 = 1$$

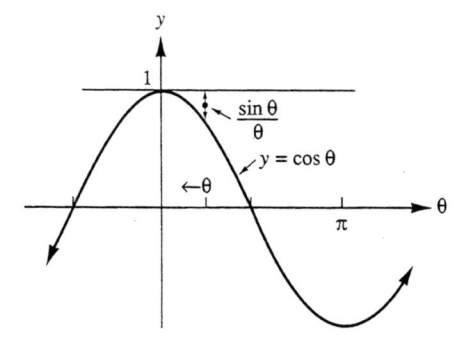

That is, $\qquad\qquad\qquad \lim_{\theta\to 0}\dfrac{\sin\theta}{\theta} = 1$

Figure 11.43

Before going further, and with the graph of $y = \cos x$ (Figure 11.43) before us, we also note that $\lim_{h\to 0}(\cos h - 1)/h = 0$. This is true because

$$\frac{\cos h - 1}{h} = \frac{\cos h - \cos 0}{h} = \frac{\cos(0 + h) - \cos 0}{h}$$

so that the limit of this quotient defines the derivative of $y = \cos x$ at $x = 0$. The value of this derivative is the slope of the tangent line to the graph at $x = 0$; in Figure 11.43 we see that this slope is 0. Hence $\lim_{h\to 0}(\cos h - 1)/h = 0$.

11.3 Differentiation and Integration of Trigonometric Functions   623

Differentiation of the Sine and Cosine

Using the limit values we just obtained we can now prove two fundamental formulas.

Theorem 11.1

$$D_x \sin x = \cos x$$
$$D_x \cos x = -\sin x$$

Proof. By definition,

$$D_x \sin x = \lim_{h \to 0} \frac{\sin(x + h) - \sin x}{h}$$

From the identity $\sin(\theta + \gamma) = \sin \theta \cos \gamma + \cos \theta \sin \gamma$ of Section 11.2, we have

$$\frac{\sin(x + h) - \sin x}{h} = \frac{\sin x \cos h + \cos x \sin h - \sin x}{h}$$

$$= \sin x \left(\frac{\cos h - 1}{h}\right) + \cos x \frac{\sin h}{h} \qquad (7)$$

Hence, because x is constant and only h varies in the limit, we have, applying the limit to Equation 7,

$$D_x \sin x = (\sin x) \lim_{h \to 0} \left(\frac{\cos h - 1}{h}\right) + (\cos x) \lim_{h \to 0} \left(\frac{\sin h}{h}\right)$$

Since $\lim_{h \to 0} \sin h/h = 1$ and $\lim_{h \to 0} (\cos h - 1)/h = 0$, we conclude that

$$D_x \sin x = \sin x \cdot 0 + \cos x \cdot 1 = \cos x$$

The remaining formula $D_x \cos x = -\sin x$ is left as an exercise.

EXAMPLE 1

Find $D_x(x^2 \cdot \sin x)$.

Solution According to the formula for differentiating the *product* $x^2 \cdot \sin x$, we have

$$D_x x^2 \sin x = x^2 D_x \sin x + (\sin x) D_x x^2 = x^2 \cos x + 2x \sin x \quad \blacksquare$$

EXAMPLE 2

Find $D_x(\sin x \cdot \cos x)$.

Solution By the same product formula,

$$D_x \sin x \cos x = \sin x \, D_x \cos x + \cos x \, D_x \sin x$$

$$= \sin x(-\sin x) + \cos x(\cos x)$$

$$= -\sin^2 x + \cos^2 x \quad \blacksquare$$

EXAMPLE 3

Find $D_x(e^x/\cos x)$.

Solution By the formula for differentiating a quotient, we have

$$D_x \frac{e^x}{\cos x} = \frac{\cos x \, D_x e^x - e^x D_x \cos x}{(\cos x)^2}$$

$$= \frac{(\cos x)(e^x) - e^x(-\sin x)}{\cos^2 x}$$

$$= \frac{e^x(\cos x + \sin x)}{\cos^2 x} \quad \blacksquare$$

A slightly more difficult form of differentiation of trigonometric functions involves composition. These problems take two forms (for the sine function):

$$D_x \sin g(x) \quad \text{or} \quad D_x f(\sin x)$$

Both of these derivatives are found by applying the chain rule: $D_x f(g(x)) = f'(g(x))g'(x)$.

EXAMPLE 4

Find $D_x \sin(x^2 + 1)$.

Solution Using the chain rule with $f(x) = \sin x$ and $g(x) = x^2 + 1$, we have [since $D_x f(g(x)) = f'(g(x))g'(x)$]

$$D_x \sin(x^2 + 1) = \cos(x^2 + 1) \cdot D_x(x^2 + 1) = 2x \cos(x^2 + 1) \quad \blacksquare$$

EXAMPLE 5

Find $D_x \sqrt{\sin x}$.

Solution Using the chain rule with $f(x) = x^{1/2}$ and $g(x) = \sin x$, we have

$$D_x \sqrt{\sin x} = \frac{1}{2}(\sin x)^{-1/2} D_x(\sin x) = \frac{\cos x}{2\sqrt{\sin x}} \quad \blacksquare$$

EXAMPLE 6

Find $D_x \cos(\ln x)$.

Solution Using the chain rule with $f(x) = \cos x$ and $g(x) = \ln x$, we have

$$D_x \cos(\ln x) = -\sin(\ln x) D_x \ln x = \frac{-\sin(\ln x)}{x} \quad \blacksquare$$

EXAMPLE 7

Find $D_x e^{\cos x}$.

Solution Here $f(x) = e^x$ [and $f'(x) = e^x$], and we have

$$D_x e^{\cos x} = e^{\cos x} \cdot D_x \cos x$$

$$= e^{\cos x}(-\sin x)$$

$$= -(\sin x)e^{\cos x} \quad \blacksquare$$

EXAMPLE 8

Find $D_x \ln|\cos x + x|$ (for those x for which $\cos x + x \neq 0$ and $\ln|\cos x + x|$ has meaning).

Solution With $f(x) = \ln x$ and $g(x) = \cos x + x$, we have

$$D_x \ln|\cos x + x| = \frac{1}{\cos x + x} D_x(\cos x + x)$$

$$= \frac{-\sin x + 1}{\cos x + x} \blacksquare$$

Integration of Sine and Cosine

The rules for differentiating sine and cosine give rise to corresponding indefinite integrals:

$$\int \sin x \, dx = -\cos x + C$$

$$\int \cos x \, dx = \sin x + C$$

EXAMPLE 9

Find $\int \cos 3x \, dx$.

Solution Integrate by substitution (Section 4.3). Let $u = 3x$, so that $du = 3 \, dx$. Then $\frac{1}{3} du = dx$ and

$$\int \cos 3x \, dx = \int \tfrac{1}{3} \cos u \, du = \tfrac{1}{3} \sin u + C = \tfrac{1}{3} \sin 3x + C \blacksquare$$

Try differentiating this solution to check its validity.

EXAMPLE 10

Find $\int \sin^7 x \cos x \, dx$.

Solution Let $u = \sin x$. Then $du = \cos x \, dx$ and

$$\int \sin^7 x \cos x \, dx = \int u^7 du = \frac{u^8}{8} + C = \frac{\sin^8 x}{8} + C \blacksquare$$

EXAMPLE 11

Find $\int (\sin x) e^{\cos x} \, dx$.

Solution Let $u = \cos x$. Then $du = -\sin x \, dx$, so

$$\int (\sin x) e^{\cos x} \, dx = \int -e^u du = -e^u + C = -e^{\cos x} + C \blacksquare$$

EXAMPLE 12

Find $\int \tan x \, dx$.

Solution Since $\tan x = \sin x/\cos x$, we must find $\int (\sin x/\cos x)\, dx$. Let $u = \cos x$. Then $du = -\sin x\, dx$, so $\int (\sin x/\cos x)\, dx = \int -du/u = -\ln|u| + C = -\ln|\cos x| + C$. Therefore $\int \tan x\, dx = -\ln|\cos x| + C$. ∎

The method of *integration by parts* (Section 7.1)

$$\int fg'\, dx = \int f\, dg = fg - \int g\, df$$

is also used to find trigonometric integrals.

EXAMPLE 13

Find $\int x \sin x\, dx$.

Solution We integrate the part $\sin x$ to obtain

$$\int x \sin x\, dx = \int x\, d(-\cos x) = -x \cos x - \int (-\cos x)\, d(x)$$

$$= -x \cos x + \int \cos x\, dx = -x \cos x + \sin x + C \quad\blacksquare$$

The next example illustrates a technique that must sometimes be used in completing a trigonometric integration. The method reaches outside calculus, back into algebra, and solves a linear equation at the appropriate point.

EXAMPLE 14

Find $\int e^x \cos x\, dx$.

Solution Integrate the part e^x to obtain

$$\int e^x \cos x\, dx = \int \cos x\, d(e^x) = e^x \cos x - \int e^x d(\cos x)$$

$$= e^x \cos x + \int e^x \sin x\, dx$$

This is an interesting result. Apparently little has been accomplished. But let us continue. We have

$$\int e^x \cos x\, dx = e^x \cos x + \int e^x \sin x\, dx = e^x \cos x + \int \sin x\, d(e^x)$$

$$= e^x \cos x + e^x \sin x - \int e^x d(\sin x)$$

$$= e^x(\cos x + \sin x) - \int e^x \cos x\, dx$$

This gives us an equation of the form $A = B - A$, where $A = \int e^x \cos x\, dx$ and $B = e^x(\cos x + \sin x)$. Solving for A, we obtain $2A = B$ or $A = B/2$. This means that

$$\int e^x \cos x\, dx = \frac{e^x(\cos x + \sin x)}{2} + C \quad\blacksquare$$

You should check this answer by differentiation.

With these examples we may add the two integration formulas, $\int \sin x \, dx = -\cos x + C$ and $\int \cos x \, dx = \sin x + C$, to our short list of *irreducible integration forms* [which additionally contains $\int x^n \, dx = x^{n+1}/(n+1) + C$, $\int e^x \, dx = e^x + C$, and $\int (1/x) \, dx = \ln|x| + C$]. Notice that every integration problem we do is reduced, by substitution or integration by parts, to one of these five basic forms to complete its solution. This reduction should be your aim in the solution of an integration problem.

Exercises 11.3

Find the derivatives in Exercises 1–30.

1. $2 \sin x$
2. $\cos 2x$
3. $x \cos x$
4. $e^x \sin x$
5. $\tan x$
6. $\dfrac{\ln x}{\sin x}$
7. $\dfrac{x}{1 + \sin x}$
8. $x^2 \cos x^3$
9. $(\sin x)\sqrt{x+1}$
10. $\sin^5 x$
11. $\sin(\pi - x)$
12. $\sin^5(-x)$
13. $e^{x^2 \sin x}$
14. $\sin^2 x^3$
15. $\cos(e^x + 1)$
16. $x + \sin \pi x^2$
17. $(\sin 2x)(\cos 3x)$
18. $(\sin 3x)^2$
19. $e^{\sin x + \cos x}$
20. $\cos(e^{-x^2})$
21. $\sin^2 x + \cos^2 x$
22. $(\sin t + 2 \cos t)^3$
23. $\sqrt{\sin u + \cos u}$
24. $(\ln x)(\cos x)$
25. $\dfrac{\sin x + \cos x}{\ln x}$
26. $\sin(x^2 - x)^3$
27. $\ln(\sin 3x)$
28. $\ln(\cos e^x) + \ln x$
29. $\ln(\cos x^3)$
30. $e^{\ln(\cos x)}$

Find the integrals in Exercises 31–44.

31. $\int \sin 5x \, dx$
32. $\int \cos^2 x \sin x \, dx$
33. $\int \cos x \, e^{\sin x} \, dx$
34. $\int \sin x \sqrt{\cos x + 1} \, dx$ (Reduce to $\sqrt{u}$.)
35. $\int \dfrac{\cos x}{\sin x + 1} \, dx$
36. $\int \sin x \cos(\cos x) \, dx$
37. $\int x \cos x \, dx$ (Integrate the part $\cos x$.)
38. $\int e^x \sin x \, dx$ (Integrate the part e^x.)
39. $\int x^2 \sin 2x \, dx$
40. $\dfrac{1}{2\pi} \int_0^1 t \sin 2\pi t \, dt$
41. $\dfrac{1}{2\pi} \int_0^1 t^2 \cos 2\pi t \, dt$
42. $\int x^3 \sin x^2 \, dx$ [Hint: Write $x^3 \sin x^2$ as $x^2(x \sin x^2)$ and integrate the part $x \sin x^2$.]
43. $\int e^x \sin 2x \, dx$
44. $\int (\sin x) \ln (\cos x) \, dx$
45. Use the trigonometric identity $(\sin x + \cos x)^2 = 1 + \sin 2x$ to find $D_x(\sin x + \cos x)^2$.

Exercises 46–50 ask you to find partial derivatives (Chapter 6) of trigonometric functions of two variables.

46. Find $\dfrac{\partial}{\partial x} \sin(x^2 y)$
47. Find $\dfrac{\partial^2}{\partial x^2} \cos(x^2 y)$
48. Find $\dfrac{\partial^2}{\partial y \partial x} \sin xy$
49. Find $\dfrac{\partial}{\partial x} (\cos x)(\sin y)$
50. Find $\dfrac{\partial}{\partial y} \dfrac{\cos x}{\sin y}$

Recall from Section 11.2 (Exercises 43–46) that $\sec x = 1/\cos x$, $\csc x = 1/\sin x$, and $\cot x = 1/\tan x = \cos x/\sin x$. Use the formulas for differentiating a quotient to find the derivatives in Exercises 51–53, and use these results in turn to find the derivatives in Exercises 54–56.

51. $D_x \sec x$
52. $D_x \csc x$
53. $D_x \cot x$
54. $D_x \sec e^x$
55. $D_x \ln \cot x$
56. $D_t \csc^2 t$
57. Show that $D_x \tan x = \sec^2 x$.
58. Show that $D_x \sec x = \sec x \tan x$.
59. Find $\displaystyle\int_{\pi/4}^{\pi/2} \cot x \, dx$
60. Find $\int \sec^2 5x \, dx$

628 Chapter 11 Trigonometric Functions

61. The value of a stock t weeks after being issued is

$$S(t) = 5(1 + t + \sin 2\pi t)$$

At what rate is its value increasing at the end of $t = 1, \frac{3}{2},$ and 2 weeks?

62. A line through the origin in the xy-plane is rotated through an angle of $\theta = 0$ to $\theta = \pi/2$ radians. At what rate is its slope changing when the line is at an angle of $\theta = \pi/6, \pi/4,$ and $\pi/3$ radians? [*Hint:* The slope of the line is $\tan \theta$.]

11.4 Applications of Trigonometric Functions

This section indicates how periodic functions may be applied in probability, differential equations, and integration, and to periodic motion and change.

Because so many phenomena are periodic, the applications of trigonometric functions are manifold. At the same time, to go much below the surface of these applications is to encounter substantial mathematical complexity. In this text we will present only a short selection of basic applications and will generally apply trigonometric functions in the context of earlier models.

An Application to Probability

EXAMPLE 1

A random variable X with range between 0 and 1 has a probability density function

$$\rho(x) = 1 + \sin 2\pi x, \quad 0 \leq x \leq 1$$

Find $P\left(0 \leq X \leq \frac{1}{4}\right)$.

Solution From Section 9.3, we have

$$P\left(0 \leq X \leq \frac{1}{4}\right) = \int_0^{1/4} (1 + \sin 2\pi x) dx$$

$$= \left(x - \frac{\cos 2\pi x}{2\pi}\right)\Big]_0^{1/4} = \frac{1}{4} - \frac{1}{2\pi}\left(\cos \frac{\pi}{2} - \cos 0\right)$$

$$= \frac{1}{4} - \frac{1}{2\pi}(0 - 1) = 0.409$$

Figure 11.44 indicates that the region of highest probability for this random variable lies between 0 and $\frac{1}{2}$. We also see from the calculation that over 40% of the probability lies in the first one-fourth of the range of X.

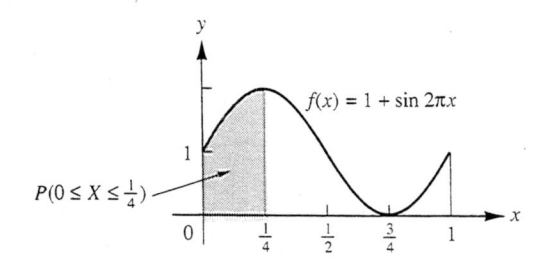

Figure 11.44

An Application to Differential Equations

Let $y(t)$ be the number of customers patronizing some product at time t over a year, $0 \leq t \leq 1$. In the absence of advertising it may be supposed that $y'(t)$, the rate at which customers come to patronize the product, is proportional to $y(t)$ ("satisfied customers are the best advertisement"). That is, $y'(t) = ky(t)$ or, equivalently, $y'/y = k$. However, if the product is, say, a sailboat or a snowmobile, then seasonal factors will intrude and the percentage rate of change, y'/y, will not be a constant k but could be some periodic function and give rise to a differential equation like that in the next example.

EXAMPLE 2

Solve the differential equation

$$\frac{y'}{y} = 0.08\left(\sin 2\pi\left(t - \frac{1}{4}\right) + 2\right)$$

with initial value $y(0) = 500$. Find $y(\frac{1}{2})$, the amount present at time $t = \frac{1}{2}$.

Solution Notice that

$$D_t \ln |y(t)| = \frac{1}{y(t)} \cdot y'(t) = \frac{y'(t)}{y(t)}$$

Therefore, applying the integral to both sides of the given differential equation (as in Section 8.1 or 8.4), we have

$$\int \frac{y'(t)}{y(t)} dt = \int 0.08\left(\sin 2\pi\left(t - \frac{1}{4}\right) + 2\right) dt$$

or
$$\ln |y(t)| = \frac{-0.08}{2\pi}\left(\cos 2\pi\left(t - \frac{1}{4}\right)\right) + 0.16t + C$$

Therefore
$$y(t) = e^{\ln y(t)} = e^{(-0.04/\pi)\cos 2\pi[t-(1/4)]+0.16t+C}$$

and
$$500 = y(0) = e^{(-0.04/\pi)\cos(-\pi/2)+C} = e^C$$

Therefore
$$y(t) = 500 e^{(-0.04/\pi)\cos 2\pi(t-1/4)+0.16t}$$

and
$$y\left(\tfrac{1}{2}\right) = 500 e^{(-0.04/\pi)\cos(\pi/2)+0.16(1/2)}$$

$$= 500 e^{0.08} \approx 542 \quad \blacksquare$$

An Application to Integration via Trigonometric Substitution

Applications of the integral can give rise to algebraic functions that cannot be integrated without using the trigonometric functions. Consider the algebraic expression $\sqrt{1 - x^2}$. If we let $x = \sin u$, then, because $\sin^2 u + \cos^2 u = 1$ we have

$$\sqrt{1 - x^2} = \sqrt{1 - \sin^2 u} = \sqrt{\cos^2 u} = \cos u \qquad (1$$

at least for $0 \leq u \leq \pi/2$, where $\cos u \geq 0$.

EXAMPLE 3

Show that

$$\int \frac{dx}{\sqrt{1-x^2}} = \int du = u + C$$

where $x = \sin u$.

Solution If we let $x = \sin u$, then $dx = \cos u\, du$. Using Equation 1 and substituting in the integral then gives

$$\int \frac{dx}{\sqrt{1-x^2}} = \int \frac{\cos u\, du}{\cos u} = \int 1\, du = u + C \quad\blacksquare$$

In Example 3, a difficult integral is reduced to a triviality by the substitution $x = \sin u$. But another difficulty remains. We would like a solution to the integral *as a function of the original variable x*, not the new variable u. It is impossible to do this without a study of the inverse trigonometric functions. We will not undertake such a study but will, at least, find a few *definite integrals*, using the following fact.

If $x = g(u)$ and if $a = g(A)$ and $b = g(B)$ and g is increasing (or g is decreasing) at every point between A and B, then

$$\int_a^b f(x)dx = \int_A^B f(g(u))g'(u)du \qquad (2)$$

In Example 3, we saw that with $f(x) = 1/\sqrt{1-x^2}$, the substitution $x = g(u) = \sin u$ actually simplifies the integral, despite the apparently complicated appearance of Equation 2. This will always be the case in the exercises we consider.

EXAMPLE 4

Find

$$\int_0^{1/\sqrt{2}} \frac{dx}{\sqrt{1-x^2}}$$

Solution With $x = \sin u$, we have $0 = \sin 0$ and $1/\sqrt{2} = \sin \pi/4$. Therefore, in Equation 2, $A = 0$ and $B = \pi/4$ and

$$\int_0^{1/\sqrt{2}} \frac{dx}{\sqrt{1-x^2}} = \int_0^{\pi/4} 1\, du = u \Big]_0^{\pi/4} = \frac{\pi}{4}$$

Moreover, this substitution is valid because $\sin u$ is increasing at every point u between $A = 0$ and $B = \pi/4$, as indicated in Figure 11.45.

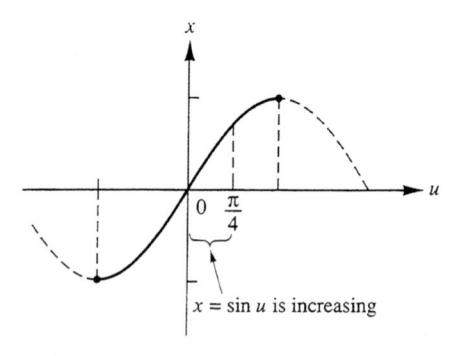

Figure 11.45

$x = \sin u$ is increasing

EXAMPLE 5

Find $\int_0^1 1/(1 + x^2)\, dx$.

Solution If we divide by $\cos^2 u$ in the identity $\sin^2 u + \cos^2 u = 1$, we obtain the identity $1 + \tan^2 u = 1/\cos^2 u$. Let $x = \tan u$. Then

$$\frac{1}{1 + x^2} = \frac{1}{1 + \tan^2 u} = \frac{1}{\frac{1}{\cos^2 u}} = \cos^2 u$$

Additionally, from Section 11.3, $dx = (1/\cos^2 u)du$, and we have

$$\int \frac{1}{1 + x^2}\, dx = \int \cos^2 u \left(\frac{1}{\cos^2 u}\, du\right) = \int 1\, du \qquad (3)$$

Now $\tan 0 = 0$ and $\tan \pi/4 = 1$ and, moreover, $\tan u$ is increasing at every point u between $A = 0$ and $B = \pi/4$ (see Figure 11.41). Therefore, from Equations 2 and 3,

$$\int_0^1 \frac{dx}{1 + x^2} = \int_0^{\pi/4} 1\, du = u \Big]_0^{\pi/4} = \frac{\pi}{4}$$

Remark. You may recall that in Section 7.5 we estimated the value of this integral using the trapezoidal rule and obtained a value $0.7828 \simeq \pi/4 = 0.7854$.

A complete discussion of this topic is impossible without the use of inverse trigonometric functions. Let us at least point out that a little more generality is possible. For example, the substitution $x = 2 \sin u$ makes $\sqrt{4 - x^2} = \sqrt{4 - 4\sin^2 u} = \sqrt{4\cos^2 u} = 2\cos u$. Thus, for example, the integral $\int_0^1 dx/\sqrt{4 - x^2}$ may be found by similar means.

The Differential Equation $y'' = -\omega^2 y$

Beginning with the function $y(t) = \sin t$, we find that $y'(t) = \cos t$ and $y''(t) = -\sin t$. That is, y satisfies the second-order differential equation $y'' = -y$. This equation characterizes basic periodic behavior; its importance may be compared to the equation $y' = ky$ characterizing exponential growth. Since a second derivative is involved, however, the solution of the general equation of this form gives rise to two unknown constants, which are to be determined by initial conditions.

632   Chapter 11   Trigonometric Functions

The solution to the general second-order differential equation

$$y'' = -\omega^2 y$$

where $\omega \geq 0$, is a function of the form

$$y(t) = A \sin \omega(t - B) \qquad (4)$$

where A and B are constants that can be determined from initial conditions imposed on $y(0)$ and $y'(0)$.

To verify this assertion, observe that, no matter what values A and B have,

$$y'(t) = A \omega \cos \omega(t - B)$$

and therefore

$$y''(t) = -A \cdot \omega^2 \cdot \sin \omega(t - B) = -\omega^2[A \sin \omega(t - B)] = -\omega^2 y(t)$$

From Section 11.2, we realize that the number A determines the amplitude of the solution, ω determines the frequency, and B determines a "phase shift" in the solution. [This allows a value for $y(0)$ other than zero itself. That is, without the constant B, we would *always* have $y(0) = 0$.]

EXAMPLE 6
Find the solution of $y'' = -4y$, for $0 \leq t \leq \pi$ and satisfying the initial conditions $y(0) = 3$ and $y'(0) = 0$.

Solution Here $y'' = -\omega^2 y$, where $-\omega^2 = -4$ and we take $\omega = 2$. Thus from Equation 4, $y(t) = A \sin 2(t - B)$. Now

$$3 = y(0) = A \sin 2(0 - B) = A \sin(-2B) = -A \sin 2B \qquad (5)$$

and, since $y'(t) = 2A \cos 2(t - B)$, then

$$0 = y'(0) = 2A \cos 2(0 - B) = 2A \cos(-2B) = 2A \cos 2B \qquad (6)$$

From Equation 6, $A = 0$ or $2B = \pi/2$. However, $A = 0$ makes Equation 5 impossible, so we only have the solution $2B = \pi/2$ or $B = \pi/4$. Substituting this in Equation 5, we have

$$3 = -A \sin 2\left(\frac{\pi}{4}\right) = -A \sin \frac{\pi}{2} = -A \cdot 1 = -A \quad \text{or} \quad A = -3$$

Therefore $y(t) = -3 \sin 2[t - (\pi/4)]$ is the solution. ∎

Remark. Had we asked for a solution on $[0, 2\pi]$, then both $2B = \pi/2$ and $2B = 3\pi/2$ would be possible. This second choice makes $B = 3\pi/4$ and yields $-3 = A \sin 2(3\pi/4) = -A$ or $A = +3$. Thus $y(t) = 3 \sin 2[t - (3\pi/4)]$ as well. Does this mean that two solutions are possible if the domain is large enough? No. Using the fact that $-\sin(x - \pi) = \sin x$, we can easily check that $-3 \sin 2[t - (\pi/4)] = 3 \sin 2[t - (3\pi/4)]$, so that these solutions are identical. In effect either $2B = \pi/2$ or $2B = 3\pi/2$ yields the same solution, and thus *any one* value of B in [0, 2π] suffices to yield a solution. This is generally true and will be used in the next discussion.

Periodic Motion: The Vibrating Spring

There are no truly simple examples of how the differential equation $y'' = -\omega^2 y$ arises in applications, but one of the simplest examples of periodic behavior is that of a spring to which a weight is first attached, then stretched further, and finally released (see Figure 11.46). The spring rebounds and raises the weight, which in turn pulls the spring back, causing a periodic, or oscillatory, motion of the spring and weight. This behavior, in the ideal (that is, ignoring friction and other damping forces), may be exactly described by a solution of an equation of the form $y'' = -\omega^2 y$.

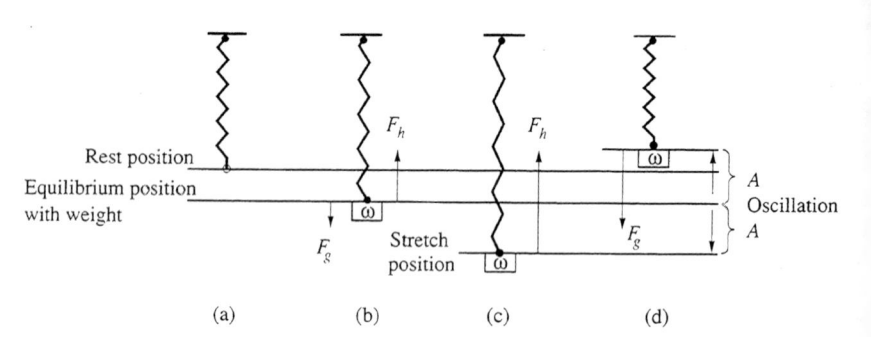

Figure 11.46

(a)   (b)   (c)   (d)

Let $y(t)$ be the position of the spring at time t, to which a weight of ω pounds has been attached [Figure 11.46(b)]. According to Hooke's law, the spring exerts a force F_h against the weight so as to restore the spring to its original length, and this force is directly proportional to the length $y(t)$ that the spring has been stretched. That is, $F_h = -k \cdot y(t)$. The number k, which is called the **spring constant**, is different for different springs.

At the same time, gravitational force F_g acts to pull the spring down according to Newton's law: F_g = (mass) × (acceleration) = $my''(t)$. Assuming no other forces (such as friction) are acting on the spring, we have that at any moment, $F_g = F_h$ or

$$my''(t) = -k \cdot y(t)$$

or
$$y'' = -\omega^2 y \quad \text{where } \omega = \sqrt{k/m}$$

Hence, from Equation 4, the position of the weight on the spring at time t is given by

$$y(t) = A \sin \omega(t - B) = A \sin \sqrt{k/m}\, (t - B) \tag{7}$$

EXAMPLE 7

An 8-lb weight stretches a spring 18 in. The spring is then stretched an additional 5 in. and released. Determine its position and the amplitude and frequency of its motion thereafter.

Solution The mass m of the 8-lb weight is $m = w/g = 8/32 = 0.25$, since the acceleration due to gravity is 32 ft/sec^2. The spring is stretched 18 in. by this weight. Since weight is the force on this mass due to gravity, $F_g = F_h$ at a distance of 18 in. Thus, converting inches to feet, $F_g = -8$ and $F_h = -k\left(\frac{18}{12}\right)$ so that $-k\left(\frac{18}{12}\right) = -8$ or $k = 5.33$. Consequently, $\omega = \sqrt{k/m} = \sqrt{5.33/0.25} \approx 4.62$ for this *particular* spring. (Note that the constants k, m, and ω are all determined without regard to any motion of the spring and depend only on the particular weight involved and how far it stretches the spring.)

According to our earlier discussion, the motion of the spring satisfies $y'' = -\omega^2 y$, where $y(t)$ is the position of the spring at time 0 and $\omega = \sqrt{k/m}$. In this example, $y(0) = -\frac{5}{12}$ ft. Since the spring is simply released at time 0, it is not moving at that moment and so $y'(0) = 0$.

Consequently, $y(t) = A \sin \omega(t - B)$ and $y'(t) = A\omega \cos \omega(t - B)$ [from Equation 7], and at time $t = 0$,

$$-\frac{5}{12} = y(0) = A \sin(-\omega B) \tag{8}$$

and
$$0 = y'(0) = A \cdot \omega \cdot \cos(-\omega B) \tag{9}$$

From Equation 9, $-\omega B = -\pi/2$ or $B = \pi/2\omega$. Consequently, from Equation 8,

$$-\frac{5}{12} = A \sin\left((-\omega)\left(-\frac{\pi}{2\omega}\right)\right) = A \sin\left(-\frac{\pi}{2}\right) = -A$$

Therefore $A = \frac{5}{12}$ and at any time t, $y(t) = \left(\frac{5}{12}\right)\sin \omega[t - (\pi/2\omega)]$ ft, with $\omega = 4.62$, by substitution for A and B in the general solution. ∎

The graph of this solution (Figure 11.47) shows what we might intuitively expect from the outset. The spring oscillates for a distance of ± 5 in. from its equilibrium position (its resting position with the weight attached), beginning 5 in. *below* the *equilibrium* position and rising to 5 in. above this position before falling back.

$y(t) = \frac{5}{12}\sin \omega(t - B)$ feet

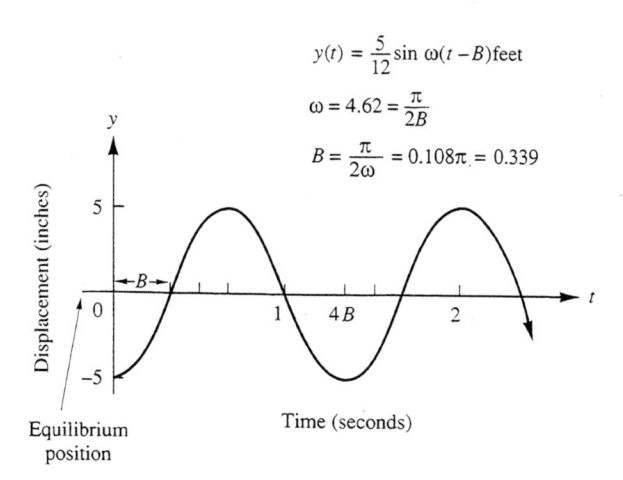

$\omega = 4.62 = \frac{\pi}{2B}$

$B = \frac{\pi}{2\omega} = 0.108\pi = 0.339$

Figure 11.47

The Lotka-Volterra Predator—Prey Model

One of the earliest applications of mathematics to ecology was made by the Austrian mathematician A. J. Lotka and the Italian mathematician Vito Volterra, who modeled the interaction of a predator on its prey as a two-compartment system, indicated in Figure 11.48. A formula description of the model is beyond the text, but we will give a qualitative description (Section 8.3) of how the model evolves over time.

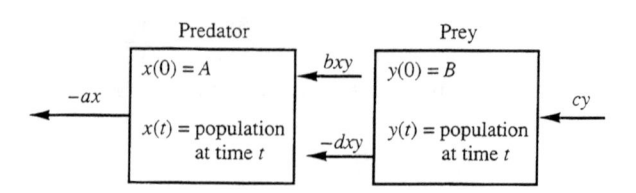

Figure 11.48

In the Lotka-Volterra model, the predator population $x(t)$ has a natural birthrate minus death rate constant a acting in proportion to its size, and the population $x(t)$ is hypothesized to increase further in a constant proportion b to the *total number of possible contacts* $x(t)y(t)$ between predator and prey. Thus, by rate-in less rate-out analysis (Section 8.5),

$$x'(t) = -ax(t) + bx(t)y(t)$$

Similarly, the prey population has a natural growth rate c but suffers from predation at a rate in constant proportion d to the number of contacts $x(t)y(t)$. Thus

$$y'(t) = cy(t) - dx(t)y(t)$$

Let us rewrite this pair of equations as

$$x' = bx\left(y - \frac{a}{b}\right) \tag{10}$$

$$y' = dy\left(\frac{c}{d} - x\right) \tag{11}$$

and attempt, using the ideas of Section 8.3, to graph the solution to this system.

Figure 11.49 illustrates the possible course of events, where the initial populations are as indicated. The key idea is this: From Equation 10, when $y(t) < a/b$, then $x'(t) < 0$ and x is decreasing; since $x(t) > c/d$, then from Equation 11, $y'(t) < 0$ and y is also decreasing. Once x decreases below c/d however, then (from Equation 11), $y'(t) > 0$ and y begins to increase and continues beyond a/b, at which time x begins to increase again. Each crossing by x or y of the horizontal line at height c/d or a/b corresponds to a *change in the direction of population growth* for y or x. And a pattern of nature is once again seen, for abstract reasons, to be natural.

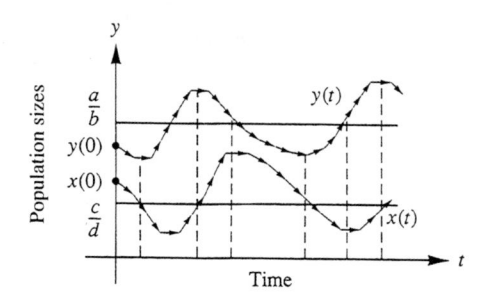

Figure 11.49

Exercises 11.4

1. A random variable X has a probability density function

$$f(x) = \frac{1 - \frac{1}{2}\cos 2x}{\pi}$$

on the interval $[0, \pi]$. Graph this function and find the following:

   a. $P(0 \leq X \leq \pi)$
   b. $P\left(\frac{\pi}{4} \leq X \leq \frac{3\pi}{4}\right)$
   c. $P\left(X \geq \frac{3\pi}{4}\right)$

2. Find the mean μ of the random variable in Example 1.

Solve the differential equations in Exercises 3–6 by the method of separation of variables.

3. $y' = y \sin t$, $y(0) = 2$
4. $y' = (1 - y)\cos t$, $y(0) = 0$
5. $y' = y \tan t$, $y(0) = 1$
6. $y' = y^2 \sin t$, $y(0) = 4$

Evaluate the integrals in Exercises 7–12 using a trigonometric substitution. Be sure that the substitution function is increasing (or decreasing) on the appropriate interval.

7. $\int_0^{1/2} \frac{1}{\sqrt{1 - x^2}} \, dx$ $\left(\text{Recall: } \sin\frac{\pi}{6} = \frac{1}{2}.\right)$
8. $\int_{-1/2}^{1/2} \frac{1}{\sqrt{1 - x^2}} \, dx$
9. $\int_{-1}^{1} \frac{1}{1 + x^2} \, dx$
10. $\int_0^1 \frac{1}{\sqrt{4 - x^2}} \, dx$
11. $\int_{\sqrt{2}}^{2} \frac{dx}{x\sqrt{x^2 - 1}}$ $\left[\text{Hint: Let } x = \frac{1}{\cos u}.\right]$
12. $\int_0^3 \frac{1}{9 + x^2} \, dx$

Find the solution to each differential equation with the initial conditions given in Exercises 13–16.

13. $y'' = -9y$; $y(0) = -2$; $y'(0) = 0$
14. $y'' = -4y$; $y\left(\frac{\pi}{2}\right) = 0$; $y'(0) = 2$
15. $y'' = -16y$; $y(0) = 0$; $y'(\pi) = 1$
16. $y'' = -2y$; $y\left(\frac{\pi}{2}\right) = 1$; $y'(2\pi) = 0$

17. When a 16-lb weight is attached to a spring, the spring stretches 6 in. Calculate the spring constant k.

18. By Hooke's law, $F_h = 0.5D$ for a certain spring when the spring is stretched a distance of D in. How far will the spring stretch when a 20-lb weight is attached to it?

19. A spring stretches 6 in. when a 3-lb weight is attached to it. If the spring is stretched an additional 5 in. and then released, describe its position at time t by a function $y(t)$.

20. Suppose that when a 6-kg mass is attached to a spring, it stretches 1 m. Describe the motion when the spring is pulled down an additional 6 cm and then released. (Here $g = 9.8$ m/sec^2.)

21. Graph $f(t) = e^{-0.5t}\sin 2\pi t$. This function can be used to model a *damped oscillation*—an oscillation that, due to friction, slowly slows down. Assume that this function models a vibrating spring subject to

damping. What is the period (the time length of one oscillation) of the spring's motion?

22. Explain why the number B is as indicated in Figure 11.47.

23. Please refer to the Lotka-Volterra model, where the predator population obeys the differential equation
$$x'(t) = -ax(t) + bx(t)y(t) = x(t)[by(t) - a]$$
Assuming that $x(t) > 0$, we may write this as
$$by(t) - a = \frac{x'(t)}{x(t)} = D_t \ln x(t)$$
Hence
$$y(t) = \frac{a}{b} + \frac{1}{b} D_t \ln x(t) \qquad (12)$$
Let T be the period of the rising and falling predator population $x(t)$; thus $x(T) = x(0)$. Using Equation 12, find $(1/T) \int_0^T y(t)\,dt$, the average population of the prey over this cycle. Interpret your answer in ordinary language.

The remaining exercises draw on application techniques from many different sections of the text.

24. Find the average value of the function $f(x) = 2 + \sin x$ on the interval $[0, \pi]$; on the interval $[0, 2\pi]$. (See Section 4.5.)

25. The cash flow to a business is
$$C(x) = 100\left(\sin \frac{2\pi}{30}x - \frac{1}{2} \cos \frac{2\pi}{30}x \right) \text{ dollars}$$
per day on day x of a 30-day month.

a. Find the days on which maximum and minimum cash flow occurs. [Hint: See Section 2.5.]

b. Find the total amount of money earned by this business during the 30-day period. [Hint: See Section 4.5.]

26. Running over rough terrain for 15 min to get to a telephone and call for help for a friend who has been hurt in an accident, Jim ran at the rate of
$$f(t) = \sin \frac{\pi}{15}t - \frac{1}{4} \cos \frac{\pi}{5}t \text{ mi/hr}$$
after t minutes. How far did Jim run? Sketch a graph of the function f.

27. The number of people living along a densely populated street is $P(t) = 2{,}000(1 + \cos 8\pi t)$ people/mi, t mi from point A. How many people live along the first 2 mi of the street from point A? (See Section 7.2.)

28. An urban area initially contains 100,000 people. New arrivals to the population occur at the rate of $A(t) = 3{,}000\left(1 + \frac{1}{2} \sin 2\pi t\right)$ individuals per year at time t. Only the proportion e^{-t} of the population living in this area at any time is still present t years later. What is the population of this urban area 5 years later? [Hint: See Section 7.4. Integrate $\int e^{-t} \sin \beta t \, dt$ by parts.]

29. A supermarket buys 200 tons of tomatoes a year. Due to seasonal factors, especially heat and cold, the concentration of rotten tomatoes per ton received is given by $R(t) = 120 + 40 \sin 2\pi t$ pounds per ton, over a full year $0 \le t \le 1$. Find the total amount A of rotten tomatoes bought by the supermarket during the year. [Hint: See Section 7.3.]

30. Let $y(t)$ be the population of quail living on an acre of Missouri farmland. Due to seasonal factors the mortality rate $y'(t)/y(t)$ of the population varies periodically and is given by
$$y'(t)/y(t) = -0.02 \sin 2\pi\left(t + \tfrac{1}{4}\right) + 0.05$$
over a 1-year period $0 \le t \le 1$ (where $t = 0$ is January 1). Additionally, $y(0) = 35$. Find $y\left(\tfrac{3}{4}\right)$, the quail population in September; $y\left(\tfrac{11}{12}\right)$, the population in November.

Chapter 11 Summary

1. The radian measure of an angle θ is the ratio A/R of arc length A to radius R in any circle in which θ is inscribed. If $R = 1$, then $\theta = A$. The radian measure of common angles ($90° \sim \pi/2$, $45° \sim \pi/4$, and so on) should be known.

2. In a circle of radius 1, the sine of an angle θ is the (vertical) y-coordinate of the point (x, y) at which the radial line determined by θ intersects the circle. The cosine of θ is the (horizontal) x-coordinate of this point. That is,
$$\sin \theta = y \quad \text{and} \quad \cos \theta = x$$

The values of these functions range between -1 and 1 as θ varies between 0 and 2π. The tangent of θ, which is defined by

$$\tan \theta = \frac{y}{x} = \frac{\sin \theta}{\cos \theta}$$

is the slope of the radial line determined by θ and varies between $-\infty$ and $+\infty$.

3. The graph of $y = C + A \sin \omega(X - B)$ is periodic, of *period* $2\pi/\omega$ (that is, repeating its pattern along any interval of length $2\pi/\omega$), with *amplitude* A alternating above and below the horizontal line $y = C$ and intersecting this line at $X = B$. The *frequency* of this graph (how often the graph repeats its pattern in an interval of length 2π) is the number ω.

4. From $\lim_{\theta \to 0} \sin \theta/\theta = 1$, we obtain

$$D_x \sin x = \cos x \quad \text{and} \quad D_x \cos x = -\sin x$$

with equivalent integration formulas,

$$\int \sin x \, dx = -\cos x + C \quad \text{and} \quad \int \cos x \, dx = \sin x + C$$

5. The trigonometric substitutions $x = \sin u$, $\tan u$, and $\cos u$ may be used to find certain definite integrals in which the algebraic forms $\sqrt{1 - x^2}$, $1 + x^2$, and $x\sqrt{x^2 - 1}$ occur.

6. The trigonometric functions can be used to model periodic change and in particular occur as the solution to the differential equation $y'' = -\omega^2 y$ with solution $y(t) = A \sin \omega(t - B)$, where A and B are determined from knowledge of $y(0)$ and $y'(0)$.

Chapter 11 Summary Exercises

What is the radian measure of the angle θ in Figure 11.50(a, b)?

(a)   (b)

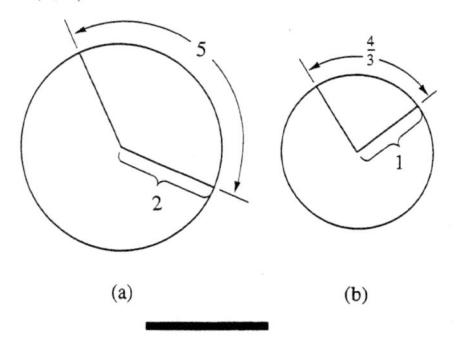

Figure 11.50

2. Discuss the relationship between degree and radian measure of an angle θ in terms of the proportion of the circle subtended by the angle θ.

3. Determine the radian measure of an angle of $30°$, $60°$, $125°$, $-35°$, $200°$, $-140°$, $420°$, and $750°$.

4. Determine the degree measure of an angle of $\pi/3$, $2\pi/5$, $3\pi/4$, $-\pi/4$, $-\pi/6$, 2, 1.5π, $-7\pi/2$, and $8\pi/3$ radians.

5. Make a table of values of $\sin \theta$ for $\theta = 0$, $\pi/3$, $\pi/4$, $\pi/6$, $\pi/2$, π, $5\pi/4$, and $7\pi/3$ radians.

6. Make a table of values for $\tan \theta$ for $\theta = 0$, $\pi/3$, $\pi/4$, $2\pi/3$, $3\pi/4$, π, and $7\pi/4$ radians.

7. Find an angle x for which $\cos 2x \neq 2 \cos x$.

8. Evaluate $\sin(x+y)$ for $x = \pi/2$ and $y = \pi/4$.

9. Evaluate $3\cos 2x$ for $x = \pi/3, \pi/4, -\pi/6$, and $-3\pi/4$.

Graph the functions in Exercises 10–13 on $[0, 2\pi]$.

10. $y = 2\sin x + 1$

11. $y = 3\cos\dfrac{x}{2}$

12. $y = \frac{1}{3}\sin 3x + 2$

13. $y = x + \cos x$

Write a formula for a function $f(x) = A\sin\omega(x - B)$ on $[0, 2\pi]$ whose amplitude, frequency, and value $f(0)$ are given, respectively, in Exercises 14 and 15.

14. $\frac{1}{2}$; 2; $f(0) = 0$

15. 2; 1; $f(0) = 2$

16. Graph $y = -\frac{1}{2}\sin 4x$ on $0 \le x \le \pi$.

17. Graph $y = 2 + 3\sin 6\pi t$ on $0 \le t \le 1$.

18. Graph $y = 1 - \frac{1}{2}\cos 4\pi t$ on $0 \le t \le 2$.

Find the derivative of each expression in Exercises 19–28.

19. $3\cos x$

20. $5\sin 2\pi t$

21. $2\sin x^2$

22. $\dfrac{\cos x}{\sin x}$

23. $\dfrac{\sin x}{x}$

24. $e^{-x}\cos x^3$

25. $\ln(2 + \sin x)$

26. $\sin e^{x^2}$

27. $\sin 2\pi\left(x - \dfrac{\pi}{2}\right)$

28. $3x\cos\left(\dfrac{\pi}{2} - x\right)$

Find the integrals in Exercises 29–36.

29. $\displaystyle\int_0^1 \sin 2x\, dx$

30. $\displaystyle\int_0^1 \cos \pi t\, dt$

31. $\displaystyle\int_0^{2\pi} \sin x \cos x\, dx$

32. $\displaystyle\int \sin x\, e^{\cos x}\, dx$

33. $\displaystyle\int x \sin 2x\, dx$

34. $\displaystyle\int_0^\pi x^2 \cos x\, dx$

35. $\displaystyle\int \dfrac{\cos x}{\sin x}\, dx$

36. $\displaystyle\int_0^1 t \sin 2\pi t\, dt$

37. Solve the IVP $y'' = -4y$; $y(0) = 0$; $y'(0) = 0$.

38. Solve the IVP $y'' = -12y$; $y(0) = 0$; $y'(0) = -1$.

39. Solve the IVP $y'/y = 2\sin 6\pi t$; $y(0) = 10$.

40. A 2-lb weight stretches a spring 9 in. The spring is pulled down an additional 3 in. and then released. Model its motion by the solution of an appropriate differential equation.

Evaluate the integrals in Exercises 41 and 42.

41. $\displaystyle\int_{-4}^4 \dfrac{dx}{4 + x^2}$

42. $\displaystyle\int_0^1 \dfrac{dx}{\sqrt{1 - \left(\dfrac{x}{4}\right)^2}}$

Appendix

Further Development

A.1 Continuity and Differentiation

The two concepts of continuity and differentiability are central concepts of the calculus. If either is not present in a model of a system, the calculus usually cannot be applied in its standard form, if at all. Of these two properties, differentiability is the stronger because any function that has a derivative must necessarily be continuous. This is the content of the following fundamental and completely general result of this section.

Theorem A.1 If the function $y = f(x)$ has a derivative at the point $x = a$, then the function f *must* also be continuous at $x = a$.

This theorem has an important meaning in applications. It says that if a natural phenomenon is described by a discontinuous function, then at points of discontinuity, we will *not* be able to study it via its rate of change.

We can see why the theorem is true, as follows: To know that f is continuous at a number $x = a$, we have to determine that $f(a + h)$ is approximately equal to $f(a)$ when h is a very small number. If f is known to be differentiable at a, we can conclude that

$$\frac{f(a + h) - f(a)}{h} \simeq f'(a) \quad \text{when } h \text{ is very small}$$

Therefore $\quad f(a + h) - f(a) \simeq h \cdot f'(a)$

Now, if h is very small, then $h \cdot f'(a)$ is approximately zero and therefore

$$f(a + h) - f(a) \simeq 0$$

Consequently, $f(a + h) \simeq f(a)$ and therefore f is continuous.

The converse of this theorem is *not true*. That is, it is possible for a function to be continuous and yet fail to have a derivative. Let us discuss this phenomenon in a familiar setting. Recall that a function is used as a mathematical model for a system outside mathematics. Our example will show not only that a continuous function may fail to have a derivative but that calculus may also be used to test for the appropriateness of a mathematical model.

EXAMPLE 1

Take a ball and throw it at the floor 3 ft in front of you. Model the flight and rebound of the ball by the graph of the function

$$f(x) = |x - 3| \qquad 0 \le x \le 5$$

That is, let

$$f(x) = \text{height of the ball at distance } x \text{ from you} = |x - 3|$$

Is this a "good model" at the point $x = 3$, where the ball hits the floor?

Solution This model is suggested by Figure A.1 and our knowledge of the graph of the absolute value function (see Section 0.4).

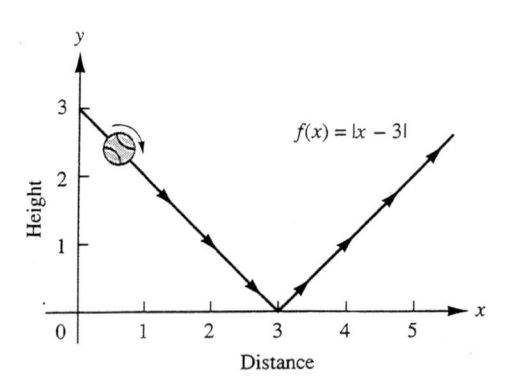

Figure A.1

Here the difference quotient

$$\frac{f(x + h) - f(x)}{h} = \frac{\text{change in height}}{\text{change in distance}}$$

measures change in vertical distance (height) relative to change in horizontal distance at point x. Thus, for example,

$$f'(3) = \text{rate of change in height per foot } at \text{ a distance of 3 ft}$$

Let us see what the computation of $f'(3)$ reveals. We have

$$\frac{f(3 + h) - f(3)}{h} = \frac{|(3 + h) - 3| - |3 - 3|}{h} = \frac{|h|}{h}$$

Now we have to be a bit careful. If we average height versus distance in the positive direction (to the right of the point $x = 3$ of impact), then $h > 0$ and

$$\frac{f(3+h) - f(3)}{h} = \frac{|h|}{h} = \frac{h}{h} = 1$$

But if we average height versus distance in the negative direction (to the left of the point of impact), then $h < 0$ and

$$\frac{f(3+h) - f(3)}{h} = \frac{|h|}{h} = \frac{-h}{h} = -1$$

(Note that these numbers, $+1$ and -1, conform to the slope of the graph in Figure A.1.)

Therefore, the limit

$$f'(3) = \lim_{h \to 0} \frac{f(3+h) - f(3)}{h}$$

$$= \lim_{h \to 0} \begin{cases} +1 & h > 0 \\ -1 & h < 0 \end{cases}$$

does not exist! The function f is *not differentiable* at $x = 3$.

In conclusion, this is not a "good model" because it does not allow us to measure the rate of change of the bouncing ball at its point of impact. ∎

If you believe that at the point of impact the vertical versus horizontal speed of an actual bouncing ball is zero, then you must question our choice of a mathematical model. The function $f(x) = |x - 3|$ does not incorporate fine details at the point of impact into the mathematical model. Thus calculus can serve as a test for the defects, or the appropriateness of a mathematical model, played against one's sense of a situation. At the same time, a model like this one could be useful, for example, in modeling the reflection of light off a mirrored surface.

Beyond this the function f in Example 1 is *continuous but not differentiable*. Thus, not all functions, even continuous ones, necessarily have derivatives. The problem with this function is that the slope of tangents to its graph changes too abruptly, from -1 to $+1$, where its graph abruptly turns a corner.

Exercises A.1

Show that $f(x) = |x|$ is continuous but not differentiable at $x = 0$.

2. Graph $f(x) = x^{1/3}$ and discuss its continuity and differentiability at $x = 0$. [*Hint:* Evaluate the difference quotient for $h = \pm 10^{-6}$.]

A.2 Functions with Equal Derivatives and L'Hôpital's Rule

The applicability of calculus to the world outside mathematics resides in its ability to derive a description of a system based on observation of how it is changing. How can we be sure that other descriptions of the system are not possible? This concern is answered by the following fundamental theorem whose application to real systems is most apparent in Chapters 4, 5, and 8.

Theorem A.2

If $y = f(x)$ and $y = g(x)$ are two functions with equal derivatives on an open interval—that is, if

$$f'(x) = g'(x) \quad \text{for all } x$$

then $\quad f(x) = g(x) + C \quad$ where C is constant on the interval

Geometrically, the hypothesis says that the graphs of $f(x)$ and $g(x)$ have the same "shape"; that is, their tangent lines have the same slope [since $f'(x) = g'(x)$] at each point. The conclusion says that the graph of one function can therefore be moved by the addition of the constant C to coincide with the graph of the other (see Figure A.2).

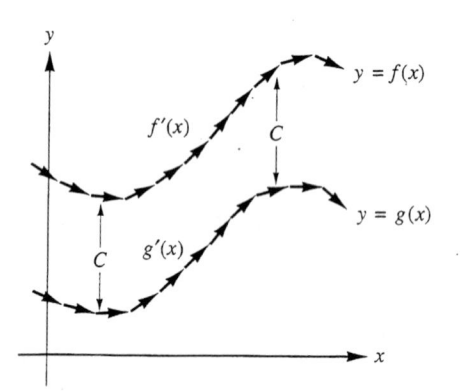

Figure A.2

The Mean Value Theorem

The proof of Theorem A.2 relies on the fundamental fact of differential calculus called the **mean value theorem**.

Consider Figure A.2. Notice that the slope of the line between the points P and Q is the number

$$\frac{f(b) - f(a)}{b - a}$$

644   Appendix   Further Development

Notice also that there are (two) tangent lines to the graph of $y = f(x)$ that appear to have this same slope. These are indicated at the points $x = z_1$ and $x = z_2$ in Figure A.3.

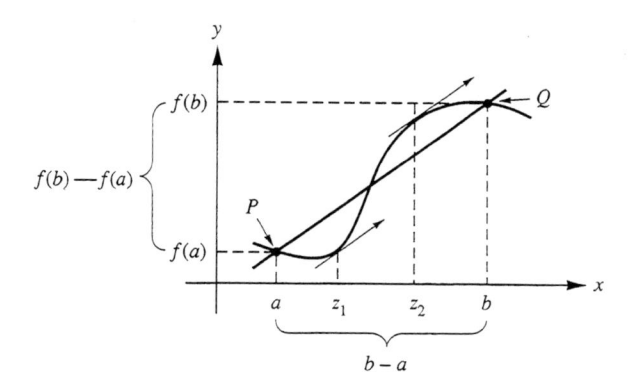

Figure A.3

The mean value theorem ensures that there will always be *at least one* such point. If we think of the graph of $y = f(x)$ as the silhouette of a hillside, the mean value theorem says that if we try to climb the hill from point P to point Q, we cannot do so without, at some point along the path, being forced to climb a slope *equal to the overall slope* from point P to point Q. This is in part why mountain roads or trails use a third dimension to "switchback" so that the driver or hiker need not climb at the full rate of the overall grade. Here is the mathematical statement of these ideas.

The Mean Value Theorem

If $y = f(x)$ is differentiable on the interval $a < x < b$ and continuous on the whole interval $[a, b]$, then there is some point z, with $a < z < b$, such that

$$\frac{f(b) - f(a)}{b - a} = f'(z)$$

Since f' measures the slope of tangents to $y = f(x)$, the point z represents a point above which the slope of the graph coincides with the overall slope from $(a, f(a))$ to $(b, f(b))$. The theorem does not say *which* point, but that, it turns out, is not important in its use. We will not prove the mean value theorem, but we will use it to prove Theorem A.2.

To prove Theorem A.2, we first define a new function

$$h(x) = f(x) - g(x)$$

Now, since the given condition in Theorem A.2 is that $f(x)$ and $g(x)$ have equal derivatives, we have that

$$h'(x) = f'(x) - g'(x) = 0 \quad \text{for all } x$$

A.2 Functions with Equal Derivatives and L'Hôpital's Rule

Let us use the mean value theorem to see that $h(x)$ must be constant for all x. Suppose we have two different values, say, a and b, of x. Then, by the mean value theorem,

$$\frac{h(b) - h(a)}{b - a} = h'(z)$$

for some number z. But $h'(z) = 0$. Multiplying both sides by $(b - a)$ yields

$$h(b) - h(a) = 0 \cdot (b - a) = 0$$

That is, $h(b) = h(a)$. Since a and b were *any two* values of x, all values of $h(x)$ must be the same. Therefore $h(x) = C$, a constant value. But this means that $C = f(x) - g(x)$ or $f(x) = g(x) + C$. This proves Theorem A.2.

L'Hôpital's Rule

We conclude with a second application of the mean value theorem that yields a powerful method for computing otherwise difficult limits. This method is named after the seventeenth-century French mathematician Guillaume L'Hôpital, who wrote the first calculus book. Recall that in a quotient limit problem

$$\lim_{x \to a} \frac{f(x)}{g(x)}$$

when both limits are zero, that is, when

$$\lim_{x \to a} f(x) = 0 \text{ and } \lim_{x \to a} g(x) = 0$$

we *cannot* compute the limit directly as

$$\frac{\lim_{x \to a} f(x)}{\lim_{x \to a} g(x)}$$

since this has the *indeterminate form* $\frac{0}{0}$. **L'Hôpital's rule** gives a way to compute such limits.

L'Hôpital's Rule

If $\lim_{x \to a} f(x) = \lim_{x \to a} g(x) = 0$ and if

$$\lim_{x \to a} \frac{f'(x)}{g'(x)} \quad \text{exists}$$

then

$$\lim_{x \to a} \frac{f(x)}{g(x)} = \lim_{x \to a} \frac{f'(x)}{g'(x)}$$

For example, the limit problem
$$\lim_{x \to 1} \frac{(x-1)^3}{x^2-1}$$
is of the form $\frac{0}{0}$. Therefore, by L'Hôpital's rule,
$$\lim_{x \to 1} \frac{(x-1)^3}{x^2-1} = \lim_{x \to 1} \frac{3(x-1)^2}{2x} = \frac{3 \cdot 0^2}{2} = 0$$
Thus L'Hôpital's rule helps us avoid the algebraic factorization usually needed in solving such a problem.

EXAMPLE 1

Compute
$$\lim_{x \to 2} \frac{x^5 - 32}{x^4 - 16}$$

Solution Since $\lim_{x \to 2}(x^5 - 32) = 0$ and $\lim_{x \to 2}(x^4 - 16) = 0$, we differentiate each function and use L'Hôpital's rule to write
$$\lim_{x \to 2} \frac{x^5 - 32}{x^4 - 16} = \lim_{x \to 2} \frac{5x^4}{4x^3} = \lim_{x \to 2} \frac{5}{4}x = \frac{5}{2} \blacksquare$$

A proof of L'Hôpital's rule requires a more general mean value theorem than the one we have stated, but we can use it to see why L'Hôpital's rule ought to be true. To compute $\lim_{x \to a} f(x)/g(x)$, imagine that x is a number b that is very close to a and that $f(a)$ and $g(a)$ are both zero. By the mean value theorem

$$\frac{f(b)}{g(b)} = \frac{f(b) - f(a)}{g(b) - g(a)} = \frac{\frac{f(b) - f(a)}{b - a}}{\frac{g(b) - g(a)}{b - a}} = \frac{f'(z_1)}{g'(z_2)}$$

for some z_1 and z_2 *between* a and b. But if a and b are very close to one another, so are z_1 and z_2. Thus $z_1 \simeq z_2$ and
$$\frac{f(b)}{g(b)} \simeq \frac{f'(z)}{g'(z)}$$
for some z between a and b. In the limit, then, with $b \to a$, we expect that
$$\lim_{x \to a} \frac{f(x)}{g(x)} = \lim_{z \to a} \frac{f'(z)}{g'(z)}$$
as L'Hôpital's rule guarantees.

L'Hôpital's rule also applies to limits "at ∞" in the same way.

EXAMPLE 2

Evaluate
$$\lim_{x \to \infty} \frac{x - 1}{x^2 + x}$$

Solution Here both limits $\lim_{x\to\infty}(x-1)$ and $\lim_{x\to\infty}(x^2+x)$ are "infinite." We can use L'Hôpital's rule as before and write, upon differentiation,

$$\lim_{x\to\infty}\frac{x-1}{x^2+x} = \lim_{x\to\infty}\frac{1}{2x+1} = 0 \quad \blacksquare$$

Remark. You can see that this problem is like the indeterminate form $\frac{0}{0}$ in that

$$\lim_{x\to\infty}\frac{x-1}{x^2+x} = \lim_{x\to\infty}\frac{\frac{1}{x^2+x}}{\frac{1}{x-1}}$$

has the form $\frac{0}{0}$. Hence it is not surprising that L'Hôpital's rule should apply in this case as well.

Here is the most general statement of L'Hôpital's rule.

If the number a is a real number, or $\pm\infty$, and if both limits

$$\lim_{x\to a} f(x) \quad \text{and} \quad \lim_{x\to a} g(x)$$

are *simultaneously* 0 or $\pm\infty$, then

$$\lim_{x\to a}\frac{f(x)}{g(x)} = \lim_{x\to a}\frac{f'(x)}{g'(x)}$$

when the limit on the right exists.

EXAMPLE 3

Show that an application of L'Hôpital's rule fails to give the correct answer to

$$\lim_{x\to 1}\frac{x-1}{x^2+1}$$

and explain why.

Solution Since $\lim_{x\to 1} x^2+1 = 2 \neq 0$ and $\lim_{x\to 1} x-1 = 0$, we have a correct answer of

$$\lim_{x\to 1}\frac{x-1}{x^2+1} = \frac{0}{2} = 0$$

But if instead we apply L'Hôpital's rule, we have

$$\lim_{x\to 1}\frac{x-1}{x^2+1} = \lim_{x\to 1}\frac{1}{2x} = \frac{1}{2}$$

This answer is incorrect; L'Hôpital's rule should not be used, because the hypothesis that *both* numerator and denominator *be zero* (or infinite) is not satisfied in this problem. $\blacksquare$

Exercises A.2

1. Consider the function $f(x) = x^2$ on the interval $a = 0$ to $b = 1$. Find a point z between 0 and 1 such that
$$\frac{f(1) - f(0)}{1 - 0} = f'(z)$$
Draw a graph illustrating this exercise.

2. Let $f(x) = 3x^2 - x^3 + 2$, $a = 0$, $b = 2$. Verify the mean value theorem for this function. How many points z exist? Draw a good graph illustrating this exercise.

3. Let $f(0) = 2$, $g(0) = 5$, and suppose that $g'(x) = f'(x)$ and that both equal x^2. Find a constant C such that $f(x) = g(x) + C$.

4. The mean value theorem is known to be logically equivalent to the truism that "between any two hills there is a valley and between any two valleys there is a hill." The mathematical version of this statement is known as **Rolle's theorem**, illustrated in Figure A.4 and stated as follows: Suppose $y = f(x)$ is a continuous function on the interval $a \leq x \leq b$, and that $f'(x)$ exists for all x, $a < x < b$. If $f(a) = f(b)$, then there must exist some point z between a and b such that $f'(z) = 0$, that is, where f has a horizontal tangent.

   a. Suppose that $f(a) = f(b)$. Apply the mean value theorem in such a case, and conclude that there is a z such that $f'(z) = 0$.

   b. Please refer to Figure A.3. Explain how, if you were to stand vertically on the line PQ, this figure, which explains the mean value theorem, would look like a version of Rolle's theorem.

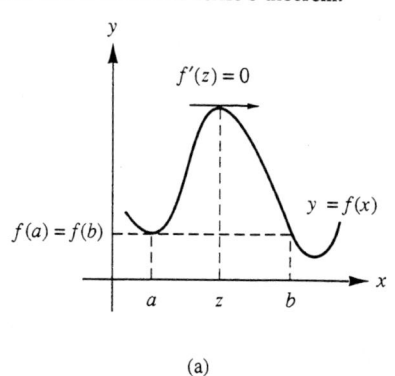

c. Consider the function $f(x) = x^2 - 2x + 1$. Notice that $f(0) = f(2)$. Find a point z between $a = 0$ and $b = 2$ such that $f'(z) = 0$. Draw a graph to illustrate your conclusion.

Compute the limits in Exercises 5–18.

5. $\lim_{x \to 3} \dfrac{x^2 - 9}{x^3 - 27}$

6. $\lim_{x \to 2} \dfrac{5x^2 - 20}{x - 2}$

7. $\lim_{x \to -1} \dfrac{x + 1}{x^5 + 1}$

8. $\lim_{x \to 1} \dfrac{x^5 - 2x + 1}{x^3 + x - 2}$

9. $\lim_{x \to 0} \dfrac{x\sqrt{x + 1}}{x(x - 1)}$

10. $\lim_{x \to \infty} \dfrac{x + 1}{x - 29}$

11. $\lim_{x \to -\infty} \dfrac{2x - 1}{x + 7}$

12. $\lim_{x \to -\infty} \dfrac{x^3 + x}{x^2 + 2x}$

13. $\lim_{x \to 1} \dfrac{x^3 - 1}{x + 1}$

14. $\lim_{x \to \infty} \dfrac{\sqrt{x + 1}}{\sqrt{x - 4}}$

15. $\lim_{x \to \infty} \dfrac{(x + 1)^2}{(x - 1)^3}$

16. $\lim_{x \to \infty} \dfrac{\frac{1}{x + 1}}{\frac{1}{\sqrt{x - 1}}}$

17. $\lim_{x \to 2} \dfrac{x + 2}{x^2 - 4}$

18. $\lim_{x \to -\infty} \dfrac{\sqrt{x^2 + 1}}{x}$

19. Show that the mean value theorem fails to be true for $f(x) = |x|$ for $a = -1$, $b = 2$.

20. Evaluate $\lim_{x \to \infty} [x - \sqrt{x(x + 1)}]$ by writing
$$x - \sqrt{x(x + 1)} = [x - \sqrt{x(x + 1)}]\left[\frac{x + \sqrt{x(x + 1)}}{x + \sqrt{x(x + 1)}}\right]$$
and simplifying.

(a)

(b)

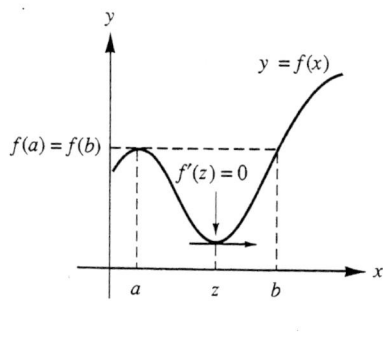

Figure A.4

A.3 Logarithmic Scaling and Best-Fit Polynomial Curves

The logarithmic nature of the response of life forms to stimuli in their environment, revealed via Weber's experiments as discussed in Section 5.3, is particularly found in the medical and life sciences. The need to represent such logarithmic relationships in the more familiar form of a linear relationship gives rise to the concept of a "logarithmic scaling" of the coordinate axis along which the variables of interest are measured. This, in turn, is closely related to how we apply the techniques of best-fit linear approximation by least squares, discussed in Section 6.4, to the problem of fitting appropriate polynomial curves to a given set of data.

Logarithmic Scaling

Consider a variable y that is an *exponential function* of x; that is, $y = a \cdot b^x$, where a and b are constant, as in Weber's model. If we apply the algebraic laws of logarithms to both sides of the equation, we obtain

$$\ln y = \ln a \cdot b^x = \ln a + \ln b^x$$

or

$$\ln y = \ln a + x \ln b$$

Let $A = \ln a$ and $B = \ln b$. Then whereas

$$y = a \cdot b^x$$

is an exponential function of x, $\ln y$ is a *linear function* of x:

$$\ln y = A + Bx$$

since A and B are constants. Now let

$$Y = \ln y \quad \text{and} \quad X = x$$

The last equation then becomes an ordinary linear equation

$$Y = BX + A$$

Its graph must therefore be a straight line in an XY-coordinate system.

How can we relate the original variables x and y to this straight line? Since $X = x$, we may simply regard the horizontal axis as the x-axis. Here is a new idea. Since $Y = \ln y$, we should replace the y-axis with a differently scaled axis representing the values of $\ln y$. That is, we should mark off the vertical axis with a logarithmic scale, as indicated in Figure A.5(a), derived from A.5(b).

Using this scaling, we may visualize the exponential relation $y = ab^x$ as linear relation. Let us see how in specific terms.

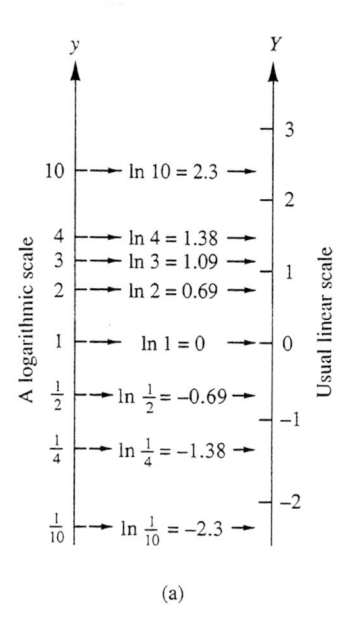

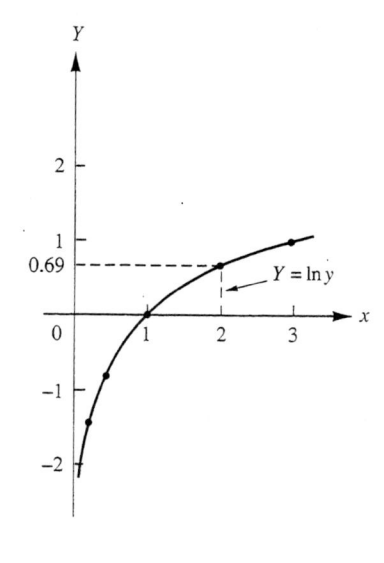

Figure A.5   (a)   (b)

EXAMPLE 1

Graph the equation

$$y = \frac{3^x}{18}$$

as a linear equation using logarithmic scaling.

Solution If we take logarithms of both sides, the equation becomes

$$\ln y = x \ln 3 - \ln 18$$

Since $\ln 3 \simeq 1.10$ and $\ln 18 \simeq 2.9$, this becomes

$$\ln y = 1.10x - 2.9$$

With $Y = \ln y$ and $X = x$ and $Y = 1.10X - 2.9$, we have the graph indicated in Figure A.6.

Let us compare this graph to values of the original equation $y = 3^x/18$. Notice that when $x = 2$, then $y = \frac{9}{18} = 0.5$; when $x = 4$, then $y = \frac{9}{2} = 4.5$. Now notice that these particular relationships are accurately indicated by the graph in Figure A.6 (page 652) of

$$Y = 1.10X - 2.9$$

when the y-axis is given a logarithmic scale. ∎

A.3 Logarithmic Scaling and Best-Fit Polynomial Curves

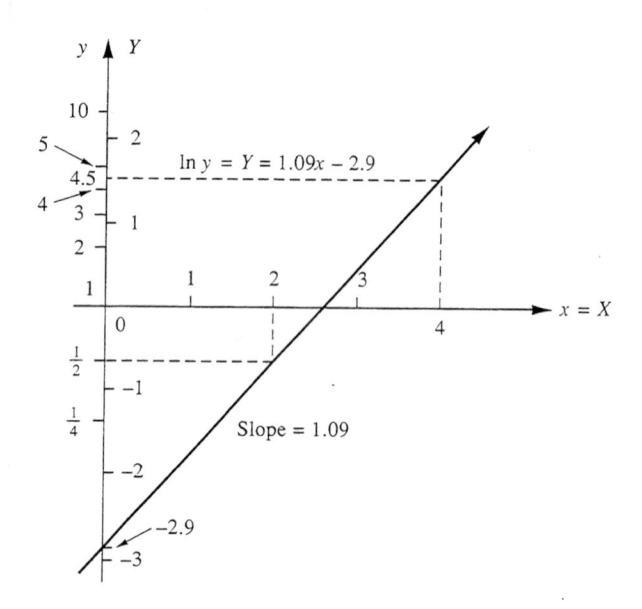

Figure A.6

Best-Fit Polynomial Curves

That a polynomial may be logarithmically transformed into a linear function allows us to adapt the best-fit techniques of Section 6.4 to the task of fitting a polynomial curve $y = Cx^k$ to a given set of data $(x_1, y_1), \ldots, (x_n, y_n)$. If we reason as before, the equation $y = Cx^k$ is equivalent to

$$\ln y = \ln(Cx^k) = \ln C + \ln x^k$$

or

$$\ln y = k \ln x + \ln C$$

If we now let $Y = \ln y$, $X = \ln x$, and $B = \ln C$, we obtain the *linear* equation

$$Y = kX + B \tag{1}$$

Suppose we are given data (x_p, y_p) and wish to determine the numbers C and k so as to obtain a best-fit polynomial curve $y = Cx^k$ to this data by the method of least squares of Section 6.4. If we simply replace the given data by the transformed data $(X_p, Y_p) = (\ln x_p, \ln y_p)$, we may determine k and B in Equation 1, as in Section 6.4. An inverse transform then determines C as well.

EXAMPLE 2

The table of values and graph in Figure A.7(a) show the average monthly temperature in degrees above freezing in Louisville, Kentucky (observed over the years 1940–1970), in the months January, March, and May, respectively. Model this rise in temperature by a polynomial curve, and predict what the April average is likely to be.

Month	Average Degrees above 32°
1 (January)	1
3 (March)	12
5 (May)	33

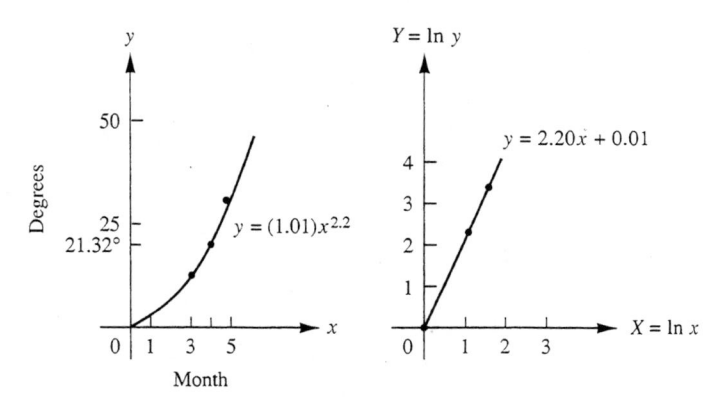

Figure A.7

(a) Original data

(b) Transformed data

Solution We apply a logarithmic transform to both sets of data, obtaining Table A.1. We now forget about the original (x, y) data and instead find the slope $A = k$ and intercept B that best fit the data pair $X = \ln x$, $Y = \ln y$ as in Section 6.4. Note how, in Figure A.7(b), the transformed data appear to fit a straight line fairly well.

Table A.1

x	y	$X = \ln x$	$Y = \ln y$	XY	X^2	
1	1	0	0	0	0	
3	12	1.1	2.5	2.75	1.21	
5	33	1.6	3.5	5.60	2.56	
		2.7	6.0	8.35	3.77	
		ΣX	ΣY	ΣXY	ΣX^2	$n = 3$

We have

$$A = \frac{3(8.35) - (2.7)(6)}{3(3.77) - (2.7)^2} = \frac{8.85}{4.02} \simeq 2.20$$

and

$$B = \frac{6(3.77) - (8.35)(2.7)}{4.02} = \frac{0.075}{4.02} \simeq 0.01$$

Consequently, $Y = 2.2X + 0.01$. We must now invert this relation to obtain the desired polynomial function $y = Cx^k$. By replacing Y by $\ln y$ and X by $\ln x$, we have

$$\ln y = (2.2)\ln x + 0.01$$

Therefore

$$y = e^{\ln y} = e^{(2.2)\ln x + 0.01} = e^{0.01} e^{\ln x^{2.2}}$$

or

$$y = (1.01)x^{2.2}$$

That is, the desired exponent is $k = 2.2$.

Consequently, the April (when $x = 4$) average temperature above freezing consistent with this data and a best-fit polynomial approximation is $y = 1.01(4)^{2.2} = 21.32°$.

The observed average temperature (1940–1970) is known to be 56°, and thus our result ($21.32 + 32 \approx 53°$) is within about 3° of the true value. ∎

How can the original data be related to the linear graph in Figure A.7? Recalling the initial discussion of this section, we see that the original data may be directly related to this linear graph by rescaling *both* horizontal and vertical axes with a logarithmic scale.

The methods presented in Section 6.4 extend to two other curves that arise naturally in applications. We will not deal with such data and functional forms in detail since the basic linear method, with only a suitable transformation of the data, is again used to treat these. But we do wish to point out the appropriate transforms in each case. Functions of the form

$$y = \frac{ax}{1 + bx} \tag{2}$$

often arise and well match the data that appear as in Figure A.8.

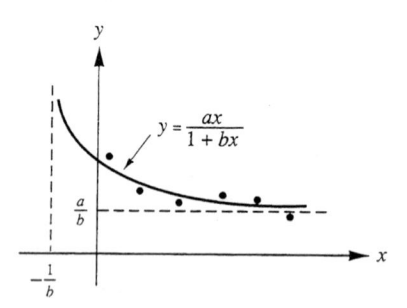

Figure A.8

In such a case we algebraically invert Equation 2 and obtain

$$\frac{1}{y} = \frac{1 + bx}{ax} = \left(\frac{1}{a}\right)\left(\frac{1}{x}\right) + \frac{b}{a}$$

If we now let $Y = 1/y$, $X = 1/x$, $A = 1/a$, and $B = b/a$, this equation takes on the linear form

$$Y = AX + B$$

Consequently, if linear regression is applied to the transformed data, $X = 1/x$ and $Y = 1/y$, a suitable curve $y = ax/(1 + bx)$ may be obtained.

A singularly useful functional form is the equation

$$Y = \frac{M}{1 + Ce^{kt}} \tag{3}$$

whose graph is the *logistic curve* (Figure A.9) modeling growth to a maximum capacity M.

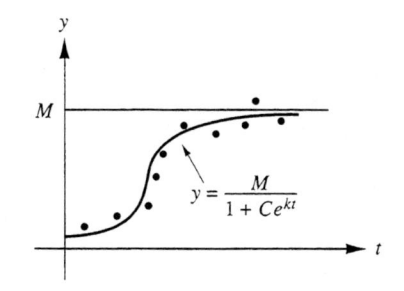

Figure A.9

In this case we rewrite Equation 3 as

$$1 + Ce^{kt} = \frac{M}{y} \quad \text{or} \quad Ce^{kt} = \frac{M-y}{y}$$

and then apply a logarithmic transformation to obtain

$$\ln Ce^{kt} = \ln\left(\frac{M-y}{y}\right)$$

or

$$Y = \ln\left(\frac{M-y}{y}\right) = \ln C + \ln e^{kt} = kt + \ln C$$

to obtain the linear equation $Y = kt + B$, where $B = \ln C$. Consequently, given data that suggest the logistic curve, we apply linear regression to the transformed data $X = t$ and $Y = \ln[(M-y)/y]$ to determine the best-fit logistic curve. Data for the case of world marathon records (Exercise 11) realistically fit a logistic curve.

Exercises A.3

Graph each of the exponential equations in Exercises 1–4 as a linear equation using logarithmic scaling. Evaluate each function at $x = 1$ and $x = 0$, and compare these values of y with the value indicated by the graph.

1. $y = (0.05)2^x$
2. $y = \frac{1}{10} 5^{-x}$
3. $y = 0.2\left(\frac{1}{3}\right)^x$
4. $y = (0.15)3^{2x}$

5. We see in Weber's model of human perception (Section 5.3) that one's perception of the strength of a stimulus is given by $P(S) = A \ln S + B$, where $A = 1/\ln(1 + r)$, $B = (-\ln S_0)/\ln(1 + r)$, S_0 is an initial stimulus, and $r = \Delta S/S$ is the constant (percentage) change needed in the level of a stimulus to perceive a higher level of stimulus. Studies have shown that the average person perceives a rising level of income if $r = 0.25$. Let $S_0 = \$10,000$ be a person's initial level of income. Graph this person's perception of rising affluence, as a straight line, using a logarithmically scaled horizontal S-axis representing rising income.

6. What does the graph in Exercise 5 tell you about one's sense of rising affluence at high versus low levels of income?

Find the best-fit polynomial curve $y = Cx^k$ for the data sets in Exercises 7–10, and graph both the data and this curve.

7. (0, 0), (1, 3), (2, 4), (3, 5)
8. (1, 1), (2, 15), (3, 30)
9. (1, 0.3), (4, 5), (5, 8), (6, 12)
10. (1, −3), (2, 15), (4, −50)

Challenge Problem

11. Actual data on world record marathon times by women runners during the years 1972–1984 are given in Table A.2. A graph of the data suggests that $M = 2.32$ hr (2 hr, 19 min) is a good guess at the human limit for women marathon runners and that the record time appears to follow a logistic curve $y = M/(1 + Ce^{kt})$. Determine values of C and k that best fit this data, assuming that $M = 2.32$. Predict when the record time will reach $0.99M$.

Table A.2

Year	Time (hr/min/sec)
1972	2:47:0
1975	2:38:5
1979	2:27:5
1980	2:26:0
1984	2:23:0

A.4 Hubbert's Curve: A Model for Resource Depletion

In 1956, M. King Hubbert, a geologist for Shell Oil Co., speaking to the Production Division of the American Petroleum Institute, predicted that between 1966 and 1971 U.S. domestic oil production rates would reach a peak and continue to decline thereafter. Hubbert's analysis was indirect, widely ignored, and correct. In 1970, the peak did occur, the "energy crisis" ensued, and the United States has been importing more than half its oil ever since.[*]

The remarkable thing about Hubbert's prediction was that in 1956 other respected sources had estimated total U.S. oil reserves to be between 150 and 200 billion barrels (bbl), and in 1956 only 52 bbl had been discovered. Let us briefly study Hubbert's remarkably simple analysis, which is based on the *shape* of the normal curve of Section 9.5.

Hubbert began by observing that, whatever the total amount of available oil it is finite. Furthermore, driven by consumption, the discovery of oil would inevitably speed up as more and more is demanded until it becomes harder and more expensive to find. The rate of discovery of new oil should therefore follow the shape of the curve indicated in Figure A.10.

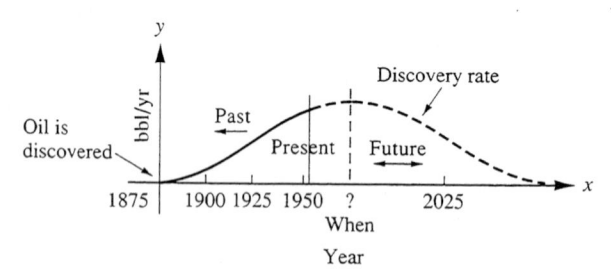

Figure A.10

The only real question was when the inevitable peak would occur. Here is how Hubbert demonstrated that the peak must occur sooner than anyone (in 1956) expected.

---
[*]See "Hubbert's Curve," *Country Journal* 7, no. 11 (1980): 56–61.

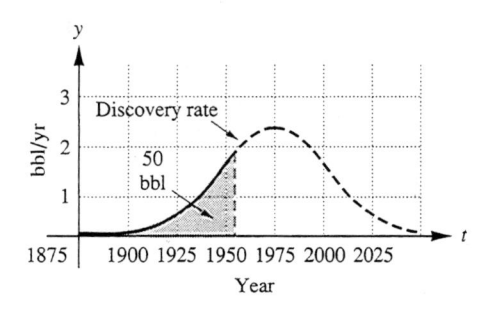

Figure A.11

Consider Figure A.11. Each square represents a rate of discovery of 1 bbl/yr, for 25 years; therefore the area of each square represents 25 billion barrels of oil. Assume that only 150 bbl of oil are available for discovery; this is equivalent to six squares. The area beneath the curve of the *rate* of discovery is of course the total amount of oil ever to be discovered. We merely have to arrange—as Hubbert did—the shape of the discovery curve so as to cover a total of six squares. This shape was of course known to Hubbert *up to* 1956, as indicated in Figure A.11. Notice then that the curve, when extended beyond 1956, is forced to be no higher than that of a curve that bounds six squares (or 150 bbl of oil!). The peak of the curve cannot be much beyond the period 1965–1970, and thus followed Hubbert's prediction, even though at the time of his prediction only 52 bbl, or only a third of the total oil reserves, had been found!

The production and consumption of a finite resource can often be reasonably assumed to follow the shape of a normal curve. For example, the rate of harvest of a ripening crop can be expected to rise slowly to a peak and decline thereafter. Let us see how calculus can be used to predict the time of a peak rate.

EXAMPLE 1

Suppose that a new product is introduced into a market, where it is estimated that a total of 150,000 customers will ultimately buy the product. The rate of new customers (in thousands per day) buying the product is assumed to have the shape of a normal curve given by

$$C(t) = \frac{A}{\sigma\sqrt{2\pi}} e^{-1/2[(t-\mu)/\sigma]^2}$$

where A is a scaling factor (see Figure A.12).

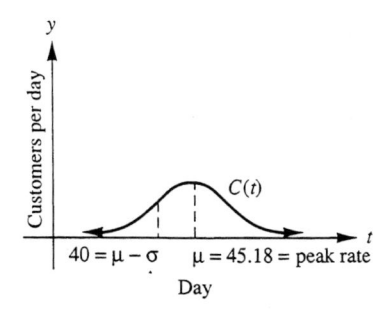

Figure A.12

A.4 Hubbert's Curve: A Model for Resource Depletion

The company marketing the product keeps a record of the values of $C(t)$ (the rate of arrival of new customers) on successive days after the product is introduced. They notice that the number of new customers first begins to rise at a slower rate 40 days after the product is introduced and that this rate is $C(40) = 7{,}000$ customers/day. Use an analysis of the shape of a normal curve to

a. Determine A.
b. Predict when the peak rate of new customers attracted to the product will be reached.
c. Determine how many total customers bought the product by day $t = 40$.
d. Predict when, after passing the peak rate, only 1,000 new customers per day will be attracted to this product.

Solution

a. Since $y = C(t)$ is the rate of new customers for the product, then

$$150{,}000 = \int_{-\infty}^{\infty} C(t)\, dt$$

is the total number of possible customers. Therefore

$$150{,}000 = \frac{A}{\sigma\sqrt{2\pi}} \int_{-\infty}^{\infty} e^{-1/2[(t-\mu)/\sigma]^2}\, dt = A \cdot 1 = A$$

since the area under the normal curve is one unit. Thus

$$C(t) = \frac{150{,}000}{\sigma\sqrt{2\pi}} e^{-1/2[(t-\mu)/\sigma]^2}$$

b. We wish to determine the location of μ, the day on which the peak rate occurs.

We are given that on day $t = 40$ the rate $C(t)$ begins to rise less rapidly, indicating that $t = 40$ is the location of $\mu - \sigma$, the left inflection point on the normal curve (see Section 9.5 and Exercises 21 and 22 in that section). Therefore $\mu - \sigma = 40$.

Now, at the same time, $C(40) = 7{,}000$, and

$$C(\mu - \sigma) = \frac{150{,}000}{\sigma\sqrt{2\pi}} e^{-1/2[((\mu-\sigma)-\mu)/\sigma]^2} = \frac{150{,}000}{\sigma\sqrt{2\pi}} e^{-1/2}$$

Therefore, $(150{,}000/\sigma\sqrt{2\pi})e^{-1/2} = 7{,}000$ since $C(\mu - \sigma) = C(40)$. Consequently, $\sigma = (150{,}000/7{,}000\sqrt{2\pi})e^{-1/2} = 5.18$. Therefore the peak rate occurs at

$$\mu = 40 + \sigma = 45.18 \text{ days}$$

after introduction of the product.

c. From Section 9.5, we know that $\frac{1}{2}(1 - 0.6826) = 0.1587$ of the area of the normal curve occurs to the left of the left inflection point $\mu - \sigma$. Therefore, on day $t = 40$, where this inflection occurs, $0.1587(150{,}000) = 23{,}805$ customers have been attracted to the product, a relatively small proportion of the total market.

d. Since $C(t) = \dfrac{150{,}000}{5.18\sqrt{2\pi}} e^{-1/2[(t-45.18)/5.18]^2}$ at any time t

the rate will be 1,000 customers when $C(t) = 1{,}000$ or when

$$e^{-1/2[(t-45.18)/5.18]^2} = \dfrac{5.18\sqrt{2\pi}}{150{,}000} \cdot 1{,}000$$

$$-1/2\left(\dfrac{t-45.18}{5.18}\right)^2 = \ln 0.0865 = -2.447$$

$$\left(\dfrac{t-45.18}{5.18}\right) = \pm\sqrt{4.89} = \pm 2.21$$

or when $\quad t = 45.18 + 11.45 = 56.63$ days ∎

Exercises A.4

1. As noted, Hubbert predicted that the peak rate of oil discovery would be reached between 1966 and 1971. The difference in dates was due to a difference of opinion among other experts as to the total amount of oil available, with estimates varying from 150 bbl up to 200 bbl. With a sheet of graph paper having a 10 by 10 mesh per square unit, let one unit on the vertical axis represent 1 bbl/year and one unit on the horizontal axis represent 25 years. Use the data indicated in Figure A.11 (up to 1950) to sketch the shape of a normal curve covering an area equivalent to

   a. 150 bbl **b.** 200 bbl

   c. How far apart do the peaks on the normal curve occur in each case? What does this suggest?

2. While Hubbert's predictions were largely ignored in their own time, a few individuals and institutions, especially in government, did pay some attention. These efforts, however, were directed at proving Hubbert wrong. Since his analysis was correct, his critics attempted to discredit the assumptions on which his conclusions were based. Since it was difficult to refute the normal curve hypothesis, reports were issued increasing the "known reserves" of oil yet to be discovered. Indeed these reserves doubled and tripled over the next few years, at least on paper! By 1975, these estimates had returned to the earlier figures between 150 and 200 bbl. As in Exercise 1, use a sheet of graph paper to answer these questions.

   a. If known reserves were in fact double (to 400 bbl) those that Hubbert used (based on the existing studies, not his own), when would the discovery rate peak?

   b. What if they were tripled?

   c. What do (a) and (b) suggest?

3. The harvesting of certain crops such as grapes or cherries tends to follow the pattern of a normal curve, with a small percentage of fruit ripening at the beginning and end of the harvest and most ripening at midseason.

   A vineyard expects to harvest 50 tons of grapes during late September and October. Using a daily tally of the rate of harvest, the owner notices that the rate first begins to rise more slowly (an inflection point) on October 10, when 2.5 tons are harvested. Assuming that the harvest follows the shape of a normal curve, determine

   a. When the peak harvest rate will be reached.

   b. The total amount that should have been harvested by October 10.

4. Answer parts (a) and (b) of Exercise 3 if the vineyard underestimated the harvest by 15%.

A.5 The Newton-Raphson Algorithm

A repeated theme of calculus is this: Approximate the complex by something simpler. The Newton-Raphson method of root approximation applies this theme to the problem of finding the roots of an equation like $x + 2 - x^3 = 0$. Here we wish to find a particular value $\bar{x}$ of x that makes this a true equation. The **Newton-Raphson algorithm** tells us how to find such a *root* $\bar{x}$ by a sequence of approximate roots x_n. Unlike the sequences we have seen in Section 10.1, the sequence of approximate roots x_n will not have an apparent pattern in terms of the counter n.

Let $f(x) = x + 2 - x^3$. Notice that $f(1) = 2 > 0$ and $f(2) = 2 + 2 - 8 = -4 < 0$. Since f is continuous, this means that somewhere between $x = 1$ and $x = 2$ there must be a value $\bar{x}$ such that $f(\bar{x}) = 0$, as indicated in Figure A.13. The Newton-Raphson method of root approximation is an *algorithm* that allows us to construct a sequence $x_1, x_2, x_3, \ldots$ of ever better approximations to the unknown root $\bar{x}$ so that finally $\bar{x} = \lim_{n \to \infty} x_n$.

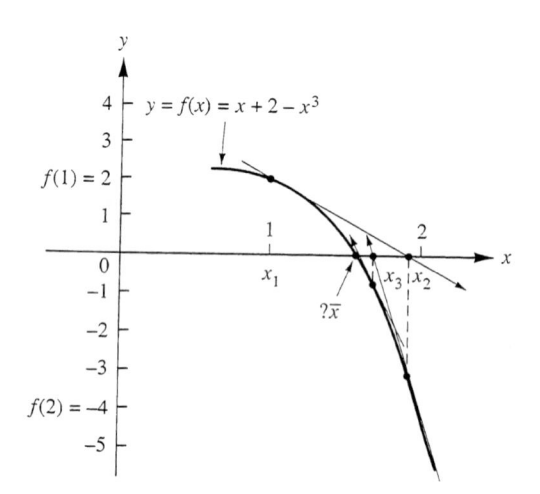

Figure A.13

The idea of the method is illustrated in Figure A.13. We first find the tangent line to the graph of f above $x_1 = 1$, where we have a known value: $f(x_1) = f(1) = 2$. We next find where this tangent has a root (at x_2). We then find second tangent, at $(x_2, f(x_2))$, and then where this tangent has a root, at x_3 closer to the unknown root $\bar{x}$ of the function f. A third such tangent yields better approximation (x_4), as Figure A.13 suggests. The sequence of approximations tends to converge very quickly to the root $\bar{x}$ (when it converges at all) and in most cases only a few approximations are actually needed to obtain good estimate of $\bar{x}$.

Let us see how to put this concept into computational form. Given an nth approximation x_n to the unknown root $\bar{x}$, we wish to find a formula for the next (better) approximation x_{n+1}. In Figure A.14, we see that x_{n+1} can be taken t

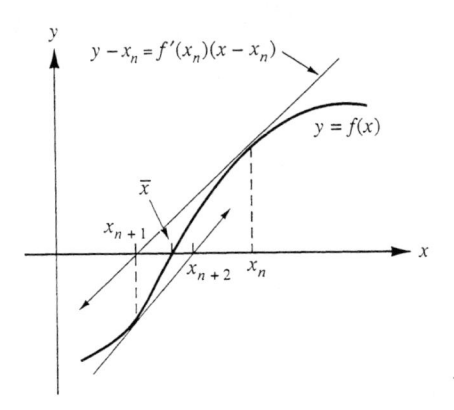

Figure A.14

be the root (x-intercept) of the equation of the tangent line to the graph of f at the point $(x_n, f(x_n))$. This equation is given in the *point–slope* form

$$y - f(x_n) = f'(x_n)(x - x_n)$$

and has a root when $y = 0$. Substituting $y = 0$ gives

$$0 - f(x_n) = f'(x_n)(x - x_n) \quad \text{or} \quad -\frac{f(x_n)}{f'(x_n)} = x - x_n$$

or

$$x = x_n - \frac{f(x_n)}{f'(x_n)} \quad \text{if } f'(x_n) \neq 0$$

Therefore we take

$$x_{n+1} = x_n - \frac{f(x_n)}{f'(x_n)}$$

to be the next best approximation to the root $\bar{x}$. We of course can only do so if $f'(x_n) \neq 0$. [If $f'(x_n) = 0$, the tangent line is horizontal and so has no intercept; thus this method of finding x_{n+1} cannot be applied.] In summary:

The Newton-Raphson Algorithm

If x_n is an approximation to a root x of $f(x) = 0$, then a (usually) better approximation is given by

$$x_{n+1} = x_n - \frac{f(x_n)}{f'(x_n)} \tag{1}$$

if $f'(x_n) \neq 0$. If $\bar{x} = \lim_{n \to \infty} x_n$ exists, then $f(\bar{x}) = 0$ [provided that $f'(\bar{x}) \neq 0$ and f' is continuous].

EXAMPLE 1

Use the Newton-Raphson algorithm to find a root of $f(x) = x + 2 - x^3$ between the two values $f(1) = 2 > 0$ and $f(2) = -4 < 0$.

Solution First, $f'(x) = 1 - 3x^2$. The Newton-Raphson algorithm gives us a formula for moving from one root approximation x_n to a next, hopefully better, approximation x_{n+1}. We have

$$x_{n+1} = x_n - \frac{f(x_n)}{f'(x_n)} = x_n - \frac{x_n + 2 - x_n^3}{1 - 3x_n^2}$$

or

$$x_{n+1} = \frac{x_n - 3x_n^3 - x_n - 2 + x_n^3}{1 - 3x_n^2}$$

$$= \frac{2x_n^3 + 2}{3x_n^2 - 1} = 2\left(\frac{x_n^3 + 1}{3x_n^2 - 1}\right) \qquad (2)$$

Let us take $x_1 = 1$ as a first approximation. The next approximation is

$$x_2 = 2\left(\frac{x_1^3 + 1}{3x_1^2 - 1}\right) = 2\left(\frac{1^3 + 1}{3 \cdot 1^2 - 1}\right) = 2$$

Using this value $x_2 = 2$ again in Equation 2 gives

$$x_3 = 2\left(\frac{2^3 + 1}{3 \cdot 2^2 - 1}\right) = \frac{18}{11} = 1.6363636$$

Applying Equation 2 to $x_3 = 1.6363636$ gives

$$x_4 = 2\left(\frac{\left(\frac{18}{11}\right)^3 + 1}{3\left(\frac{18}{11}\right)^2 - 1}\right) = 1.530392$$

so

$$x_5 = 2\left(\frac{(1.530392)^3 + 1}{3(1.530392)^2 - 1}\right) = 1.5214415$$

$$x_6 = 2\left(\frac{(1.5214415)^3 + 1}{3(1.5214415)^2 - 1}\right) = 1.5213797$$

and $x_7 = 1.5213797$ again!

A good approximation to the root between $x = 1$ and $x = 2$ is $\bar{x} = 1.5213797$. Note that after only six approximations we have three-digit accuracy (compare x_5 and x_6) and then that $x_7 = x_6$ (so far as our calculator is concerned), bringing us to six-digit accuracy in seven steps.

Let us, in this example, check $\bar{x}$ as a root. We have $f(\bar{x}) = f(1.5213797) = 1.5213797 + 2 - (1.5213797)^3 = -0.000000027 \approx 0$, a very good approximation. ∎

The Newton-Raphson algorithm can also be used to solve nonalgebraic equations.

EXAMPLE 2

Find a root of the equation $e^x = x + 2$.

Solution Let $f(x) = e^x - x - 2$. Notice that $f(0) = e^0 - 0 - 2 = -1 < 0$ and $f(2) = e^2 - 2 - 2 > 2^2 - 4 = 0$. Since f is continuous, there must be a root $\bar{x}$ between $x = 0$ and $x = 2$.

Now $f'(x) = e^x - 1$. Notice that $f'(0) = 0$ so we cannot begin with $x_1 = 0$ as a first approximation. Let us use, say, $x_1 = 1$ as a first approximation to $\bar{x}$. We first find that, in general,

$$x_{n+1} = x_n - \frac{f(x_n)}{f'(x_n)} = x_n - \frac{e^{x_n} - x_n - 2}{e^{x_n} - 1}$$

$$= \frac{x_n e^{x_n} - x_n - e^{x_n} + x_n + 2}{e^{x_n} - 1}$$

or
$$x_{n+1} = \frac{e^{x_n}(x_n - 1) + 2}{e^{x_n} - 1} \tag{3}$$

Using a calculator and Equation 3, we obtain, beginning with $x_1 = 1$,

$$x_2 = \frac{e^1(1 - 1) + 2}{e^1 - 1} = 1.1639534$$

$$x_3 = \frac{e^{1.1639534}(1.1639534 - 1) + 2}{e^{1.1464211} - 1} = \frac{2.5250721}{2.2025694} = 1.1464211$$

$$x_4 = \frac{e^{1.1464211}(1.1464211 - 1) + 2}{e^{1.1464211} - 1} = \frac{2.4607741}{2.1469104} = 1.1461932$$

We thus have three-digit accuracy within three computations. Thus $\bar{x} \simeq 1.1462$. Indeed

$$f(1.1462) = 0.0000145 \simeq 0 \quad \blacksquare$$

Remarks.
1. Accuracy of the approximation is of course limited by our means of calculation. Using a calculator with an eight-digit limit on accuracy works very well.
2. Exercise 22 completes the discussion of this example.

There is no guarantee that, in general, the Newton-Raphson algorithm will converge to a root $\bar{x}$. Figure A.15 illustrates how the algorithm may cycle rather than converge. The figure shows that $x_1 = x_3$ and, consequently, $x_2 = x_4$. For the same reason, $x_5 = x_3$. Therefore $\lim_{n\to\infty} x_n$ cannot exist or equal $\bar{x}$.

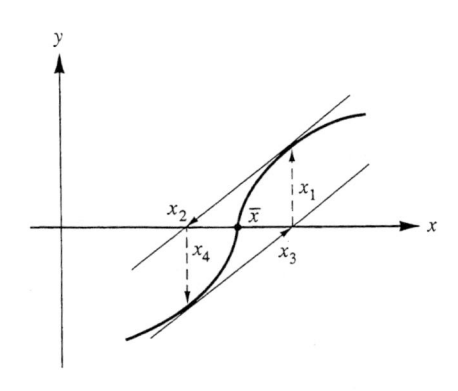

Figure A.15

Exercises A.5

In Exercises 1–6, use the Newton-Raphson algorithm to compute x_5, given x_1 and $f(x)$.

1. $f(x) = x^2 - 9$, $x_1 = 2$
2. $f(x) = x^2 - 4$, $x_1 = 1$
3. $f(x) = x - e^{-x}$, $x_1 = 0$
4. $f(x) = x + \ln x$, $x_1 = \frac{1}{2}$
5. $f(x) = x^3 + 4x + 10$, $x_1 = -1$
6. $f(x) = \sqrt{x} - 2$, $x_1 = 1$

Find a root with three-digit accuracy, and sketch the graph between the indicated values of x for each function given in Exercises 7–10.

7. $f(x) = x^2 + 2x - 7$, where $f(1) = -4$ and $f(2) = 1$
8. $g(x) = x^3 + x - 1$, where $g(0) = -1$ and $g(1) = 1$
9. $f(x) = x^2 - e^{-x^2}$, where $f(0) = -1$ and $f(1) = 0.632$
10. $h(x) = x^4 - 4x^3 - 8x^2 + 4$, where $h(0) = 4$ and $h(1) = -7$

An nth-degree polynomial has n roots. Some may be complex numbers, others may be repeating roots. Find all roots to three-digit accuracy for the polynomials in Exercises 11–14 (none of which has complex roots), and sketch the graph.

11. $2x^3 - 4x + 1 = 0$ 12. $x^2 + 2x - 1 = 0$
13. $4x^3 - 4x - 1 = 0$
14. $x^4 - 4x^3 + 4x^2 - \frac{1}{2} = 0$
15. Find two approximate solutions to $e^{-x} = x^2$.
16. Find two approximate solutions to $x \ln x = -\frac{1}{4}$.
17. Find x_{n+k} in the Newton-Raphson algorithm when it happens that x_n is exactly a root of $f(x) = 0$, and $f'(x_n) \neq 0$.
18. Apply the Newton-Raphson algorithm to $f(x) = mx + b$, $b \neq -m$, with $x_1 = 1$.

Challenge Problems

19. What happens when you apply the Newton-Raphson algorithm to $f(x) = x^3 - x + 1$, where $f(1) = 1 > 0$, $f(-2) = -5$, and $x_1 = 1$ is chosen as a first approximation? What if $x_1 = 0$ is chosen as a first approximation? $x_1 = -1$?

20. Apply the algorithm to $f(x) = x^2 - a$, and show that $x_{n+1} = x_n/2 + a/2x_n$. Let $b = \lim_{n \to \infty} x_n$. Show that $b = \sqrt{a}$.

21. Show that if $\lim_{n \to \infty} x_n$ exists and equals, say, a, then $f(a) = 0$ if $f'(a) \neq 0$ and f' is continuous. [Hint: Note that $\lim_{n \to \infty} x_{n+1} = \lim_{n \to \infty} x_n = a$, and apply this to the Newton-Raphson algorithm.]

22. Show that $f(x) = e^x - x - 2$ has only one positive root by showing that f is a strictly increasing function for $x > 0$. Find a second, negative root of $e^x = x + 2$.

23. This application of the Newton-Raphson algorithm is related to Exercise 18 of Section 7.2. An environmental specialist is trying to model the dispersal of particulate matter in a circular urban area of radius 2 mi centered about a source that emits 500 tons of pollutants over the course of a year. The specialist hypothesizes two possible models for the density of matter falling to earth x mi from the pollutant source. These are

$$f(x) = 250 e^{-kx^2} \text{ tons/mi}^2 \quad \text{in model I}$$

and $\quad g(x) = \dfrac{250}{kx^2 + 1}$ tons/mi^2 $\quad$ in model II

According to the methods of Section 7.2, this mean that in model I the unknown dispersal constant k must satisfy the equation

$$\int_0^2 2\pi x (250 e^{-kx^2}) \, dx = 500$$

while in model II

$$\int_0^2 2\pi x \left(\frac{250}{kx^2 + 1} \right) dx = 500$$

a. Solve for the dispersal constant k in each model.
b. Graph the density functions f and g.
c. 1.25 mi from the source, soil measurements indicate an actual dispersal density of 30 tons/mi^2. Which model is most consistent with this observation?
d. Suppose instead that soil samples from a 150,000 square-foot children's playground 1 mi from the pollutant source show that there are approximately 600 lb of pollutants on the playground. Which model is more consistent with these data?

664 Appendix Further Development

Answers to Odd-Numbered Exercises

CHAPTER 0

Exercises 0.1

1. $x \to (20 - 2x)(10 - 2x)x$; 144, 192, 96, 174.2 in.^{3}
3. 1 week
5. a. 74.28 min; 145.71 min b. 16.5 min; 22.18 min; 37.33 min; 204 min c. It would be necessary to divide by 0, a mathematical impossibility.
7. a. $\frac{3}{2}$ b. $\frac{60 - 2t}{40 - t}$ c. $\frac{10}{7}, \frac{6}{5}, \frac{2}{3}$
9. Let x = distance from bottom of ladder to wall
    y = distance from top of ladder to ground
    a. $y = \sqrt{25 - x^2}$ b. 4 ft
11. Let x be the position between points A and C; V = vertical speed.
$$V = \begin{cases} 0 \text{ ft/sec, if } x \text{ is between } A \text{ and } B \\ 20 \text{ ft/sec at } x = B \text{ (a guess!)} \\ \frac{6}{5} \text{ ft/sec, if } x \text{ is between } B \text{ and } C \end{cases}$$
13. 13.89 sec; 14.27 sec; 15.65 sec; no; yes

Exercises 0.2

1. $3, 1, -1, 5, 2\sqrt{2} + 1, 2, 2a + 1, 2a + 3, 2a + 2h + 1$
3. $0, -1, 0, 3, 1, -\frac{3}{4}, a^2 - 1, a^2 + 2a, (a + h)^2 - 1$
5. $-1, -\frac{1}{3}, 0, -3, -(5 + 4\sqrt{2})/7, -\frac{3}{5}, (a + 1)/(a - 3), (a + 2)/(a - 2), (a + h + 1)/(a + h - 3)$
7. $4, 1, 0, 9, (\sqrt{2} + 1)^2, \frac{9}{4}, (a + 1)^2, (a + 2)^2, (a + h + 1)^2$
9. $(-\infty, +\infty)$
11. $w \neq -1$
13. $t \neq 0, -1, 2$
15. $0 = A(0), 24 = A(1), 46 = A(2), 66 = A(3), 84 = A(4)$ or many other values
17. $x = -1$
19. $z = \frac{5}{3}$
21. $s = (-3 \pm \sqrt{17})/2$
23. Not in range
25. $0, 1, \frac{1}{2}, 1, \sqrt{2} - 1, 3, a$
27. 4; 20; not possible
29. a. $f(t) = (120 - 4t)/(40 - 2t)$ b. $f(0) = 3; f(10) = 4; f(15) = 6$; the relative difficulty of finding 2 washers initially; after 10 min; after 15 min c. $0 \leq t < 20$; after 20 minutes no washers would be left and no ratio is possible
31. a. The number of bicycles produced by the end of the 2nd day of the year b. The number of bicycles produced by the end of the 10th day of the year c. The number of bicycles produced by the end of the ath day of the year d. The number of bicycles produced by the end of the $(5 + h)$th day of the year e. The number of bicycles produced from the end of the 29th day to the end of business h days later f. The number of bicycles produced by the end of the 30th day of the year g. The number of bicycles produced by the end of the 1st year h. The number of bicycles produced by the end of the $(a + 3)$rd day of the year i. The number of bicycles produced on the 30th day j. The average number of bicycles produced from the end of the 29th day to the end of the 34th day: the average of production on days 30 through 34
33. a. $h(0) = 100$ ft; $h(1) = 84$ ft; $h(2) = 36$ ft b. $t = 2.5$ sec c. 2 sec after beginning its fall

Exercises 0.3

1. $y = -2x + 3$

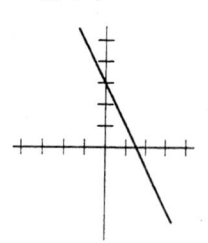

3. $y = -x$

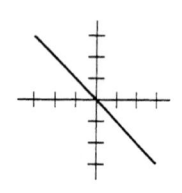

5. $y = 1$

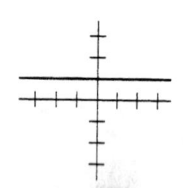

7. $y = \frac{1}{2}(x - 1)$

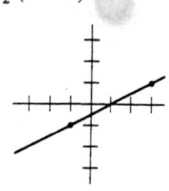

9. $m = 3, (0, -3)$

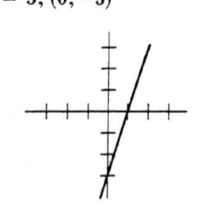

11. $m = -\frac{1}{2}, (0,2)$

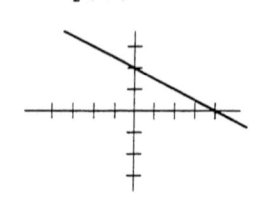

13. $m = 0, (0,2)$

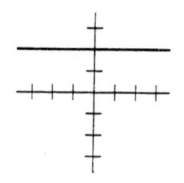

15. $m = 1, (0,-3)$

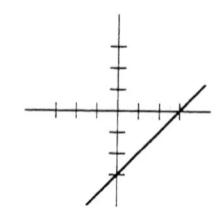

17. $g(x) = -3$

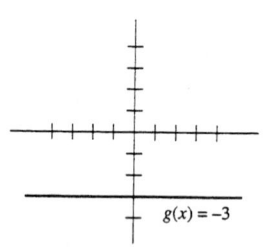

19. $\frac{1}{10}, \frac{1}{2}, \frac{37}{40}, \frac{39}{40}; t = 1.8$ hr. It takes an ever-longer amount of time to learn the remaining details.

21. $C(x) = 22x + 30{,}000$; 22 is the slope and 30,000 is the y-intercept

23. $A = 0.08P$; 0.08

25. a. $H(x) = 0.05(80 - x)$
    b. $H(x) = 0.05(80 - x)$
    c. $H(10) = 3.5, H(25) = 2.75, H(40) = 2$. Why pay more to be overheated?

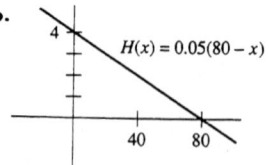

Exercises 0.4

1. $f(x) = 2x + 1$

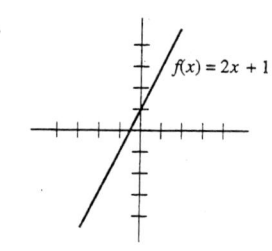

3. $f(x) = -|x|$

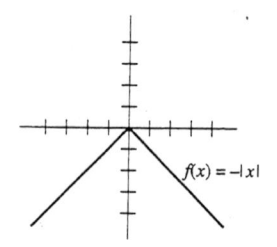

5. $h(x) = |x - 2|$

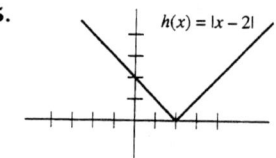

7. $w(x) = \frac{1}{x^2}$

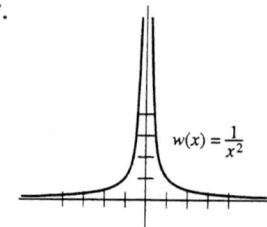

9. $s(x) = \sqrt{x}$

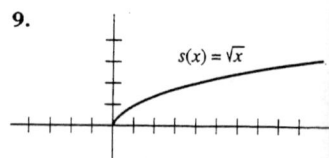

11. $y = 3^x$

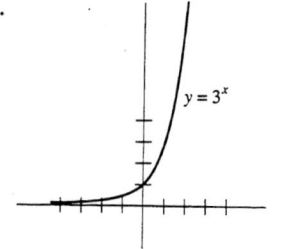

13.

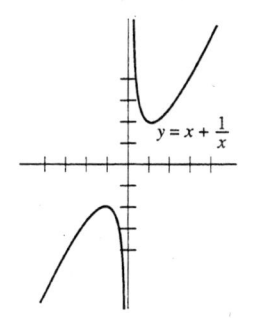

[Graph: $y = x + \frac{1}{x}$]

15.

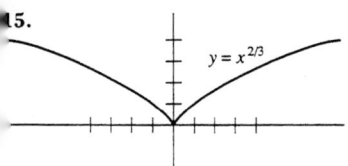

[Graph: $y = x^{2/3}$]

17.

[Graph: $y = x(x-1)(x+2)$]

19.

$f(x) = \begin{cases} 1, 0 \le x \le 2 \\ x - 1, 2 < x \le 3 \\ (x-3)^2 + 2, 3 < x \le 4 \end{cases}$

21.

[Graph: $V(x) = (10 - 2x)(10 - 2x)x$]

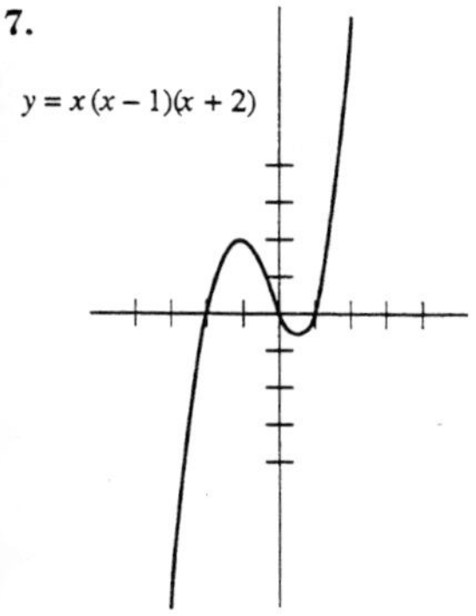

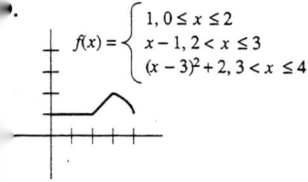

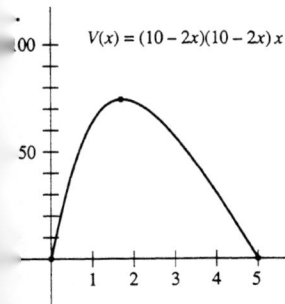

near $x = 2$. (The precise answer is $x = \frac{5}{3}$.)

23.

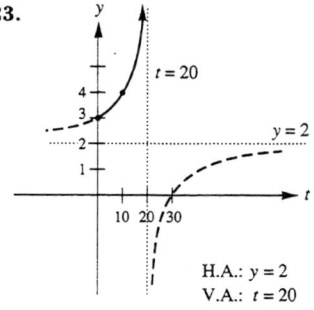

[Graph: $t = 20$, $y = 2$, H.A.: $y = 2$, V.A.: $t = 20$]

25.

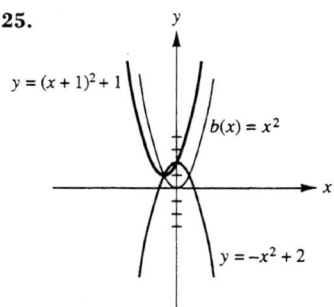

[Graph: $y = (x+1)^2 + 1$, $b(x) = x^2$, $y = -x^2 + 2$]

27.

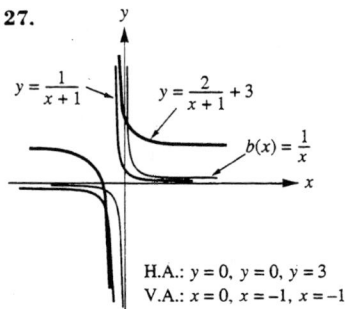

[Graph: $y = \frac{1}{x+1}$, $y = \frac{2}{x+1} + 3$, $b(x) = \frac{1}{x}$, H.A.: $y = 0, y = 0, y = 3$, V.A.: $x = 0, x = -1, x = -1$]

29.

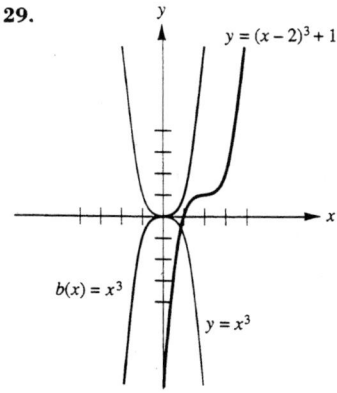

[Graph: $y = (x-2)^3 + 1$, $b(x) = x^3$, $y = x^3$]

31.

[Graph: $b(x) = 2^x$, $y = 2^{x-1}$, $y = 2^{x+1} - 3$]

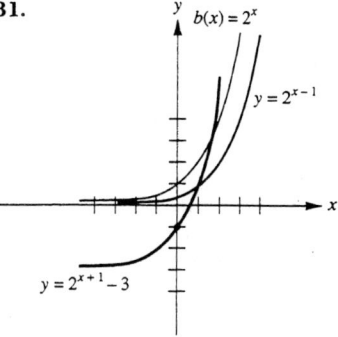

33. $y = (x - 2)^2 + 1$
35. $x = 2, -2$
37. $x = (-3 \pm \sqrt{13})/2$
39. $x = \pm \sqrt{5}/5$
41. $x = 0$
43. 2.665, 2.0665, 1.0083, 0.5035, 0.999993
45. Some values of x and $f(x)$ on the interval $[0, 2]$

x	$f(x)$
0.00	1.00
0.11	1.06
0.74	1.12
1.05	1.08
1.26	1.07
1.79	1.23
2.00	1.42

Exercises 0.5

1. a. $3 - 2x^2$ b. $-4x - 4x^2$
   c. -5 d. $3 - 2(a + 1)^2$
3. a. $x - 1$ b. $\sqrt{x^2 - 1}$ c. 1
   d. a
5. a. $\frac{1+x}{1-x}$ b. $\frac{x-1}{x+1}$ c. -3
   d. $\frac{2+a}{-a}$
7. $f(z) = 2z^{13}; g(x) = x + 1$; no
9. $f(x) = (x^2 + 1)^{1/3}; g(x) = x - 1$; no
11. $f(x) = 2^{x^2}; g(x) = x - 1$; no
13. a. $\frac{1+x^2}{x}$ b. $\frac{1}{x^2}$
15. a. $\frac{1}{1+x}$ b. $\frac{x}{1-x}$
17. a. $\frac{x}{x^2 + 3x + 2}$ b. $-\frac{x+1}{x+2}$

Answers to Odd-Numbered Exercises   667

19. $3{,}000; 5{,}500; \dfrac{s + 10{,}000}{20};$
$\dfrac{s + h + 10{,}000}{20}; B(g(w))$ is the number of bass for w spiders; $g(B(x))$ has no meaning

21. $6; 8; \tfrac{1}{2}z; \tfrac{1}{2}(z + 4);$ the number of points won when attempting t deep backhand returns; $p(w(x))$ has no meaning

23. a. 900 b. 48.2 c. $\dfrac{x}{100} + 3 + \dfrac{900}{x} = \dfrac{C(x)}{x}$

25. a. 472 b. $R(x) = -(x^2/50) + 24x$ c. $P(x) = 21x - (3x^2/100) - 900$ d. Approx. 46 e. $A_C(x) = (x/100) + 3 + (900/x)$, $= A_R(x) - (x/50) + 24$; 46 and 654 items

Exercises 0.6

1. a. The number of students who have caught the flu by noon of the 5th day b. The number of students who have caught the flu from the end of the 2nd day through the end of the 4th day c. The average number of students who caught the flu from the end of day 4 through the end of day 6 d. The average number of students (per day) who have caught the flu by the end of the 5th day e. The number of students who caught the flu the first 20 minutes of the 5th day (20 min = $\tfrac{1}{24} \cdot 8$ hr) f. The average rate of change per day in the number of students catching the flu during the 1st h minutes of the 5th day g. The average rate of change per day in the number of students catching the flu during the last h minutes of the 4th day h. The number of students who caught the flu between the end of day t and the end of day $t + h$

3. a. 5 b. 1 c. 4 and 6; 3 and 7; 2 and 8

5. a. $f(6) - f(5)$ b. $\dfrac{f(6) - f(0)}{6}$
   c. $\dfrac{f(6) - f(4)}{2}$ d. $\dfrac{f(t + 1) - f(t)}{1}$
   e. $\dfrac{f(4.5) - f(4)}{0.5}$

7. a. $\dfrac{f(3) + f(4) + f(5)}{3}$
   b. 258 (approximately)
   c. $(f(3) + f(3.5) + f(4) + f(4.5) + f(5.0) + f(5.5))/6$ d. 316
   e. 333; the area under the graph between 3 and 6

9. a. 0.8, 0.5; 0.1, 0.2; 14.3%, 66.7%; 1st b. How many more (or fewer) cars were sold in June than May; this number as a percentage of total sales in May: thus the percent change in sales from May to June c. Percent change (as a decimal) in sales from one month to the next

Chapter 0 Summary Exercises

1. $9, 0, 4 + 2\sqrt{3}, 4 - 4h + h^2,$ $a^4 + 2a^2 + 1$
3. $\tfrac{3}{5}, 0, \tfrac{1}{2}, (-2h + h^2)/(2 - 2h + h^2), (a^4 - 1)/(a^4 + 1)$
5. $3, -6, 3\sqrt{3} - 5, -h^3 + 3h^2 - 3h - 4, a^6 - 5$
7. $t \geq 1$
9. $z \neq 2, -1$
11. $x \geq 1$
13. $x = 1$
15. $u = \tfrac{5}{3}$
17. $z = -2$
19. 3
21. $(2 + x)/(2 - x)$ for $x \neq 0, 2$
23. $\tfrac{5}{3}$
25. $(2 + a^2)/(2 - a^2)$
27. $m = 2$, x-intercept $\left(\tfrac{3}{2}, 0\right)$, y-intercept $(0, -3)$

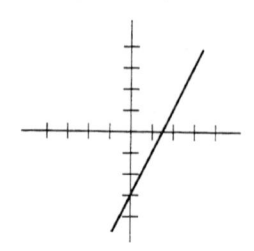

29. $m = \tfrac{1}{3}$, x-intercept $(4, 0)$, y-intercept $\left(0, -\tfrac{4}{3}\right)$

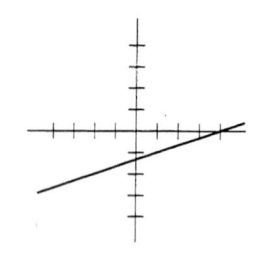

31. $y = -2x + 1$

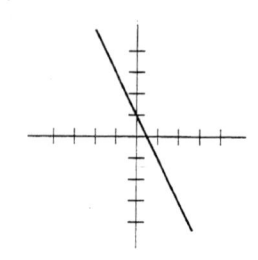

33. $y = \sqrt{2}x + 2$

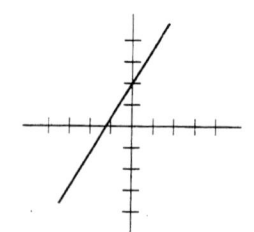

35. $y = \tfrac{1}{2}x - 1$

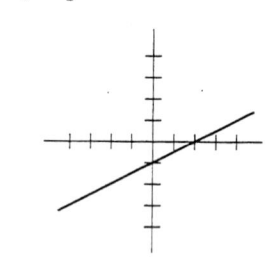

37. $y = x^2 - 2x + 1$

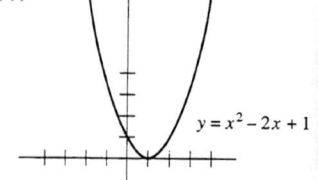

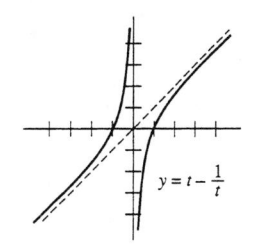

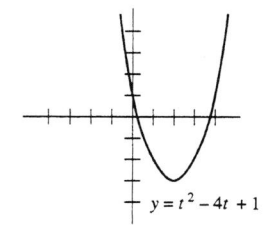

5. a. $9\frac{3}{4}$ yd/sec; $11\frac{1}{2}$ yd/sec; 12 yd/sec; 8 yd/sec; 10 yd/sec; 6 yd/sec

b.

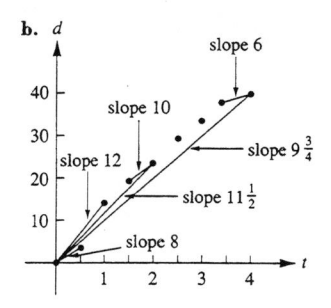

[Graph showing d vs t with slopes labeled: slope 6, slope 10, slope 12, slope $9\frac{3}{4}$, slope $11\frac{1}{2}$, slope 8; d-axis values 10, 20, 30, 40; t-axis values 1, 2, 3, 4]

7. Check for when the rate of sales growth begins to decrease.

9. a. $f(x) = x^9$
Estimate of the slope of the tangent at $a = 0$, beginning at $a + h = 0.1$

Increment h in a	Secant slope
0.100000	0.000000
0.050000	0.000000
0.010000	0.000000
0.005000	0.000000

b. $f(x) = x^9$
Estimate of the slope of the tangent at $a = 0.5$, beginning at $a + h = 0.6$

Increment h in a	Secant slope
0.100000	0.081246
0.050000	0.053045
0.010000	0.038104
0.005000	0.036596

c. $f(x) = x^9$
Estimate of the slope of the tangent at $a = 1$, beginning at $a + h = 1.1$

Increment h in a	Secant slope
0.100000	13.579480
0.050000	11.026540
0.010000	9.368515
0.005000	9.182119

d. $f(x) = x^9$
Estimate of the slope of the tangent at $a = 1.5$, beginning at $a + h = 1.6$

Increment h in a	Secant slope
0.100000	302.761100
0.050000	263.930500
0.010000	236.906800
0.005000	233.758600

11. a. $f(x) = 2^x$
Estimate of the slope of the tangent at $a = 0$, beginning at $a + h = 0.1$

Increment h in a	Secant slope
0.100000	0.717734
0.050000	0.705299
0.010000	0.695550
0.005000	0.694346

b. $f(x) = 2^x$
Estimate of the slope of the tangent at $a = 1$, beginning at $a + h = 1.1$

Increment h in a	Secant slope
0.100000	1.435466
0.050000	1.410589
0.010000	1.391101
0.005000	1.388645

c. $f(x) = 2^x$
Estimate of the slope of the tangent at $a = 2$, beginning at $a + h = 2.1$

Increment h in a	Secant slope
0.100000	2.870922
0.050000	2.821179
0.010000	2.782107
0.005000	2.777195

Exercises 1.2

1. $\lim_{x \to b} f(x) = 3$

3. $\lim_{x \to b} f(x) = 2.6$

5. $\lim_{x \to b} f(x)$ does not exist

7. 1.00003, 1.00005, 1.00007; 0.9991, 0.99915, 0.9992

9. a. Likely 2   **b.** Likely 0.75   **c.** May not exist

11. -6

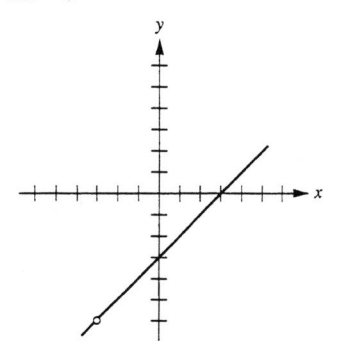

[Graph of a line in the xy-plane]

13. 3

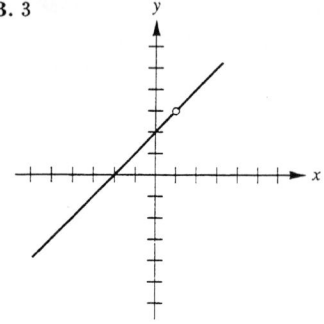

15. ≈0.693
17. 1
19. 2.7183

Exercises 1.3

1. 575
3. 1
5. 0
7. 896
9. Does not exist
11. 0
13. −1
15. 0
17. $2^7 = 128$
19. 2
21. −2
23. Does not exist
25. $\frac{1}{6}$
27. 0
29. Properties 2 and 3
31.

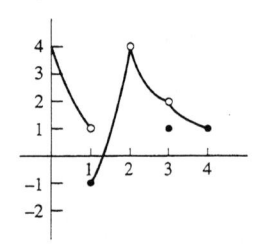

33. Continuous
35. Not continuous at $x = 4$
37.

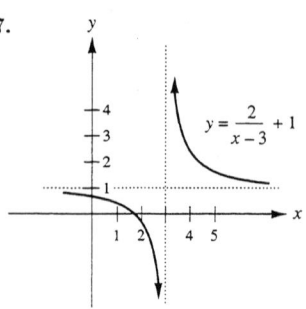

39. $\frac{2}{3}$
41. 0
43. 3
45. Does not exist
47. 0
49.

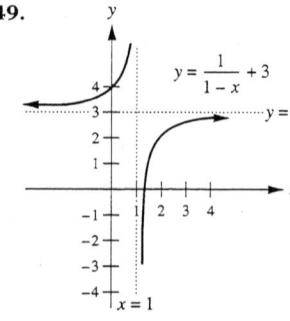

51. a. 2, 2.25, 2.37, 2.488, 2.594, 2.705, 2.718, 2.718 (to 3 decimal places) b. The limit is the same; let $x = \frac{1}{n}$
53. 0
55. a. −1 b. −1 c. $g(0) = -1$ d. $g(0) = -1$ fills in the "hole" in the graph of g.
57. 20; 20; yes, the hypotenuse would equal $10\sqrt{2} \approx 14.14$. While the stair appears to approximate the hypotenuse, it does not do so (in length). The discrepancy is that while two curves can be very "close" to one another in space, one curve can be much, much longer than the other.

Exercises 1.4

1. a. 4 b. $2a$ c. $-2, 1, 2\sqrt{2}$, 2.042 d. Where $f' > 0$, the graph of f rises; where $f' < 0$, the graph falls.

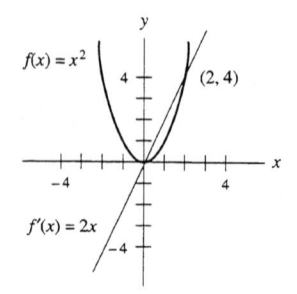

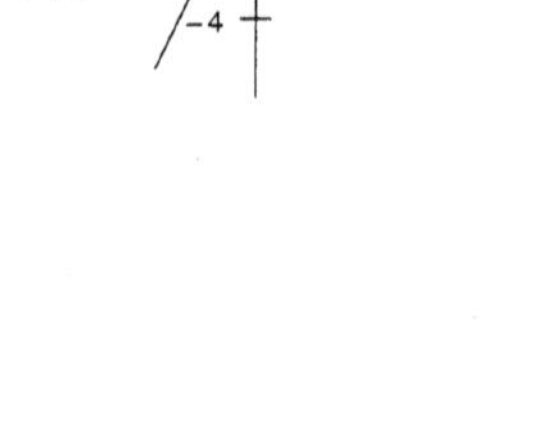

3. $f'(a) = \lim_{h \to 0} \frac{f(a + h) - f(a)}{h} = -\frac{1}{a^2}$

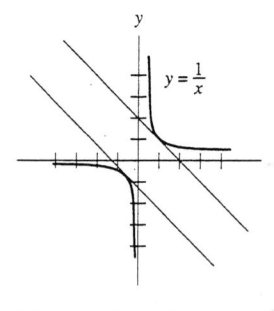

5. $f'(a) = 0$. Every horizontal has slope 0. The derivative of constant function is zero.
7. Tangent line: $y = 2x - 1$

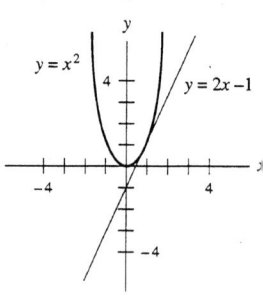

9. Tangent line: $y = \sqrt{7}$

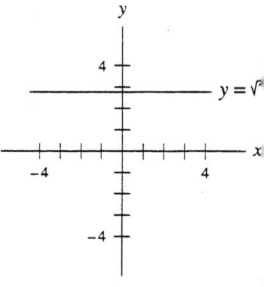

11. The light beam is always in the direction of the tangent. No, since (1.6, 0.2) is not a point the tangent line to the curve a (1, 1)
13. Equation of tangent line at $\left[\frac{1}{2}, g\left(\frac{1}{2}\right)\right]$ is given by $y = 0.70 + 0.3535$. Hence, along this tangent line, at $t = 1$, $y = 1.0605$ is beyond the scale.

670 Answers to Odd-Numbered Exercises

15. $f(x) = 2 \cdot x$
Estimate of $f'(a)$ at $a = 0$. Begin approximation at 0.5.

Increment h in a	Difference quotient
0.500000	0.828427
0.250000	0.756828
0.050000	0.705299
0.025000	0.699186

17. $f(x) = 2 \cdot (-x \cdot 2)$
Estimate of $f'(a)$ at $a = 0$. Begin approximation at 0.5.

Increment h in a	Difference quotient
0.500000	−0.318207
0.250000	−0.169587
0.050000	−0.034628
0.025000	−0.017326

Exercises 1.5

1. $R'(5)$ = rate at which revenue is changing per item sold when 5 items are sold
$R'(x)$ = rate at which the revenue is changing per item sold when x items are sold
3. $R'(5)$ = number of people per day hearing the rumor on the 5th day
$R'(x)$ = number of people per day hearing the rumor on the xth day
5. $V'(5)$ = rate at which the speed is changing per second (acceleration) at 5 sec
$V'(t)$ = rate at which the speed is changing per second (acceleration) at t sec
$P'(5)$ = kill rate per mg of penicillin at 5 mg dose
$P'(x)$ = kill rate per mg of penicillin at x mg dose
$B'(5)$ = people per mile 5 mi from trailhead
$B'(x)$ = people per mile x mi from trailhead
a. $c(x) = 50{,}000 + 10x$ b. $10,
$10   c. Yes   d. $−0.04
e. Cost per item is constant in (b); the formula in (d) reflects an "economy of scale" in which cost per item declines with increasing production.

13. 2 ft/sec
15. a. 80 ft   b. $t = 2$ c. 64 ft
    d. $t = 5$ e. 64 ft/sec
    f. 96 ft/sec   g. 128 ft high, speed = 32; 80 ft high, speed = 64
17. a. $V'(r) = 4\pi r^2$ b. $S'(r) = 8\pi r$
19. $t = \frac{2}{3}$; String's sales will increase faster
21. $\frac{1}{8}$ ft

Chapter 1 Summary Exercises

1. 3
3. $2\sqrt{7}$
5. 0
7. $\frac{1}{4}$
9. $\sqrt{10}$
11. Continuous
13. Not continuous
15.

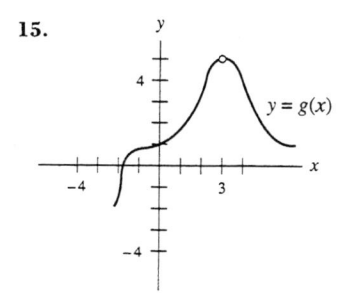

17. $4y = 4x − 1$; $m = 1$
19. $100y = −x + 20$; $m = -\frac{1}{100}$
21. $\left(0, \frac{2}{5}\right)$
23. $f'(15)$ is the rate of change in the cost to produce one item when 15 items are produced
25. $f'(15)$ is the population change rate in year 15
27. a. $\frac{1}{16}$ yr   b. 1 yr
29. a. 3 sec   b. 160 ft   c. Falling; 144 ft   d. 32 ft/sec   e. Approx. 6.16 sec   f. Approx. 101.2 ft/sec
31. 4
33. $+\infty$
35. $x = -1$
37. None
39. The limit at $+\infty$ is -1

CHAPTER 2

Exercises 2.1

1. $3x^2$
3. $\frac{1}{3}x^{-2/3}$
5. $\frac{1}{2}t^{-3/2}$
7. $2x - 3x^2$
9. $\frac{1}{2}u^2 + \frac{3}{2}u^{1/2}$
11. $1 + 2x$
13. $-12x^{-4} + 12x - 2$
15. $6z^2 + 6z^5$
17. $2v - 6$
19. $2x + 1$
21. $x^3 + x^{-4}$
23. a. $4x^3, 12x^2, 24x, 0, 0$ b. $-\frac{1}{x^2}, \frac{2}{x^3}, -\frac{6}{x^4}, \frac{-120}{x^6}, \frac{(-1)^n n!}{x^{n+1}}$
25. $6 - 24x^{-5}$
27. $-\frac{1}{4}x^{-3/2} - \frac{2}{3}x^{-5/3}$
29. a. $f'(0) = 0$, $f''(0)$ does not exist

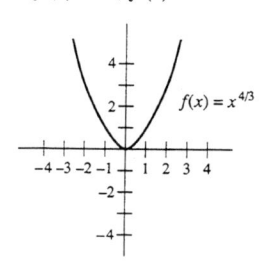

b. $f'(0)$ and $f''(0)$ do not exist

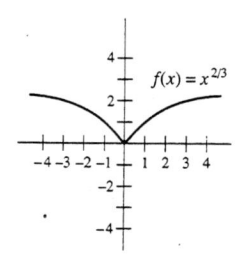

31. Jim; Bob; no, since $B\left(\frac{3}{2}\right)$ seems inhuman; yes, since $J\left(\frac{3}{2}\right)$ is physically reasonable

Exercises 2.2

1. $x^3 + x$; ( )^5; 5( )^4
3. $1 - x^4$; ( )$^{1/3}$; $\frac{1}{3}$( )$^{-2/3}$
5. $\sqrt{x + 1} + 2$; ( )$^{-3}$; $-3($)$^{-4}$
7. $-8(1 - x)^7$
9. $18(1 + 2t)^8$
11. $\left(\frac{1+z}{4}\right)^3$
13. $-6x(x^2 + 1)^{-4}$
15. $3(1 - 2x)(x - x^2)^2$
17. $\frac{5}{2}t^4(t^5 + 4)^{-1/2}$

Answers to Odd-Numbered Exercises

19. $\frac{7}{6}(3x^2 + 2x)(x^3 + x^2)^{1/6}$
21. $-\frac{3}{2}(1 - r)^{-1/2}(\sqrt{1-r} + 2)^2$
23. $-10(1 + 2x)^{-6}$
25. $-12u^2(2 + u^3)^{-2}$
27. $\frac{1}{4}(s + 1)^{-1/2}(1 + \sqrt{s+1})^{-1/2}$
29. $(1 + x^6)(3x^2)$
31. $1 + (1 + x)^2$
33. $1/(x - 1)$
35. $6; 5/2; 5$
37. $\frac{1}{2}, \frac{1}{3}, \frac{7}{12}$
39. a. $50\pi t; 100\pi$ b. $25\pi; 25\pi$
    c. $150\pi(t - 1)^2[(t - 1)^3 + 1]$; 300π
41. a. $1{,}507{,}772{,}228$ dollars/day
    b. $250t^{-1/2} + 750(500\sqrt{t} + 1{,}000)^2 t^{-1/2}$

Exercises 2.3

1. $-1 + 2x - 3x^2$
3. 0
5. $-\dfrac{z + 2}{z^3}$
7. $\dfrac{-2}{(x - 1)^2}$
9. $\dfrac{-2t}{(t^2 + 2)^2}$
11. $\dfrac{7x^6}{12} + \dfrac{5x^4}{12} - \dfrac{3x^2}{2}$
13. $\frac{8}{3}x + 3x^{1/2} + \frac{11}{3}x^{5/6} + 4x^{1/3}$
15. Composition of functions; chain rule; quotient rule
17. Composition of functions; chain rule; quotient rule
19. Product of functions; product rule; chain rule
21. Composition of functions; chain rule; chain rule
23. $\dfrac{-4(x + 1)}{(x - 1)^3}$
25. $\dfrac{2v^2 + 1}{\sqrt{v^2 + 1}}$
27. $(x^3 - x)^3 \left(\dfrac{x^2}{2} + 1\right)^{-6} (x^4 + 15x^2 - 4)$
29. $\dfrac{2x^4 + 7x^2 - 1}{(x^2 + 1)^2}$
31. $\dfrac{1}{\sqrt{t}(1 - \sqrt{t})^2}$
33. $\dfrac{5x(x^2 + 1)^{3/2}(1 - 3x - 2x^3)}{(x^3 + 1)^6}$

35. $(x^8 - 2x^7 - 4x^6 - 9x^4 + 6x^3 - x^2 + 2x - 1)/(1 - x + x^3 - x^5)^2$
37. $-\dfrac{(u^4 + 2u^3 + 6u^2 + 2u + 1)}{(u^3 - 1)^2}$
39. $(x^2 + 1)^{-1/2}(x + 2)^2(5x^3 + 4x^2 + 4x + 2)$
41. $-\dfrac{1}{x^2}; 1$
43. $-\dfrac{6x}{(x^2 + 1)^4}; -6x(x^2 + 1)^{-4}$
45. a. $\dfrac{x}{x + 1}$ b. $\dfrac{x}{x + 1} - \sqrt{x^2 + 21}$
    c. $-\dfrac{109}{121}$
47. a. $\dfrac{145 - 8x}{30 - 2x}$ b. $\frac{1}{2}$ "frustration units" per min

Exercises 2.4

1. a.

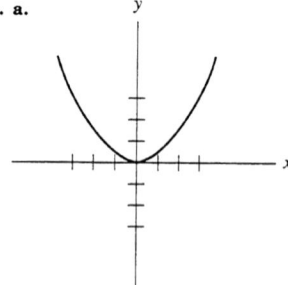

b.

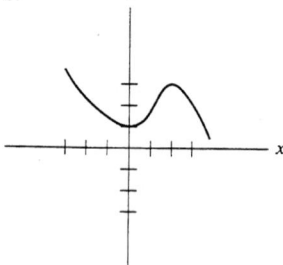

3. a.

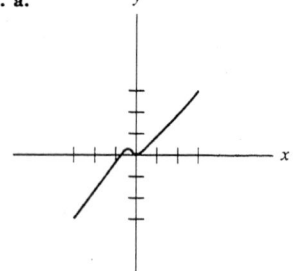

b.

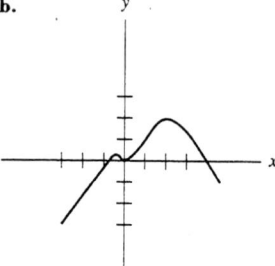

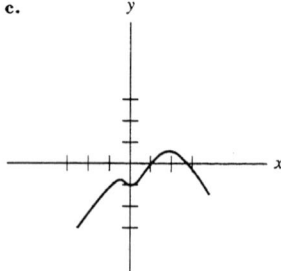

5. a.

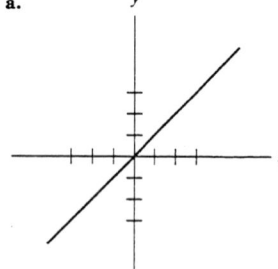

b.

672 Answers to Odd-Numbered Exercises

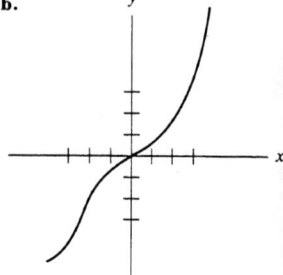

7. f is increasing [decreasing] at all x in $(0, 2)[(-2, 0)]$.
9. $f'(x) = 3x^2 > 0$ for all x except $x = 0 : f'(0) = 0$
11. Second graph
13. First graph

15.

17.

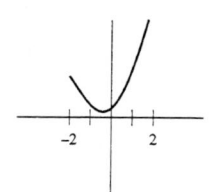

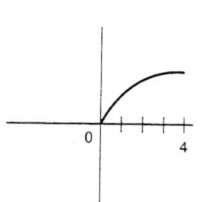

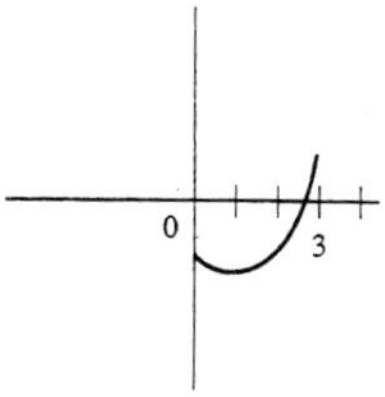

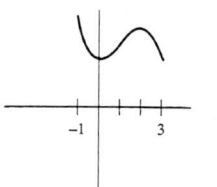

Decreasing on, and at each point in: $(-1, 0)$ and $(2, 3)$
Increasing on, and at each point in: $(0, 2)$

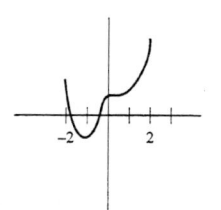

Increasing at each point in: $(-1, 0)$ and $(0, 2)$
Increasing on $[-1, 2]$
Decreasing on, and at each point in: $(-2, -1)$

25.

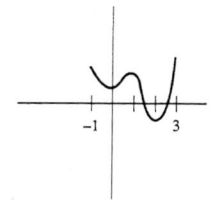

Increasing on, and at each point in: $(0, 1)$ and $(2, 3)$
Decreasing on, and at each point in: $(-1, 0)$ and $(1, 2)$

27.

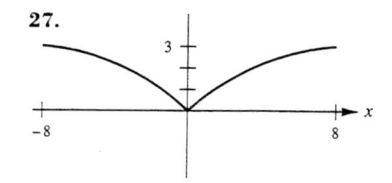

29. $f'(x)$ is increasing; $g'(x)$ is decreasing.
31. Compare to the graph. It is *very* similar.

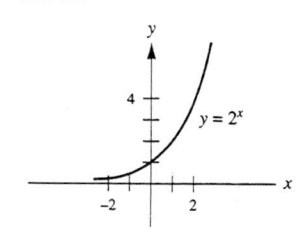

33.

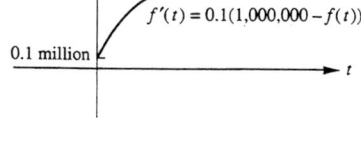

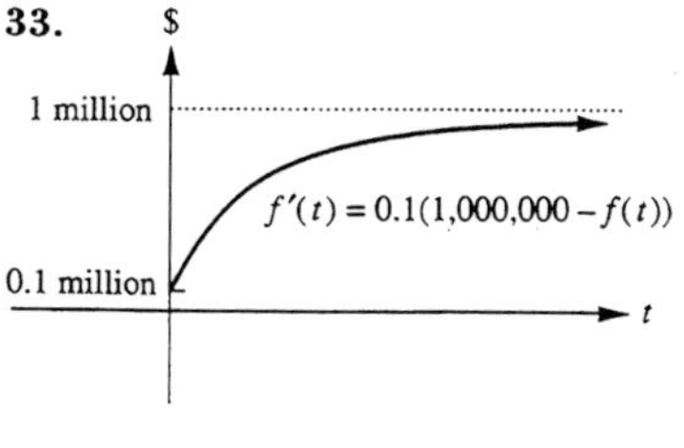

Exercises 2.5

1. $x < 1$, $x > \frac{7}{2}$; $1 < x < \frac{7}{2}$; $x = 1$, $\frac{7}{2}$; local max at $x = 1$; local min at $x = \frac{7}{2}$
3. $x < 2$, $x > \frac{13}{2}$; $5 < x < \frac{13}{2}$; $[2, 5]$ and $x = \frac{13}{2}$; local max at $[2, 5]$; local min at $x = \frac{13}{2}$

5. The "worst" graph must still have a similar shape and can have no other peaks or valleys.

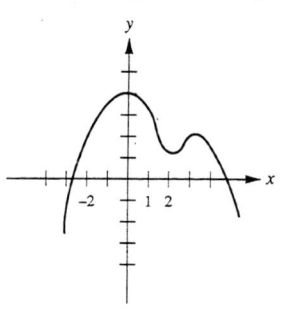

7. (1) $x > \frac{5}{2}$; $x < \frac{5}{2}$; $x = \frac{5}{2}$
   (2) Always positive (3) $x > 6$; $x < 2, 5 < x < 6$; $x = 6$ and $[2, 5]$ (4) $2 < x < 3$; $x < 2$, $x > 3$; $x = 2, 3$
9. The "worst" graph must be of similar shape and can have no other peaks, valleys, or level points.

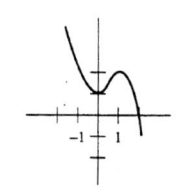

11. Local min at $x = -1$

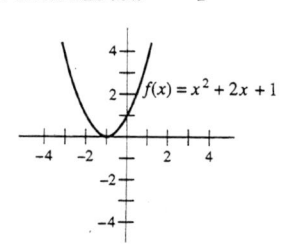

13. Local min at $x = \sqrt{2/3}$; local max at $x = -\sqrt{2/3}$

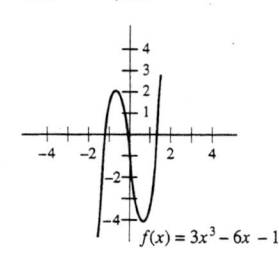

Answers to Odd-Numbered Exercises   673

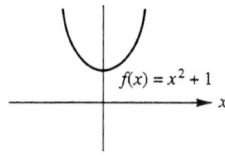

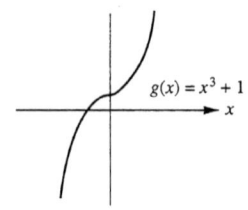

15. Local min at $x = \pm 1$; local max at $x = 0$

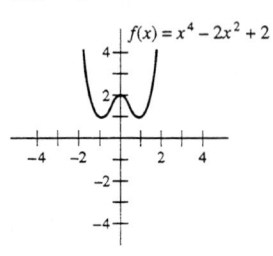

$f(x) = x^4 - 2x^2 + 2$

17. Local min at $x = 1$; local max at $x = -2$

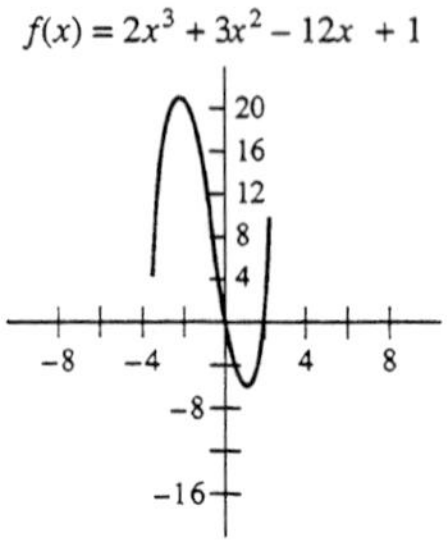

$f(x) = 2x^3 + 3x^2 - 12x + 1$

19. Local min at $x = 1$

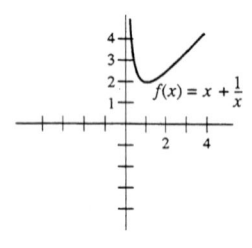

$f(x) = x + \dfrac{1}{x}$

21. Local min at $x = -1$; local max at $x = 1$

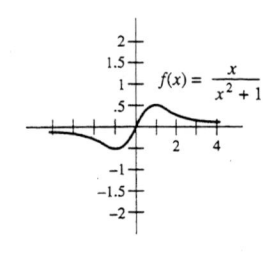

$f(x) = \dfrac{x}{x^2 + 1}$

23. a. In each case the derivative is zero only at $\bar{x} = 0$.
b. Conclusive only for f
c.

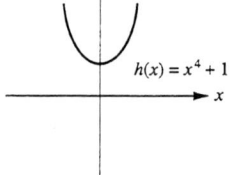

$f(x) = x^2 + 1$

$g(x) = x^3 + 1$

$h(x) = x^4 + 1$

d. g

25. Local min at $x = 2$

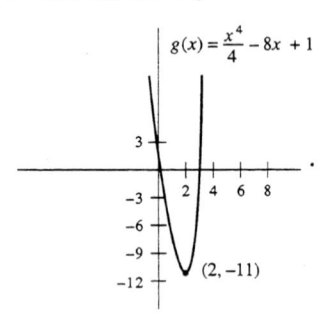

$g(x) = \dfrac{x^4}{4} - 8x + 1$

$(2, -11)$

27. Local min at $x = 1$

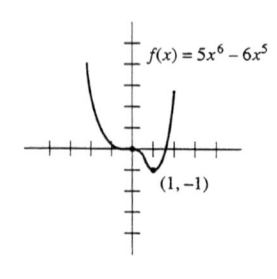

$f(x) = 5x^6 - 6x^5$

$(1, -1)$

29. Local min at $x = \pm 2$; local max at $x = 0$

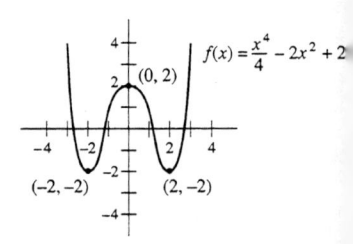

$f(x) = \dfrac{x^4}{4} - 2x^2 + 2$

$(0, 2)$, $(-2, -2)$, $(2, -2)$

31. Local min at $x = 0$

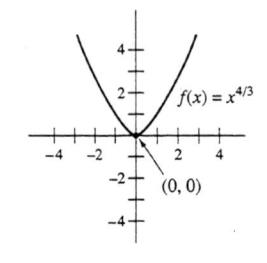

$f(x) = x^{4/3}$

$(0, 0)$

33. Local min at $x = 3$

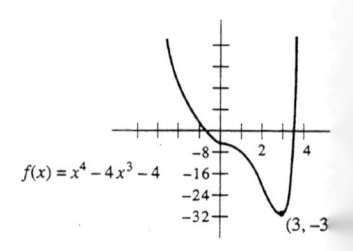

$f(x) = x^4 - 4x^3 - 4$

$(3, -3)$

35. $x = 2$; $x > 2$; $x < 2$

$f(x) = (x - 2)^3$

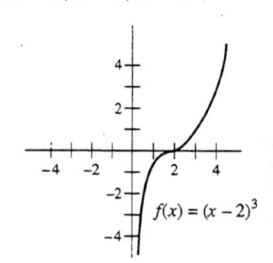

37. $x = 0$; $x > 0$; $x < 0$

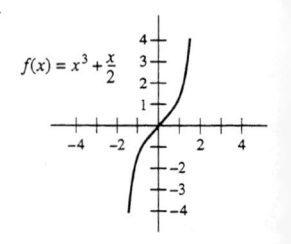

$f(x) = x^3 + \dfrac{x}{2}$

674 Answers to Odd-Numbered Exercises

39. $x = -\frac{1}{3}; x > -\frac{1}{3}; x < -\frac{1}{3}$

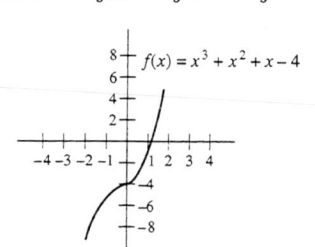

[graph: $f(x) = x^3 + x^2 + x - 4$]

41. $x = 0; x < 0; x > 0$

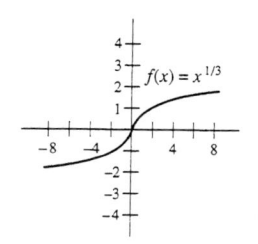

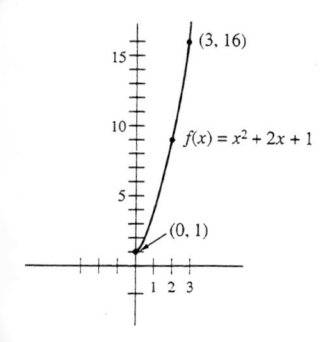

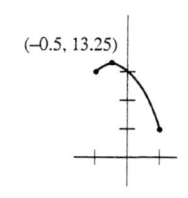

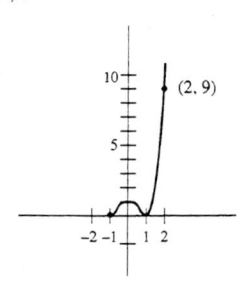

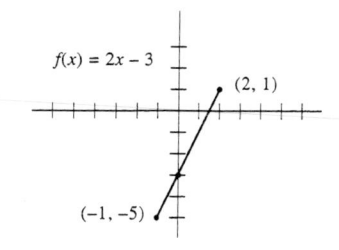

[graph: $f(x) = x^{1/3}$]

43. 16; 1

[graph: $f(x) = x^2 + 2x + 1$, points (3, 16), (0, 1)]

45. $\frac{13}{4}$; 1

[graph: (−0.5, 13.25)]

47. 9; 0

[graph: (2, 9)]

49. 1; −5

[graph: $f(x) = 2x - 3$, points (2, 1), (−1, −5)]

51. $1 + \sqrt[3]{4}$; 1

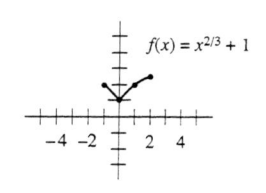

[graph: $f(x) = x^{2/3} + 1$]

53. It is an increasing function. To insure a maximum value the interval needs to be closed.

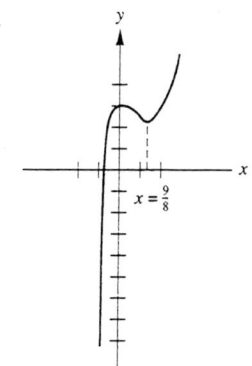

55.

[graph with $x = \frac{9}{8}$]

Values of x and $f(x)$ on the interval $[-1, 2]$, step 7.692308E-02

x	$f(x)$
−1.00	−8.22
−0.92	−5.70
−0.54	1.24
−0.08	2.98
−0.00	3.00
0.08	2.99
0.46	2.80
1.00	2.72
1.08	2.72
1.54	2.76
2.00	3.93

Exercises 2.6

1. 17.5, 17.5
3. No
5. $\frac{5}{2}, \frac{5}{2}$, 10
7. a. 13×13 b. $\frac{25}{2} \times \frac{25}{2}$
9. a. Bend entire wire into a circle
  b. Circle of radius $\frac{10}{\pi + 4}$, square of side $\frac{20}{\pi + 4}$
11. $x = 1 - \frac{\sqrt{2}}{2}$
13. $(5^{2/3} + 4^{2/3})^{3/2}$
15. $r = \frac{\sqrt[3]{35}}{2}; h = \frac{4\sqrt[3]{35}}{\pi}$
17. $l = 19.876; w = 10.841$
19. a. $4.37 \times 4.37 \times 6.55$ b. yes, same dimensions as in (a)
21. 28,985; every 44 days
23. 34.64 tons
25. $247.50 for 124 members is the mathematical solution. In practice, since 125 members at $245/member yields more revenue than 120 members at $255, the club will charge $245.
27. 22 or 23
29. 72 or 73; in multiples of five, 70 or 75 yields equal (but smaller) revenue.
31. $x = 62.3$ using $C(x) = 20x + 60\sqrt{50^2 + (x - 80)^2}$
33. $D(x) = -\frac{x}{450} + \frac{44}{3}$
35. $9.63; 131 pizzas
37. 20
39. a. $P(x) = 1.3x - \frac{x^2}{10,000} - 1,500$ b. 4.85 cents c. No, only 0.075 cent of the 0.15 cent increase should be passed on in order to maximize overall profit.

Chapter 2 Summary Exercises

1. $1 - 2x$
3. $1/x^2$
5. $6x^2 + 6x + 1$
7. $\frac{4}{(2x + 3)^2}$
9. $\frac{x^3 + 2x}{(x^2 + 1)^{3/2}}$
11. $6t(1 + t^2)^2$
13. $x^3 + 1$

15. $-\dfrac{2}{(w-1)^2}$

17. 20

19. $(x+1)^3(42x^2+24x+22)$

21. $\dfrac{4(t^2+1)^3(t^2-2t-1)}{(t-1)^5}$

23. $m=-1/\sqrt{3}$; $3y=-\sqrt{3}t+2\sqrt{3}$

25. $p=M/2$

27. $1,991.60 per item

29.

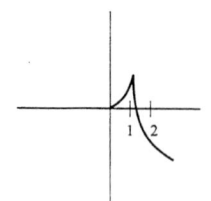

31.

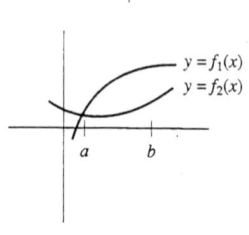

33.

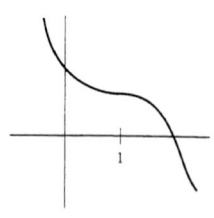

35. $f(a)=f(g)=f(i)=0$
$f(x)<0$ for $x<a$, $g<x<i$
$f(x)>0$ for $a<x<g$, $x>i$
$f'(x)>0$ for $x<d$, $-e<x<f$, $x>h$
$f'(x)<0$ for $d<x<e$, $f<x<h$

37.

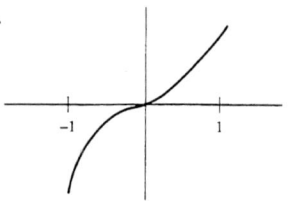

39.

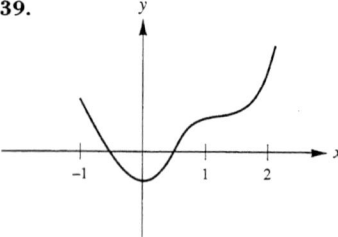

41.

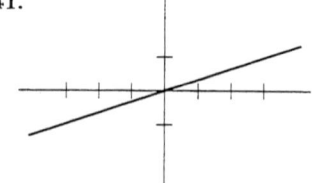

43. Many possible, for example:

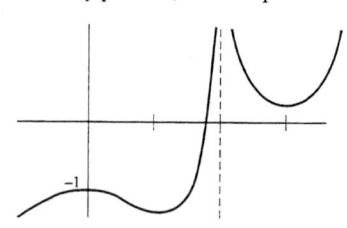

45. Local max at $x=-\sqrt{3}/3$; local min at $x=\sqrt{3}/3$; concave up for $x>0$; concave down for $x<0$

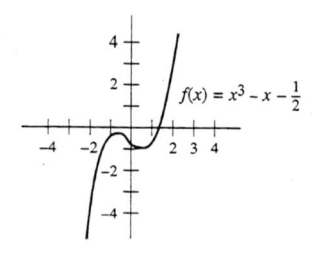

47. Local max at $t=1$; local min at $t=3$; concave up for $t>2$; concave down for $t<2$

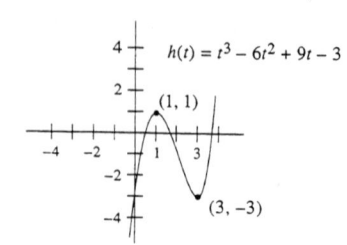

49. Local max at $x=-2$; always concave down

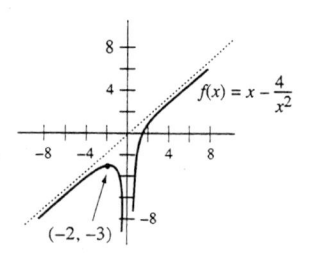

51. Local min at $x=0$

53. No local extrema

55. $f(-1)=-\dfrac{31}{2}$ is minimum; $f(3)=\dfrac{177}{2}$ is maximum

57. $f\left(-\dfrac{3}{2}\right)=-\dfrac{5}{4}$ is minimum; $f(1)=5$ is maximum

59. $\sqrt{200}$ ft from the corner on each wall

61. 253

63. a. $100x-5x^2$ b. $98x-5x^2$
   c. 10 items

65. a. $18.35; 400   b. With 400 members, charge $3.25/hr.

67. $x=6.31$ mi from the dirtier plant

CHAPTER 3

Exercises 3.1

1. $-\dfrac{x}{y}$

3. $-\dfrac{2y}{x}$

5. $\dfrac{3x^2-2y}{2x+2y}$

7. $\dfrac{2x+3y^2}{3y^2-6xy}$

9. $\dfrac{2x^{1/2}y^{1/2}-y}{x-2x^{1/2}y^{1/2}}$

11. $-\dfrac{(y^2-1)^{1/2}}{y}$

13. 0

15. $y=2$

17. $3y=x+3$

19. $4y=-x+1$

21. $y=8x+16$

23. $3y=-4x+25$; $3y=4x+2$

25. $-\dfrac{600y}{(1+12y^2)^3}$

27. $-\dfrac{C}{40,000}$

676   Answers to Odd-Numbered Exercises

Exercises 3.2

1. $Dty = -\dfrac{D_t x}{2y}$

3. $Dty = \dfrac{(y-2)D_t x}{3-x}$

5. a. 25 b. $-\dfrac{8x}{3}$ c. $-\dfrac{40}{3}$

7. a. 18.3 ft b. 0.66 ft/sec c. $\dfrac{2y}{x}$

9. a. 0.9 yd/sec b. $\dfrac{25}{8}$ yd/sec
   c. $\dfrac{10}{3}$ yd/sec; $\dfrac{50}{3}$ yd/sec; $\dfrac{50}{27}$ yd/sec

11. a. $\dfrac{27}{2\pi}$ ft/min b. $\dfrac{6}{\pi}$ ft/min
    c. $\dfrac{6}{\pi}$ ft/min

13. 1.64 ft/sec

15. −$0.328 per case per year

17. a. 392.5 b. $116.7 + 0.012xy + 0.036y^2$

19. −0.14 unit/week; −1.125 unit/week

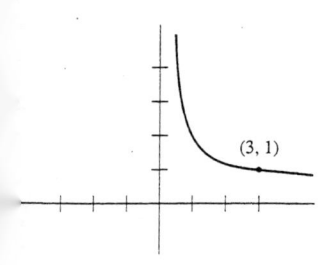

$D_t y = \dfrac{0.2y^2}{x^2} - \dfrac{0.3y^2}{R^2}$ in general;
0.625 Ω/sec at $x = 4$, $y = 2$
0.0139 km/hr; 0.1326 km/hr
0.56 ft/sec; 0 ft/sec

Exercises 3.3

5; $\dfrac{29}{4}$; $\dfrac{9}{4}$; 2

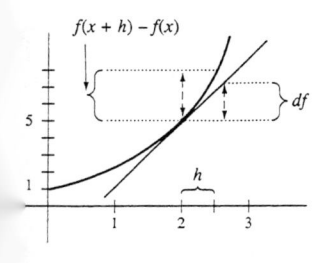

3. 0.12
5. 0.02
7. 0.8
9. $\dfrac{1}{25}$ in.; 0.125%; 0.25%
11. $\dfrac{1}{2\pi} \approx 0.159$ ft ≈ 1.9 in.
13. $f(4) + df(4; 0.1) = 2.025$
15. $f(3) + df(3; -0.2) = 2.9$
17. $\dfrac{59}{90}$
19. a. ±1.3 mi/hr b. 55.6 mi/hr; ±0.05 mi/hr
21. 672 rev. new; 695 rev. worn

Chapter 3 Summary Exercises

1. $\dfrac{2(x-1)}{3y^2}$

3. $\dfrac{2x^{1/2}(y-1)+1}{2x^{1/2}(1-x)}$

5. $3y = x + 1$
7. $4y = -3x + 5$
9. $48y = -x + 24$
11. a. 50π ft^3/min; 200π ft^3/min b. 0.14 ft/min
13. 0.4 thousand dollars/day
15. Use the standard formulae for volume and surface area of a sphere and compare the two rates of change.
17. 0.999
19. 0.3875
21. 384 in^3

CHAPTER 4

Exercises 4.1

1. 3; 3
3. 2; 2
5. $\dfrac{1}{8}$
7. a. 1 b. 1 c. 1.1
9. b. 0.285 c. 0.3325
11. No, both $\dfrac{7}{8}$ and 1 are chosen from the same subinterval.
13. Yes
15. [0, 1], $f(x) = (1-x)x$, $\Delta x = \dfrac{1}{4}$
17. $\int_0^{15} f(x)\,dx$ is the population of the city starting at the center of the city and going east 15 mi along Main Street.
19. $\dfrac{1}{3}$

21. a. $f(x) = x\cdot 3$ on the interval $[-1, 1]$
Riemann sums by evaluation of f at the midpoint of each subinterval

Number of subintervals	Riemann sum
5	0.000000
10	0.000000
15	0.000000
20	0.000000
25	0.000000

b. $f(x) = x\cdot 3$ on the interval $[-1, 2]$
Riemann sums by evaluation of f at the midpoint of each subinterval

Number of subintervals	Riemann sum
5	3.615001
10	3.716251
15	3.735000
20	3.741564
25	3.744599

c. $f(x) = x\cdot 3$ on the interval $[-2, 1]$
Riemann sums by evaluation of f at the midpoint of each subinterval

Number of subintervals	Riemann sum
5	−3.615001
10	−3.716250
15	−3.735000
20	−3.741562
25	−3.744599

Exercises 4.2

1. a. $\int_1^4 3\,dx = 9$ b. 9
3. 24
5. $\dfrac{26}{3}$
7. $\dfrac{16}{3}$
9. $-\dfrac{1}{12}$
11. $4\sqrt{2} - 1$
13. −2
15. $\dfrac{73}{100}$

Answers to Odd-Numbered Exercises 677

17. 1st integral = $\frac{3}{2}$; quotient of integrals = $\frac{5}{3}$

19. Remember that the area up to 2 contributes to $A(3)$.

21. a. $5 - \frac{5}{x(x+30)}$ b. $\frac{674}{135} \approx \$4.99$

c. $\frac{10(x+15)}{x^2(x+30)^2}$

Exercises 4.3

1. $u = 1 + x^2$; $du = 2x\,dx$
3. $u = z^2 + 3$; $du = 2z\,dz$
5. $u = t^3 - 1$; $du = 3t^2\,dt$
7. $\frac{2}{3}(1 + x^2)^{3/2} + C$
9. $\frac{2}{3}(x^3 - 4)^{3/2} + C$
11. $\frac{1}{9}(3t^2 + 1)^{3/2} + C$
13. $-\frac{1}{8}(1 - x^2)^4 + C$
15. $\frac{1}{6}(2x^4 - 1)^6 + C$
17. $-\frac{1}{2}(x^2 + 1)^{-2} + C$
19. $\frac{2}{9}(3y + 2)^{1/2} + C$
21. $\frac{1}{6}(x^3 - x^2 + x)^{-6} + C$
23. $\frac{1}{8}(u^2 + 2u)^4 + C$
25. $-\frac{(1-x)^{100}}{100} + \frac{(1-x)^{101}}{101} + C$
27. $\frac{2}{7}(x + 1)^{7/2} - \frac{4}{5}(x + 1)^{5/2} + \frac{2}{3}(x + 1)^{3/2} + C$
29. $\frac{1}{3}x^6 + \frac{1}{5}x^5 + x^4 + \frac{2}{3}x^3 + x^2 + x + C$
31. $\sqrt{2} - 1$
33. $\frac{6}{25}$
35. $F(x) = \frac{(x^2 - 5)^4}{8} + C$
37. $F(v) = (v^2 + 1)^{1/2} + C$
39. $F(y) = \frac{2}{5}(1 + y)^{5/2} - \frac{2}{3}(1 + y)^{3/2} + C$

Exercises 4.4

1. a. $f(x) = \frac{x^2}{2} + x + C$

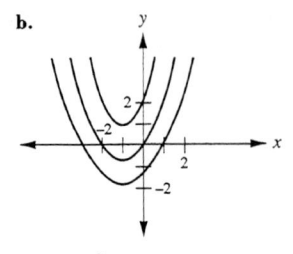

b.

c. $f(x) = \frac{x^2}{2} + x + \frac{3}{2}$

3. a. $f(x) = -\frac{1}{3}(1 - x^2)^{3/2} + C$

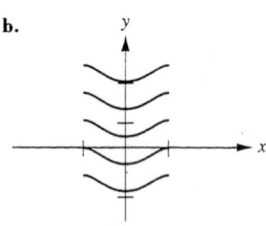

b.

c. $f(x) = -\frac{1}{3}(1 - x^2)^{3/2} + \frac{7}{3}$

5. $y(t) = \frac{t^2}{2} - \frac{(1-t)^3}{3} + \frac{4}{3}$
7. $y(t) = -\frac{2}{3}(1 - t)^{3/2} + \frac{11}{3}$
9. $y(t) = -1/(t^2 + 1) + 2$
11. $(3{,}000\sqrt{5} + 12{,}000)$ dollars
13. Approx. 2,125
15. Joe Piston; approx. 2.4 yd; approx. 50.9 yd/sec
17. 3 sec; 36 ft
19. 18.5 days
21. Approx 12 sec; 625 ft; approx. 184 ft/sec
23. Approx. $9\frac{1}{2}$ days later

Exercises 4.5

1. $\frac{10}{3}$
3. $\frac{4}{3}$
5. 2
7. 2.02
9. $\frac{4}{3}$
11. 0.066
13. $25 + \frac{2}{3}\sqrt{10}$ lb/tree
15. $A(x) = \pi x$; $V = 2\pi$

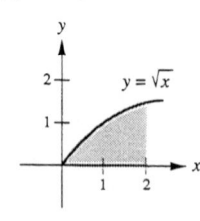

17. $A(x) = \pi(1 - x^2)$; $V = \frac{4\pi}{3}$

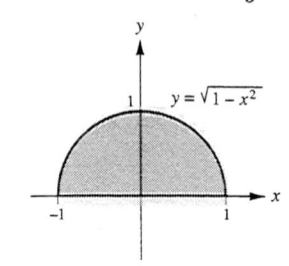

19. $A(x) = \pi(x - 1)^4$; $V = \frac{33\pi}{5}$

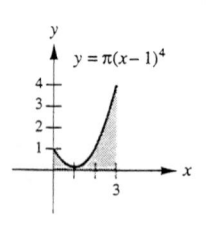

21. $A(x) = \pi(2x^3 - 9x^2 + 12x)^2$; $V = \frac{1{,}216\pi}{35}$

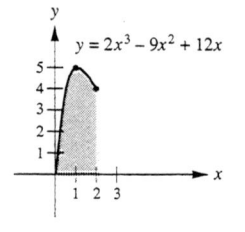

23. Imagine a Riemann sum defined on the interval $[0, k]$. Each Δx-thick slice of the figure perpendicular to the vertical line of length k will have cross-sectional area ab.

25. 200 in.^{3}
27. 15.27 ft^3 yields approx. 5,516,000 BTU's.
29. 845; 396
31. 66.33; 50
33. $281,250,000; $1,875,000
35. $\frac{120}{7}$ mg
37. 20 ft/sec; 200 ft
39. 900,000 in.^{3}/hr
41. 0.9 ton

Chapter 4 Summary Exercises

1. 4
3. $\int_0^{45} f(t)\,dt$ represents the distance in feet from your dorm to class.
5. 1
7. $\frac{14}{3}$
9. $\frac{7}{8}$
11. $-\frac{3}{8}$
13. $\frac{127}{14}$
15. $u = 1 - x^3$; $du = -3x^2\,dx$
17. $-\frac{1}{3}(1 - x^2)^{3/2} + C$
19. $-\frac{1}{6}(1 - x^4)^{3/2} + C$

21. $\frac{1}{5}(t^2 + 1)^{5/2} + C$

23. $\frac{1}{3}(x^2 + a^2)^{3/2} + C$

25. $\int f(x)dx$ represents all functions $F(x)$ for which $F'(x) = f(x)$. Since the derivative of a constant is zero, adding C is necessary to give a complete answer. Adding C has no effect, since

$[F(x) + C]\Big|_a^b = [F(b) + C] - [F(a) + C] = F(b) - F(a)$.

27. $y = -\frac{t^2}{2} + 3$

29. $y = \frac{1}{3}(1 + t^2)^{3/2} + \frac{2}{3}$

31. $\frac{(x + 1)^{15}}{15} - \frac{(x + 1)^{14}}{14} + C$

33. $\frac{2[2\sqrt{7} - 1]}{3}$

35. a. $\frac{15}{8}$ sec b. -2 ft/sec c. 0
    d. $\frac{15}{16}$ sec e. 3.1 sec
    f. 68.8 ft/sec g. -32 ft^2/sec

37. 3,018 items. The other answer (496,981) cannot occur until long after revenue reaches a peak.

39. $\frac{91}{3}$ square units

41. $\frac{2\pi}{3}$ cubic units

43. 750; 250

45. $f(x_k)\Delta x$ approximates the number of pounds of fly ash on a strip Δx feet long, located x_k feet from the smokestack; $\int_0^{500} f(x)dx$ represents the total amount of fly ash on the ground up to 500 feet north of the stack.

47. 153.75 tons

CHAPTER 5

Exercises 5.1

1. a. 128, $\frac{1}{8}$, 1.4, 0.71, 1, 2.8, 11.2, $\frac{1}{1,024}$ b. 2.6, 8.56

3. a. 4.096, 10.49, 1.6, 7.4, 2.7, 7.4, 7.4, 1, 1.81, 10.1, 18.49, 7.4, 54.76 b. 4.1, 22.2

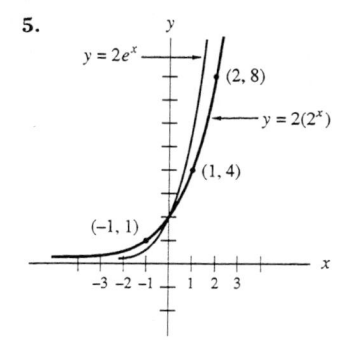

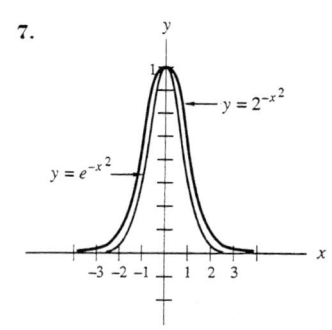

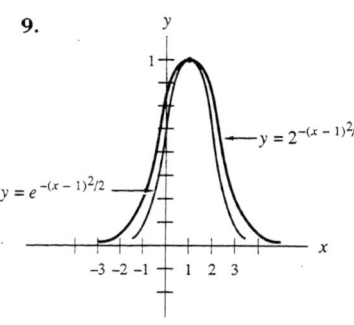

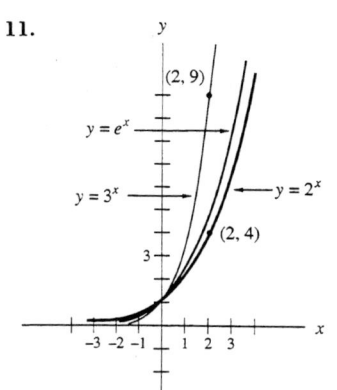

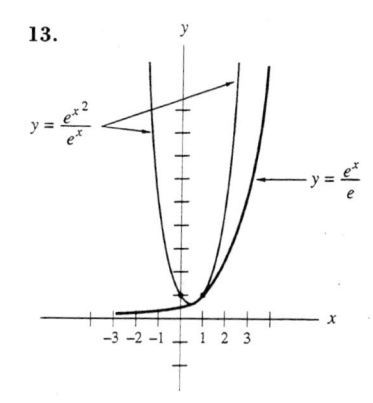

5.

7.

9.

11.

13.

15. 19.219, 0.6058, 2.111, 4.073; 20.086, 0.6065, 2.117, 4.1133

17. $e^x - 1$

19. $1/(e^x - 1)$

21. $e^{(x-1)}$

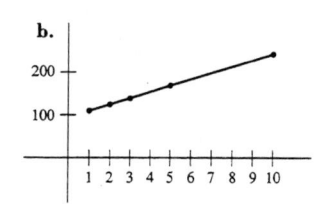

23. a. $110, $110.51; $121, $122.14; $133.10, $134.99; $161.05, $164.87; $259.37, $271.83

    b.

    c. Here growth "feeds on itself" over a specific interval of time (1 year).

25. $xe^x(x + 2)$

27. $5e^x$

29. $-e^{-t}$

31. e^x

33. $e - 1$

35. $e - \frac{2}{3}$

37. $(e - 1)/e$

39. No max; $f(-1) = -1/e$ is min

41. $f(1) = 1/e$ is max; no min

43. $f''(x) = e^x > 0$ for all x

45. $e^x(x + 5)$

47. For example:

h	$D(h) = (\exp(h) - 1)/h$
0.1000000	1.0517100
0.0769462	1.0394800
0.0513308	1.0261110
0.0257154	1.0129700
0.0001000	1.0001660

Exercises 5.2

1. $f(x) = e^x$, $g(x) = 2x$
3. $f(x) = \sqrt{x}$, $g(x) = e^x + 1$
5. $f(x) = e^x + 3$, $g(x) = (x-1)/x$
7. $5e^{5x}$
9. $xe^{x^2/2}$
11. $\dfrac{e^x}{2(e^x+1)^{1/2}}$
13. $(-t)e^{(1-t^2)}$
15. $e^x + e^{-x}$
17. $(1 + 2x)e^{(x+x^2)}$
19. $2xe^{x^2}$
21. $-\dfrac{e^{1/x}}{x^2}$
23. $(2x^2 + 1)e^{x^2} + 1$
25. $(2xe^{x^2})e^{e^{x^2}}$
27. $2e^{2x} - 2e^{-2x}$
29. $e^{3x} + e^{2x} + e^x + 1$
31. $\dfrac{e^x(2x^3 - x^2 - 2x) + 2x - 2x^3}{e^{x^2}}$
33. $g(1) = 1$ is local min
35. $g(0) = 0$ is local min
37. $f(-3) = 27/e^3$ is local max
39. No local extrema
41. $(e^x - y)/x$
43. e^{-y}
45. 1
47. e to any power is positive.
49. $\tfrac{1}{3}(e^3 - 1)$
51. $(e^3 - 1)/2e^4$
53. $\tfrac{2}{3}(e^x + 1)^{3/2} + C$
55. $-e^{1/x} + C$
57. $e^{1/2} - 1$
59. $(\pi/2)(e^2 - 1)$
61. 5.6 mL; yes
63. Approx. 752 fish; the number of fish killed decreases rapidly as you move away from the spill.
65. 1.019 million

Exercises 5.3

1. 1.61
3. -0.288

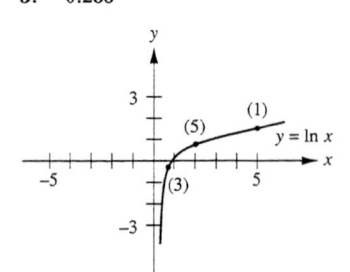

$y = \ln x$

5. 0.693
7. $\ln 2$
9. $\ln(x + 1)$
11. $\ln 2$
13. $(-2)\ln 2$
15. $\ln a$
17. $(\ln 6)/\ln 2$

19. a.

$y = \dfrac{1}{x}$

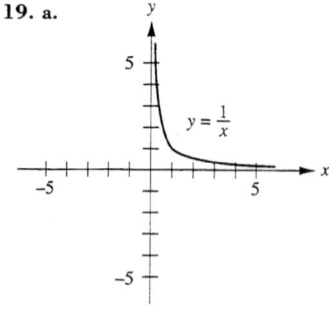

b. $\displaystyle\int_1^a \dfrac{1}{x}\, dx = \ln x \Big|_1^a = \ln a - \ln 1 = \ln a$

21. $x = \pm 2$
23. $x = e^{-4} - 1$
25. $x = -\tfrac{1}{2}\ln 5$
27. $x = \ln 4$
29. $x = \pm\sqrt{4 - e^{-3}}$
31. $\ln y = \ln A + k \ln x$
33. $S(t) = \dfrac{e^C M e^{kt}}{1 + e^C e^{kt}}$; $S(3) \approx 89.6$
35. $50e^{0.15} \approx 58.1$
37. $A = 100$; $k = \ln(2.5)$
39. 125

41.

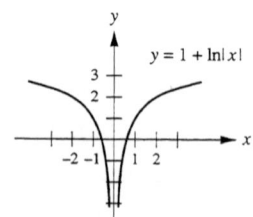

$y = 1 + \ln|x|$

43.

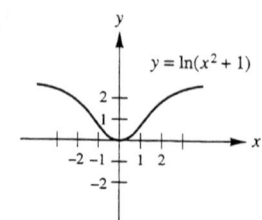

$y = \ln(x^2 + 1)$

45.

$y = \ln\left(\dfrac{1}{1+x}\right)$

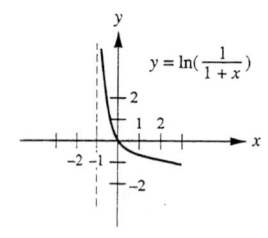

47. $y'' = -1/x^2 < 0$
49. Yes; no
51. a. \$3,750; \$8,750 b. \$12,500; \$15,625; \$19,531.25; \$30,517.58; \$59,604.65; \$93,132.26
c. $n = \dfrac{\ln(S_n) - \ln(10{,}000)}{\ln 1.25}$
d. $D_S n = \dfrac{1}{S_n \ln 1.25}$ i. $\dfrac{1}{12{,}500}$; $\dfrac{1}{31{,}250}$; $\dfrac{1}{50{,}000}$ ii. The higher the income level the lower the perception of a \$1,000 raise
53. a. Approx. 0.83 min
b. Approx. 0.2025 mg

Exercises 5.4

1. $\dfrac{1}{x+3}$
3. $\dfrac{2x+1}{x^2+x}$
5. $\dfrac{6x^{5/2} + 1}{2x^{7/2} + 2x}$
7. $2(x - \ln x)\left(1 - \dfrac{1}{x}\right)$
9. $\dfrac{1 - \ln x}{x^2}$
11. $\dfrac{1}{x \ln x}$
13. 1
15. $\dfrac{1}{x}$
17. $\ln t$
19. $-e^{-x \ln x}(1 + \ln x)$
21. $\dfrac{(x+1)\ln(x+1) - x\ln x}{x(x+1)[\ln(x+1)]^2}$
23. $\dfrac{1}{3x[7 + \ln(3x)]^{2/3}}$
25. $\ln(x^2) = 2 \ln x$
27. 16 new ants; 15.7 ants/min
29. $\tfrac{1}{2}$
31. 2.52
33. $\ln|x^3 + 9| + C$

35. $\dfrac{[\ln(x+1)]^2}{2} + C$

37. $\tfrac{1}{2}\ln|\ln x| + C$

39. $\tfrac{1}{2}(2x+3)[\ln(2x+3) - 1] + C$

41. $(\ln x)[\ln(\ln x) - 1] + C$

43. Local min at $x = e^{-1}$

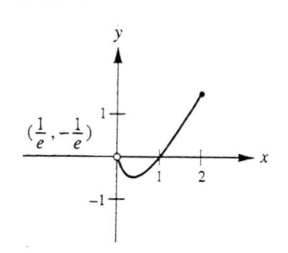

$(\tfrac{1}{e}, -\tfrac{1}{e})$

5. Maximum at $x = e$
7. Minimum at $x = e$
9. For $x > e (x < e)$, $y = \ln(\ln x)$ grows more slowly (rapidly)
11. $p = \tfrac{1}{2}$

Exercises 5.5

1. $65,026,052,250
3. 5.8 years; 8.7 years
5. $t = \ln 2/k$
7. 6.67 mg; 12.9 mg; 20.3 mg
9. 536%
11. $290.46
13. 17.33 years
15. 31.9 weeks
17. 13.86 years; 21.97 years
19. 30.4%
21. 5.35 years
23. Tree farm is best; $5,126.21 vs. $3,793.65
25. 28.2 days; 15%
27. 65 years
29. a. 26.6 days b. 9 years
31. 20.3 years
33. a. Let $1/n = r/m$.
    b. Amount compounded quarterly at 8% 1 yr
    c. Amount after 1 yr compounded daily d. As $m \to \infty$, the amount (1 year) $\to 500e^r$.

Chapter 5 Summary Exercises

... $+ e^{2x}$

9. $1 + \ln 3$
11. 15
13. $e^{-x}(1-x)$
15. 1
17. $e^{-x}(x-2)$
19. $e^t - e^{-t}$
21. $e^x\left[\ln x + \dfrac{1}{x}\right]$
23. $e^{-x}[(1/x - \ln x]$
25. $1 + \ln x$
27. $\dfrac{2x - 1 - x^2}{1 + x^2}$
29. $\dfrac{1}{2x\sqrt{\ln x}}$, $x > 1$
31. $-\tfrac{2}{3}(1 - e^x)^{3/2} + C$
33. $\ln 2$
35. $\sqrt{e} - 1$
37. $-\ln|1 - x| + C$
39. $\ln|x^2 - x| + C$
41. $\dfrac{(1 + e^x)^3}{3} + C$
43. $\dfrac{x^2}{2} + C$
45. $x + C$
47. $f(2) = 6/e$ is max
49. $f(1/e) = -2/e$ is min
51. No max or min
53. No; worth $911.06
55. 62 days
57. $3,351.60
59. 36 trees; 24%

CHAPTER 6

Exercises 6.1

1. $-8, -4, 1$
3. $0, -1, 1$
5. $e^{-1}, 0, 17$
7.

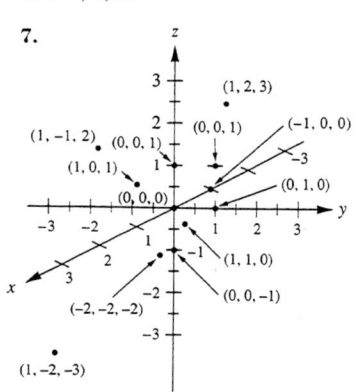

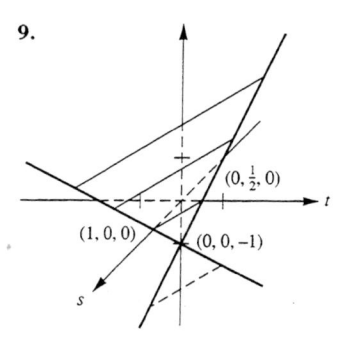

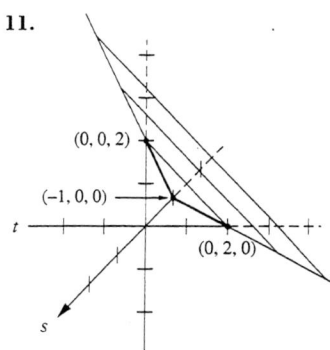

9.

11.

13. Slope $= 2$

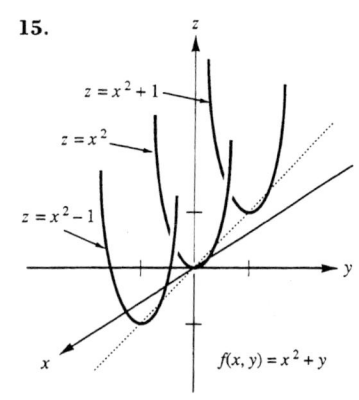

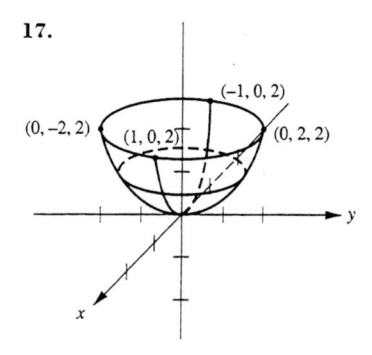

15.

17.

Answers to Odd-Numbered Exercises 681

19. $f(x, y) = \ln y$

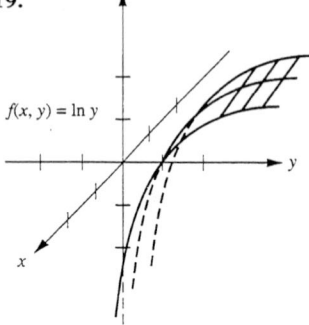

21. $f(x, y) = y^2 - x^2 + 3$

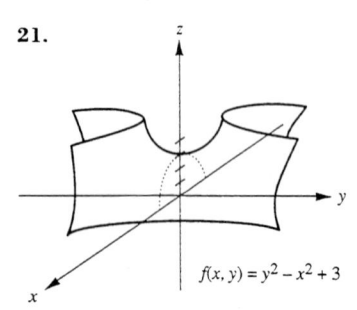

23. $f(x, y) = 3 - (x^2 + y^2)$, $(0, 0, 3)$

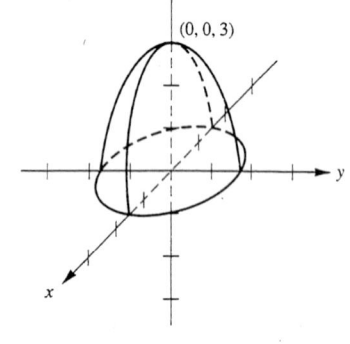

25. $f(x, y) = \frac{x}{y}$, $x, y > 0$

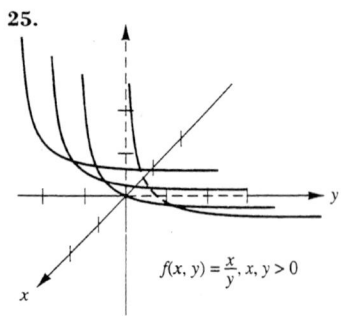

27. $(0, 0, 100)$

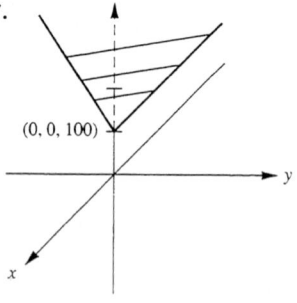

29. $f(2, 3)$ is the revenue when coffee sells for $2/pound and tea is $3/pound; the revenue at x dollars/pound for coffee and $3/pound for tea; the revenue at $3/pound for coffee and y dollars/pound for tea

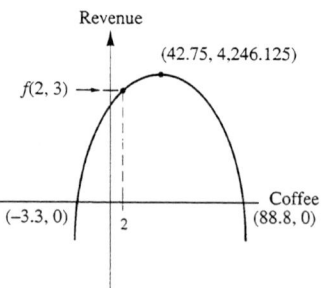

Revenue, $f(2, 3)$, $(42.75, 4,246.125)$, $(-3.3, 0)$, $(88.8, 0)$, Coffee

31.

x	y	$f(x, y)$
-1.00	0.00	0.00
1.00	0.00	0.00
-1.00	0.40	0.40
-0.60	0.40	0.14
-0.20	0.40	0.02
0.20	0.40	0.02
0.60	0.40	0.14
1.00	0.40	0.40
-1.00	0.80	0.80
1.00	0.80	0.80
-1.00	1.20	1.20
1.00	1.20	1.20
-1.00	1.60	1.60
-0.60	1.60	0.58
-0.20	1.60	0.06
0.20	1.60	0.06
0.60	1.60	0.58
1.00	1.60	1.60
-1.00	2.00	2.00
1.00	2.00	2.00

33.

x	y	$f(x, y)$
-10.00	0.00	-99.00
10.00	0.00	-99.00
-10.00	0.40	-107.16
-6.00	0.40	-39.96
-2.00	0.40	-4.76
2.00	0.40	-1.56
6.00	0.40	-30.36
10.00	0.40	-91.16
-10.00	0.80	-115.64
10.00	0.80	-83.64
-10.00	1.20	-124.44
10.00	1.20	-76.44
-10.00	1.60	-133.56
-6.00	1.60	-56.76
-2.00	1.60	-11.96
2.00	1.60	0.84
6.00	1.60	-18.36
10.00	1.60	-69.56
-10.00	2.00	-143.0
10.00	2.00	-63.0

Exercises 6.2

1. $2, 3$
3. $7, -8$
5. $\frac{1}{2}, \frac{1}{2}$
7. $\partial f/\partial x = 3(x - y)^2$; $\partial f/\partial y = -3(x - y)^2$
9. $\partial f/\partial x = (x + 1)e^{x+y}$; $\partial f/\partial y = xe^{x+y}$
11. $\partial f/\partial x = 5y(xy + 1)^4$; $\partial f/\partial y = 5x(xy + 1)^4$
13. $f_x = \dfrac{4xy^2}{(x^2 + y^2)^2}$; $f_y = -\dfrac{4x^2y}{(x^2 + y^2)^2}$
15. $f_x = \dfrac{e^x}{e^x + e^y}$; $f_y = \dfrac{e^y}{e^x + e^y}$
17. $f_{ss}(1, -1) = -\frac{1}{2}$; $f_{st}(1, -1) = 0$; $f_{tt}(1, -1) = \frac{1}{2}$
19. $f_u = (u + 1)v\, e^{u+v}$; $f_v = (v + 1)u\, e^{u+v}$; $f_{uu} = v(u + 2)e^{u+v}$; $f_{vv} = u(v + 2)e^{u+v}$; $f_{uv} = (v + 1)e^{u+v}(u + 1)$; $f_{vu} = (u + 1)e^{u+v}(v + 1)$
21. $f_x = 2/x$; $f_y = 1/y$
23. a. $P(x, y) = 15x + 7y$ **b.** $P_x = 15$; $P_y = 7$

25. $R_x(50, 4{,}000) = 52$; $C_x(50, 4{,}000) = 30$; $P_x(50, 4{,}000) = 22$
  $R_y(50, 4{,}000) = 0.725$; $C_y(50, 4{,}000) = 0.35$; $P_y(50, 4{,}000) = 0.375$
27. $R_x(6, 5) = 144$; $R_y(6, 5) = 214$

Exercises 6.3

1. a. $(0, 0)$ b. $D(0, 0) = 4$ c. $f(0, 0) = 1$ is local max
3. a. $(0, 2)$ b. $D(0, 2) = -4$
5. a. $(0, 1)$ b. $D(0, 1) = -3$
7. a. $(1, 2)$ b. $D(1, 2) = 4$ c. $f(1, 2) = 0$ is local min
9. a. $(\pm 1, -\ln 2)$ b. $D(\pm 1, -\ln 2) = -1$ c. No max or min
11. a. $\left(-\tfrac{1}{5}, \tfrac{2}{5}\right)$ b. $D\left(-\tfrac{1}{5}, \tfrac{2}{5}\right) = -5$
13. a. $(0, 1)$ b. $D(0, 1) = -1$
15. a. $(0, 0), \left(\tfrac{12}{53}, -\tfrac{36}{53}\right), \left(0, -\tfrac{2}{3}\right)$
  b. $D(0, 0) = 0$, $D\left(\tfrac{12}{53}, -\tfrac{36}{53}\right) > 0$; $D\left(0, -\tfrac{2}{3}\right) < 0$ c. $f\left(\tfrac{12}{53}, -\tfrac{36}{53}\right)$ is local min
17. a. $(0, 0), (-1, -1)$ b. $D(0, 0) = -1$; $D(-1, -1) = e^{-4}$ c. $f(-1, -1) = e^{-2}$ is local max
19. a. $x = 0, y \neq 0$ b. $D(0, y) = 0$
21. a.

  b. 0; 0 c. 0; 0 d. $f_x(0.2, y) = \dfrac{-0.2}{(0.96 - y^2)^{1/2}}$
23. $x = 3{,}500, y = 5{,}600$; maximum profit $= \$6{,}130$
25. 3.5 by 5.2 by 3.5 in.; minimum cost ≈ 22 cents
27. Use $R(x, y) = xp + yq - C(x, y)$.
29. 2 square units

Exercises 6.4

1. $y = \tfrac{13}{10}x - \tfrac{1}{2}$
3. $y = -\tfrac{45}{59}x + \tfrac{168}{59}$
5. $y = 0.11e^{0.945x}$
7. $y = 14.56e^{-0.294x}$
9. Omit $(x, y) = (0, 85)$; $y = -5.6x + 102.6$; approx. 83%
11. Omit first 3 years; approx. 2.07 bbl; $y = -0.0568x + 3.208$, with $x = 0$ being 1970.
13. Approx. 0.3°F ($y = -1.4x + 21.3$)
15. Approx. 1,993 ($y_1 = 0.9e^{0.10x}$ imports and $y_2 = 0.64x + 2.96$ exports)

Exercises 6.5

1. $f(2, 0) = 4$ is min
3. $f\left(\tfrac{1}{2}, \tfrac{1}{2}\right) = \tfrac{1}{2}$ is min
5. $h\left(\tfrac{1}{2}, \tfrac{1}{2}\right) = \tfrac{1}{2}$ is max
7. $f(\pm 1, 0) = 1$ is max
9.

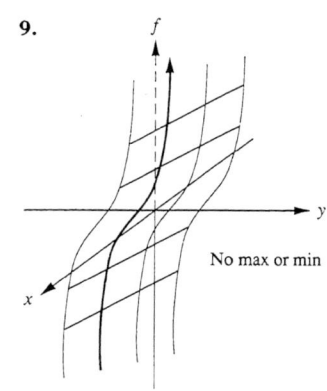

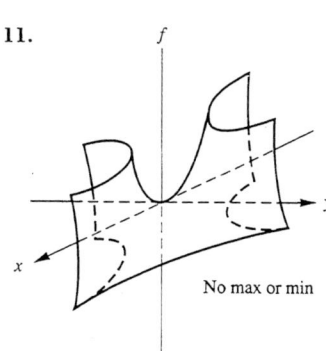

No max or min

11.

No max or min

13. Minima at $(0, \pm 2\sqrt{2})$, $(\pm 2\sqrt{2}, 0)$; maxima at $(-2, \pm 2)$, $(2, \pm 2)$
15. $f(2, 4) = 13$ is max; $f(-2, -4) = -7$ is min
17. 9 ft by 18 ft
19. $x \approx 1.2, y \approx 0.8$
21. $x, y = 10{,}000$
23. Minimum cost $= C(94, 31) = \$9{,}350$
25. $L_x = P_x + \lambda a$; $L_y = P_y + \lambda b$

Exercises 6.6

1. 47.55
3. 1.925
5. 8.617
7. 0.075
9. 3.9
11. 3.875 ft^2
13. 3.51 ft^3; 0.175 ft^3; 4.98%
15. 12.13
17. 100 units
19. 43%

Exercises 6.7

1. $\tfrac{1}{2}x^2$
3. $\tfrac{8}{3} + 2x$
5. $\tfrac{3}{2}e^{-y}$
7. 1
9. $\tfrac{5}{12}$
11. $\tfrac{4}{3}$
13. $(e - 1)/2e$
15. -4
17. $\dfrac{2}{3} + \dfrac{e^6}{3} - e^2$
19. $\tfrac{32}{3}$ units
21. 8 grams; \$98.80

Chapter 6 Summary Exercises

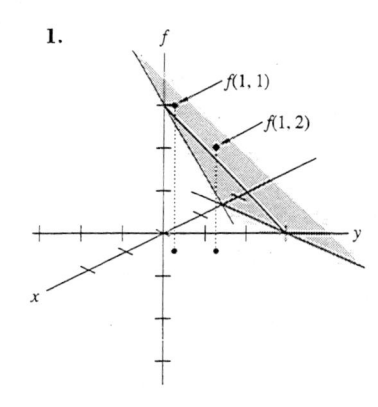

1.

3. An x-section is obtained by fixing the value of x and viewing f as a function of one variable, y, and is the two-dimensional graph of this function. For a y-section, hold y fixed. The graph of $z = f(x,y)$ is made up of infinitely many x- and y-sections.
5. $1 + 2y; 2x + 3; 0; 0; 2; 2$
7. $2xy^3; 3x^2y^2; 2y^3; 6x^2y; 6xy^2; 6xy^2$
9. $\frac{1}{2}(x+y)^{-1/2}; \frac{1}{2}(x+y)^{-1/2};$
   $-\frac{1}{4}(x+y)^{-3/2}; -\frac{1}{4}(x+y)^{-3/2};$
   $-\frac{1}{4}(x+y)^{3/2}; -\frac{1}{4}(x+y)^{-3/2}$
11. 16.5; 263.9
13. $(0, 0), (-1, 1)$
15. $(1, 1); D(1, 1) < 0$
17. Minus; minus
19. $1,667 on newspaper ads and $2,000 on television ads
21. $25 \text{ yd}^2; 2.563 \text{ yd}^2$
23. 53 returns/thousand
25. $f\left(\pm\frac{\sqrt{2}}{2}, \pm\frac{\sqrt{2}}{2}\right) = \frac{1}{2}$ is a max;
    $f\left(\pm\frac{\sqrt{2}}{2}, \mp\frac{\sqrt{2}}{2}\right) = -\frac{1}{2}$ is a min
27. $y = 1.554e^{-0.345x}$
29. 3.55
31. $\frac{9}{2}x; \frac{135}{4}$
33. $\frac{9}{4}$
35. $\frac{20}{3}$
37. Approx. 19,550 trees

CHAPTER 7

Exercises 7.1

1. $\frac{3}{2}x^2 + C$
3. $\frac{1}{9}x^9 + C$
5. $h(x) = \frac{1}{3}x, k(x) = e^{3x}; h(x) = e^{3x}, k(x) = \frac{x^2}{2}$
7. $h(x) = 2x, k(x) = \frac{(x+1)^6}{6};$
   $h(x) = (x+1)^5, k(x) = x^2$
9. $\frac{1}{9}e^{3x}(3x - 1) + C$
11. $-\frac{1}{x}(\ln x + 1) + C$
13. $\frac{2}{9}x^{3/2}[3 \ln x - 2] + C$
15. $e^x(x^2 - 2x + 2) + C$
17. $\frac{x^2}{4}[2 \ln (3x) - 1] + C$
19. $-(x + 2)e^{-x} + C$

21. $\frac{e^x}{x+1} + C$
23. $2e^{\sqrt{x}}(\sqrt{x} - 1) + C$
25. $-\frac{3}{4}e^{(1-2x)}[2x + 1] + C$
27. $1 - \frac{2}{e}$
29. $3 \ln 3 - 2$
31. $\frac{e^2 - 1}{3e}$
33. $(\ln 4)[\ln(\ln 4) - 1]$
35. $420e^{-0.05} - 2400e^{-5}$
37. $\frac{1}{3}(\ln 4 - 1)$
39. $\ln |x + \sqrt{x^2 + 9}| + C$
41. $-\frac{1}{2\sqrt{3}} \ln \left|\frac{y + \sqrt{3}}{y - \sqrt{3}}\right| + C$
43. $\frac{1}{9}(6 - 7 \ln 7)$
45. $\frac{x}{2}\sqrt{x^2 - 1} + \frac{1}{2} \ln |x + \sqrt{x^2 - 1}| + C$
47. $\ln \left|\frac{y}{1-y}\right| + C$

Exercises 7.2

1. $\frac{1}{4} - \frac{21}{4}e^{-20}$
3.

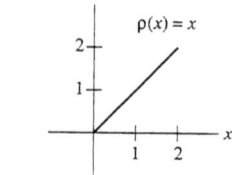

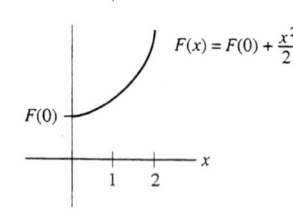

5.

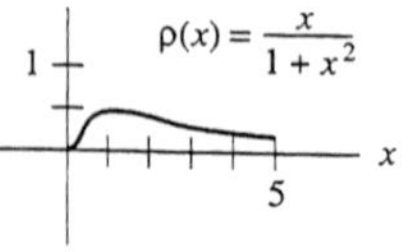

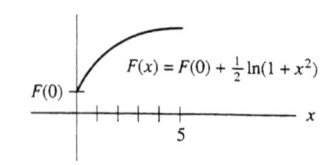

7.

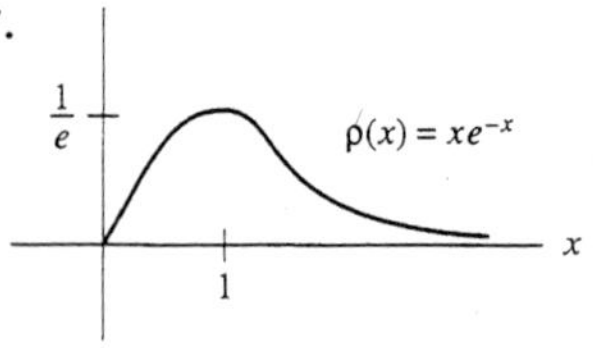

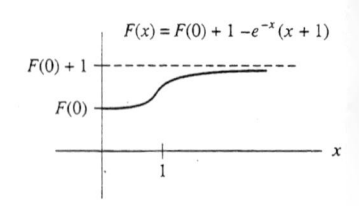

9. a. 62.5; 200   b. 2.25
11. a. 28.3 mg   b. $250(1 - e^{-0.02x})$
    c. 4.17 cm
    d.

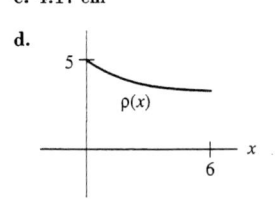

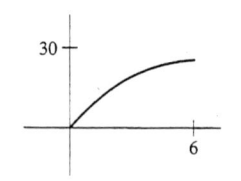

13. 26.04 ft^3
15. 7.235×10^{14} rems
17. 5.85

Exercises 7.3

1. 51.8 L/sec
3. 6.1 L/sec
5. 588.4
7.

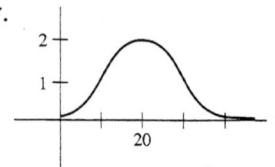

0.017 L/sec
9. 5659.6 gal/hr; 6433.6 gal/hr

11.

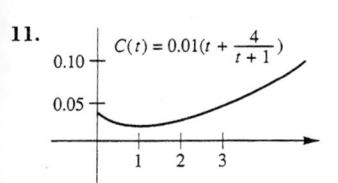

$C(t) = 0.01(t + \frac{4}{t+1})$

59 defective cars

Exercises 7.4

1. 239
3. 395.5
5. 21.4
7. $2,590.26
9. $26,580.83; $33,959.64
11. $73,539.16
13. 15.545
15. 679.85
17. a. 79 b. 45

c. $\dfrac{6}{6 + 4.526[0.47e^{-0.47T} - e^{-0.47T} + 1]}$

19. $1,239.42
21. $9,395,380.10

Exercises 7.5

1. 0.34375
3. 0.749
5. 0.6938
7. 0.7847
9. 0.11 miles
11. $71,182.16
13. $13,163.68
15. 14,375 lbs
17. 38,265

Chapter 7 Summary Exercises

1. $\rho(x) = 30e^{-0.3x}$
3. $\rho(x) = -0.00015(x - 200)^2 + 6$
5. Approximately $[\rho(300)](4) = 180$
7. 44.626/ft; ants per ft of length;
9. 6.13
11. 0.37 l/sec
13. $\approx 137{,}626$
15. ≈ 69.29 units
17. 200
19. 1,741.83
21. $17,293,294
23. 33

CHAPTER 8

Exercises 8.1

1, 3, 5. Verify that equality results upon substitution of the function in each side of the differential equation.

7, 9. Verify that equality results upon substitution of the function in each side of the differential equation and verify the function value at the given point.

11. $C = -2$
13. $C = 4$
17. a. Approximately 71
    b. Approximately 24.5 days
19. 1.24 mL
21. a. $B(6) \approx 671$ bass
    b. $B(6) \approx 671$ bass

Exercises 8.2

1. $y(t) = 5 - 4e^{-3t}$
3. $A = 2$
5. $A(t) = \dfrac{15}{5 - 2e^{-6t}}$
7. $y(t) = \dfrac{5}{1 + 4e^{10t}}$
9. $t = 0.096$
11. Impossible
13. a. $\approx 60.26°$ b. 28 minutes
    c. No
15. a. $A' = 0.12[150{,}000 - A]$
    b. $A(t) = 150{,}000[1 - e^{-0.12t}]$
    c. $A(5) \approx \$67{,}678$; $A(10) \approx \$104{,}821$
17. a. $G' = k(M - G)$, where $M = R/k$ b. $G(t) = R/k - Ce^{-kt}$, where $C = R/k - G(0)$ c. This is the long run concentration of glucose.

19. a. $A' = 0.25A(1 - A)$; $A(0) = 0.15$
    b. $A(t) = \dfrac{1}{1 + \frac{17}{3}e^{-0.25t}}$, $C = -\dfrac{17}{3}$
    c. Approximately 6.94 days; approximately 12.48 days; approximately 15.73 days
    d.

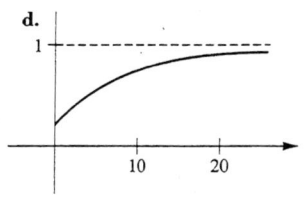

21. $S' = KS(A/N - S)$
23. With $\dfrac{A'}{A} = k\left[\dfrac{M - A}{M}\right]$ use $k = K/M$
25. a. Combine $N'/N = k$ and $N(0) = U = \alpha R$ with $P = N(T)$.
    b.

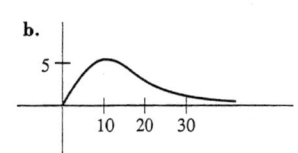

c. The maximum of $f(x) = axe^{-bx}$ occurs at $x = 1/b$ using the first derivative test.   d. $R = 14.81$, $P = 4.905$; $P = (ek\beta)^{-1}$ e. Improved control on any one of the constants α, β, or k has the same effect on efficiency.

Exercises 8.3

1. $y(t) = 5$
3. $y(t) = 1$; $y(t) = -2$
5. There is no constant solution.
7.

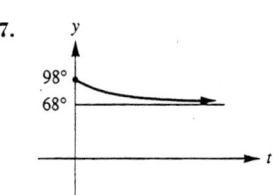

9.

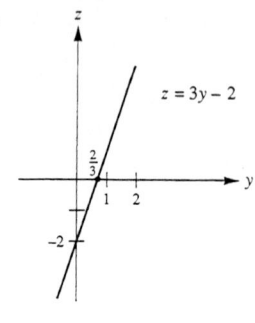

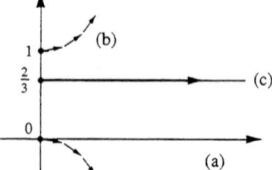

$z = 3y - 2$

11.

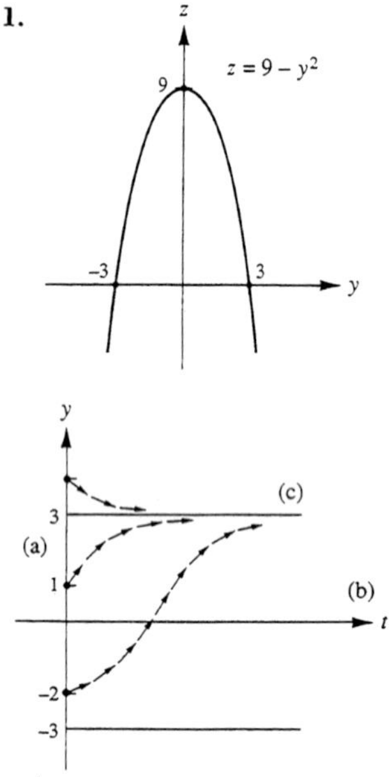

$z = 9 - y^2$

13.

$z = 1 + y^2$

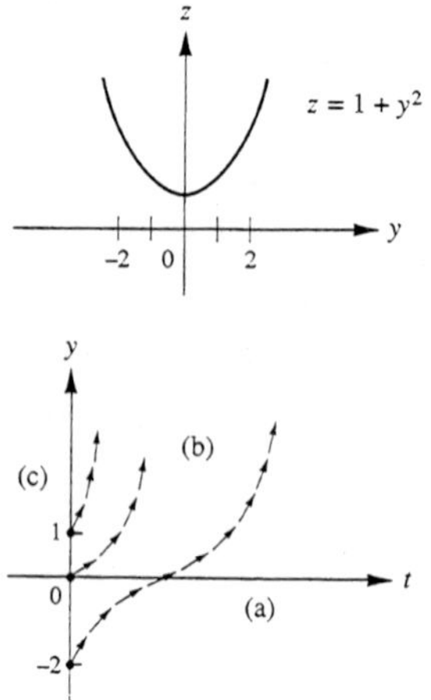

15.

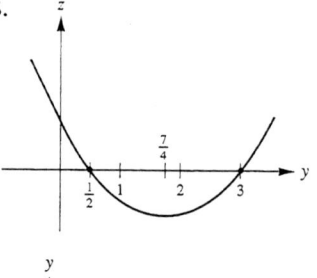

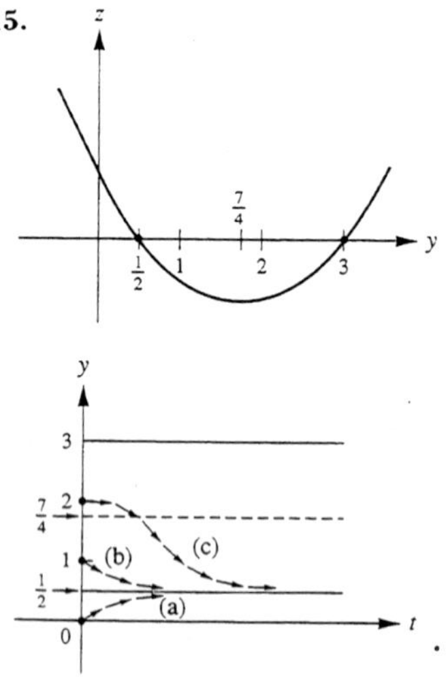

17.

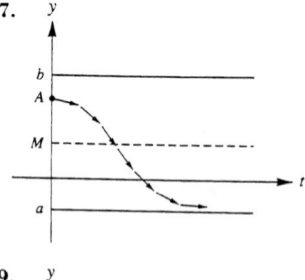

19.

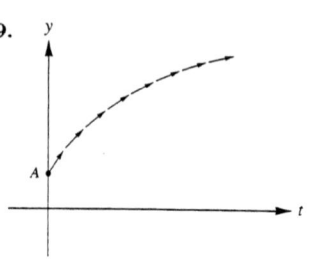

21.

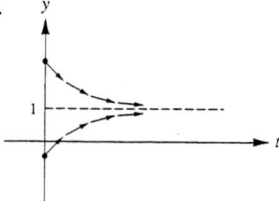

23.

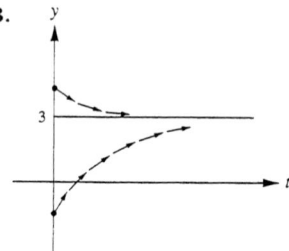

25.

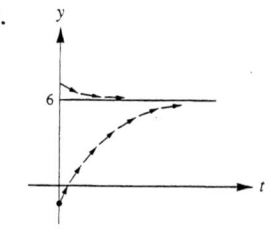

27.

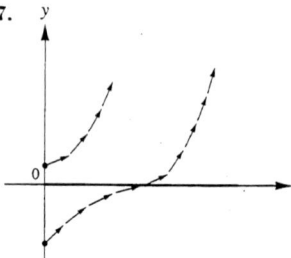

29.

686 Answers to Odd-Numbered Exercises

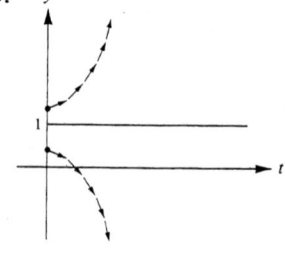

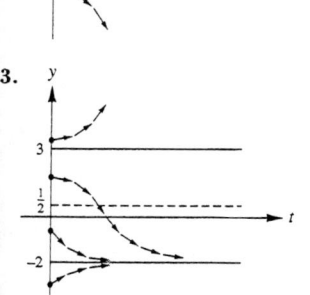

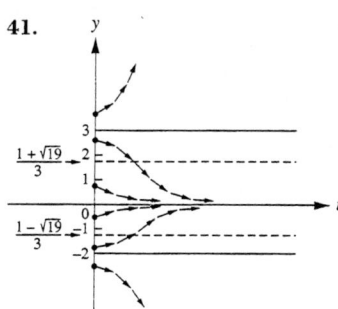

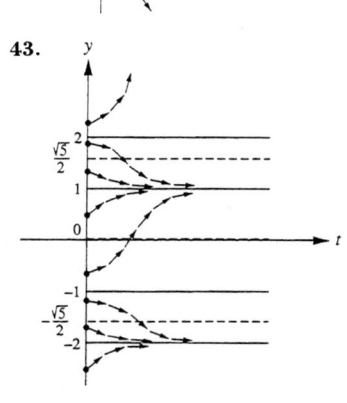

31. [graph: horizontal asymptotes at $y=5$ and $y=0$; dashed line at $y=\frac{5}{2}$]

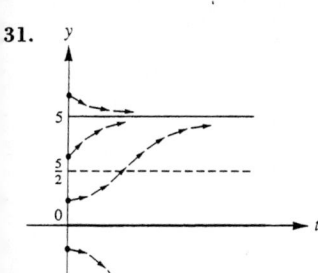

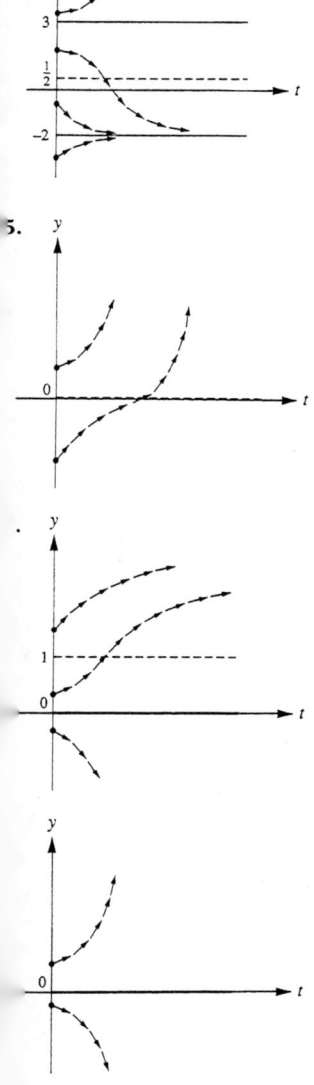

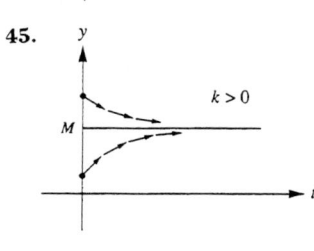

41. [graph: horizontal lines at $y=3$, $y=\frac{1+\sqrt{19}}{3}$, dashed, $y=\frac{1-\sqrt{19}}{3}$, $y=-2$]

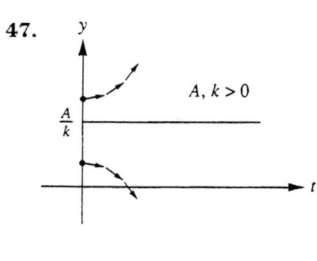

49. a. $g'(y) = kM - 2ky$; $g''(y) = -2k$ **b.** $g''(M/2) = -2k > 0$ if $k < 0$

c. [graph, $k > 0$, asymptotes at M and $\frac{M}{2}$ dashed]

33. [graph: $y=3$, dashed $y=\frac{1}{2}$, $y=-2$]

43. [graph: $y=2$, dashed $y=\frac{\sqrt{5}}{2}$, $y=1$, $y=-1$, dashed $y=-\frac{\sqrt{5}}{2}$, $y=-2$]

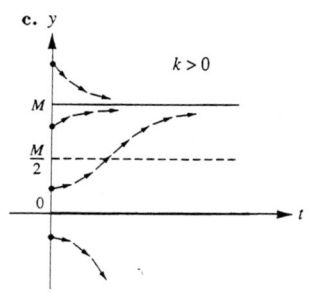

d. [graph, $k < 0$, lines at M and $\frac{M}{2}$ dashed]

35. [graph]

45. [graph, $k > 0$, line at M]

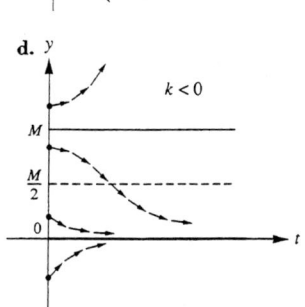

51. [graph of parallel lines, slope 2]

47. [graph, $A, k > 0$, line at $\frac{A}{k}$]

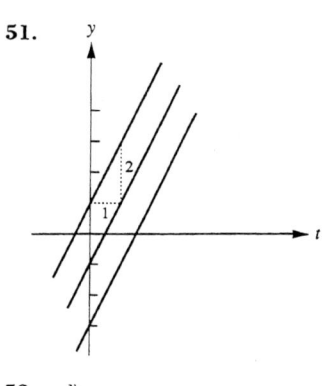

53. [graph: horizontal line at 6,250 and curve at 5,000]

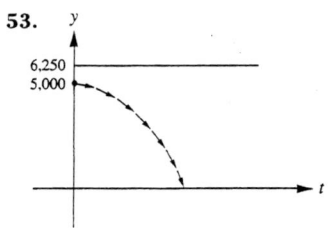

55.

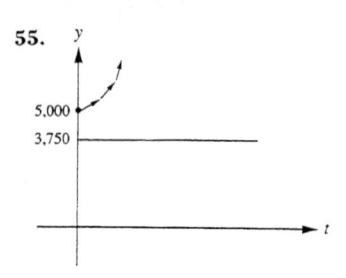

57. When its length is $L_\infty/2$

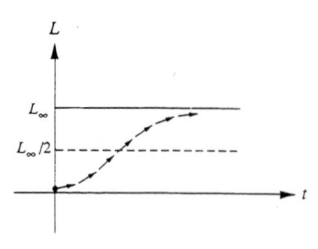

59. $p' = (\alpha + \beta)\left(\frac{\alpha}{\alpha+\beta} - p\right)$

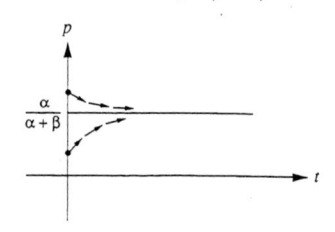

Exercises 8.4

1. $y = Ce^{t^2/2}$
3. $y = 1 - Ce^{-t}$
5. $y = 1 - Ce^{-e^t}$
7. $y = Ce^t - 3$
9. $y^2 = \frac{2}{3}t^3 + C$
11. $u^3 = t^3 + C$
13. $y^2 = 2\ln(t) + C$
15. $y = Ce^{t^3/3}$
17. $u = Ce^{(-1/2)e^{-t^2}}$
19. $A = Cxe^{bx}$
21. $y = -7e^{(1/2)t} + 8$
23. $y = 3$
25. $y = \frac{2}{3} + \frac{1}{3}e^{(-3/2)t^2}$
27. $y = -1 - (t^2 + 2t + 9)^{1/2}$
29. $y^{3/2} = \frac{9}{8}t^{4/3} + 1$
31. $y = \frac{4}{\sqrt[3]{2}}(t^3 + 2)^{1/3} - 1$
33. $y = [68 - e^{-t}(2t^2 + 4t + 4)]^{1/2}$
35. $y = t^t e^{1-t}$
37. $y = \dfrac{10^6}{50{,}000\,t + 1}$

41. 60%
43. $y = M - Ce^{-kt}$

Exercises 8.5

1. a. $y(t) = \frac{1}{2} + Ce^{-2t}$ b. $C = -.25$ c. $.50$ d. $y = \frac{1}{2}$ e. $t = 0$
3. a. $y(t) = 200 + Ce^{-.04t}$ b. $C = 0$ c. $y(5) = 200$ d. $y = 200$ e. No solution
5. 11 years
7. $4725.60
9. a. $277,777.78 b. $231,861.42
11. $16.65 more interest paid for 1.1% monthly interest
13. a. $R(t) = 700$; $R(t) = 1050$; $R(t) = 1190$; 850 million b. $R(t)$ would drop to 952; $R(t)$ would rise to 1008 c. $R(t)$ would rise to 1050
15. b; by $14.01 per month
17. a. $\frac{1}{15}$ g/ft³ b. $\frac{1}{3}$ g/min c. $\frac{2}{3}$ g d. $\frac{1}{6}$ g e. $\frac{1}{18}$ g
19. a. 68 words b. 150 words
21. 109.9 years
23. 13.86 minutes
25. a. 1409 b. 72 turtles/year
27. a. 4.9 weeks b. Clean more frequently
29. a. $y' = 100 - 0.20y$ b. $y(t) = 500(1 - e^{0.20t})$ c. 8.05 hours d. 1.44 hours
31. a. 22.15 cc/hr b. 5 cc/hr c. No
33. a. $w'(t) = 0$; $w(0) = 10^7$ b. $w(t) = 10^7$ c. $y'(t) = 10{,}000 - 0.03\,y(t)$; $y(0) = 0$ d. $y(t) = \frac{10^5}{3}(1 - e^{-0.03t})$ e. y

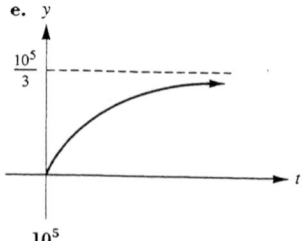

f. $\frac{10^5}{3}$ g. 76.75 years h. 0.3%
i. 0.09 gallons

Exercises 8.6

1. 3
3. Increasing
5. $e^{t^2/2}$; $e^2 \approx 7.389$; 3.28
7. $\frac{19}{16}$
9. 2.101
11. 93.826
13. $g(t, y) = t*y + 1$
    $y(1) \approx 2.836571$; 2.941893, 2.998988 for $N = 8, 16, 32$ subintervals, respectively
15. $g(t, y) = t*y + 1$
    $y(2) \approx -5.417817$, -5.874378, -6.143287 for $N = 8, 16, 32$ subintervals, respectively
17. a. 1704.18; 1575.45; $g(t, y) = -0.005*y + 100*t*(\exp(-0.$
    b. 1845.3; 1670.7; 1300.4

Exercises 8.7

1. $y(t) = (-e^{-t} + C)e^{2t}$
3. $y(t) = Ce^t - t - 1$
5. $y(t) = Ce^{t^2/2}$
7. $y(t) = t^2 + Ct$
9. $y(t) = 1 + t + 2e^t$
11. $y(t) = \frac{t}{2}(t^2 + 9)$
13. a. $y(t) = 800e^{0.03t}$; $y(10) = \$1{,}079.89$ b. $y(t) = \dfrac{e^{0.03t}}{0.03}[-500e^{-0.03t} + 524]$; $y(10) = \$6910.87$ c. $y(t) = \dfrac{e^{0.03t}}{0.09}[572 - 500e^{-0.09t}]$; $y(10) = \$5530.15$
15. $a(2) = 1.35$ mL; $C(2) = 1.094$ mL
17. a. with $y = 1/u$, we have $y' = -u'/u^2$, and substitution yields $u' = -\alpha u - \beta$. b. $u(t) = (\beta(t) + C)e^{A(t)}$, where $A(t) = \int -\alpha(t)\,dt$ and $\beta(t) = \int -\beta(t)e^{A(t)}\,dt$
c. $y(t) = \dfrac{e^{-A(t)}}{\beta(t) + C}$ d. $y(t) = \dfrac{0.65e^{0.05t}}{C + e^{0.65t}}$

Exercises 8.8

1. Substitute these functions in the equations.

3. $x(t) = C_1 e^t + C_2 e^{3t}$
   $y(t) = C_1 e^t - C_2 e^{3t}$
5. $x(t) = C_1 e^{2t} + C_2 e^{-3t}$
   $y(t) = C_1 e^{2t} - \frac{1}{4} C_2 e^{-3t}$
7. $x(t) = C_1 e^t + C_2 e^{3t}$
   $y(t) = -C_1 e^t + C_2 e^{3t}$
9. $x(t) = -te^t$
   $y(t) = (1+t)e^t$
11. $x(t) = 1.437 e^{2.019t} - 0.437 e^{-1.287t}$
    $y(t) = 0.239 e^{2.019t} + 0.761 e^{-1.287t}$
13. 0.467
15. $x(t) = \frac{15}{4} + \frac{25}{4} e^{-8t}$
    $y(t) = \frac{25}{4} - \frac{25}{4} e^{-8t}$; never
17. $x(t) = 20 + 30 e^{(-5/12)t}$
    $y(t) = 45 - 45 e^{(-5/12)t}$
19. $r_1 = -0.6469$, $r_2 = -0.0462$

Chapter 8 Summary Exercises

1. Verify by substitution.
3. $C = -6$
5. $y(t) = 2e^{-t^2/2}$
7. $\frac{y'}{y} = k\left(\frac{500 - y}{500}\right)$, the logistic equation
9.

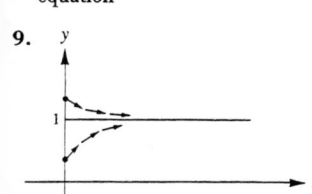

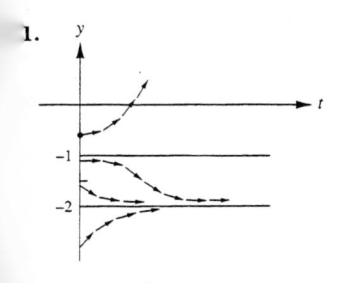

11.

. $y(t) = e^{(t-1)/t}$
. $y(t) = 7 - e^{-t}$
. $y(t) = 1 + t + \ln t$
. The logistic equation has the form $y' = ky(m - y)$. Its solutions are best illustrated by a graph. It models percentage rate of growth in proportion to percentage of remaining capacity.

21. $y(t) = \pm\sqrt{t^2 + 2t + C}$
23. \$53.92 per month

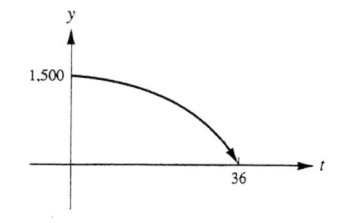

25. $t \approx 4.7$ years
27. $t \approx 8.959$ after starting the procedure
29. $y(3) \approx -17/16$
31. $y(t) = [-2e^{-t^2/4}(t^2 + 4) + 9]e^{t^2/4}$
33. a. 50 tons per year are added; F/V is the proportion of $A(t)$ that is lost per year.
    b. $A(t) = \frac{50}{0.03}[1 - e^{-0.03t}]$;
       $A(20) = 752$ tons   c. $B'(t) = (50 - 0.05t) - 0.03B(t)$; $B(0) = 752$ d. $B(10) = 986.8$
    e. 32.16 years
35. $u(t) = C_1 e^{0.707t} + C_2 e^{-0.707t}$
    $v(t) = 0.414 C_1 e^{0.707t} - 2.414 C_2 e^{-0.707t}$
37. $x(t) = \frac{1}{2} e^{22t} - \frac{3}{2} e^{26t}$
    $y(t) = -\frac{1}{2} e^{22t} - \frac{1}{2} e^{26t}$

CHAPTER 9

Exercises 9.1

1. 0
3. 0
5. $-\infty$
7. ∞
9. ∞
11. 1
13. a. ∞ b. ∞ c. 1000
15. 8
17. $\frac{1}{3}$
19. $2e^{-1/2}$
21. e^4
23. $\left(\frac{3}{2}\right)(e^{10})$
25. ∞
27. $\frac{1}{4}$
29. ∞

31. 1
33. e
35. $\int_1^\infty \pi\left(\frac{1}{x}\right)^2 dx = \pi$
37. Does not exist.
39. 200 lbs

Exercises 9.2

1. a. The probability that X is larger than 0 and smaller than 9
   b. The probability that X is less than or equal to 0   c. The probability that X is greater than or equal to 8   d. The probability that X is greater than or equal to 3.5 and less than or equal to 5
   e. The probability that X is less than or equal to 5 minus the probability that X is less than or equal to 3   f. The probability that X is less than or equal to 2 plus the probability that X is greater than or equal to 4
3. a. $\frac{1}{4}$ b. $\frac{1}{4}$ c. $\frac{3}{4}$ d. $\frac{3}{4}$

$$F(t) = \begin{cases} 0, & t < 1 \\ \frac{1}{4}, & 1 \le t < 2 \\ \frac{1}{2}, & 2 \le t < 3 \\ \frac{3}{4}, & 3 \le t < 4 \\ 1, & t \ge 4 \end{cases}$$

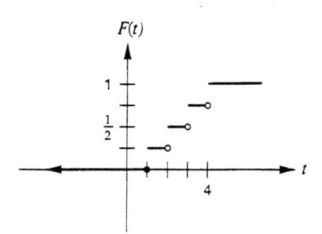

   f. $F(3) - F(1)$
5. $\frac{1}{2}$; 1;

$$F(t) = \begin{cases} 0, & t < 0 \\ \frac{1}{2}, & 0 \le t < 1 \\ 1, & t \ge 1 \end{cases}$$

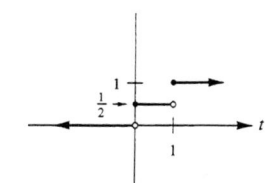

7. a. $\frac{1}{2}$; $\frac{1}{4}$; 0; $\frac{1}{4}$, $\frac{3}{8}$, $\frac{1}{4}$ b. $\frac{1}{4}$; $\frac{7}{32}$

c. $F(t) = \begin{cases} 0, t < 0 \\ \left(\frac{1}{2}\right)t, 0 \le t \le 2 \\ 1, t > 2 \end{cases}$

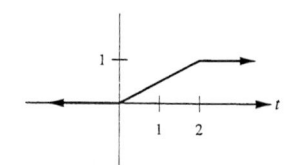

9. $F(t) = \begin{cases} 0, t < 0 \\ t, 0 \le t \le 1 \\ 1, t > 1 \end{cases}$

$\rho(t) = \begin{cases} 1, 0 \le t \le 1 \\ 0, \text{elsewhere} \end{cases}$

11. $F(t) = \begin{cases} 0, t < 1 \\ \left(\frac{1}{3}\right)(t-1), 1 \le t \le 4 \\ 1, t > 4 \end{cases}$

$\rho(t) = \begin{cases} \frac{1}{3}, 1 \le t \le 4 \\ 0, \text{elsewhere} \end{cases}$

13. a. $\frac{1}{3}$ b. $\frac{1}{6}$ c. $\frac{1}{2}$

15. a. 0.20 b. 0.1 c. 0.05
  d. 0.25 e. 0.007 f. 0.1
  g. 0.01

17. a. (0,0,0,0), (0,0,0,1), (0,0,1,0),
  (0,0,1,1), (0,1,0,0), (0,1,0,1)
  (0,1,1,0), (0,1,1,1), (1,0,0,0),
  (1,0,0,1), (1,0,1,0), (1,0,1,1)
  (1,1,0,0), (1,1,0,1), (1,1,1,0),
  (1,1,1,1)
  b. To each outcome there is associated one and only one sum.
  c. 6; $\frac{3}{8}$ d. Since $(-1)^{wk} = 1$ when a tail is flipped and -1 when a head is flipped. e. $\frac{3}{8}$; 0, $\frac{1}{16}$

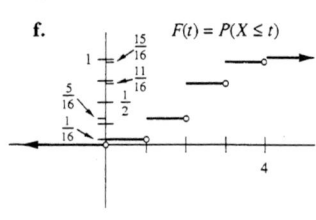

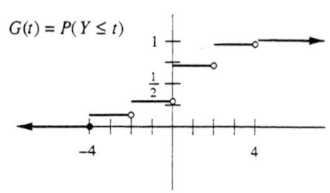

f.

$G(t) = P(Y \le t)$

Exercises 9.3

1. $C = \frac{1}{2}$;

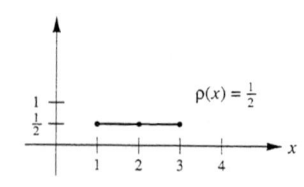

3. $C = 6$;

5. $\frac{1}{4}$
7. $\frac{7}{27}$
9. $F(t) = \begin{cases} 0, t < 1 \\ \frac{t}{2} - \frac{1}{2}, 1 \le t \le 3 \\ 1, t > 3 \end{cases}$

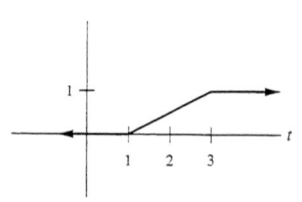

11. $F(t) = \begin{cases} 0, t < 0 \\ 3t^2 - 2t^3, 0 \le t \le 1 \\ 1, t > 1 \end{cases}$

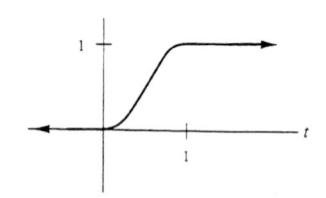

13. a.

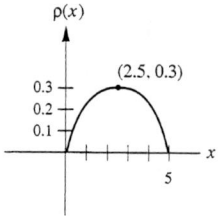

b. $2 \le X \le 3$
c. 0.248; 0.544; 0.296; 0.104
d. $\frac{1}{2}$; 0.648 e. For example,

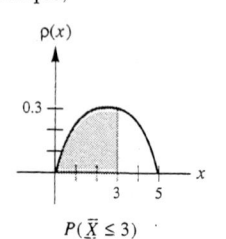

$P(\bar{X} \le 3)$

f. 0.144 vs. 0.1435
15. $e^{-3} = 0.05$
17. $1 - 3e^{-2} \approx 0.593$
19. a. 0.02752 b. 0.200 c. 0.403
  d. 0.387 e. Yes

Exercises 9.4

1. 2.5; 0.866

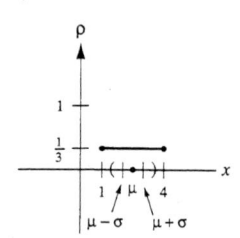

3. 1.5; 0.671

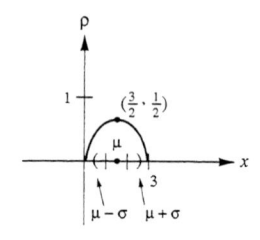

690 Answers to Odd-Numbered Exercises

5. $\dfrac{2}{\ln 3}$; 0.57

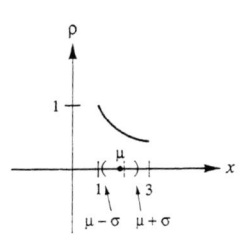

[graph: ρ vs x, curve decreasing, with μ at 1, point at 3, μ−σ and μ+σ marked]

7. 0.93; 0.835

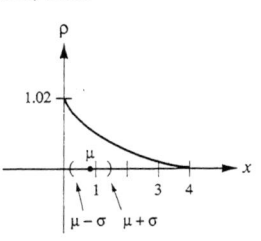

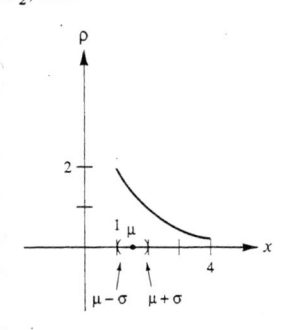

[graph: ρ vs x, 1.02 marked, curve decreasing, μ at 1, 3, μ−σ and μ+σ]

9. $\dfrac{3}{2}$; 0.50

[graph: ρ vs x, 2 marked, curve decreasing, μ at 1, 4, μ−σ and μ+σ]

1.50; 0.471

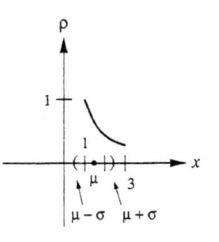

[graph: ρ vs x, curve decreasing, μ at 1, 3, μ−σ and μ+σ]

0; $\sqrt{2}/2$

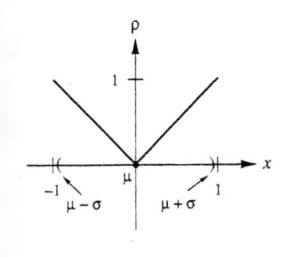

[graph: ρ vs x, V-shape, μ at 0, −1, 1, μ−σ and μ+σ]

15. $\mu = 2$; $\sigma = 2$
17. $\mu = 1$; $\sigma = \sqrt{1/2}$
19. $M = \dfrac{5}{2}$
21. $M = 2\ln 2$
23. $F(t) = \begin{cases} 1 - e^{-kt}, & t \geq 0 \\ 0, & t < 0 \end{cases}$

$1 - e^{-2k}$; $e^{-3k} - e^{-5k}$; e^{-6k}; e^{-1}

25. $F(t) = \begin{cases} 1 - e^{-t/65}, & t \geq 0 \\ 0, & t < 0 \end{cases}$

0.341; 0.292; yes

27. 51%
29. 60 tons
31. 15; 0.1988
33. 0.041

Exercises 9.5

1. $\dfrac{1}{\sqrt{2\pi}} \int_0^{1.5} e^{-z^2/2}\,dz = 0.4332$

3. $\dfrac{1}{\sqrt{2\pi}} \int_{0.5}^{1.5} e^{-z^2/2}\,dz = 0.2417$

5. $\dfrac{1}{\sqrt{2\pi}} \int_{-1}^{-0.5} e^{-z^2/2}\,dz = 0.1498$

7. $\dfrac{1}{\sqrt{2\pi}} \int_{-\infty}^{-0.55} e^{-z^2/2}\,dz = 0.2912$

9. $\dfrac{1}{\sqrt{2\pi}} \int_{0.33}^{\infty} e^{-z^2/2}\,dz = 0.3707$

11. $\dfrac{2}{\sqrt{2\pi}} \int_1^{\infty} e^{-z^2/2}\,dz = 0.3174$

13. The possible outcomes can be represented by (x, y, u, v), where each letter can be H or T; there are then 16 possibilities since $16 = 2^4$. Only 1 outcome has $x = y = u = v = H$; 4 outcomes with exactly one of x or y or u or $v = T$ are possible, and this event coincides with 3 H's, etc.

15. 0.3779
17. 0.2486
19. 0.3707
21. Compute $f''(x) = -\dfrac{e^{-x^2/2\sigma^2}}{\sigma^2 \sqrt{2\pi}}\left(\dfrac{-x^2}{\sigma^2} + 1\right)$

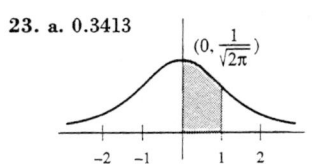

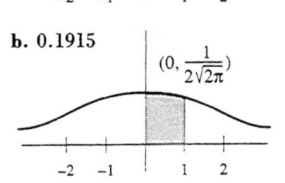

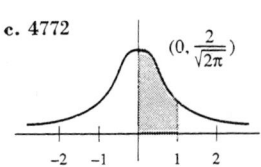

23. a. 0.3413

[graph: normal curve, shaded 0 to 1, peak at $(0, \tfrac{1}{\sqrt{2\pi}})$]

b. 0.1915

[graph: normal curve, shaded, peak at $(0, \tfrac{1}{2\sqrt{2\pi}})$]

c. 4772

[graph: normal curve, shaded, peak at $(0, \tfrac{2}{\sqrt{2\pi}})$]

25. Integrate by parts using $\int (-x\sigma^2)\,d(e^{-x^2/2\sigma^2})$

27. 0.27; 0.338
29. a. 0.1915 b. 0.6915 c. 0.1587
    d. 0 e. 0.5328 f. 0.1915
    g. 0.0179
31. 0.5; 0.1498; 0.0919; 0.0062
33. 34.13%; 14.98%; 567

Chapter 9 Summary Exercises

1. $\dfrac{x^2}{2}\ln x - \dfrac{x^2}{4} + C$
3. $e - 2$
5. $\dfrac{1}{2}e^{x^2} + C$
7. 2
9. $\dfrac{1}{e}$
11. a. 0.2 b. $\dfrac{4}{7}$
13. $C = \dfrac{3}{4}$
15. a. $\mu = \dfrac{1}{3}$, $\infty = \dfrac{1}{3}$ b. $1 - e^{-6}$; $1 - e^{-1}$; $1 - e^{-1}$
17. $F(t) = \begin{cases} 1 - e^{-2t}, & t \geq 0 \\ 0, & t < 0 \end{cases}$

$1 - e^{-8}$; e^{-10}

19. 0.8185
21. $\rho(x) = \dfrac{1}{5}e^{-x/5}$, $x \geq 0$

23. a.

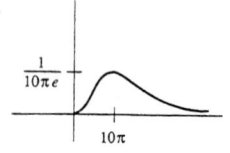

b. 0.17   c. 0.075   d. $\mu = 20$;
   $\sigma = 94$ e. ≈ 3.2 times
25. 0.2476; 0.0475; 0.6306

CHAPTER 10

Exercises 10.1

1. 1, 3, 7, 63, 2047
3. $\frac{1}{3}, \frac{1}{9}, \frac{1}{27}, \frac{1}{729}, \frac{1}{177147}$
5. $e^{-1}, 2e^{-2}, 3e^{-3}, 6e^{-6}, 11e^{-11}$
7. 2 ln(ln 2), 3 ln(ln 3), 4 ln(ln 4), 7 ln(ln 7), 12 ln(ln 12)
9. $\frac{1}{4}, \frac{1}{5}, \frac{1}{6}, \frac{1}{9}, \frac{1}{14}$
11. $a_n = 2n$
13. $a_n = e^{(-1)^n}$
15. $a_n = (-1)^{n+1}/n$
17. $a_{n+1} = a_n + 5(0.1)^{m+1}$ where $a_1 = 0.5$
19. $1, \frac{6}{5}, \frac{6}{5}, \frac{20}{17}, \frac{15}{13}$
21. 2
23. 0
25. 0
27. $\frac{1}{2}$
29. Does not exist
31. 0
33. 0
35. 0
37. 2
39. 0
41. $+\infty$

Exercises 10.2

1. $y_n = 2\left(\frac{3}{2}\right)^n$; 3, 4.5, 6.75, 15.1875, 76.88672
3. $y_n = 0$; 0, 0, 0, 0, 0
5. $y_n = 3n + 2$; 5, 8, 11, 17, 29
7. $y_n = 2^{1-n}$; $1, \frac{1}{2}, \frac{1}{4}, \frac{1}{16}, \frac{1}{256}$
9. $y_n = 2(1.25)^n$; 2.5, 3.125, 3.90625, 6.103513, 14.90116
11. $y_n = 2 - \left(\frac{1}{2}\right)^n$; 1.5, 1.875, 1.984375, 1.999023
13. $y_n = 4(1 - 2^n)$; $-4, -28, -252, -4092$
15. $y_n = 1$; 1, 1, 1, 1

17. $y_n = -1 + 2\left(-\frac{1}{2}\right)^n$; $-2, \frac{5}{4}$, $-\frac{31}{32}, -\frac{511}{512}$
19. $y_n = -\frac{1}{2} + (-1)^{n+1}; \frac{1}{2}, \frac{1}{2}, -\frac{3}{2}, -\frac{3}{2}$
21. $y' = .087y$; $1090.90, $1,190.06, $1,544.96, $2,386.91:
    $y_{n+1} - y_n = .09y_n$; $1090, $1,188.10, $1,538.62, $2,367.37
23. 37,010
25. $953.98 owed; $753.98 interest paid
27. 151, 1681, 8 yrs, 222/yr
29. $y_n = 14.31 - 4.31(0.3012)^n$; approaches 14.31 over time
31. 44
33.

Mature	Young	Total (pairs)
50	50	100
50	100	150
100	150	250
150	250	400
250	400	650

Note that each total is the sum of the previous two.
35. -1
37. 2, -1
39. 0
41. M

Exercises 10.3

1. $1, \frac{4}{3}, \frac{13}{9}, \frac{364}{243}$
3. 2, 4, 6, 12
5. $\frac{1}{2}, \frac{2}{3}, \frac{3}{4}, \frac{6}{7}$
7. $3, \frac{9}{2}, \frac{21}{4}, \frac{189}{32}$
9. 3, 15, 120, 5050; $S_n = n(n+1)/2$; no
11. $1 + \frac{1}{3} + \frac{1}{9} + \frac{1}{27} + \cdots$; $S = \frac{3}{2}$
13. $1 + 0 + 0 + 0 + \cdots$; $S = 1$
15. $1 - 0.6 + 0.36 - 0.216 + \cdots$; $S = 0.625$
17. 1
19. $\frac{1}{6}$
21. $\frac{5}{99}$
23. $\frac{1}{3}$
25. $\frac{5}{99}$
27. $\frac{1}{2}$
29. 20; 5
31. $R = 0$; converges
33. $R = \frac{5}{4}$; diverges
35. $R = \infty$; diverges
37. $R = 0$; converges

39. $R = 0$; converges
41. Diverges
43. Diverges
45. Diverges
47. $\frac{11}{3}$
49. $-0.090909; -0.0909090909$; $-\frac{1}{11}$
51. Diverges; compare to $\Sigma\, 1/n$
53. Converges; geometric with $r < 1$
    $1/e$
55. Converges; compare to $\Sigma\, 1/n^2$
57. Diverges; compare to $\Sigma\, 1/n$
59. Converges; compare to $\Sigma\, 1/2^n$
61. Converges; compare $1/(2^n - n)$ to $1/n^2$ for $n \geq 5$
63. Estimation of the sum of an infinite series

N-TH PARTIAL SUM S(N)

s(100) = 5.187378
s(200) = 5.878032
s(300) = 6.282666
s(400) = 6.569931
s(500) = 6.792825

N-TH TERM A(N)

a(100) = .01
a(200) = .005
a(300) = .0033
a(400) = .0025
a(500) = .002

Estimation of the sum of an infinite series

N-TH PARTIAL SUM S(N)

s(5000) = 9.094514
s(10000) = 9.787613

N-TH TERM A(N)

a(5000) = .0002
a(10000) = .0001

Exercises 10.4

1. $a_n = \begin{cases} 0 & n \text{ even} \\ -1 & n \text{ odd} \end{cases}$; $r = 1$; $(-1, 1)$
3. $a_n = \begin{cases} 0 & n \text{ even} \\ -1/n! & n \text{ odd} \end{cases}$; $r = \infty$; $(-\infty, +\infty)$
5. $a_n = (-1)^n/n!$; $r = +\infty$; $(-\infty, +\infty)$
7. $a_n = n^n$; $r = 0$; $x = 0$

9. $1 + x^2 + x^4 + x^6 + \cdots$; $(-1, 1)$
11. $x + x^2 + x^3 + \cdots$; $(-1, 1)$
13. $1 - x + x^2 - x^3 + x^4 - x^5 + \cdots$; $(-1, 1)$
15. $x - x^2 + \dfrac{x^3}{2!} - \dfrac{x^4}{3!} + \cdots$; $(-\infty, +\infty)$
17. $\dfrac{1}{1 + x^2}$
19. $e^{x/2}$
21. $S(x) = \ln(1 + x)$; $S\left(\tfrac{1}{2}\right) = \ln\left(\tfrac{3}{2}\right)$
23. $S(x) = (1 - x)^{-2}$; $S\left(\tfrac{1}{2}\right) = 4$
25. $S(x) = -\tfrac{1}{2}\ln(1 - x^2)$; $S\left(\tfrac{1}{2}\right) = \tfrac{1}{2}\ln\left(\tfrac{4}{3}\right)$
27. $6!$; $-25!$
29. -6; -25
31. $x - \dfrac{x^2}{2} + \dfrac{x^3}{3} - \dfrac{x^4}{4} + \cdots + \dfrac{(-1)^{n+1}x^n}{n} + \cdots$
33. $a_0 = 1$, $a_n = 2$
35. $a_n = n + 1$
37. $a_0 = 1$; $a_1 = \tfrac{1}{2}$; $a_n = \dfrac{(-1)^{n+1} 1 \cdot 3 \cdot 5 \cdots (2n - 3)}{2^n n!}$, $n \geq 2$

39. 1.424
41. 0.463

Chapter 10 Summary Exercises

1. $-2.303, -1.609, -0.693, -0.223$
3. $a_n = 2 - \dfrac{1}{2^{n-1}}$
5. 0
7. $-\tfrac{1}{5}$
9. 1
11. 0
13. $y_n = 2\left(\tfrac{5}{2}\right)^n$; 5, 12.5, 31.25
15. $y_n = 0$; 0, 0, 0
17. $y_n = \tfrac{1}{2}[1 + 3^{n+2}]$; 14, 41, 122
19. $y_n = \tfrac{20}{7}((0.3)^n - 1)$; $-2, -2.6, -2.78$
21. -1
23. $y_n = 800 - 650(1.01)^n$; $y_{24} = -25.3$
25. $\tfrac{3}{4}, \tfrac{11}{12}, \tfrac{49}{40}$
27. $\tfrac{7}{2}, \tfrac{29}{6}, \tfrac{169}{20}$
29. $\tfrac{4}{5}$; $S_5 = \tfrac{205}{256}$

31. $\tfrac{4}{9}$
33. Converges absolutely
35. Converges absolutely
37. Converges
39. Converges absolutely
41. Converges absolutely
43. $r = 1$; $(-1, 1)$
45. $r = +\infty$; $(-\infty, +\infty)$
47. $1 + (2x) + (2x)^2 + \cdots + (2x)^n + \cdots$
49. $x - x^2 + x^3 - x^4 + \cdots + (-1)^{n+1} x^{n+1} + \cdots$
51. $S(x) = \dfrac{1}{1 + 4x^2}$
53. $S(x) = -\ln(1 - x)$
55. $x - \dfrac{x^2}{2} + \dfrac{4x^3}{3} - \dfrac{x^4}{4} + \dfrac{x^5}{5} - \dfrac{x^6}{6} + \cdots$
57. $x - x^2 + \dfrac{x^3}{2!} - \dfrac{x^4}{3!} + \cdots + \dfrac{(-1)^{n+1} x^n}{(n - 1)!} + \cdots$
59. $S(x) = \dfrac{1}{(1 + x)^2}$

CHAPTER 11

Exercises 11.1

1. $\pi/8$
3. $\pi/5$
5. $\pi/9$
7. $5\pi/6$
9. $4\pi/3$
11. $5\pi/2$
13. $20°$
15. $210°$
17. $225°$
19. $450°$
21. a. $5\pi/3$ b. $5\pi/4$ c. $13\pi/12$ d. $7\pi/4$
23. 750 yds
25. 1 radian
27. 1.5 radians

Exercises 11.2

θ	0	$\pi/6$	$\pi/4$	$\pi/3$	$\pi/2$	$2\pi/3$	$3\pi/4$	$5\pi/6$	$5\pi/4$	$3\pi/2$	$11\pi/6$	2π
$\cos \theta$	1	$\sqrt{3}/2$	$\sqrt{2}/2$	$1/2$	0	$-1/2$	$-\dfrac{\sqrt{2}}{2}$	$-\dfrac{\sqrt{3}}{2}$	$-\dfrac{\sqrt{2}}{2}$	0	$\dfrac{\sqrt{3}}{2}$	1

5. a. .89 b. $\alpha = \pi/9$
. The y-coordinate of the radial line determined by $-\theta$ is the same as that for θ except for a change of sign.

7. Verify by substitution.
9. $0, \sqrt{3}, 1, \sqrt{3}/3, -1, 1, \sqrt{3}/3$

11.

x	0	$\pi/3$	$\pi/4$	$\pi/6$	$\pi/2$	$3\pi/4$	π	$7\pi/6$	$3\pi/2$	2π
$\sin 2x$	0	$\sqrt{3}/2$	1	$\sqrt{3}/2$	0	-1	0	$\sqrt{3}/2$	0	0
$\sin 4x$	0	$-\sqrt{3}/2$	0	$\sqrt{3}/2$	0	0	0	$\sqrt{3}/2$	0	0

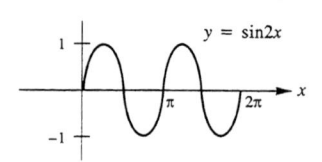

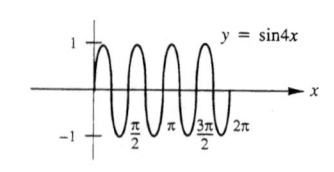

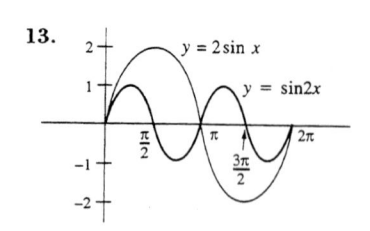

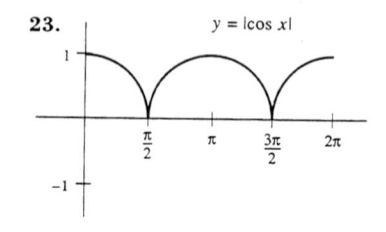

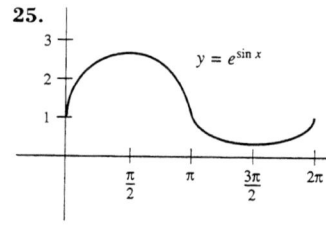

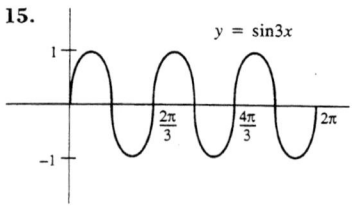

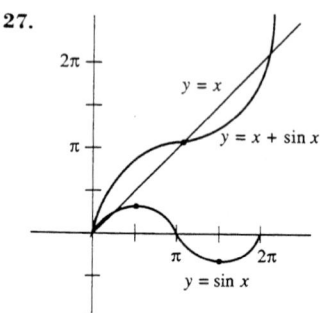

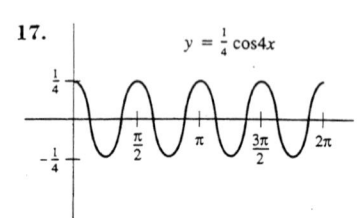

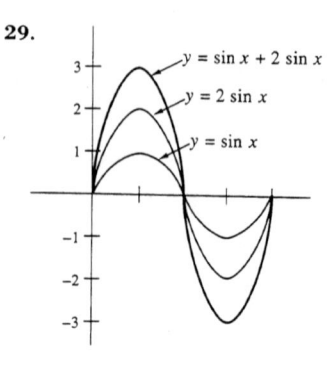

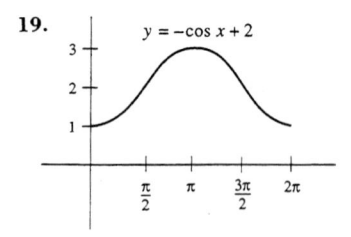

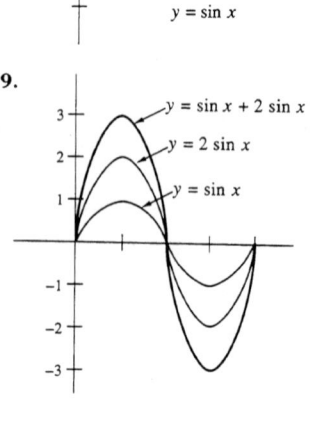

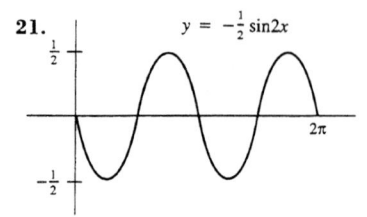

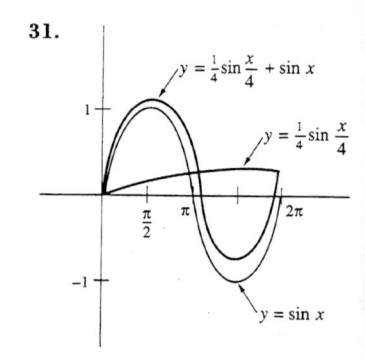

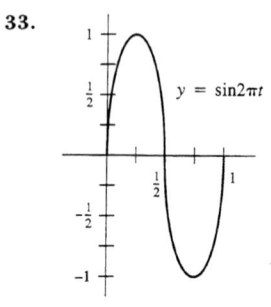

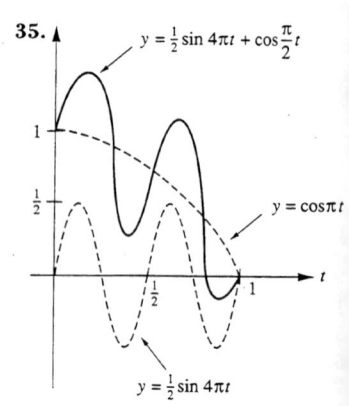

13. ; it is a true equation only when $x = k\pi$

694   Answers to Odd-Numbered Exercises

37. a.

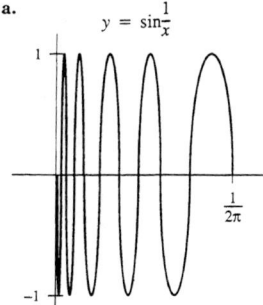

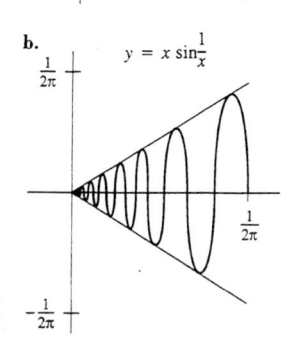

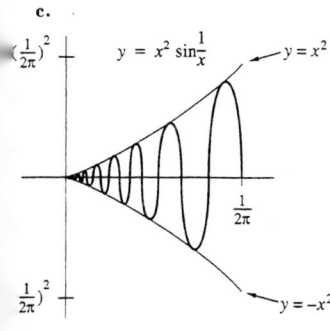

$y = \sin\frac{1}{x}$

b. $y = x\sin\frac{1}{x}$

c. $y = x^2 \sin\frac{1}{x}$; $y = x^2$; $y = -x^2$

$y = \left|\sin\left(\frac{\pi t}{80}\right)\right|$, t in seconds

a. $y = 10 + 0.5x$, x in days
b. $y = 10 + 0.5x - 0.25 \sin 2\pi x$

$-1, 1, 1, 2, \sqrt{2}$; $\sec\frac{\pi}{2} = \frac{1}{\cos(\pi/2)}$ and $\cos\frac{\pi}{2} = 0$; $\csc 0 = \frac{1}{\sin 0}$ and $\sin 0 = 0$

Exercises 11.3

1. $2 \cos x$
3. $\cos x - x \sin x$
5. $\sec^2 x$
7. $\dfrac{1 + \sin x - x \cos x}{(1 + \sin x)^2}$
9. $\dfrac{\sin x + 2(x+1)\cos x}{2\sqrt{x+1}}$
11. $-\cos(\pi - x)$
13. $xe^{x^2 \sin x}(x \cos x + 2 \sin x)$
15. $-e^x \sin(e^x + 1)$
17. $-3 \sin(2x)\sin(3x) + 2 \cos(3x)\cos(2x)$
19. $e^{\sin x + \cos x}(\cos x - \sin x)$
21. 0
23. $\dfrac{\cos u - \sin u}{2\sqrt{\sin u + \cos u}}$
25. $\dfrac{(x \ln x)(\cos x - \sin x) - (\sin x + \cos x)}{x(\ln x)^2}$
27. $3 \cot(3x)$
29. $-3x^2(\tan(x^3))$
31. $-\frac{1}{5}\cos(5x) + C$
33. $e^{\sin x} + C$
35. $\ln|1 + \sin x| + C$
37. $x \sin x + \cos x + C$
39. $\dfrac{x}{2}\sin(2x) + \left(\dfrac{1 - 2x^2}{4}\right)\cos(2x)$
41. $\dfrac{1}{4\pi^3}$
43. $-\dfrac{e^x}{5}[2 \cos(2x) - \sin(2x)]$
45. $2 \cos(2x)$
47. $-2y[2x^2 y \cos(x^2 y) + \sin(x^2 y)]$
49. $-(\sin x)(\sin y)$
51. $\sec x \tan x$
53. $-\csc^2 x$
55. $-\dfrac{1}{\sin x \cos x}$
59. $\frac{1}{2}\ln 2$
61. $5(1 + 2\pi)$; $5(1 - 2\pi)$; $5(1 + 2\pi)$

Exercises 11.4

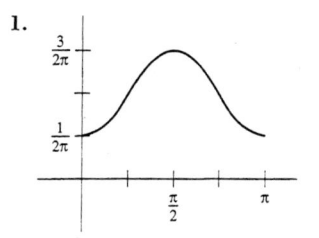

1.

a. 1 b. $\dfrac{1}{2}\left(1 + \dfrac{1}{\pi}\right)$ c. $\dfrac{1}{4}\left(1 - \dfrac{1}{\pi}\right)$

3. $y = 2e^{(1 - \cos t)}$
5. $y = \sec t$
7. $\pi/6$
9. $\pi/2$
11. $\pi/12$
13. $y(t) = 2 \sin(3t - \pi/2)$
15. $y(t) = \frac{1}{4}\sin(4t)$
17. 32 lbs/ft
19. $y(t) = \frac{5}{12}\sin(8t - \pi/2)$
21. y

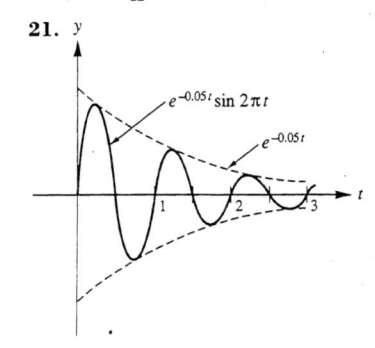

23. a/b
25. a. tenth day; twenty-fifth day
    b. 0
27. 4000
29. 24,000 lbs

Chapter 11 Summary Exercises

1. a. $5/2$ b. $\frac{4}{3}$
3. $\pi/6$; $\pi/3$; $25\pi/36$; $-7\pi/36$; $10\pi/9$; $-7\pi/9$; $7\pi/3$; $25\pi/6$

Answers to Odd-Numbered Exercises   695

5.

θ	0	π/3	π/4	π/6	π/2	π	5π/4	7π/3
sin θ	0	$\sqrt{3}/2$	$\sqrt{2}/2$	1/2	1	0	$-\sqrt{2}/2$	$\sqrt{3}/2$

7. Any x except $1.655 + 2k\pi$ or $4.63 + 2k\pi$
9. $-\frac{3}{2}; 0; \frac{3}{2}; 0$

11. [graph of $y = 3\cos(\frac{x}{2})$, passing through $(2\pi, -3)$]

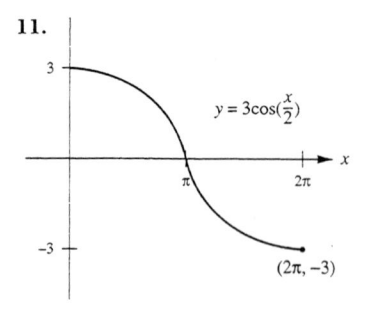

13. [graph of $y = x + \cos x$, passing through $(2\pi, 1 + 2\pi)$]

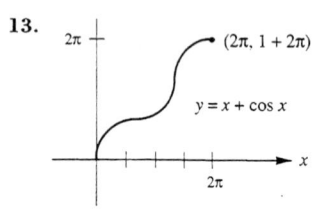

15. [graph of $f(x) = 2\sin 2\pi(x + \frac{1}{4})$]

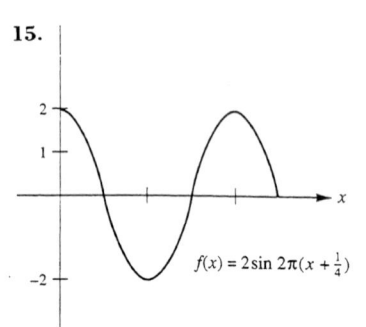

17. [graph of $y = 2 + 3\sin(6\pi t)$, passing through $(1, 2)$]

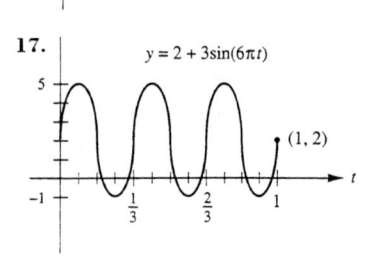

19. $-3 \sin x$
21. $4x \cos(x^2)$
23. $(x \cos x - \sin x)/x^2$
25. $\dfrac{\cos x}{2 + \sin x}$

27. $2\pi \cos[2\pi(x - \pi/2)]$
29. $-\dfrac{\cos(2x)}{2} + C$
31. 0
33. $\frac{1}{4}[\sin(2x) - 2x \cos(2x)] + C$
35. $\ln|\sin x| + C$
37. $y = 0$
39. $y = 10e^{[1 - \cos(6\pi t)]/3\pi}$
41. $\tan^{-1}(2)$

APPENDIX

Exercises A.1

1. $\lim\limits_{h \to 0} \dfrac{|h|}{h}$ does not exist since $\dfrac{|h|}{h} = \pm 1$ for $h > 0, < 0$.

Exercises A.2

1. $z = \frac{1}{2}$ [graph with Slope = 1, point $(\frac{1}{2}, \frac{1}{4})$]

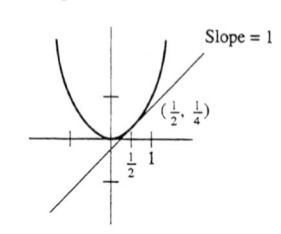

3. $C = -3$
5. $\frac{2}{9}$
7. $\frac{1}{5}$
9. -1
11. 2
13. 0
15. 0
17. Undefined
19. Since $\dfrac{f(2) - f(-1)}{2 - (-1)} = \dfrac{1}{3}$ and $f'(x) = \pm 1$ for any x where f' exists

Exercises A.3

1. [graph: $y(1) = 0.1$, $y(0) = 0.05$, points $(4.33, 0)$, $(0, -3)$, $\ln y = Y = 0.693x - 3$]

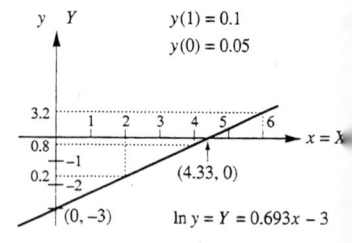

3. [graph: $y(1) = 0.067$, $y(0) = 0.2$, points $(-1.46, 0)$, $(0, -1.61)$, $\ln y = Y = -1.1x - 1.6$]

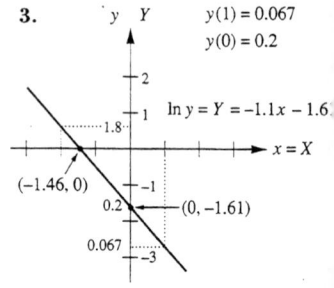

5. [graph of $P(s) = 4.48 \ln s - 41.28$, points $(9.21, 0)$, $(0, -41.3)$]

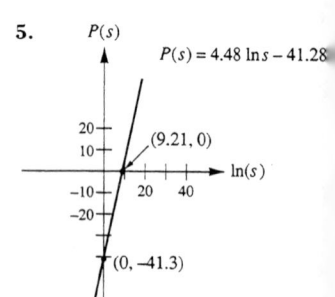

7. [graph of $y = 2.98x^{0.457}$]

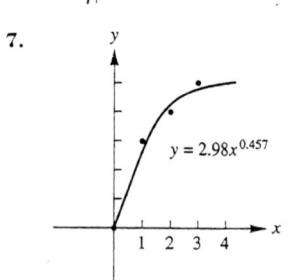

9. [graph of $y = 0.298x^{2.048}$]

696 Answers to Odd-Numbered Exercises

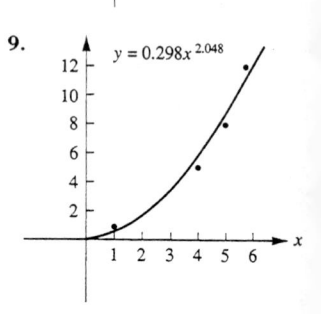

11. $y = \dfrac{2.32}{1 - 0.176e^{-0.16t}}$; 1990

Exercises A.4

1. a. Peak is near 1972. b. Peak is near 1987. c. About 15 years; thus even a 33% increase in actual reserves makes little difference in where the peak rate will occur.
3. a. October 12 b. 7.935 tons

Exercises A.5

1. $x_5 = 3$
3. $x_5 = .56714329$
5. $x_5 = -1.556773265$
7. $x = 1.828$

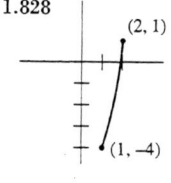

9. $x = .753089165$

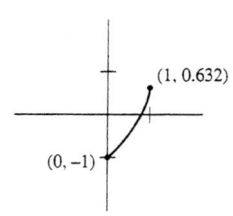

11. $-1.525; 0.258; 1.267$

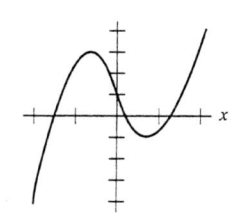

13. $-0.837; -0.269; 1.107$

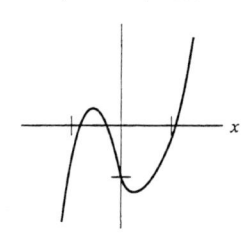

15. $0.815; -1.429$
17. $x_{n+k} = x_n$
19. Diverges; diverges; converges to -1.324717957
21. After taking the limit one has
$$a = a - \dfrac{f(a)}{f'(a)}; \text{ hence } f(a) = 0.$$
23. a. $k = 1.567828$ for $f(x)$; $k = 4.685192$ for $g(x)$
    b. $f(x) = 250e^{-1.568x^2}$
    $g(x) = \dfrac{250}{4.6852x^2 + 1}$
    c. $g(x)$ d. $f(x)$

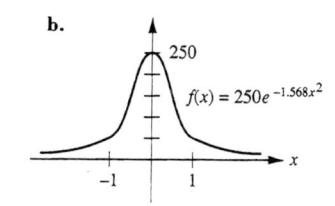

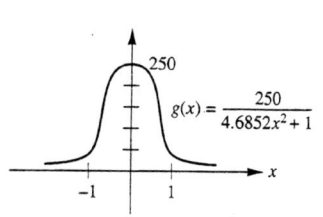

Index

Absolute convergence, 590
Absolute value, 20
  function, 31
Accumulation, 213–214, 253–254, 399, 415
Advertising, 95–96
Agriculture, 163–164, 312–313, 505–507
Algae growth, 5–6, 235–236
Algorithm, 660
Allocation of resources, 160–164, 364–365
Allometric relationship, 444
Alternating series, 589
Amortization, 477–478
Amplitude, 615
Animal husbandry, 575–577
Antiderivative, 225
Antidifferentiation, 233
Approximation, 58
  Euler's method, 489, 494
  principle (AP), 243, 384
  Riemann sum, 210
Area
  beneath a graph, 51, 207–208
  between two curves, 243–244
Arithmetic sequence, 573
Asymptote, 32
  horizontal, 32, 81
  vertical, 32, 80
Autonomous differential equation, 455
Autone-evolution graphing, 462
Auxiliary algebraic equation, 457
Average amount present, 51
Average cost, 45
Average rate of change, 48–49, 95

Average value, 543
  of a function, 244–246
Axial line, 247

Bacteria growth, 49, 571–572
Blood flow, 374–375, 411–412
Box volume, 10

Cancellation, 82–83
Carbon-dating, 316
Carrying capacity, 449
Cartesian coordinate system, 22–23
Cash flow, 98
Cell incubation, 484–485
Center of symmetry, 35
Chain rule, 112, 122
Change-of-state equation, 186
Circle, 10
Closed interval, 20
Cobb–Douglas production function, 338, 364
Coefficients of power series, 593
Comparison test, 591
Compartment mixing, 410
  accumulation, 415
  departure, 419
  first-order difference equation, 574–575
  Lotka-Volterra predator–prey model, 636
  multiple input/output rates, 485
  one-compartment mixing processes, 473
  rate-in less rate-out analysis, 473

  two-compartment mixing processes, 502
  variable rates, 496
Completing the square, 36
Composition of functions, 40
Concavity, 142–143, 146
Concentration, 401, 410, 481–484, 498–501, 577–578
Cone, 192
Constant
  function, 18
  derivative, 111
  multiple rule, 109–110
  proportionality, 29
  rule, 111
  solutions, 455
Constraint, 360
  equation, 155, 361
Construction, 159–161, 556–557
Consumers' surplus, 250–251
Consumption, 585
Continuous compound interest, 309–310, 417–418, 475–476
Continuous function, 74, 93, 641–643
  limit, 75
Continuous random variable, 529
Continuum, 67
Convergence
  interval of, 595
  power series, 593–595
  radius of, 595
  of a sequence, 566
  of a series, 584
    absolute, 590
    comparison test, 591
    ratio test, 587–588, 594

699

Coordinates, 23
Cosine, 610, 618–619
   differentiation, 624
   integration, 626
Cost
   average, 45
   fixed, 166, 167–168
   inventory, 161–163
   marginal, 96, 167
   production, 27–28, 166–168, 235, 505–507
   variable, 166
Critical point, 139, 341
Cumulative distribution, 401
Cylinder, 192

Decay constant, 307–308
Decomposition, 43
Decreasing function, 128–131
Definite integral, 206
   area, 207
   definition of, 208
   Riemann sum, 210
Demand, 37–38, 45–46, 165–166
   function, 45, 165
Density, 399
   function, 533, 534
      exponential, 545–546
      standard normal, 551
      gamma, 548
Dependent variable, 5
Derivative
   chain rule, 112, 122
   constant multiple rule, 109–110
   constant rule, 111
   first
      test for extrema, 140
   of a function, 86
   higher-order, 110–111, 336–338
   notation, 108–109
   partial, 323, 331
   power rule, 108
   product rule, 119, 122
   quotient rule, 119, 122
   second, 110
      test for concavity, 143
      test for extrema, 144, 344
      test for inflection points, 147
   sum rule, 109–110
   tangent line, 89
   third, 110
Deterministic model, 239, 311, 523
Difference equation, 570
   of arithmetic growth, 573
   first-order, 571
      compartment mixing, 574–575

   of geometric growth, 572
   logistic, 579
   solution, 571
Difference quotient, 95, 401
Differentiable, 87, 641–643
Differential, 196, 371
   equation, 133, 436, 453
      autonomous, 455
      change to remaining capacity, 447
      exponential change, 304–307, 438–440
      logistic, 449
      notation, 442
      qualitative solution, 462
      separation of variables, 465, 467–468
      system of, 503
      trigonometric function application, 630
   estimate, 199, 200, 372–373
Differentiation, 91, 93, 124
   exponential function, 269–270, 277
   implicit, 179
   invariant function, 262–264
   partial, 323, 331
      higher-order, 336–338
   sine and cosine, 624
Discontinuity, 74
Discrete change, 570
Discrete random variable, 529
Discriminant, 343
Distribution, 528
   uniform, 529
Distributive law, 28
Divergence of a series, 584
   comparison test, 591
   harmonic, 586
   ratio test, 587–588
Divergent improper integral, 516
Domain of a function, 14, 15–16, 44, 266
Drug concentration, 499–501, 577–578
D-test for extrema, 345
Dummy variable, 98

e, 264. See also Exponential function
Ecology, 636
Endpoint extrema, 150
Environmental impact, 311–312, 441–442
Epidemics, 450–451, 452
Equalizing tax, 252–253
Equation
   auxiliary algebraic, 457
   change-of-state, 186

   constraint, 155, 361
difference, 570
   of arithmetic growth, 573
   first-order, 571
   of geometric growth, 572
   logistic, 579
   solution, 571
differential, 133, 304–307, 436, 453
   autonomous, 455
   logistic, 449
   qualitative solution, 462
   separation of variables, 465, 467–468
   system of, 503
   trigonometric function application, 630
logistic, 451
   difference, 579
polynomial, 38
   roots, 152
quadratic, 38
   roots, 151–152, 240
simultaneous, 349
state, 186
Equilibrium price, 37
Euler's method, 489, 494
Expected value, 541, 542
Exponential change, 5, 303–304
   differential equation, 133, 304–307
   time-dependent, 438–440
Exponential decay, 307–309
   half-life, 310
Exponential function, 262, 264
   defined, 265
   differentiation, 269–270, 277
   domain, 266
   graph, 267–269
   integration, 270, 279–280
   properties, 266–267, 272–273
   range, 266
   values, 273
Exponential growth, 32, 134, 268, 303–304
   doubling-time, 309
Exponentially distributed random variables, 545
Exponents, laws of, 274
Extreme points. See Local extrema

Factorization, 82–83
Family of antiderivatives, 225
Family of functions, 131, 437
Fick's law for passive diffusion, 507
First derivative test for extrema, 140
First-order difference equation, 571
   compartment mixing, 574–575

Fish farming, 348
Flow rate, 409–411
Forecasting, 354–355
Forestry, 48–50, 51–52
Form, mathematical, 43–44
　outer, 114
Fractional power, 152
Frequency, 615
　histogram, 549
Function
　absolute value, 31
　average value, 244–246
　composite, 40
　　domain, 44
　　simplification, 43–44
　constant, 18
　　derivative, 111
　continuous, 74
　critical point, 139, 341
　decreasing, 128–131
　definition of, 13
　derivative, 86
　differentiable, 87
　differentiation invariant, 262–264
　domain, 14, 15–16, 44
　exponential, 262, 264
　　defined, 265
　　density, 545–546
　　differentiation, 269–270, 276–277
　　domain, 266
　　graph, 267–269
　　integration, 270, 279–280
　　properties, 266–267, 272–273
　　range, 266
　　values, 273
　family, 131, 437
　graph
　　definition of, 23
　　exponential, 267–269
　　linear, 24
　　logarithmic, 286
　　nonlinear, 30–33
　　of several variables, 323–329
　growth, 442
　image, 14
　increasing, 128–131
　limit, 68, 72
　linear, 24
　logarithmic
　　definition of, 285
　　differentiation, 292–294, 297–300
　　domain, 286
　　graph, 286
　　integration, 300–302
　　inversion formulas, 288

　　properties, 287–288
　　range, 286
　　values, 295
　notation, 13–14, 48–49, 108–109, 442–443
　objective, 154
　periodic, 610–612
　piecewise constant, 18–19
　probability density, 533, 534
　　standard normal, 551
　range, 14, 17
　several variables, 323
　Taylor-MacLaurin series, 598
　trigonometric, 610–612
　value, 14
　　estimate of unknown, 199–201
Fundamental Theorem of Calculus, 218

Gamma densities, 548
Genetics, 6–8
Geology, 406–407
Geometric
　sequence, 572
　series, 584
Gompertz growth functions, 442
Graph
　center of symmetry, 35
　frequency histogram, 549
　function, 23–24
　　exponential, 267–269
　　logarithmic, 286
　　of several variables, 323–329
　linear, 24
　nonlinear, 30–33
　reflection, 34
　translation, 34
Growth constant, 305
　functions, 442
　See also Rate of change

Harmonic series, 586
Heat loss, 337–338
Higher-order derivatives, 110–111
Horizontal asymptote, 32
H-test for Lagrangian optimization, 363
Hubbert's curve, 656
Hyperbola, 31
Hyperbolic paraboloid, 328

Identities, 613
Image of a function, 14
Implicit background variable, 185

Implicit differentiation, 179
Improper integral, 515
　definition of, 516
　divergent, 516
Increasing function, 128–131
Indefinite integral, 225
Independent variable, 5
Infinite series, 582
　alternating, 589
　comparison test, 591
　convergent, 584, 593–595
　definition of, 584
　divergent, 584
　geometric, 584
　harmonic, 586
　notation, 586–587
　nth partial sum, 582
　power series, 593
　　convergent, 593–595
　　Taylor-MacLaurin series, 598
　properties, 587
　ratio test, 587–588
Infinity, 20, 81–82
Inflation, 308–309, 310
Inflection points, 146
Inherent error, 197–199
Initial condition, 437
Initial value problem (IVP), 437
Inner part, 114
Integral
　definite, 206
　　area, 207
　　definition of, 208
　　Riemann sum, 210
　double, 380
　improper, 515, 516
　indefinite, 225
　iterated, 381
　sign, 208
　table, 397–398
Integrand, 208, 226
Integration, 205, 206
　approximate, 424–427
　exponential function, 270, 279–280
　irreducible forms, 628
　limits of, 214
　by parts, 391
　sine and cosine, 626
　by substitution, 227, 229, 279–280
　trigonometric function application, 630–632
Interest rates, 42–43, 114
Intermediate value theorem, 131
Interval, 20
　of convergence, 595
Invariant, 262
　exponential function, 270

Index   701

Inventory, 161–163, 573–574
Investment
    growth, 133–134, 306–307, 309–310, 573
    rate of, 281–282
    return curve, 133–134
Irrational number, 20

Lagrange multipliers, 361–362, 376
Least squares, 351–352
L'Hôpital's rule, 646
Limit
    evaluation of, 78–79
    of a function, 68, 72
    at infinity, 81–82
    of integration, 214
    properties, 77–78, 568
    of a sequence, 566
    value, 68, 72
Linear
    function, 24
    regression, 351, 353
    systems, 503
ln x, 285. *See also* Natural logarithm
Loans, 477–478
Local extrema, 138, 341
Local maximum value, 138, 341
Local minimum value, 138, 341
Logarithm. *See* Natural logarithm
Logarithmic scaling, 650
Logistic
    curve, 654
    equation, 451
        difference, 579
        differential, 449
Lotka-Volterra predator–prey model, 636

Manufacturing, 148–150, 190–191
Marginal cost, 96, 167
Marginal productivity, 365–366
Marginal profit, 96, 168
Marginal rates of change, 338
Marginal revenue, 167
Marketing, 281, 471, 519–520, 657–659
Mathematical analysis, 57
Mathematical model, 3
    deterministic, 239, 311, 523
    random, 523
Maximum and minimum values, 138, 140, 144, 149, 341
Mean value, 543
    theorem, 644, 645

Medicine, 374–375, 407–408, 411–412, 450–451, 452, 499–501
Method of sections, 325
Minimax point, 329
Mortality rate, 441–442
Multiplier effect, 585

Natural logarithm
    defined, 285
    differentiation, 292–294, 297–300
    domain, 286
    graph, 286
    integration, 300–302
    inversion formulas, 288
    properties, 287–288
    range, 286
    values, 295
Newton-Raphson algorithm, 660–661
Newton's law of heating/cooling, 445
Nondeterministic model, 523
Normally distributed random variable, 554
Notation
    derivative, 108–109
    differential equation, 442–443
    function, 13–14, 48–49
    Fundamental Theorem of Calculus, 220
    infinite series, 586–587
nth partial sum, 582
nth term, 564

Objective function, 154
Oil production, 656–657
One-compartment mixing processes, 473
    multiple input/output rates, 485
Open interval, 20
Optimization, 154, 156, 341
    constrained, 361
Origin, 22
Outer form, 114

Packaging, 158–159, 346–347
Parabola, 30
Paraboloid, 327
Parameters, 536
Partial derivative, 323, 331
    higher-order, 336–338
Partial sum, 582
Passive diffusion, 507
Percentage error, 198
Percentage rate of change, 439

Periodic functions, 610–612
Periodic motion, 634–635
Piecewise constant function, 18–19
Point–slope form, 27
Polluters' surplus, 251–252
Pollution, 189–190, 213–214, 251–254, 282–283, 441–442, 498–499
Polynomial, 38
    curve, 652
    equation roots, 152
Population, 401–402, 403–404, 419–420, 441
Power rule, 108
Power series, 593
    coefficients, 593
    convergence, 593–595
        interval, 595
        radius, 595
        ratio test, 594
    Taylor-MacLaurin series, 598
Present value, 313–314, 420–423, 429
Pricing, 165–166
Probabilistic model, 523
Probability
    definition of, 524
    density function, 533, 534
        standard normal, 551
    trigonometric function application, 629
Producers' surplus, 249–250
Product rule, 119, 122
Production, 163–164
    average, 384–385
    costs, 27–28, 166–168, 235, 505–5
    function, 338
    maximization, 364–365, 376–377
Productivity, 190–191
    marginal, 365–366
Profit, 137, 164–168
    marginal, 96, 168
Proportion, 399
Proportional reasoning, 6
Proportionality constant, 29
Protein fixing, 482–484
Pythagorean theorem, 191

Quadratic equation, 38
    formula, 38
    roots, 151–152, 240
Qualitative solution, 462
Quotient rule, 119, 122

Radial line, 605
Radian measure, 605

702   Index

Radioactive decay, 316
Radius of convergence, 595
Random effects, 239–240
Random variable, 523
  continuous, 529
  discrete, 529
  distribution, 528
  expected value, 541, 542
  exponentially distributed, 545
  normally distributed, 554
  standard deviation, 543
  standard normal, 551
  variance, 543
Range management, 575–577
Range of a function, 14, 17, 266
Rate-in less rate-out analysis, 473
Rate of change
  average, 48–49, 95
  and maximum capacity, 448–449
  Newton's law of heating/cooling, 445
  percentage, 439
  and remaining capacity, 447
Rate of decay, 315
Rate of growth. See Rate of change
Ratio test, 587–588, 594
Rational number, 20
Real number system, 19–20
Rectangle, 10
Rectangular coordinate system, 22–23
Reflection, 34
Regression. See Least squares
Related rates, 185
Relative rate of change, 439
Remaining capacity, 447
Resource
  allocation, 160–164, 364–365, 478–479
  depletion, 475–476, 478–479, 656–657
  return on investment, 133–134
  revenue, 164–168, 201
Riemann sum, 210
Rolle's theorem, 649
Roots of an equation, 38, 151–152, 240, 660–661

Sales growth, 58–61, 91–93
Sample space, 527
Secant line, 59

Second derivative, 110
  test for concavity, 143
  test for extrema, 144, 344
  test for inflection points, 147
Sedimentation, 480
Separation of variables, 465, 467–468
Sequence, 67, 564
  arithmetic, 573
  convergent, 566
  geometric, 572
  limit, 566
    properties, 568
  nth term, 564
Series. See Infinite series
Simplification, 43–44
Simulation, 8–9
Simultaneous equations, 349
Sine, 610, 618–619
  differentiation, 624
  integration, 626
Slope, 24
Slope–intercept form, 25
Soil management, 505–507
Solid of revolution, 248
Sphere, 191
Spoilage, 413
Spring constant, 634
Square, completing the, 36
Standard deviation, 543
Standard normal density function, 551
  table of values, 558
Standard normal random variable, 551
State, mathematical, 9–10, 106
  equation, 186
  steady, 520
Substitution, integration by, 227, 229, 279–280
Sum of an infinite series, 582
Sum rule, 109–110
Supply and demand, 37–38
Surge–dissipation curve, 280–283
Symmetry, center of, 35
System, 3, 9–10
  of differential equations, 503

Tangent
  function, 619–620
  line, 59
Taylor-MacLaurin series, 598
Terminus, 605

Theory of the firm, 164
Third derivative, 110
Time-evolution graphing, 462
Toxicity, 441–442
Transfer coefficients, 505
Translation, 34, 613
Trapezoidal rule, 427
Triangle
  area of, 191
  right, 10, 191
  similar, 619
Trigonometric functions, 610–612
Trigonometric identities, 613
t-section, 325
Two-compartment mixing processes, 502

Uniform distribution, 529

Variable, 3
  dependent, 5
  implicit background, 185
  independent, 5
  random, 523
    continuous, 529
    discrete, 529
    distribution, 528
    expected value, 541, 542
    exponentially distributed, 545
    normally distributed, 554
    standard deviation, 543
    standard normal, 551
    variance, 543
  separation, 465, 467–468
Variance, 543
Vertical asymptote, 32
Volume of a solid, 246–249

Water management, 480
Weighted average, 541
Whole number, 20, 569

x-axis, 22
x-section, 325

y-axis, 22